AF893228

Our Earth's Changing Land

International Geographical Union
COMMISSION ON LAND USE/COVER CHANGE

This work was supported by the International Geographical Union's Commission on Land Use/Cover Change (IGU-LUCC)

The Commission involves a network of geographers with research interests in land use/cover change. Its objectives are:

- To promote geographical research on land use/cover changes, at scales ranging from the local to the global
- To stimulate the production and the use of land-use information bases of both the present and the past
- To coordinate the comparative study and the model study of land use/cover changes and their driving forces in different regions.

Contents

List of Entries

Guide to Related Topics

Analysis of Land Change

Aerial Photography
Auto-Correlation
Cadastre
Change Detection
Conversion
Cumulative Change
Decision-Making
Disequilibrium Dynamics
Driving Forces
Feedback
Georeferencing
Hot Spot Identification
Human Appropriation of Net Primary Production
Initial Conditions
Land Cover Classification System
Land Evaluation
Land-Use History
Land-Use System
Land-Use Transition
Mediating Factor
Meta-Analysis
Metabolism
Mode of Interaction
Multiplier Effect
Paleo Perspectives
Pattern Metrics
Pattern to Process
Pixelizing the Social
Proximate Causes
Remote Sensing
Scale
Socializing the Pixel
Spatial Diffusion
Syndrome
Systemic Change
Thermal Band Analysis
Tradeoffs
Transition Matrix
Validation

Biomass Burning

Advanced Very High Resolution Radiometer
Along Track Scanning Radiometers World Fire Atlas
Alternatives to Slash-and-Burn (ASB) Programme
Amazonia
Atmosphere-Land Interlinkages
Biodiversity
Biomass
Boreal Zone
Brazilian Cerrado
Canada
Carbon Cycle
Climate Impacts
Cumulative Change
Degradation Narrative
Disequilibrium Dynamics
Ecological Colonization
Fire
Global Burnt Area 2000
Global Observation of Forest and Land Cover Dynamics
GLOBCARBON
Grassland
Great Plains
Hot Spots of Land-Cover Change
Human Impact on Terrestrial Ecosystems (HITE) Initiative
LANDFIRE Project
Modification
Mountains
Nutrient Cycle
Paleo Perspectives
Pasture
Pristine Myth
Remote Sensing
Russia
Savannization
Secondary Vegetation
Slash-and-Burn Agriculture
Southern Africa
Southern Cone Region
Taiga
Threshold

Tropical Humid Forest
Turnover
United States of America
Vegetation

Case Studies

Amazonia
Boreal Zone
Brazilian Cerrado
Canada
China
Coastal Zone
East-Central Europe
Europe
Great Plains
Himalayas
Japan
Latin America
Meta-Analysis
Miombo
Mountains
River Basin
Russia
South Africa
South America
Southeast Asia, Mainland
Southeast Europe
Southern Africa
Southern Cone Region
Sudan-Sahel
Taiga
United States of America
West-Central Europe
Western Europe

Causes

Access
Agricultural Intensification
Agriculture
Agro-Industry
Arena of Land Conflict
Cattle Ranching
Colonization
Consumption
Contract Farming
Conversion
Cultural Factors
Cumulative Change
Decision-Making
Deforestation
Degradation Narrative
Desertification
Driving Forces
Economic Growth
Economic Liberalization
Economic Livelihood
Economic Restructuring
Endogenous Factor
Exogenous Factor
Extensification
Fire
Forest Degradation
Forest Transition
Global Environmental Change
Globalization
Green Revolution
Industrialization
Institutions
Investments
Land Abandonment
Land Concentration
Land Degradation
Land Fragmentation
Land Rehabilitation
Land Rent
Land Rights
Land Tenure
Land-Use Policies
Land-Use Transition
Leapfrogging
Mineral Extraction
Mode of Interaction
Modification
Pasture
Population Dynamics
Poverty
Property
Proximate Causes
Reforestation
Regrowth
Remittances Landscape
Slash-and-Burn Agriculture
Soil Degradation
Soil Erosion
Subsistence Agriculture
Suburbanization
Swidden Cultivation
Syndrome
Systemic Change
Technological-Scientific Revolution
Tragedy of Enclosure
Tragedy of the Commons
Transhumance
Urbanization
Urban Sprawl

Conventions and Environmental Agreements

Agenda 21
Conservation
Cropland Change
Forest
Institutions
Kyoto Protocol
Public Policy
Land Evaluation
Land-Use Planning
Land-Use Policies
Millennium Ecosystem Assessment
Ramsar Convention on Wetlands
United Nations Convention on Biological Diversity
United Nations Convention to Combat Desertification
United Nations Environment Programme
United Nations Food and Agriculture Organization

Cropland Change

Agricultural Intensification
Agricultural Revolution
Agriculture
Agrodiversity
Agro-Industry
BIOME 300 Project
Biotechnology
Cash Crops
Cassava
Coffee
Colonization
Consultative Group on International Agricultural Research
Consumption
Contract Farming
Conversion
Cotton
Cumulative Change
Decision-Making
Desertification
Disintensification
Domestication
Economic Livelihood
Eutrophication
Extensification
Fallow
Food Crops
Food Security
Globalization
Green Revolution
Hot Spots of Land-Cover Change
Human Appropriation of Net Primary Production
Initial Conditions
International Food Policy Research Institute
Land Abandonment
Land Concentration
Land Degradation
Land Rent
Land Rights
Land Tenure
Landholding
Land-Use History
Land-Use Legacy
Land-Use Policies
Land-Use System
Maize
Modification
Monoculture
Nutrient Cycle
Property
Proximate Causes
Remittances Landscape
Rice
Slash-and-Burn Agriculture
Soil Degradation
Soil Erosion
Spatial Diffusion
Species Extinction
Subsistence Agriculture
Swidden Cultivation
Tobacco
Urban-Rural Fringe
Wheat

Data

Aerial Photography
BIOME 300 Project
Cadastre
Center for International Earth Science Information Network
Center for Sustainability and the Global Environment
Change Detection
Continuous Data
Coordination of Information on the Environment (CORINE) Database
Discrete Data
Earth Resources Observation Systems (EROS) Data Center
Forest
Geographical Information System

Dryland Change

Forest Change

Vegetation
World Agroforestry Centre

Impacts

African Trypanosomiasis
Agrodiversity
Albedo
Atmosphere-Land Interlinkages
Biodiversity
Biodiversity Novelties
Biomass
Carbon Cycle
Carbon Sequestration
Cascades
Climate Impacts
Cumulative Change
Ecological Footprint
Ecosystem Services
Environmental Change
Eutrophication
Exotic Species
Feedback
Fire
Global Environmental Change
GLOBCARBON
Greenhouse Gases
Holocene
Human Appropriation of Net Primary Production
Human Health
Human Immunodeficiency Virus (HIV)/ Acquired Immunodeficiency Syndrome (AIDS)
Land Fragmentation
Land Quality
Malaria
Metabolism
Nutrient Cycle
People at Risk
Salinization
Savannization
Species Extinction
Systemic Change
Threshold
Tradeoffs
Water-Land Interlinkages
Watershed

Journals

Agriculture, Ecosystems & Environment
Ambio, A Journal of the Human Environment
BioScience
Ecology and Society
Environment
Frontiers in Ecology and the Environment
Global Environmental Change
Journal of Arid Environments
Journal of Land Use Science
Land Degradation & Development
Land Use Policy

Land-Cover Conversion and Modification

Agricultural Frontier
Agricultural Intensification
Anthropocene
Arc of Deforestation
BIOME 300 Project
Change Detection
Conversion
Cumulative Change
Deforestation
Desertification
Ecological Colonization
Eutrophication
Extensification
Fallow
Fire
Forest Degradation
Forest Transition
Holocene
Hot Spot Identification
Hot Spots of Land-Cover Change
Human Appropriation of Net Primary Production
Land Abandonment
Land Degradation
Land Rehabilitation
Land-Use Transition
Land-Use/Cover Change (LUCC) Project
Leapfrogging
Mineral Extraction
Modification
Monoculture
NASA Land-Cover/Land-Use Change (LCLUC) Program
Pasture
Peatlands
Proximate Causes
Reforestation
Regrowth
Savannization
Secondary Vegetation
Siltation
Soil Degradation
Soil Erosion

Suburbanization
Sustainable Land Use
Urban Sprawl
Urbanization
Vegetation
Watershed
Wetlands

Land-Cover Products

AFRICOVER
Along Track Scanning Radiometers (ATSR) World Fire Atlas
BIOME 300 Project
Coordination of Information on the Environment (CORINE) Database
Earth Resources Observation Systems (EROS) Data Center
European Remote Sensing (ERS-1/-2) Satellites
Global Burnt Area 2000
Global Land Cover Map of the Year 2000
Global Land Cover Network
Global Terrestrial Observing System
GLOBCARBON
GLOBCOVER
GLOBICE
GLOBSCAR
GLOBWETLANDS
History Database of the Global Environment
Hot Spots of Land-Cover Change
IGBP-DIS Global 1-Km Land Cover Data Set
Sahel Land Cover
Tropical Ecosystem Environment Observations by Satellite (TREES) Project
United States Geological Survey (USGS) Program

Land-Use Dynamics

Access
Agricultural Frontier
Agricultural Intensification
Agricultural Revolution
Agrodiversity
Agroforestry
Agro-Industry
Arena of Land Conflict
Biomass
BIOME 300 Project
Biotechnology
Cattle Ranching
Colonization
Common-Pool Resources
Community Involvement
Conservation
Consumption
Contract Farming
Conversion
Corporate Strategies
Cotton
Cultural Factors
Cumulative Change
Decentralization
Decision-Making
Deforestation
Degradation Narrative
Desertification
Disequilibrium Dynamics
Disintensification
Domestication
Driving Forces
Ecological Colonization
Economic Growth
Economic Liberalization
Economic Livelihood
Economic Restructuring
Exotic Species
Extensification
Fallow
Feedback
Fire
Food Crops
Food Security
Forest Anomalies
Forest Degradation
Forest Transition
Forestry
Globalization
Green Revolution
Green Wall Project
Human Appropriation of Net Primary Production
Human Impact on Terrestrial Ecosystems (HITE) Initiative
Indigenous Knowledge
Industrial Revolution
Industrialization
Initial Conditions
Institutions
Investments
Land Abandonment
Land Concentration
Land Privatization
Land Reform

Models and Scenarios

People

Programs, Projects, and Organizations

Earth System Science Partnership
George Perkins Marsh Institute
Global Environmental Change and Food Systems (GECAFS) Project
Global Land Cover Network
Global Land Project
Global Observation of Forest and Land Cover Dynamics
Global Terrestrial Observing System
Green Wall Project
Human Impact on Terrestrial Ecosystems (HITE) Initiative
Integrated Land Ecosystem-Atmosphere Processes Study
Intergovernmental Panel on Climate Change
International Food Policy Research Institute
International Geographical Union (IGU) Commission on Land-Use and Land-Cover Change
International Geosphere-Biosphere Programme
International Human Dimensions Programme on Global Environmental Change
International Institute for Applied Systems Analysis
Joint Research Centre (JRC) of the European Commission
Land Degradation Assessment in Drylands (LADA) Project
LANDFIRE Project
Land-Ocean Interactions in the Coastal Zone (LOICZ) Project
Land-Use/Cover Change (LUCC) Project
Large-Scale Biosphere-Atmosphere Experiment in Amazonia
Millennium Ecosystem Assessment
NASA Land-Cover/Land-Use Change (LCLUC) Program
National Institute of Public Health and the Environment
Past Global Changes (PAGES) Project
Population-Environment Research Network
Project on People, Land Management and Ecosystem Conservation
Tropical Ecosystem Environment Observations by Satellite (TREES) Project
United Nations Environment Programme
United Nations Food and Agriculture Organization
United States Geological Survey (USGS) Program
World Agroforestry Centre

Rangeland Change

Access
Agricultural Frontier
Amazonia
Arc of Deforestation
Cattle Ranching
Common-Pool Resources
Conversion
Desertification
Disequilibrium Dynamics
Grassland
Great Plains
Hot Spots of Land-Cover Change
Land Concentration
Land Degradation Assessment in Drylands (LADA) Project
Large-Scale Biosphere-Atmosphere Experiment in Amazonia
Modification
Pastoral Mobility
Pasture
Resilience
Savannization
South America
Southern Africa
Sudan-Sahel
Tragedy of the Commons
Transhumance
United Nations Convention to Combat Desertification

Satellite Imagery

Advanced Very High Resolution Radiometer
AFRICOVER
Airborne Light Detection and Ranging
Albedo
Along Track Scanning Radiometers (ATSR) World Fire Atlas
Center for International Earth Science Information Network
Center for Sustainability and the Global Environment
Center for the Study of Institutions, Population and Environmental Change
Change Detection
Continuous Data
Coordination of Information on the Environment (CORINE) Database

Discrete Data
Earth Resources Observation Systems (EROS) Data Center
ENVISAT ASAR
European Remote Sensing (ERS-1/-2) Satellites
Georeferencing
Global Environmental Change and Food Systems (GECAFS) Project
Global Land Cover Network
Global Observation of Forest and Land Cover Dynamics
Global Positioning System
Global Terrestrial Observing System
Hot Spot Identification
IKONOS
JERS Radar Mosaics
Joint Research Centre (JRC) of the European Commission
Land Cover Classification System
Landsat
Medium Resolution Imaging Spectrometer
Moderate Resolution Imaging Spectroradiometer
NASA Land-Cover/Land-Use Change (LCLUC) Program
Normalized Difference Vegetation Index
Pattern Metrics
Pattern to Process
Pixel
Pixelizing the Social
Remote Sensing
Socializing the Pixel
SPOT Vegetation
TERRA-ASTER
Thermal Band Analysis
Transition Matrix
Tropical Dry Forest
Tropical Ecosystem Environment Observations by Satellite (TREES) Project
United States Geological Survey (USGS) Program
Validation

Urban-Industrial Change

Anthropocene
Atmosphere-Land Interlinkages
Carbon Cycle
Cascades
Consumption
Corporate Strategies
Economic Growth
Economic Restructuring
Forest Transition
Globalization
Greenhouse Gases
Industrial Revolution
Industrialization
Institutions
Japan
Kyoto Protocol
Land-Use Transition
Metabolism
Mineral Extraction
Population Dynamics
Remittances Landscape
Suburbanization
Syndrome
Systemic Change
Technological-Scientific Revolution
Urbanization
Urban Sprawl
Urban-Rural Fringe
Western Europe

Preface

Our Earth's Changing Land: An Encyclopedia of Land-Use and Land-Cover Change serves as a reference tool for university, college, and high school students, as well as others—including scientists and educated laypeople—who want to gain a better understanding of land change as a forcing function in global environmental change. There is a long history of studies of human impacts upon the earth's surface and, equally, of environmental changes that impact humans. However, it is remarkable that our understanding has intensified over the last decade, involving contributions from both the social and natural sciences. Concerns emerged in the mid-1970s with the realization that land surface processes influenced climate. It was recognized that land-cover change modified the exchange between the land surface and the atmosphere, which impacted regional climates. In the early 1980s, terrestrial ecosystems were highlighted as sources and sinks of carbon, which underscored the impact of land-use/cover change on the global climate via the carbon cycle. Later on, the contribution of local evapotranspiration to the water cycle, driven by land-use/cover change, highlighted yet another considerable climatic impact. A much broader range of impacts of land-use/cover change on ecosystem goods and services was further identified. Of primary concern are impacts on biotic diversity worldwide, soil degradation, and the ability of biological systems to support human needs. Land-use/cover changes also determine, in part, the vulnerability of places and people to climatic, economic, and sociopolitical perturbations. When aggregated globally, land-use/cover changes significantly affect central aspects of earth's systemic functioning.

Today, there is increasing awareness that some ancient societies—such as the Anasazi in the U.S. Southwest, the Mayas in Mesoamerica, and Easter Island in Polynesia—collapsed due to specific conditions in land-use/cover change, most notably deforestation. Likewise, it has been recognized that, especially in irrigated dryland zones, the hydrological consequences of land-use/cover change are most negative and occasionally illustrate modern or industrial society's capacity to transform large coupled human-environment systems rapidly, such as in the disastrous case of the Aral Sea basin in Central Asia. All impacts are not negative, though, as many forms of land-use/cover change are associated with continuing increases in food and fiber production, resource-use efficiency, wealth, and human well-being. Therefore, a careful consideration of "tradeoffs," a crucial entry in this book, is needed between impacts that are perceived to be "negative" and those highlighted as "positive." This reference tool aims at providing immediate and informed key access to such considerations.

There is good indication that a great deal of the intensified knowledge of land-use/cover change has been due to the International Geosphere-Biosphere

Programme (IGBP) and the International Human Dimensions Programme (IHDP) on Global Environmental Change, both established by the International Council of Scientific Unions (ICSU). In 1995, the programs for the first time jointly initiated and put on track the Land-Use/Cover Change (LUCC) project, which operated until 2005. This encyclopedia presents crucial terms and topics as they emerged over one decade of land-use/cover change research under the umbrella of the LUCC project. Most importantly, a difference has been made between land cover and land use that pervades the logic of most of the entries: land cover relates to the measurable properties of the biophysical state of the earth's surface (and its immediate subsurface); while land use involves both the manner in which biophysical attributes of land are manipulated and the purpose or intent of that manipulation. The latter turns land-use/cover change into a veritable issue for public policy (another entry), and is a reminder that the use of land is a highly political activity and sometimes—if not in most cases—an arena of conflict.

While it is still frustratingly difficult to separate out the effects of "natural noise" from human impact in terrestrial "global environmental change," biophysical and social factors often combine to create a particular effect that has been the heart of, and also the reason for, the notable success of LUCC research in the past. It came as no surprise that dealing with factors that drove land-use/cover changes led to the initiation in 2005 of further projects on global land change, global land-atmosphere interactions, and global land-ocean interactions, all of them jointly sponsored by IGBP and IHDP under the umbrella of the Earth System Science Partnership (ESS-P) (and all of them covered by entries in the encyclopedia).

Apart from the study of interlinked human and natural factors, an important feature of the recent, intensified research on land-use/cover change has been that these studies have become increasingly global in scope. The subtitle of LUCC's final synthesis book (*Local Processes, Global Impacts*), edited by Éric Lambin and me, points this out. This book can serve as a major source of information for those wishing to dive even deeper into the concerns of land-use/cover change (see the bibliography for this book). The global notion is especially true for some land use-related climate impacts, but it is also true for the ramifications of some processes, such as agricultural intensification, that link human and natural systems increasingly on the global level. In the late twentieth century, global change truly led to remapping the world, and globalization (see entry) has become a unifying theme, because its many processes amplify or attenuate land-use/cover change by removing regional barriers and strengthening global (at the expense of local to national) connections.

Given the importance of coupled social-biophysical phenomena at a whole range of temporal and spatial scales (see entry), and given the rapidity of intellectual thoughts since LUCC started in 1995, several other strains of LUCC or LUCC-related research have developed, all of them related to the vision of LUCC's science plan: the LUCC group of the International Geographical Union (IGU), ESS-P activities under the auspices of the Global Land Project (GLP), the U.S. National Aeronautics and Space Administration's (NASA) Land Cover Land Use Change (LCLUC) program, LUCC

China, and the Global Observation of Forest and Land Cover Dynamics (GOFC-GOLD) group—just to name the big five. *Our Earth's Changing Land* profited tremendously from the nearly 100 contributors with backgrounds in at least one of these five programs. In particular, I cannot forget the immediate commitment and enthusiasm of so many scientists from IGU—LUCC, LUCC-endorsed, and otherwise affiliated projects. These colleagues sometimes truly persuaded me to take up and continue with editorial work, since it was a particular encyclopedia on land change that was so desperately needed in teaching and informing a larger group of interested people. I am grateful to all of them—and I won most of them for contributions to this work.

The encyclopedia covers key terms and topics of land change in an alphabetical listing. For those who wish to get started, the entry "Land Change as a Forcing Function in Global Environmental Change" is recommended as an introductory reading. There are a few longer entries (more than 3,500 words) that deal with four major, broad issues of the earth's changing land: the causation and history of land use ("Land-Use Change in a Global, Historical Perspective"), the dynamics of terrestrial surface cover and related issues of detecting them ("Land-Cover Change"), the dynamics of land change and how to model it ("Modeling"), and some selected impacts of land-use/cover change ("Climate Impacts" and "Water-Land Interlinkages"). The first three are located in the chapter titled "Introductory Essays." These anchor entries include numerous cross-references (indicated by bold letters) that point to other entries that deepen our understanding of land-use dynamics, land-cover change, modeling, and impacts. In addition, almost every entry is accompanied by a bibliographical listing called "Further Reading," which provides the reader with sources for more up-to-date research and study. The Further Reading section also provides the reader with Web Sites where suitable. Most of the more than 300 entries are medium-length (about 1,000 to 2,500 words) and are designed to deepen the reader's understanding of land-change issues raised in the longer entries. The encyclopedia is not limited to definitions and topics of discussion but also includes biographies, histories, and abundant examples; that is, profiles of people, organizations, and journals central to the study of terrestrial land change. These are found in shorter entries (of fewer than 1,000 words).

Conception and execution of the encyclopedia owes a great deal to Kevin Downing, who contacted the LUCC project of IGBP and IHDP as early as 2000 for gathering inputs into a work already provisionally titled *Our Earth's Changing Land*, and who finally convinced me to get started. As editor and contributor alike, I know all too well my indebtedness to various collaborators and supporters, in particular Éric Lambin, the chair of the LUCC project, for drawing lots of inspiration from jointly published work (also acknowledging the support of the Belgian Federal Science Policy Office). Web services such as the forest policy expert listserver of the Center for International Forestry Research, managed by David Kaimowitz, delivered up-to-date information. The publication project has benefited from numerous comments by scientists who cannot all be named here, but I am aware, for example, that John Townshend, the chair of GOFC-GOLD, reviewed and approved the entry written on his group (see above). Special thanks go to

Heike, Milena, and Klara for their understanding and support, in particular during the last work-intensive year before completion of the encyclopedia.

Over the last decade, a huge and still diverse field of issues and research has opened, and *Our Earth's Changing Land* hopefully informs and guides those who seek to better understand and assess the tradeoffs inherent in our earth's changing land.

Helmut Geist
Erfurt, Germany
April 2005

Introductory Essays

Land Change as a Forcing Function in Global Environmental Change

Human-induced changes in the land surface of the earth, intentional or not, are pivotal to a wide range of changes in the earth's structure and function; also known as **global environmental change**. Change in land cover (biophysical conditions of the land) contributes significantly to greenhouse gas emissions, especially carbon dioxide, methane, and water vapor; and it directly affects the earth's heat balance through changes in **albedo** (reflectivity) and the hydrologic cycle, including local and regional precipitation impacts. These changes also fragment ecosystems and **watersheds** and trigger **biodiversity** losses, species invasions, **soil erosion**, and sedimentation. Land use (the intent to which land covers are managed), in turn, usurps significant proportions of net primary productivity, releases large amounts of nitrogen through fertilizer use, and draws down surface and subsurface waterstocks. Likewise, urban-industrial land use not only generates extra-urban heat regions but also leads to large-scale tropospheric pollution that threatens crop production and **human health**. Recognition of these varied and significant forcing functions has elevated land-change science to the forefront of global-change, earth-system, and sustainability themes—attested to by such international science efforts as the **International Geosphere-Biosphere Programme**, the **International Human Dimensions Programme on Global Environmental Change**, the **Millennium Ecosystems Assessment**, and **DIVERSITAS**.

CUMULATIVE AND SYSTEMIC LAND CHANGES

The most ancient global-scale forcing from land use led to the extinction of megafauna at the end of the Pleistocene era. The mastery of **fire** for the hunt and the repeated ecosystem changes it triggered across the globe apparently joined climate and other forcing functions to eliminate the mastodon, saber-toothed tiger, camelids, and giant sloth, among other "giant" fauna. This linkage is increasingly documented in the prehistoric record (e.g., 46,000 years ago in Australia), supported by modeling exercises indicating that the megafauna die-off is difficult to explain without the human fire component.

Fire also played a role in expansion of cultivated lands at the expense of **forest** and **grasslands** for grazing at the expense of wooded savanna. Indeed, the scale of this activity has been so large throughout human history that the impacts from human burning constitute the first **cumulative change** to reach systemic status (affecting biogeochemical cycling), having significantly

The classical photograph of the earth (showing Africa, the Arabian peninsula and the south polar ice cap) as taken on December 7, 1972, and as seen by the Apollo 17 crew traveling toward the moon.

PHOTO: Apollo. CREDITS: Image Science and Analysis Laboratory, NASA-Johnson Space Center, 18 Mar. 2005 (Earth from Space—Image Information). http://earth.jsc.nasa.gov/sseop.efs

affected atmospheric carbon dioxide before the twentieth century, with some claims pushing this impact back thousands of years ago. Today, **deforestation** and **biomass** burning are about one-fifth of the total flux in carbon.

By the end of the twentieth century, irrigation, livestock, and fertilizer use joined deforestation as forcing functions with systemic consequences. Increases in irrigation, especially for wet **rice** cultivation; ruminant livestock (**pasture** and feedlots); biomass burning; landfills; and **wetland** reclamation account for more than half of the current emissions of methane to the atmosphere, which now stand at 1,700 parts per billion compared to 670 parts per billion in pre-industrial times. The production of nitrogen fertilizer (especially ammonia) is the largest human-driven nitrogen (N) flux, followed by the cultivation of leguminous crops (e.g., soybeans). Biomass burning emits N-gases (e.g., NO_x and N_2O) to the atmosphere, and burning by humans is estimated to add about 10 Tg N^{yr-1}.

As important as these and other systemic consequences are, land change is an even more powerful forcing function in altering the states of the earth, in many cases accumulating to global-scale impacts. Land-use changes disrupt ecosystem functions through fragmentation and edge effects on biota; soil loss by wind and water erosion; water withdrawals from aquifers, wetlands, lakes, and rivers; and nutrient release, especially nitrogen, in water bodies, in turn creating hypoxia (dead zones in water). In addition, these changes may affect regional climate. For example, land change, especially tropical deforestation, is claimed by some to be leading to a "mass extinction" of biota. About 12.5 percent of the known 270,000 plant species may be threatened by extinction, an estimate significantly larger than the ambient extinction rates and generated largely by human-induced land changes. In another example, recent empirical and modeling work indicate that regional-scale land changes in some environments may reduce regional rainfall significantly, as in the case of southern Florida with massive urban growth and water withdrawals from the adjacent Everglades, primarily for agriculture. In most cases examined to date, the degradation or loss of native **vegetation**, if sufficiently expansive, has lowered regional precipitation beyond that accounted for by global climate change. Global change science has thus begun to unravel various land-climate myths, such as the idea that "rainfall follows the plow." Overall, the range, complexity, and consequences of land change are such that the notion of earth system services (e.g., ecosystem and other services) has gained currency, as has the means of valuing those services that are otherwise not accounted for by common market indicators (e.g., loss of pollinators in agricultural systems).

Land as a Forcing Function in Human History: Examples

Land change has played an important role in the history of human-environment relationships, characterized by increasing manipulation and "control" of nature, largely with the provisioning of food and resources. The inequalities of, and waste in, production and **consumption** that have marked all higher levels of social order notwithstanding, this history is one in which humankind has sought to increase the production of food, fiber, and fuel (historically, biomass fuel) while simultaneously lessening the risk of production shortfalls from natural and other hazards. For the most part, this history is marked by attempts to overcome the vagaries (e.g., drought) and limits (e.g., soil nutrients) of nature through the substitutes of labor and technology. In terms of the sheer number of our species, these efforts have been hugely successful on the global scale and over the long haul. Many examples exist of high-performing cultivation sustained over centuries, even millennia (e.g., wet rice in **China** and irrigation in Java). Human history is littered, however, with collapses of production systems triggered by inappropriate land uses, often combined with climate and political-economic change. Increasingly, the successes and failures of **land-use systems** partly reside in external forcing functions—those foreign to the location of impacts—that change the **access** and governance of those systems.

Ecosystems have long been burned to support food production. In parts of Africa, for example, pastoralists expanded and maintained savannas through annual burning to increase new vegetative growth for their livestock, leading to co-evolved landscapes. As these systems became increasingly agropastoral in kind, early dry season burning expanded to reduce the risk of catastrophic fires that could occur at the end of the dry season when large biomass buildup was ripe for ignition from lightning. The scale of these activities in western Africa reached such proportions in the middle to late twentieth century (coinciding with extensive drought and **land degradation**) that some international programs pointed to burning as the base cause of **desertification**. Supported by international donor agencies, African states sought to reduce landscape burning. The impacts of this regulation not only affected local production capacity, but they increased the exposure of settlements and farms to high-intensity fires and challenged the integrity of landscapes that required frequent burning.

Humankind has long used irrigation successfully to enhance food production, especially in regions of minimal or erratic precipitation. Irrigation systems in arid environments are particularly sensitive; and if not used and maintained correctly, they can rapidly degrade. The degradation of these systems in Mesopotamia, for example, is linked to population and state collapses. Large-scale and inappropriate irrigation in the Tigris-Euphrates watershed triggered significant **salinization** and waterlogging as early as 1700 BC, leading to the abandonment of formerly cultivated lands, a process also linked to diminished levels of state power and population, until such time that new irrigation systems were constructed. The synergy of the relationship consequences between land use and society is well established, if not always clear in its causal pathways: from land degradation to society disruption, or from societal disruption to land degradation. Such problems persist throughout the world today, despite advances in technology. At least one estimate holds that up to 37 percent of the total area brought into irrigation over the last 300 years has become unusable.

The forcing functions of land use may have broader regional impacts than the Mesopotamian irrigation example. The collapse and depopulation of the classic period lowland Maya of the southern Yucatán peninsular region is a case in point. By the ninth-century AD, the lowland Maya had denuded much of their tropical forest homelands for food, fiber, and fuel; and their agricultural systems required intensive labor inputs to retain soil fertility and to fight crop pest-diseases, and perhaps invasive species of fern. A large Mayan population was confronted with an extended winter dry season, which, in a karstic landscape, placed a premium on water storage. Production and occupation in these conditions would have been constrained significantly in the face of increasing aridity; these are precisely the desiccating conditions that may have taken place in the climate system, perhaps reinforced by local land-atmosphere dynamics owing to the denuding of regional forests. Such regionwide landscape changes triggering local climate change or reinforcing changes underway in the large climate system have been demonstrated recently among global change scientists.

Perhaps the most dramatic cumulative land change of the twentieth century is that of the "death" of the Aral Sea in Central Asia. Formerly the fourth-largest freshwater ecosystem in the world, this inland water body desiccated from the incremental buildup of irrigated land along the only two rivers feeding it. Begun in the 1960s, more than 5 million hectares of irrigation constructed by the late 1980s effectively cut off discharge to the sea, which lost 60 percent of its volume and 50 percent of its area, and more than doubled its salinity. The ecosystem, which once sustained a lucrative fishing industry, is dead for all practical purposes. In addition, inappropriate use of fertilizers and pesticides has contaminated local water, generating significant human health problems. The Aral Sea catastrophe is one of the twentieth century's few examples of the near-collapse of a regional-scale human-environment system.

MAJOR LAND CHANGES WORLDWIDE

Perhaps 50 percent of the ice-free surface of the earth has incurred significant human-induced land-cover change, involving change from one cover state to another (e.g., forest to cultivated lands) or substantial alterations in the conditions of an existing state (e.g., co-evolved grasslands). The land uses triggering these changes have been estimated to usurp as much as 50 percent of earth's net primary productivity, although the large error boundaries in the estimates warrant considerable caution.

No land change has received more attention of late than tropical deforestation. While open to various charges of Western or Northern Hemisphere biases—after all, long-term, significant deforestation took place in those areas—attention to tropical deforestation follows from special characteristics and functions of tropical forest **biomes**: the amount of biodiversity and biomass contained in them, and the critical role of tropical forests in the maintenance of the Walker circulation and thus the transfer of energy from the tropics poleward, a driver of global climate. Recent annual rates of tropical deforestation are high but variable, owing to varying spatiotemporal dynamics that create forest losses and recovery, and to different measures and metrics used to define forests and their change. Estimates from satellite images, which provide internal consistency in their measures and metrics, provide deforestation (total amount cut) rates of 1.3–4.4 percent and net change (deforestation minus **regrowth**) rates of 0.3–3.6 percent. The process of tropical deforestation is tied into a myriad of factors often residing outside the control of the local inhabitants and, in part, tied in complex ways to the forestation of much of the mid-latitudes.

Second only to tropical deforestation in importance, and intimately tied to it, are the expansions of **agriculture**—on the order of a 400 percent increase in area since the beginning of plant **domestication**—and **agricultural intensification**. Much deforestation is currently undertaken for the expansion of livestock—in the Americas often involving the introduction of African grasses—and cultivation in the tropical world. In contrast, marginal agricultural lands in the world of the "haves" are being taken out of production, fostering major forestation in some regions, such as the northeastern **United States of America** and **western Europe**, while fossil fuel–based inputs intensify production on better lands. This intensification triggers significant impacts through the release of nitrogen from fertilizer throughout watersheds and ecosystems, and the release of the **greenhouse gas** methane from increased irrigation and ruminant livestock.

Land degradation remains an acute problem globally. In its broadest sense, land degradation is the drawdown in the biophysical stocks of the land, and it affects ecosystem function and thus land-use production. These stocks include biomass, soil, soil nutrients, and moisture capacity, among other factors; and they affect regional precipitation, vegetation regeneration, biodiversity, and a large array of **ecosystem services**. For example, agriculture and other land uses have depleted the pool of soil organic carbon by 60 percent and 75 percent in the temperate and tropical regions, respectively. Most of this carbon is released to the atmosphere, contributing to global warming, while reducing agricultural productivity and increasing costs to farmers to offset these loses. Land degradation also involves loss of soil to wind and water erosion, with its concomitant impacts on runoff, sediment, and aquatic systems. Alternatively, water withdrawal for cultivation literally prevents the flow of several major rivers around the world from reaching the sea, affecting coastal marine ecosystems significantly.

The growth in the worldwide agglomeration of settlements—**urbanization** and megalopolises—increasingly becomes a land change of environmental consequences. The total area covered by the impervious surfaces of settlement is small compared to such land changes as deforestation or agriculture. The urban **ecological footprint** is large, however, demanding considerable space elsewhere for production and consumption to support the agglomeration in the first place. Some of the environmental consequences that have been registered of late include watershed-scale flooding, witnessed in central Europe in 2002, and extra-urban "heat islands" that affect leafing and leaf loss in deciduous forest zones. Looming as a larger footprint, however, is regionwide pollution of the troposphere, leading to human and crop health problems. Three heavily fertilized and industrialized regions—the eastern United States, western Europe, and eastern China—that make up 23 percent of the world's land area and produce about 60 percent of global food stocks produce a majority of NO_x emissions, giving rise to photochemical smog, including O_3 (ozone), in the troposphere. Ozone at this level in the atmosphere is harmful to plant production, potentially reducing the regional productivity of agriculture. At least one estimate for China indicates that crop production is reduced 5–30 percent from optimal yields due to haze from air pollution alone.

THE CAUSES OF LAND CHANGE

These few examples illustrate the wide array of forcing functions that land change entertains on the earth system. While the examples focus on global and large regional-scale consequences, in fact land change is pivotal at all spatiotemporal **scales** of analysis, from the ecosystem over the decades to the earth system over the millennia. For the most part, land change is undertaken incrementally at the micro-spatiotemporal scale, complete with ecosystem-scale and regional-scale consequences and **feedbacks** on the land use, while the totality of similar land changes across the world (e.g., global deforestation for agriculture) ultimately contributes to earth system changes (e.g., climate change) with global feedbacks. Compounding these scalar dynamics are the influences of **exogenous factors** or forces (e.g., markets, state policies, international treaties, absentee landlords) that affect land-use outcomes.

The complexity of these linkages has led the research community to address land change in terms of proximate and distal or underlying "drivers" of change. **Proximate causes** or drivers are those most immediately linked to activity engendering the land change in question: for example, cutting and burning the biomass of the forest to create cropland; draining a wetland for urban settlement; or annual burning of grasslands to reduce growth of woody species and rejuvenate grasses for livestock. Distal drivers or underlying **driving forces** usually operate through nested linkages (e.g., national markets, middlemen, and farmers; state policies, local land **institutions**, and farmers; international corporations, absentee landlords, and land managers) in which decision factors flow from afar to the place of use. In the Aral Sea case, the decision to invest in irrigation (proximate cause) was made by those in control of the former Soviet state (distal cause), who passed on the responsibility to construct and control the system to engineers and bureaucrats, the last of whom established the rules of access to irrigation and the metrics of rewards for agricultural production that contributed to the malfunctioning of the human-environment system. The complexity and linkages involved have proven an obstacle to the development of land-change theory to guide in forcing function models.

See also: Atmosphere-Land Interlinkages; Boreal Zone; Carbon Cycle; Change Detection; Climate Impacts; Coastal Zone; Colonization; Decision-Making; Degradation Narrative; Global Land Project; Hot Spots of Land-Cover Change; Industrialization; Integrated Model to Assess the Global Environment; Kyoto Protocol; Land Abandonment; Land-Use History; Land-Use/Cover Change (LUCC) Project; Marsh, George Perkins; Millennium Ecosystem Assessment; Modeling; Nutrient Cycle; Past Global Changes (PAGES) Project; People at Risk; Property; Reforestation; Remote Sensing; Species Extinction; Systemic Change; Tropical Humid Forest; Water-Land Interlinkages.

Further Reading

Foley, Jonathan A., DeFries, Ruth, Asner, Gregory P., Barford, Carol, Bonan, Gordon, Carpenter, Stephen R., Chapin, F. Stuart, Coe, Michael T., Daily, Gretchen C., Gibbs, Holly K., et al., "Global Consequences of Land Use," *Science* 309 (2005): 570–74;

Gutman, Garik, Janetos, Anthony C., Justice, Chris O., Moran, Emilio F., Mustard, John F., Rindfuss, Ronald R., Skole, David, Turner, B. L. II, and Mark A. Cochrane, eds., *Land Change Science: Observing, Monitoring and Understanding Trajectories of Change on the Earth's Surface*, Dordrecht, NL, Boston, London: Kluwer Academic Publishers, 2004;

Johnson, Douglas L., and Lawrence A. Lewis, *Land Degradation: Creation and Destruction*, Oxford, UK: Blackwell, 1995;

Kalnay, Eugenia, and Ming Cai, "Impact of Urbanization and Land-Use on Climate," *Nature* 423 (2003): 528–31;

Kasperson, Jeanne X., Kasperson, Roger E., and B. L. Turner, II, eds., *Regions at Risk: Comparisons of Threatened Environments*, Tokyo: United Nations University Press, 1995;

Lambin, Éric F. and Helmut Geist, eds., *Land Use and Land Cover Change: Local Processes, Global Impacts*. Berlin: Springer-Verlag, 2006;

Perera, Judith, "A Sea Turns to Dust," *New Scientist* 140 (1993): 24–27;

Ramankutty, Navin, and Jonathan A. Foley, "Estimating Historical Changes in Global Land Cover: Croplands from 1700 to 1992," *Global Biogeochemical Cycles* 13 (1999): 997–1027;

Redman, Charles L., *Human Impact on Ancient Environments*, Tucson, AZ: University of Arizona Press, 1999;

Steffen, Will, Sanderson, Angelina, Tyson, Peter, Jäger, Jill, Matson, Pamela, Moore, Berrien III, Oldfield, Frank, et al., eds., *Global Change and the Earth System: A Planet under Pressure*, Berlin: Springer-Verlag, 2004;

Turner, B. L. II, and Susannah McCandless, "How Humankind Came to Rival Nature: A Brief History of the Human-Environment Condition and the Lessons Learned," in *Earth System Analysis for Sustainability*, eds. Clark, William C., Crutzen, Paul, and Schellnhuber, H. John, Cambridge, MA: MIT Press, 2005, pp. 227–43;

Watson, Robert T., Noble, Ian R., Bolin, Bert, Ravindranath, N. H., Verardo, D. J., and D. J. Dokken, *Land Use, Land-Use Change and Forestry: Special Report of the IPCC*, NY: Cambridge University Press, 2001.

B. L. TURNER II

Land-Use Change in a Global, Historical Perspective

Land is necessary for mankind to survive, and humans have been using and changing their surrounding landscape as long as they have inhabited the planet. The pace and rate of change, however, have increased dramatically over the last centuries. Johan Goudsblom and Bert de Vries distinguish several periods in mankind's evolution—from using **fire** as a tool, to the period of "agrarization" with the **domestication** of plants and animals, and later to the so-called **Anthropocene**, with almost complete control of humans over their environment. One cannot fully understand current land use in the world without looking back to the past. What processes lie underneath the different land use practices all over the world, and in what stage of development are they? Will the world be one big cropland in the next few decades to feed all earth's people, or are we somewhere else on the agricultural expansion timeline? In order to address these questions, one has to first look a bit further back in time than the last 300 years, because land-use changes already had occurred long before the start of the **Industrial Revolution** at the beginning of the eighteenth century.

The scale and intensity of the land-use changes have not always been the same. Obviously, there is a strong relationship with the total number of people living in a region. Global population numbers have been relatively low for a long time. The total global estimates for 1700 range from 410–679 million people, with a best guess of 587 million. Then, in large parts of **Europe**, there was a great, sustained fall in mortality accompanied by high birth rates. This resulted in a large excess of births, thus producing a high natural increase of the population. Changes in **agriculture** and industrialism triggered emigration within Europe and migration toward the New World.

This out-migration to the New World proved to have considerable consequences for the landscape of the new lands. Although in some regions historical population densities already had been relatively high for a long time period, a lot of continents were virtually empty in the seventeenth century. This often resulted in massive **conversion** and/or removal of the native vegetation cover in the need for agricultural land.

THE BEGINNING: FROM HUNTERS AND GATHERERS TO THE FIRST FARMERS (10,000–500 BC)

The period from roughly 10,000 BC to 500 BC is known as the Neolithic period. Caused by climate change and/or overpopulation, humans were forced to find better ways to improve or secure their food supply. They evolved toward a more sedentary existence and adapted new crops and some domestication of animals as well. The ox-drawn plow was introduced and horses were domesticated by 4000 BC, innovations in metallurgy boosted the use of tools (Bronze and Iron ages), and water buffaloes were domesticated by 3000 BC. Irrigation became widespread in Eurasia between 4000 and 2000 BC. By 2000 BC, the major **food crops** of today were already cultivated. This Neolithic **land-use transition** from (mobile) hunting/gathering to (sedentary) farming led to an increasing population over time, albeit a very modest increase over thousands of years. Most of the agricultural activities seemed to have developed along mountain ridges, hills, and river valleys. Examples of these are the Iranian Plateau and its edges with traces of farming from 10,000–4000 BC, regions in northern **China** (Yellow and Yangtze rivers), the Indus and Ganges river basins in India, and parts of Central America.

THE CLASSICAL PERIOD (500 BC–AD 500)

The Mediterranean forests during the Roman era (500 BC–AD 500) were the main source of fuel, shipbuilding, and war material. They had to be cleared before agriculture could be practiced. **Deforestation** started around the city-states and spread throughout the Greek colonies, to Africa, and to other European Roman provinces. Forests were also the main fuels for domestic and industrial purposes (more than 90 percent of the total wood demand was for fuel), mostly in the form of charcoal. Because charcoal could be more easily transported, it allowed people to exploit more distant forests instead of nearby town/village forests, which were overexploited quite soon as town population numbers increased. When forests were exhausted near villages, people turned to nearby rivers and seaports as means of transport for wood products from more distant forests. Inaccessible forested **mountain** areas were likely to remain intact.

Mineral extraction was another activity that led to increasing deforestation. For example, in ancient Cyprus, copper and silver smelting required considerable amounts of wood, which at first was abundant in the densely forested plains. This led to an expansion of cultivated area, especially since the forests were not protected. As noted by Strabo of Amasia in AD 19, "They allowed anyone who was willing and able to fell trees to adopt the land that was thus won as their own property, without having to pay any taxes."

THE MIDDLE AGES (AD 500–1500)

Unlike the progress and expansion of agriculture in Europe during the classical period, the Middle Ages were quite the opposite. Also known as the "Dark Middle Ages," this period is synonymous with stagnation of several aspects of life. Religion left a heavy mark on daily life, and progress was not encouraged. On top of this, catastrophic events like the outbreak of epidemic diseases, a deteriorating climate (little ice ages), and economic recessions (due to numerous wars) had an opposite effect on forests. Population numbers declined staggeringly in many parts of the world; for example, 25–30 percent of people in central Europe died due to periods of the Black Plague, and the introduction of smallpox in the Americas by the first colonists resulted in regional mortality rates of 50–90 percent. Michael Williams also estimated that between one-fifth and a quarter of all settlements were abandoned across Europe, which lessened pressure on the forests so they could advance again. Nevertheless, even in the Middle Ages, forest exploitation was already highly planned and regulated by local authorities in Europe. Certain trees were favored (oaks were sometimes not used for wood but for acorns as food for pigs), and over-exploitation was avoided as much as possible by means of controlled grazing and/or strict laws.

Forests were under pressure in other parts of the world during this time period. In China, for example, scarce information is available about large-scale deforestation during the northern Sung period (AD 910–1126). The iron and steel industry was already flourishing during that time period (with estimated production numbers similar to those of total **western Europe** at the beginning of the eighteenth century). Thus, it is not hard to imagine the consequences this ferocious need for fuelwood must have had on the local forests. By the year 1300, Sung production had declined by half, but whether it was caused by shortage of fuelwood or other factors (such as Mongul invasions) remains unclear.

William Ruddiman argues that based on a wide array of evidence, the population density in China around AD 1 was 4.5 times higher than that of England at the Domesday survey in 1086. Knowing the high degree of deforestation in England at that time, one cannot avoid the conclusion that deforestation levels in China and other highly populated regions in Eurasia around AD 1 must have been at least comparable to those in England. Also in Central America, the landscape was already influenced by 500 BC by human agricultural activities. Local tribes had used the forest for centuries for hunting, collecting medicinal products and building materials, and small-scale agriculture. It resembled the shifting cultivation of the tropical forest of today. Original forestland was also cleared for limited agriculture and fuel needs. When the first Europeans arrived, relatively important sections of forestland had already been changed to more open, parklike vegetation.

THE BIG CHANGE (1500–PRESENT DAY)

Notwithstanding the severe local changes in land use in some parts of the world over the past millennia, the really big changes in global land use started with the launch of the Industrial Revolution at the end of the seventeenth century. By the end of the fifteenth century, the emergence of a global economy had already enhanced migration throughout Europe and later into the New World. New politics (imperialism) supported by religion and technology (new weaponry and improved ways of transport) enabled the old economies of Europe to explore and exploit similar natural environments as at home (i.e., temperate forests in Asiatic Russia and North America), and later also more exotic regions. The nineteenth and early twentieth centuries were the high point of large agricultural plantations in Asia and **Latin America**. Earlier successes in the Caribbean with sugar and **cotton** in North America created confidence and finances to look further into the New World. European owners mobilized local labor to clear lands in **tropical humid forests** in order to grow, for example, rubber in Malaysia, cacao in Africa, tea in India, **coffee** in Brazil, and bananas in Central America.

So, worldwide, except for some already densely populated Asian regions, the patterns of land-use and land-cover change followed the patterns of European settlement and economic development. With European settlement, intensified land-use activities rapidly spread through the colonized areas.

Europe

Around AD 1600, European society was essentially agrarian. Most of the people were engaged in **subsistence agriculture** and the production of food. A small increase in population created pressure in the countryside, and food shortages resulted in migration to the towns. In many regions, most of the suitable land was now occupied by agriculture, and efforts were made to improve yields through enclosures and enhanced fertility. The result of this was that in many countries hardly any **forest** was left. Lots of wood was needed in the prosperous mining and metal-smelting industries, as well as in the glass and salt production sectors. However, already during the sixteenth and seventeenth centuries, an intensive trade existed between central Europe and some western European ports. The European river-trading towns were major import centers for shipped goods from the Baltic region and their hinterland. Poland, Hungary, and Russia were important providers of timber and grain for many western European countries. This lessened the pressure on the local forests.

The first changes occurred in the United Kingdom during 1600–1750. Subsistence farming gave way to industry and was driven by innovation and technology, marking the start of the Industrial Revolution. The landscape changed dramatically by the improvement of the transportation infrastructure. The formation of railways and new roads provided the expanding towns with fresh food and raw materials. Also, an emerging world market introduced a more diverse diet to the people. Increasingly, however, the countryside was becoming a remote and distant environment to an industrialized society engaged primarily in trade. This process continued into the twentieth century, and by the 1960s, agriculture was transformed in many European countries to a highly mechanized sector. Improved technology, pesticide use, and ample fertilizer use increased food production in a huge way. Specialization took place, and farm size increased. However, by the end of the twentieth century in many highly industrialized countries, consumer demand for a more sustainable way of agricultural production had risen.

Kees Klein Goldewijk estimated the total cropland area in Western Europe in 1700 to be around 82 million hectares. Eastern Europe was comparable with **Russia** at that time, that is, 32 million hectares and 31 million hectares, respectively. Michael Williams presents estimates for the expansion of cropland in Russia for 1860 that range widely from 49 to 95 million hectares

(Kees Klein Goldewijk: 65 million in 1850), 113–208 million hectares in 1900 (Kees Klein Goldewijk: 92 million), 109–259 million hectares in 1930, and 196–369 million hectares in 1960 (Kees Klein Goldewijk: 128 million in 1950).

Asia and Australia

Asia is unique in many ways concerning **land-use history**. Parts of Asia have been (densely) populated for thousands of years, yet a more or less sustainable way of agriculture in the form of paddy **rice** cultivation flourished for a long time. Until the middle of the nineteenth century, this wet-rice cultivation remained uninfluenced by external forces. With the arrival of European colonists, rice became an export product, and new methods of irrigation farming were introduced (massive dams, lined canals). **Agricultural intensification** and commercial cropping became common practice, replacing shifting cultivation in the hills. Land became scarce, as did fuelwood. Land clearance uphill resulted in **soil erosion**, frequent landslides, and flash floods. Also, formerly sparsely populated areas in French Indochina, Burma, and Thailand were converted to rice production. Pasture is not a big issue in Asia, livestock is not very important, the area is often scarce, and all land is needed for permanent rice/**wheat** cultivation, which leaves no room for growing fodder.

John Richards and Elizabeth Flint presented a thirteen-country study (India, Sri Lanka, Bangladesh, Myanmar, Thailand, Laos, Cambodia, Vietnam, Malaysia, Brunei, Singapore, Indonesia, and the Philippines) for the period 1880–1980. In this whole area, forest/woodland and **wetlands** declined over these hundred years by 47 percent (131 million hectares). At the same time, the cultivated area increased by 106 million hectares, which is nearly a doubling since 1880. Thus, 81 percent of the lost forest and wetland **vegetation** appears to have been converted to agricultural land.

Malaysia experienced a tripling of the annual cropping area over a hundred-year period from 1880 to 1980. Rubber was planted in huge quantities, as well as oil palm plantations. The result is that more than 80 percent of the agricultural land is now occupied by perennial crops. Virtually all of the forest/woodland vegetation outside the protected areas on mainland Malaysia is now logged. Nowadays commercial timber exploitation for export in Sabah and Sarawak is responsible for massive deforestation on the island of Kalimantan.

Land use in Indonesia has two different faces. The islands of Java and Bali are very densely populated and supported by intensive paddy rice cultivation (sawah), upland field crops, and home gardens. Because of its rich population history, it was already largely deforested by 1880. Other islands such as Sumatra, Sulawesi, and Kalimantan were relatively sparsely populated and limitedly exploited before 1920. Commercial logging, especially in Kalimantan, started during the 1960s, helped with financial injections from abroad (Japan). Deforestation exploded during the 1980s when agricultural fires from **cash crop** farmers and many small slash-and-burn families went out of control, with an estimated loss of 3.5 million hectares of forests.

India, Bangladesh, and Sri Lanka experienced a staggering population growth over the last three centuries. Starting with about 150 million people around the year 1700, the numbers first slowly increased to 212 million in

1850, then accelerated to 427 million in 1950, and virtually exploded after 1950, reaching 1.166 billion in 2000, with still further increases. This has had severe impacts on agriculture. When technology is limited, the first response is often expansion of cropland. Sometimes intensification was achieved by introducing multiple cropping practices. Pressure on remaining forests increased, although India has a long history of protective forest laws. The individual right to exploit forests for fuel was functionally liberalized rather than restricted, causing a steady transformation of the existing woodland/forests to a lower **biomass** content. The enormous livestock population (183 million heads in 1880 and 377 million heads in 1980) added extra pressure on the remaining forest/woodlands. Degradation of biomass resources from overgrazing is a widespread phenomenon, especially in the drier regions.

China also has a long agricultural history. Michael Williams presents population estimates for China for the Ming dynasty in 1400, with 65–80 million people and a cultivated area of 25 million hectares. In 1770, this was 270 million people and 63 million hectares. Illustrative is the example of tea, as it was recognized as an economic opportunity by Europeans in the late seventeenth century. China opened up its borders and thus created an international market. The Dutch and British East Indian companies opened trade routes to the Manchu area. John Richards estimated the cultivated Manchu area in 1600 to be 33 (±7) million hectares, reaching 81 (±3) million hectares around the 1870s. Similar expansion of agricultural settlement and land clearance was strongly encouraged by the administration in Hunan province. During the late eighteenth century, some 3.2 million hectares were under cultivation, a doubling in less than forty years, according to John Richards. In the eighteenth century, intensification and commercial cropping became common practice, replacing shifting cultivation in the hills. Land became scarce, as did fuelwood. Land clearance uphill resulted in erosion, frequent landslides, and flash floods. After World War II, the cultivated area grew to 105 million hectares. It is estimated to be currently around 149–153 million hectares, according to the **United Nations Food and Agriculture Organization** and the China State Statistics Bureau, respectively.

Australia was already inhabited for thousands of years before the first European settlers arrived around 1788 in the southeastern part. However, the landscape was more or less untouched by the native population because of its nomadic lifestyle. The arriving Europeans introduced their agricultural habits and livestock, and immediately disputes over land ownership began. Expansion of the agricultural area took place mostly in the fertile and rainfed regions of the southeastern parts, a bit to the northeast, and in the southwestern Perth area. Estimates of cultivated area are 0.4–9.7 million hectares in 1860, 3–16 million in 1900, 10–22 million in 1930, and 12–35 million in 1960. Still, vast areas of inland Australia are hardly used for any agricultural purposes, although ranching exists in large parts in some regions. Impacts are very limited, though, because of the very low densities of livestock.

The Americas

The first European settlers arrived in the Americas at the beginning of the seventeenth century, initially along North America's eastern seashores, and gradually others followed. They started clearing quite extensive areas of forestland for timber exploitation, cultivation, and grazing purposes. The indigenous population was decimated by diseases introduced by the settlers, so that much of the land was depopulated and constituted **agricultural frontier** zones. Throughout the following decades, more and more pioneers arrived in the New World: Michael Williams estimates 0.25 million people in 1700, 1.6 million in 1760, 3.9 million in 1790, and 7 million in 1810. The cropland area in the **United States of America** increased from 3 million hectares in 1700 to 50 million in 1850. Furthermore, policy measures like the Homestead Act of 1862 (providing each settler 160 acres of free government land for cultivation over at least five years) led to a rapid settlement of public lands in the next decades. The total cropland area reached 179 million hectares in 1920 and 193 million hectares in 1950, decreasing slowly again to the 1920 level of 179 million hectares in 2000.

Not all the increase of cropland was at the expense of forests. A considerable area was converted from land other than forests. Old cultivated land from the Native Americans was used, as were vast areas of pristine prairie in the Midwest converted to cropland and **pasture**. Estimates for pasture range from 2.5 million hectares in 1700, 11.3 million in 1800, and 145 million in 1900, reaching a peak of 265 million hectares in 1960, and again decreasing toward the end of the twentieth century to 234 million hectares.

The forests of Latin America were already influenced centuries before the Europeans came. Highly civilized cultures in Middle America (Mayas, Aztecs, Olmecs, etc.) deforested some areas, but one after another collapsed, so that many areas were abandoned and left for **regrowth** again. The Spanish conquistadors, after arriving at the start of the sixteenth century, introduced diseases that took their toll on the indigenous population (sometimes killing up to 90 percent). This lessened the pressure on natural resources for quite a while. Therefore, it is generally accepted that serious land conversion occurred at the end of the nineteenth century only. Following the example of the successful sugar cane plantations in Cuba, many other plantations (such as rubber, coffee, and bananas) were established in the Latin American continent. In 1700, around 15.5 million hectares were converted to cropland (2.5 million in Central America and 13 million in **South America**), slowly increasing to 16.5 million in 1800 and 32 million in 1900. The expansion accelerated during the twentieth century: 79 million hectares in 1950 and 168 million hectares in 2000. However, even more forests were cleared for **cattle ranching**. Large areas were converted to pasture, especially in Brazil. Estimates for Latin America range from 37–92 million hectares in 1700, 149 million hectares in 1800, 193–202 million hectares in 1850, and 277 million hectares in 1900. The total pasture area increased rapidly in the twentieth century to 439 million hectares in 1950, 534 million in 1970, and 615 million in 2000.

Africa

There is not as much information on deforestation in the African continent as compared to other parts of the world. Some estimates of historical forest areas are listed by Alexander Mather for the Ivory Coast (14.5 million hectares in 1900 and 3.9 million hectares in 1980), Liberia (6.5 million hectares in 1920 and 2 million hectares in 1980), and Ghana (9.8 million hectares in 1920 and 1.7 million hectares in 1980). Intensive forest clearing is currently going on in Gabon and in the Congo basin. Clearance for plantations (cacao, coffee) occurred at a level that is generally not as high as in Southeast Asia.

A decline of the local iron industry during the colonial period in Togo in West Africa is an example of the reduced local demand for charcoal, which in its turn diminished the pressure on the remaining forest. But especially after World War II, the increasing population and livestock numbers in sub-Saharan Africa took their toll on the remaining woodlands and forests. The expansion of cropland (driven by population growth), illegal and destructive logging operations, overgrazing, and droughts caused a decrease of the total forest/woodland area of 735 million hectares in 1961 to 713 million hectares in 1994, a loss of 22 million hectares in just three decades.

WHERE ARE WE NOW?

The face of the earth has changed dramatically during the last three centuries. With Europe and parts of Asia already transformed into highly cultivated landscapes at the beginning of the eighteenth century, large parts of the world were colonized and brought under cultivation with often-devastating consequences for the original land cover. It started with massive deforestation and conversion of savanna/**grasslands** in the more temperate regions (United States, Australia, Argentina) and gradually shifted into the tropical regions (**Amazonia**, Central America, Africa, and Asia). Although the loss of forests and natural grasslands has been enormous, there has also been a substantial gain of especially forested land. In North America, parts of Europe, and **Japan**, the total forest area has increased during the last century. This reversal, the so-called **forest transition**, is marked first by a gradual or sudden depletion of forest resources, then later by slow or radical changes in economic structures or organization. Hopefully, this transition can also be encouraged in some tropical regions, in order to avoid the near-extinction of native forest, like some developed countries experienced in the past.

See also: Agricultural Revolution; Arena of Land Conflict; BIOME 300 Project; Colonization; Consumption; Driving Forces; Food Crops; Great Plains; Holocene; Hot Spots of Land-Cover Change; Human Appropriation of Net Primary Production; Industrialization; Land Abandonment; Land Rights; Population Dynamics; Pristine Myth; Proximate Causes; River Basin; Southeast Asia, Mainland.

Further Reading

Goudsblom, Johan, and Bert De Vries, eds., *Mappae Mundi, Humans and their Habitats in a Long-Term Socio-Ecological Perspective*, Amsterdam: Amsterdam University Press, 2002;

Houghton, Richard A., Lefkowitz, David S., and David L. Skole, "Changes in the Landscape of Latin America between 1850 and 1985, I: Progressive Loss of Forests," *Forest Ecology and Management* 38 (1991): 143–72;

Klein Goldewijk, Kees, "Estimating Global Land Use Change over the Past 300 Years: The HYDE Database," *Global Biogeochemical Cycles* 15 (2001): 417–33;

Mather, Alexander S., *Global Forest Resources*, London: Belhaven Press, 1990;

Ramankutty, Navin, and Jonathan A. Foley, "Characterizing Patterns of Global Land Use: An Analysis of Global Croplands Data," *Global Biogeochemical Cycles* 12 (1998): 667–85;

Ramankutty, Navin, and Jonathan A. Foley, "Estimating Historical Changes in Global Land Cover: Croplands from 1700 to 1992," *Global Biogeochemical Cycles* 13 (1999): 997–1027;

Redman, Charles L., *Human Impact on Ancient Environments*, Tucson, AZ: University of Arizona Press, 1999;

Richards, John F., "Land Transformation," in *The Earth as Transformed by Human Action: Global and Regional Changes in the Biosphere over the Past 300 Years*, eds. B. L. Turner II, William C. Clark, Robert W. Kates, John F. Richards, Jessica T. Mathews, and William B. Meyer, Cambridge, UK: Cambridge University Press, 1990, pp. 163–78;

Richards, John F., and Elizabeth P. Flint, *Historic Land Use and Carbon Estimates for South and Southeast Asia 1880–1980* (Publication No. 4174), Oak Ridge, TN: Carbon Dioxide Information Analysis Center, Oak Ridge National Laboratory, Environmental Sciences Division, 1994;

Ruddiman, William F., "The Anthropogenic Greenhouse Era Began Thousands of Years Ago," *Climatic Change* 61 (2003): 261–93;

Strabo 14.6.5 [AD 19], Strabo of Amasia, Greek historian and geographer, cited in The History of Cyprus in 90 Centuries, www.heureka.clara.net/sunrise/history.htm; UK Agriculture, Countryside [Online, May 2004], The UK Agriculture Web Site, www.ukagriculture.com;

United Nations Food and Agriculture Organization, FAOSTAT Database [Online, May 2004], The FAO Web Site, www.fao.org;

Williams, Michael, "Dark Ages and Dark Areas: Global Deforestation in the Deep Past," *Journal of Historical Geography* 26 (2000): 28–46;

Williams, Michael, *Deforesting the Earth, from Prehistory to Global Crisis*, London: University of Chicago Press, 2003;

Williams, Michael, "Forests," in *The Earth as Transformed by Human Action: Global and Regional Changes in the Biosphere over the Past 300 Years*, eds. B. L. Turner II, William C. Clark, Robert W. Kates, John F. Richards, Jessica T. Mathews, and William B. Meyer, Cambridge, UK: Cambridge University Press, 1990, pp. 179–202;

Wolman, Gordon, and F. G. A. Fournier, eds., *Land Transformation in Agriculture (SCOPE 32)*, Chichester, UK, New York: John Wiley & Sons, 1987.

KEES KLEIN GOLDEWIJK

Land-Cover Change

During the 1990s and the first half of this decade, significant research efforts have been geared toward better understanding land-use and land-cover changes at various geographical **scales**. Land use, and by extension land-use change, has been mapped in many regions and countries during the twentieth century; an early example being the Land Utilization Survey of the United Kingdom, which was coordinated by Sir Dudley Stamp in the 1920s and 1930s. However, such efforts have been piecemeal: not all countries or world regions have been mapped, and generally there has been no common approach in terms of analytical methodologies or land-use/land-cover classification systems. A major advance in adopting a common methodology and classification was the Europe-wide **Coordination of Information**

on the Environment (CORINE) database, though even in that case Great Britain produced a separate product.

As **global environmental change** grew to become the dominant research paradigm in the 1980s and 1990s, demand was created for global data sets of land use and land cover (and changes in land use and land cover) to meet the needs of the growing global science community in the following areas: global and regional climate modeling, biogeochemical cycling, water budgeting, **biodiversity** studies (specifically to analyze **species extinction** rates and evaluate the impacts of invasive species) and, more generally, in the arena of sustainable development. The few suitable global products that did exist until recent advances—for example, global land-use data for nine world regions by Richard Houghton—did not meet the requirements of spatially explicit **modeling**; that is, that they should be geo-located data and produced for regular intervals in time. In the context of global land-cover change driven by human activities, research has focused on two time scales: the last three centuries (1700–1990) and the last two decades (1981–2000).

The **History Database of the Global Environment**, produced by the Dutch **National Institute of Public Health and the Environment** (RIVM), facilitated the reconstruction of past land use (i.e., cropland, **pasture**, and natural **vegetation** types) using historical human population data. Similar research at the **Center for Sustainability and the Global Environment** (SAGE) at the University of Wisconsin–Madison has reconstructed historical distributions of cropland and potential **vegetation**. Together these are milestones in the search for geo-located global databases and maps at regular intervals in time. They both used multiple sources to create the databases and then produced land-use maps going back to 1700. Their research, carried out under the **Past Global Changes (PAGES) project** and the **Land-Use/Cover Change (LUCC) project**, was later combined with paleo-environmental research to create the **BIOME 300 project** or initiative. One aim of this was to reconstruct past land-cover products at a spatial resolution of 0.5° latitude and longitude for the last 300 years (i.e., back to 1700) at 50- to 100-year time slices in the earlier years within this time frame, and ten to twenty-five years in later years.

The global statistics on land-use change calculated from the RIVM and SAGE reconstructions reveal that the cropland area in 1700 was 3–4 million km^2 and that this had increased to 15–18 million km^2 by 1990, an increase of 450–500 percent. Most of the 12–14 million km^2 cropland that has appeared during the last three centuries has been at the expense of **forest** and woodland. The global extent of pasture increased from 4–5 million km^2 in 1700 to 31–33 million km^2 in 1990. This increase of 560–675 percent was mostly accomplished by the **conversion** of natural **grassland** to pasture.

At the beginning of the eighteenth century, the largest areas of pasture were in the following areas:

1. Australia, in an arc extending from South Australia to New South Wales
2. A discontinuous belt centered approximately on latitude 50°N extending from Mongolia westward to modern-day Kazakhstan
3. The Asir Mountains of the southwestern Arabian Peninsula
4. A discontinuous area in northwestern **Europe** comprising northern France, England, and southern Ireland

5. Southern Spain and northern Morocco
6. A further discontinuous area comprising relatively small areas of pasture in **southern Africa** (modern-day **South Africa**, Mozambique, Botswana, Zimbabwe, and Namibia)
7. Three areas in the **Southern Cone region** of **South America** adjacent to the cities of Buenos Aires and São Paulo, and in Entre Rios province in Argentina

The main areas of cropland were in a belt immediately to the south of the **Himalaya** and Hindu Kush mountains extending from modern-day Bangladesh to Afghanistan; around major African cities such as Kano, Lagos, and Khartoum; and in a dispersed cluster of areas in northern Eurasia that extended from the British Isles to western **Russia**. Western Russia and the plains of eastern **China** formed two large areas of mixed cropland and pasture. A century later (1800) the spatial distribution of cropland and pasture showed elements of both consolidation and slight expansion, but very few new areas had been converted from natural vegetation to **agriculture**. Expansion of existing areas was noticeable in the pasturelands of southeastern Australia, Central Asia, and Argentina. The mosaic of cropland and pasture in western Russia had consolidated and extended southward to the Caucasus, and that in eastern China had expanded. The areas of natural vegetation that had been converted during the nineteenth century included the croplands of southern India, around Mexico City, adjacent to the northeast coastline of the **United States of America**, and along the St. Lawrence River on either side of the U.S.-**Canada** border. New pasture had appeared in the **Sudan-Sahel** in what is now modern-day Chad, western Sudan, and western Senegal and Mauritania.

Present-day patterns of land use were more or less in place by 1950. Most of the areas of cropland and pasture outlined above expanded significantly between 1800 and 1950. For example, much of interior Australia was pasture by 1950; in South Asia the croplands of southern India and the Indo-Gangetic plains had expanded and merged; the mosaic of cropland and pasture in the east China plains had expanded to the north, south, and west; most of western Saudi Arabia was now pasture; and the croplands around African cities such as Kano, Lagos, Addis Ababa, and Khartoum continued to grow. Consolidation and expansion of croplands and pasture in northern Eurasia meant that a contiguous area of agriculture extended from the USSR-Poland border eastward to the Mongolia-China border with two southward extensions, one through the Caucasus to western Iran and the other into Afghanistan. With the exception of the western Iberian Peninsula and central and northern Scandinavia, Europe was a mosaic of pasture and cropland by this time. Most of southern Africa, with the exceptions of the Kalahari and Namib deserts, was pasture by 1950, and this type of land use extended along the Atlantic seaboard of Angola and northeastward along the African dry corridor into Kenya and Somalia. The areas of pasture that had existed in the Sahel 150 years earlier now merged to form a belt of pastureland extending from the Atlantic coast to the Horn of Africa. All of northern Argentina was now either cropland or pasture, with Buenos Aires province having switched from pasture to cropland, and the pasturelands around São Paulo expanding northward.

A significant number of new agricultural areas had appeared by 1950. One of the most important was Java, where most of the forest had been converted to cropland since 1800—the first major evidence of tropical forest conversion. Areas of cropland had been created along the Mediterranean littoral in the Near East and western Turkey, a large pastoral area was centered on the Atlas Mountains, and significant pockets of cropland had appeared along the West African coast. Much of the North American prairie grasslands had been converted to either cropland or pasture. This area extended from southern Canada into eastern Mexico, as far south as Mexico City, westward to the Rockies, and eastward into Texas, the Mississippi Valley, and the Corn Belt of the Midwest. A mosaic of cropland and pasture ran through Middle America southward into the **Andes**.

A relatively simplistic global-scale analysis of these data indicates that the main areas where natural vegetation has been converted to agricultural land in the past 300 years have been those areas with high concentrations of people; and, as the world's population has grown, more areas of productive cropland and pasture have been created. **Meta-analyses** of many case studies of tropical **deforestation** and **desertification** have shown that the land conversion-population link is too simplistic and that a political ecology of underlying **driving forces** and **proximate causes** of land-use and land-cover change exists. Nonetheless, the BIOME 300 land-cover reconstructions clearly show very old agricultural areas with high human population, cultivation, and stock densities (e.g., the Indo-Gangetic plains, eastern China, and northwestern Europe). They also show unambiguously that land-use and land-cover change followed European settlement on other continents. What is not clear visually from the BIOME 300 maps are the land-cover changes directly attributable to urban growth because of the scale constraints that map generalization at a large scale has produced. Though zones of cropland and pasture can be seen in close proximity to cities like Lagos, Kano, and Khartoum, well-established agricultural areas like the Nile Valley and the Tigris-Euphrates valleys are omitted. BIOME 300 products also show that land-use and land-cover change have followed more-or-less exponential trajectories. In **Latin America**, for example, little natural vegetation (grasslands, dry woodlands, or humid forest) was converted to cropland or pasture until the second half of the nineteenth century. Since then the amount of natural vegetation that has been converted has increased dramatically and, in some parts of the continent, conversion continues apace. It is this seemingly incessant conversion of the remaining natural vegetation at very high rates that has worried the global science community, development experts, and policymakers because of its environmental and socio-economic repercussions. A pertinent question at this point is to ask, "If rates of conversion are exponential, what recent land-cover changes have occurred, given that the current global pattern of land cover was more or less in place by 1950?"

This question has been answered by Erika Lepers and colleagues who, in the context of the **Millennium Ecosystem Assessment**, have produced a rapid land-cover change product that focuses on land-use conversions in the last two decades using evidence from **remote sensing** analyses, census data (forty-nine data sets in total), expert opinions, and global consultation with key informants. The results of the analysis show that deforestation remains an important issue, especially humid forest loss in **Amazonia** and Southeast Asia, where much forest has been converted to cropland. Weakened state control during the 1990s has enabled loggers to rapidly expand their activities in the **boreal zone** of Siberia. Cropland areas have expanded on many continents, but especially so in Bangladesh, along the Indus Valley, in the Middle East and Central Asia, throughout the Great Lakes region of East Africa, along the **arc of deforestation** to the south of the Amazon basin, and in the **Great Plains** of North America. However, the agricultural heartland of eastern China has experienced rapid loss of cropland in the last two decades, as has the southeastern United States of America, where cropland converted after European settlement is being lost. Many of the most populated areas are in the tropics—especially the Indo-Gangetic plains, the plains and northern plateau of China, and Java; while concentrations of cities with populations greater than 750,000 people are mainly grouped along the east coast of the United States of America, in **Western Europe**, and in India and east Asia. Intriguingly, evidence from this product has failed to lend support to the widely held view that the African Sahel is a hot spot of desertification.

Despite the rigors of the approaches adopted in developing the rapid land-cover change product, the results must be interpreted in the light of a number of caveats. The majority of recent land-cover change studies have focused on the pan-tropical belt, with boreal and temperate forests being relatively neglected. As a consequence, more data gaps exist for higher-latitude forests than tropical forests, thereby skewing comparisons between these and tropical forests. The possible dichotomy in the land-cover dynamics for forests that this creates is accentuated by the fact that most research on land-cover dynamics in the tropics is based on remotely sensed data, while those for temperate forest analysis are generally based on census and statistical databases. Further pitfalls in determining land-cover change in forested areas lie in the abundance of forest definitions (over ninety) available. The reliability of our understanding of land-cover change in drylands (i.e., dryland degradation or desertification) is less than that for forests. This is mainly because the data sets are in worse shape: there is a lack of reliable data, not least because they are heterogeneous in terms of data acquisition methods and the indicators used, and because it is difficult to distinguish between human-induced trends in land use and **land degradation** and climate-driven variability. Population census data—the key data source for studies of **urbanization**—are relatively accurate; but there are too few data on urban areas or shapes of urban agglomerations, let alone linkages between these key spatial properties for land-cover change studies and population data. The most reliable data used in creating the rapid land-cover change product was that for croplands. The timing of the rapid land-cover change product after the massive efforts among scientists the world over to investigate land-use/cover change has highlighted major research deficiencies. The absence of reliable spatially explicit data sets meant that regional-to-global scale estimates of afforestation, **reforestation**, and pasture were not made for the rapid land-cover change product. Data gaps in more reliable data sets mean that our knowledge of land use and land cover in Southern Hemisphere dry woodlands (e.g., the **miombo** woodlands of southern Africa and the Chaco woodlands in central South America) and boreal forests is not only poor, but we may be missing significant land-use and land-cover change in areas with data gaps.

It is apparent from the two major research efforts summarized above that efforts to monitor land-use change by scientists during the past two decades have been underpinned by the development of regional and global land-use and land-cover maps from remotely sensed data. In fact, Erika Lepers and colleagues make the case for the use of remote sensing over statistical data in land-cover change analysis. Many of the earlier maps were derived from **advanced very high resolution radiometer** (AVHRR) imagery; though coarse spatial resolution data from the **along track scanning radiometers (ATSR)**, **SPOT vegetation**, and the **moderate resolution imaging spectroradiometer** (MODIS) are now used as well. In addition, pan-tropical forest-cover change mapping has been accomplished using **Landsat** MSS, TM, and ETM data; for example, by the Global Land Cover Facility at the University of Maryland. However, many of the maps that have been produced are not comparable (thereby creating uncertainties in land-cover change trajectories and measurements) because different producers have developed separate land-use and land-cover classification systems. One of the early efforts to develop a standard land-cover classification on a global scale was the **IGBP-DIS global 1-km land cover data set** (IGBP-DISCover). This comprises nineteen classes: evergreen needleleaf and broadleaf forests, deciduous needleleaf and broadleaf forests, mixed forest, closed and open shrublands, woody savannas, savannas, grasslands, permanent **wetlands**, croplands, cropland and natural vegetation mosaics, barren and sparsely vegetated areas, urban/built-up areas, snow and ice, and water bodies. This classification was linked to the widespread use of AVHHR data for continental and global-scale land cover mapping. It has subsequently been adopted as the classification system for one of the three MODIS land-cover products, along with land-cover maps based on the fourteen-class University of Maryland land-cover classification system and the LAI/FPAR (leaf area index/fractional photosynthetically active radiation) product. Neither has it been restricted to mapping from one-kilometer AVHRR data; for example, the land cover map of Asia is produced at eight-kilometer resolution using IGBP-DISCover. However, this classification, along with all other categorical land use and land cover systems, is subject to revision or, at worst, rejection when applied to land-cover mapping projects in new areas, with different data inputs and when the purposes of mapping change. The consensus is that classification schemes need to meet two main aims—maximum flexibility and the highest level of standardization possible—if they are to be adopted universally. Several of the many global land-cover classification schemes that have been developed have not been universally accepted because they do not meet these aims, though in some cases it is apparent that political decisions and independence of research teams override scientific considerations.

As a consequence, recent calls for standardization in global and regional-scale land-cover classification have focused on the **Land Cover Classification System** (LCCS) developed by the **United Nations Food and Agriculture Organization** (FAO). This is a hierarchical system in which the highest level contains eight major land-cover classes: cultivated and managed terrestrial areas, natural and semi-natural terrestrial vegetation, natural and semi-natural aquatic vegetation, cultivated aquatic and regularly flooded areas, artificial surfaces and associated areas, bare areas, natural and artificial water bodies, and snow and ice. End users are able to customize classes below the top level. Because it adopts an *a priori* classification method, the use of LCCS enables comparisons to be made regardless of the data sources used, the economic sector, or the country; it is multipurpose rather than single-purpose; and it minimizes the problems generated by having to deal with a large number of pre-defined classes. Classification criteria (classifiers)—pre-selected independent diagnostic attributes—are pre-defined, rather than classes being pre-defined as in most classification systems, for example, IGBP-DISCover. The number of classifiers employed by an end user determines the level of detail in the map. The LCCS has been adopted for recent major land-cover mapping projects such as the **AFRICOVER** initiative and the **Global Land Cover Map of the Year 2000**.

Another major issue in land-cover change mapping encompasses the geographical scale-spatial resolution-map generalization nexus. Land-cover and land-cover change maps have been produced at various scales. Given the nature of the data inputs from remote sensing data and statistics for administrative and/or political units, the spatial resolution of the data in comparison to the spatial properties of the real land-cover changes can be critical. Land-cover change can be inferred from statistics for an administrative unit, for example, a comparison of the areas of forest in the unit for two or more time periods, but it is unlikely that the rates of change (i.e., quantification of the inference about land-cover change) can be applied with any reasonable level of accuracy to the entire administrative unit. In fact, the probability of error associated with the rate of land-cover change increases with increasing size of the unit. In reality, only small areas within a unit will be undergoing change in a period between two dates (and then at rates greater than the average for the entire unit), but areas where no land-cover change has occurred within the unit will also be indicated as areas of change in generalized maps and are errors of comission. The same holds true for small areas of change that are detected by coarse spatial resolution remote-sensing data, though it is often possible to investigate these in detail with finer spatial resolution imagery. A further problem is created by errors of comission. This mainly applies to land-cover change products derived from remote-sensing data where small areas of land use (e.g., small cities, peri-urban fringes, narrow belts of irrigation along rivers, oasis agriculture, and small pockets of agriculture common in high **mountain** zones) are either not detected by sensors or are lost in map generalization. As a consequence, urbanization and irrigation farming are underrepresented in many global and continental-scale land-cover maps and in global land-cover change statistics. There are ways around this problem that have not been well exploited by the global land-cover change community so far. These include multi-sensor approaches to land-cover mapping (to exploit a wider range of spectral properties and different spatial resolution simultaneously) and innovative methods of cartographic representation.

In summary, the general consensus on recent land-cover change needs to be weighed against future actions to improve land-cover change analyses. It is clear from recent land-cover change that different trajectories exist for different groups of broad land-cover types. For example, many cropland areas are reverting back to forest and non-forest (e.g., recreational) uses in temperate regions such as the southeastern United States and under set-aside policies for agricultural land in Europe, while in many parts of the tropics forest continues to be converted to cropland and pasture. Urban areas continue to expand onto croplands in many very old agricultural regions, such as in Egypt, South Asia, and China (where the early twenty-first-century growth of Shanghai is a spectacular example); while in many parts of North America, **suburbanization** had led to cities expanding into woodlands and desert—the growth of Houston and Phoenix being prime examples.

Despite the improvements that have been made in our understanding of land-cover change, further progress can be made; and the priority list of tasks before the global land-cover change community are as follows:

1. Accurate assessments of land-cover change products need to be carried out using finer-spatial-resolution remotely sensed data in areas of actual land-use change to complement assessments carried out at coarser spatial resolutions.
2. Hierarchical, standardized land classification systems (e.g., the LCCS) need to be adopted at all scales to ensure compatibility and comparability.
3. Along with adopting a standardized classification system, alternative representations of land-cover change need to be developed, particularly derivates of land cover that can be used more directly by policymakers, scientists, and social scientists in particular (examples include fractional tree cover and the MODIS LAI/FPAR product).
4. Existing data gaps need to be filled with detailed case studies of past and recent change.
5. Operational **monitoring** of areas that are not **hot spots of land-cover change** at the present time need to be undertaken so that future land-cover change can be detected an early stage; this is a potentially large undertaking, which needs to be focused through the lens of coarse-resolution remote-sensing data and then through the application of fine scale analysis and/or predicting areas of future rapid change with land-cover change models.
6. The focus in land-cover change on vegetation cover and its properties needs to expand to the requirements of environmental models and economic analyses of land productivity, and needs to routinely and systematically include soil, surface, and near sub-surface hydrological properties.
7. New empirical work is required to define and measure dryland degradation, and to link the key parameters of urbanization, that is, demographic metrics, urban plan forms, and urban areas.
8. Systematic observations of poorly measured processes of land-cover change are required urgently.

See also: Change Detection; Colonization; Cumulative Change; Extensification; Georeferencing; Global Land Cover Network; Global Observation of Forest and Land Cover Dynamics; Global Terrestrial Observing System; Hot Spot Identification; Land-Use History; Modification; Paleo Perspectives; Pixelizing the Social; Population Dynamics; Pristine Myth; Socializing the Pixel; Southeast Asia, Mainland.

Further Reading

Belward, Alan S., Estes, John E., and K. D. Klein, "The IGBP Global 1-Km Land Cover Data Set DISCover: A Project Overview," *Photogrammetric Engineering and Remote Sensing* 65 (1999): 1013–20;

Comber, Alexis, Fisher, Peter F., and Richard Wadsworth, "Actor-Network Theory: A Suitable Framework to Understand How Land Cover Mapping Projects Develop?" *Land Use Policy* 20 (2003): 299–309;

De Fries, Ruth S., Hansen, Mathew C., Townshend, John R. G., Janetos, Anthony C., and Tom R. Loveland, "A New Fractional Global 1-Km Dataset of Percentage Tree Cover Derived from Remote Sensing," *Global Change Biology* 6 (2003): 247–54;

Di Gregorio, Antonio, Louisa, H. and J. M. Jansen, *Land Cover Classification System*, Rome: Food and Agricultural Organization, 2000;

Houghton, Richard A., "The Annual Net Flux of Carbon to the Atmosphere from Changes in Land Use 1850–1990," *Tellus* 51B (1999): 298–313;

Klein Goldewijk, Kees, "Estimating Land Use Change over the Past 300 Years: The HYDE Database," *Global Biogeochemical Cycles* 12 (2000): 667–83;

Lambin, Éric F., Geist, Helmut J., and Erika Lepers, "Dynamics of Land-Use and Land-Cover in Tropical Regions," *Annual Review of Environment and Resources* 28 (2003): 205–41;

Lepers, Erika, Lambin, Éric F., Janetos, Anthony C., DeFries, Ruth, Achard, Frédéric, Ramankutty, Navin, and Robert J. Scholes, "A Synthesis of Rapid Land-Cover Change Information for the 1981 to 2000 period," *BioScience* 55 (2005): 115–25;

Ramankutty, Navin, and Jonathan A. Foley, "Estimating Historical Changes in Global Land Cover: Croplands from 1700 to 1992," *Global Biogeochemical Cycles* 13 (1999): 997–1027;

Ramankutty, Navin, Graumlich, Lisa, Achard, Frédéric, Alves, Diogenes, Chhabra, Abha, DeFries, Ruth, Foley, Jonathan, Geist, Helmut, Houghton, Richard, Klein Goldewijk, et al., "Global Land Cover Change: Recent Progress, Remaining Challenges," in *Land Use and Land Cover Change: Local Processes, Global Impacts*, eds. Éric F. Lambin and Helmut Geist, Berlin: Springer-Verlag, 2006;

Waisanen, R., and N. Bliss, "Changes in Population and Agricultural Land in Conterminous United States Counties, 1790 to 1997," *Global Biogeochemical Cycles* 16 (2002): 1137–56.

ANDREW MILLINGTON

A

Access. Demarcates a range between impedances to and full availability of land. Individual and societal land use are directly related to access to land and the forces controlling access. Political, historic, physiographic, and economic forces define boundaries, restrictions, protections, and access. In turn, access to land shapes and reflects values, perspectives, and land uses today. Access to land refers to both tangible and intangible accessibility. Tangible accessibility includes the ability to navigate to a land parcel via existing transportation infrastructure as well as forging **agricultural frontier** and settlement frontier environments. Intangible accessibility has to do with **property** rights and **land tenure** policies in an area, as well as financial mechanisms (e.g., mortgages) available to assist in land procurement.

Much of accessibility **modeling** involves understanding both the seasonal and longer-term impedances to land. For example, in seasonally flooded environments, many people are unable to migrate in or out or to reach their lands or nearby village centers during high-flooding stages. Some river-based areas experience greater accessibility during flooding, using boats as their primary means of movement. Other road-based areas become landlocked when floodwaters make for impassable conditions. Other seasonal impediments include snow and ice, forcing vertical seasonal migration among high-elevation residents in areas such as the **Andes**. Longer-term physical impediments may be economic or political in origin, such as the closing of a border due to social conflict, embargo, or military action.

In the broader sense, access to land refers to the ability to derive benefits from land and its associated resources. For example, local people may have physical access to a **forest** located next to their village, but they may not possess any property rights in the forest. Even if they hold a property right (be it ownership, right to use, etc.), they may only capture a small share of the economic benefits derived from extracted forest resources. The rural poor in many developing countries rely on farming for subsistence. Inadequate access to land is often cited as a reason for environmental or **land degradation** in regions where resources such as land are scarce. In many developing countries where access to land is scarce, squatters have succeeded at illegally obtaining land for their own use. This can become a difficult issue for local governments and law enforcement to deal with, especially when there are a large number of trespassers. Boundaries are also used to protect tracts of land for the purpose of **conservation**, that is, reserving floral and faunal species in a particular area. When access to land is limited, these areas can be subject to poaching to provide resources for local peoples.

Access in the broader sense, therefore, is closely related to property and **land rights**. In areas where land title or property require a substantial cost in addition to labor (e.g., land clearing), access to land can also be blocked by inaccessibility to financial **institutions** to support purchase or acquisition of title. Many developing countries either do not have such institutional aids available to them, or the interest rates are prohibitively high (30–50 percent rather than 5–8 percent).

See also: **Degradation Narrative; Exogenous Factor; Land Privatization; Land-Use Policies; People at Risk; Poverty; Subsistence Agriculture; Transhumance.**

Further Reading

Kitay, Michael G., *Land Acquisition in Developing Countries: Policies and Procedures of the Public Sector*, Boston: Lincoln Institute of Land Policy, 1985; Ribot, Jesse C., and Nancy L. Peluso, "A Theory of Access," *Rural Sociology* 68 (2003): 153–81; Verburg, Peter H., Overmars, Koen P., and Nol Witte, "Accessibility and Land-Use Patterns at the Forest Fringe in the Northeastern Part of the Philippines," *The Geographical Journal* 170, no. 3 (2004): 238–55.

KELLEY A. CREWS-MEYER, AMY L. NORMAN

Advanced Very High Resolution Radiometer. A series of satellite sensors operated by the American National Oceanic and Atmospheric Administration (NOAA) that have provided global data sets at a spatial resolution of one kilometer from 1970 to the present day. Image data is acquired in five spectral bands, 0.58–0.68 μm (red), 0.725–1.1 μm (near infrared), 3.55–3.93 μm (middle infrared), 10.3–11.3 μm (thermal infrared), and 11.5–12.5 μm (thermal infrared). The advanced very high resolution radiometer (AVHRR) has been used for many applications including meteorology, oceanography, and terrestrial sciences. The AVHRR data set has been used to create global land-cover products through the **IGBP-DIS global 1-km land cover data set** DISCover project. The **normalized difference vegetation index** has been used extensively for mapping land-use/cover change. The thermal bands on AVHRR have been used to detect **fires**.

See also: **Change Detection; Global Land Cover Map of the Year 2000; GLOBCARBON; Remote Sensing.**

Further Reading

Goward, Samuel N., Dye, Dennis, Kerber, Arlene, and Virginia Kalb, "Comparison of North and South American Biomass from AVHRR Observations," *Geocarto International* 2 (1987): 27–39; Tucker, Compton J., and Piers J. Sellers, "Satellite Remote Sensing of Primary Production," *International Journal of Remote Sensing* 7 (1986): 1395–1416.

KEVIN J. TANSEY

Aerial Photography. A type of airborne **remote sensing** that takes photographs of the earth's surface from a platform flying above the surface, but not in orbit. The platform (aircraft, kites, or balloons) carries a photographic device (a camera) to capture a section of the earth's surface. The devices come in various sizes and collect reflected energy (radiance) from the ground. Depending on the objective of the data to be collected, a wide range of films, from black-and-white to color and infrared, can register specific segments of the electromagnetic spectrum.

The first known aerial photography was taken by Gaspard Felix Tournachon above the French city of Bièvre in 1858 using a balloon carrying a photographic device. The first aerial photographs obtained from airplanes were taken in 1908 by Wilbur Wright. In World War I, aerial photography became an increasingly important tool for military reconnaissance, with over a million photographs taken. French aerial units were developing and printing as many as 10,000 photographs each night, and Germany had 2,000 aerial cameras in use. In 1934, the American Society of Photogrammetry was founded (it is now called the American Society of Photogrammetry and Remote Sensing) as a professional organization to advance the field of aerial photography.

Similar to other remote sensing data, key properties of aerial photographs are the spatial resolution and the **scale**, which are determined by the characteristics of the film, lens, and the platform (flying height). Today, most analog aerial photographs are scanned for storage and processing in digital form, but digital cameras are increasingly used. A main advantage of aerial photography is the spatial resolution, which can be substantially higher than the resolution of satellite imagery, dependent on the flight height as well as the film and camera characteristics.

Aerial photographs are one of the most versatile and economical forms of remote sensing. The use of aerial photography as a means of examining the earth's surface is well established and still widespread for areas that are not overly large. Aerial photography is especially common for smaller areas but is often replaced by the use of satellite sensors for larger areas like regions or countries, as costs and data load of aerial photography are becoming increasingly prohibitive.

Aerial photography is the basis for a wide range of spatial analyses and has a long tradition in land and forest use management as well as for **change detection** and **monitoring**. The first land-cover maps using remote sensing techniques were based on a visual interpretation of aerial photographs in combination with topographic information and field observations. Until today, many land-cover maps, particularly for smaller areas, are still derived from aerial photography. Land-use modelers interpret aerial photographs to derive land-cover data and land-use information as well as other features like roads or populated areas. Aerial photography is also essential to compile land-cover and land-use information for periods prior to the existence of satellite imagery in the 1970s. Other typical applications of aerial photography include cadastral mapping, species mapping, urban planning, and habitat analysis. Almost all topographic maps on medium and large scales are based on aerial photography.

See also: **Airborne Light Detection and Ranging; Cadastre; Land-Use Planning; Pattern Metrics; Pixel.**

Further Reading

DeMers, Michael N., *Fundamentals of Geographic Information Systems*, 2nd ed., New York: John Wiley & Sons, 2000; Lillesand, Thomas M., Kiefer, Ralph W., and Jonathan W. Chipman, *Remote Sensing and Image Interpretation*, 5th ed., New York: John Wiley & Sons, 2004.

DANIEL MÜLLER

African Trypanosomiasis. A vector-borne parasite causing disease in people, livestock, and wildlife throughout much of sub-Saharan Africa. The trypanosome parasites (*Trypanosoma spp.*) are predominantly transmitted by the bite of an infected tsetse fly (*Glossina spp.*). Some trypanosome species are infectious to humans and cause sleeping sickness, a fatal disease if untreated. Sleeping sickness remains a public health concern in many infected countries, including Uganda, the Democratic Republic of the Congo, Angola, and Sudan, where sporadic outbreaks occur. Trypanosome parasites are also infectious to a variety of animal species, including cattle, wild pigs, wild ungulates, buffalo, and some reptiles. The impact of trypanosomiasis on livestock, particularly cattle, has been a significant economic burden in infected regions.

African trypanosomiasis is both *effected by* land-use/cover patterns and *affects* land-use/cover change. The disease is affected by land use predominantly by the role of land cover in driving the habitat of the tsetse fly vector. As the parasite is transmitted by an insect vector, distributions and patterns of land cover affect the available habitat of the various tsetse species. Stephen Leak's comprehensive book on tsetse biology and ecology provides a thorough review of the determinants of tsetse distributions and ecology. There are over twenty-five tsetse species and subspecies, which are generally classified into three main groups, the *fusca, palpalis,* and *morsitans* tsetse species. As noted by Stephen Leak, the climate and **vegetation** of the tropical rain forest control habitat for the *fusca* and *palpalis* groups of tsetse species, and woodland distributions generally control the habitat of the *morsitans* group. Distributions of specific vegetation covers and species can directly drive the distribution of the tsetse fly, and thus the potential for parasite infestation and disease risk. Early observations in Nigeria, for example, suggested that the distribution of the plant *Mimosa asperata* was almost synonymous with the presence of the fly. Similarly, in Uganda, increased growth of the bush *Lantana camara* has been associated with disease outbreak. In this latter example in southeastern Uganda, changes in land cover from **cotton** and **coffee** plantations to uncultivated *Lantana* bush (resulting from social and economic upheaval) are believed to have created ideal tsetse habitat, leading to a substantial sleeping sickness outbreak in the 1980s.

Population density has also been associated with tsetse distributions, with high densities believed to suppress tsetse presence. This is due to human activity reducing the host and reservoir availability, as well as vegetation cover available for habitat. For example, Anthony Jordan notes that the expansion of the road network in Nigeria has been associated with a reduction in the abundance of tsetse. Robin Reid and colleagues suggest that increasing human populations may lead to a reduction in the densities of savanna tsetse species, but not the riverine species. These processes of land-use and cover change drive the availability of hosts and habitat for the tsetse fly vector, which, in turn, drives the distribution of trypanosomiasis risk in sub-Saharan Africa.

The close association between land cover in sub-Saharan Africa and the tsetse vector of trypanosomiasis has resulted in the reciprocal process of African trypanosomiasis *affecting* land-use/cover patterns in affected regions. The continental distribution of cattle in sub-Saharan Africa is inversely correlated with the distribution of the tsetse fly. Stephen Leak estimates that 46 percent of tropical sub-Saharan Africa is infected with the tsetse fly; this has a direct effect on the suitability of land for livestock. Livestock farmers in sub-Saharan Africa have generally responded to the risk of disease by avoiding tsetse-infested regions wherever possible. For example, the Fulani of West Africa graze their cattle on the edge of the southern tsetse belt during the dry season, yet are forced to move northward as the wet season encourages northern spread of the tsetse distribution. This association results from reluctance of livestock farmers to settle in areas where there is significant risk of cattle infection with trypanosomiasis and its resulting economic impacts. In many of these areas, livestock are a significant cultural and economic asset and are central to poor livelihoods. The presence of the tsetse fly and the trypanosome parasites across large regions of sub-Saharan Africa is, therefore, considered to be a significant constraint to livestock and economic development. This constraint to agricultural development, and its associated influence on land use in sub-Saharan Africa, has led to occasional conservationist arguments that the tsetse fly (and therein the risk of African trypanosomiasis) has been the savior of African resources, particularly vegetation cover and wildlife populations.

In association with the impact of trypanosomiasis risk on land use in Africa is the implementation of tsetse control methods that influence land cover. Previous techniques, used particularly during the colonial period, often focused on removal of vegetation as a habitat for tsetse and elimination of game as a parasite reservoir. These techniques, however, have largely been replaced by insecticides via localized and aerial spraying. During the significant animal epizootics and human epidemics of the colonial period, for example, widespread abandonment and evacuation of land near rivers and shores occurred around Lake Victoria in Uganda. This abandonment of land cultivation and occupation resulted in increased bush growth, causing further expansion of tsetse-infested lands. Proceeding re-occupation of these lands may have contributed to more recent epidemics of disease. These examples illustrate the reciprocal relationship that exists between African trypanosomiasis and land use/cover.

See also: **Conservation; Human Health, Human Immunodeficiency Virus (HIV)/ Acquired Immunodeficiency Syndrome (AIDS); Land Quality; Pastoral Mobility.**

Further Reading

Jordan, Anthony M., *Trypanosomiasis Control and African Rural Development,* Harlow, UK: Longman, 1986; Jordan, Anthony M., "Trypanosomiasis Control and Land Use in Africa," in *Outlook on Agriculture* 10 (1979): 123–9; Leak, Stephen G. A., *Tsetse Biology and Ecology: Their Role in the Epidemiology and Control of Trypanosomosis,* Wallingford, UK, Nairobi, Kenya: CABI Publishing, International Livestock Research Institute, 1999; Perry, Brian D., Randolph, Thomas F., McDermott, John J., and Philip K. Thornton; *Investing in Animal Health Research to Alleviate Poverty,* Nairobi, Kenya: International Livestock Research Institute, 2002; Reid, Robin S., Kruska, Russ L., Deichmann, Uwe, Thornton, Philip K., and Stephen G. A. Leak, "Human Population Growth and the Extinction of the Tsetse Fly," in *Agriculture, Ecosystems, and Environment* 77 (2000): 227–36; Reid, Robin S., Wilson, Cathy J., Kruska, Russ L., and Woudyalew A. Mulatu, "Impacts of Tsetse Control and Land-Use on Vegetative Structure and Tree

Species Composition in South-Western Ethiopia," *Journal of Applied Ecology* 34 (1997): 731–47.

LEA BERRANG FORD

AFRICOVER. Even though a large amount of **remote sensing** data for the assessment of natural resources is now available, and technologies exist for its storage, analysis, and integration, the actual situation in Africa shows a severe shortage of countrywide, as well as regional and sub-regional, quantitative and qualitative information on **vegetation** cover and current land use. This has proven to be the major limiting factor in proper planning, development, and sustainable management of renewable natural resources in Africa.

The AFRICOVER project was developed by the **United Nations Food and Agriculture Organization** (FAO) in response to a number of national requests for assistance to develop reliable and georeferenced information on natural resources required at sub-national, national, and regional levels (e.g., for early warning, **agriculture**, planning, **watershed** catchment management, **biodiversity**, and climate change modeling). The main aim of the AFRICOVER initiative is therefore to assist in improving national and sub-regional capacities in the establishment, update, and operational use of geographic, referential land-cover maps and geodatabases.

The initiative is divided into five sub-regional modules. The first operational component, AFRICOVER East Africa, was initiated in 1995 through the financial support of the Italian government and involved ten countries: Burundi, the Democratic Republic of Congo, Egypt, Eritrea, Kenya, Rwanda, Somalia, Sudan, Tanzania, and Uganda.

For these and other countries, digital georeferenced databases on land cover and a geographic referential (geodesy, toponymy, roads, hydrography) at a 1:250,000–1:200,000 scale (1:100,000 for small countries and specific areas in Africa) have been created and made available to users. The land-cover data is mainly derived from visual interpretation of high-resolution satellite images digitally enhanced. The processing and the interpretation of imagery have been facilitated by the development and use of the vector-based editing system GeoVIS and the integrated image processing interpretation mapping tool AIMS (AFRICOVER Interpretation and Mapping System). The interpretation and classification are based on the FAO **Land Cover Classification System** (LCCS), which was conceived by an international working group and was developed by several FAO services, with the support of a number of national and international agencies.

The cartography is carried out by national institutions and agencies, with the assistance of international experts. FAO provides assistance in project preparation and implementation, including technical, institutional, and financial arrangements. In addition, the AFRICOVER project initiates national and regional collaboration and ensures standardization and harmonization of methods, procedures, tools, information, and data. This standardization facilitates the integration of national and regional data and assists the distribution of the final products, including databases and maps.

The success of AFRICOVER has triggered the creation of the **Global Land Cover Network** initiative, a global partnership led by FAO and the **United Nations Environment Programme** with the objective of developing harmonized global land-cover data required by a wide user community.

See also: **Georeferencing; Global Terrestrial Observing System.**

Further Reading

AFRICOVER [Online, December 2004], The Africover Web Site, www.africover.org.

REUBEN SESSA

Agenda 21. A major outcome from the United Nations Summit on Sustainable Development and the Environment (colloquially known as the Earth Summit) held in Rio de Janeiro in 1992. The Earth Summit itself was prompted by the publication of *Our Common Future*, also known as the Brundtland Report, in 1987. Agenda 21 is both a report on the state of planet earth as of 1992 and a global strategy for the future actions required globally to attain the goals of sustainable development. It contains forty chapters covering most aspects of the world's environment and development issues. Vitally, it recognizes that to achieve sustainable development, many stakeholders need to be involved, not just national governments. For example, it recognizes important roles for non-governmental organizations (NGOs), businesses, trade unions, women's organizations, youth groups, and local councils. In fact, Local Agenda 21s (LA21s)—the implementation of Agenda 21 at the local scale—have been a success story, with 6,416 LA21s planned or underway in 113 countries by 2002. Agenda 21 is not a legally binding document, but rather a blueprint for action. Conventions that have been negotiated during and after the Earth Summit (e.g., the United Nations Framework Convention on Climate Change and its **Kyoto Protocol**, the **United Nations Convention on Biological Diversity**, and the **United Nations Convention to Combat Desertification**) represent global, legal instruments that bind countries to specific actions.

See also: **Alternatives to Slash-and-Burn Programme, Global Land Cover Network; Public Policy; Sustainable Land Use.**

Further Reading

Brundtland, Gro Harlem, *Our Common Future: The World Commission on Environment and Development,* Oxford: Oxford University Press, 1987; Dodds, Felix, ed., *The Way Forward: Beyond Agenda 21,* London: Earthscan, 1996.

ANDREW MILLINGTON

Agent-Based Model. A computer representation of a system that comprises multiple, interacting actors (i.e., agents). In a land-use/cover change context, agents can include landowners, farmers, collectives, migrants, management agencies, and/or policymaking bodies, all of whom are involved in **decision-making** or take actions that affect land-use patterns and processes. By simulating the individual actions of many diverse actors, and measuring the resulting system behavior and outcomes over time (e.g., the changes in patterns of land cover), ABMs can be useful tools for studying the effects on land-use/cover processes that operate at multiple **scales** and organizational levels and their effects.

Agents are discrete entities that are defined in terms of both their attributes and their behaviors. The attributes of agents can be continuous measurements, like age or wealth; or discrete categories, like farmers or members of a particular interest group. The actions of agents can be scheduled to take place synchronously (i.e., every agent performs actions at each discrete time step) or asynchronously (i.e., agent actions are scheduled with reference to a clock or to the actions of other agents). The behaviors of agents can vary from completely reactive—that is, agents only perform actions when triggered to do so by some external stimulus (e.g., actions of another agent)—to goal-directed (e.g., through seeking of a particular goal). For example, a farmer agent could be programmed to plant corn every spring (i.e., a relatively reactive agent) or choose whether to plant and which crop to maximize return on **investment**, and plant at the time that is expected to produce the highest yield (i.e., a goal-directed agent). Agent-based models for land change research are nearly always spatially explicit, which means that the agents and/or their actions are referenced to particular locations on the earth's surface. For this reason, many agent-based models have either direct or indirect interaction with **geographical information systems**.

ABMs belong to a category of models known as discrete event simulations, which run with some set of starting conditions over some period of time, allowing the programmed agents to carry out their actions until some specified stopping criteria are satisfied, usually indicated by either a certain amount of time or a specified system state. As such, ABMs are similar to discrete-event models used in a wide range of fields, including individual-based models (IBMs) and gap models, which have been used in biology and ecology to simulate the behavior of animals and plants, respectively.

Advantages of Agent-Based Models

Agent-based models have a number of strengths that contrast with traditional methods for **modeling** land-use/cover change. Important among these strengths is the ability to represent agent behavior explicitly, while at the same time providing a range of options for representing behavior. Traditional land-use models derive from rational utility theory, in which a land-use pattern is explained in terms of actors who maximize utility obtained from various land uses at various locations (in some models the agent decision is what land use to engage in at a location; in others it is where to locate a particular land use). For example, rational utility theory is the basis for models of agricultural land-use patterns around a city center such as the one of **Johann Heinrich von Thünen**, and later variants of this model. In such traditional models, homogeneity is assumed in both the decision-making approach and in the factors that influence the utility calculation for each member of the population. Further, all agents are assumed to have perfect information about alternatives and about future utility, and to maximize utility in all situations.

Research in the behavioral sciences has produced a range of alternative behavioral models that can be implemented and evaluated with agent-based models. Some examples are as follows:

1. Heterogeneity in agent behavior can be represented by drawing parameters of a utility function from a statistical distribution, or using alternative decision-making approaches for different agent types (e.g., residents evaluate esthetics and distance to work, and farmers evaluate crop price and yield).
2. Rationality can be bounded by imposing limits on the amount of effort agents use to search for and/or evaluate alternatives.
3. Alternatives to utility maximization can be included by, for example, using satisfying behavior, in which agents select alternatives that are "good enough" using heuristics to determine agent choices (e.g., based on simple psychological models).
4. Randomness in environmental conditions, information availability, or decision outcomes can be incorporated through inclusion of stochastic processes.

In addition to their richer behavioral representations, ABMs inherit several advantages from their computer-simulation heritage. Because an ABM is a dynamic system, the model can incorporate positive and negative feedbacks, such that the behavior of an agent has an influence on the subsequent behavior of other agents. These feedbacks can be used to represent the endogeneity of various **driving forces** of land-use and **land-cover change**. For example, if the existing amount of developed land is a variable that predicts how much new land will be developed, then developed land area is an endogenously coupled variable. By representing the sequential effects of land development at each time-step in the model, a simulation model can explicitly represent this endogeneity. Other simulation models, like **cellular automata**, share this ability. An important characteristic that distinguishes ABMs from cellular automata (CA) is that, whereas CA usually have a fixed interaction topology (i.e., which neighbors a cell interacts with is fixed by the cellular geometry), interactions in the ABM can be dynamically changed as the model runs, because they are defined at the agent level, rather than in terms of the partitioning of space.

Purpose and Use of Agent-Based Models

Like other modeling approaches, the use of ABMs might be summarized in two broad purposes: explanatory and prognostic. The explanatory uses of models involve their use as a means to codify understanding about processes, evaluate the implications of assumptions, and evaluate the efficacy of this understanding relative to observations. Prognostic uses of models involve their use for extrapolation of trends, evaluation of **scenarios**, and prediction of future states. Whether explanatory or prognostic, many projects that use ABMs begin with exploratory uses of the models to understand system processes and dynamics before building truly explanatory or prognostic models.

In the context of land-use/cover change research that focuses on explanation and understanding land-related decision-making processes and their implications, ABMs serve as microcosms, or **land-use systems** *in silica*, within which such land-related processes can be evaluated. Such studies build on the tradition of complex systems research that has been referred to as "generative social science," an approach elaborated by Joshua Epstein. The goal of such studies is to understand the emergence of patterns, trends, or other characteristics observable in society or geography by generating them from models built to represent the actors but not, explicitly, the collective outcomes.

For example, it is a common goal of land-use/cover change studies to characterize the spatial patterns of landscapes through the calculation of metrics that describe the composition and configuration of land cover in a particular area. Such **pattern metrics** might be used to observe that **forests** are becoming more fragmented, that urban settlements are becoming more dispersed, or that **grasslands** are declining. A goal of agent-based modeling in these contexts could be to program plausible agent-level behaviors and interactions that, when run in a simulation, produce similar trends in the patterns as those observed through analysis of maps. Such a model represents a candidate explanation for the emergence of the observed patterns. The challenges in such applications are to ensure that the agent-level representations are plausible, and to test alternative models to identify the range of microbehaviors that can produce a given macrobehavior. If that range is large, then the set of candidate explanations is, too.

Prognostic uses of ABMs in land change are most likely to be successful in the context of scenario development and policy analysis. In such applications, an ABM whose mechanisms are well supported by empirical understanding and data can be used to evaluate the possible effects on model outcomes of changes in **initial conditions**, in information available to agents, in constraints or incentives to agent behavior, and so on. Such analyses can be used to understand how system dynamics and outcomes can be affected by changing contexts. An obvious application of such approaches is to evaluate the effects of system changes over which some agency or group has some control (e.g., policy analysis), to evaluate the range of system effects a particular change in policy might be expected to have.

Positive **feedbacks**, common in many such systems, can create system behavior referred to as path dependence, in which the path a process takes is very sensitive to both initial conditions and small variations in stochastic processes. For this reason, predictive modeling with agent-based models, or any models of systems that contain positive feedbacks, can be challenging. Some land-use processes do contain feedbacks. A classic example is the development of certain industrial regions (e.g., Silicon Valley in the United States) in which the returns to locating in a region increase as more and more facilities are located there. Because the location of such a large agglomeration of industries of a given type can be very sensitive to some early decisions that are hard to predict, the land-use system itself can be hard to predict. The impulse to calibrate a model to fit existing data can result in an overfit model, in which the model is overly constrained so that it fits the data but is insufficiently general to represent the full range of possible outcomes from the system.

Building Agent-Based Models

The process of building agent-based models starts with conceptual modeling, in which the basic questions or goals, the elements of the system (i.e., agents and other features), the relevant behaviors of the agents, and the measurable outcomes of interest are identified. Depending on the goals of a particular model application, a model may involve the use of designed or empirically grounded agents. If designed, the agents are endowed with characteristics and behaviors that represent, though often simplified, conditions for testing specific hypotheses about general cases. If empirically grounded, the agents are intended to represent real individuals or organizations for which data about decision-making processes are available. Similarly, the environment within which the agents act can be designed, that is, given characteristics that are simplified to focus on specific attributes, or empirically grounded to represent specific places. In practice, a study may start with simple models, often with designed agents and environments, to evaluate and understand the dynamics of the system. Where such understanding is more mature, and where a project has prognostic goals related to specific cases, a study would use empirically grounded agents and environments.

Following conceptual modeling, the computer programs must be designed in ways that represent the conceptual model with fidelity and that are computationally efficient. The design of an ABM program follows the framework of object-oriented programming. As such, classes of objects are defined, some of which are agents and some of which can include other computational objects—for example, an object that observes and reports the outcomes of the models. The defining characteristics of object-orientation include encapsulation, in which the attributes and methods of objects are bundled together so that the actions of objects and agents can be run independently; and inheritance, in which a sub-class of objects (or agents) inherits the attributes and methods of its superclass. Inheritance makes it possible to efficiently define all agents as a class with a set of attributes (e.g., type and color), as well as types of agents as a sub-class with different sets of additional attributes and methods for each type.

Once a program has been designed, the software must be written. ABMs use object-oriented programming languages, like C++, Java, or Objective-C. There are several tools available to assist in building ABMs. Many ABMs are written using sets of software libraries that provide pre-defined routines specifically designed for use in agent-based modeling. Perhaps the most prominent and early of these libraries was SWARM, originally developed at the Santa Fe Institute and programmed with Objective-C but later supporting models written in Objective-C or Java. Recent years have witnessed the development of several more libraries, including REPAST, ASCAPE, and MASON, all written with Java. Several systems based on Smalltalk have been developed and include CORMAS, which has a natural resource focus, and SDML. The object-oriented programming paradigm of ABMs allows for the incorporation of a range of other software libraries. Importantly for land change research, libraries of geographic information systems (GIS), like GEOTOOLS, can enhance the functionality of ABMs through provision of data management and spatial analytical functionality.

In addition to software libraries, several packages provide routines and software environments that simplify the programming of ABMs and the building of visualizations based on these models. Notable among these packages are STARLOGO, its derivative NETLOGO, and AGENTSHEETS. While these packages are comparatively much easier to use than the Java-based libraries, they also provide more limited functionality.

Once a model has been coded, it must be verified by running the model to test that the specific programmed behaviors perform as expected. These tests can be posed as hypotheses about the behavior of the model under a range of different parameter settings. The hypotheses should be ones for which the answer is known, given the conceptual model as designed. If the coded model returns different results, then the modeler should further evaluate the model for logical errors and programming mistakes. Given that complex systems can often return counterintuitive results, however, care must be taken to determine whether the unexpected result is an error or a feature of the system. Only after the various methods programmed in the model have been tested in this way should the model be considered sufficient for scientific application.

Calibration and Validation

Some of the hardest issues to deal with in building ABMs are related to calibration, that is, setting model structure and parameter values in ways that accurately reflect a real-world situation, and **validation**, that is, confirming that the model behaves substantially similarly to the way the target system behaves. Calibration typically requires data on the microlevel processes that the model is programmed to represent. These data can be acquired through surveys, statistical analysis of empirical data (e.g., through hedonic price analysis), or experiments specifically designed to elicit decision-making strategies and factors. Validation usually involves comparison of model outcomes, often at the macrolevel, with comparable outcomes in the real world (e.g., land-cover patterns). Because the models are dynamic, these comparisons often require multitemporal and spatially explicit data to fully evaluate that a model reflects dynamics accurately. Given the path dependence that can be present in land-use systems, comparison of a model that has multiple possible outcomes with a real world that has only one outcome presents new analytical challenges for agent modelers.

See also: **Endogenous Factor; Pattern to Process.**

Further Reading

AGENTSHEETS [Online, July 2004], Software Package Providing Routines and Environments That Simplify ABM Programming and the Building of Visualizations Based on ABMs, Agentsheets Web Site, www.agentsheets.com; ASCAPE [Online, July 2004], Agent-Based Modeling Software Library Web Site, www.brook.edu/es/dynamics/models/ascape; Brown, Daniel G., Page, Scott E., Riolo, Rick, Zellner, Moira, and William Rand, "Path Dependence and the Validation of Agent-Based Spatial Models of Land Use," *International Journal of Geographical Information Science*, in press; CORMAS [Online, July 2004], Agent-Based Modeling Software Library Web Site, cormas.cirad.fr/indexeng.htm; Epstein, Joshua M., "Agent-Based Computational Models and Generative Social Science," *Complexity* 4 (1999): 41–60; GEOTOOLS [Online, July 2004],

Geographic Information System Software Library Web Site, www.geotools.org; Gilbert, Nigel, and Steven Bankes, "Platforms and Methods for Agent-Based Modeling," *Proceedings of the National Academy of Sciences USA* 99 (Suppl. 3) (2002): 7197–8; Gimblett, Randy H., *Integrating Geographic Information Systems and Agent-Based Modeling Techniques for Simulating Social and Ecological Processes,* New York: Oxford University Press, 2002; Janssen, Marco A., ed., *Complexity and Ecosystem Management: The Theory and Practice of Multi-Agent Systems,* New York: Edward Elgar, 2003; MASON [Online, July 2004], Agent-Based Modeling Software Library Web Site, cs.gmu.edu/~eclab/projects/mason; NETLOGO [Online, July 2004], Software Package Providing Routines and Environments That Simplify ABM Programming and the Building of Visualizations Based on ABMs, ccl.northwestern.edu/netlogo; Parker, Dawn C., Berger, Thomas, and Steven Manson, ed., *Agent-Based Models of Land-Use and Land-Cover Change: Report and Review of an International Workshop, Irvine, CA, USA, October 4–7* (LUCC Report Series No. 6), Bloomington, IN: LUCC Focus 1 Office, 2002; Parker, Dawn C., Manson, Steven M., Janssen, Marco A., Hoffman, Matthew J., and Peter Deadman, "Multi-Agent Systems for the Simulation of Land-Use and Land-Cover Change: A Review," *Annals of the Association of American Geographers* 93 (2003): 314–37; REPAST [Online, July 2004], Agent-Based Modeling Software Library Web Site, repast.sourceforge.net; SDML [Online, July 2004], Agent-Based Modeling Software Library Web Site, sdml.cfpm.org; STARLOGO [Online, July 2004], Software Package Providing Routines and Environments That Simplify ABM Programming and the Building of Visualizations Based on ABMs, www.media.mit.edu/starlogo; SWARM [Online, July 2004], Agent-Based Modeling Software Library Web Site wiki.swarm.org.

DANIEL G. BROWN

Agricultural Frontier. The modern expression of frontier development corresponds to a nation-state's attempts to integrate some areas considered either unoccupied or economically unproductive into the mainstream economy within the nation's territorial boundaries. This effort of the nation-states to achieve territorial occupation is what entails a process of frontier development, which involves both economic and political aims. The former is intended to articulate the frontier economies into broader circuits of capital, labor, and goods exchange; and the latter seeks to include the frontier areas within the political domain of the nation-state. In this vein, the process of frontier occupation has been an opportunity for several nation-states to benefit economically from the previously unutilized natural resources, but also it has posited several challenges for them in terms of the emergent distribution of benefits and sustainability of natural resource use.

The social relationships in the frontier are defined by a continuous tension between cooperation and individualistic behaviors in an atmosphere of the state's weak presence, and tensions among different stakeholders fighting for multiple resources (i.e., land, **forests**, minerals, biodiversity **conservation**, etc.). Furthermore, this tension constitutes the arena in which short- versus long-term profit maximization expectations coexist, and this strongly influences the way stakeholders access and use the natural resources, which in turn affects the ecological sustainability of such natural resources and hence the natural habitats. Finally, individual choices counterbalance the efforts of the state to regulate human interventions, and this makes it more difficult to harmonize individual versus social interests in the use of the natural resources as a result of the frontier expansion. All those factors define to a large extent the human-environment relationships in the frontier.

Despite the classical view that largely relegated primitive accumulation and economic rents to the early stages of capitalism, both are relevant features in the frontier. The frontier's expansion is a process of original accumulation of capital largely as a result of the appropriation of natural resources, and enlarged by institutional rents provided by the nation-states. It is hard to imagine a frontier without speculation. The economic rents are captured without significant **investment** of capital and with little use of labor, because the new landowners obtain an important portion of their earning from the income flowing from that ownership rather than from the fruits of their labor, management skills, management investment, or higher productivity. This is not a fair competition because who runs first is often who claims legitimacy over the land and other resources, or who is able to enforce such right by legal or non-legal means. Most of the resources acquired initially are based on speculation, which can be sold or mined, or used in long-term agricultural or forest production.

The academic debate around frontier development acquired relevance after Frederick J. Turner delivered his seminal paper to the American Historical Association in 1893. In his essay, he analyzed the influence of the westward movement of land occupation in the formation of American society. This famous paper marked a starting point in the debate over the implications that the frontier evolution has had on **economic growth** and social development. Though Frederick Turner's frontier paradigm did not have any echo in further frontier analysis, it allowed for the subject to gain visibility in broader academia. The analysis of the frontier as a subject has since then captured large attention across disciplinary boundaries, and different paradigms have been applied to understand their underlying dynamics. The most important of such views are addressed below.

There is some comparative research focus on explaining the frontier's evolution, particularly those frontiers that had somewhat similar features, like the land occupation movements taking place in North America, **Canada**, Australia, and **South Africa's** colonization, or those other processes of frontier occupation prompted by Spanish and Portuguese **colonization** in **South America**. While the former have been labeled as frontiers of exclusion because they systematically negate pre-existing local societies to be integrated in the frontier, the second have represented a frontier of assimilation of ethnic societies along the process of the frontier's formation. Furthermore, the frontier approach has been extensively applied to tropical frontiers in which complex interactions take place because of the existence of poor soils, land speculation, presence of complex ecologies, struggles among different actors, and violence and conflict.

Theoretical Explanations

Nowadays, no single theory exists to understand the evolution of the frontiers, although diverse explanations have been provided from disparate disciplinary perspectives to understand the **driving forces** of frontier expansion. Most of the approaches look at the frontier as a process of unoccupied places and unexploited resource appropriation driven by the self-interest of individuals who push the frontier further, supported by broader forces such as government policies, market development, **access**, and technological change. To tackle such issues, both temporal and spatial perspectives have been used to support the different views.

The conventional perspectives, when looking at the underlying economic logic of frontier development, often adopt a Ricardian rent theory, and the theory of crops allocation by **Johann Heinrich von Thünen**. According to David Ricardo, the profits obtained from a piece of land depend on the rent of the land, which is determined by its location and quality. In turn, Thünen's interpretation mentions that the intensity of production is higher closer than farther from the city. When looking at the competition among crops, this theory suggests that crops with higher profits will be produced near the markets in lands whose rents are comparatively higher. In this regard, the location of crops along the frontier is determined by the relationships between costs and revenues—in short, by its profitability. Nevertheless, the spatial dynamics of contraction or expansion of some crops will depend on a range of biophysical factors (land suitability, rainfall patterns, etc.), market factors (e.g., price elasticity), and technological changes (e.g., introduction of adapted seed varieties).

The frontier analysis has been strongly shaped by neoclassical economic perspectives. In this view, the agricultural frontier(s) are seen as the expansion of available land for agricultural production, which enlarge the contribution of **agriculture** to food supply in the domestic markets, and commodities production for international trade. This has collateral benefits for the economy in that it generates rural employment and attracts unemployed people from other sectors of the economy or from less productive rural areas. The expansion of available land for production has an effect on reducing land prices, which make investments more attractive in such areas but at the same time less profitable due to higher transportation costs. The agricultural frontiers evolve because economic agents look to maximize their profits, although some have mixed objectives of subsistence and capital accumulation. It is likely that some process of frontier expansion is linked to expansion of international demand of some commodities, which tend to stimulate the process of land occupation depending of the opportunity costs of incorporating additional croplands into production.

The Marxian perspective, which inspired the further development of political economy and political ecology approaches, emphasized that the process of frontier evolution constituted a process of primitive accumulation of capital, which was mediated by power relationships. This process of accumulation thus takes place when some influential social groups obtain access to resources in the frontier, from which they can make a profit, benefiting not only from the economic rents of using such resources but also from the additional benefits they receive through capturing institutional rents. The latter constitute the incomes that favored social groups obtain from public sources without having to provide products or services of comparable market value in return.

The political economy and political ecology views are the two relevant perspectives in understanding the dynamics of frontier evolution. Political economy focuses on the frontier as a realm of social struggles among social groups in their attempts to accumulate capital, and reproduces the social system legitimizing the capitalist social relationships. Political ecology, in turn, assesses the implications of social, economic, and political dynamics on nature appropriation and transformation. There is abundant literature about the way in which local elites and dominant groups, favored by government policies and political patronage, have enlarged their benefits from frontier occupation in the tropics. Anthropological and cultural ecology views have drawn attention to the fact that the frontiers were actually places occupied by local societies, which underwent violent processes of occupation to fight against domination. To the extent that the expansion of the frontiers has implicitly been supported by the premise of nation-states' territorial consolidation, this expansion has involved a dimension of cultural homogenization and economic integration through the inclusion of local ethnic societies to the flows of trade, and articulating them to the national markets.

Finally, there are some voices in frontier expansion analysis, mainly when looking at tropical frontiers, that highlight the influence of biophysical factors such as soil quality and rainfall patterns in the process of agricultural expansion. This view, in practice, appeals to a sort of naive environmental determinism. It argues that soils and rainfall can operate as physical barriers that may limit agricultural frontier expansion because climate is one of the key factors in the relative failure of agriculture in the tropics; hence, primary crops display low levels of productivity. This perspective forgets that agricultural research, mainly through technological adaptation of seed varieties, has largely contributed to expanding the frontier borders beyond what was previously expected.

The previous review suggests that the different approaches discussed above are not necessarily contradictory, because the social, economic, political, and ecological dimensions need to be integrated for one to understand the dynamics of frontier evolution.

Typology

There are different types of frontiers, most importantly mining, timber, and agricultural frontiers. In practice there are strong interactions among the different type of frontiers, since some precede others in time. Hence, the last frontier depends very much on the type of natural resources around which the frontier evolves, because the original process of accumulation in the frontier tends to determine its further economic development. Agricultural frontiers differ from extractive frontiers because they promote a larger distribution of settlements and a massive land occupation due to their contribution to food-staple supply for the incoming flows of migrants, and due to food provision for the rest of the society. Furthermore, agricultural uses demand larger intervention on the original landscapes, which can have sustained effects on altering the natural habitats with accumulative global effects. The fact that the implications of human interventions in the tropical landscapes have larger global environmental effects has enlarged the attention over tropical agricultural frontiers.

There are two types of agricultural frontier in tropical landscapes. The first are the *populist frontiers* that are dominated by small-scale agriculturalists, often settled in the frontier in part due to the stimulus of colonization programs through pre-determined patterns of land occupation—the most known is the fish-bone pattern by which lots were distributed along both sides of a main road. The second are the *entrepreneurial frontiers* dominated by medium- and large-scale landholders who have occupied (through legal or semi-legal means) large tracts of land to implant agricultural activities. Hence, much of the state land-distribution decisions, or the process of land occupation in frontier areas, have determined the further development of the frontier, because even though there is some land concentration in colonization areas, it is likely that they will not experience massive processes of **land concentration** to evolve toward entrepreneurial frontiers, or for entrepreneurial frontiers to transform into populist ones. One is likely to find a sort of hybrid frontier, due to a mixed pattern of occupation.

The populist frontiers have not opened important opportunities for land occupation to large contingents of migrants displaced from different zones to the tropical regions—that is, the **Andes** in Peru and in Bolivia and the northeast in Brazil—motivated by expectations of improving their well-being. The states often made use of colonization programs to motivate such population movements in the 1970s and 1980s, mainly as a way to ameliorate land pressures in other areas, to reduce rural underemployment, and to increase the supply of agricultural goods. The colonization programs had little success because they were costly and promoted settlements in areas little-suited to agriculture, and because agriculturalists had little knowledge of the new local environments. A large rate of agriculturalist desertion was one of the main outcomes of such settlements. This situation has changed dramatically over time, because as new waves of migrants arrive to the frontier, there is an accumulation of learned experiences of the most suitable agricultural practices; and the settlements get closer to the markets, which makes it easier for small farmers to combine subsistence with **cash crops**. Several studies carried out in colonization areas show that people tend to enhance their livelihoods in the frontier areas, although there is a wide range of different situations.

In some cases, much land-frontier penetration is the responsibility of the rural poor's incursion into the frontier in search of enhanced livelihoods. Nevertheless, it has been widely reported, particularly in the Amazon frontiers, that after some period of time, smallholders tend to sell the land to either move deep into the forest or to abandon the frontier. This has been labeled the *hollow frontier* or sell-out effect. Common wisdom explaining this effect is that smallholders decide to abandon their plots because they cannot overcome adverse health, market, and technological conditions that make their lives difficult at the frontier. Alternative explanations mention that smallholders prefer to sell their land when the land value increases along with the frontier development. People migrating to the frontier are those with the lowest human capital and, hence, low opportunity costs. As the frontier matures and homesteading is no longer required to secure **property** rights, speculation becomes possible. Then, people with higher capital and higher opportunity costs bid current owners off the land. The sell-out effect, however, does not apply in every case, as conventional wisdom assumes. There is new evidence showing that as the frontier evolves, small farmers tend to consolidate on it.

The entrepreneurial frontiers have evolved in **Amazonia** in part as result of the significant incentives—that is, tax holidays and subsidized credit—that the region's governments, particularly the Brazilian government, spent on attracting large companies to invest in livestock production in the frontier areas in the 1970s and 1980s. Most of the companies who went to the remote frontier areas were motivated by some sort of speculative interests to obtain short-term gains through institutional rents and then selling the land after it had increased its original value. Hence, a large-scale pattern of land occupation was promoted through the incentive system. The initial opening of the frontier attracted a large number of medium-scale ranchers interested in benefiting from the road expansion and market development already brought by the state-supported projects. The entrepreneurial frontiers have had different outcomes over time. Broadly speaking, after the initial speculative process was over, there was an important process of livestock-production modernization through the adoption of semi-intensive production systems that were more concerned with reaching higher profits with beef cattle production than getting speculative gains over land.

Land Change and Legacy

Frontier expansion increases the amount of land available for agriculture. Yet this does not constitute an automatic relationship. On one hand, some frontier areas may not be suitable for agriculture due to different factors (soil composition, weather conditions, and terrain structure, etc.). On other hand, the soils in land frontier areas can be exhausted in situations in which mining nutrients persist, and then abandoned rapidly. In this regard, **land abandonment** extends into the tropical frontiers for two different reasons. First, small agriculturalists use shifting cultivation, known as **slash-and-burn agriculture**, by which they leave the cultivated land **fallow** after some short period of cultivation to allow for soil-nutrient restoration. Second, medium- and large-scale ranchers abandon some areas after the dynamics of **soil degradation** take place, leading to a drastic reduction in pasture productivity to support cattle herds' reproduction.

The phenomenon of abandonment is not confined to the tropics, and besides technological practices, some economic factors contribute to it as well. Factors influencing producers' profits, such as increases in the prices of agricultural inputs, and reduction of prices on agricultural goods, also affect the increment or abandonment of areas or the expansion of one crop to the detriment of others. For instance, in the Amazon, soybeans extend over areas already established as **pasture** for beef cattle production, producing an effect of land-use substitution from pasture to grain production. In parallel, this produces an effect of land displacement in the sense that in the frontier, soybean expansion is pushing pastures further away over forestlands. Most explanations of the dynamics of land-use substitution use an economic rationality that depends on the different land uses' profitability.

A characteristic of the frontier's evolution is that it is, at least in its earlier stages, prompted by pioneer individuals looking to enlarge their self-interest in the absence or weakness of the state. In the history of frontier development, the states often arrived later in their attempts to expand the ramifications of services provision and to capture a portion of the rents from private earnings. The absence of the state has, in some cases, led to exacerbation of land struggles among competing social groups, chaotic occupation of the frontier, non-compliance with environmental regulations (if any), and the emergence of illegal economies, in several cases supported by corruption. This situation favors the emergence of strong local elites counterbalancing the powers of the central state.

The weaknesses of the central state **institutions** have inspired some to propose that frontiers of governance be built. For some, this means expanding the capacity of the central state to monitor and enforce social (i.e., labor standards), financial (i.e., land and forest fees collection), and environmental regulations (i.e., forest management and forest clearing norms). Others, in contrast, support the idea that market mechanisms would be more efficient at incentivizing the agents to comply with environmental regulations, which, in turn, might stimulate frontier governance.

The legacy of agricultural frontier expansion is diverse, and the implications in the tropical environments are more complex as expansion of agriculture competes with **forestry** uses, which deplete forest resources and endanger natural habitats and **biodiversity**. The benefits that agricultural frontier brings to the national economies are several. The latter, in some cases, do not offset the ecological costs of clearing forest resources, but in others have helped to consolidate prosperous **agro-industries** that contribute to improved national earnings and support regional economic growth. There are situations, however, in which the economic benefits of frontier development have been captured by small groups of people and have enlarged social disparities. In other cases, small farmers settled in the frontiers have enhanced their livelihoods but, in others, they still face large difficulties in making their livings. The final outcome in the frontier development, as noted previously, depends on diverse economic, social, and ecological factors. The different disciplinary perspectives explain each piece of these dynamics.

See also: **Arc of Deforestation; Arena of Land Conflict; Brazilian Cerrado; Cattle Ranching; Cumulative Change; Deforestation; Economic Livelihood; Extensification; Land Rent; Mineral Extraction; Population Dynamics; Pristine Myth; Subsistence Agriculture; Tropical Dry Forest; Tropical Humid Forest.**

Further Reading

Schneider, Robert R., *Government and the Economy on the Amazon Frontier,* (Environmental Paper No. 11), Washington, D.C.: World Bank, 1995; Turner, Frederick J., *The Frontier in American History,* New York: Henry Holt, 1920.

PABLO PACHECO

Agricultural Intensification. Aims to raise crop yields per unit area and per unit time, in other words, to increase productivity. To achieve this goal, usually the inputs, labor, or capital (in terms of fertilizers, pesticides, energy, or technology) are increased. To raise crop yields, a broad range of methods are being applied, often in combinations, including breeding, irrigation, organic and inorganic fertilization, green manure and cover crops, pest and weed management, multicropping, crop rotation, and the reduction of fallow periods.

History

Intensification of land use has a long history, probably starting with irrigation farming in the plains of Mesopotamia's Euphrates and Tigris river regions (today's Iraq and Syria) and Egypt around 5500 BC, not counting the **domestification** of **rice** in east Asia, of **wheat** in Mesopotamia, and of **maize** in Mexico one to three millennia earlier. Not all historical intensification efforts have been successful or sustainable in the long run, for example, **salinization** of soils from irrigation in ancient India or Mesopotamia, a problem that seems to be persistent even today. The major **driving forces** behind intensification have been and still are population growth, changing life styles, and **consumption**, causing increasing demands and markets for food, fiber, and animal products. In the past, growing demands have often been met by expanding croplands and grazing areas. **Extensification** may be followed by intensification, when all virgin land has been acquired, but has often occurred simultaneously. **Amazonia** and Indonesia may serve as examples for recent simultaneous extensification and intensification, planting new land with improved **coffee** or cocoa varieties and using an increasing number of agrochemicals. Nowadays, in many parts of the world, additional land suitable for agriculture is becoming scarce or is no longer available (e.g., **China**, **Europe**), which means that intensification is the only option to meet increasing demands or persist in (international) agricultural markets.

While organic amendments in the form of mulch or dung have been used in **agriculture** for millennia, the history of chemical fertilizers started only in the 1840s, when the German chemist Justus von Liebig proposed the use of chemical fertilizers to replenish the chemical substances removed from the fields. Agricultural productivity increased considerably, as his ideas soon were integrated into agricultural practice first by large-scale farmers in Germany and elsewhere. Already a few years later, in 1849, commercial mixed fertilizers were available in the United States. For a second time, the use of chemical fertilizers increased considerably after they became available as cheap by-products of steel production in the late nineteenth century and again after 1913 when the German chemists Fritz Haber and Carl Bosch developed a methodology to derive ammonia from molecular nitrogen, which in turn again advanced the intensification of land use.

During the last four decades, agricultural production has doubled, but it is increasingly lacking behind fertilizer consumption, which has risen 600 percent in the case of nitrogen and 250 percent in the case of phosphate fertilizers. Increased productivity was achieved by using high-yielding varieties, which are more efficient than previous cultivars in producing plant **biomass** and grains, but require higher amounts of fertilizer and water, in combination with fertilizers and other agrochemicals and extended irrigated areas, though the rate of increase has slowed down since the 1980s. At the global scale, cereal yields in the past four decades have more than doubled. However, two regions, namely Oceania and sub-Saharan Africa, barely achieved any yield increases in that period, while other developing countries raised agricultural productivity at a rate similar to that of the developed countries. These different developments can partly be explained by the so-called Green Revolution.

The **Green Revolution** (GR) started in Mexico in the late 1960s with the introduction of high-yielding wheat cultivars, in combination with optimized water and nutrient management. After considerable yield increases in Central America, GR technologies spread to **South America** and Southeast Asia and were extended to maize and rice in subsequent years. Critics argue that the high-yielding crop varieties needed (and still need) additional inputs of fertilizers and other agrochemicals, which were (and still are) either not available to or not affordable for many small-scale farmers, and thus rather increase than decrease social and income disparities in poorer countries. Additionally, GR intensification was criticized to be restricted to only a few crops like maize, rice, and wheat cultivars. Subsequently, agricultural research organizations (especially the eighteen institutes of the **Consultative Group on International Agricultural Research**) extended their research to a broader range of cereals and to legumes, roots, and tubers.

During the 1990s, the **International Institute for Applied Systems Analysis** in collaboration with the **United Nations Food and Agriculture Organization** (FAO) developed the global agro-ecological zones approach, a model that provides, among other information, estimates about maximum attainable crop yields for major crops under various management conditions, including fertilizer and irrigation. Comparing present yield levels with maximum attainable yields, one can derive a potential for further intensification. Actual cereal yields of Egyptian, French, or Dutch farmers at six to eight tons per hectare almost reach or even slightly surpass calculated maximum attainable yield levels in their agro-ecological zone already; while in other African, Asian, or European countries, considerable yield increases can potentially be achieved.

Since World War II, huge amounts of money have been supplied by international donors like the World Bank and many national development agencies to advance agricultural technology and boost crop yields in less developed countries via loans or grants, either directly for agricultural projects (e.g., irrigation in China) or indirectly via supporting improvements in infrastructure, education, and so on.

During the last decades, increasing concern has been raised about the sustainability of intensified agricultural land use, because these farming or **land-use systems** usually require inputs of chemical fertilizers, pesticides, and fossil energy. Many of these inputs can have adverse effects on the health of farmers and consumers, as well as potentially contaminating systems and organisms adjacent to the fields, such as (natural) ecosystems, animals, soil, water, and air. An ongoing debate exists about whether or not highly intensive agricultural land use can be sustainable. Present trends include developing integrated production systems, reducing the adverse impacts of industrialized agriculture on society and the environment by partly replacing the input of capital (energy, fertilizer, and pesticides) by labor (working hours per unit crop, and knowledge), and better adapting production systems to local conditions. Furthermore, organic farming, which has demonstrated its beneficial environmental effects, is gaining increased popularity both in industrialized and less industrialized countries.

Means and Methods

Fallow Cycles

The use of **fallows** originates in the experience of farmers that soils need to "rest" for a certain period to restore their fertility and to decrease the abundance of pests and diseases. The relative frequency of fallow cycles of two different agricultural systems are considered here: cycles in shifting cultivation systems and cycles in permanent agriculture.

Shifting cultivation, the most extensive form of agriculture besides hunting and gathering—also related to **swidden cultivation** and **slash-and-burn agriculture**—is associated with bush or forest fallows of usually ten to more than 100 years. Assuming that no inputs but labor are available in these systems of **subsistence agriculture**, intensification can be achieved in prolonging the cropping cycle, shortening the fallow cycle, or increasing the variety of crops grown on the field (multiple cropping). Note that initially increased productivity may be reduced in subsequent years by nutrient mining or the spreading of grasses and weeds, supported by prolonged cropping cycles or shortened fallow periods.

In permanent agriculture, bare fallows are included in many farming systems around the world, like the three-year rotation winter cereals—summer cereals or root crops—fallow, which has been practiced successfully in Europe for centuries. As an intensifying measure, the soil can be planted with cover crops (often nitrogen-fixing legumes) during the fallow period, which can either be harvested (e.g., rapeseed) or used as fodder or mulch (e.g., clover or lupin). As in shifting cultivation, the cropping period may be extended, as it happened in Europe in the sixth century when farmers switched from a two- to a three-year cycle, resulting in a three-, four-, or five-year rotation cycle in which 33 percent, 25 percent, or 20 percent of the fields were fallow.

Multiple Cropping

The term multiple cropping means growing two or more crops in the same field in one year at the same time (also called mixed cropping), or one after the other. Thus, in all farming systems growing more than one crop per year, the cropping intensity is greater than one. Under mixed cropping conditions, often crops, which are complementary in the use of resources (rooting zone, demands for light and nutrients) are planted. Crop yields of each crop are frequently lower than under sole-cropping conditions, while total yields are usually higher.

Double cropping is the most frequent form of multiple cropping, practiced around the globe wherever growing periods and water availability support two crops. Relay cropping, a special form of double cropping, means planting a second crop while the first is still growing. It has been established in regions where growth periods are too short for two crops to be planted subsequently, for example, in the temperate zone in North America, which is limited by frost, or in Ethiopia, which is limited by the length of the rainy season. Triple cropping is practiced in some parts of the tropics like Indonesia, where in parts of the country sufficient water is available to support crop growth year round. In drier places, the second or third crop is either optional in wet years or can only be grown if irrigation water is available.

Irrigation

Irrigation is among the oldest methods used to increase agricultural outputs and was already used more than 4,000 years ago in India. Irrigation technology ranges from simple rainfed irrigation, which can still be found in paddy rice fields in Southeast Asia, to highly sophisticated drip irrigation in orange plantations in Israel and elsewhere. Most irrigated land is more productive than rainfed land, and in some regions, such as the Nile Valley, the Middle East, and dry regions in India and Pakistan, it is essential to crop production. The threefold increase in irrigated areas worldwide contributed considerably to the yield increases reported above. However, fresh water has to be considered one of the most limited natural resources, of which agricultural water users consume more than two-thirds of the freshwater used globally. In many regions, water resources are already exploited to their limits, which means that farmers must increase the efficiency of water use if they are to raise productivity.

Fertilizers

Fertilizers are among the major means of intensification. Here we distinguish between organic and chemical fertilizers.

Organic fertilizers include mulch that originates either from the same field—like cover crops or straw or other crop residues—or is taken from other plots, which may in turn lose a certain amount of nutrients with the exported biomass used as mulch. Dung from various animals including cattle, horses, sheep, goats, camels, pigs, and chickens is widely used as fertilizer both in solid form and as liquid manure. Nutrient contents and application rates of organic fertilizers vary widely but usually stay below the rates of chemical fertilizers. They may in many (extensive) systems be the only addition of nutrients if no other inputs are available or affordable, but may also be part of sophisticated integrated nutrient management systems used in countries of the European Union (EU) and elsewhere. Organic fertilizers like mulch or cover crops support organic carbon levels in soils, which improve water and nutrient retention capacities and other favorable properties.

Chemical (commercial) fertilizers are available in a broad range of either pure substances like potassium-nitrate or various mixed fertilizers usually containing at least two of the most frequently applied elements: nitrogen, phosphorus, and potassium. Together with calcium and magnesium, the three elements mentioned before are called "plant macronutrients." They are usually the elements limiting productivity, if not supplied in amounts adequate for (optimal) crop growth either by the soils or added as fertilizers. Note that yield increases do not increase linearly with addition of fertilizers, but plant growth reaches a limit, caused by physiological limitations, a fact that is described by Mitscherlich's law. Understanding the response of crops to fertilizer amendments also helps one understand why increased global fertilizer consumption is not matched by yield increases, and also explains why farmers usually do not opt for maximum attainable crop yields, because the increase in costs (energy, fertilizer, labor, machinery) prevails the benefits.

Environmental problems caused by wrong or excess fertilizer use, high costs, and the growing concern for sustainable agriculture have helped to promote both organic farming and "integrated plant nutrient systems" by the FAO and others, aiming at reducing fertilizer rates to the amounts needed by the crop under location and management-specific conditions.

Pesticides

The purpose of pesticide application is to increase productivity via reducing losses of crop yields either before or after the harvest. Potential losses can be caused by plant diseases, animal pests like insects or nematodes, or weeds competing with the crop for space, water, and nutrients. Major consumers (and producers) are farmers in the European Union, **Japan**, and North America, but increasing amounts are applied in less developed countries. The plethora of pests is combated with a wide range of toxic substances, most of which are also harmful or even toxic for humans and other non-target organisms and have caused many environmental problems. Agricultural research organizations and many agricultural extension services nowadays favor "integrated pest management." This approach aims at reducing yield losses while minimizing the use of chemical pesticides and simultaneously promotes biological methods like supporting natural enemies of pests and diseases.

Plant Breeding

According to carbon dating, plant breeding—including genetically modified organisms (GMOs)—started with the domestication of the ancestors of wheat and barley in the Middle East in the eighth millennium BC, **millet** and rice in China and Southeast Asia by 5500 BC, and maize and squash in Mexico about 8000 BC. Legumes found in Thessaly and Macedonia have been dated as early as 6000 BC. Since then, farmers and scientists have continuously improved the properties of crop plants with respect to crop yields, nutrients, starch or protein contents, disease resistance, and tolerance toward environmental constraints such as scarcity of water and nutrients, acid or alkaline soils, salinity, extreme temperatures, and other factors. Crop breeding has resulted in thousands of varieties, providing a certain level of yield in almost any environment. This setting changed drastically, when high-yielding varieties (HYVs) were bred starting only decades ago; these were soon covering major parts of the world's most suitable agricultural land, displacing locally adapted varieties. The importance of this process lies in the partly irreversible loss of important genetic resources (genetic erosion), both of crop plants and of adjacent flora and fauna (e.g., birds), because the use of HYVs or GMOs usually comes hand in hand with other intensification methods like fertilizer and pesticide application, mechanization, and so on, resulting in a general loss of **biodiversity** of the agro-ecosystem. Furthermore, **food security** is affected because the genetically uniform crops (HYVs or GMOs), often grown in huge **monocultures**, are all susceptible to the same diseases or pests, and in addition do often have a much higher water demand than local varieties, which may cause the crops to fail (completely) in drier years.

Mechanization

Huge combines used in precision farming and stirred by **global positioning systems** are just some of the most visible results of agricultural mechanization, which started millennia ago with the invention of the sickle, scythe, hoe, plow, and flail. Milestones in the course of mechanization of field preparation and harvesting were the introduction of plows—the use of animal traction—and the replacement of animals by first steam and later diesel engines. Mechanization is partly driven by shortages of agricultural labor forces during labor-intensive periods like harvests. On the other hand, during continued mechanization, for example, since the mid-nineteenth century when the first steam engines came into use in agriculture in Europe and North America, a huge amount of labor was substituted by capital, a process that had severe implications for the labor market. According to FAO data, the number of tractors and harvester-threshers globally roughly doubled from the 1960s to 1980s and has fluctuated around the same level since the late 1980s. During the 1990s, the decrease in the amount of agricultural machinery in the industrialized countries reflected an increase in farm size and a continued trend toward larger equipment. In countries with less industrialized agriculture, where only 10 percent of the world's tractors are located, the trend toward increased mechanization continues, though regionally at quite different rates.

See also: **Agricultural Revolution; Agrodiversity; Biotechnology; Boserup, Ester; Brazilian Cerrado; Carson, Rachel Louise; Conservation; Cotton; Cumulative Change; Decision-Making; Degradation Narrative; Disintensification; East-Central Europe; Human Health; Land Rehabilitation; Land Rent; Mediating Factor; Nutrient Cycle; Population Dynamics; Private Property; Reforestation; Sustainable Land Use; Technological-Scientific Revolution.**

Further Reading

Bruinsma, Jelle, ed., *World Agriculture Towards 2015/2030: An FAO Perspective*, London: Earthscan, 2003; Carswell, Grace, "Agricultural Intensification and Rural Sustainable Livelihoods: A 'Think Piece'," (IDS Working Papers No. 64), Brighton, UK: University of Sussex, 1997; International Federation of Organic Agriculture Movements [Online, December 2004], The IFOAM Web Site, www.ifoam.org; International Institute for Applied Systems Analysis, Global Agro-Ecological Zones [Online, December 2004], The IIASA GAEZ Web Site, www.iiasa. ac.at; Pesticide Action Network [Online, December 2004], The PAN-UK Web Site, www.pan-uk.org; Stoate, C., Boatman, N. D., Borralho, R. J., Rio Carvalho, C., de Snoo, G. R., and P. Eden, "Ecological Impacts of Arable Intensification in Europe," *Journal of Environmental Management* 63 (2001): 337–65; United Nations Food and Agriculture Organization, FAO Statistical Databases [Online, December 2004], The FAOSTAT Web Site, apps.fao.org; United States Department of Agriculture, A History of American Agriculture 1776–1990 [Online, December 2004], the USDA Web Site www.usda.gov; University of Reading, Environmental Challenges in Farm Management [Online, December 2004], The University of Reading Web Site, www.ecifm.rdg.ac.uk.

JÖRG A. PRIESS

Agricultural Revolution. The improvement and increase of agricultural production by revolutionary changes in technique, technology, and equipment of **agriculture** in the context of the fall of the feudal mode of production and the development of capitalism. It is characterized above all by a transition from the triennial (farming under **fallow**) to the alternate systems, by cultivating fodder plants such as clover, alfalfa, and especially root plants such as potatoes and sugar beet, which caused great greed for new plots of arable lands. Agricultural production increases have been achieved above all by improvements in drilling and sowing procedures; by the onset and increasing use of natural and later synthetic fertilizers in order—together with deep tillage—to restore at least or increase soil fertility; by the breeding of cattle in stables; the improvement of tillage tools; the use of machines to prepare the soil to sow and (to care for and harvest the products) to trash grain; and, finally, by the application of science (pedology, physiology of plants, agrobiology, and agrochemistry such as Thaer's humus theory and Liebig's mineral theory from the 1840s). Its historical result was the start of fundamental changes in land use and land cover. The broader term agrarian revolution involves also political, legal, and social changes in agriculture.

The gradual pace of the Agricultural Revolution (AR) in **Europe** generally went almost parallel or rather with some delay to the **Industrial Revolution**. Sociopolitical processes such as enclosures in Britain, the Homestead Act and abolition of slavery in the United States, and the 1848–1849 revolutions in Europe—abolition of manorial labor and creation of competitive labor markets, land kept in tenure of farmers (copyholders) came under their ownership, and so on—accelerated them considerably. The AR in such a sense began in the middle of the eighteenth century and culminated in the second half of the nineteenth century at big estates and later at smaller farms. This process occurred typically by phase shift between different countries in gradients from west to east.

The AR was accompanied by changes in the use and area of especially agricultural land categories; that is, by an extensive expansion of arable land to the detriment of **pastures** and meadows, followed by the almost complete fall of triennial farming by the end of the nineteenth century; and the **conversion** of permanent **grasslands** in fertile lowlands into arable land. Many lakes and ponds were drained to acquire more arable land for the growing of sugar beets, turnip, **wheat**, barley, and potatoes, and in hilly regions for oats, rye, and potatoes. After a previous, almost permanent decrease of **forest**, a period of their steady growth began, which lasted until recent times. The total area of historically created arable land was, after the culmination of the AR, relatively stabilized, and its acreage began to be reduced by plots of less fertile soils or with worse locations to markets, which needed higher inputs of capital to reach competitive yields. These plots have been changed mainly into permanent grasslands and forests.

The total extinction of fallow farming by the end of the nineteenth century brought about an extension of the cultivated arable lands area by about one-third (compared to the end of the eighteenth century). Together with the new arable lands, these areas have been permanently "naked" and exposed to the natural and anthropogenic impacts of decrease in landscape **biodiversity** and large-scale **soil erosion** by wind and water, the effects of which are reinforced by plants such as corn, potatoes, sugar beets, **cotton**, and **tobacco**.

The AR linked agriculture with the agricultural industry (e.g., sugar, distilling, milk, and brewing industries) and with agricultural engineering. These links were responsible for introducing an industrial character into agriculture.

See also: **Agricultural Intensification; Agro-Industry; Biotechnology; Cadastre; Carson, Rachel Louise; Consultative Group on International Agricultural Research; Green Revolution; Land Reform; Land Tenure; Technological-Scientific Revolution.**

Further Reading

Chambers, Jonathan D., *The Agricultural Revolution, 1750–1880,* New York: Shocken Books, 1966; Jeleček, Leoš, "Changes in the Production and Techniques in the Agriculture of Bohemia 1870–1945," in *Agriculture in the Industrial State,* eds. Michael A. Havinden and Edward J. T. Collins, Reading, UK: University of Reading, Rural History Centre, 1995, pp. 126–45; Kerridge, Eric, *The Agricultural Revolution,* New York: A.M. Kelley, 1968.

LEOŠ JELEČEK

Agriculture. A source of food and fiber, which also shapes the rural environment and is a key element in global land-use/cover change. A considerable part of the land surface is managed by farmers, and agriculture accounts for around 40 percent of the total land resource that is available in countries of the Organisation for Economic Cooperation and Development, on average. Historically, agriculture has been the key factor of **economic growth** and prosperity in most countries. Today, less than 5 percent of the labor force in many developed countries is employed in agriculture, and agriculture's share of gross domestic product (GDP) shows a declining trend, but it is more significant in economic terms when indirect and related economic activities—for example, provision of (intermediate) input, transport and processing of food, and retailing—are taken into account. In contrast, the share of agriculture in GDP and employment remains high in countries with limited trading opportunities.

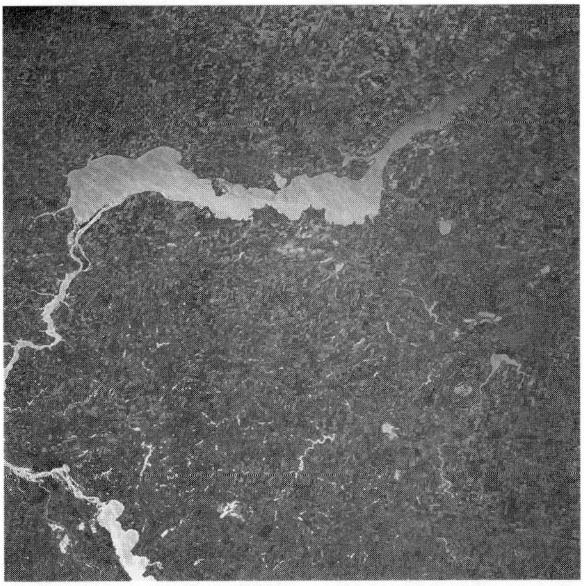

Agricultural region (wheat, corn, potatoes, melons) surrounding the Kakhovsk Reservoir of Dnepr River in Ukraine in July 1996.

PHOTO: Shuttle, Mir Space Station. CREDITS: Image Science and Analysis Laboratory, NASA-Johnson Space Center, 18 Mar. 2005 (Earth from Space—Image Information). http://earth.jsc.nasa.gov/sseop.efs

The production of food by farmers has substantial side effects or externalities to the physical and natural environment, including loss of **biodiversity**, harmful effects of the use of agrochemicals (e.g., fertilizers and crop protection products) on the quality of water and soils, and excess amounts of nutrients applied on the soils. Such externalities illustrate the size of the problem, and multifunctionality applied to agriculture means a system

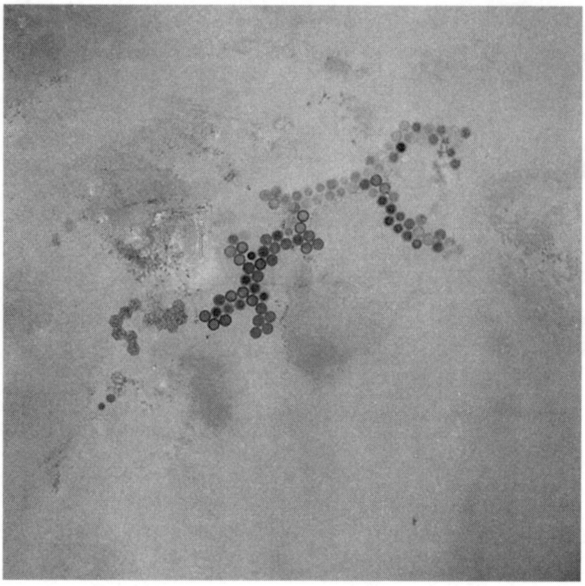

Pivot irrigation agriculture in the Al Kufrah Oasis in southeastern Libya, October 1988 (circular fields in the desert of ca. one kilometer in diameter, with dark color indicating current cultivation).

PHOTO: Space Suttle. CREDITS: Image Science and Analysis Laboratory, NASA-Johnson Space Center, 18 Mar. 2005 (Earth from Space—Image Information). http://earth.jsc.nasa.gov/sseop.efs

that does more than the production of food and fiber. There are important public goods (quality of water, soil, and air, and landscape values) that offer opportunities to generate income. Strengthening the relationship between agriculture and sustainable development would improve the provision of public goods and services. Sustainability is a main societal objective, incorporating economic, environmental, and social dimensions of agriculture, and also considering the supply of human, natural, physical, and social capital.

Diverse Agricultural Systems

Farming types are very diverse, ranging from small-scaled farms that mainly operate for subsistence in agrarian societies with limited trading opportunities, to larger farms whose main emphasis is the provision of food. Specialization in agriculture tends to rise with an increase in welfare, and agricultural systems remain diverse, with efficiency gains to be achieved by increasing the production per unit of land and/or labor. Improvements in efficiency from fixed inputs like land or labor depend on the factors that are most limiting. Resources like land and labor are used more efficiently over time due to technological progress.

In contrast to other exporting countries on the globe, average farm size in **Europe** is rather small—fewer than twenty hectares—and mixed farming types remain an important feature of landscapes. European landscapes are largely cultural and are heavily influenced by management of the land from farming and **forestry**. Around 90 percent of agricultural land in Australia is used for grazing cattle and sheep for meat and wool production, and only 5 percent is **pasture** grass and 5 percent is cropped.

Main Trends

Land use for agriculture has been falling in most OECD countries but is being used more intensively over time and supplying an increase of production per unit of land area. The pressure of population and food demand for the intensity of land use is observed in large parts of the globe. India, for example, accounting for about a sixth of the world's population and only about 2 percent of its land surface, has faced increasing pressure on land use. Food production doubled over the past two decades by an increase of yields rather than expansion of cultivated area. Similar trends to intensify production are also observed in other parts of the world. In Central America, for example, agricultural production increased considerably over the past decades, which was largely due to **agricultural intensification** rather than the expansion of agricultural land or increases in labor input. Two dominant trends in current farming practices are intensification, concentration of production and specialization in some areas, and marginalization and **land abandonment** in others. They both involve a move away from traditional forms of low-input, labor-intensive crop and livestock production, which have long characterized most agriculture.

First, intensification and specialization involve the development of capital-intensive and geographically specialized farming, which is mainly observed in regions where agriculture is most productive. Competitive advantages may arise in some regions because of better biophysical conditions, more rationalized farm structures, the integration of primary production with food processing industries, and well-developed farm extension services. Here, sustainable farming practices emerge to better respond to changes in consumer demand (in terms of quality and diversity of food) and meeting environmental constraints. Ambitions for **sustainable land use** are linked to the long-term economic viability of agriculture and strengthening the competitive position on export markets.

Second, marginalization and large-scale abandonment of agricultural land tend to occur in remote areas with unfavorable economic or social conditions, or on less fertile land where traditional extensive agriculture is threatened by its inability to compete effectively with intensive production in other regions. Abandonment, **land degradation**, and economic decline threaten regions where harsh natural conditions, poor soils, and long distances to markets increase the costs of agricultural production, and where rural populations show a declining trend. Ambitions for sustainable agriculture could be linked to strengthen multifunctionality in an effort to cope with marginalization in agriculture.

Control of External Effects

Agriculture is by far the most important activity shaping rural landscapes, affecting water quality as well as carbon stocks and nutrient flows in soils. Proper land management practices that ensure the productivity of the land also are vital to the maintenance of primary production. Environmental and rural concerns are a high public priority. Increasingly, modern society values the resources that arise as joint outputs with primary production, including water supply of good quality and sufficient amounts, habitats and wildlife, land settlements and their associated cultural artifacts, landscapes, and open spaces. What are the main external effects from agriculture, both beneficial and benign?

Agriculture is a major user of land and water; and farming practices can have harmful effects on soil, air, and water quality. Agriculture is a dominant activity using the land, and agriculture also is a major user of the available water resources. Agriculture may account for more than half of total water consumption, especially in cases of irrigation. Societal debate on nitrates and pesticides in water that started in the late 1980s has given incentives to better control the environmental effects of farming practices, especially in regions with intensive farming practices (with high and/or improper use of nutrients or pesticides). **Soil degradation** from salinity and erosion are key processes that are a widespread concern in regions where irrigation farming accounts for a high share of water use and excessive extraction of groundwater levels.

The public sense of urgency regarding harmful effects of hog production on the environment is high and largely focused on water pollution, noise, and odor; and more recently also on ammonia and **greenhouse gas** emissions. Since then, the interest has moved toward a more targeted and rationalized use of inputs. This has been shown to be a proper strategy to control external effects of agriculture on the environment. Mandatory measures are introduced to enhance farm management practices that better respect the environment. In addition and more recently, environmental quality measures are linked with food safety aspects. Constraints are imposed on individual farmers, either directly through mandatory measures like standards or through regulations that affect farm practices in an indirect way (e.g., standards for water quality putting constraints on farming practices). The farming community increasingly responds to the societal demands to alter production methods applied in agriculture. Such societal demands might be reflected by rules on the use of inputs, put either by the food processing industry and food retailers or by **public policy**. Rules include conditions regarding the use of agrochemicals like nutrients and pesticides, and codes of good agricultural practice are important in clarifying the responsibilities of managing environmental resources by farmers.

The provision of public benefits from agriculture (e.g., management of biodiversity, improving the balance of water availability with local needs, nutrient recycling, and the protection of biodiversity and wildlife) can help to increase the value of market activities. The support of such non-commodity outputs requires public support when markets fail or are non-existent. Public policies with voluntary standards for compensation eligibility are widely initiated by public agencies, as are standards developed by food processors, retailers, or other markets.

Public Support

Agriculture is a highly supported sector. Traditionally, the main emphasis has been given in most of the developed countries to the markets of agricultural products and the income formation of the producers. However, parallel to the pressure for market **access** and the abolishment of export subsidies, there also has been a slow downward trend in total support since the mid-1980s. In addition, major changes have taken place in the introduction of payments that are conditional to environmental constraints. Public support needs to respond to public concerns; and issues like the sustainable use of natural resources, protection of the environment, development of rural areas, animal welfare, alleviation of **poverty** and **food security** are put on the agenda of reforming agricultural or **land-use policies**. The ambition of European agricultural policy is to enhance a sustainable and viable agricultural sector. This is supported by policies that acknowledge the wide diversity of farming systems.

With the introduction of cross-compliance, market and price support measures for dairy products, beef, sheep, and cereals are important to provide incentives for sustainability in agriculture. Cross-compliance is an instrument introduced to reinforce legislative standards related to environment, food safety, animal welfare, and nature management. It is a basis for expressing social responsibility of the agricultural sector that provides food, and it has a supplementary role in managing the rural countryside. Cross-compliance is part of the process of integrating environmental, food safety, welfare, and nature concerns in the European Union common agricultural policy (CAP), but is essentially meant to maintain the status quo and not meant to promote the provision of public goods beyond what is legally required. Being part of the first pillar of the CAP, it implies that direct payments might be withdrawn in part when farmers do not respect the requirements. Of the eighteen pieces of legislation, five are environmental and will be applicable from January 1, 2005, including the birds and habitats directives. Rather than giving positive signals to farmers, cross-compliance is an instrument suitable to reversing farming practices that are harmful for the environment and nature.

In addition, the public increasingly demands healthy and safe food. The agrifood sector plays a vital role in meeting environmental requirements, producing safe food, and improving animal welfare as well as **human health** and animal health. Retailers and the food processing industry, for example, are demanding better and audited farming systems in response to changed consumer demands. In doing so, they promote sustainable practices in agriculture. Therefore, agriculture must respond to and work with others in the agrifood chain. Public-private partnerships may be the way forward for meeting societal demands on the agricultural sector. The incorporation of environmental concerns in marketing strategies from retailers could change farming practices and also contribute to reducing efforts needed for meeting public policy objectives.

Agro-Food Sector and Sustainability

Major structural changes are taking place in the agro-food sector. Processes of concentration and internationalization have given food retailers substantial market power vis-à-vis their suppliers. This in turn has triggered a process of consolidation among the food processing industry, wholesalers, and even farmers. Firms participating in a production and distribution chain for agricultural and food products—farmers, processing industry, wholesalers, and retailers—are increasingly working together to gain efficiencies in logistics and information exchange and to set up quality monitoring and control systems throughout the food chain.

Consumers in developed countries have become more concerned about the quality of food products and also about the quality of production and processing methods applied on the farm and in the manufacturing plant. Such consumer concerns relate to food safety and quality, environmental sustainability, and ethically appropriate methods of production. As a result, farmers, the food processing industry, and retailers have initiated efforts to guarantee safe products produced in a sustainable way. The environmental issue has even become part of the competition strategy of farmers, the food processing industry, and retailers.

Food retailers have become particularly concerned about the quality of fresh produce because they either sell top-quality products under private labels or they advertise their company as being an environmentally conscious food supplier. Not only is fresh produce like fruits and vegetables increasingly sold under private labels; but chilled foods, ready-to-eat meals, prepared vegetables, and fruit salads are popular products within the own-brand strategy. For private label products, retailers take responsibility for quality, because it is their brand that is at risk if quality flaws appear.

These structural changes in food processing and food retailing lead to more elaborate quality control systems throughout the whole agro-food chain. Quality control at the point of purchase is no longer sufficient, as some quality characteristics cannot easily be measured and as the cultivation methods used on the farm have become part of the quality characteristics of the final product. The food processing industry and retailers set strict requirements for sustainable cultivation practices by their suppliers. Quality monitoring and control systems also give the food processing industry and retailers more insight into the primary production parameters, and thus more options for (re)directing cultivation decisions. Once measurable sustainable agriculture indicators have been established, it becomes possible to select and reward suppliers on the basis of their score on these indicators.

Future Perpectives

Different factors drive long-term food demand, and different factors create pressure to increase food production. Global population may further increase from the current six billion people to eight or nine billion people within the next couple of decades. Also, per capita income is foreseen to increase by a factor of three or more by 2050. In addition, the share of animal calories in diets may well increase from about 15 percent today to about a third in 2050. Such factors all increase environmental pressure on land and water. Technological progress remains an important driving factor in increasing productivity in agriculture and allowing further intensification of production.

See also: **Agricultural Frontier; Agricultural Revolution; Agrodiversity; Agro-Industry; Biomass; Biotechnology; Carbon Cycle; Cash Crops; Coffee; Colonization; Conservation; Consumption; Contract Farming; Cotton; Desertification; Domestication; Driving Forces; Ecological Colonization; Economic Livelihood; Economic Restructuring; Environmental Change; Fallow; Food Crops; Forest Transition; Green Revolution; Indigenous Knowledge; Investments; Land Evaluation; Land Fragmentation; Landholding; Land Quality; Land Reform; Land Rent; Land-Use System; Metabolism; Monoculture; Nutrient Cycle; Pristine Myth; Private Property; Proximate Causes; Remittances Landscape; Subsistence Agriculture; Swidden Cultivation; Technological-Scientific Revolution; Turnover; United Nations Food and Agriculture Organization; Water-Land Interlinkages.**

Further Reading

Brouwer, Floor, and D. E. Ervin, eds., *Public Concerns, Environmental Standards and Agricultural Trade*, Wallingford, UK: CABI Publishing, 2002; Organisation for Economic Cooperation and Development, *Agriculture, Trade and the Environment: The Pig Sector*, Paris: Organisation for Economic Cooperation and Development, 2003;

Pretty, J., *Agri-Culture: Reconnecting People, Land and Nature,* London: Earthscan, 2002; Sharma, Neeraj, "Population as a Driver of Future Land Use Pattern in India," in *Rural Lands, Agriculture and Climate beyond 2015: Usage and Management Responses,* eds. Floor Brouwer and Bruce McCarl, Dordrecht, NL: Kluwer, 2005, in press.

FLOOR BROUWER

Agriculture, Ecosystems & Environment. An international journal for scientific research on the interactions among **agriculture**, agro-ecosystems, and the environment, with a focus on agro-ecosystem sustainability in as wide an international or comparative context as possible. Since 1974, *Agriculture, Ecosystems & Environment* has covered, among others, topics such as farming systems changes or the dynamics of **land-use systems**, influence of agricultural production methods on the environment, agro-biodiversity, and issues of **sustainable land use**. The journal addresses scientists in agriculture, food production, **agroforestry**, ecology, the environment, and earth and resource management, as well as administrators and policymakers in these fields.

See also: **Agricultural Intensification; Agrodiversity.**

Further Reading

Elsevier Science Ltd. [Online, January 2005], The Elsevier Web Site, www.elsevier.com.

HELMUT GEIST

Agrodiversity. The many ways in which farmers use the natural diversity of the environment for production, including crop choice (or the management of species in land use), soil and water management practices, and marketing arrangements, typically as a feature of whole fields, farms, communities, and landscapes. **Harold Brookfield** and Michael Stocking divide agrodiversity into four principal elements or types. Biophysical diversity means the diversity of the natural environment (mainly features of soils, hydrology, and microclimate), which controls the intrinsic quality of the natural resource base. This allows the base to be exploited by agricultural or **land-use systems**, including ecosystem **resilience**. Management diversity includes all methods (biological, chemical, and physical) of managing land, water, and biota for crop and livestock production and the maintenance of soil fertility and structure. Farmers' local knowledge, constantly modified by new information, leads to the formation of management diversity. Agro-biodiversity is shorthand for biological diversity or **biodiversity** on land used for agricultural purposes. Typically, an attribute of an organism or site, the diversity of crops, semi-domesticates, and wild species is particularly important: the manner in which these are used to sustain or increase production, reduce risk, and enhance **conservation**. Finally, and underpinning management diversity, organizational diversity includes diversity in the manner in which farms are owned and operated, and in the use of resource endowments, such as labor, gender, household features, and off-farm employment. All categories of agrodiversity are closely interrelated, they are shaped by **mediating factors**, and they play out differently at various time and spatial **scales**. For example, crop choice and type of conservation practices often differ between poor and rich farmers, thus affecting the pattern of management diversity, and feeding back to enlarge differences in natural **land quality**.

See also: **Agriculture; Biodiversity Novelties; Biomass; Biotechnology; Cash Crops; Coffee; Cotton; Domestication; Food Crops; Participatory Land Management; People at Risk; Project on People, Land Management and Ecosystem Conservation; Turnover; World Agroforestry Centre.**

Further Reading

Brookfield, Harold, *Exploring Agrodiversity*, New York: Columbia University Press, 2001; Brookfield, Harold, Parsons, Helen, and Muriel Brookfield, eds., *Agrodiversity: Learning from Farmers across the World*, Tokyo: United Nations University Press, 2003; Wilson, Edward O., *The Diversity of Life*, Cambridge, MA: Belknap Press of Harvard University Press, 1992; Xu, Jianchu, Mikesell, Stephen, eds., *Landscapes of Diversity: Indigenous Knowledge, Sustainable Livelihoods and Resource Governance in Montane Mainland Southeast Asia*, Kunming, China: Yunnan Science and Technology Press, 2003.

HELMUT GEIST

Agroforestry. A collective name and abbreviation for—as well as the science about—agrosilvicultural systems (involving **agriculture** and silviculture, or **forestry**), silvopastoral systems (involving forestry and **pasture**), and agrosilvopastoral **land-use systems** (involving agriculture, forestry, and pasture). Such integrated farming and forest systems embrace the interactions of livestock; woody perennials (trees and shrubs); herbaceous plants (crops and pasture); and people's cultural patterns, **indigenous knowledge**, and traditional land management techniques.

Among all land uses, agroforestry systems possibly achieve the highest degrees of **agrodiversity** worldwide. They can be centered upon **forests**, the composition of which is deliberately manipulated. Alternatively, they can be centered upon cropping systems, livestock systems, or mixed livestock-cropping systems; or based on the production of annual and/or perennial plants, trees grown for multiple purposes (e.g., shade, timber, fruit crops, leaf fodder, medicinal products, resin, latex), and domestic livestock. **Swidden cultivation** in tropical Southeast Asia, for example, provides many examples of traditional forest communities that have created complex and highly productive agroforestry systems—in sharp contrast to prevailing views that shifting cultivation is the major cause of tropical **deforestation**.

A famous example of the agrosilvicultural variant of agroforestry is the *taungya* system of Myanmar, which—in its modern manifestation in the form of forest villages in Thailand—comprises teak, **maize**, and **rice**. Numerous other combinations exist, such as trees on cropland in the traditional parklands or savannas of the **Sudan-Sahel** in West Africa. There, millet and sorghum are grown with *Faidherbia (Acacia) albida* on predominantly infertile soils, deficient in phosphorus and nitrogen. However, crops within a 510-meter radius have double or triple the yields of crops growing in the open, since the trees have an ability to capture nutrients and fix nitrogen, which are then cycled to the crops. The trees drop their leaves at the beginning of the rainy season, and the leaves rapidly decay, releasing their nutrients. The trees also improve the microclimate in these semi-arid regions. In addition to improving crop growth, the trees provide wood fuel and valuable fodder, and pods and shade for animals during the dry season. As in humid forest zones, dryland cases, which show the integration of farming with trees and animal husbandry, tend to contradict **degradation narratives** such as the prevailing views of **desertification** (e.g., population growth). In densely populated village areas of the Nigerian shrub-**grassland** zone, Michael Mortimore and colleagues found that recent land-use/cover changes over a decade were not synonymous with **land degradation** in the sense of lost plant biomass productivity. Rather, data suggest that the transformation of woodland or shrubland into farmland added economic value if—not only marketable crops, but also—the value of all **biomass** produced from farms were considered in assessment, that is, including subsistence crops, fodder, construction and craft materials, fuel, or medicine, with a considerable part of these goods and services stemming from the management of trees on farms.

Agroforestry systems are not limited to the tropical zone but exist in the subtropical and temperate zones as well. For example, long-term uses of trees have been incorporated into sustainable agricultural systems such as the sugar maple-based land-use system in southern **Canada** and northeastern **United States of America**, and the modern paulownia-wheat intercropping system in **China**. Different from **monocultures**, the integration of various land-use elements in agroforestry generates high ecological **resilience**, controls **soil erosion**, promotes **nutrient cycling**, and maintains soil organic matter. Provided that **biodiversity** is maintained or even increased, the mosaic pattern of land use created by agroforestry means a reproductively viable and long-term **sustainable land use**. This insight has triggered experimental research—along with organic farming—under the umbrella of the **World Agroforestry Centre**.

See also: **Alternatives to Slash-and-Burn (ASB) Programme; Community Involvement; Consultative Group on International Agricultural Research; Cultural Factors; Decision-Making; Non-Timber Forest Products; Regrowth; Secondary Vegetation; Tradeoffs.**

Further Reading

Hladik, Annette, "Agroforestry," in *Encyclopedia of Global Environmental Change, Vol 3: Causes and Consequences of Global Environmental Change,* eds. Ted Munn and Ian Douglas, Baffins Lane, Chichester, UK: John Wiley & Sons, 2002, pp. 172–8; Mead, Donald J., Agroforestry [Online, March 2005], The Encyclopedia of Life Support Systems Web Site, www.eolss.net; Mortimore, Michael, Harris, Frances M. A., and Beryl Turner, "Implications of Land Use Change for the Production of Plant Biomass in Densely Populated Sahelo-Sudanian Shrub-Grasslands in North-East Nigeria," *Global Ecology and Biogeography Letters* 8 (1999): 243–56.

HELMUT GEIST

Agro-Industry. A synonym for agribusiness or a large commercial, agricultural enterprise or sector; connoting a particular business configuration

Almost entirely by mechanical means, "truck crops" such as beans and peas have been grown on thousands of hectars on Seabrooks Farms (own operation as well as contract farming), between Bridgeton and Vineland in New Jersey in 1936 (for its own large cannery for rapid freezing and lots for the market of Philadelphia, Baltimore and Washington).
PHOTO: Edwin Rosskam. [Courtesy Library of Congress]

in which a single corporation (usually transnational) coordinates industrial or commercial activities through subsidiary companies in the spheres of manufacture and distribution of farm supplies; production operations of farms (contract farming); and storage, processing, and distribution of farm commodities and items made from them. The process of agro-industrialization thus comprises at least three related sets of changes: growth of commercial, off-farm agro-processing, distribution, and input provision activities; institutional and organizational change in the relations between farms and firms both upstream and downstream (such as a marked increase in vertical integration and contract farming); and related changes in product composition, technologies, and sectoral and market structures. The concepts of agro-industry and, more particularly, agro-food chains, agro-food complexes, and agro-food regimes are complementary in the sense that they shed light on various institutional aspects of the contemporary system of food production.

The concepts relate to what Karl Kautsky earlier called the "agrarian question," that is, the various forms in which capitalist relations transform the agrarian sector and the political alliances, struggles, and compromises that emerge around different trajectories of change, including land-use/cover change and its impact on human well-being and biogeochemical cycles. According to Kautsky, agro-industry is rooted in the increasing application of science, technology, and capital to the food processing, farm input, and farm finance systems. Historically, industry—or more properly agro-industrial capital—was the motor of agricultural development, but the peculiarities of **agriculture** (its biological character and rhythms), coupled with the capacity for family farms to survive through self-exploitation, has hindered some tendencies such as the development of classical agrarian capitalism.

Contemporarily, there are several links between, for example, **economic liberalization** and **land degradation**, especially in the developing world. However, the actual and potential environmental effects of agro-industrialization upon natural environments have been sparsely documented to date. Christopher Barrett and colleagues propose to categorize the distinctive characteristics and multiple prospective effects—in the context of **globalization** and international development—into three classes: direct effects on farming and upstream supplier industries, direct effects on downstream processing and distribution in the commercial agro-food channel, and indirect effects due to income growth and induced structural changes.

As for the first situation, there is evidence that retail consumers in markets of the developed world increasingly value the esthetic appeal of agricultural produce, thus inducing farmers in developing countries, especially on large **landholdings**, to use chemical fungicides, insecticides, and pesticides in greater volumes than ever before, in particular for cut flowers and **cash crops** such as **cotton** and **tobacco**. On the other hand, there is decent indication that vertical coordination may also result in inducements to adopt environmentally friendly land-use practices in order to get premium prices, such as for organically grown **coffee** paid by primarily wealthier and informed consumers in developed countries. Likewise related to the direct effects on farming and upstream supplier industries are technologies to introduce genetically modified organisms shaped by private firms. However, ecological risks remain poorly understood, and firms do not have to take social costs and benefits into account.

In the **United States of America**, the agro-industrial complex developed in particular during the second half of the twentieth century. While the first half of the century had been characterized as the golden age of American agriculture due to high demand and high prices for agricultural products (and a continuous appreciation of land), the problem of surplus production continued to exert pressure on farmers after World War II. The farmers successfully managed to increase output per hectare by utilizing pesticides, chemical fertilizers, and technological innovations for rural development. Since the end of the twentieth century, U.S. farmers have had to meet rising and very sophisticated consumer demands while facing fierce competition in a global economy and a global supply of goods and products. In the course of globalization, U.S. agriculture turned to an agro-industrial mode of production: in 1940, 6 million farms, with an average of sixty-seven hectares, existed in the United States. At the end of the 1990s, only 2.2 million rural corporations and businesses, with an average of 190 hectares, existed, many of them in the hands of shareholders. During this period of time, employment in the agricultural sector decreased dramatically, that is, from 12.5 million to 1.2 million workers employed by farmers and corporations. In other words, while agricultural production more than doubled during the last fifty years, more than two-thirds of all farms and corporations have disappeared. Today, 150,000 U.S. farmers and corporations produce the bulk of food and raw materials. These producers are ranked as the most powerful ones worldwide, although they are only a minority of all self-employed within the agricultural sector in the United States (and a diminutive minority of all self-employed on a global scale).

See also: **Agricultural Intensification; Agricultural Revolution; Biotechnology; Cash Crops; Corporate Strategies; Food Crops; Industrialization; Monoculture; Technological-Scientific Revolution; Tragedy of Enclosure.**

Further Reading

Barbier, Edward B., "Links between Economic Liberalization and Rural Resource Degradation in the Developing Regions," *Agricultural Economics* 23 (2000): 299–310; Barrett, Christopher B., Barbier, Edward B., and Thomas Reardon, "Agroindustrialization, Globalization, and International Development: The Environmental Implications," *Environment and Development Economics* 6 (2001): 419–33; Clay, Jason, *World Agriculture and the Environment: A Commodity-by-Commodity Guide to Impacts and Practices*, Washington, D.C.: Island Press, 2004; Fold, Niels, and Bill Pritchard, *Cross-Continental Agro-Food Chains: Structures, Actors and Dynamics in the Global Food System*, London: Routledge, 2005; Goodman, D., and Michael J. Watts, eds. *Globalising Food: Agrarian Questions and Global Restructuring*, London: Routledge, 1997; Kautsky, Karl, *The Agrarian Question*, London: Zwan, 1899/1988; Whatmore, Sarah, "From Farming to Agribusiness: The Global Agro-Food System," in *Geographies of Global Change: Remapping the World in the Late Twentieth Century*, eds. R. J. Johnston, Taylor, Peter J., and Michael J. Watts, Oxford, UK: Blackwell, 1995, pp. 36–49.

HEIKE GRIMM, HELMUT GEIST

Airborne Light Detection and Ranging. Also known as airborne laser terrain mapping, airborne light detection and ranging (LIDAR) is a fairly new technology that holds much promise for the scientific community, as it provides accurate elevation and positioning of topography and vegetated surfaces. In contrast to traditional **remote sensing** systems that acquire information about the terrain and landscape on a horizontal plane, LIDAR systems provide information on the vertical structure of a landscape. LIDAR determines range information by recording the time interval between a transmitted laser pulse and its reflected pulse, usually in the near-infrared portion of the electromagnetic spectrum. An accurate (to within fifteen centimeters) three-dimensional location for each laser spot on the ground is determined by combining the range information with precise positioning from a **global positioning system** and an inertial motion unit. The irregular sampling of LIDAR pulses are often posted to a regular grid, creating a digital surface model. In many applications, the LIDAR data are filtered to remove points that are reflections from **vegetation** or man-made features, thus creating a model of the terrain, or a digital elevation model. LIDAR applications also include support of hydrological modeling, vegetation structural analysis, and urban-density **modeling**.

See also: **Change Detection.**

Further Reading

Airborne 1 [Online, October 2004], The Airborne Laser Mapping Background Web Site, www.airbornelasermapping.com; Rees, William G., *Physical Principles of Remote Sensing*, 2nd ed., New York: Cambridge University Press, 2001.

KELLEY A. CREWS-MEYER, AMY L. NEUENSCHWANDER

Albedo. The ratio between the amount of light reflected by an object and the amount of incident light. In simple terms, it is a measure of the reflectivity or intrinsic brightness of an object. Albedo is normally expressed as a fraction between 0 and 1. A perfect reflecting surface would have an albedo of 1.0, and a perfect absorbing surface would have an albedo of 0.0. The albedo of the earth's surface depends on the type of land cover. Typical albedo values of general land-cover types are the following: **forests**, 0.05–0.2; sand/desert, 0.2–0.3; fresh snow, 0.75–0.95; and grass, 0.15–0.25. The albedo of water depends on the roughness of the surface and the solar incidence angle; it is 0.03–0.1 at high sun angles and in calm conditions, and 0.7 at low sun angles and in rough conditions. A change in albedo, detected from an airborne or spaceborne sensor, can in certain cases be interpreted as a change in land cover or land use.

See also: **Climate Impacts; Desertification; Disequilibrium Dynamics; Remote Sensing.**

Further Reading

Betts, Richard A., "Offset of the Potential Carbon Sink from Boreal Forestation by Decreases in Surface Albedo," *Nature* 408 (2000): 187–90.

KEVIN J. TANSEY

Along Track Scanning Radiometers (ATSR) World Fire Atlas. A database of monthly global **fire** maps produced by the series of along track scanning radiometers (ATSR) onboard successive European Space Agency **European remote sensing (ERS-1/-2)** and ENVISAT satellites. Data are available through the Internet from 1995 to the present day. Satellite images are acquired during darkness to reduce solar artifacts, in four thermal bands at one-kilometer resolution. This ensures almost complete global coverage on a daily basis. Two algorithms are used to compute the locations of active fires. The product has been used to compute the emissions of **greenhouse gases** and aerosols from **biomass** burning and to explore the impacts on tropical ozone levels.

See also: **Atmosphere-Land Interlinkages; Global Land Cover Map of the Year 2000; GLOBSCAR; Remote Sensing.**

Further Reading

European Space Agency [Online, August 2004], ATSR World Fire Atlas Web Site, shark1.esrin.esa.it/ionia/FIRE/AF/ATSR; Schultz, Martin, "On the Use of ATSR Fire Count Data to Estimate the Seasonal and Inter-Annual Variability of Biomass Burning Emissions," *Atmospheric Chemistry and Physics* 2 (2002): 387–95.

KEVIN J. TANSEY

Alternatives to Slash-and-Burn (ASB) Programme. A systemwide program of the **Consultative Group on International Agricultural Research** (CGIAR) and conceived from recommendations at the 1992 Rio Earth Summit (**Agenda 21**, Chapter 11, Combating Deforestation), ASB has operated since 1992. The **World Agroforestry Centre**, or International Center for Research in Agroforestry (ICRAF), convenes the ASB program and hosts its global coordination office in Nairobi, Kenya. As a partnership of over eighty institutions around the world, including research institutes, non-governmental organizations, universities, community organizations, and farmers' groups, ASB operates as a multidisciplinary consortium for research, development, and capacity building. ASB applies an integrated natural resource management approach to analysis and action through long-term engagement with local communities and policymakers.

ASB undertakes participatory research and development of technological, institutional, and policy innovations to raise productivity and income of poor rural households in the humid tropics without increasing **deforestation** or undermining essential environmental services. **Poverty** reduction in the humid tropics depends on finding ways to raise productivity of labor and land, often through **agricultural intensification** of smallholder production activities. Although there are some opportunities to reduce poverty while conserving tropical forests, tropical deforestation typically involves **tradeoffs** among the concerns of poor households, national development objectives, and the environment. ASB partners work with households to understand their problems and opportunities. Similarly, consultations with local and national policymakers bring in their distinctive insights. In this way, participatory research and policy consultations guide the iterative process necessary to identify, develop, and implement combinations of policy, institutional, and technological options that are workable and relevant.

ASB benchmark sites have been established in the Amazon of Brazil and Peru, the Congo basin forest of Cameroon, the island of Sumatra in Indonesia, the northern mountains of Thailand, and the island of Mindanao in the Philippines. These are areas (roughly 10^2–10^3 km^2) of long-term study and engagement by partners with households, communities, and policymakers at various levels. All benchmark sites are in the humid tropical and subtropical broadleaf forest biome, as mapped by the World Wide Fund for Nature (WWF). The most biologically diverse terrestrial **biome** by far, **conversion** of these forests leads to the greatest species loss per unit area of any land-cover change.

ASB's multi-site network helps to ensure that analyses of local and national perspectives and the search for alternatives are grounded in reality. ASB focuses on the landscape mosaics (comprising both **forests** and **agriculture**) where global environmental problems and poverty coincide. Of course, the rural poor are not alone in using the slash-and-burn method to clear forests, and their activities often are not the major cause of tropical deforestation. This method is used by many actors—government agencies and private companies, large and small-scale farmers, rich and poor—who convert forests to other uses. Nor is it accurate to equate **slash-and-burn agriculture** with permanent forest conversion and unsustainable land use. **Swidden cultivation**, or traditional shifting cultivation, which has been practiced in the tropics by generations of local people, is sustainable as long as population densities are low enough to allow long **fallow** rotations. But traditional shifting cultivation tends to disappear as rural population densities and market integration increase. The resulting land uses may not be economically or environmentally sustainable because of **soil degradation**, nutrient depletion, and loss of other ecosystem functions. Alternatively, **access** to markets may make forest-derived land uses so profitable that they attract an inflow of migrants, which, in turn, accelerates forest conversion. Understanding the factors at work in a given situation is a crucial first step if policymakers are to introduce effective measures to curb deforestation, and to do so in ways that reduce poverty.

ASB researchers have developed and tested innovative indicators of above- and below-ground **biodiversity**, carbon stocks and **greenhouse gas** emissions, agronomic sustainability, returns to labor and other determinants of adoptability by smallholders, and national policymakers' concerns. These methods have been applied to a range of land uses at ASB benchmark sites. The results link global environmental benefits to **sustainable land use** alternatives. ASB is extending its analyses to include hydrological, ecological, and other environmental services at the **watershed** or landscape scale.

To weigh up tradeoffs across various objectives, ASB researchers developed a tool known as the ASB matrix. In the ASB matrix, natural forest and forest-derived land uses can be compared with reference to various indicators reflecting the objectives of different interest groups. At each site, researchers have evaluated **land-use systems** both as they are currently practiced and in the alternative forms that could be possible through policy, institutional, and technological innovations. To enable results to be compared across sites, the systems specific to each site are grouped according to broad categories, ranging from **agroforests** to **grasslands** and **pastures**.

ASB analyses show that no forest-derived system is a perfect substitute for the global environmental benefits of forest **conservation**. Moreover, under a wide range of circumstances, intensification can accelerate the deforestation process. But the results also indicate that many smallholder land uses are agronomically sustainable. A key insight from this work is that these (locally) sustainable options differ significantly in their environmental impacts and their profitability and adoptability by poor households. These results indicate that a middle path of development can be created—involving smallholder tree-based systems and community-based forest resource management—that could attain an attractive balance between the environment and development. Whether or not this balance can be achieved depends on a range of policy and institutional innovations, including means to effectively protect natural forests and to compensate households for foregone opportunities.

ASB is conducting a crosscutting assessment titled "Forest and Agroecosystem Tradeoffs in the Humid Tropics" for the **Millennium Ecosystem Assessment** (MEA). This multiscale assessment, which matches ASB's ecoregional focus, aims to synthesize results across ASB benchmark sites and place these results within the broader context of other scientific evidence.

Translating ASB's work on tradeoffs among global, national, and local objectives into impacts depends on effective dissemination of information to stakeholders in forms they can use. The ASB Policybriefs series is designed to deliver relevant, concise reading to key people whose decisions will make a difference to poverty reduction and environmental protection in the humid tropics. The ASB Voices series conveys for a broad audience insights and perspectives from people's real-life experiences and challenges in the humid tropics.

An ASB Website in English, with mini-sites in Spanish, French, and Portuguese, includes a searchable database of more than 780 ASB publications. Key recent publications include two edited collections of ASB results. The work of Cheryl Palm and colleagues covers the development of the ASB approach and presents results from a decade of thematic research, analysis of site-specific alternatives, national perspectives, and cross-site comparisons. Thomas Tomich and colleagues report results from efforts to extend the ASB analysis to the landscape/watershed scale, including work on smoke pollution, biodiversity functions, and watershed functions.

See also: **Agricultural Frontier; Carbon Cycle; Center for International Forestry Research; Fire; Forest Degradation; Hot Spots of Land-Cover Change; Land-Use Policies; Participatory Land Management; Project on People, Land Management and Ecosystem Conservation; Proximate Causes; Public Policy; Tropical Humid Forest.**

Further Reading

Alternatives to Slash-and-Burn Programme, Innovations to Reduce Poverty and Conserve Tropical Forests [Online, November 2004], The ASB Web Site www.asb.cgiar.org; Palm, Cheryl A., Vosti, S. A., Sanchez, P. A., and P. J. Ericksen, eds. *Slash and Burn Agriculture: The Search for Alternatives*, New York: Columbia University Press, 2005; Tomich, Thomas P., van Noordwijk, Meine, and David E. Thomas, eds., "Environmental Services and Land Use Change: Bridging the Gap between Policy and Research in Southeast Asia," a Special Issue of *Agriculture, Ecosystems and Environment* 104 (2004).

THOMAS P. TOMICH

Amazonia. The equatorial Amazon corresponds to the Amazon basin in **South America**. Although it is often related to what is labeled Brazilian Legal Amazon (BLA), the two correspond to two different areas. While the former refers exclusively to the rainforest area, the latter corresponds to the Brazilian government's legal definition of the Amazon basin of Brazil, by government decree of 1953. According to the decree, the BLA covers an area of 5 million square kilometers embracing the six "north" states (Acre, Amapá, Amazonas, Pará, Roraima, and Rondônia), and part of three others (Tocantins, north of the 130° parallel; Mato Grosso, north of the 160° parallel; and Maranhão, west of the 440° meridian). In 1977, the entire state of Mato Grosso was included in the BLA through the complementary law N° 31 (art. 45). The main reason the Brazilian government created the Legal Amazon was to define an area for the administration of economic development, rather than delimiting a region according to a uniform ecosystem. That is the reason why the BLA includes, besides forestland, extensive areas of natural savanna (*cerrado*) to the east and south, and a transition zone between closed forest and *cerrado.*

The equatorial Amazon embraces the portion of tropical forest of the Amazon Basin. There is no agreement, however, about the extent of this area. The **United Nations Food and Agriculture Organization** (FAO) estimated it to be about 356 million hectares in 1981, whereas the Brazilian Institute of Statistics (IBGE) located it in 379 million hectares seven years later. In turn, the National Institute of Spatial Research (INPE) estimated in 1998 that the tropical moist forest area was equal to 419 million hectares. The latter is assumed as the official estimate used by INPE to annually estimate the official rates of deforestation in the Brazilian Amazon.

The terms "tropical moist rainforest" and "evergreen tropical forest" are often used to describe the Amazon forest. This type of forest is commonly characterized as evergreen or partly evergreen located in the lowland tropics, generally around 800 meters or less above the sea level and in frost-free areas, in areas of high and uniform levels of precipitation normally in the 2,000–4,000 millimeter range, and with high and uniform mean temperatures. The tropical **forests** are closed, multi-storied, and comprise a diversity of species. The majority of plants are woody tall trees with an average height of over thirty meters, along with associated herbs, climbers, epiphytes, stranglers, saprophytes, and parasites. The aboveground **biomass** estimated for dense forests of the Brazilian Amazon is about 298 mg ha^{-1}, in a range between 175 and 397 mg ha^{-1}.

It is widely known that underneath the luxurious forest, soil quality is generally poor. In the tropics, soils have, over millions of years, been affected by high temperatures and heavy rainfalls, thus leaching nutrients from the soil. As a result, they are clay soils with a low-nutrient status. High soil acidity and high content levels of aluminum and manganese are common, as are poor levels of nitrogen and phosphorus. Most of the physical and chemical characteristics of tropical soils constrain the development of agriculture, which is even more limited by heavy rainfall and high temperatures, contributing to reduced agricultural production. This does not necessarily mean that all tropical soils are unsuitable for **agriculture**; besides, many of the soil limitations can be managed by implementing already-known technologies.

The Amazon holds the largest continuous tropical forest in the world, although it has lost the highest amount of forest among all tropical countries. **Deforestation** in the Amazon contributes to **global environmental change** because of its globally cumulative impacts (i.e., loss of **biodiversity**) and their links to globally systemic **environmental change** (i.e., climate change).

The Amazon sustains about 60 percent of the world's remaining tropical rainforests and stores billions of metric tons of carbon, which are emitted as **greenhouse gases** when forest is cleared to convert it to other land uses. Tropical deforestation is the most important **land-cover change** as a major component of the **carbon cycle**. Recent assessments suggest that unaccounted parts of forests, including large areas of **regrowth**, constitute much of the missing carbon sink. The Amazon has been labeled a heat engine because it drives large-scale patterns of atmospheric circulation and rainfall. These global links suggest that land-use changes in the Amazon have complex effects on distant regions of the world, though most of their impacts are local. In this regard, it has been argued that more than 50 percent of the rains falling in the Amazon are recycled back into the atmosphere through a process of plant evapotranspiration. This pattern of rainfall contributes to maintaining a humid climate that prevents flooding and **soil erosion**.

According to the INPE, total deforestation in the BLA grew from 15.2 million hectares in 1978 to 41.5 million hectares in 1990, and was about 65.4 million hectares in 2003. That number includes only complete removal of natural forest. The effects of human intervention on forest condition have some additional impacts. Some scholars found that while 6 percent of the closed forest canopy has been cleared, 15 percent of the forested Amazon was affected by forest fragmentation and edge effects. Other researchers expanded such figures, arguing that the cryptic deforestation from logging crews severely damaged the forests from 1 to 1.5 million hectares per year. In the light of this evidence, some researchers argue that the area of the Amazon forest affected by surface **fire** each year may be similar in scale to the area affected by deforestation. Much deforestation in the Amazon takes place in the labeled "**arc of deforestation**" located at the eastern and southern flanks of the Brazilian Amazon.

Since forest may regenerate after abandonment, not all of these areas remain deforested. Although the area of secondary forest succession is not accurately known, some estimates mention that it would be significant. It has been reported that roughly 30 percent of the deforested area in the Amazon is regenerated forest, and more than a half of this forest it is estimated to be less than five years of age. Much of the secondary succession, however, may be temporary **fallow**, and it is not properly forest. The estimates of the amount of forest regeneration, however, depend to a large extent on the definition adopted to classify forest succession.

Most of the human impacts in the Amazon's ecology have taken place in the last five-decade period. Nevertheless, there are ambiguous evidences of ancient occupation in the region. While some argue that dense and relatively complex settlement patterns dominated some areas of the Amazon, others mention that there is not enough evidence to support such assertions, and these arguments cannot be easily extrapolated from archeological information gathered in isolated locations. Despite that, the relative abundance of "dark earth" areas (soils with high organic matter) scattered all over the Amazon, along with the increasingly accepted hypothesis that they would have resulted from intensive human intervention for agricultural uses, reinforces the notion that the Amazon was an intervened landscape around 500 years ago.

Indeed, when Europeans entered the Amazon in the middle of the sixteenth century, they found it densely inhabited from the Japurá River eastward to the mouth of the Amazon. The areas closer to the main rivers were more populated; less is known about settlement patterns and population densities away from the major rivers. In this light, it has been suggested that rather than shifting cultivation as the dominant agricultural system, the landscape was dominated by semi-permanent fields intermingled with fruit orchards, managed fallows, house gardens, and brief bush fallows, by thousands of people located in semi-permanent villages surrounded by zones of forests manipulated by hunting and gathering activities. Such a complex system of integrated land use could have created pockets of dark earth and used them to support relatively dense settlements. In this view, the tropical rainforest was less much pristine than we believe it is nowadays. The **agricultural intensification** is a phenomenon driven by population concentration within a restricted area. Hence, it could have happened that when population pressures were removed, labor-intensive production techniques started to relax after the dissemination of foreign diseases in the second half of the sixteenth century following the Amazon's conquest.

The dynamics of the Amazon occupation were reinvigorated about 150 years ago along with the boom of **non-timber forest products**, and pressures for land occupation centered on mining activities, timber, and livestock production. The history of the Amazon region as a whole is that of plenty of boom and bust events resulting from attempts to occupy the territory through short-term enterprises that target natural resources extraction. The rubber boom at the turn of the last century and its decline after World War I, followed by several cycles of expansion and decline of Brazil nut production, then mining accompanied by gold rushes in several regions of the Brazilian Amazon, were the main extractive economies; they had large impacts on the region's land occupation. To a diverse extent, they have stimulated the expansion of urban settlements and villages supplying the labor force and commercial services demanded by such extractive economies. In a multitude of cases, the extractive economies' collapse has often been followed by depopulation. Nevertheless, in many places such economies have contributed to developing settlements, commercial networks, and infrastructures that have had a key role in further economic regional development.

Most of the deforested land has resulted in **pasture**. This land use embraces about 70 percent of the total forest removal, and the rest is distributed between agricultural areas and areas left as either temporary fallows or abandoned areas. It is still controversial as to who is the most important agent responsible for forest removal. While some argue that medium- and large-cattle ranchers are the main agents driving forest **conversion** to pasture, others still argue that small landholders share that responsibility. The process of cattle expansion has in part been the product of exhaustion of valuable timber species, after which forestlands are converted to pasture as a way to justify land ownership. Furthermore, selling timber became an important source of capital for cattle producers to use to invest in cattle ranch formation, and it also allowed some timber extractors to become cattle ranchers once the valuable timber was extracted.

Several scholars, after some assessments of the dynamics of livestock growth in the region and because of the lack of cattle herd and pasture management practices, considered that the process of land occupation was dominated by a broader logic of land fertility mining by which ranchers were interested only in taking advantage of land productivity with short-term perspectives on profits generation. The latter led to extensive **soil degradation**. Consequently, large areas of forest were converted to pasture, scarcely considering ecological and economic criteria and using pasture varieties—available at that time—inappropriate for the region conditions. This system of **cattle ranching** failed because it was not sustainable in the long term. Yet, some better cattle herd and pasture practices have been adopted over time, as well as pasture seed varieties appropriate for the region (*Brachiaria Brizanta*), which have helped to improve not only pasture productivity but also the financial performance of cattle ranching. Cattle production will continue exerting pressure on the forest because it is the easiest way to occupy the land, it helps to save some revenues on physical assets, and it demands a small labor force.

It is worth mentioning that only a small proportion of the total BLA is occupied by agricultural establishments. According to IBGE, in 1995–1996, the areas declared as agricultural establishments reached 24 percent of the whole Amazon. The areas under some sort of protection, including protected areas and indigenous territories, are 29 percent of the region; and about 47 percent of the Amazon's territory remains as public land. Due to this fact, it is not a surprise that timber extraction constitutes one of the main land uses in the Brazilian Amazon.

There are more than seventy forest frontiers in the Brazilian Amazon, mostly in the southern and western portions of the BLA and distributed in the states of Pará, Mato Grosso, and Rondônia, in which there is better connection with the main road infrastructure. In 1998, it was estimated that about 28 million m^3 were logged from the different forest frontiers, making a total of 11 million m^3 of sawed timber. The total number of timber companies reached 2,570 in the same year, most of which were sawmills and only a few processing plants. According to the Brazilian legal framework, most of the timber logged in the Amazon requires the approval of a management plan, but not all forest operations comply with such regulations. It is estimated that about half of the timber logged in the Amazon comes from illegal sources. An unknown portion of timber originates in areas where forest is eliminated and land is devoted to agricultural uses. Only a small number of private landholders comply with legal regulations regarding the approval of forest clearing plans by the government environmental agency. By law, private landholders should maintain a minimum of 80 percent of their areas as forest reserves. This is hard to enforce, particularly in older frontiers, because strong incentives for forest clearing persist.

The growing trend of converting forest to pasture has motivated some researchers to maintain that deforestation is going to continue taking place as an inevitable outcome of regional economic development in the Brazilian Amazon. This pressure is expanding with the arrival of soybeans as a new factor pushing further the expansion of cattle herds into the forest edges. Soybeans tend to convert degraded pasture areas to croplands, which in turn could be motivating new forest clearings for pasture areas to sustain a growing cattle population. According to official estimates, the soybean area in the BLA states increased from 1.6 million hectares in 1990 to 4.2 million hectares in 2002, much of this explosive growth taking place in the northern state of Mato Grosso. Due to attractive prices in the international markets, this has become one of the main commodities of Brazilian **agro-industry**. Yet, the bad shape of roads, along with the lack of productive capacity for grain storage, constitutes important constraints for the growing of grain production in the Brazilian Amazon. As a way to overcome such constraints, there are ongoing proposals for road construction through public-private partnerships.

Currently, the main dilemma in the Brazilian Amazon is how to provide forest **ecosystem services** by ameliorating forest removal, while at the same time expanding competitive economic sectors. Nothing suggests that this dilemma will be resolved in the near future. On one hand, the government is spending a lot of effort to regulate the process of Amazon land occupation through a program of economic and ecological zoning, and issuing new **forestry** regulations and a law of environmental crimes. On the other hand, agricultural development is a priority as well. Hence, credit programs are incentivizing development in the region, and programs of animal health control are carried out to assure the **access** of cattle production into the international market. The struggles between **conservation** and **economic growth** will still be difficult to reconcile in the future.

See also: **Agricultural Frontier; Brazilian Cerrado; Climate Impacts; Decision-Making; Extensification; Hot Spots of Land-Cover Change; Mineral Extraction; Pristine Myth; Savannization; Secondary Vegetation; Slash-and-Burn Agriculture; Systemic Change.**

Further Reading

Denevan, William, *Cultivated Landscapes of Native Amazonia and the Andes*, Oxford: Oxford University Press, 2001; Laurance, William F., Powell, George, and Lara Hansen, "A Precarious Future for Amazonia," *Trends in Ecology & Evolution* 17 (2002): 251–2; Moran, Emilio F., Brondizio, Eduardo, Mausel, Paul, and You Wu, "Integrating Amazonian Vegetation, Land-Use, and Satellite Data," *Bioscience* 44 (1994): 329–38; Skole, David L., Chomentowski, Walter H., Salas, William A., and Antonio D. Nobre, "Physical and Human Dimensions of Deforestation in Amazonia," *BioScience* 44 (1994): 314–22.

PABLO PACHECO

Ambio, A Journal of the Human Environment. Founded in 1972, the year of the first United Nations Conference on the Environment in Stockholm, Sweden, *Ambio* is a multidisciplinary, non-profit publication of the Royal Swedish Academy of Sciences. The journal addresses the scientific, social, economic, and cultural factors that influence the condition of the human environment. The journal has English- and Chinese-language versions and is published in eight issues per year; it is widely recognized as an important international forum for debate on these issues. Among the detailed subjects are ecology, environmental economics, geology, geochemistry, geophysics, paleontology, hydrology, earth sciences, meteorology, and (physical) geography, among others.

Further Reading

Royal Swedish Academy of Sciences, *Ambio: A Journal of the Human Environment* [Online, December 2004], The Ambio Web Site, www.ambio.kva.se.

HELMUT GEIST

Andes. The world's longest **mountain** chain, extending from some of the northernmost parts of **South America** in the Caribbean Sea (11° N latitude) to nearly its southern projection (56° S) into the cold waters of the southern oceans. This includes high elevations in Venezuela, Colombia, Ecuador, Peru, Bolivia, Argentina, and Chile. There are several important environmental gradients that control the kinds of land cover found in particular landscapes and regions, including the latitude, the elevation, the change in elevation from place to place (both topography and relief), and the resulting continua of climatic patterns. Climate change will be especially important in affecting the land cover of the high elevations over the next several decades. In terms of land use, some parts of the Andes have for millennia been considered

Mount Aconcagua in the Chilean Andes Mountains, Austral Summer 1995/1995.

PHOTO: Philip Hall. CREDITS: NOAA Central Library and the Office of the CIO/HPCC (corp 3012, NOAA Corps Collection).

home to farmers who modified their landscapes with irrigation farming systems, terraces, and a cornucopia of domesticated plants. Current drivers of land-use change include the demands for water and other goods destined for urban populations in growing cities, ever-more elaborate interconnections due to **globalization** of commerce and information, and needs for products from tree plantations, industrialized **agriculture**, and **mineral extraction**.

The climate systems range from tropical to subtropical until about 30° S, and temperate systems are found along the remainder of the north-to-south extent of the Andes. Tropical environments have little temperature fluctuation in the course of the year, although precipitation often changes dramatically, with recognizable dry and rainy seasons. The Intertropical Convergence Zone is a zone of low pressures where the tropical trade winds converge and large thunderstorms can form; dry seasons occur when the Zone shifts away seasonally. The Andes are so tall, however, that the trade winds themselves are altered in strength and direction, with areas facing into the trade winds among the wettest places in the world (particularly the Chocó area of the western Colombian coast, and along much of the eastern Andean flanks that face into the Amazon basin); here are found biodiverse **tropical humid forests** with abundant epiphytes and frequent fog. In turn, the parts of the Andes that do not face the trade winds can be so dry as to be classified as semi-arid, arid, or hyperarid, with shrubland, succulent, and ephemeral desert **vegetation**.

The subtropical portions of the Andes have similar climatic conditions except that temperatures do change during the course of the year, creating cool winters, especially as cold air masses move northward from the Southern Cone of Chile and Argentina during the austral winter, in the months from

May to October when westerly wind patterns are frequent. The native vegetation types are more seasonal, with deciduous trees dominating in some southerly **forests**. There is a narrow band of Mediterranean-type climates from 29° to 38° S in Chile where hot summers are characterized by drought conditions, producing a unique type of shrubland where plant species prosper that grow during the moister but cooler part of the year.

The middle to higher latitudes of the Andes have cool-to-cold winters and persistent westerly air flows. Thus, the most humid temperate forests are on the western side of the Andes from about 38° to 55° S; these are tall forests dominated by southern beeches (*Nothofagus* spp.). In turn, the eastern flanks of the Andes at those latitudes grade from humid forests into seasonal woodlands and finally into **grasslands** and deserts. These parts of the Andes in Chile and Argentina have extensive areas of modified land cover due to **forestry** and range **land-use systems**. Often tree plantations are considered more economically productive, and as a result many native Andean forests have been cleared and then replanted to Monterey pine (*Pinus radiata*). Shrublands and grasslands are used for grazing livestock, particularly cattle and sheep. Most land-cover **conversion** began with the arrival of European colonists in the mid-1800s, with industrial forestry becoming especially prevalent from the 1970s to the present. The largest Argentinean and Chilean cities are in lowland areas, connected to the mountains by roads and dependent on the uplands for supplying water to urban populations. Some highland areas, such as national parks and nature reserves, have been placed in protected areas for protection of upper **watersheds** and for tourism, for example skiing.

There are valuable mineral deposits located along the length of the Andes. Often these are extracted in large open-pit mines that are visible in satellite images. Unlike the middle and high latitudes, the subtropical and tropical Andes have large highland cities, some of which predate the arrival of European settlers. Other tropical highland cities were founded during the early colonial period, when nearby mines for extracting gold and silver fomented the development of cities.

Many areas of the tropical and subtropical Andes that are humid enough in theory to support native forests are in fact covered by shrublands. Andean forests from Venezuela to Bolivia can be found restricted to isolated hillsides or protected ravines. Thus, it appears that deforestation has occurred, some of it associated with European **colonization** and some perhaps predating those changes in land use. Large parts of the tropical Andes, especially in Ecuador, Peru, and northern Bolivia, are home to some ancient land-use practices that predate European colonization, for both the planting of crops and the grazing of native domesticated animals like llamas and alpacas. **Domestication** began at least several millennia ago, so native crops such as potatoes now have thousands of land races, providing farmers with much genetic diversity for planting in difficult environments.

There can be soil-related (edaphic) changes even within a single valley due to underlying shifts in bedrock. Adding further to the spatial heterogeneity that farmers must deal with are microsite differences in climate due to slight changes in both elevation and exposure to clouds and winds. This mosaic of biophysical factors results in multiple ecological zones, often arrayed in complex landscape mosaics. A typical Andean farmer traditionally grows crops in widely separated fields to increase chances of successful harvests. Most farmers also are part of family and community-based social networks wherein they share or coordinate **access** to communal lands, especially for grazing of livestock including cattle, sheep, alpacas, and goats. They also share labor for harvesting and for the maintenance of irrigation systems and trails or roads.

These highland landscapes show fairly subtle and fine-grained shifts in land-use/cover patterns: fields are fallowed, often for several years, then plowed and planted; forest patches are dissected for the establishment of new fields as populations grow; new forested patches are created by the planting of pines or eucalypts; irrigation projects intensify usage in nearby fields; and shrublands and grasslands are converted into **pasture** through burning. **Driving forces** of land-use change can be due to changing markets and accessibility, plus the economic inputs from overseas remittances.

Land-use changes that are dramatic and conspicuous are occurring in the more humid, forested parts of the tropical and subtropical Andes that are being rapidly converted by **deforestation**. Roads were built as part of determined national and international programs to colonize the humid tropical lowlands in the 1960s. Access to previously forested lands was provided to colonists by the construction of penetration routes, the founding of settlements, and massive government subsidies. **Land-cover change** has been rapid, with forests modified by extraction of valuable timber, cut for agriculture, and burned to provide rough pasture.

The high elevations of the Andes show effects of global climate changes in temperatures and humidity. Many icecaps are smaller than they were in the 1970s, although this could be due to warming temperatures or declining precipitation. These changes of ice cover are of concern both because they can increase risks from natural hazards and because the high elevations serve as sources of water for adjacent lowlands. The protected areas set up in the last forty years may not adequately protect flora and fauna if climate changes cause even larger shifts in land cover over the next several decades. Sustainable practices in the landscapes and countries of Andean South America will require further research and knowledge of land-use dynamics and associated land-cover types.

See also: **Agricultural Frontier; Agricultural Intensification; Agro-Industry; Arc of Deforestation; Biodiversity; Conservation; Exotic Species; Fallow; Latin America; Monoculture; Remittances Landscape; Southern Cone Region; Transhumance; Urbanization.**

Further Reading

Denevan, William M., *Cultivated Landscapes of Native Amazonia and the Andes*, Oxford, UK: Oxford University Press, 2001; Lentz, David L., *Imperfect Balance: Landscape Transformations in the Pre-Columbian Americas*, New York: Columbia University Press, 2000; Mayer, Enrique, *The Articulated Peasant: Household Economies in the Andes*, Boulder, CO: Westview Press, 2002; Young, Kenneth R., "Deforestation in Landscapes with Humid Forests in the Central Andes: Patterns and Processes," in *Nature's Geography: New Lessons for Conservation in Developing Countries*, eds. Karl S. Zimmerer and Kenneth R. Young, Madison, WI: University of Wisconsin Press, 1998, pp. 75–99.

KENNETH R. YOUNG

Anthropocene. This term was coined by Paul Crutzen and Eugene Stoermer to highlight the extent to which human activities have come to dominate the earth system from the late eighteenth century onward. The increasing role of human activities in the functioning of the earth system may be exemplified by the accelerated use of fossil fuels, which generate steeply rising emissions and an unprecedented gradient of change in atmospheric carbon dioxide (CO_2) concentrations; the impact of human activities on key global biogeochemical cycles, for example that of nitrogen; and major changes at the earth's surface as a result of **deforestation**, land reclamation, irrigation farming, and **soil degradation**.

See also: **Agricultural Intensification; Carbon Cycle; Cumulative Change; Driving Forces; Nutrient Cycle; Proximate Causes; Systemic Change.**

Further Reading

Crutzen, Paul J., and Eugene Stoermer, "The 'Anthropocene'," *International Geosphere-Biosphere Programme Global Change Newsletter* 41 (2001): 12–3.

FRANK OLDFIELD

Anthropological Center for Training and Research on Global Environmental Change. A training and research center for academics and practitioners at Indiana University in Bloomington, as well as a nurturing environment in which university faculty and students use and learn innovative research methods and curricula addressing the human dimensions of **global environmental change**. From its inception in 1992, the Anthropological Center for Training and Research on Global Environmental Change (ACT) has developed strong links with the disciplines of anthropology, geography, political science, demography, environmental science, agronomy, economics, and forestry. Four major ACT topics relate to the human dimension of **environmental change**: human ecology, population and the environment, land-use and land-cover change, and **deforestation**—as well as secondary successional processes. Since 1998, ACT researchers have helped lead the **Large-Scale Biosphere-Atmosphere Experiment in Amazonia**, and from 1999 to 2005, ACT was the home of the Focus 1 office of the **Land-Use/Cover Change (LUCC) project**.

See also: **Institutions; Population Dynamics; Secondary Vegetation.**

Further Reading

Anthropological Center for Training and Research on Global Environmental Change [Online, November 2004], The ACT Web Site, www.indiana.edu/~act.

HELMUT GEIST

Arc of Deforestation. Area to the south and east of the flanks of the Brazilian Legal Amazon in which **deforestation** is occurring faster than average (rate) and within which most of the deforestation taking place in the

Rapid destruction of tropical humid forest in Rondônia, western Brazil, in August 1992 (dark areas show remaining forest canopy).

PHOTO: Space Shuttle. CREDITS: Image Science and Analysis Laboratory, NASA-Johnson Space Center, 18 Mar. 2005 (Earth from Space—Image Information). http://earth.jsc.nasa.gov/sseop.efs

Brazilian Amazon occurs (magnitude). Deforestation along this arc represents the advance of cattle, and recently soybean frontier, from the center-west of the country into the southern part of the Amazon. Deforestation in Brazil is a phenomenon concentrated in few geographical areas, since it has followed the roads associated with the national development planning. First, it is located along the Belém-Brasilia highway, the Trans-Amazon highway, the Cuiabá-Santarem highway, and the Cuiabá-Porto Velho highway connecting Rondônia to the south of Brazil and Manaus. It is along the Belém-Brasilia and the Cuiabá-Porto Velho highways that the highest rates of deforestation have been observed.

The Brazilian Amazon holds the largest continuous tropical forest in the world, but it has also been losing the highest amount of tropical forest among all tropical countries. It constitutes an area of about 5 million square kilometers, 4 million of which are **forest**. In 2003, the total gross deforestation estimated in the Brazilian Amazon was equivalent to 654,000 square kilometers. The spatial distribution of deforestation has been relatively well explored due to the explosion of **remote sensing** analysis and adoption of **geographical information systems** since the last decade. Brazil is the country that has the most complete series of deforestation rates, due to a system of annual deforestation **monitoring** in pace since the late 1980s. This work is carried out by INPE, the Brazilian Agency of Spatial Research.

The INPE's estimate of deforestation considers a binomial classification, separating forest from non-forest. Yet, that constrains our understanding of the land-use dynamics that take place within the arc of deforestation, and in the Brazilian Amazon as a whole. Fortunately, several land-use assessments undertaken within the arc of deforestation, as well as information gathered from survey censuses carried out by the Brazilian Institute of Geography and Statistics, make it possible to differentiate different emergent land uses from forest **conversion**. According to them, in 1995–1996, there was an overwhelming influence of **pasture** resulting from forest conversion to agricultural uses, since around 70 percent of all cleared forest area remained as pasture. Beginning in the late 1990s, however, there was an increasing expansion of soybean plantations, particularly expanding northward from the *Cerrado* area into northern Mato Grosso and southern Pará.

The soybean frontier growth leads to a land-use substitution effect, since pasture areas are converted to soybean growth due to its comparatively higher profitability. That produces a land-displacement effect by which pastures and cattle herds move deeper into the frontier, fueling more deforestation. This is not a linear relationship because a large portion of soybean growth may establish itself over low-productivity degraded pastures, and hence it does not necessarily influence further increase of pastures, and mixed systems may also develop. The latter exemplifies that competition among agricultural land uses is taking place in the arc of deforestation, which has contradictory effects on forest **conservation**. This also reflects that socioeconomic dynamics are quite different in kind in the earlier occupied areas with respect to the fronts where pasture competes with forestland use.

The arc of deforestation reflects the imprint in the forest of **agricultural frontier** expansion. Nowadays, it is quite common to find references to the arc of deforestation area in the literature about deforestation in the Brazilian Amazon. Yet those references have to be taken with some caution while analyzing the human-environment relationships at the finest scales, because diverse interactions are actually taking place within this area, which result in differentiated patterns of deforestation. The latter means that a closer look into the arc of deforestation will reveal a multiplicity of social, economic, and ecological interactions taking place within this area. Broadly speaking, differentiated dynamics can be found in the areas of southern Pará, northern Mato Grosso, Rondônia, and Acre. These four zones account for most of the arc of deforestation.

The southern Pará is one of the oldest frontiers, which began to be occupied in the early 1960s through large land states establishing extensive systems of **cattle ranching** over public state lands, stimulated by government programs of fiscal incentives and subsidized credit with slight changes until the mid-1980s. Deforestation in Pará is one of the largest in the Brazilian Amazon (222,000 km^2 by 2003). The road network connecting this area to the Belém-Brasilia highway and a railroad built to export the mining production through Sao Luis in Maranhaõ, facilitated the **access** of small farmers to the area, escalating the struggles between smallholders, landless people, and ranchers to enlarge their access to the land, and securing land ownership. Most of the incentives for cattle production were removed in the late 1980s, but ranchers still benefited from cheap credit to sustain their beef cattle production. The improvement of road infrastructure and the comparative advantages of the region for beef production (cheap, suitable lands for pasture production, and cheap labor force) have stimulated increasing investments in slaughterhouses and meatpacking plants that motivated incursion into more semi-intensive systems of production. In parallel, land struggles continue to fuel a persistent movement of ranch invasions.

The state of Mato Grosso in the southern portion of the arc of deforestation has been one of the most active frontiers of livestock and agricultural expansion in the Brazilian Amazon. Accumulated deforestation has expanded quickly in Mato Grosso, achieving 169,000 km^2 by 2003. This is partly a result of the pressures of the agricultural and cattle frontier expansion that originated in the center-west states of Brazil. Furthermore, the expansion's proximity to the southern states of Brazil, which constitute the largest markets of the country, has made this region one of the most dynamic centers for the growth of **agro-industry**. This state embraces a large portion of the **Brazilian Cerrado**, which has been occupied for grain production, mainly soybeans. The opportunities for soybean production since the early 1990s, due to larger international demand for this product, along with the national currency devaluation, have motivated the occupation of the main proximate areas, some already occupied with pastures for beef cattle production. These factors have given this portion of the arc of deforestation one of the highest rates of deforestation ever seen.

The state of Rondônia in the western portion of the arc of deforestation has contrasting characteristics, since it constitutes one of the areas in which the federal government stimulated small farmers' settlements as part of a broader strategy to occupy the Amazon region in the 1970s. Since then, large contingents of population have been settled in this area, stimulated by the likelihood of obtaining cheap land and some services for the agricultural production. The intensive **colonization** accelerated even more in the early 1980s, after the road linking Rondônia with the south of Brazil was paved. The state government of Rondônia has actively promoted agricultural and livestock production through rural credit programs. Farm establishments statewide ballooned in the last two decades. In contrast, inflows of migrants to Acre were lower due to the fact that the links of the state to the southern markets and even Rondônia were more tenuous and further away (the stretch of BR-364 between Porto Velho and Acre was paved in the early 1990s).

Government policy and discussions on Acre have often centered around a developmental strategy involving forest products. The state of Acre shows a contrasting reality because the dominance of extractive economies in this area is based primarily on rubber extraction. In this part of the Amazon, rubber tappers resisted cattle ranchers' invasions. They also lobbied the federal government to create extractive reserves in the late 1980s and early 1990s. The system of **non-timber forest products** extraction has proved to be more environmentally sound, and it favors social equity. Nevertheless, pressures on deforestation persist due to the implantation of some areas in pasture.

See also: **Amazonia; Arena of Land Conflict; Public Policy.**

Further Reading

Alves, Diogenes, "Space-Time Dynamics of Deforestation in Brazilian Amazonia," *International Journal of Remote Sensing* 23 (2002): 2903–8; Browder, John O., and Brian J. Godfrey, *Rainforest Cities, Urbanization, Development and Globalization of the Brazilian Amazon,* New York: Columbia University Press, 1997; Kaimowitz, David, "Amazon Deforestation Revisited," *Latin American Research Review* 37 (2002): 221–35.

PABLO PACHECO

Arena of Land Conflict. Land conflict is the social conflict that arises over the **access** to or use of land or land-based resources. Land conflicts arise over different proximate issues, take different forms, and develop and are mediated within different geographical and institutional settings. In short, land conflict is "played out" in different ways and settings. These can be referred to as "arenas" of land conflict. An understanding of the arenas of land conflict within particular socioecological environments is necessary to develop effective strategies to better manage land conflict.

Land conflict may result from competing claims to land or land resources. Examples of such conflicts include those that arise between local communities and the state as a result of the establishment of nature protection areas or sites of multinational **investment**, conflicts within communities over boundaries between agricultural fields, and conflicts within families over inheritance of land. Conflicts can result from competing uses of the same land or land resource—especially under common property situations—conflicts over the impact of logging or mining on the productivity of natural and agricultural areas; crop damage caused by livestock; and farming by one group on the hunting or gathering activities of another group using the same land base. Land conflict can involve many different types of competing interests held by social groups with quite different historical and contemporary relationships.

Competing interests necessarily will develop as human societies utilize land resources. Competing interests over land can result in quite different forms of conflict including social tension/avoidance, informal mediation, legal sanctions, and violence. The forms that conflict can take not only reflect the intensity of the competing interests but the power relations among disputants and the institutional context through which the conflict is mediated. When there are large inequities of **decision-making** authority among disputants, competing interests are most likely to be shown through social tension/avoidance, and possibly punctuated by flare-ups of violence. The forms of conflict are also affected by a community's, district's, or state's institutional capacities to manage conflict. A major trend within rural areas of the world has been the erosion of the local **institutions** that manage conflict resulting from the extension of political authority of colonial and postcolonial states, the increase of individualism through the influence of western media and markets; and the migrations of people to and from rural areas. This, coupled with the limited legal and administrative capacities to manage land conflicts of many governments, has led to an increased potential of competing interests over land to escalate into serious conflict.

The "arena of land conflict" concept provides a good basis for understanding the different ways in which land conflicts develop, escalate, and are mediated. Still, its basic premise—that social conflicts arise out of conflicts over land that is scarce either physically or socially—must be seriously examined. Many conflicts in rural areas of the developing world relate in some ways to land—land is a major form of wealth, holds significant social and cultural meaning, and is the major substrate for productive activities. Therefore, social conflict in rural areas, no matter its cause, most often involves land in some way. In using the "arena of land conflict" concept, it is important that conservationists, developers, and scientists first verify that social conflicts are over land rather than that they simply implicate land. This important distinction has often not been adequately addressed.

See also: **Blaikie, Piers; Common-Pool Resources; Conservation; Decentralization; Degradation Narrative; Desertification; Economic Liberalization; Hot Spot Identification; Indigenous Knowledge; Land Degradation; Land Reform; Land Rehabilitation; Land Rights; Land Tenure; Mediating Factor; Mineral Extraction; Participatory Land Management; Pastoral Mobility.**

Further Reading

Bennett, Olivia, ed., *Greenwar: Environment and Conflict*, London: Panos Institute, 1991; Peluso, Nancy L., and Michael Watts, eds., *Violent Environments*, Ithaca, NY: Cornell University Press, 2001.

MATTHEW D. TURNER

Atmosphere-Land Interlinkages. The exchange of energy, trace gases, momentum, and particles between the land surface and the atmosphere depends on their continuous production and/or consumption through biochemical processes, and on the physical transport across surface-air boundaries. Consequently, all land-atmosphere exchange processes respond to long-term climate trends as well as to shorter-term variation. But relatively small alterations in their magnitude in response to climate forcing will influence climate in turn through a number of important biogeochemical **feedbacks**: directly, for example, by mediating the surface's energy balance and **albedo**; or indirectly through the radiative forcing potential of emitted trace gases, their tropospheric chemistry, and cloud-formation properties. Relatively slow changes in **vegetation** cover but also extreme and/or episodic events like **fire** set additional constraints in the surface-atmosphere system as they affect vegetation patterns and cause dynamic changes in ecosystem structure and species composition.

The terrestrial biota thus have had a major influence on the development of contemporary atmospheric conditions, and they still do. But human influence via land use, land-use change, and fossil fuel burning has resulted in major alterations of the naturally evolved synergies, particularly in the last 200–250 years or so. The ensuing increases in atmospheric **greenhouse gases** and aerosol load can partially be attributed to the twentieth century's observed climatic changes being the consequence of human activities. But a range of key processes in the land-atmosphere system is still only imperfectly understood. Further progress will hinge on the simultaneous study of the interactive biogeochemical cycles through field and **modeling** experiments, as given in the following four examples.

First, the best-studied trace gas exchanged between land surfaces and the atmosphere is carbon dioxide, or CO_2. Its atmospheric concentration has increased by more than 30 percent over the last 150 years, due to fossil fuel burning and following land-use/cover change. Large uncertainties are associated with the contributions of the latter to regional and global budgets, because the translation of vegetation changes into net CO_2 fluxes to or from the atmosphere is non-trivial. Time lags between increased (or decreased) carbon assimilation and productivity and respiratory CO_2 loss from the terrestrial biosphere add considerable complications; in many cases it is the respiratory part of the budget where mechanistic understanding is particularly lacking.

Second, coupled climate-ecosystem model simulations have compellingly demonstrated the potential magnitude of terrestrial biosphere CO_2 feedback in the climate system, but obtaining a comprehensive understanding of the related processes that control the atmosphere-land fluxes of CO_2 on the regional to global scale is a top research priority. Increasing evidence also suggests tight coupling of inert and reactive trace gas exchange: emission and deposition of gaseous substances and particles are directly or indirectly controlled by feedback loops involving water (e.g., available soil water, presence/absence of stomatal control) and thus the surface energy balance.

Third, methane is one of the most potent contributors to the atmospheric greenhouse effect and plays an important role in tropospheric chemistry. Microbial processes in natural **wetlands** have always been a major source of atmospheric methane and are considered even today to represent a chief part of total emissions. Methane emissions do not contribute drastically to the carbon budget of wetlands, which normally act as net carbon sinks. However, due to their strong greenhouse radiative potential, methane emissions will result in wetlands having a positive radiative forcing. Due to its ability to act as a sink for OH radicals, methane will indirectly participate in atmospheric chemistry and aerosol dynamics, and its global emissions are therefore not only of importance for the radiative forcing in the earth's energy budget but also of significance for the oxidative capacity of the atmosphere.

Fourth, aerosols represent the largest uncertainty in climate forcing. Their so-called indirect effects, that is, changes induced in clouds and on precipitation, represent a chief uncertainty in regional and global climate issues. Organic aerosol particles emitted by the terrestrial biota (as well as by anthropogenic sources), are effective cloud condensation nuclei. The formation of aerosol particles and their subsequent growth to cloud condensation nucleus size have been observed at remote continental sites, and their connection to biospheric processes has been noted. The exact aerosol formation processes not fully understood depend on the presence of organic vapors. A feedback loop linking the type of vegetation, its photosynthetic activity, and the capacity to emit volatile organic substances, which contribute to aerosol growth, can thus be proposed.

These few examples indicate how a more detailed quantification of the processes and feedbacks associated with vertical, bi-directional exchange and transformation of energy, momentum, and mass is needed. Progress in understanding the land-atmosphere system requires that measurements conducted at a multitude of **scales** be combined with regional and global models in physically and biologically accurate ways.

See also: **Carbon Cycle; Climate Impacts; Integrated Land Ecosystem-Atmosphere Processes Study; Kyoto Protocol.**

Further Reading

Foley, Jonathan A., Costa, Marcos H., Delire, Christine, Ramankutty, Navin, and Peter Snyder, "Green Surprise? How Terrestrial Ecosystems Could Affect Earth's Climate," *Frontiers in Ecology and Environment* 1 (2003): 38–44; Intergovernmental Panel on Climate Change, IPCC Third Assessment Report: Climate Change 2001: The Scientific Bases [Online, November 2004], The IPCC Web Site, www.ipcc.ch; Monson, Russell K., and Elisabeth A. Holland, "Biospheric Trace Gas Fluxes and Their Control over Tropospheric Chemistry," *Annual Review of Ecology & Systematics* 32 (2001): 547–76; Scholes, Mary C., Matrai, Paty A., Andreae, Meinrat O., Smith, K. A., and M. R. Manning, "Biosphere-Atmosphere Interactions," in *Atmospheric Chemistry in a Changing World: An*

Integration of a Decade of Tropospheric Chemistry Research, eds. Guy P. Brasseur, Ronald G. Prinn, and Alexander A. P. Pszenny, New York: Springer, 2003, pp. 19–71.

ALMUT ARNETH

Auto-Correlation. The degree of similarity between values of a variable and values of the same variable collected in a different place (in the case of spatial auto-correlation) and/or time (in the case of temporal auto-correlation). The amount of land in **forests** in a given year is often very similar to the amount in the previous year(s) (i.e., temporal auto-correlation). Similarly, the amount of land in forests in one county is often similar to the amount in neighboring counties (i.e., spatial auto-correlation).

A basic assumption in many statistical analyses, like regression analysis, is that all observations are independent, that is, not auto-correlated. Such analyses must, therefore, use sampling or estimation methods that acknowledge the existence of auto-correlation and account for their effects on the statistical tests.

In most cases, auto-correlation is positive, such that the value at a given time or place is positively correlated with the value at previous times or nearby places. Measures of auto-correlation, like Moran's I for spatial auto-correlation, can be calculated for data recorded with temporal or spatial sampling intervals of some size. Auto-correlation, if it exists, is usually present over a finite number of lags (the number of intervals before or away from a given observation). To examine over how many lags auto-correlation exists, researchers can compute correlograms, which plot a measure of auto-correlation against the number of lags. In spatial analysis, such tools are sometimes used with **pattern metrics** and in the analysis of **scale** of variables.

Further Reading

Chatfield, Chris, *The Analysis of Time Series: An Introduction,* 5th ed., New York: Chapman and Hall, 1996; Odland, John, *Spatial Autocorrelation,* Beverly Hills, CA: Sage Publications, 1988.

DANIEL G. BROWN

B

Biodiversity. The variety of species of plants, animals, and other organisms, as well as the genes they contain and the ecosystems of which they are part. Hence, biodiversity or biological diversity comprises three fundamental levels of organization: genetics, organisms, and ecosystems; including not only the compositional aspects at each hierarchical level, but also structural and functional aspects.

As for biodiversity, many trends in distributional patterns are well established. The generality of the latitudinal gradient is the most evident of these: species diversity increases from high to low latitudes in most groups that have been well analyzed (plants, beetles, butterflies, frogs, and birds, for example). The species richness of mammals (excluding bats) for natural communities in the New World shows the same overall latitudinal gradient. It indicates that the regional increase of species richness toward the equator (gamma diversity) appears to be a consequence of great increases both in local species richness (alpha diversity) and in species differentiation among communities (beta diversity). In general, alpha diversity means the number of species per unit area or locality as a measure of the biological richness of a region; beta diversity is a measure of the changes in species composition from one area to an adjacent area in dealing with species diversity; and gamma diversity is a comparative term used for all the species in a given region (such as tropical America) or in a **biome** (such as the short-grass prairies of North America).

Biodiversity inside an area can be estimated by taking into consideration species richness, specific diversity, and functional diversity, each emphasizing distinct perspectives. Species richness, also called diversity, refers to the number of species within an area, giving identical importance to each species; specific diversity quantifies the species in an area, adjusting the sampling effect and the specific abundance; functional diversity refers to the evaluation of the functional richness aspects.

Genetic diversity is the intraspecific genetic variation among populations physically unconnected and also among individuals of the same population. This diversity maintains the population's reproductive vigor, increases resistance against diseases, and favors the capacity to adapt to changing environmental conditions. Genetic diversity is affected by the reproductive behavior of the individuals of the population. The sexual reproduction recombines genes, originating individuals with unique combinations of genes and chromosomes. At small scales, landscapes are organized by areas regarded as ecosystems or communities. The diversity among areas is measured as beta diversity. Systems diversity is assessed as the richness of ecological systems in a certain region or landscape.

Ecosystem diversity includes the distinct biological communities and their association with the environment (ecosystem). Therefore, it represents the joint answer of species to different environmental conditions. The communities assure the functioning of the ecosystems and a number of **ecosystem services** that are vital for humanity, from flood control to soil protection, water quality, pollination, and so on.

At a global scale, organisms are not uniformly distributed. Their spatial distribution is well illustrated by a global classification scheme that includes large biogeographic zones, biomes, eco-regions, ocean domains, and a variety of smaller **scales**: landscapes, ecosystems, communities, and associations. In terrestrial ecosystems, communities can be categorized in relation to the physical environment (eco-region), the physiognomic type dominant (biome), and the faunistic/floristic zone (biogeographic) in which they occur. In marine ecosystems, communities are characterized according to the physical environment and faunistic zone (biogeographic).

Biodiversity has recently emerged as an issue of both scientific and political concern primarily because of an increase in extinction rates caused by human activities. During earth's history, the reasons for **species extinction** were probably modifications of the physical environment after impacts such as meteorites or volcanic activity. Contemporary extinction of species is mainly a result of anthropogenic factors, namely land use. Human **population dynamics** and **economic growth** cause the **conversion** of vast natural areas to settlement, **agriculture**, and **forestry**, resulting in habitat destruction, degradation, and fragmentation (i.e., the reduction of the area of the original habitat available to a particular species, decreasing its potential for dispersal and **colonization**), which are among the most important causes of species' decline and extinction. The loss of biodiversity is unique among the major anthropogenic changes because it is irreversible.

The global recognition of the alarming loss of biodiversity and the acceptance of its value resulted in the **United Nations Convention on Biological Diversity**. An unknown number of species have become extinct since the first *homo sapiens*. We do not know how many species do exist today, but several studies estimate 10–30 million different species in the world, of which we know around 1.7 million. From these, 1.4 million are insects, the huge group of arthropods that dominates the world's diversity.

Pressures on natural systems are increasing, and we will probably never have the time to learn everything we could about the diversity of species in the world. With particular emphasis after 1980, the recognition that species may play an important role in ecosystems functioning, as well as the fast-growing interest in the preservation of biodiversity, has driven ecologists to focus on the potential importance of biodiversity for the adequate functioning of ecosystems. Ecosystem functioning represents the assemblage of processes such as primary production, decomposition, **nutrient cycling**, and their interactions.

Species perform diverse ecological functions. A species may regulate biogeochemical cycles, modify disturbance regimes, or change the physical environment. Other species regulate ecological processes indirectly through trophic interactions such as predation or parasitism, or functional interactions such as seed dispersal and pollination. Both empirical and theoretical research in ecology suggests that the processes that maintain ecosystem functioning could be adversely affected by the loss of diversity, but debate continues as to whether this has been adequately demonstrated.

Major Causes of Biodiversity Loss

Habitat degradation and destruction rank high among the major causes of biodiversity loss. The size and distribution of the human population over the near future will be a crucial factor in determining the loss of biodiversity, pushing animal and plant populations past critical **thresholds** of tolerance and renewal. In the biodiversity hot spots, which are priority areas in **conservation** efforts as they have high biological diversity, high levels of endemism, and are currently threatened by anthropogenic impacts such as habitat destruction and species extinction, the human population growth rate between 1995 and 2000 was estimated to be 1.8 percent yr^{-1}, substantially larger than the average world growth rate (1.3 percent yr^{-1}), and even larger than the developing countries' growth rate (1.6 percent yr^{-1}). This obviously indicates that human population expansion will be the main leading force affecting biodiversity conservation in the future. Habitat conservation is affected by **proximate causes** of land-use/cover change (human demands for housing, arable land, freshwater, and industries), spreading **land degradation**, and climate change, consequently threatening ecosystem functioning and raising the risk of further extinctions. In particular, rainfall variability and human disturbances rather than density dependence or competition are **driving forces** in dry areas, which include 45 percent of the land surface and support 50 percent of the world's livestock. There, overgrazing and **fire** together with episodic drought result in accelerating land degradation and **desertification**. The way that these factors, however, affected the north and south Mediterranean environments are different. During the second half of the twentieth century, the evolving populations in north and south Mediterranean countries showed different trends. Population in the Maghreb has increased 300 percent, leading to an overexploitation of rangelands and to desertification, while in northern countries the increase is only 30 percent. However, the relocation of population to the coastal border in northern Mediterranean countries, along with the **agricultural intensification**, are leading to increased land degradation at both shores. On the other hand, **land abandonment** in the interior is in some cases allowing soils and **vegetation** to recover. In many cases, soils are so exhausted and degraded that their recovery will not be possible in the near future.

Habitat fragmentation is further considered to be one of the greatest threats to biological diversity. Populations inhabiting a landscape composed of patches of suitable habitats will be subjected to local extinction events, and that regional persistence will depend on their abilities to disperse and colonize those patches. The ability of individual species to persist in spatially and temporally changing landscapes depends on their ability to locate and exploit suitable habitats. The relative connectivity of similar patches and the permeability of species inhabiting the landscape play a central role in species distribution and abundance across landscapes. Thus, the persistence of species on fragmented landscapes caused by an anthropogenic impact will depend on the dispersal and colonization abilities (species with smaller seeds disperse easier than those with larger seeds) and on patch area—distance between patches and spatial configuration. In metapopulations, which are populations perceived to exist as a series of subpopulations linked by intermigration, single populations are connected by low levels of dispersal across a network of occupied and unoccupied habitat patches. Individual populations are prone to extinction, but metapopulations persist, provided that extinction rates are exceeded by rates of recolonization from populations nearby. Patches should be close enough so that colonization of empty habitats is rapid and mortality during migration is low. These circumstances are most likely to be found near the core of a species distribution, although some very substantial and persistent metapopulations can be found where extensive patch networks survive close to range margins. The metapopulation theory indicates that, when resources are limiting, conservation efforts should concentrate on the best patches of habitat within the core of species' range. Marginal habitat quality and greater isolation mean that conserving populations at the edges of species' ranges can be extremely difficult and may make only a limited contribution to the persistence of metapopulations as a whole.

In fragmented landscapes, conservation actions have to be implemented in the human-dominated matrix. What happens in these habitats beyond reserves is the key to the future of biodiversity, for two reasons. First, the wider matrix is important in its own right. In most parts of the world, protected areas are unlikely ever to cover much more than 5 percent of the land surface. If this non-reserved landscape is thoroughly hostile (so that very few species remain in it), and if reserves are distributed at random with respect to biological diversity, then species extinction is easily predictable even if no species are lost from protected areas. The condition of the matrix will also have a serious effect on the status of biological communities within reserves. Parks are not islands, and their destiny is closely tied up with processes occurring beyond their boundaries. Marginal and degraded landscapes have great importance to global conservation. Habitat fragments that seem small to vertebrate ecologists may support strong and persistent local populations of many insect species, as they do for some plants.

Evidence is now accumulating that ongoing anthropogenic climate change is already leading to the differential loss of populations at the warmer margins of species' ranges. This problem is compounded by the effects of habitat fragmentation. Whereas in the past, individuals dispersing from a marginal habitat in a warming (or cooling) world may have found more suitable conditions close by, in the twenty-first century the likelihood is that they will be separated from the nearest patch of a more appropriate habitat by a considerable expansion of agricultural or other unsuitable habitats. Fragmentation thus greatly reduces the probability of successful dispersal and establishment. Artificial and natural corridors between remaining habitat patches may become increasingly important in favoring range shifts through highly fragmented landscapes, although it is suspected that large protected areas, having a series of climatically discrete habitats, may be of even greater value. In facing major **environmental changes**, species that fail either to relocate their ranges or to adapt accordingly elsewhere simply go extinct. How far we can maintain the potential for dynamic responses such as adaptive evolution and range shifts, despite population reductions and habitat fragmentation, may be a key determinant of the impact of the sorts of new and rapidly developing anthropogenic challenges. Species restricted to isolated habitat fragments and reserves must rely either on their limited physiological tolerances, or on evolutionary adaptation *in situ,* to survive quick global warming in the coming centuries.

Conservation of Biodiversity

As human population continues to increase, species extinctions also increase, and the number of species lost or at risk of extinction increases in spite of the conservation efforts developed in some countries. At the same time, it is obvious that the present network of protected areas preserves only a limited proportion of biodiversity at risk, and there is a scant scope for bringing many more wild lands or wilderness areas under protection within the foreseeable future because of rising human numbers and their demands on protected natural environments. Moreover, most protected areas are proving unable to preserve more than a bit of their biodiversity in the long run.

One of the most productive priorities in conservation planning is to pin down "hot spots" of biodiversity, these being areas where exceptional concentrations of endemic species face an exceptional threat of habitat destruction. At least 20 percent of plant species are confined to 0.5 percent of the earth's land surface, and in eighteen areas where habitat destruction is moving apace. Twenty-five hot spots, comprising less than 2 percent of earth's land surface, contain the sole remaining habitats of 34 percent of the four tetrapod groups of vertebrate species and 46 percent of higher plant species. Equally important, they harbor 60 percent of all species, including non-endemic, and at least 65 percent of Red Data Book species. The World Conservation Union's Red Data Book is produced to assess the conservation status of species, subspecies, varieties, and even selected subpopulations on a global scale, and it provides the most objective, scientifically based information on the current status of globally threatened biodiversity. The twenty-five areas are expected to lose most, if not virtually all, of their former habitats within the first half of the twenty-first century at most.

Conservation efforts traditionally focused on certain "charismatic" or flagship species—beautiful endangered animals such as gorillas and pandas. However, the benefits of such species approach are not very auspicious. It seems reasonable to assume that concentration of too many efforts in this strategy may have led to an overlap of activity, forgetting the ecosystem in which the species lives. Although it is important to keep the one-habitat species concern, the switch to whole-ecosystem practice seems crucial. This transition has to be done even if it represents a high risk for conservation due to the limited number of case studies and the lack of scientific data.

The ecosystem approach is necessary in managing the earth's resources because it addresses the interactions that link biotic systems, of which humans are an integral part, with the physical systems on which they depend. This applies on the scale of the earth as a whole, a continent, or a farmer's field. An ecosystem approach is critical to resource management as we tackle the sustainable use of resources in an era of increasing human population and **consumption** and large, fast changes in the global environment.

Management plans to preserve biodiversity need urgent implementation. However, these plans will never succeed without **community involvement**. This involvement will depend on our ability to show that ecological sustainability is crucial for the economic and social sustainability of all countries, in particular in those extremely dependent on tourism and where the scarcity of natural resources is frequent. Nature conservancy must be understood as a tool to maintain diversity of our natural heritage through the promotion of economic, social, and cultural activities that at the same time enhance diversity and local development.

See also: **Agricultural Revolution; Agrodiversity; Biodiversity Novelties; Biomass; Biotechnology; Brazilian Cerrado; Cadastre; Carson, Rachel Louise; Climate Impacts; Deforestation; Degradation Narrative; DIVERSITAS; Ecological Colonization; Exotic Species; Forest Degradation; Grassland; Hot Spot Identification; Hot Spots of Land-Cover Change; Integrated Model to Assess the Global Environment; Land-Use History; Resilience; Savannization; Taiga; Tropical Dry Forest; Turnover.**

Further Reading

Cervigni, Rafaello, *Biodiversity in the Balance: Land Use, National Development and Global Welfare*, Cheltenham, UK, and Northampton, MA: Edward Elgar, 2001; Chapin, F. Stuart, Matson, Pamela, and Hal Mooney, *Principles of Terrestrial Ecosystem Ecology*, Berlin, Heidelberg: Springer, 2002; Dirzo, Rodolfo, and Peter H. Raven, "Global State of Biodiversity and Loss," *Annual Review of Environment and Resources* 28 (2003): 137–67; Levin, S. A., ed., *Encyclopedia of Biodiversity*, San Diego, CA: Academic Publishers, 2001; Mace, Georgina M., Balmford, Andrew, and Joshua R. Ginsberg, *Conservation in a Changing World*, Cambridge, UK: Cambridge University Press, 1998; Magurran, Anne E., and Robert M. May, *Evolution of Biological Diversity*, Oxford, UK: Oxford University Press, 1999; Myers, Norman, "The Biodiversity Challenge: Expanded Hot-Spots Analysis," *Environmentalist* 10 (1990): 243–56; Pimm, Stuart L., Russell, George J., Gittleman, John L., and Thomas M. Brooks, "The Future of Biodiversity," *Science* 269 (1995): 347–50; Wilson, Edward O. *The Diversity of Life*, Cambridge, MA: Belknap Press, Harvard University Press, 1992; Xu, Jianchu, and Andreas Wilkes, "Biodiversity Impact Analysis in Northwest Yunnan, Southwest China," *Biodiversity and Conservation* 13 (2004): 959–83.

HELENA FREITAS

Biodiversity Novelties. New discoveries of species in the order of around 15,000–20,000 new species described annually. Rodolfo Dirzo and Peter Raven note that it should not come as a surprise that strikingly novel and wholly unexpected new taxa continue to be discovered. They note that perhaps one in ten species of the earth have been identified yet. Even for relatively well-known areas like the United States, the rate of discovery suggests that no more than a third to half of the estimated total of 500,000–750,000 species of organisms have yet been discovered. Some of the biodiversity novelties are of evident economic importance, such as a species of **maize** (*Zea diploperennis*) from the state of Jalisco in western Mexico. Discovered only five years ago on a disturbed area about the size of a football field, this maize species is interfertile with cultivated maize but unique in its perennial habit. Human activities have greatly accelerated the rate of **species extinction** far beyond the natural rate. At least one species per day suffers the fate of extinction, and a minimum of 100,000 species have died out within the past 100 years. For the past 300 years, recorded extinctions for a few groups of organisms reveal rates at least several hundred times the rate expected on the geological record. Of all the species that have lived on earth since life first appeared, only about one in 1,000 is still living today.

See also: **Agrodiversity; Biodiversity; Exotic Species; United Nations Convention on Biological Diversity.**

Further Reading

Dirzo, Rodolfo, and Peter H. Raven, "Global State of Biodiversity and Loss," *Annual Review of Environment and Resources* 28 (2003): 137–67; Iltis, H. H., Doebley, J. F., Guzmán, R., and B. Pazy, "*Zea diploperennis* (Graminae): A New Teosinte from Mexico," *Science* 203 (1979): 186–8.

HELMUT GEIST

Biomass. Sum of organic non-fossil material of biological origin. Plant biomass, terrestrial and marine, is formed through photosynthesis, a complex biochemical process in which solar energy is fixed by chemical reduction of carbon dioxide to organic carbon. In the most productive ecosystems, plants convert about 5 percent of the incident solar energy into biomass energy (gross primary production). A part of this energy is lost through plant respiration, and the remainder is termed net primary production (NPP), that is, the total amount of trophic energy available in an ecosystem for heterotrophs. Biomass produced by plants either enters heterotrophic food chains (grazing, detritivores) consisting of several stages (such as herbivores, then carnivores of first, second, etc., order) with a minimum energy loss of 90 percent in each transfer step; or, biomass not consumed by heterotrophs accumulates within an ecosystem (in non-equilibrium stages) and builds up a stock, the standing crop of biomass. Biomass can be measured in the mass of fresh weight or dry weight. Vegetable biomass has an energy content of 17–21 MJ/kg dry matter, animal biomass around 23 MJ/kg dry matter. The carbon content of dry matter biomass is about 42–55 percent. Biomass is indispensable for humans as food and serves important other purposes such as clothing, construction material, provision of technical energy, and so on. Combustion of biomass (such as fuel wood) probably was the first sort of technical energy conversion by humans. Due to its carbon content, biomass plays a fundamental role in the global **carbon cycle**.

According to Robert Whittaker and Gene Likens, the earth's total biomass stock is estimated to be around 1,800 10^{15} grams, only 0.2 percent of which is marine biomass. The NPP of the earth's ecosystems amounts to about 170 10^{15} grams/year, of which a third originates from marine and two-thirds from terrestrial ecosystems.

Human use of biomass, or socio-economic **metabolism**, is an important driver of land use. Land use alters biomass stocks, biomass production, and biomass **turnover** of ecosystems to an extent that has already become globally significant. In pre-industrial societies, biomass was the almost exclusive source of socio-economic energy. Hunter-gatherers used biomass by and large without deliberately altering ecosystem patterns and processes (an uncontrolled solar energy system). In contrast, agrarian societies pervasively alter ecosystems in order to maximize useful biomass output (a controlled solar energy system), but their energy system remains area-dependent. Even in industrial societies, despite their heavy reliance on fossil fuels, biomass still accounts for a considerable fraction of their material and energy inputs.

In recent years, the use of biomass as a renewable energy source gained much attention in the sustainability debate, because under certain circumstances, biomass can be an almost carbon-neutral energy resource. In exploiting such options, however, it is important to consider possible impacts of increased use of biomass on ecosystems and not to exceed their regenerative capacity.

See also: **Agriculture; Agrodiversity; Biodiversity; Biotechnology; Colonization; Disequilibrium Dynamics Driving Forces; Ecological Colonization; Fire; Forestry; Human Appropriation of Net Primary Production; Industrialization; Proximate Causes; Threshold.**

Further Reading

Smil, Vaclav, *General Energetics: Energy in the Biosphere and Civilization*, Manitoba, New York: John Wiley & Sons, 1991; Vitousek, Peter M., Mooney, Harold A., Lubchenco, Jane, and Jerry M. Melillo "Human Domination of Earth's Ecosystems," *Science* 277 (1997): 494–9; Whittaker, Robert H., and Gene E. Likens "The Biosphere and Man," in *Primary Productivity of the Biosphere*, eds. Helmut Lieth and Robert H. Whittaker (Ecological Studies 14), Berlin: Springer, 1975, pp. 305–28.

KARLHEINZ ERB

Biome. Synonym for major life zone; large ecosystem; or large plant, animal, and soil community; classified according to predominant features of **vegetation** and classified by organisms that have adapted to that particular ecosystem. With regard to the land surface, typical terrestrial biomes are tropical **forest**, coniferous forest (or **taiga**), temperate forest, tundra, shrubland, desert, **grassland**, steppe, and savanna. Types and further subdivisions in any biome list—such as tropical forests subdivided into **tropical humid forest** and **tropical dry forest**—are arbitrary, depending on the list's specific purpose. Biomes often span several continents, thus normally consisting of a multitude of much smaller ecosystems that reflect **initial conditions** of the environment as well as interactions and mainly climate-induced gradients of local environmental factors. Usually, biomes are not defined by land-use features, but can serve as indicators for land management practices—in **agriculture** and **forestry**, mainly—that would best fit given ecosystem properties, for example, in biome models. The so-called biome paradox refers to the consideration in climate change **modeling** and **scenario** construction of important **feedback** effects between major zones in the past behavior of the earth system. For example, there is evidence for warmer winters in the early to mid-**Holocene** in mid- to high latitudes of the Northern Hemisphere (taiga-tundra feedback effect). In the past, biome shifts—simple shifts of intact biomes—have been interpreted as ecosystem responses to changing climate. However, there is increasing recognition of a need to move away from simple cause-effect chains with single drivers. Instead, multiple and interacting biophysical and human stresses on particular ecosystems need to be considered, including land-use/cover change.

See also: **Biodiversity; BIOME 300 Project; Brazilian Cerrado; Land Quality; LandSHIFT Model; Paleo Perspectives; Sustainable Land Use.**

Further Reading

Cramer, Wolfgang, "Biome Models," in *Encyclopedia of Global Environmental Change, Volume 2, The Earth System: Biological and Ecological Dimensions of Global Environmental Change*, eds. Ted Munn, Harold Mooney, and Josep G. Canadell, Baffins Lane, Chichester, UK: John Wiley & Sons, 2002, pp. 166–71; Woodward, Susan, *Biomes of Earth: Terrestrial, Aquatic, and Human-Dominated,* Westport, CT: Greenwood Press, 2003.

HELMUT GEIST

BIOME 300 Project. A recent effort to reconstruct historical changes in permanent cropland at a global scale from the last 300 years, jointly done by the **Past Global Changes (PAGES) project** and the **Land-Use/Cover Change (LUCC) project**. In March 2000, a group of about forty interested participants came together in Bern, Switzerland, highlighting two preliminary efforts of the U.S. **Center for Sustainability and the Global Environment** (SAGE) and the Dutch **National Institute of Public Health and the Environment** (RIVM), namely the **History Database of the Global Environment**. All these efforts compiled various contemporary and historical statistical inventories on agricultural land—such as census data, tax records, and land surveys—and applied different spatial analysis techniques to reconstruct historical maps of agricultural areas. The data sets were created at a spatial resolution of 0.5° latitude and longitude, and at an annual resolution from 1700 to 1990, by spatializing data based on the **IGBP-DIS global 1-km land cover data set** derived by **remote sensing**. A fast-track product of the historical land-cover database was subsequently widely distributed using CD-ROMs.

See also: **Cadastre; Land-Use History.**

Further Reading

Goldewijk, Kees Klein, "Estimating Global Land Use Change over the Past 300 Years: The HYDE Database," *Global Biogeochemical Cycles* 15 (2001): 417–34; Ramankutty, Navin, and Jonathan A. Foley, "Estimating Historical Changes in Global Land Cover: Croplands from 1700 to 1992," *Global Biogeochemical Cycles* 13 (1999): 997–1027.

HELMUT GEIST

BioScience. A monthly journal of the American Institute of Biological Sciences (AIBS) with content written and edited for accessibility to researchers, educators, and students alike. Since 1964, articles have provided overviews of current research in biology, accompanied by essays and discussion sections on education, **public policy**, history, and the conceptual underpinnings of the biological sciences. Recent special sections have addressed ecological boundaries, **conservation**, acid rain, and the U.S. Long-Term Ecological Research (LTER) Network.

Further Reading

The American Institute of Biological Sciences [Online, December 2004], The BioScience Web Site, www.aibs.org/bioscience.

HELMUT GEIST

Biotechnology. A wide array of technological applications that use biological systems, living organisms, and parts thereof to alter or modify living or non-living materials for human use. In the broadest sense, biotechnology includes conventional practices of plant and animal breeding, fermentation, and the use of yeast to leaven bread. Recent scientific advances have refined the definition to specify molecular and reproductive technologies that alter the genetic integrity of living organisms to produce traits or characteristics not attainable through conventional methods. Recombinant DNA transfer from one species to another is the basis of transgenic technology, which results in the creation of a genetically modified organism, or GMO. Technologies that modify genetic composition of plants and animals are dramatically restructuring food and agricultural commodity production. Consequently, biotechnology will shape new processes and patterns of land-use/cover change in regions where the economy supports advanced **agro-industry**.

Biotechnology may contribute to further **agricultural intensification** and **modification** as **agriculture** and **forestry** are coupled with energy and pharmaceutical sectors to produce more non-food or non-feed crop products. Alternative bioenergy sources, such as fuelwood, charcoal, and energy crops (e.g., sugar cane, sweet sorghum, rapeseed, and corn), produce heat, ethanol, biodiesel, bioelectricity, and biogas. Genetic engineering offers new avenues to increase plants' **biomass** production and improve burning efficiencies. These products are important alternatives to hydrocarbons and offer mechanized farmers new avenues to modify and intensify land use. Another developing sector is "biopharming," or molecular farming, the cultivation of genetically modified plants, such as corn, soybeans, and **rice**, to produce pharmaceuticals such as antibodies, contraceptives, growth hormones, protein drugs, edible vaccines, and industrial enzymes. While biopharming will be limited to developed countries for the short term, it holds the potential to transform land use by exponentially increasing the value of a commodity produced per unit of land.

Biotechnology may conserve **forests** as well as undermine **conservation** efforts. Currently, genetically modified trees hold the potential to alter forestry practices and plantation cultivation. Transgenic technologies, which transfer recombinant DNA from one species to another, produce trees with important traits not available through natural plant selection. In addition to resistance to insects, diseases, or environmental stresses, genetically modified trees could grow faster, produce better-quality fiber, and sequester larger quantities of carbon, a trait that some believe could help mitigate global warming. Advocates argue that the world's increasing demand for wood products could be met by higher-yield and more productive transgenic forest resources because these GM plantations may relieve logging pressure on natural forests. However, GM technologies may support **extensification** of agricultural production. Commercial genetically modified crops are developed to resist herbicides, insects, and other environmental stress. These same traits may provide the qualities required to open new lands previously limited to agricultural production.

Reliance on genetic resources as the raw materials for use in the pharmaceutical and agriculture industries also offers new pathways to increase the economic value of ecosystems and traditional farming systems rich in **biodiversity**. Debt-for-nature swaps, *in situ* conservation programs, and bioprospecting agreements between suppliers (countries or communities) and corporations represent important economic incentives that may promote ecosystem preservation and conservation of crop germplasm, principally land races of ancestral crop varieties and wild crop progenitors. **Ecosystem service** valuation, which relies on international trade and intellectual property rights agreements, however, is not without critics. Arguments against the privatization of genetic material required in this neo-liberal model of conservation cite unequal power relations and possibilities for exploitation of communities who steward genetic resources.

See also: **Agricultural Revolution; Agrodiversity; Brazilian Cerrado; Carbon Sequestration Industrial Revolution.**

Further Reading

Kloppenberg, Jack, *First the Seed: The Political Economy of Plant Biotechnology, 1492–2000,* Cambridge, UK: Cambridge University Press, 1988; Rifkin, Jeremy, *The Biotech Century: How Genetic Commerce Will Change the World,* London: Phoenix, 1999; Shiva, Vandana, *Biopiracy,* Boston: South End Press, 1997; Tokar, Brian, *Redesigning Life? The Worldwide Challenge to Genetic Engineering,* New York: Zed Press, 2001.

WENDY JEPSON

Blaikie, Piers (1947–). Professor emeritus and professorial fellow in the School of Development Studies, University of East Anglia, Norwich, United Kingdom, who dedicated himself to the study of environment-society relations in the developing world, political ecology and environmental policy in South Asia, Southeast Asia, central Africa, and **Southern Africa**. Piers Blaikie made a significant contribution to the debate on environmental change, particularly **land degradation**, through a seminal work co-authored by **Harold Brookfield**. They demanded careful studies of land management at the local level and demonstrated that study of **environmental change** needed to be carried out at multiple and linked **scales** of analysis.

See also: **Arena of Land Conflict; Degradation Narrative; Human Immunodeficiency Virus (HIV)/Acquired Immunodeficiency Syndrome (AIDS).**

Further Reading

Blaikie, Piers M., "A Review of Political Ecology: Issues, Epistemology and Analytical Narrative," *Zeitschrift für Wirtschaftsgeographie* 43 (1999): 131–47; Blaikie, Piers M., and Harold Brookfield, *Land Degradation and Society*, London, New York: Routledge, 1987; Wisner, Ben, Blaikie, Piers M., Cannon, Terry, and Ian Davis, *At Risk: Natural Hazards, People's Vulnerability, and Disasters*, 7th ed., London: Routledge, 2003.

HELMUT GEIST

Boreal Zone. Circumpolar northern coniferous **biome**, often synonymously used with the Russian term **taiga**. About two-thirds of the boreal zone is in Eurasia, and the appearance of its forests is like that of North American boreal plant communities. Many genera and species are in common; the general physical canopy structure is similar. The coniferous canopy is one-layered, and the ground layer is dominated by low shrubs underlain by mosses and lichens with varying density and thickness.

Vegetation

The taiga growth forms are adapted to the cold and the physiological drought of winter and the short-growing season: the conical shape of the coniferous trees promotes shedding of snow; needleleafs reduce transpiration and have a thick wax coating to protected against wind evaporation; the evergreen foliage allows photosynthesis as soon as temperatures permit; and the dark-needle green absorbs maximum heat.

The primary succession in the boreal zone can be described as fen > swamp > bog > musked > treed bog > **forest**; secondary succession (after disturbances such as flooding or **fire**) as bare surface > annual herbs > perennial herbs and shrubs > treed bog > forest. In boreal forests, disturbances are frequent and may occur every few decades, which is different from human-induced disturbances (as discussed below).

Edaphic conditions result in sometimes extensive, persistent patches of **vegetation** other than spruce and fir: bogs (muskegs) occur in poorly drained, glacial depressions; pine forests occur on sandy outwash plains and former dune areas; larch forests claim the thin, waterlogged substrate in level areas underlain with permafrost (in Alaska stands of *Larix larichina* are localized phenomena; in Siberia east of the Yenisey River, extreme continentality and nearly continuous permafrost create vast areas dominated by *Larix dihurica*).

The northern border, or transition zone between forest and the treeless tundra, coincides roughly with the climatic frontal zone existing between the arctic and southern air masses in summer. The southern limit cannot be distinctively defined, since it varies with regional climate conditions. The southern transition can consist of broadleaf deciduous or steppe **grassland** communities. Alpine expressions of the boreal biome, or oroboreal regions, exist in higher elevations of both continents.

Soils

With regard to soil development, podzolization occurs as a result of the acid soil solution produced under needleleaf trees. The main soil order associated with the taiga is the relatively infertile spodosol. Characteristic of this is the leached, acid, grayish white horizon near the surface. In the acidic environment of a podzol, the populations of soil biota are low, and little mixing of the soil occurs. The vegetation is shallow-rooted because decomposition occurs most rapidly in the growing season and the trees capture nutrients as they are released.

Climate System

Dominated by the conflict between cold polar and warm tropical air masses and by the movement of frontal cyclones, the cyclonic climate regions lie in a broad belt between 35° and 70° latitude of the Northern Hemisphere. The subarctic cyclonic climate, or taiga, lies to the north of the humid continental climates, between 50° and 70° north latitude. It covers much of Alaska and northern **Canada** in North America and much of Scandinavia, **Russia**, the Baltic countries, Belarus, and Manchuria in Eurasia; here are the source regions of the cold continental polar air masses. The forest distribution is limited in the north by the position of July's 15° Celsius isotherm and in the south by the 18° Celsius isotherm. Departures occur in regions under montane or oceanic influences.

The boreal zone is a major carbon storage area and so plays an important role in regulating the earth's climate. The boreal region is a "hot spot" of the global climate system. Small perturbations in the energy balance may be strongly amplified through physical and biogeochemical **feedback** mechanisms. Boreal forests contain about 18 percent of the earth's total aboveground **biomass** (the dry weight of organic matter). Siberia's taiga alone represents 57 percent of the earth's coniferous wood volume. Boreal soils store 12 percent of the total world soil organic carbon. Recent research suggests that carbon compensation rates are saturating in the boreal zone, thus contributing to the net increase of carbon into the atmosphere with possible enforced feedback mechanisms with the **greenhouse gas** (GHG) effect. Key feedback mechanisms are the compensating effects of an increased growing season versus increased respiration in a warming world, changes in fire frequency, changes in **wetland** extent and emissions of methane, and melting of permafrost. In the European Commission (EC) Framework Programme 5 (FP5) Project TCOS-Siberia (Terrestrial Carbon Observations Siberia), elements of an *in situ* observing network have been established in Siberia and western Russia, complementing similar efforts in Canada (AIES), Alaska (NOAA-CMDL, or National Oceanic and Atmospheric Administration–Climate Monitoring and Diagnostics Laboratory), and eastern Siberia (NIES-Japan, or National Institute for Environmental Studies). This network is still too sparse, and most records are too short to determine regional (1,000 kilometer, ten-year average) GHG budgets, which are needed for a comprehensive assessment. The contradictions among carbon storage values given by different authors are striking. One cause for this disagreement seems to be the difficulty of allowing for disturbances.

The major types of disturbances in the boreal zone are fires and biotic (insect damage and diseases) and abiotic disturbances (extreme climate, human use). Fire regimes in the boreal biome have been linked to climate via large-scale atmospheric circulation patterns. There is evidence of acceleration of fire regimes in vast regions of northern Eurasia during the last two decades. Boreal fires also contribute to the interannual variations in atmospheric carbon concentrations in the Northern Hemisphere. The second GCOS Adequacy Report (Global Climate Observing System) identifies the need for a long-term fire database. Several burned area products are available: **Global Burnt Area 2000**, **GLOBSCAR**, and **GLOBCARBON**. Fire frequency/active fire detection is available in the **Along Track Scanning Radiometers (ATSR) World Fire Atlas**, World Fire Web, and from **moderate resolution imaging spectroradiometer** (MODIS), but cross-sensor comparisons show large discrepancies.

Vegetation and snow determine the radiative properties of the land surface at northern latitudes to an extent that influences global climate. The snow-vegetation feedback is one of the major feedbacks operating between the land surface and atmosphere and is thought to have played an amplifying role in past glaciations and deglaciations. In these temperature-limited environments, climate change is expected to alter the composition, extent, and properties of vegetation and snow earlier and more strongly than elsewhere. **Remote sensing**–based global vegetation and snow products are available from a range of sources, such as MODIS, **advanced very high resolution radiometer** (AVHRR), Special Sensor Microwave Imager SSM/I, and **SPOT Vegetation**.

Permafrost underlies more than 25 percent of the world's land area and is primarily controlled by climatic factors. Climate change **scenarios** indicate that anthropogenic warming will be most pronounced at northern latitudes. This could cause the disappearance of up to 25 percent of the present terrestrial permafrost. Large increases of methane (CH_4) and carbon dioxide (CO_2) emissions are therefore associated with **land degradation** of permafrost, representing a positive feedback to **global environmental change**. Because permafrost is highly sensitive to long-term warming, it is also seen as a valuable indicator for observing and forecasting **environmental changes**. Permafrost is surveyed by the Global Terrestrial Network for Permafrost (GTN-P) network, which is relatively sparse in Eurasia. Permafrost state and dynamics are influenced by climate, but also very much by local geographical and ecological conditions.

Human Impact and Sustainable Use

In the nineteenth century, travelers like the writer Anton Chekhov experienced the taiga as a green, endless monster. In the twentieth century, the **perception** of the boreal forests has changed dramatically. Canadians and Americans saw the taiga as an economically barren land only useful for excessive exploration. In the Scandinavian countries, deep appreciation of the taiga is part of their societies' identity and culture. However, misguided **forestry** experiments to increase wood production have caused great environmental problems also in these countries. Public awareness has already produced "remedies," for example, Sweden's National Forest Act. The antithesis of viewing the forests as a worthless land is found among the aboriginal cultures on both continents. Here, harmony between nature and people is coded in spiritually based rules of behavior.

A different perception developed in Russia, where Siberia is the land of opportunities based on gas, oil, diamonds, hydropower, and forests. Despite the view of "Mother Siberia," Russia has dangerously mismanaged boreal resources—and conditions did not improve after the breakup of the Soviet Union (USSR). Modern forest management led to significant changes with strong fire influences. Fires that develop around clearcuts and roads fundamentally alter the taiga landscape: fire frequency is increased, fire refuges disappear, and the forest mosaic is simplified. The least architecturally complex forests are formed on the most severely disturbed parts (abandoned agricultural lands or clearcuts), with a very even canopy of secondary growth trees, most often birch and gray alder. Forests with an absolute domination of pioneer deciduous species are especially typical of the large, landscape-sized clearcuts of the 1950s through 1980s. On January 22, 1997, the Russian Duma adopted the new Russian Federation Forest Code. Lack of law enforcement, resulting from bureaucracy problems, is one of the reasons for the sad reputation of Russia's (illegally logged) forest products. Mass public violation due to **poverty** and pure criminal activities are others.

The twenty-first century forces several countries to shift gears toward a sustainable use of their resources. In 1999, Canada's Senate Subcommittee on Boreal Forests released the report "Competing Realities: The Boreal Forest at Risk," which supported the Canadian Ministry of Natural Resources's commitment to building a sustainable economy, with an ecological framework document as the basis for governmental actions with a National Indicator and Reporting Office. The European Union (EU) has the following policy instruments: EU Strategy for Sustainable Development; EU Forestry Strategy; Integrated Product Policy–Life Cycle Thinking; EU Eco-Labeling Scheme; EU Action Plan for Forest Law Enforcement, Governance, and Trade (FLEGT); and Combating Climate Change. Legally binding conventions exist, yet only on an international level on **biodiversity** or the climate convention. Influential organizations such as the **United Nations Environment Programme** (UNEP) with its World Conservation Monitoring Center (WCMC) have launched initiatives to improve forest management and **conservation**. Its Forest Programme cooperates with the World Conservation Union (IUCN) to produce the Internet-based "Conservation Atlas of Temperate and Boreal Forests." IUCN also fosters a Forest Conservation Programme in Russia. The Timber Committee of the UN's Economic Commission for Europe (UNECE), together with the European Forestry Commission of the **United Nations Food and Agriculture Organization**, published the Temperate and Boreal Forest Resource Assessment (TBFRA) 2000.

On November 18, 2004, Russia deposited its instrument of ratification with the United Nations. This marked the start of the ninety-day countdown to the entry in force of the **Kyoto Protocol**, an international and legally binding agreement to reduce greenhouse gas emissions worldwide. New mechanisms for forest management will develop. For example, emissions trading is likely to increase forest resources and wood production, especially in developing and transition economies. Sinks will act as an economic subsidy for wood production, which impacts volume and price supply.

Large-Scale Projects

Recognizing the importance of the boreal zone, large-scale projects investigate scientific topics such as ecosystem processes, greenhouse gas exchanges, mapping, and inventory issues. The following section introduces a selection of projects, without claiming complete coverage of all activities.

The Boreal Ecosystem-Atmosphere Study (BOREAS) is a large-scale, international, interdisciplinary experiment in the northern boreal forests of Canada. BOREAS had eighty-five science teams organized into six disciplinary groups: Airborne Fluxes and Meteorology (AFM), Hydrology (HYD), Remote Sensing Science (RSS), Terrestrial Ecology (TE), Tower Fluxes (TF), and Trace Gas Biogeochemistry (TGB). The 1994 and 1996 BOREAS field campaigns put over 300 scientists and aircrew into the field, supported by eleven research aircraft.

The **International Geosphere-Biosphere Programme** (IGBP) runs so-called high-latitude transects: the Far East Siberia Transect (FEST), the East Siberia Transect (EST), the Scandinavian Transect (ScanTran), and Alaska Transect. They span significant variation and co-variation of several environmental variables.

The **International Institute for Applied Systems Analysis** (IIASA) has the Forestry Program (FOR): its overall theme is global change and forests, which addresses the question of how to manage the forest sector to harmonize geo- and biospheric functions with socioeconomic development.

Japanese Aerospace Exploration Agency (JAXA)'s Kyoto & Carbon Initiative is an international collaborative project led by JAXA Earth Observation Research Center (EORC). The Initiative provides data and information to support the Kyoto Protocol and international efforts for a Terrestrial Carbon Observing (TCO) system, as outlined by the Integrated Global Observing Strategy Partnership (IGOS-P).

The Northern Eurasia Earth Science Partnership Initiative (NEESPI) is an international program of coordinated research on the state and dynamics of terrestrial ecosystems in northern Eurasia to enhance scientific knowledge and develop predictive capabilities to support informed **decision-making** and practical applications.

The Reducing Canada's Vulnerability to Climate Change (RCVCC) program of Natural Resources Canada's earth sciences sector is part of Canada's agreement to sign the Kyoto Protocol. The goal of the RCVCC program is to lessen the **vulnerability** of Canadians, their infrastructure, and communities to climate change via research. Research being conducted by the portfolio of program projects includes the **monitoring** of Canada's landscape and coastal areas through satellite observation and *in situ* measurements; paleo reconstructions for understanding historical climate and landscape evolution; spatially explicit, process-based **modeling** of landscape dynamics; and simulations and assessments of landscape changes in response to climatic, socio-economic, and policy scenarios.

The EC FP5 project SIBERIA-II (Multisensor Concepts for Greenhouse Gas Accounting of Northern Eurasia) demonstrates the viability of full carbon accounting (including greenhouse gases CO_2, CO, CH_4, N_2O, and NO_x) on a regional basis for more than 200 million hectares by employing earth observation instruments, detailed existing databases of field information, and dynamic global vegetation models.

See also: **Atmosphere-Land Interlinkages; Carbon Cycle; Climate Impacts; European Remote Sensing (ERS-1/-2) Satellites; Global Terrestrial Observing System; Hot Spots of Land-Cover Change; Land Abandonment; Mineral Extraction; Paleo Perspectives; Public Policy; Reforestation; Regrowth; Secondary Vegetation; Sustainable Land Use; Systemic Change.**

Further Reading

Canada's Senate Subcommittee 1999 Report, Competing Realities: The Boreal Forest at Risk [Online, December 2004], The Parliament of Canada's Web Site, www.parl.gc.ca/36/1/parlbus/commbus/senate/com-e/bore-e/rep-e/rep09jun99-e.htm; Eurostat, Forest and Environment, Sustainable Management of EU Forests, News Release, N° 121/2000, 24 October 2000 [Online, December 2004], The Eurostat Web Site, europa.eu.int/comm/eurostat/; Forest.ru, General Forest Characteristics of the Region [Online, December 2004], The Russian NGOs Forest Club Web Site, www.forest.ru/eng/publications; Henry, John David, *Canada's Boreal Forest*, Washington, D.C.: Smithsonian Institution Press, 2002; International Geosphere Biosphere Programme, IGBP High Latitude Transects [Online, December 2004], The IGBP Web Site, picea.sel.uaf.edu/projects/igbpdesc.html; Natural Resources Canada, Reducing Canada's Vulnerability to Climate Change [Online, December 2004], The RCVCC Web Site, rcvcc.nrcan.gc.ca/index_e.cfm; United Nations Economic Commission for Europe, Forest Resources Assessment [Online, December 2004], The UNECE Web Site, www.unece.org/trade/timber/fra/welcome.htm; World Meteorological Organisation, The Second Report on the Adequacy of the Global Observing Systems for Climate in Support of the UNFCCC, GCOS–82, WMO/TD No. 1143, April 2003.

CHRISTIANE SCHMULLIUS

Boserup, Ester (1910–1999). Writer, consultant, and scientist (development economics) whose works continue to impact the evolution of thoughts in anthropology, demography, economics, and sociology, in particular about the role of human **population dynamics** in land **extensification** and, more so, gradual land-use intensification, that is, the interrelationships among economic, demographic, and technical change in general. Ester Boserup began her career as a government planner in Denmark during World War II and then turned, via the United Nations, into a consultant concerned with contemporary problems of the developing world. She examined and found deep flaws not only in the assumed **driving forces** of land-use change, but also in the early gender-blind approaches to **economic growth** or development as a whole. One postulate, for example, has been that land tends to be used more intensively as it becomes scarcer in relation to rising population densities, with **land degradation** being a possible but not very likely outcome.

See also: **Land-Use History; Technological-Scientific Revolution.**

Further Reading

Boserup, Ester (posthum), "Technological Society and Its Relation to Global Environmental Change," in *Encyclopedia of Global Environmental Change, Volume 5: Social and Economic Dimensions of Global Environmental Change*, eds. Ted Munn and Peter Timmermann, Baffins Lane, Chichester, UK: John Wiley & Sons, 2002, pp. 86–96; University of Copenhagen, Museum Tusculanum, Ester Boserup, 1910–1999 [Online, December 2004], The Ester Boserup Web Site, www.mtp-dk/authors/boserup-e.

HELMUT GEIST

Brazilian Cerrado. Tropical savanna covers 22 percent of Brazil's territory (1.78 million km^2) and is located primarily on the central plateau of the Brazilian shield, an area between 24° and 4° south latitudes. The Cerrado ranks twelfth on a list of global "hot spots," a designation that identifies a region with high levels of plant endemism and increasing levels of land conversion. Only 5.2 percent (92,280 km^2) of the Cerrado is protected by a **conservation** regime. **Cattle ranching** and mechanized commercial agriculture, including **cotton**, **rice**, **maize**, and soybeans, have expanded rapidly in the Cerrado during the past three decades. Although the Cerrado is a major agricultural and cattle region, available **biome**-scale land-cover loss estimates are imprecise, varying between 40 and 80 percent. Recent research indicates that while the original land-cover loss is high, there are important trends in vegetation **regrowth** and **agricultural intensification**.

The Cerrado's undulating topography is interrupted by tributaries of the Amazon, Paraná-Paraguay, and São Francisco rivers. The region varies in elevation from 300 to 3,000 meters. Soils are deep, porous, highly acidic, and high in aluminum saturation but deficient in base cations, phosphorous, and nitrogen. They are chiefly formed by oxisols (46 percent), inceptisols (20 percent), and ultisols-alfisols (26 percent). The Cerrado has Köppen Aw climate with an annual precipitation between 900 and 1,800 millimeters, and average annual temperatures range between 18° and 28° C. Ninety percent of rain falls during the wet season between October and March.

Cerrado **vegetation** is a mix of grasses, shrubs, woody plants, and fire-resistant twisted trees with thick, corky bark and sclerophyllous leaves. It varies in height and density, ranging from an open herbaceous land cover (*campo*) to orchard-like woodland savanna (*cerradão*) that approaches a closed canopy of twelve to fifteen meters. Seasonal savannas are present to a lesser extent in the Cerrado region; and **forests**, including riparian and mesophytic, represent a small but important land cover. Physiognomic differences are determined by complex biophysical relationships at varying temporal and spatial **scales**. At the global scale, climate influences overall vegetation distribution. However, soil fertility, nutrient status, **fire** regime, and drainage characteristics influence vegetation structure at a landscape scale.

The Brazilian Cerrado is a repository of **biodiversity**, claiming an estimated 160,000 species of plants, fungi, and fauna. While large areas of the Cerrado are insufficiently inventoried or underrepresented in data sets, the best scientific measurements estimate that 44 percent of vascular plants, 9.3 percent of mammals, and 3.5 percent of avifauna are endemic species. Species are not distributed evenly throughout the Cerrado. Whereas it is unusual to find more than 120 species of trees and larger shrubs per hectare, there are continuous changes in botanical composition from one area to another and evidence of extreme floristic heterogeneity. Botanists also have identified a strong biogeographic pattern that allows for the designation of five distinct botanical sub-regions within the Cerrado. The mosaic landscape of **grassland**, scrub, and woodland environments creates complex habitats for fauna. Cerrado mammal and bird species composition is dependent upon the relationship among open landscapes, forest edges, and closed canopy forests during dry seasons and fire events.

Agricultural **colonization** projects, state development policy, and agricultural technology contribute to Cerrado **conversion** and **modification**. Between 1970 and 1990, cooperatives and private companies allotted over 900,000 hectares of Cerrado to colonists in the state of Mato Grosso, areas that are currently at the forefront of Brazil's soybean production. In the mid-1970s, the Brazilian government funneled resources into two showcase development programs to encourage mechanized **agriculture** in the Cerrado. The first program, POLOCENTRO (Programa para o Desenvolvimento do Cerrado, 1975–1982), improved infrastructure throughout the Cerrado, provided financial support for agricultural research, and offered cheap rural credit to mechanized farmers, financing 3,373 agricultural projects on approximately 1.5 million hectares. The second program, PRODECER (Programa Cooperativo Nipo-Brasileiro para o Desenvolvimento do Cerrado, 1976), was a bilateral initiative between Brazil and **Japan** to encourage agricultural modernization. The program targeted commercial farmers in smaller agricultural colonization projects. While PRODECER's relatively modest operations did not directly impact large Cerrado areas, the development initiative stimulated expansion of diverse **agricultural frontiers** in the states of Minas Gerais (1979–1990), Goiás (1983–1990), Bahia (1983–1990), and Tocantins and Maranhão (1994–2001). PRODECER's long-term dedication of human and financial resources to new agronomic research and technologies contributed greatly to land-cover modification beyond the designated project areas.

Over the past thirty years, Brazilian scientists have developed a suite of soil conservation and agronomic techniques, such as crop-pasture rotation, no-till agriculture, and the heavy application of lime and fertilizers, to increase and intensify production on the Cerrado's nutrient-deficient soils. Scientists also developed strategies to integrate annual cultivation with livestock operations through pasture-crop rotation schemes. In addition, Brazilian researchers selected maize and soybean varieties to tolerate the high aluminum toxicity and low pH of Cerrado soils. Scientists also bred soybean varieties, which were originally sensitive to the photoperiods of temperate regions, to tolerate the day lengths characteristic of low latitudes.

The Cerrado soybean boom pushed Brazil into its current position as the world's second-leading producer. Areas under soybean cultivation in the 1970s were concentrated in the southern states of Rio Grande do Sul and

Paraná. Since 1976, Cerrado regions converted to soybean production have contributed more area to the total land under soybean cultivation. For example, in 1976 Mato Grosso did not produce soybeans, but in 2003, the state planted 22 percent (4.27 million hectares) of Brazil's soybean crop. In 2003, farmers on the Cerrado's most dynamic frontier in western Bahia and southern Maranhão planted 1.1 million hectares of soybeans. Livestock remains an important economic activity in the Cerrado as it supports a large proportion of the national herd that sustains Brazil's beef exports.

See also: **Amazonia; Arc of Deforestation; Biotechnology; South America; Tropical Dry Forest.**

Further Reading

Jepson, Wendy. "A Disappearing Biome? Reconsidering Land-Cover Change in the Brazilian Savanna," *The Geographical Journal* 171 (2005): 99–111, Oliveira, Paulo, and Robert J. Marquis, eds., *The Cerrados of Brazil: Ecology and Natural History of a Neotropical Savanna,* New York: Columbia University Press, 2002; Sano, Sueli Matiko, and Semíramis Pedrosa de Almeida, eds., *Cerrado: Ambiente e Flora,* Brasília, BR: EMBRAPA-Cerrados, 1998; Silva, Jose Maria Cardoso da, and John M. Bates, "Biogeographic Patterns and Conservation in the South American *Cerrado*: A Tropical Savanna Hotspot," *BioScience* 52 (2002): 225–34.

WENDY JEPSON

Brookfield, Harold Chilingworth (1926–). Professor emeritus in the Department of Anthropology, Research School of Pacific and Asian Studies at the Australian National University, and principal scientific coordinator of the **Project on People, Land Management and Ecosystem Conservation** (PLEC) from 1992 to 2002, now PLEC's senior adviser. Eric Waddell titled him a "magnificent loner" whose writings on the borderlands among disciplines (geography, anthropology, sociology, agronomy, botany, history, development studies) are concerned with matters of common sense, ordinary people, and reality. Harold Brookfield dedicated himself to the study of rural societies in developing countries and, more specifically, to the dynamic relationships between land and people. He proposed a more grounded perspective on development studies, explored issues of **agrodiversity** and **land degradation**, and made a significant contribution to the debate on **environmental change**, through a seminal work co-authored by **Piers Blaikie**. According to Brookfield, fieldwork and "local study with comparative method," the title of one of his articles (and, in his view, one of his best publications) make it possible to "see the wood for the trees" and generate theoretical discussion out of practical work.

Further Reading

Blaikie, Piers M., and Harold Brookfield, *Land Degradation and Society,* London, New York: Routledge, 1987; Brookfield, Harold, *Exploring Agrodiversity,* New York: Columbia University Press, 2001; Brookfield, Harold C., "Local Study and Comparative Method: An Example from Central New Guinea," *Annals of the Association of American Geographers* 52

(1962): 242–52; Brookfield, Harold, Parsons, Helen, and Muriel Brookfield, eds., *Agrodiversity: Learning from Farmers across the World*, Tokyo: United Nations University Press, 2003; Waddell, Eric, The 1997 Robert McC. Netting Award: Harold Chilingworth Brookfield [Online, November 2004], The Cultural Ecology Specialty Group Web Site, www.cwu.edu/~geograph/Brookfield.html.

HELMUT GEIST

C

Cadastre. List of persons or real estates made by a state administration primarily for tax purposes. The term derives from Latin and means "inventory" (*caput* = head, *capitastrum* = inventory after heads, later meaning "according to any unit"). In a narrower sense, this term indicates comprehensive and systematic written and cartographic descriptions and inventory of individual plots of land, elaborated to detect their acreage, land use, evidence of their **property**, and to register and chart their land use as property changes. The cadastre contains data on pricing of land plots, which as a rule express their natural fertility. These data serve as a basis for the land tax assessment of each plot of land and also for fixing of so-called official prices of land (which are generally lower than market prices). Written and cartographic data must be identical.

Cadastral maps are maps of properties and identify the property of owners (state, nobility, church, corporations, peasants, etc.) as registered in written documents and with some additional data like names of owners, areas of plots, their fertility, tax class, etc. A cadastre is therefore an important source of exact numerical and cartographic information on the state of land use to date, when counted or surveyed. Whereas modern cadastre has existed for about 300 years, it represents an inestimable database for the study of long-term land-use/cover changes in small territorial units. These units can be joined according to different administrative or geographical units (such as historical lands, provinces, districts, counties, or catchment basins, mountains ranges, etc.).

The so-called "stabile (Franciscan) cadastre," founded in the first half of the nineteenth century in the Austrian part of the Habsburg Monarchy, distinguished about fifty land-use categories in its beginning, such as arable land (including arable land with fruit trees, wood, alternate pastures, or fallow land, etc.), meadows, and **forests** (deciduous, coniferous, mixed, high/low stand forest, clearings, shrubs, etc.). All monarchy descendent states continued in such cadastral systems from 1918 until recently.

Land-use change research at higher regional levels generally uses archive-written cadastral data on the micro-regional level, rather than the old or new cadastral maps or large-scale maps created by field observations of current land use. Cadastral parishes, or units, territories, etc., that usually surround each settlement (village, township, territorial parts of cities) serve as basic territorial units of cadastre. They are characterized by data on land-use categories, but at the same time represent sort of a "black box," inside of which it is impossible to observe land-use structure and its changes. For studying land change in such detail, it is necessary to use old and new cadastral maps or maps of smaller **scale**.

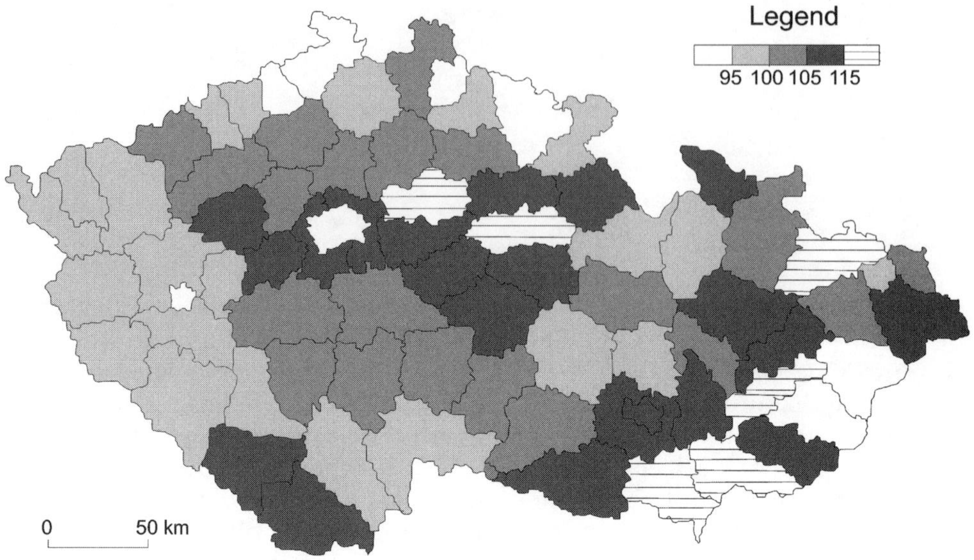

Czechia: Percentage change of arable land 1845–1948 (1845 = 100).

The figure titled "Czechia: Percentage changes of arable land, 1845–1948" shows the high informative value of cartograms elaborated according to comparable cadastral units of an average area of about 790 hectares (3,200 acres), demonstrating not only changes of arable land area within some historical period, but also its regional differentiations on the territory in Czechia.

Cadastral data regarding land-use categories are of use also in landscape ecology, including the survey of at least simplified changes of **biodiversity** at the landscape level. These are expressed, for example, by the ratio of the area of land-use categories with a high intensity of anthropogenic impact—for example, arable land, built-up areas, and categories with a lower intensity of anthropogenic stress, such as forests, meadows, etc. They can demonstrate simplified quantification of the human impact on the landscape in different regions and historical periods. Because of the consequences, for example, of **agriculture**, **industrialization**, and the **Agricultural Revolution**, cadastral maps have to be innovated, thus enabling the observation of historical development of the character and aspects of all landscapes types—that is, tracing **land-use history**.

Cadastral maps are based on exact trigonometric surveys that have been and are under permanent control. Their topography is very detailed and includes place names and names of parts of cadastral units (many of them express their use or character in the past: compare *-crest, -grove,* [Holly]-*wood,* etc.). They also provide information on spot objects in the landscape, such as castles, churches, mills, sawmills, factories, etc. Cadastres, which are organized or sponsored by the state, usually cover and contain data on smaller territory (district, county, parish) and then data according to higher administrative units (provinces, lands, states). As regards the private sector, its cadastre dealt only with the property of manors and great estates of landlords, and is considered to have been the forerunner of the state cadastre.

Like other activities of human beings, the technique of mapping land property was known and used to a limited extent in ancient Mesopotamia and Egypt from about the late third millennium BC. Ancient Greece and mainly the Roman Empire used cadastral maps (e.g., incised in bronze) and registers of state land from the second century BC as well.

The nobility began to secure its private rights to land property (in early feudalism, usually gained by purchase or rather as feud from state represented by king or emperor for services to them) in **Europe** around AD 1000 by recording it into official registers (land books), such as the Doomsday (Domesday) Book in England in the eleventh century. The hereditary tenures of land (or farms) by serfs and their incumbencies (pledges) to the feudal lords were registered in special books called *urbars*. Land farmed by serfs was called rustical land, and land of feudal lords was called dominical land. The land of kings or nobility was not burdened by any land tax until the middle of the eighteenth century (e.g., in the case of Austria). Many medium-scale maps of manors drawn for noblemen from the sixteenth to the eighteenth centuries show the field patterns and some gross data on their land use. They also served as inventories of their land property. These maps have limited possibilities of use compared to later cadastral maps because of their relatively small scale, which is not suitable for detailed study.

The idea of cadastral mapping sponsored or organized by the state was born in eighteenth-century Europe in the Enlightenment era. It was influenced by the struggle between a gradually centralized state and nobility aiming to preserve the old no-tax character of its land property. The birth and development of capitalism urgently required all land to be converted into a free commodity on free markets. Such a commodity had to be liberated from feudal survivals, priced, and then taxed, i.e., therefore exactly measured and mapped.

States with increasing bureaucracy and the necessity of permanent armies required much larger state budgets. The direct taxes, such as those from farmed land, were the most rapid way to its reinforcement. This approach corresponded with the economic philosophy of physiocratism, the leading economic theory at that time. It meant a shift in the use of cadastre and cadastral maps by the state, that is, by the government as one of tool of land management and consolidation of the nation-state and capitalism.

Tax assessment was usually based on land plots. Each plot got its number (valid until recent days) and then, after the complicated pricing (based on net income, for example), it got another number expressing its soil value, according to which tax assessment of land was calculated. Cadastral maps were elaborated for each cadastral unit with delineated boundaries of all plots. For all the surveying and mapping of enclosures and tithes in England, very early cadastral mapping in the Netherlands, the so-called "ancient cadastre" of France, the plats together with the notes of the United States Federal Land Survey, and the Austrian Stabile Cadastre are typical for their long duration, therefore offering valuable historical sources for the historical study of land use.

Farmers and land owners in communist countries, in which agriculture was more or less collectivized or nationalized, had to put their land *gratis* into a cooperative and cultivate it collectively. The state wanted therefore only to know who farmed the land, not who owned it. The cadastre was then degraded into the records and maps of users of land, not of their owners and the land property changes. Ongoing updates of land property changes became obsolete; it was not coincident with reality any longer. The restoration of **private property** rights followed by restitution of land properties, that is, the former owners of land or their descendants who obtained their land back, have been therefore complicated and tedious since 1989. The reconstruction of cadastre in post-communist countries into its former quality needs time and has not been fully accomplished thus far.

See also: **BIOME 300 Project; Change Detection; Land Privatization; Land Reform; Land Rights; Land Tenure.**

Further Reading

Clawson, Marion, *The Bureau of Land Management*, New York: Praeger Publishing, 1971; Kain, Roger J. P., and Elizabeth Baigent, *The Cadastral Map in the Service of the State: A History of Property Mapping*, Chicago: University of Chicago Press, 1992; Kain, Roger J. P., and Hugh C. Prince, *The Tithe Surveys of England and Wales*, Cambridge, UK, and New York: Cambridge University Press, 1985; Schwartz, Seymour I., and Ralph H. Ehrenberg, *The Mapping of America*, New York: H. M. Abrams, 1980; U.S. Bureau of Land Management, *Manual of Instructions for the Survey of the Public Lands of the United States*, Washington, DC: Department of the Interior, 1947.

LEOŠ JELEČEK

Canada. Canada and **Russia** together contain almost half of the world's remaining intact **forest**, making the boreal forest or **taiga** a highly significant ecosystem on a global scale. While **deforestation** in the tropics has received much media attention, deforestation rates in some parts of the **boreal zone** have actually exceeded those in **Amazonia**, reaching rates of 0.91 percent per year. The cumulative impacts of **land-cover changes** due to human land-use and global climate changes will likely reduce the **biodiversity** of the boreal forest.

Cumulative impact is a synergistic phenomenon in Canada, in which the total impact on the ecosystem is greater than the sum of the individual disturbances. In Canada, cumulative impacts are currently of a lower magnitude—but accelerating, creating a need for **conservation** of the remaining intact boreal habitat. In the Canadian landscape, cumulative impacts are a result of the combined impacts of forestry, **agriculture**, climate change, changes in the fire regime, and oil and gas exploration and development (cutting lines for seismic exploration, pipelines, heavy oil extraction).

Forestry is a major economic factor in Canada, and changes in the boreal forest due to climate change could have serious impacts on the industry. Planning is essential for continued boreal forestry. The forestry industry is beginning to realize that mimicking natural patterns in forest management is essential in maintaining the integrity of the boreal forests. Practices such as green-tree retention and fire management are being investigated as possible means of maintaining both biodiversity and long-term forest productivity. However, it has been shown that forestry, even when carefully planned to mimic fire patterns, cannot fully reproduce the early post-fire stage in the boreal region.

Mechanical forestry causes significant alterations to the post-disturbance soils and plant communities, with the effect of eliminating the early post-fire stage from managed areas. Preservation of the quality of the whole landscape is very important.

Because forests are intensively used for commercial purposes, different types of forest inventory systems have been developed to produce information on forest resources, health of forests, and biodiversity of forests at national and regional levels. Nowadays, these systems also utilize satellite **remote sensing** in various countries. As the cumulative impacts of human and natural disturbances alter the boreal region, it is important to be able to monitor the changes in the biophysical land cover. This includes assessment of current conditions and quantification of changes in landscape composition and patterns.

One of the most distinct characteristics of the Canadian boreal region is the complex and dynamic nature of its internal spatial arrangement. The boreal region consists of an intricate mosaic of forests and **wetlands**, driven by factors such as temperature, hydrology, substrate, insect outbreaks, and fire. While dominated by just a few tree species, the spatial heterogeneity of the boreal forest provides a wide variety of habitat types. This mosaic pattern along with the boreal forest's location along bird migration routes allows it to support one of the highest levels of bird species diversity in North America. The boreal region also hosts a large diversity of fungi, plants, insects, and animals.

Fire is an important factor in boreal forest succession, as fire and insect outbreaks are the primary drivers of the pattern and species compositions present. An average of 2.4 million hectares of boreal forest are burned annually in Canada, and 22 million hectares globally. The fire cycle in the boreal forest ranges from 50–270 years, averaging 110–155 years. The majority of area is burned in a few large fires. Eighty-five percent of fires are less than four hectares in size, but some fires can be up to 200,000 hectares. Fire impacts ground-level thermal characteristics and the **nutrient cycle**. Many species depend on fire for their continued presence in the boreal landscape. Post-fire succession depends on many factors, including burn severity and pre-fire species composition. Post-fire **colonization** in the boreal may occur by several strategies, mainly wind-dispersed seeds, serotinous (fire-opened) cones, fire tolerance (in which individuals are not killed by fire), or vegetative reproduction (root and stump sprouting after a burn). Fires burn different areas at different times and at different severities, leaving behind a complex mosaic of vegetation types and age classes. Species-level adaptations to fire were in place by the Pliocene (12 million years BP), meaning that fire was already established as a dominant process in the boreal **biome** as the boreal forest re-colonized landscapes exposed by retreating glaciers.

The boreal forest plays a major role in the global climate and carbon budget, and an understanding of this role is crucial to our analysis of global warming. The boreal region is a significant carbon pool, with the total amount of carbon stored estimated to be between 559 and 709 pg (1 pg = 10^{15}, g = 1 gt, or gigaton). The carbon is stored in above-ground **biomass**, soils, peat, and **vegetation**. There is a significant flux of carbon among these pools and the atmosphere through fire, respiration/photosynthesis, and decomposition. The overall source/sink status depends on the relative impact of changes in forested areas, carbon uptake by trees, carbon storage in soils, decomposition rates, fire regime, and peat/permafrost dynamics. The boreal forest may currently be a sink for atmospheric carbon; however, this sink will become saturated over time due to reduced forest area and loss of permafrost.

The boreal ecosystem is particularly sensitive to global warming due to its location and ecology. In fact, the largest changes in mean annual temperature have been recorded in the boreal. Since 1900, the global mean temperature has increased by 0.6 ± 0.2°C, whereas western Canada has increased by as much as 1.5°C, primarily manifested as warmer winter lows. On a global scale, the largest temperature increases have been in the continental regions of the Northern Hemisphere, primarily in northwest Canada and northwest and northeast Russia, with anomalies as high as +4°C. This means that large regions of the boreal forest will likely experience much greater warming than the global mean. The responses to global warming will manifest as changing fire regimes, northward migration of the boreal biome, and changes in permafrost distribution. Some areas that are currently tundra may become forested at the northern fringe of the boreal, but at the same time in the south, forest areas may be lost, becoming **grassland**. The response of the boreal forest to global warming is likely to be non-linear, making precise estimates of the impact difficult.

The most obvious and direct impact that humans are having on the boreal ecosystem is the **conversion** of natural habitats to human use and the fragmentation of the remaining natural habitat. This is mainly happening through the conversion of natural lands (forest or grasslands) to agriculture or **pasture**, urban and road network expansion, alteration of forest composition by forestry or other resource extraction, and cutting of seismic lines for oil and gas exploration.

In the Canadian boreal landscape, habitat loss and fragmentation are primarily caused by forestry and oil exploration, with agricultural disturbance generally limited to the southern fringe. The total forest cover remains relatively constant with forestry, but there are qualitative changes such as alteration of the age/class distribution. Agriculture reduces the amount of forest cover more permanently, as well as introducing community alterations and higher predation rates at agriculture edges, which do not occur at clearcut edges.

Recent studies indicate that habitat loss is the most critical factor in the loss of species, and more specifically that below certain levels of remaining habitat, species begin to go locally extinct. This relationship has been shown both theoretically and in the field. Some authors have modeled theoretical responses to habitat loss and found that habitat loss is the key factor in **species extinction** (much more important than fragmentation), often with a distinct threshold of around 20 percent remaining habitat. Other authors have found that fragmentation effects can be mediated by the total amount of habitat in the area.

Once the **threshold** for habitat is reached, fragmentation begins to play a role in the loss of species. The effects of fragmentation depend on landscape factors such as patch size and shape, connectivity, time since fragmentation, amount of remaining habitat, and species-specific biological factors. As well as the direct impacts of fragmentation on a given species, there are also indirect impacts such as the effects of changing community dynamics as the increased proportion and/or effectiveness of generalist predators in fragmented landscapes, for example.

Time since fragmentation has been shown to be important both theoretically and empirically, meaning that multitemporal remote sensing analysis is critical. Models of habitat fragmentation effects indicate delayed responses to habitat loss. Field studies of boreal birds have also shown sensitivity to the amount of time passed since fragmentation. This is related to the concept of extinction lag, the delay between destruction of a species habitat and its extinction. The length of lag depends on the amount of habitat. The extinction will be slower if the amount of remaining habitat is close to the critical threshold.

See also: **Atmosphere-Land Interlinkages; Carbon Cycle; Cumulative Change; Great Plains; Hot Spots of Land-Cover Change; Malaria; Mineral Extraction; Monitoring; Peatland; Suburbanization; Urban-Rural Fringe.**

Further Reading

Bayne, Erin M., and K. A. Hobson, "Comparing the Effects of Landscape Fragmentation by Forestry and Agriculture on Predation of Artificial Nests," *Conservation Biology* 11 (1997): 1418–29; Boutin, Stan, and D. Hebert, "Landscape Ecology and Forest Management: Developing an Effective Partnership," *Ecological Applications* 12 (2002): 390–7; Schmiegelow, Fiona K. A., and M. Mönkkönen, "Habitat Loss and Fragmentation in Dynamic Landscapes: Avian Perspectives from the Boreal Forest," *Ecological Applications* 12 (2002): 375–89.

ARTURO SÁNCHEZ-AZOFEIFA

Carbon Cycle. The global carbon cycle is one of the major biogeochemical cycles and of particular importance to human life on earth. Carbon dioxide (CO_2) is cycled among the atmosphere, the oceans, and the terrestrial biosphere in the form of inorganic and organic compounds. The largest natural carbon exchanges occur between the atmosphere and the terrestrial biosphere and between the atmosphere and ocean surface waters. The CO_2 content of the atmosphere is a result of carbon cycle **feedbacks** between the atmosphere and the terrestrial and oceanic carbon reservoirs. Ocean uptake of carbon is limited by the solubility of CO_2 in seawater; the exchange from and to terrestrial ecosystems is controlled by photosynthesis and respiration. These processes again are largely controlled by temperature.

Feedbacks, Interaction, and Human Perturbation

Carbon cycle feedbacks act on different time **scales**. Equilibrium in the exchange of carbon between the atmosphere and carbonate rocks takes place on a million-year time scale while the exchange between the terrestrial biosphere and the atmosphere is balanced within decades to centuries. The temporal as well as the spatial dynamics of the carbon cycle are therefore closely coupled with variations in ecosystem processes such as **nutrient cycling**, stability, succession, and **biodiversity**.

During the last decades, the human perturbation of the global carbon cycle has become more and more important. Atmospheric CO_2 concentration has risen by about 32 percent during the last two centuries, and it is common scientific understanding that this increase stems to a large degree from human activities, of which fossil fuel combustion is the most important one.

Atmospheric CO_2 is, however, increasing at only about half the rate of fossil fuel emissions; the rest of the CO_2 emitted either dissolves in seawater and mixes into the deep ocean or is taken up by terrestrial ecosystems. This imbalance in the global carbon budget is the so-called "missing sink," because the accumulation of carbon has not been observed. This biosphere sink is sensitive to climate events and human activities. In certain years almost all the fossil fuel emissions are reabsorbed, whereas in other years the sink capacity of the earth's surface is almost zero. A large fraction of these oscillations have been associated on the one hand with natural phenomena such as El Niño effects and major **biomass**-burning events and on the other hand with land-use changes.

Depletion and recovery of ecosystem carbon stocks due to land-use changes are highly asymmetric. While depletion can be nearly instantaneous, for example, by fire, recovery may take decades or centuries. Effects of land-use change on soil carbon stocks can be detected even after centuries.

Six types of land-use change are known to have particular impacts on the change of carbon stock: **conversion** of natural ecosystems to permanent croplands, conversion of natural ecosystems to **slash-and-burn agriculture**, conversion of natural ecosystems to **pasture**, abandonment of croplands, abandonment of pastures, and conversion of natural forests to tree plantations and managed forests.

Terrestrial Carbon Stocks

The oceans, **vegetation**, and soils are significant reservoirs of carbon. The largest carbon reservoir is the ocean, with roughly 39,000 gigatons of carbon. With about 2,500 gigatons of carbon, terrestrial ecosystems contain about three and a half times as much carbon as the atmosphere. Within terrestrial ecosystems, the amount of 2,000 gigatons carbon stored globally in soils is about four times higher than that in vegetation (see table). There are large differences in carbon stocks of soils and vegetation according to the vegetation type and climatic zone. Soils are major carbon pools in all terrestrial ecosystems. Vegetation-related carbon stocks are largest in forest ecosystems, making up about 46 percent of the carbon stored in all terrestrial ecosystems on 28 percent of the land surface. **Forests** contain 77 percent of the global carbon stocks stored in vegetation biomass and roughly 39 percent of global soil carbon. Forests of the **boreal zone** are the largest carbon reservoirs of all forest ecosystems, although more than 50 percent of vegetation carbon stock is contained in tropical forests. This is because boreal forest soils show a particular high carbon density (344 tons of carbon per hectare). Taking into account that tropical forests and boreal forests make up only about 12 percent and 10 percent of the terrestrial surface, respectively, the particular importance of these ecosystems as carbon pools becomes obvious.

Temperate **grasslands** and savannas cover approximately 23 percent of the earth's land surface and contain some 26 percent of the earth's terrestrial

Global carbon stocks in vegetation and top 1 m of soils (based on GACGC, 1998)

Biome	Area (10^6 km^2)	Share (%)	Carbon Stocks (gt C) Vegetation	Soils	Total	Fraction of Total Carbon Vegetation	Soils	Carbon Density (tC/ha) Vegetation	Soils
Tropical forests	17.6	11.6	212	216	428	45.5	10.7	120	123
Temperate forests	10.4	6.9	59	100	159	12.7	5.0	57	96
Boreal forests	13.7	9.1	88	471	559	18.9	23.4	64	344
Tropical savannas	22.5	14.9	66	264	330	14.2	13.1	29	117
Temperate grasslands	12.5	8.3	9	295	304	1.9	14.7	7	236
Deserts and semideserts	45.5	30.1	8	191	199	1.7	9.5	2	42
Tundra	9.5	6.3	6	121	127	1.3	6.0	6	127
Wetlands	3.5	2.3	15	225	240	3.2	11.2	43	643
Croplands	16	10.6	3	128	131	0.6	6.4	2	80
Total	151.2	100.0	466	2,011	2,477	100.0	100.0	31	133

carbon stocks, mainly in their soils. The soil carbon pool per unit of area is two to four times greater in temperate grasslands than in the savannas.

In four other **biome** types—deserts and semi-deserts, tundra, wetlands, and croplands—the carbon stocks are particularly dominated by soil carbon, which makes up at least 93 percent of the total carbon stored in these biome types. **Wetlands** are the biome type with the highest carbon density of all terrestrial ecosystems. Compared to the average carbon density of croplands, global wetlands contain eight times more carbon. The share of tropical wetlands in global wetland areas ranges between 10 and 30 percent, excluding **rice**-growing areas. However, the carbon stocks of tropical wetlands are of a magnitude similar to those of the northern hemisphere's wetlands. This is due to their stocks per unit area, which are several times larger, both in the vegetation biomass and the soil compartment.

Terrestrial Carbon Fluxes

The dynamics of global carbon fluxes, which are associated with different pools, can be split up into three major categories: the short-term carbon uptake, the medium-term carbon storage, and the long-term carbon storage. Photosynthesis of plants is a key process of the global terrestrial carbon cycle. Via photosynthesis, plants take up atmospheric carbon dioxide. This uptake is called gross primary productivity (GPP). However, plants use about 50 percent of the assimilated carbon for maintenance and therefore release CO_2 via respiration. The result of this process is the net primary productivity (NPP) of plants, which globally makes up about 60 gigatons of carbon per year and is equivalent to the short-term carbon uptake (days to years). Forest ecosystems account for 80 percent of this exchange with the atmosphere.

During the life cycle of plants, the major fraction of carbon sequestered by photosynthesis is released as CO_2 due to respiration of organisms when decomposing plant litter, while the remaining fraction enters the soil organic carbon pool. The resulting carbon fluxes in ecosystems, which comprise both carbon fixation (due to photosynthesis) and plant respiration as well as microbial respiration (due to decomposition of organic material), are called net ecosystem productivity (NEP). The NEP is about 8 percent of GPP, which globally sums up to a flux of ten gigatons of carbon per year. The NEP describes fluxes relevant for medium-term carbon storage (decades).

To cover the long-term carbon storage of ecosystems (centuries), one has to consider natural disturbances, for example, large forest fires as well as further decomposition of soil organic carbon. The related carbon fluxes are called net biome productivity (NBP). The NBP accounts for less than 1 percent of GPP, which is about one gigaton of carbon per year.

Land Use and Its Effects

Prior to the nineteenth century, humans had only a modest influence on terrestrial carbon storage—through **fire**, fuel use, **deforestation**, and agricultural activities. With the beginning of the **Industrial Revolution**, human activities started to interfere more severely with the global carbon cycle. Between 1850 and 1980, anthropogenic emissions summed up to more than 100 gigatons of carbon. Until the present, the bulk originated from fossil fuel combustion, but land-use changes are assessed to represent about one-third of these anthropogenic carbon emissions. Until the late nineteenth century, most of the emission-relevant land-use activities such as forest clearing and wetland drainage took place in regions of the temperate zone. In the twentieth century, the area of temperate forests stabilized and the conversion of tropical forests became the primary source of land-use-related carbon emissions.

Conversion of natural vegetation to **agriculture** is a major source of CO_2 due to increased decomposition of soil organic matter caused by different kinds of cultivation as well as due to losses of biomass and annual harvesting. During the long history of arable farming, about 750 million hectares of forest have been converted to agricultural land. Globally, this expansion of cropland due to deforestation has caused a loss of roughly 100 gigatons of carbon contained in aboveground biomass and a further twenty-five gigaton loss of carbon stemming from different kinds of soil cultivation (e.g., tillage). Currently, the conversion of forests to agricultural land is taking place mainly in the tropics. During the 1990s, about 90 percent of the 12 million hectares of forest area that were annually converted was utilized as agricultural land. The emissions from tropical land-use change were assessed to account for 1.6 gigatons of carbon per year. About half of tropical forest areas were in turn cleared in shifting cultivation for **subsistence agriculture** or the self-provisioning of the local population. However, this utilization of former forest area as agricultural land is not only caused by agricultural expansion. There are several **proximate causes** and underlying **driving forces** that trigger tropical deforestation in a synergetic way. With that respect, agricultural utilization is often interconnected with infrastructure extension as well as with wood extraction.

Grassland conversion to cropland is assessed to have affected globally about 660 million hectares of natural grasslands. Within this conversion type, the reduction in soil carbon is most important. The carbon losses are calculated to amount to roughly seventy-seven gigatons. Drainage and subsequent cultivation of wetland areas result in particularly high soil carbon losses. Up to 50 percent of the stored soil carbon could be lost during the first five years of agricultural utilization. These losses of soil carbon are at least ten times higher than natural carbon accumulation in soils.

Conversion of primary forests to secondary forests in the tropics results in 30–80 percent lower carbon stocks because old natural forests store large quantities of carbon in above-ground biomass. The conversion of temperate natural forests to managed forests is combined with a loss of 40–50 percent of carbon in aboveground biomass. Even higher is the loss of aboveground carbon if natural forests are converted into fast-growing forest plantations. Thus, conversion of primary natural forests into secondary natural forests as well as managed forests results in a decrease of carbon in both the vegetation and soil pools. However, managed forests are a source of wood products, which are again a carbon pool. How the carbon pool of different wood products relates to the losses of carbon due to forest conversion is still a matter of debate and is of particular importance for the implementation of terrestrial sinks in the **Kyoto Protocol**.

The options for afforestation range from fast-growing species, for example, for energy or paper production, to mixed forest stands managed to follow natural vegetation patterns. In terms of carbon, the first option means a low carbon stock in aboveground biomass while carbon fluxes—growth rates—are high. Therefore, cumulative effects, such as those due to substitution of fossil fuels when used for energy production, are of particular interest. The second option has considerably lower annual growth rates but the potential to store much more carbon in both aboveground biomass and soils.

The use of high-yielding plant varieties, residue management, and reduced tillage can reduce losses and enhance uptake within agricultural soils. The **Intergovernmental Panel on Climate Change** estimated that appropriate management practices could cumulatively increase carbon sinks by 24–43 gigatons of carbon over fifty years. Energy efficiency improvements and production of energy from dedicated crops and residues would result in further cumulative carbon storage of 16–68 gigatons of carbon over fifty years.

See also: **Anthropocene; Atmosphere-Land Interlinkages; Carbon Sequestration; Forestry; Global Burnt Area 2000; GLOBCARBON; Human Appropriation of Net Primary Production; Paleo Perspectives; Reforestation; Secondary Vegetation; Water-Land Interlinkages.**

Further Reading

German Advisory Council on Global Change, *The Accounting of Biological Sinks and Sources under the Kyoto Protocol: A Step Forwards or Backwards for Global Environmental Protection?* Bremerhaven, GE: GACGC, 1998; Houghton, Richard A., "The Annual Net Flux of Carbon to the Atmosphere from Changes in Land Use 1850–1990," *Tellus* 51B (1999): 298–313; Intergovernmental Panel on Climate Change, *Climate Change: The Scientific Basis*, Cambridge, UK: Cambridge University Press, 1996; Intergovernmental Panel on Climate Change, Climate Change 2000: The Scientific Basis [Online, November 2004], The IPCC Web Site, www.ipcc.ch; Intergovernmental Panel on Climate Change, *Land Use, Land Use Change, and Forestry*, Cambridge, UK: Cambridge University Press, 2000; Lal, Rattan, Kimble, John M., Levin, Elissa, and Bobby A. Stewart, *Soils and Global Change*, Boca Raton, FL: CRC Press, 1998; Lambin, Éric F., Geist, Helmut J., and Erika Lepers, "Dynamics of Land-Use and Land-Cover Change in Tropical Regions," *Annual Review of Environment and Resources* 28 (2003): 205–41;

Nilsson, Sten, and Wolfgang Schopfhauser, "The Carbon-Sequestration Potential of a Global Afforestation Program," *Climatic Change* 30 (1997): 267–93; Schimel, David S., "Terrestrial Ecosystems and the Global Carbon Cycle," *Global Change Biology* 1 (1995): 77–91; Steffen, Will, Sanderson, Angelina, Tyson, Peter D., Jäger, Jill, Matson, Pamela A., Moore III, Berrien, Oldfield, Frank, et al., *Global Change and the Earth System: A Planet under Pressure*, Berlin: Springer Verlag, 2004.

GERALD BUSCH

Carbon Sequestration. The withdrawal of carbon from the atmospheric pool and its secure medium to long-term storage in aquatic (largely oceanic), terrestrial, and geological sinks by natural processes and human activities. Industrial activities to sequester carbon are often referred to as carbon capture and storage (CCS), for example, the Sleipner oilfield CO_2 injection project in the North Sea. Land-cover and land-use changes have a great potential for terrestrial carbon sequestration (CS) in the next few decades. CS is controlled by a dynamic equilibrium among atmospheric, aquatic, and terrestrial carbon pools. Enhanced CS in the terrestrial ecosystem is based on the rebuilding of carbon stocks in living **biomass**, dead organic matter, and soils, which have been depleted in the past or are still being lost by **deforestation**, degradation of **vegetation** cover, or **soil degradation** due to destructive agricultural practices. Measures to enhance terrestrial CS include **conversion** to land-use types with more living biomass (e.g., degraded land to **grassland**, grassland to **forest**) and numerous land management strategies improving soil organic matter content (e.g., **conservation** tillage or mulch farming on cropland). Rebuilding depleted carbon stocks in soils has beneficial side effects on soil fertility, soil **resilience**, and ecological functions of soils.

See also: **Carbon Cycle; Ecosystem Services; Environmental Change; Integrated Model; Kyoto Protocol; Reforestation; Sustainable Land Use.**

Further Reading

Intergovernmental Panel on Climate Change, *Land Use, Land-Use Change and Forestry* (IPCC Special Report), Cambridge, UK: Cambridge University Press, 2000; Robert, Michel, *Soil Carbon Sequestration for Improved Land Management* (World Soil Resources Reports No. 96), Rome: United Nations Food and Agriculture Organization, 2001.

JUSSI BAADE

Carson, Rachel Louise (1907–1964). Scientist (biology, ecology), teacher, writer, and early environmental activist in the United States who shaped modern-day researchers' thinking and launched environmental awareness, especially with regard to **conservation** and the detrimental consequences of **agricultural intensification**. *Silent Spring*, a best-selling book, is her lasting legacy. Since financial constraints prevented Carson's pursuit of a doctorate, she assumed teaching positions at the Universities of Maryland and Johns Hopkins while continuing her zoological field research. In the 1930s, she became a government marine biologist and then joined the U.S. Fish and Wildlife Service from 1936 to 1952. It was while working there that

Carson began learning of significant declines in certain bird and mammal species, inspiring her to begin researching the causes of these declines. Her job also helped her to sharpen her skill at putting research results into reports significant to scientists and laypeople alike. She began a popular writing career that gave her commercial success and fame (e.g., National Book Award); she was also popular because she knew how to provide the larger picture while comprehending the details. The commercial success of her unofficial writing enabled Carson to resign from her government job and to more fully dedicate herself to writing about the growing threats to marine and terrestrial ecosystems. In particular, she collected evidence of the damage caused by fertilizer and (chemical) pesticides that were widely applied after the end of the Second World War. Her work on the subject was originally published in the *New Yorker* magazine and then in book form as *Silent Spring* in 1962. This best-seller triggered a storm of controversy, especially with a view upon banning the use of dichloro-diphenyl-trichloro-ethane (DDT) and other chemicals, linked in her work with widespread damage to wildlife and **human health**. The belief had been promoted by the food and chemical industry that the use of active ingredients or potent poisons was essential to **agriculture** and daily life. Rachel Carson, however, with her reputation for scientific integrity and independence from industry interests, was able to demonstrate to the international public the flaws in these arguments. She succeeded both in establishing a new public understanding of the ecology of the earth and in demonstrating how these new chemical inputs damaged crucial elements of **ecosystem services** and functioning. Unmarried and childless—the industry lobby argued that she could not legitimately claim concern for future generations—Carson received a merciless attack, with the charges highly personal, but defended her well-researched work with appearances before Congressional committees (see photo).

Rachel Carson speaking in 1963 before US Senate Government Operations subcommittee studying pesticide spraying.

PHOTO: United Press International Photo. [Courtesy Library of Congress]

Even more than forty years after its publication, *Silent Spring* remains a vitally important book, especially for anticipating and accurately mapping the more subtle health risks of chemical pesticides associated with land-cover **modification**, including endocrine disruption and xeno-estrogenic effects. The book is considered to be one of the most powerful works questioning the attitudes and practices of modern (industrial) society as it relates to the human-environment condition. It galvanized people and their governments in various countries, and many innovative new laws and agencies followed. In the United States, these include, among others, the Environmental Defense

Fund, programs of the Food and Drug Administration, the National Academy of Sciences, and the U.S. President's Office of Science and Technology. In 1972, Congress enacted legislation banning the further use of DDT in the United States, and also began regulating use of other chemicals that could potentially harm the environment. DDT can still be manufactured and sold outside the United States, especially in tropical countries, but many other countries issued similar bans.

See also: **Agricultural Revolution; Agro-Industry; Biodiversity; Green Revolution; Industrial Revolution; People at Risk; Public Policy; Species Extinction; Technological-Scientific Revolution.**

Further Reading

Carson, Rachel L., *Silent Spring*, Boston: Houghton Mifflin, 1962; Lear, Linda, Life and Legacy of Rachel Carson [Online, December 2004], The Rachel Carson Web Site, www.rachelcarson.org.

HELMUT GEIST

Cascades. Energy and material flows in ecosystems and society are organized in several levels or stages. In ecosystems, trophic (food) energy flows from primary producers (e.g., plants) to herbivores, carnivores, and finally decomposers. The trophic cascade hypothesis claims that the amount of energy flowing through each trophic level is influenced by the other levels. Its "top-down" version holds that predators control their prey (i.e., ultimately, top predators might regulate primary producers), while adherents of "bottom up" regulation believe that organisms are limited by resources available from the level below (i.e., primary productivity would limit all other trophic levels). Human societies extract materials and energy from their environment, use and transform them in various stages of production and **consumption**, and ultimately hand wastes or emissions back to nature. Industrial ecologists try to minimize both resource extraction and waste generation through a strategy of "cascade utilization" that aims to maximize the number of times a resource can be made useful for society through recycling and re-use, energetic use of wastes and residues, or utilization of waste energy. Cascade utilization of **biomass** is relevant for minimizing socio-economic land demand.

See also: **Food Crops; Human Appropriation of Net Primary Production; Metabolism; Vulnerability.**

Further Reading

Carpenter, Stephen R., Kitchell, James F., and James R. Hodgson, "Cascading Trophic Interactions and Lake Productivity," *BioScience* 35 (1985): 634–9; Haberl, Helmut, and Susanne Geissler, "Cascade Utilization of Biomass: How to Cope with Ecological Limits to Biomass Use," *Ecological Engineering* 16 (Supplement, 2000): 111–21; Schmitz, Oswald J., Hambäck, Peter A., and Andrew P. Beckerman, "Trophic Cascades in Terrestrial Systems: A Review of the Effects of Carnivore Removals on Plants," *American Naturalist* 155 (2000): 141–53.

HELMUT HABERL

Cash Crops. Agricultural products produced for sale either on domestic or on export markets. Cash crops may be **food crops** (e.g., soybeans, **maize**, **wheat**, **rice**, sugar cane) or non-food crops (e.g., **cotton**, **coffee**, oil palm, cocoa, **tobacco**, tea, rubber). Cash crops can be differentiated from crops that are not sold on markets, such as those fed to livestock (fodder) or grown for home consumption (subsistence crops). Both food crops and non-food crops can enhance the cash income of farm households, if surplus production is marketed and not entirely for home **consumption**.

Cultivation and Production

The most important cash crops are also the main food staples, such as maize, wheat, and rice. Harvested areas of selected food and non-food cash crops are depicted in the figure titled "Harvested area of major cash crops" and show a major increase in the extent of harvested area of soybeans by a factor of 3.5 to 0.84 million km^2. Most of the soybean production is sold as soy meal for animal feed. Apart from the major cereals, soybeans are by far the most important cash crop in terms of area harvested, followed by cotton and sugar. Almost all the cash crops depicted in this figure show increases in the extent of their harvested areas.

Land-Use Change

Various cases are documented across the world where the expansion of areas under cash cropping is associated with the **conversion** of forested land into commercial agricultural cultivation. Prominent examples are the expansion of coffee cultivation in Vietnam, Indonesia, and Brazil; banana plantations in

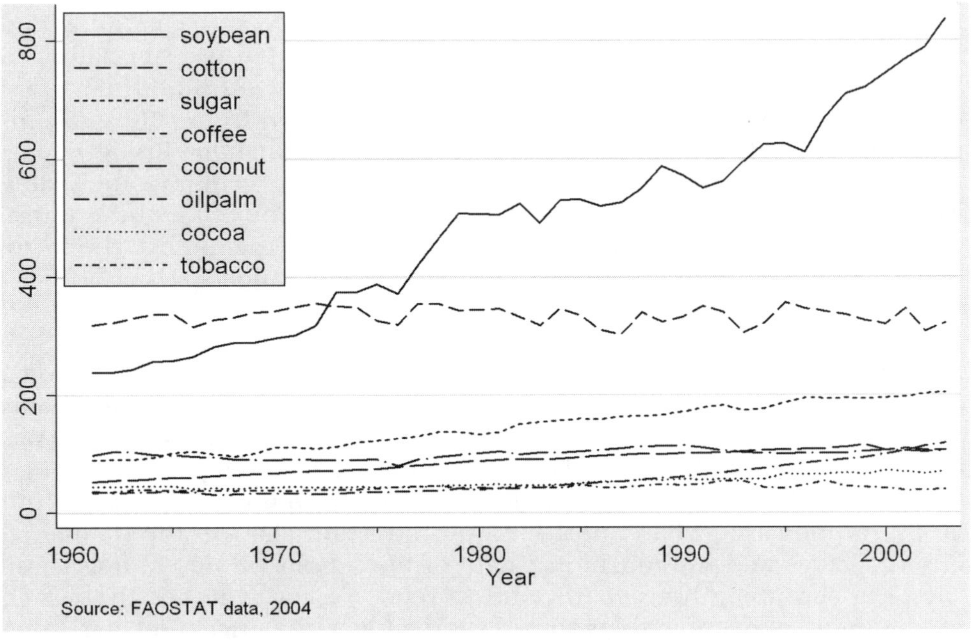

Harvested area of major cash crops (1,000 sqkm).

Latin America; the conversion of tropical forests for oil-palm plantation in Indonesia, Malaysia, and Papua New Guinea; the spread of cocoa in several countries of West Africa; and rapidly increasing soybean cultivation in the Amazonian rainforest and in Argentina. This expansion of often large-scale and monocultural agricultural production for cash cropping is identified as one of the major causes of tropical **deforestation** in lowland and upland areas. The creation of large-scale commercial plantations is pushing many small farmers into marginal and less-productive land with detrimental consequences for the natural resource base, like the acceleration of forest conversion and rapid **soil erosion**.

Higher applications of agricultural inputs like labor and capital for cash crop production increase yields and farm income on the same amount of land. In that way, they have the potential to reduce the pressure to expand agricultural production into uncultivated areas. The additional income can also enable farmers to invest in the sustainable use of farm resources. However, more intensive cash crop production might also increase the demand for irrigation water, affect soil conditions adversely, and, due to higher agricultural profitability, boost the demand for agricultural land, thereby providing an incentive to convert land into agricultural cultivation. The widespread development of monocultural cash cropping also has detrimental consequences for **biodiversity**, both of natural plant and animal species as well as on the genetic diversity of agricultural crops.

Agricultural Development

The term "cash crop" is often used in the sense of export crop. However, cash crops may also be sold on domestic markets. The promotion and subsidization of cash crop cultivation are two of the most important development activities in the past three decades. Cash crops are often subsidized by governments and international agencies to increase foreign exchange earnings and households' cash income. This has often caused a rapid transformation of local economies due to increasing integration into markets and the world economy. This transition involves a shift of production from cultivation for subsistence needs to higher degrees of market orientation. Rising export earnings and increases in farm incomes from the marketing of high-value cash crops may facilitate a country's economic development and the transition into an industrialized economy. Technological progress like the introduction of high-yielding crop varieties, improved irrigation systems, artificial fertilizers, and mechanization enables farmers to shift from subsistence food crops to cash crops. Therefore, **agricultural intensification** is often accompanied by an increasing share of marketed agricultural production. The income a household derives from marketing cash crops can facilitate **investments** in agricultural production, education, food, health, or housing, in that way improving people's living standards and reducing **poverty**.

There are also potential economic risks involved in the cultivation of cash crops. An increasing share of cash crop production changes the traditional crop structure and the nutritional base of households in developing countries. Cash-cropping households tend to rely on a few commodities, which render them more dependent on and vulnerable to the prices and marketing potential of these commodities. A high dependence on cash crops might come at the expense of a decreasing household **food security**, if cash income from crop production is declining due to, for example, climatic factors or decreases in the market price.

Prices for cash crops are set on global commodity markets. Farmers, regions, and countries that rely on few cash crops to generate income or export earnings are vulnerable to price fluctuations and market shocks caused by bumper harvests elsewhere or changing consumer demand. Rising prices for cash crops make agricultural production more profitable, induce an expansion of agricultural area, and might thereby stimulate the clearing of **forest** areas.

See also: **Agricultural Frontier; Agriculture; Agro-Industry; Amazonia; Brazilian Cerrado; Cassava; Economic Growth; Economic Livelihood; Exogenous Factor; Himalayas; Landholding; Monoculture; Population Dynamics; Vulnerability.**

Further Reading

Barbier, Edward B., *Cash Crops, Food Crops and Agricultural Sustainability*, London: International Institute for Environment and Development, 1987; Clay, Jason, *World Agriculture and the Environment: A Commodity-by-Commodity Guide to Impacts and Practices*. Washington, D.C.: Island Press, 2004; Ruthenberg, Hans, *Farming Systems in the Tropics*, 3rd ed., Oxford, UK: Clarendon Press, 1980; United Nations Food and Agriculture Organization, FAOSTAT Data [Online, August 2004], The FAO Web Site, faostat.fao.org.

DANIEL MÜLLER

Cassava. A perennial root crop (tuber) and woody shrub that grows from one to three meters tall, the only widely cultivated food staple in the world that is poisonous to consume prior to processing (due to high cyanide levels), and nonetheless a basic **food crop** in developing countries of the tropics, lending itself to conditions of **poverty** alleviation for the rural as well as urban poor. Among staple food crops, cassava—also called manioc and yucca—is the greatest producer of carbohydrates per hectare. Both the starchy roots, as the main crop, and the leaves are used for human **consumption** and animal feed. To a minor degree, cassava also has non-food and industrial uses in the form of alcohol.

As an arable crop, cassava is grown either for subsistence or cash cropping, and it employs little technology only. In most cases, it is interplanted with other crops and directly consumed by local producers or sold in local markets (which makes it hard to track the total global area cultivated or the volume produced). Monocropped cassava over relatively extensive areas of land is rare and limited to Brazil and Thailand, notably.

Land-use systems that involve cassava are as dynamic as any other agricultural system, reacting to environmental stimuli as well as socio-economic factors and technological change. For example, remote influences from **urbanization**—such as in the vicinity of Ibadan, Nigeria—create market opportunities for food crops, which trigger land-use/cover change in the surroundings of the city. Following the removal of **forest** cover, major crops such as cassava, maize, and yams (but also vegetables and fruits) are intercropped on permanent or shifting plots. It has been found that at the city edge, the density of trees is substantially higher on land under permanent cultivation (such as cassava) than it is in areas nearby under shifting cultivation.

As a food crop, cassava has some inherent characteristics that make it attractive to small farmers throughout the world in terms of **food security** and maintenance of **economic livelihood** opportunities: it is available all year round and cheap for poor people to buy; it produces more calories on less land and without inputs than almost any other crop; it is relatively resistant to drought, pests, and diseases; it is relatively tolerant of low soil fertility; and roots can remain in the ground for several years after they mature. In particular, impoverished small farmers tend to integrate cassava into diversified land-use systems as an adaptive measure to combat declining soil fertility, but also to take advantage of new economic opportunities created by the market. In the forest and savanna frontier zones of Ghana, for example, early maturing cassava has now replaced traditionally grown, longer-maturing crops.

Among the impacts of cassava cultivation are habitat **conversion** on lower value, more marginal lands, which often have a high **biodiversity** value. Also, the cassava plant does not produce enough **vegetation** to cover the soil well, even if it is interplanted with other crops. Thus, the production of cassava can result in considerable **soil erosion** during the entire life of the plant. Because few producers can afford agricultural inputs for mitigation or long **fallow**, and because little else grows on marginal soils, the erosion often continues well after the root crop is harvested. Standard soil **conservation** techniques and intercropping, particularly leguminous trees with cassava, have great potential to decrease soil losses and conserve water in the ground.

Domesticated by Indians in the Amazon, cassava was taken to many different parts of the world with the arrival of Europeans. Compared to other permanent arable, staple crops such as **maize**, **wheat**, and **rice**, cassava holds a lower share in global arable land, only 4 percent of the world's total harvested area (together with potatoes), or 16.8 million hectares of cassava planted in 2000. Likewise, and due to its poor overall nutritional value, cassava plays a less-prominent role in the diets of most people in developing countries. Still, it constitutes the fourth most important food crop. Nigeria, Brazil, Thailand, Indonesia, and the Democratic Republic of Congo are the top five cassava-producing countries, and Thailand has always been the largest cassava exporter globally, accounting for 85 percent of all global exports. Most of the international cassava trade continues to depend heavily on imports to **Japan** and **Europe** (mainly, the Netherlands, Spain, and Belgium), where cassava is used for industrial purposes and mixed with vegetable meals as animal feed.

See also: **Cash Crops; Domestication; Driving Forces; Land Degradation; Subsistence Agriculture; Tragedy of Enclosure.**

Further Reading

Bryant, Raymond L., and Sinéad Bailey, *Third World Political Ecology*, London, New York: Routledge, 1997; Clay, Jason, *World Agriculture and the Environment: A Commodity-by-Commodity Guide to Impacts and Practices*, Washington, D.C.: Island Press, 2004; Leff,

Billie, Ramankutty, Navin, and Jonathan A. Foley, "Geographic Distribution of Major Crops in the World," *Global Biogeochemical Cycles* 17 (2004); Mannion, Antoinette, *Dynamic World: Land-Cover and Land-Use Change*, London: Arnold, New York: Oxford University Press, 2002.

HELMUT GEIST

Cattle Ranching. An extended economic activity in the tropics, which has been identified as one of the main **proximate causes** of **deforestation** in **Latin America**, since **pasture** constitutes the main land use in the tropics following **forest** conversion. There have been numerous studies seeking to identify the causes of cattle ranching development in the Amazon. Although cattle ranching is dominant in large-scale **landholdings**, it is not unusual to find that small-scale landholders are also devoted to cattle raising. Some of the factors influencing cattle adoption are as follows: the land-extensive nature of beef production; low labor and supervision requirements; transportability and limited risk; limited use of purchased inputs; biological and economic flexibility; low costs of production and banking services; indirect financial benefits such as wealth maintenance; and secure source of cash flow. Furthermore, cattle can assist farmers in slowing spontaneous forest

Cattle being rounded up near Eagle Pass in Maverick County, Texas, in March 1939.
PHOTO: Russell Lee. [Courtesy Library of Congress]

regeneration; smallholders are left with significant cleared land that they cannot use for anything but cattle.

Cattle ranching has played a crucial role in the occupation of the **agricultural frontiers** because it allows squatters to demand **land rights** with little initial investment, although it leads to extensive and low-productive cattle-ranching systems. There is little incentive to intensify cattle-ranching systems in contexts in which there is easy **access** to cheap land and a scarcity of labor, which are common factors in tropical frontiers. Therefore, benefits of cattle production do not depend on gains in productivity but on the amount of land and pasture, and cattle herd size. Yet, in contexts in which land becomes scarce—above all in consolidated frontier areas—and markets come to demand quality, beef producers begin to develop semi-intensive production systems. That primarily consists of improving cattle herd and pasture management through **investments** in better-quality pastures, separating pasture areas with fences, and buying higher-quality herds, among others.

In the Brazilian Amazon, common wisdom suggests that cattle ranching is a low-productive and unprofitable activity, and hence it has grown only as a result of institutional rents (i.e., fiscal incentives and subsidized credit) captured by corporate ranching. This view, however, has been challenged by the fact that cattle ranching has continued growing even after subsidies and incentives were eliminated by the Brazilian government in the late 1980s. This has led to the argument that factors other than government policies explain the persistent expansion of cattle ranching over forestland. In this direction, recent cost-benefit analysis undertaken in several areas of the eastern Amazon suggests that cattle ranching is in fact a profitable activity. This by itself is a strong factor that helps to explain why cattle ranching is still growing in the Amazon.

In 2002, Brazil contained the world's second-largest cattle herd, with 185 million heads, and became the main beef exporter in the world, followed by the **United States of America** and Australia in 2003. The cattle herds were traditionally placed in the rural areas closer to the main consumer centers of the south (i.e., São Paulo and Rio de Janeiro), and the northwest; they have expanded northward following the occupation of public lands. Since more than a decade ago the cattle population in the south has been close to stagnation, but its growth has been explosive in the Amazon, with annual rates of growth closer to 6 percent. In the Amazon, cattle herds grew in size from 26 million in 1990 to 57 million heads in 2003 in the Brazilian Amazon states.

There is a diversity of situations within the frontier regarding cattle beef production. In the older frontiers, where land becomes a scarcer resource, there is better road infrastructure and more developed markets; some moves toward intensification are evident. In contrast, in the new frontier areas, cattle ranching still constitutes a means to justify land ownership, and longer distances to markets do not justify developing any other activity than extensive cattle ranching. In parallel, in the consolidated frontier areas, there is more investor interest in developing more productive chains to integrate beef production more effectively with more demanding markets. This leads to a growing trend of investing in slaughterhouses and meatpacking plants to better get to more distant markets.

Most cattle ranchers in the frontier areas have looked at the domestic markets. In most of the Amazon countries, cattle production has been targeted to supply the local and national demand for beef, and little has been exported. Therefore, it practically constitutes a non-tradable sector. In the early 1990s, much of the meat production was consumed within the region itself. The Amazon region did not even produce enough beef to feed its own population until 1991. This situation has started to change recently because there is a larger trade of beef products (natural and processed) originating in Amazonian countries, specifically Brazil. The latter does not deny the fact that domestic consumption, much of which is determined by the level of urban incomes, continues to be one of the main factors influencing beef production levels in this region.

Brazilian beef exports have begun to increase since the mid-1990s as a result of the confluence of several factors such as currency devaluation, mad-cow disease influence on **consumption** preferences, and increasing demand for Brazilian cattle beef in international markets. Although most of those exports have originated in the south of the country, the explosive growth of the cattle population in the north suggest that it is filling the gaps created in the domestic market of beef. That means that the increasing expansion of Brazilian beef exports constitutes a **driving force** for cattle ranching expansion in the Amazon, and this directly fuels forest **conversion** to pasture. This fact has restated the debate about the "hamburger connection." This term, coined by Norman Myers in the early 1980s, describes how the rapid growth of beef exports from Central America directed to fast food chains in the United States was driving deforestation. In the current years, this term can be applied to Brazil regarding European, Russian, and Middle East markets.

See also: **African Trypanosomiasis; Agricultural Intensification; Amazonia; Arc of Deforestation; Brazilian Cerrado; Corporate Strategies; Cumulative Change; Domestication; Great Plains; South Africa; South America; Southern Africa; Southern Cone Region.**

Further Reading

Faminow, Merle D., *Cattle, Deforestation, and Development in the Amazon: An Economic, Agronomic, and Environmental Perspective*, Wallingford, Oxon, UK, and New York: CAB International, 1998; Hecht, Susanna B., "The Logic of Livestock and Deforestation in Amazonia," *BioScience* 43 (1993): 687-95; Kaimowitz, David, and Arild Angelsen, *Will Livestock Intensification Help Save Latin America's Tropical Forests?* Bogor, Indonesia: Center for International Forestry Research, 2001; Margulis, Sergio, *Causes of Deforestation in the Brazilian Amazon*, Washington, DC: World Bank, 2003; Myers, Norman, "The Hamburger Connection: How Central America's Forests Become North America's Hamburgers," *Ambio* 10 (1981): 3–8.

PABLO PACHECO

Cellular Automaton. A tessellation of identical finite-state automata that corresponds to an idealization of a physical phenomenon in which space and time are discrete. It can be seen as a discrete model of spatiotemporal dynamics. Cellular automata (CA) are composed of identical cells geometrically located on a one-, two-, or three-dimensional interconnected grid. Each cell of the grid corresponds to a finite-state automaton with a finite set of inputs, a state transition function defined on input-state pairs, and a cell neighborhood. The neighborhood is the set of cells influencing a particular cell, and it usually corresponds to those geometrically near it. In a two-dimensional cellular automaton, the neighborhood of a given cell typically consists of the four or eight adjacent cells (Moore and von Neuman neighborhoods, respectively). Cellular automata function on a discrete time base: after each constant time step, all cells of the grid undergo a simultaneous state transition by concurrently applying the transition function, after receiving inputs from neighbors. Processing local conditions in each cell in such a parallel mode can result in a surprisingly overall complex behavior. The most popular example of emergent behavior is Conway's game of life, which became widely known after being described by Martin Gardner. John von Neumann and Stanislaw Ulam introduced cellular automata in the early 1960s; after that there were several periods of interest related to computer hardware advances. Stephen Wolfram and others claim that cellular automata represent the basic form of information processing in the universe. However, others have pointed out that the finite state and parallel synchronous nature of the state transitions are unlikely to provide useful descriptions of many spatially explicit phenomena. Recently, there has been great interest in the use of cellular automata for **modeling** and simulation of landscape dynamics based on knowledge of local processes supported by distributed data. Cellular spaces are widely used for representing the spatial variability of geographical and environmental factors, especially within **geographical information systems**, and can be directly used in a cellular automaton. Advances in high-performance computing technology and in newer formalisms such as parallel discrete event systems specifications are converging to obviate most of the computational limitations associated with the use of cellular spaces for modeling complex spatiotemporal phenomena.

See also: **Agent-Based Model; Change Detection.**

Further Reading

Gardner, Martin, "Mathematical Games: The Fantastic Combinations of John Conway's New Solitary Game 'Life'," *Scientific American* 223 (1970): 120–3; Wolfram, Stephen, A New Kind of Science [Online, December 2004], The History of CA Web Site, www.wolframscience.com/reference/notes/876b.

MARIA JOSÉ P. DE VASCONCELOS

Center for International Earth Science Information Network. Established in 1989 as an independent non-governmental organization, the Center for International Earth Science Information Network (CIESIN) became a unit of the Earth Institute at Columbia University in 1998. CIESIN's mission is to improve the data and information base for understanding interactions between humans and the environment. This mission places

CIESIN at the intersection of social, natural, health, and information sciences. To accomplish its mission, CIESIN creates distributed data and information resources, provides users with new types of interactive analysis and visualization capabilities, and conducts research on human interactions with the environment as well as on the management of data and information.

Among the data products CIESIN produces are the Gridded Population of the World, a globally consistent grid of human population density useful for **land-cover change** research; the Urban Points and Extents data sets (from the Global Rural-Urban Mapping Project) with associated population figures for all urban areas with populations greater than 5,000; and the Human Footprint and Last of the Wild data sets depicting the degree of anthropogenic impact on natural landscapes and the remaining wilderness areas by ecosystem type. Its information resources include Thematic Guides to **Global Environmental Change**, including two recent guides on land use, land-cover change, and social science applications of **remote sensing**, and reports such as the Environmental Sustainability Index.

CIESIN manages the Socioeconomic Data and Applications Center (SEDAC), one of the Distributed Active Archive Centers of the U.S. National Aeronautics and Space Administration (NASA), and has received funding from numerous agencies including the **United States Geological Survey (USGS) program**, the National Oceanic and Atmospheric Administration (NOAA), the United Nations Development Program, and the World Bank.

See also: **Population Dynamics; Population-Environment Research Network; Urbanization.**

Further Reading

Columbia University, The Center for International Earth Science Information Network [Online, December 2004], The CIESIN Web Site, www.ciesin.columbia.edu.

ALEXANDER DE SHERBININ

Center for International Forestry Research. An international collaborative research and global knowledge institution that is committed to the **conservation** of **forests** and the improvement of the livelihoods of people in the tropics. The Center for International Forestry Research (CIFOR) runs a headquarters in Bogor, Indonesia, and regional offices in Brazil, Cameroon, and Zimbabwe. It reaches out for local communities and small farmers to gain their rightful share of forest resources, while increasing the production and value of forest products. CIFOR was created in 1993 by the **Consultative Group on International Agricultural Research**, responding to a call by the 1992 Earth Summit in Rio de Janeiro for more authoritative **forestry** knowledge. CIFOR's main task is to address the need to save tropical forests on the one hand while ensuring that they still provide timber, food, and a living for those who depend on them, on the other hand. For the more than 300 researchers who are associated with about fifty organizations, priority areas are spread over more than thirty countries, including **tropical humid forests** in Southeast Asia, central West Africa, and the Congo basin, and **tropical dry**

forests such as the **miombo** woodlands in East and **southern Africa**. In 1997, CIFOR established POLEX, an electronic forest policy expert listserver.

See also: **Alternatives to Slash-and-Burn (ASB) Programme; Economic Livelihood; Tradeoffs.**

Further Reading

Center for International Forestry Research [Online, February 2005], The CIFOR Web Site, www.cifor.cgiar.org.

HELMUT GEIST

Center for the Study of Institutions, Population and Environmental Change. A multidisciplinary research center at Indiana University-Bloomington, funded by the National Science Foundation (NSF) Human Dimensions of Global Change Program. Researchers of the Center for the Study of Institutions, Population and Environmental Change (CIPEC) study **deforestation**, **reforestation**, and **forest degradation** processes as mediated by factors such as institutional arrangements, land suitability, and demographics. The following questions frame the overall research agenda at CIPEC:

1. How are regional and global political and economic processes linked to human behaviors at household and community levels?
2. How can macro-scale physical and biological processes observed and modeled at a global scale be linked to meso- and micro-human organizational and **decision-making** processes?
3. How do institutional arrangements influence the impact of human **driving forces**, such as population density and transportation networks, on ecosystems and global change?

CIPEC conducts interdisciplinary cross-site analysis among field sites in North America, Central America, **South America**, Asia, and Africa with an emphasis on the detection of **land-cover change** through **remote sensing** techniques and the spatial integration of social and biophysical data sources.

See also: **Agent-Based Model; Anthropological Center for Training and Research on Global Environmental Change; Community Involvement; Cultural Factors; Environmental Change; Forest; Forest Anomalies; Institutions; Mediating Factor; Modeling; Population Dynamics; Regrowth; Secondary Vegetation.**

Further Reading

Center for the Study of Institutions, Population and Environmental Change [Online, December 2004], The CIPEC Web Site, www.indiana.edu/~cipec; Moran, Emilio, Elinor Ostrom, and J. C. Randolph, Ecological Systems and Multitier Human Organization, in *UNESCO Encyclopedia of Life Support Systems*, Oxford, UK: EOLSS Publishers, in press.

THOMAS P. EVANS

Center for Sustainability and the Global Environment. Dedicated to the study of pressing global problems stemming from interactions among human activities, natural resources, and the environment, the Center for Sustainability and the Global Environment (SAGE) at the University of Wisconsin-Madison conducts cutting-edge research on critical global environmental problems and disseminates that knowledge through instruction and outreach. SAGE aims to improve our understanding of the global environment and how it is being affected by human activities. To achieve this goal, it is developing maps, computer models, and data sets to describe the behavior of earth's terrestrial ecosystems, hydrological systems, and climate. SAGE has developed a new generation of global ecosystem (IBIS) and terrestrial hydrology (HYDRA) models, as well as fully coupled climate-vegetation models (GENESIS-IBIS and CCM3-IBIS). In addition, new data sets were created for global ecosystems, historical land use, and **land-cover change**, as well as specific to the Amazon basin.

See also: **Amazonia; BIOME 300 Project; Environmental Change.**

Further Reading

University of Wisconsin–Madison, Center for Sustainability and the Global Environment [Online, December 2004], The SAGE Web Site, www.sage.wisc.edu.

KEES KLEIN GOLDEWIJK

Century Ecosystem Model. A generalized ecosystems model that has been used extensively during the last fifteen years to simulate the impact of land use and **environmental changes** on ecosystem properties. The important land-use changes considered include **conversion** of **grasslands**, savanna, and **forest** systems to agricultural cropping systems, conversion of forests to grazing systems, use of different agricultural cropping rotation systems and grassland grazing management practices, and use of alternative management practices for harvesting trees in forest systems. Century has been used to simulate the impact of environmental changes such as changing atmospheric CO_2 levels, and climatic factors such as solar radiation, atmospheric humidity, air temperature, and precipitation. The model is a generalized plant/soil ecosystem model that is parameterized for all of the major natural arid land, grasslands, savanna, and forest systems and the important agricultural cropping systems used in the world. The model simulates plant production for trees, agricultural crops, and grasslands; soil **nutrient cycling** (nitrogen and phosphorus); soil carbon dynamics; and leaching of organic and inorganic carbon, nitrogen, and phosphorus compounds (dissolved organic carbon, nitrogen, and phosphorus, NH_4, NO_3, and PO_4). There are two versions of the Century model: one with monthly time steps (CENTURY) and a second with daily time steps (DayCent). The DayCent version of the Century model is also capable of simulating daily nitrogen trace gas fluxes (N_2O, N_2, and NOx) and CH_4 uptake by the plant soil system, and is thus capable of simulating full **greenhouse gas** budgets for managed and unmanaged ecosystems.

The daily and monthly versions of the Century model use a similar structure and include plant production, nutrient cycling, soil organic matter, and nitrogen trace gas flux submodels. The plant production submodel simulates growth and death of plants in grassland, cropland, savanna, and forest ecosystems and is capable of simulating tree/grass or tree/crop competition in savanna systems. Plant growth is controlled by environmental factors such as atmospheric CO_2 levels, soil moisture, and temperature and available nitrogen and phosphorus for plant growth. The soil organic matter submodel simulates carbon dynamics and decomposition of dead plant material and three soil organic matter pools with different **turnover** times (active: 0.1–1.0 yr^{-1}, slow: 10–50 yr^{-1}, and passive: 1,000–2,000 yr^{-1}). Decomposition of the dead plant and soil organic matter pools results in the mineralization of nitrogen and phosphorus, which are potentially available for plant growth. The nutrient cycling submodel simulates the nitrogen (N) and phosphorus (P) dynamics for the organic and inorganic N and P pools, the mineralization of inorganic N and P from organic matter pools, and the amount of inorganic N and P available for plant growth. The DayCent version of Century includes daily soil water and temperature submodels and a nitrogen trace gas flux submodel that simulates NOx and N_2O fluxes from nitrification and N_2 and N_2O gas fluxes from denitrification.

See also: **Carbon Cycle; Climate Impacts; Ecosystem Services; Great Plains; LandSHIFT Model; Modeling.**

Further Reading

Colorado State University, Natural Resource Ecology Laboratory [Online, December 2004], The Century Ecosystem Model Web Site, www:nrel.colostate.edu/projects/century.

WILLIAM J. PARTON JR.

Change Detection. The process of extracting the type, amount, location, and configuration of land change that varies with data type. Change detection techniques can therefore be divided into those that involve **continuous data** and those that involve thematic data, also sometimes associated with the level of automation in processing. Change detection methods have advanced greatly in the last three decades, owing to greater availability of imagery and photography time series as well as concomitant increases in computing power and speed. Change detection work has evolved to include not only extraction of change but also explanation, prediction, and simulation of change.

Non-spatially explicit change detection was once the norm, used in contexts of sound bytes of information such as “the Amazon basin has seen a 40 percent reduction in forests over ten years.” This information, while helpful or alarming depending on your view, was neither spatially explicit nor thematically explicit. That is, one was left to wonder what exactly replaced the forests in those ten years: **agriculture**, **pasture**, settlements, oilfields? Furthermore, no locational information could be provided in such a summary statistic, again leaving one to wonder about the spatial footprint of that change: was it pervasive throughout the basin, or located more in some places and less in others? In the 1970s, change detection advanced to what most simply consisted of overlay analysis, whereby the same area was compared at two or more points in time (“from-to” change detection, often associated with thematic data). More automated methods of change detection make use of continuous data, either using spectral responses (radiance or reflectance) or indices derived from those spectral responses. When comparing spectral responses directly, slight differences in atmospheric conditions can be problematic if left uncorrected. This technique, known as image differencing, quite simply calculates the spectral response difference in two or more images on a band-by-band basis. This process requires scene characteristics to be as similar as possible, either through scene normalization, scene correction, or minimally by controlling for seasonal and time of day differences through the image-selection process. A similar approach includes band ratioing, whereby image bands from different times are compared via division rather than subtraction; this method also suffers from similar drawbacks. The advantage of these approaches is that they are extremely fast and replicable; the disadvantage is that it is unclear what exactly the results indicate. That is, if the digital number changed from 45 to 51 in an 8-bit scaled **Landsat** thematic mapper band 4 (near-infrared), what would be the ecological meaning of that change, and what would even be the appropriate **threshold** at which one would call such a shift "change"? This is more of a problem when using reflectance data than when using radiant flux data, the latter of which allows for more direct comparison.

Another automated method of detecting change involves comparing indices calculated from band ratioing. Indices commonly used include the **normalized difference vegetation index**, the Soil-Adjusted Vegetation Index, and mineral-based indices such as the Iron Oxide Index and Ferrous Minerals Index. The index of choice would be calculated the same way for each input image, and then the resulting index images would be differenced. This procedure provides for more intuitively interpretable results, though it too suffers from error introduced by scene differences owing to atmospheric differences as well as differences in soil/litter background and soil moisture.

Spectral transformations provide a third, more automated approach for change detection and typically involve orthogonal transformation of three or more bands of imagery for each scene. The two most common of these are principle components analysis (a technique originally developed for data compression) and the Tasseled Cap Index (developed for Landsat MSS imagery and extended for use with other sensor systems). These techniques are more processing intensive than image differencing or index differencing but still require less time and oversight than methods associated with thematic change detection. Both transformations can be sensitive to atmospheric differences.

Change detection performed using thematic data is referred to as post-classification change detection; images are each classified into a thematic classification or attribution scheme, and then "from-to" change can be mapped and summarized in table format. Change classes give thematically explicit information regarding landscape change, such as which areas changed *from* **forest** *to* agriculture, compared to areas that changed *from* forest *to* urban and compared to areas that changed *from* barren *to* agriculture. From-to change detection is the most common type of change detection methodology used today, though it has been used explicitly only since the late 1970s. Necessarily the level of detail of the thematic attribution scheme (often referred to as an Anderson level) in part dictates how much change can possibly be observed; the more classes used, the greater the possible number of change classes with the expected value of the amount of total change to be equal to or greater than change detection using less-detailed schema. Spatial **scale** also plays an important role, as the ability to detect landscape classes generally and change classes specifically is scale-dependent. In addition to mapping, results are typically presented in a cross-tab format, with the "from" classes listed vertically and the "to" classes listed horizontally. Change classes can be read across (an old class became which new classes?) or down (a new class came from which old classes?). Diagonal elements represent areas classified the same in both observations (no change).

The protocol typically employed for post-classification or thematic change detection analysis begins with defining the area, problem or hypothesis, and classification scheme to be used. Next, remotely sensed data are selected based not only upon availability but also the data characteristics necessary for testing the particular hypothesis in that particular landscape. Such data characteristics include but are not limited to spatial resolution (typically associated with the instantaneous field of view—the area on the ground represented in a typical **pixel**), temporal resolution (or return time—the amount of time between consecutive observations of the same area on the ground), spectral resolution (the number, width, and placement of observations in the electromagnetic spectrum), radiometric resolution (bit depth, or level of precision of reading), and directional resolution (acquired in nadir or off-nadir mode, which can impact return time). Other important sensor system/image factors include signal-to-noise ratio (an indicator of quality) and geometric distortion (related to sensor design and platform altitude). Cost also plays a part in image acquisition, though the current subsidizing of many systems has lessened this burden.

Landscape considerations are equally important: phenological, crop, flooding, and precipitation cycles all impact the usability of imagery due to landscape condition and/or cloud cover. Soil moisture conditions and atmospheric conditions are also important and can be accounted for either in image selection or in data preprocessing. Preprocessing of data normally consists of dealing with atmospheric (radiometric) and geometric corrections or normalizations, whereby either both images are corrected to the "truth" as closely as possible or else are matched to each other to prevent detection of artificial change. After the change detection is performed using one of the methods discussed above, accuracy of that change detection must be assessed before reporting those results to the broader community.

Increasingly, change detection is becoming more temporally explicit, moving from pairwise comparison of thematic imagery to a longitudinal or panel approach for extracting dominant temporal patterns of landscape change (borrowing heavily from the life-course method of sociology). With larger time-series, such as those assembled using **advanced very high resolution radiometer** or **moderate resolution imaging spectroradiometer** imagery (which both have greater temporal resolution or return time), a more automated approach to longitudinal analysis is used. One such example is Fourier analysis, whereby thematic data are assessed for dominant frequencies of temporal patterns using a discrete Fourier transform. This approach is typically employed with binary data sets (e.g., fire/no **fire**, flood/no flood), though it can be extended to multiple classes as well as to continuous data (using, for example, a fast Fourier transform).

A concern with change detection involves how accuracy can be assessed for change pairs. Developments in accuracy assessment and error propagation are still evolving, but currently involve comparing a spatially random sample (often stratified by thematic class) of each image to a more accurate product such as finer-scaled imagery or field sampling data. Obviously such assessment is problematic when using archived data or working in sites that are inaccessible for physical or political reasons. Geometric co-registration of the images in the change pair also remains an important issue.

Change detection continues to evolve, and the field currently has observed growing numbers of applications in simulation and **modeling**. Empirically fitted models are more prevalent than previously, for example, Markov modeling. Statistical techniques, often used in econometrically based models, have been offered as a means of explaining landscape change as a function of predictor variables representing socioeconomic, political, biophysical, and geographic **driving forces** of change. Worth noting is the distinction between land cover and land use. Land cover refers to the tangible elements, natural or man-made in origin, on top of the earth's surface: grass, rocks, concrete, trees, water, etc. Land use refers to cover types and how they are managed by humans: agriculture, silviculture, orchards, irrigation supply, etc. Land-use information provides more nuanced information about likely trajectories of the landscape given the current environs, but is more difficult to discern remotely. Mixed classification schemes that have elements of land use attributed where possible but otherwise use land cover refer to that scheme as a land-use/land-cover (LULC) classification and the change as land-use/land-cover change (LULCC). Various research communities may refer to these concepts as the **Land-Use/Cover Change (LUCC) project**, or the **NASA Land-Cover/Land-Use Change (LCLUC) program**.

Process-based (or dynamic process) models have appeared in this arena more recently, and rather than modeling data, outcomes attempt to model landscape processes. One such example is **cellular automata** or CA modeling, a rule-based grid capable of simulating the diffusion of landscape classes (e.g., **urbanization**, **deforestation**). CA models, first popularized in simulating urbanization, have now been extended to phenomena including alpine tree-line processes, agricultural extensification and intensification, human migration and settlement, deforestation, and coca planting. Another example of process-based modeling is the use of **agent-based models** (ABM), which instead focus on modeling the processes or behaviors of the agents of change on a landscape (e.g., household **decision-making** of landscape management or migration) and then move the impacts of agent behavior accordingly on the background landscape. This approach grew out of biological work in the 1960s that noted that the behavior of some species was more properly explained at a level of aggregation above the individual (e.g., ants). More recently, integration of CA and ABM has allowed for greater success in understanding and predicting landscape change as observed in change detection, and in both back-simulating and forward-simulating landscape change **scenarios** under varying conditions. The next steps of change detection may well see further advances in modeling, as richer data sets are created for more areas and more times. Although not typically referred to as change detection, vertical or structural change detection has also undergone rapid development using radar and, more recently, **Airborne Light Detection and Ranging** (LIDAR) data. Interferometry involves creating detailed digital elevation models (DEMs) and then examining the differences in those products over time. The processing involved is completely different from working with optical imagery, but the analogy of extracting either thematic (elevation class) or continuous (elevation) data and then looking for change across time still holds. Interferometry has been used successfully in multiple areas of the United States undergoing subsidence due to population-induced groundwater drawdown. LIDAR-based DEMs, still under development, have not yet been applied in such cases per se, but offer much-improved horizontal and vertical resolution for detecting more nuanced vertical or structural change, such as change in canopy height or surface roughness over time.

See also: **Cadastre; Conversion; Cumulative Change; Environmental Change; Modification; Systemic Change.**

Further Reading

Jensen, John R., *Introductory Digital Image Processing*, 3rd ed., Upper Saddle River, NJ: Prentice Hall, 2004; Lunetta, Ross S., and Chris D. Elvidge, eds., *Remote Sensing Change Detection: Environmental Monitoring Methods and Applications*, Chelsea, MI: Ann Arbor Press, 1998; Song, Conghe, Woodcock, Curtis E., Seto, Karen C., Pax Lenney, Mary, and Scott A. Macomber, "Classification and Change Detection Using Landsat TM Data: When and How to Correct Atmospheric Effects?" *Remote Sensing of Environment* 75 (2001): 230–44; Tso, Brandt, and Paul M. Mather, *Classification Methods for Remotely Sensed Data*, New York: Taylor and Francis, 2001.

KELLEY A. CREWS-MEYER

China. The world's most populous country and one of the oldest civilizations on the globe. China has a long history of, and is profoundly affected by, human land use. This is reflected by the expansion of crop farming, because **agriculture** dominates in Chinese civilization. As early as AD 2, the total area of cultivated farmland in the country reached 38 million hectares, supporting a population of 59.6 million. It reached 52.8 million hectares in 1812 and 93.2 million hectares in 1935, accounting for some 8 percent of the country's territory. At present, China is undergoing rapid **industrialization** and **urbanization**, which substantially increase pressure on land resources and environment. The urban share of the total population has been increasing by 1 percent a year since the early 1990s, resulting in the loss of a large area of primary farmland to **urban sprawl**. In terms of the societal aims of land use, China is in a crucial moment of its history, that is, in finding a balance among **food security**, **economic growth**, and ecological **conservation**.

Present Land Use and Cover

China assumed its present size and shape in 1949. According to a recent survey of land use and land cover for the whole country (completed in 1996), the land territory of the country covers 9.51 million km^2 except Taiwan, Hong Kong, and Macao. In the total territorial land, there are 130.04

million hectares of cultivated land (13.7 percent of the entire area), 10.02 million hectares (1.0 percent) of orchards, 227.61 million hectares (23.9 percent) of **forest**, 266.06 million hectares (28.0 percent) of **pasture**, 24.08 million hectares (2.5 percent) of residential and industrial land, 5.47 million hectares (0.6 percent) of built-up transport infrastructure, 42.31 million hectares (4.5 percent) of water area, and 245.08 million hectares (25.8 percent) of unused natural land cover.

The vast country covers land with varied climates from the **temperate zone** to the tropics and from humid to arid zones. **Mountains** and plateaus occupy two-thirds of the territory and are distributed mainly in the western half. Most of the farmland and urban built-up land are distributed in the eastern half, where the terrain is flatter and the climate is more humid than in the west. The northwest is covered mainly by desert, **grassland**, and pasture.

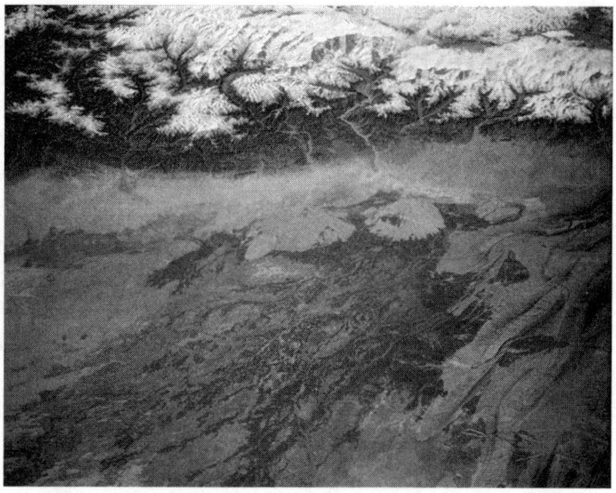

Snowcapped Kunlun and Tien Mountains and arid Tarim Basin (irrigation farming zone in dark tone) in Western Xinjang Province of Northwest China in October 1994.

PHOTO: Space Shuttle. CREDITS: Image Science and Analysis Laboratory, NASA-Johnson Space Center, 18 Mar. 2005 (Earth from Space—Image Information). http://earth.jsc.nasa.gov//sseop.efs

Land Tenure

In China, land is held by either state or village collectives. **Land tenure**, or the system of land ownership, in rural areas is complex because the real **land rights** are distributed among farmers' collectives and households. A village is usually a natural settlement in a rural area with about 300–500 households. Collectives hold formal ownership and the right to allocate land-use rights to member households. Farm households are allocated the rights to use land for a finite period and to keep or sell the produce.

In 1996, some 53 percent of land in the country was state-owned. The major land-use types of the state-owned land are unused land (37 percent of the total state-owned land), pastureland (35 percent), forestland (18 percent), and water area (6 percent). The major land-use types of the collectively owned land are forestland (30 percent of the total collectively owned land), cultivated land (29 percent), and pastureland (19 percent). Some 76 percent of the country's unused land and 67 percent of pastureland are state-owned, while 97 percent of cultivated land, 70 percent of residential and industrial land, and 57 percent of forestland are collectively owned.

All the residential and industrial land in urban areas is state-owned. In the process of urban expansion, ownership has to be changed before the collectively owned land can be occupied.

Land-Use Policy

Even the collectively owned land in China is heavily influenced by **public policies**. Governments at every level make land-use plans to control the conversion among the eight general use types. Three major public or national interests direct **land-use planning**, including food security, urban land demand, and ecological conservation. For instance, some 109 million hectares of cultivated land were zoned in 2002 as primary farmland. **Conversion** of the primary farmland to other uses is strictly controlled to guarantee national self-sufficiency in food. In fact, halting the cultivated land decline to stabilize domestic food supply has been always the government policy for many years, especially after the mid-1980s. A strategy called "maintaining a dynamic balance of cultivated land" was initiated in 1996. According to this policy, provincial governments were required to take the responsibility to maintain a dynamic balance of cultivated land in their jurisdictions. Losses of cultivated land have to be compensated by reclaiming new land and other means. Protection of primary farmland, however, is a great challenge because of the ever-growing demand for land by rapid industrialization and urbanization. For about twenty-five years, the annual growth of the gross domestic product (GDP) has averaged about 10 percent, while some 60 percent of the country's population still lives in rural areas. Lands at the **urban-rural fringe** are usually primary farmland. The pressure on these lands is mounting. On the other hand, reclamation on sloping land and arid land generates serious soil erosion and land degradation. A large flooding on the middle and lower reaches of the Yangtze River in the summer of 1998 and severe dust storms in north China in the spring of 2000 triggered countrywide concerns for ecological conservation. From 2001 onward, the government implemented an ambitious program to subsidize farmers with grain for giving up farming on sloping land and returning it to forest. Some 14.7 million hectares of cultivated land will be converted to forest according to the so-called "grain for green" program.

Change of Land Use

Thousands of years had seen expansion of cultivated farmland and shrinking natural **vegetation** area in China before the late 1950s, when cultivated area reached its peak. Reclamation speeded up after 1949 when the Communist party took the rule of the country. Some 71 million hectares of cultivated land were reclaimed from 1949 to 1996. The figure amounts to about 72 percent of the total cultivated area in 1949. Most of the land was reclaimed from **wetlands**, grassland, and forestland in the northeast and western border provinces. During the same period, grassland area decreased by 32 percent.

Expansion of urban and rural built-up land and shrinking cultivated area feature land-use changes in modern China, in response to the transition from an agriculture-dominated society to an industrial country. Only 10 percent of China's population lived in urban areas in 1949; the figure reached 40 percent in the early 2000s. In 1952, the total area of built-up land was 10.8 million hectares. It reached 29.5 million hectares in 1996, some 2.7 times that of 1952. Built-up land has still expanded at a speed of about 350,000 hectares each year during the past decade.

Concomitant to the expansion of built-up land, the country's cultivated land has been diminishing. According to official statistics, over 50 percent of the expanded built-up land was from cultivated land. Between 1978 and 2003, total cultivated land decreased by 11.4 million hectares, accounting for 8.5 percent of the total cultivated land stock. Most of the loss to built-up land occurred in coastal and central provinces, where land is relatively fertile and productivity is high.

In response to Chinese society's changing demand for agricultural products, land use varies in usage type and intensity. The acreage of fruit garden has increased fourfold during the past two decades and that of vegetable land by a factor of 2.6. At the same time, sown area of **food crops** has decreased. The proportion of food crop sowing area in the total sowing area was about 80 percent in 1983. It dropped to as low as 65 percent in 2003. The sown area of food crops like **wheat** and **rice** has decreased by over 20 percent, while that of feed crops like **maize** has increased by nearly 30 percent.

Chinese agriculture is famous for its tradition of intensity. The level of **agricultural intensification** has improved greatly during the past five decades. For instance, the area on which irrigation farming is practiced has increased by more than 80 percent, mainly in the 1950s to 1980s. Inputs of chemical fertilizers, pesticides, energy, machinery, and improved seeds have increased quickly during the recent two decades.

During the past two decades, both intensification and **extensification** of agricultural land use have taken place in China. They have shown a clear spatial disparity. The central and western provinces have still kept the increasing trend in intensification in terms of multicropping index, application of chemical fertilizer, and irrigation area. The western provinces have been weak in industrialization and have shown the strongest trend in agricultural intensification. In contrast, the east coastal provinces, where the economy is more developed and rural workers find more job opportunities in the non-agricultural sectors, experienced a decreasing trend in multicropping index and irrigated area.

Environmental Impacts of Land-Use Change

Environmental and ecological impacts of land-use and **land-cover change** are deep in magnitude and wide in spatial extent in China. Perhaps the most notorious impact is **land degradation**. China is a country with two-thirds of its territory under mountains and plateaus. According to the 1996 survey, some 28 percent of the country's cultivated farmland is sloping land with a gradient over 15 degrees. Farming on these sloping lands induces severe **soil erosion**, especially in the loess plateau. Moreover, more than 20 percent of the country's cultivated farmland is located in the arid and semi-arid northwestern part where the natural **vegetation** is steppe or desert steppe. Wind erosion and **desertification** are the results of farming in these areas. It is estimated that the total area of land degradation amounted to 116 million hectares by 1949. By the 1990s, the figure was as high as 180 million hectares.

China is also a country with intensive exploitation of water resources. In total water **consumption**, agriculture has a share of about 70 percent, and farmland irrigation has changed river hydrology dramatically. Dam construction has decreased the runoff coefficient of the rivers. As early as in the 1960s, several large rivers in the North China Plain fell dry. The Yellow River is the second-longest river in China and is also one of the longest rivers in the world. Between 1972 and 2000, there were twenty-one years with a drying-up period along its lower reaches, mainly due to human water diversion from the river. In 1997, the duration of drying-up in parts of the river's downstream courses was as long as 330 days.

Exploration of groundwater for farmland irrigation triggered the depletion of aquifers. This is a serious environmental problem in northern China. Thanks to the expansion of irrigated area and improved crop cultivars, the cropping system in the North China Plain changed from three crops per two years to two crops a year in the early 1970s. This was a dramatic increase of water consumption by crops. Together with the increase of domestic water consumption and climate change, irrigation caused overpumping of groundwater, which is now a widespread phenomenon in the northern part of the country. The results are dropping water tables and falling land levels. The average water table depth in the Hebei Plain dropped from 7.23 meters in 1983 to 11.52 in 1993. Between the mid-1970s and the mid-1990s, some 6,340 km^2 in the Hebei Plain went through a falling of land levels over 0.2 meters.

Direct drainage of wetland for farming also threatens the natural ecosystems and their environmental functions. The Three-Rivers Plain was once the largest marsh region of China. The acreage of marsh in the plain amounted to 2.44 million hectares in 1975. It declined to only 1.13 million hectares in 1990. Lake Dongting, one of the largest freshwater lakes in the country, shrank by 30 percent from the early 1950s to the late 1970s, mainly because of land reclamation for crop farming. It is also estimated that about 50 percent of the coastal wetlands have been lost to farming and built-up land during the past few decades. The shrinking wetland area is not only a threat to **biodiversity** but also raises the risk of flooding.

Environmental impacts of land-use and cover change have raised the attention of both China's government and society, especially in the 1990s. In 1998, the central government forbade logging of the natural forests on the upper reaches of the Yangtze and Yellow rivers. From 1999 onward, reclamation in the Three-Rivers Plain has also been forbidden. Conversion of land has been taken as one of the major countermeasures for ecological conservation. A program called "returning land back to the lake" was initiated in the Dongting Lake region. With the ambitious "grain for green" program, some 5.57 million hectares of cultivated farmland were returned to nature between 1998 and 2003.

See also: **Domestication; Green Revolution; Green Wall Project; Himalayas; Hot Spots of Land-Cover Change; Land Concentration; Land Privatization; Land Rehabilitation; Land-Use Policies; Paleo Perspectives; Private Property; Regrowth; Tradeoffs.**

Further Reading

Li, Xiubin, and Xiuhong Wang, "Changes of Agricultural Land Use in China: 1981–2000," *Asian Geographer* 22 (2003): 27–42; Lin, George, and Samuel Ho, "China's Land Resources and Land-Use Change: Insights from the 1996 Land Survey," *Land Use Policy* 20 (2003): 87–107; Liu, Changming, and Shifeng Zhang, "Drying up of the Yellow River: Its Impacts and Counter-Measures," *Mitigation and Adaptation Strategies for Global Change* 7 (2002): 203–14; Liu, Changming, Yu, Jingjie, and Eloise Kendy, "Groundwater Exploitation and Its Impact on the Environment in the North China Plain," *Water International* 26 (2001): 265–72; Liu, Jiyuan, Liu, Mingliang, and Dafang Zhuang,

"Study on Spatial Pattern of Land-Use Change in China during 1995–2000," *Science in China Series D* 46 (2003): 373–84; Yang, Hong, and Xiubin Li, "Cultivated Land and Food Supply in China," *Land Use Policy* 17 (2000): 73–88.

XIUBIN LI

Climate Impacts. The distribution and function of **vegetation** from previously natural ecosystems have been significantly altered by **land change as a forcing function in global environmental change**, resulting in manmade landscapes over much of the earth. The processes of land **conversion** and **modification** have accelerated in intensity and scale over the last few centuries. Field observations, atmospheric profiling, satellite **monitoring**, and climate **modeling** have been used to show that human changes in vegetation have subsequent impacts on **atmosphere-land interlinkages**, with further impacts on other components of the climate system. Complex relationships exist between land surface properties, atmospheric processes, and vegetation cover, with changes in vegetation impacting land surface and near-earth atmospheric processes by altering the fluxes of radiation, momentum, energy, and moisture among the soil, vegetation, and the atmosphere. At larger scales, changes in the land surface fluxes resulting from land-cover changes influence atmospheric processes operating over various temporal and spatial scales with numerous **feedback** mechanisms. There is further indication that the local changes in surface fluxes interact with larger-scale atmospheric conditions to impact weather and climate remotely through changes in global atmospheric circulation and moisture transport. The interactions among **land-cover change**, the atmosphere, and other components of the climate system are often classified based on the spatial scales on which they occur and the time frames over which they operate. Pavel Kabat and colleagues categorize the vertical and horizontal scales at which the land surface interacts with the atmosphere and therefore climate into three classes: the climate near the ground (micro and local), regional climate (meso), and global climate (macro). Roger Pielke and colleagues categorize the time frames that these interactions occur at into three classes: short-term interactions (minutes to a few weeks), long-term interactions (months to 100 years), and very long-term interactions (greater than 100 years). This review investigates how changes in land-surface properties from human land use at the local- to mesospatial scale impact climatic processes at the micro- to macrospatial scale, over the short- to long-term time frame to establish the more direct impacts of land-use and land-cover change on climate.

Huge boiling masses of dust that blocked out the sun were common sights in Texas during the Dust Bowl years, ca. 1935.

PHOTO: Russel Lord, "To Hold this Soil", Miscellaneous Publication No. 321, US Department of Agriculture, 1938. [Courtesy Library of Congress]

Surface Radiation Budget

The optical properties of vegetation canopies alter surface **albedo** and the effective emissivity from values that occur with bare soil or other types of vegetation. The changes in surface albedo and emissivity affect the amount of shortwave and longwave radiation absorbed, reflected, and emitted by the land surface, which in turn impacts the net surface radiation budget. The changes in radiation budget also have direct implications for the amount of energy available in the surface energy balance for heating and evaporation.

Vegetation impacts the surface albedo by changing the dynamics of shortwave radiation at the surface. These changes are a result of the optical properties of individual leaves and the light-trapping properties of collections of leaves within the canopy. The optical properties of leaves are strongly influenced by chlorophyll, which absorbs strongly at visible wavelengths (0.4–0.7 μm) to drive photosynthesis, but absorbs very little at near-infrared wavelengths (0.7–4.0 μm).

The individual leaves' difference in optical properties result in visible wavelengths of radiation being completely absorbed after being reflected only a few times within the canopy, but near-infrared wavelengths being scattered numerous times through the canopy before being absorbed or reflected to the ground or back to the sky. In the figure titled "Solar radiation dynamics," (a) illustrates the representation of leaves as scattering objects for a beam solar radiation. The additional radiation dynamics of vegetation over bare soil are shown in (b) and (c).

The impacts of land-cover change on intercepted solar radiation are the result of changes in the optical properties of the leaves and the position and orientation of the leaves within the canopy with changes in vegetation structure and density. The net impact of these changes on surface albedo of changes in vegetation is complex, dependent on changes in the density of the vegetation, changes in optical properties of the vegetation, and the color of the underlying soil.

Vegetation also affects the dynamics of longwave radiation at the surface. Roger Pielke demonstrated how the vegetation canopy acts as an intermediary layer absorbing both sky- and ground-emitted longwave radiation, as well as emitting longwave radiation both upward and downward itself. The figure

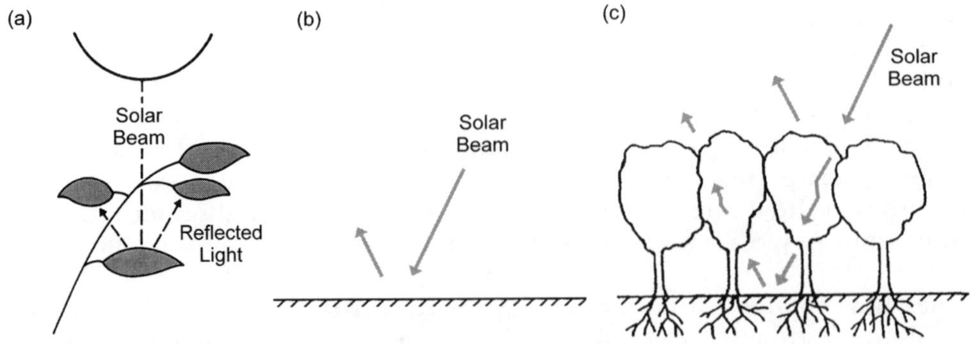

Solar radiation dynamics for (a) individual leaves (after Dickinson [1992]), (b) bare soil, and (c) dense vegetation canopy (adapted from Pielke [2001b]).

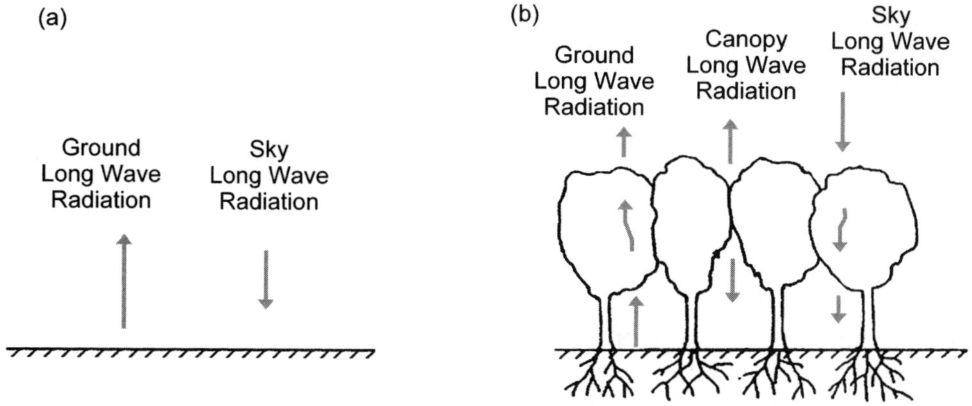

Longwave radiation dynamics for (a) bare soil and (b) dense vegetation canopy (adapted from Pielke [2001b]).

titled "Longwave radiation dynamics" illustrates the intermediary nature of a dense vegetation canopy on the transfer of longwave radiation compared to bare soil. The impacts of vegetation changes on the longwave radiation budget are therefore dependent on the soil temperature, the canopy temperature, the changes in vegetation density, and the downward flux of longwave radiation from the sky.

Effects on Surface Hydrology

The physiology and density of vegetation at the land surface modify the surface hydrology through deep-rooted vegetative transpiration and through the interception of precipitation and dew by the canopy. The figure titled "Hydrological processes" illustrates the differences in the evaporative fluxes to the atmosphere, and the hydrological cycle of dense vegetation compared to bare soil.

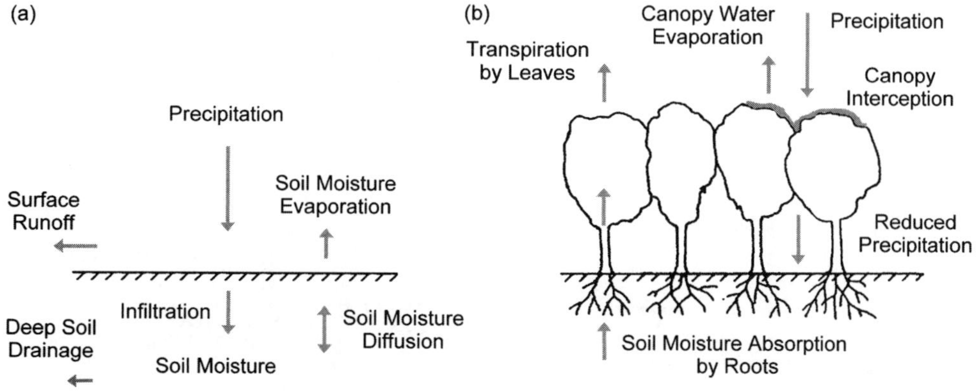

Hydrological processes for (a) bare soil and (b) dense vegetation canopy (only processes additional to bare soil hydrology shown) (adapted from Pielke [2001b]).

Land-cover changes that alter vegetation density affect the canopy leaf area and therefore the number of stomatal pores present available for transpiration, which impacts potential canopy transpiration rates. The vegetation density also affects the leaf area of the canopy for intercepting precipitation and dew, as well as the leaves' moisture storage capacity. Changes in physiology can also lead to changes in the function of stomatal pores, making the leaves more or less resistant to transpiration. Grasses with shallower roots and more efficient water utilization may result in higher resistance to transpiration and reduced access to soil moisture than the **forests** that they replace, resulting in lower potential transpiration rates. The reverse may also be true where irrigated crops replace dry shrublands.

Evaporative processes are some of the main phases of the hydrological cycle, with significant impact on the water budget for a region. The evaporative fluxes also strongly influence the partition of surface energy into sensible heat and latent heat, which impact the structure and atmospheric stability of the atmospheric boundary layer. Changes in land cover therefore can have impacts on the surface hydrology and evaporative fluxes with impacts on the atmospheric boundary layer.

Surface Energy Balance

The energy available at the surface is a major limiting factor in sensible, latent, and ground heat fluxes, with the net energy available at the surface dependent on radiative fluxes. The division of energy into each of these fluxes is dependent on the direction of energy flow and the properties of the land surface, the atmosphere, and the substrate for the transfer of moisture and energy. The presence of vegetation at the surface influences the surface radiation budget, and the fluxes of sensible and latent heat through changes in surface hydrology and turbulent mixing. The differences in surface fluxes are illustrated in the figure "Surface energy balance processes" for bare soil compared with a dense vegetation canopy.

Surface Friction

Vegetation structure and density affect the properties of surface roughness and displacement height (measures of surface frictional characteristics)

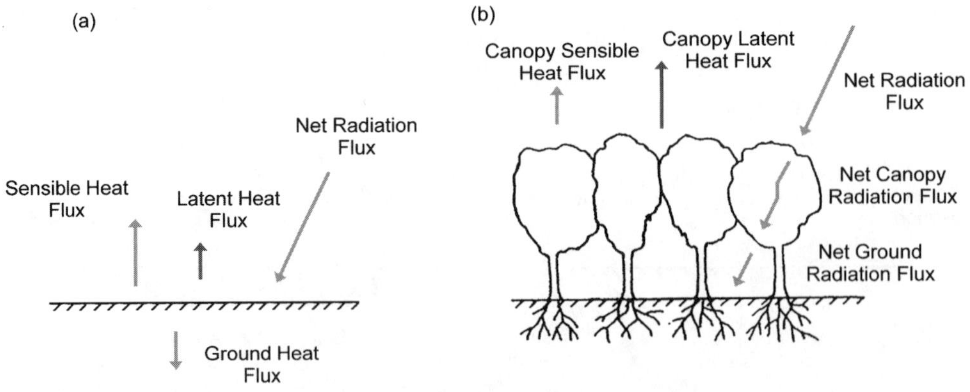

Surface energy balance processes for (a) bare soil and (b) dense vegetation canopy (only processes additional to bare soil shown) (adapted from Pielke [2001b]).

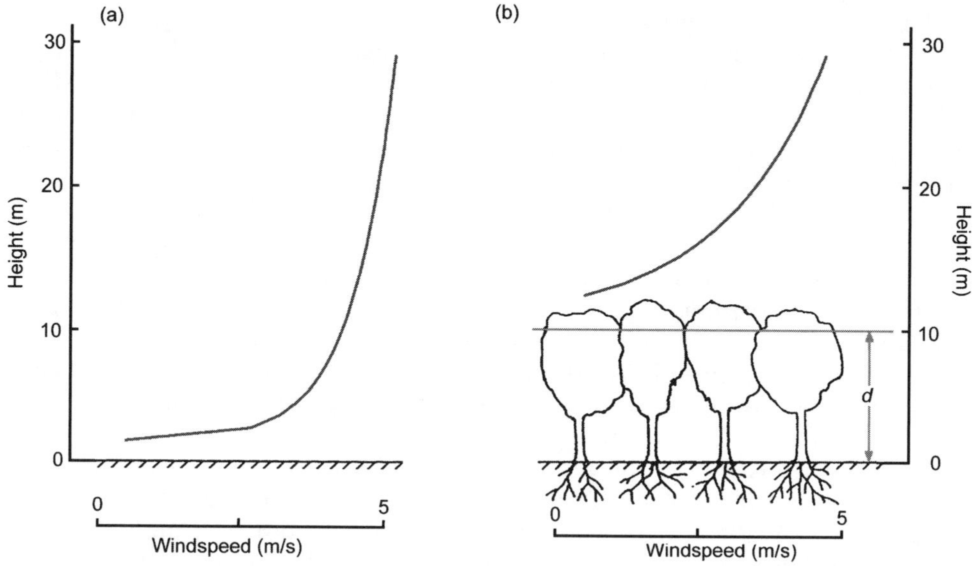

Logarithmic wind profile for a wind speed of 5 ms^{-1} at a height of 30 m for (a) bare soil and (b) dense vegetation canopy (adapted from Pielke [2001b] and Monteith and Unsworth [1990]).

by changing the distribution of roughness elements, such as trees, which in turn changes the surface friction imparted on the air moving over the surface. The figure titled "Logarithmic wind profile" demonstrates how combined measures of surface frictional effects (surface roughness and displacement heights) appropriate for bare soil and a dense vegetation canopy produce different vertical friction profiles with different wind speeds for the same ambient wind speed of 5 ms^{-1} at 30 meters.

There are simplified expressions for surface frictional effects that can be determined from canopy height and total leaf area. In these models, surface friction is strongly influenced by vegetation density and vegetation height. Taller vegetation creates larger surface friction; however, the relationship between vegetation density and frictional effects is more complex. Increasing leaf density increases surface friction to a maximum value where the canopy starts to close. After this point, increases in leaf density decrease the surface friction with turbulent mixing also reduced within the canopy air space.

Local-Scale Atmospheric Processes

Over time periods of a single day, the immediate influence of the land-cover change is restricted to a shallower zone known as the planetary or atmospheric boundary layer. The atmospheric boundary layer represents a series of idealized boundary layers, which are approximately parallel to the surface. Land-cover change affects the exchange of momentum, matter, and energy from the land surface to the atmosphere above through changing the transfer of momentum and sensible and latent heat fluxes across these boundary layers, which are highly dependent on surface friction.

The changes in physical properties of the land surface and atmospheric boundary layer influence turbulence, convective heating, and the distribution of moisture within the daytime atmospheric boundary layer. These changes impact the vertical structure of moisture and energy, static stability, and convective processes, which in turn strongly impact cloud formation and precipitation.

Turbulent heat fluxes near the ground are strongly affected by land surface characteristics for the division of radiative energy into latent and sensible heat. Over dry, bare land, radiative energy is primarily transferred as sensible heat to the atmosphere or heats the ground directly. Over wet, bare land, radiative energy is primarily transferred to the atmosphere as latent heat fluxes through evaporation. For vegetated land, however, large latent heat fluxes can continue even with a dry soil surface, as transpiration draws moisture from below the soil surface.

The characteristics of the atmosphere above dry, wet, or vegetated land surfaces are significantly different. Over dry bare soils the strong sensible heat flux produces vigorous turbulent mixing, resulting in an unstably stratified planetary boundary layer. The strong turbulent mixing and heating result in a deep boundary layer extending to heights of 2,000–3,000 meters or more above the land surface in afternoon hours. The deep and thermally unstable boundary layer also entrains dry air from the troposphere above, further reducing the average moisture content of the boundary layer.

Over wet or vegetated land surfaces, the slower heating rate limits the growth of the planetary boundary layer to typically less than 1,000 meters. Evaporation and transpiration supply additional moisture to the shallower boundary layer, which combines with significantly less entrainment of dry air to produce a moister boundary layer than develops over dry bare soil. The shallower, cooler boundary layer with more moisture is unstable to moist convection and under favorable conditions may promote deep cumulus convection and heavy precipitation. For example, the influences of land-cover changes on deep cumulus convection over the central plains of the United States are discussed by Roger Pielke and colleagues. This modeling study looked at the convective processes associated with the current-day land surface of irrigated cropping, compared with the same processes associated with the natural land surface of short prairie grass and found that the higher evaporation and transpiration rates of the current landscape produced higher convective available potential energy in the boundary layer, which in turn resulted in the development of a thunderstorm over the area during the simulation. The lower evaporation and transpiration of the natural landscape, however, only produced a shallow line of cumulus clouds under the same conditions.

Regional-Scale Atmospheric Processes

At the regional scale, Nathan Moore and Stuart Rojstaczer studied the influence of irrigation on precipitation in the Texas high plains of the United States. The study found that irrigation enhanced summer precipitation by 6–18 percent for areas up to ninety kilometers downwind, with storms of greater duration, length, and total accumulation. The study also found that the irrigation was not a significant source of the moisture in the storms; rather the cool, wet surface increased low-level instability, triggering the storms.

Lawton and colleagues investigated the impacts of lowland **deforestation** in Costa Rica on cloud immersion of adjacent tropical montane cloud forests during the dry season using satellite imagery and atmospheric modeling. The results of the study indicated that the **pastures** were relatively cloud free, while forests produced well-developed cumulus cloud fields, with the cloud base heights significantly higher over the pastures than the forests. The study found that the reduced cloud formation and higher cloud base height from deforestation reduced the amount of cloud immersion experienced by the montane cloud forests with serious impacts on the moisture-dependent ecosystem. Thomas Chase and colleagues also found weather influences in the high Rocky Mountains due to the presence of irrigated farmland in the plains below. Irrigated regions affected the daily summer mountain-plains breeze by altering temperature patterns, thereby allowing communication between the two regions. Such changes in local circulation regime would be expected to alter the transport of pollutants or atmospherically transported micronutrients and so may be of importance in many locales. Joseph Eastman and colleagues described a "biological teleconnection" where changed ecosystem characteristics affected local weather such that effects were communicated to regions distant from actual changes in surface characteristics, while Elfatih Eltahir found a strong role of surface vegetation in maintaining the regional West African monsoon circulation.

Differences in land-surface characteristics at regional scales also can produce organized circulations between areas with different surface fluxes, such as those between land and water, or irrigated crops and dry **grassland**. It has been shown that these organized circulations and associated fluxes can generate clouds, alter precipitation patterns, and induce thunderstorms.

From an investigation of the **scales** at which heterogeneous landscapes with different land surface characteristics form organized circulations using large eddy simulation, experiments have modeled the dynamics of the convective boundary layer over a range of heterogeneity lengths within the mesoscale domain. The results of these modeling studies indicate that for land surface heterogeneity with homogeneous features less than five to ten kilometers long, the vertical buoyancy caused by uneven heating is disturbed by horizontal pressure gradients caused by the same uneven heating, preventing well organized circulations from developing. With land surface heterogeneity greater than five to ten kilometers, however, the horizontal pressure gradients are too weak to disturb the vertical buoyancy, due to the distance between the regions of uneven heating; and organized circulations are able to form between the areas of buoyancy and the areas of subsidence.

Global-Scale Atmospheric Processes

At global scales, changes in land surface properties associated with changes in vegetation can have impacts on continental and global atmospheric circulation, with possible large impacts on regional and continental climate. For example, Pavel Kabat and colleagues, Roger Pielke and colleagues, and Richard Betts and colleagues provide extensive reviews of the complex relationships that exist between vegetation and other components of the climate system at the local, regional, and global scales. These reviews detail the differences in magnitude and sign that similar vegetation change investigations have identified in different geographic localities over the earth.

Early research into the impact of land-cover change on surface albedo and climate by Jule Charney found that a reduction in darker vegetation over brighter soils in the Sahel region of the southern Sahara Desert led to a decrease in the radiative heating of the surface and a relative cooling of the atmosphere above. The study suggested that atmospheric cooling resulted in a sinking motion that was superimposed on the descending branch of the Hadley cell, leading to increased stable descending air over the region, perpetuating arid conditions.

In similar research, B. Lofgren used global scale climate models to show that thermally induced overturning circulations between oceans and land could be affected by albedo at low latitudes. The study found that decreased surface albedo increased radiative heating at the surface, resulting in enhanced upward atmospheric motion over the continent. The increased upward motion caused increased low-level convergence of air over the continent and an associated low-level divergence over the surrounding oceans. This circulation pattern resulted in enhanced moisture flux from the ocean to the continent, with increased precipitation over the continent. The increased precipitation further enhanced the thermal forcing through latent heat release on condensation.

Modeling investigations into the climate impacts of tropical deforestation in **South America**, Africa, and Southeast Asia identified that the main impact of removing tropical forests was on the hydrological cycle through reduced transpiration. The reduction in atmospheric moisture combined with the increase in sensible heat flux resulted in reduced cloud formation and precipitation. The reduced cloud cover significantly increased solar radiation flux, which offset the increased albedo associated with the deforestation, to result in further increases in sensible heat flux and further reduction in precipitation and cloud cover.

Paul Dirmeyer and O. Reale investigated historical land-cover and climate changes in northern Africa from 2,000 years ago to see whether the changes in land cover could be responsible for changes in Mediterranean climate. Climate proxies from the region suggested a widespread drying trend across the Mediterranean since the Roman classical period (RCP). The study also showed that the desert areas of northern Africa corresponding with modern Egypt, Tunisia, and Algeria were significantly moister during the RCP, with wealthy agricultural economies making the area the most productive in the Roman world. To investigate whether land-cover changes associated with deforestation and cultivation across the Mediterranean may have contributed to the widespread drying, the study modeled the region's climate with a fine-resolution general circulation model using vegetation that existed in the RCP compared to the climate modeled with the vegetation representing modern-day cover. The modeling experiments showed significant changes in the climate under the two vegetation scenarios, with a northward shift in the Inter-Tropical Convergence Zone (ITCZ) and the creation of a sea-land circulation over northwestern Africa under the RCP vegetation. The changed atmospheric circulation resulted in substantially moister conditions in northern Africa, with the speculation that the vegetation of the RCP would be sustained under the wetter conditions in areas that under current-day conditions are too dry. The conclusion from the research was that clearing of the Mediterranean by human activity since the RCP may have triggered a positive climate feedback with a drift toward the dryer conditions of modern day.

Finally, there is accumulating evidence that large-scale land-cover changes, particularly in the tropics, generate remote climatic effects of global extent. Remote climate effects far from where the surface has been directly affected by land-cover changes in climate models have been discussed by Thomas Chase and colleagues and H. Zhang and colleagues, among others. Others have also noted isolated extra-tropical effects due to simulated tropical vegetation changes. Thomas Chase and colleagues also examined global circulation model (GCM) simulations of the effect of observed levels of land-cover change globally and found strong evidence of changes in global-scale circulations. Other studies again demonstrated that the remote effects of observed levels of land-cover change were prevalent in a variety of models under a range of configurations and model assumptions and that remote temperature anomalies resulting from land-cover change could be similar in magnitude as effects of the historical increase of the radiative effect of increased carbon dioxide.

Nicola Gedney and Paul Valdes, also using a GCM, specifically examined the effects of a wholesale removal of the Amazonian rainforest on remote climates and found significant evidence for a reduction in large-scale circulations generated by tropical convection and for propagating atmospheric waves that affected rainfall in Northern Hemisphere winter. David Werth and Roni Avissar, in several studies, find statistically significant remote effects due to deforestation in **Amazonia**, Central Africa, and Southeast Asia; and Ruth DeFries and colleagues examined potential impacts of future land-use changes and found regional temperature anomalies of up to 1.5°C in regions not directly affected by land-cover changes.

Conclusion

The biophysical changes in land-surface properties associated with land-cover change are shown to be complex, with impacts on the lower atmosphere through changes in the surface radiation budget, surface hydrology, and the surface energy balance. The wide range of land-cover and climate studies from different regions of the world suggest that the impacts that changes in land-surface fluxes have on the atmospheric boundary layer and larger-scale atmospheric processes also are complex, as these systems are closely coupled with the land-surface fluxes, with numerous feedback mechanisms operating at a range of spatial and temporal scales.

See also: **Biodiversity; Carbon Cycle; Cumulative Change; Desertification; Disequilibrium Dynamics; Driving Forces; Ecosystem Services; Environmental Change; Fire; Greenhouse Gases; Kyoto Protocol; People at Risk; Systemic Change.**

Further Reading

Avissar, Roni, and David Werth, "Global Hydroclimatological Teleconnections Resulting from Tropical Deforestation," *Journal of Hydrometeorology* (submitted); Bounoua, Lahouari, DeFries, Ruth, Collatz, G. James, Sellers, Piers, and H. Khan, "Effects of Land Cover Conversion on Surface Climate," *Climatic Change* 52 (2002): 29–64; Charney, Jule G., "Dynamics of Deserts and Drought in the Sahel," *Quarterly Journal of the Royal Meteorological Society* 101 (1975): 193–202; Chase, Thomas N., Pielke Sr., Roger A., Kittel, Timothy G. F., Baron, Jill S., and Thomas J. Stohlgren, "Potential Impacts on Colorado Rocky Mountain Weather and Climate Due to Land Use Changes on the Adjacent Great Plains," *Journal of Geophysical Research* 104 (1999): 16,673–90; Chase, Thomas N., Pielke, Roger A., Kittel, Timothy G. F., Zhao, M., Pitman, Andrew J., Running, Steve W., and Ramakrishnan R. Nemani, "Relative Climatic Effects of Land Cover Change and Elevated Carbon Dioxide Combined with Aerosols: A Comparison of Model Results and Observations," *Journal of Geophysical Research* 106 (2001): 31,685–91; Chase, Thomas N., Pielke, Roger A., Kittel, Timothy G. F., Nemani, Ramakrishna R., and Steven W. Running, "Simulated Impacts of Historical Land Cover Changes on Global Climate in Northern Winter," *Climate Dynamics* 16 (2000): 93–105; DeFries, Ruth S., Bounoua, Lahouari, and James G. Collatz, "Human Modification of the Landscape and Surface Climate in the Next Fifty Years," *Global Change Biology* 8 (2002): 438–58; Dickinson, Robert E., "Changes in Land Use," in *Climate System Modeling*, ed. Kevin E. Trenberth, London: Cambridge University Press, 1992, pp. 689–701; Eastman, Joseph L., Coughenour, Michael B., and Roger A. Pielke, "The Effects of CO_2 and Landscape Change Using a Coupled Plant and Meteorological Model," *Global Change Biology* 7 (2001): 797–815; Eltahir, Elfatih A. B., "Role of Vegetation in Sustaining Large-Scale Atmospheric Circulations in the Tropics," *Journal of Geophysical Research* 101 (1996): 4255–68; Gedney, Nicola, and Paul J. Valdes, "The Effect of Amazonian Deforestation on the Northern Hemisphere Circulation and Climate," *Geophysical Research Letters* 27 (2000): 3053–6; Kabat, Pavel, Claussen, Martin, Dirmeyer, Paul A., Gash, John H. C., Bravo de Guenni, Lelys, Meybeck, Michel, Pielke Sr., Roger A., Vörösmarty, Charles J., Hutjes, Ronald W. A., and Sabine Lütkemeier, *Vegetation, Water, Humans and the Climate: A New Perspective on an Interactive System*, Berlin, Heidelberg, New York: Springer, 2004; Lawton, R. O., Nair, U. S., Pielke, Roger A., and R. M. Welch, "Climate Impact of Tropical Lowland Deforestation on Nearby Montane Cloud Forests," *Science* 294 (2001): 584–7; Lofgren, B. M., "Sensitivity of Land-Ocean Circulations, Precipitation, and Soil Moisture to Perturbed Land Surface Albedo," *Journal of Climate* 8 (1995): 2521–42; Moore, Nathan, and Stuart Rojstaczer, "Irrigation's Influence on Precipitation: Texas High Plains, U.S.A.," *Geophysical Research Letters* 29 (2002): 21–4; Pielke, Roger A., "Influence of the Spatial Distribution of Vegetation and Soils on the Prediction of Cumulus Convective Rainfall," *Reviews of Geophysics* 39 (2001): 151–77; Pielke, Roger A., Avissar, Roni, Raupach, Michael R., Dolman, A. J., Zeng, X., and A. Scott Denning, "Interactions between the Atmosphere and Terrestrial Ecosystems: Influence on Weather and Climate," *Global Change Biology* 4 (1998): 461–75; Pielke, Roger A., Marland, G., Betts, Richard A., Chase, Thomas N., Eastman, Joseph L., Niles, J. O., Niyogi, D., and Steve W. Running, "The Influence of Land-Use Change and Landscape Dynamics on the Climate System: Relevance to Climate-Change Policy beyond the Radiative Effect of Greenhouse Gases," *Philosophical Transactions of the Royal Society of London*, Series A 360 (2002): 1–15; Sellers, Piers J., "Biophysical Models of Land Surface Processes," in *Climate System Modeling*, ed. Kevin E. Trenberth, London: Cambridge University Press, 1992, pp. 451–90; Werth, David, and Roni Avissar, "The Local and Global Effects of Amazon Deforestation," *Journal of Geophysical Research* 107/D20 (2002); Zhang, H., Henderson-Sellers, A., and K. McGuffie, "Impacts of Tropical Deforestation Part I: Process Analysis of Local Climatic Change," *Journal of Climate* 9 (1995): 1497–1517; Zhao, M., Pitman, Andrew J., and Thomas N. Chase, "The Impact of Land Cover Change on the Atmospheric Circulation," *Climate Dynamics* 17 (2001): 467–77.

PETER J. LAWRENCE, THOMAS N. CHASE

Coastal Zone. The transitional area where land and ocean meet, where the interaction between natural processes and human activity is most active. The rise and fall of tides, the passing of storms, and other physical processes create a region of natural process and interaction between land and sea, such as beaches, tidal flats, coastal marshes, and mangroves. Human beings have lived at the coast from ancient times, utilizing the rich and easily accessible resources. Its importance has been ever increasing to societies worldwide. The coastal zone provides space for human activities for urban development, tourism and recreation, **agriculture**, fisheries, industry, and transportation. In earlier years, resources in the coast were considered to be unlimited, and people had little environmental concern for their use. Now, few coastal zones exist without marked human impact and some form of environmental degradation. However, in recent years, sustainable environmental principles regarding the use of coastal zones have been gaining greater attention.

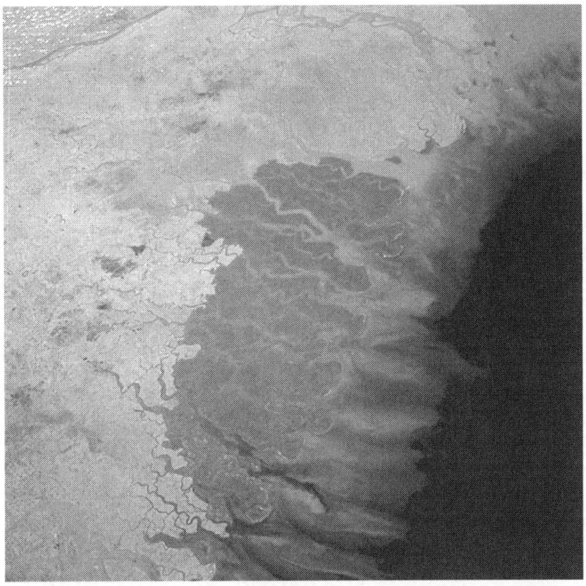

Mouths of the Ganges River (south of the confluence of the Ganges and Brahmaputra Rivers) in India and Bangladesh in November 1985, with croplands (rice, sugarcane, jute) in bright tone, and the Sundarbans, an abundant mangrove swamp and largest remaining habitat of the Bengal tiger (in dark tone).

PHOTO: Space Shuttle. CREDITS: Image Science and Analysis Laboratory, NASA-Johnson Space Center, 18 Mar. 2005 (Earth from Space—Image Information). http://earth.jsc.nasa.gov/sseop.efs

Definition

The "coastal zone" can be related to the term "coastal area." In coastal management, some authors consider that the concepts of coastal area and coastal zone have different meanings. Others assert that coastal area is a generic geographic space that has not been defined as a zone. The concept of coastal zone, on the other hand, is specific and geographically defined. It is subject to the management process and interaction between humans and their environment. According to the **United Nations Environment Programme**, the coastal area is a notion geographically broader than the coastal zone, with loosely defined borders. On the other hand, the coastal zone is that part of the coastal area concerned with a coastal management program. "Zone" implies a planning zone.

Others believe the concept of the coastal zone comes from natural sciences such as geology and geomorphology, while coastal area is more often used by social scientists. Some authors prefer to use the term "coastal area" to avoid any confusion between coastal zone and planning zone. However, many authors, even in the field of coastal management, consider coastal zone and coastal area as synonymous.

Defining the boundaries of coastal area or zone is more relevant to planners and managers than academics. A coastal zone is the narrow area between adjacent land and ocean where the two directly affect each other. The interaction between land and ocean is transitional. Therefore, the choice of boundary between the landward and seaward limits of a coastal zone depends to a large extent on why the definition is necessary. International organizations and coastal nations define the limits of a coastal zone according to the use of the particular coastal management, such as non-point sources of marine pollution management or wetland preservation managements. For example, the **Land-Ocean Interactions in the Coastal Zone (LOICZ) project** defines the coastal zone as the area from the coastal plains to the outer edges of the continental shelves.

Population and Coastal Use

Human **population dynamics** are one of the major **driving forces** of coastal land-use changes. It is estimated that 50 percent of the population of the industrialized world lives within one kilometer of the coast. Other studies report that eight of the ten largest cities in the world are located by the coast, and that the number of people in the world living within 100 kilometers of the coast has increased from 2 billion in 1990 to 2.2 billion in 1995, which is 39 percent if compared to the world population. Coastal cities have the highest rates of growth of other areas. Increasing tourism, which is one of the largest and fastest-growing sectors of the global economy, adds to the pressure on the coastal environment. This in turn brings further ecosystem changes and land-use change to the coast.

Increasing population density exerts pressure on the coastal zone. For instance, coastal **wetlands** are drained and reclaimed, flood plains adjacent to estuaries are covered, and mangroves are cut down. Developments of port and harbor, transportation systems, industries, sewage treatment plants, and waste disposal landfills destroy the wetlands. In addition, pollution from upstream degrades marine and coastal resources. Since 1900, the world has lost approximately half of its coastal wetlands through coastal development.

Coastal Ecosystems and Their Capacity

Increasing population, human-induced coastal pollution, and exploitation of coastal resources destroy coastal ecosystems. Coral reefs, mangroves, tidal wetlands, estuaries, and a variety of other habitats each provide their own distinct goods and services. They are the primary producers of fish, shellfish, and seaweed for both human and animal **consumption**.

Coastal ecosystems are already losing much of their capacity to produce fish because of overfishing and destruction of nursery habitats. As the extent of mangroves, coastal wetlands, and sea grasses declines, coastal habitats are losing their pollutant-filtering capacity. Coastal ecosystems provide the important service of maintaining water quality by filtering or degrading toxic pollutants and absorbing nutrient inputs. This capacity is easily exceeded by chemical pollutant discharges from agriculture and industry, oil spills in the ocean, oil discharges from land-based sources, excessive nutrient inputs from urban runoff, and sewage effluent.

Many of the world's coral reefs, mangroves, and wetlands are under threat by human activities. Many coastal wetlands have been converted to agricultural land, shrimp, and fishponds, or have been reclaimed. Mangroves cover about one quarter of tropical coastlines, and some 112 countries and territories have mangroves. Scientists estimate that anywhere from 5 to 85 percent of original mangrove areas in various countries have been lost, with extensive losses occurring in the past fifty years. Many coastal countries in Southeast Asia have lost half or more of their mangrove forest since the mid-1960s, largely because of industrial logging for timber and fuelwood, **conversion** to **rice** fields and other agriculture purposes, and urban development. **Eutrophication** and trampling by tourists, overfishing, and blast for ports or navigation are the sources of damage.

Integrated Coastal Zone Management

The integrated coastal zone management approach is an interdisciplinary activity. Natural and social scientists, coastal managers, and policymakers aim to manage the diverse problems of coastal areas and mitigate accelerating environmental degradation of worldwide coastal ecosystems. Historically, the national park movement influenced the development of coastal ecological management. In the late nineteenth century, they strove to protect valuable ecosystems and conserve the **biodiversity** of the protected areas or parks. The first such park in coastal marine areas was established in the 1930s. Expansion of **land-use planning** in the late nineteenth and early twentieth centuries also influenced the development of coastal area management. There were conflicting land uses by expansion of urban uses into coastal areas for new residential areas and industries as well as recreational uses.

In recent years, the issue of sustainability has become central to most coastal management efforts worldwide. In early years, however, various authorities operated independently in ecological management, resource management, engineering, or urban development of coastal areas. By late 1960s, efforts to combine the independent approaches culminated in the development of the U.S. Coastal Zone Management Act. With increasing degradation of the environment, as a consequence of **Agenda 21** of the United Nations Conference on Environment and Development in 1992, recent efforts have focused on the issue of sustainable development in coastal ecosystems. Introduced in Agenda 21, a number of international organizations including the World Bank, Organisation for Economic Cooperation and Development, United Nations Environment Programme, and the World Conservation Union of Nature have developed guidelines for coastal management that is "integrated in content." The Intergovernmental Oceanographic Commission (IOC) of the United Nations Education, Science and Culture Organization (UNESCO) defines integrated coastal area management as an interdisciplinary activity where natural and social scientists, coastal managers, and policymakers focus in the long-term on how to manage the diverse problems of coastal areas. Integrated coastal zone management is a comprehensive policy and management response option, because the coast is a distinctive system with both terrestrial and ocean characteristics. Any **land-use policies** or management of the coastal zone requires joining the different elements into some overall unified coastal management system.

Coastal Wetlands

Coastal wetlands are one of the major areas that deserve immediate integrated management approaches. Coastal wetlands, such as tidal flats, salt marshes, and mangrove swamps, are developed in estuaries, river deltas, and coasts with back barrier environments. Many coastal wetlands have already been lost or will be degraded due to land reclamation, unsustainable exploitation of wetland resources, and pollution. The **Ramsar Convention on Wetlands**, for example, already provides principles and guidelines for incorporating wetland issues into integrated coastal zone management, which intends to increase understanding and recognition of the importance and vital role played by wetlands in the coastal zones of the world. The Ramsar Convention is a global intergovernmental treaty signed in Ramsar, Iran, in 1971. The convention specifically addresses the **conservation** and wise use of coastal zone ecosystems. Ramsar defined the concept of "wise use" as sustainable utilization for the benefit of humankind in a way compatible with the maintenance of the natural properties of the ecosystem. It provides a comprehensive list of important coastal and inland wetlands in the world.

Coastal wetlands have important economic values and functions. The wetlands have been utilized for spawning, nurseries, feeding grounds for fish and shellfish, timber for building, fuelwood, reeds for thatching and weaving, and water transport. They buffer floods and storm surges, provide coast protection, and mitigate the effects of rising sea levels. The wetlands in the coast are also important for the conservation of biological diversity, particularly migratory species such as water birds. As a consequence of various development activities, many coastal wetlands are being destroyed at an alarming rate.

Two instances are illustrated here for human-induced wetland destruction and resulting environmental degradation. In the **United States of America**, about 55 percent of the area of estuaries and coastal wetlands has been destroyed or damaged during the past 200 years, primarily because of dredging and filling and waste contamination. Wetland loss in coastal Louisiana, particularly in the Mississippi delta plain, increased dramatically during the second half of the twentieth century. Net loss of coastal wetlands would be occurring without human intervention to some degree. However, a variety of human activities have caused wetland loss to accelerate greatly. Major causes of Louisiana wetland loss are related to fuel and mineral resources. Construction for ship channels to inland ports, oil and gas development and pipelines, and hydrologic modifications all contribute significantly. The accelerated wetland loss is basically a consequence of the sinking of the land and deficiencies in the buildup of soil to offset this sinking, rather than erosion of edges. The interconnected network of dredged canals in the Louisiana wetlands accounts for about 10 percent of the land losses. The dredged canals, however, are responsible indirectly for additional losses of **vegetation** by intrusion of saltwater into freshwater marshes, changes in marsh hydrology, changes in sediment dispersion, and changes in nutrient distribution. Other wetland losses in Louisiana are caused by old land reclamation projects that failed. Land development by reclamation was very active in the early 1900s, and began in the early eighteenth century. These developments eventually resulted in land loss as levees sank, organic soils decomposed, and the wetlands subsided. It is estimated that Louisiana has lost 1,900 square miles of wetlands since the 1930s and that another 700 square miles are likely to vanish by 2050.

Reclamation of tidal flats is another example of wetland destruction. A tidal flat is a broad, flat land developed in the intertidal zone along the coast. It is a source of food, purifies water, and provides stopovers for migrating water birds, as well as being an economic base for the local community. Tidal wetlands are one of the most productive ecosystems known among the different land covers. They produce large amounts of plant organic material, which support the mud flat inhabitants, such as crabs, snails, and clams. Tidal wetlands act as nurseries for many species of fish, shellfish, and prawns. These and other creatures, in turn, serve as the food base for predatory animals, primarily fishes and birds. Seasonally, many tidal flats serve as essential stopovers for migrating birds to rest, feed, and breed. Tidal flats purify pollutants discharged from the land into the sea through filtering and removing excess nutrients from the water that flows through them. All the creatures living in the tidal flat are involved in this purification process directly or indirectly. Tidal wetlands also stabilize the coastline, preventing **soil erosion** during storms.

Land reclamation of coastal wetlands is one of the major pressures induced by human activities. The polder, which is Dutch in origin, is one of the most well-known coastal zone alterations. A polder is land converted from sea and constantly defended from it thereafter. The term may be used to describe a drained marsh or a reclaimed coastal zone. Extensive areas of coastal wetland are cleared and filled for ports, residential, commercial and industrial development, and agricultural needs. Conversion of coastal wetland into agricultural fields goes back a few thousand years and is practiced on many continents. Tidal flats are one of the most desirable wetlands for agricultural land because of their usually fine sediments.

Coastal lowlands of western Korea, southern **Japan**, and southern **China** support a high-density population with rapid economic development, resulting in huge demands of land for urban uses. Many of the coastal mudflats have been reclaimed for industrial, infrastructural, agricultural, and urban development. Even large-scale modern airports in Korea and Japan are located on reclaimed land, via sea walls drying out tidal flats.

Although there are strong economic and social reasons to support each individual reclamation project, such conversion destroys ecologically rich natural resources whose productivity may one day be of even greater benefit. It is exploitation for uncertain economic gain with little concern for the consequences. For instance, a large reclamation project located south of Inchon in western Korea built a dyke to cut off the tidal flat from the ocean and to prepare a large regulating freshwater reservoir. However, the project of ten-plus years failed when the fresh reservoir was filled with toxic pollutants from adjacent industrial districts, urbanized areas, and agricultural and feeding lots. The sea dyke was eventually reopened to the ocean.

See also: **Arena of Land Conflict; Ecosystem Services; Land Rehabilitation; Mineral Extraction; Nutrient Cycle; Species Extinction; Urbanization; Water-Land Interlinkages.**

Further Reading

Bourne, Joel K. Jr., "Gone with the Water," *National Geographic*, October 2004, 88–105; Global Forum on Oceans, Coasts, and Islands, Global Web Service on Oceans, Coasts, and Islands [Online, January 2005], The Global Forum Web Site, www.globaloceans.org; Goldberg, Edward D., *Coastal Zone Space: Prelude to Conflict?* Paris: United Nations Education, Science and Culture Organization, 1994; Hwang, Manik. "Land Reclamation from Tidal Flats in Korea: A Historical Perspective," in *Land-Use Changes in Comparative Perspective*, Yukio Himiyama, Manik Hwang and Toshiaki Ichinose, eds., Enfield, New Hampshire: Science Publishers, Inc., 2004, pp. 209–17; Kay, Robert, and Jacqueline Alder, *Coastal Planning and Management*, London, New York: E & FN Spon, 1999; United Nations Atlas of the Oceans [Online, February 2005], The UN Atlas of the Oceans Web Site, www.oceansatlas.org; United Nations Education, Science and Culture Organization, Intergovernmental Oceanographic Commission, Marine Sciences and Observations for Integrated Coastal Area Management [Online, January 2005], The UNESCO IOC Web Site, ioc.unesco.org/icam; Vallega, Adalberto, *Fundamentals of Integrated Coastal Management (The Geo-Journal Library Vol. 49)*. Dordrecht, NL: Kluwer Academic Publishers, 1999; Vernberg, F. John, and W. B. Vernberg, *The Coastal Zone—Past, Present, and Future*, Columbia, SC: University of South Carolina Press, 2001; Walker, H. Jesse, "The Coastal Zone," in *The Earth as Transformed by Human Action: Global and Regional Changes in the Biosphere over the Past 300 Years*, eds. B. L. Turner II, William C. Clark, Robert W. Kates, John F. Richards, Jessica T. Mathews, and William B. Meyer, Cambridge, UK: Cambridge University Press, 1990, pp. 271–94.

MANIK HWANG

Coffee. Comes from the genus *Coffea*, and the two dominant commercial species globally are *coffea arabica* and *coffea canephora*, hereafter referred to as the arabica and robusta coffees, respectively. The arabica coffee produces a superior-tasting blend and is the dominant variety. Robusta became important in the production of instant coffee, where an inferior-tasting bean was appropriate. The popularity of such instant coffee only arose during World War II, and therefore its history is brief compared to that of arabica.

Arabica Coffee

Arabica coffee originated in Ethiopia, where the first records of its use date to AD 850, when it was used more for medicinal purposes than as a beverage. Coffee gained popularity as a beverage in the sixteenth century, and by the seventeenth century it was a dominant part of European life, with coffeehouses opening throughout **Europe**. While coffee production initially was monopolized within originating African countries, coffee is now grown across the world. Today, some of the most important coffee-producing countries are Brazil and Colombia in Latin America, Tanzania and Uganda in Africa, and Vietnam and Indonesia in Asia. Coffee currently covers over 2 million hectares in Asia, 2.9 million hectares within Africa, and over 5.9 million hectares in **Latin America** and the Caribbean. Coffee is considered the second most-valuable legal export commodity traded today in the world market (after oil), and it brings in revenue in excess of $10 billion. As such, its importance as an agricultural crop cannot be understated.

Conditions needed for growth of the most important coffee variety, arabica, differ from those for robusta. Globally, about 70 percent of the coffee produced is of the arabica variety, with the remaining 30 percent being the lower-quality robusta. Arabica is an evergreen tree, which can grow as high as seven meters, although it is usually pruned to maintain a height around two meters. This helps in picking the beans, but also such pruning enhances the yield. This evergreen has arrowhead-shaped leaves, broad and shiny, of seven to fifteen centimeters in length. These trees do have small, white flowers that cluster at the leaf bases, and it is the appearance of these flowers, at around five years in age, that indicates the trees' maturity and thus their bearing of the coffee cherries. Depending primarily on climate and soils, a tree can produce between one and twelve pounds of coffee a year. Each coffee cherry contains two coffee beans. Arabica grows best in **mountain** environments, with subtropical climates with annual temperature ranges of 17–25°C and a mean temperature around 20°C. Ideally, there should be limited diurnal or annual variability. It is a crop that is not able to tolerate frosts and is highly sensitive to moisture conditions. Arabica needs a defined dry period of around ten to twelve weeks, although if this extends too long, a poor crop will result. Outside of this dry period, the arabica needs a fairly even distribution of rainfall throughout the year, with no excessively moist periods, which may result in potentially waterlogged soils, or intense precipitation events, which may cause either the flowers or the coffee cherries to fall, depending on the timing of the intense rain events. Annual precipitation values should range between 1,200 and 1,500 millimeters per year, with an upper limit around 2,500 millimeters per year. Soils should be deep loams, with pH values around 6.0.

Robusta Coffee

Robusta was first discovered in the late nineteenth century, growing in lower altitudes in Uganda. In general, it grows at lower altitudes and also at warmer temperatures than the arabica, and it is also noteworthy that it is grown exclusively in the Eastern Hemisphere. Robusta is less sensitive to frosts and is better adapted to higher temperatures. The annual temperature range for robusta is 24–30°C, and in general it grows better at higher temperatures. Like arabica, though, it requires minimal diurnal and annual variability. Annual precipitation requirements for robusta are a minimum of 1,550 millimeters per year, and like arabica a dry season is needed for flower buds to break their dormancy. Robusta requires soil pH greater than 4.5. Robusta tends to be much hardier, higher-yielding, and more disease resistant; and it is these characteristics that make it a less-expensive agricultural crop to maintain, although it is also less desirable and so fetches a lower market price.

Once picked, the coffee bean undergoes either a dry or wet production method, in which the bean is dried in the sun or by a mechanical dryer, and the husk is removed. The huller then removes the two thin layers still remaining on the bean, and then the beans are cleaned at least once, and twice for better-quality coffees. The cleaning process removes debris and defective beans. Beans are then graded based on their taste, preparation, and quality of picking, country, altitude, and size. Next, these green beans are roasted at temperatures between 193 and 218°C, while being turned in rotating drums, which removes moisture, releases the oils, and brings out the flavor. During this roasting process all the starches are converted to sugars, 18–20 percent of the bean weight is lost, and the beans themselves actually plump up. Rapid cooling is required after roasting. Temperature and duration of roasting are critical to the flavor of the coffee, which is now ready for sale.

Landscape Impacts

While coffee was traditionally grown under a full canopy, given its origins as an understory bush in the East African rainforests, today most coffee production does result in significant landscape change. Even when coffee is grown under a canopy, the canopy is usually a secondary crop, such as banana, rather than a native **forest** cover. As such, coffee production can produce complete landscape transformation; given where coffee grows, this often occurs on steeper slopes and in areas not previously cleared for **agriculture**, and so represents new changes in land use and land cover. In addition, the more recent movement from shade coffee to sun-grown coffee removes the need for an overstory completely and thus results in a much more significant landscape change, with the native cover being removed and replaced by a **monoculture** of coffee shrubs. This has significant implications for soil stability and **soil erosion** rates in the years it takes for the coffee plants to mature and provide a canopy cover for the soil. Given that much of the coffee prefers mountainous conditions, this is potentially a huge problem, with farmers increasingly switching to sun-grown coffee.

While coffee prices have been in a major slump in recent years, causing many to temporarily abandon their coffee plantations, recent news seems positive. According to the **United Nations Food and Agriculture Organization**, there are now some indications that coffee may be recovering from its price depression, although it is still quite uncertain and the recovery is still fragile. Most coffee-producing countries are still struggling, as the fundamental problem of oversupply does still remain. Current emphasis on the production of higher-quality coffee may thus be problematic, as reduced coffee prices and reduced revenues make **investment** in the crop difficult. One current trend to switch to organic coffee production may help bolster coffee prices, although the process of certification can be problematic and expensive.

See also: **Agricultural Intensification; Agrodiversity; Cash Crops; Consumption; Land Abandonment; Reforestation.**

Further Reading

Clarke, Ronald J., and Robert Macrae, *Coffee*, New York: Elsevier Applied Science Publishers Ltd., 1988; Food and Agriculture Organization of the United Nations [Online, August 2004], The Coffee Commodity Notes Web Site, www.fao.org; Guhl, Andrés, "Coffee and Landscape Change in the Colombian Countryside 1970–2002," PhD dissertation, University of Florida, 2004; Tucker, Catherine M., Munroe, Darla K., Southworth, Jane, and Harini Nagendra, "Trading Spaces: Following the Impact of Forest-Coffee Conversions on Honduran Livelihoods," *IHDP Update* 1 (2004): 14–5; Willson, Ken C., *Coffee, Cocoa and Tea*, New York: CABI Publishing, 1999.

JANE SOUTHWORTH

Colonization. With respect to human societies, colonization can have two different meanings, political and socio-ecological. The political notion of colonization refers to the imperialistic expansion of nations, with the aim of economic exploitation of the colony's natural resources and the creation of new markets for the colonizer. The term is used mostly in context with the European colonization between the fifteenth century and the middle of the eighteenth, where European dominion was expanded to all of North and **South America** and Australia, most of Africa, and much of Asia. This colonization resulted in a global spread of European production and **consumption** patterns. With regard to **land-use change in a global, historical perspective**, colonization resulted in large-scale changes in land use and in a transfer of European land-use practices and species to **biomes** not present in **Europe**. This often had destructive and unsustainable consequences, such as **soil erosion**, excessive pesticide use, excessive fertilizer use, pests, and irreversible ecosystem damage.

The socio-ecological notion of colonization extends the notion of **ecological colonization** and applies it to human societies. Colonization in this sense refers to the increasing degree of human control over, and consequently human alteration of, natural systems. Since the Neolithic period, human societies have been not merely adapting themselves passively to the prevailing abiotic and biotic conditions, eventually changing ecosystems accidentally in the process; they actively and purposely intervene in patterns and processes in ecosystems in order to render them more useful for societal needs. Examples include the cultivation of selected plant species, the exclusion of species deemed detrimental, and changes in soil fertility and water availability. Many of these measures induce deviations from natural (dynamic equilibrium) states or pathways such as natural succession toward climax associations.

Colonization in the socio-ecological sense has been defined as a bundle of deliberate human interventions into natural processes in order to achieve socio-economic ends. Within this framework, land use can be seen as the colonization of terrestrial ecosystems. Societies expend materials, labor, and technical energy to control and change natural systems and to keep them in a desired state. An important aim of colonizing interventions into terrestrial ecosystems is the provision of materials and energy, above all in the form of **biomass**. Colonization may also be pursued for other reasons, for example, to raise the uptake of carbon dioxide (CO_2) of ecosystems or to establish defined physical conditions such as those in a building.

In the context of land-use/cover change, the socio-economic significance of colonization can be evaluated by assessing the efficiency of colonization from a societal point of view, for example, relating socio-economic expenditures to an intended benefit or resource gained. These expenditures may be in the form of materials, energy, labor, know-how, social organization, **institutions** (e.g., property rights), and the like. Indicators of colonization efficiency analyze, for example, how much energy society gains per unit of energy invested through activities such as **agriculture** or **forestry**. The ecological significance of colonization can be assessed by analyzing changes in natural systems induced by colonization, such as effects on energy flows, changes in biomass stocks, species composition, and other alterations.

See also: **Degradation Narrative; Disequilibrium Dynamics; Domestication; Driving Forces; Economic Growth; Globalization; Human Appropriation of Net Primary Production; Land Rights; Land-Use History; Pristine Myth; Turnover.**

Further Reading

Crosby, Alfred W., *Ecological Imperialism: The Biological Expansion of Europe, 900–1900*, Cambridge, UK: Cambridge University Press, 1986; Fischer-Kowalski, Marina, and Helmut Haberl, "Sustainable Development: Socio-Economic Metabolism and Colonization of Nature," *International Social Science Journal* 158 (1998): 573–87.

KARLHEINZ ERB

Common-Pool Resources. Resources that meet two criteria: first, difficulty excluding potential beneficiaries from their use (non-excludability); and second, use by one person foreclosing potential uses by others (subtractability in **consumption**). Common-pool resources include numerous resources associated with land use, such as **grasslands**, **forests**, and irrigation farming. They pose a particular challenge to management because of the free-rider problem: as resource managers cannot exclude other beneficiaries from use, they cannot fully capture the benefits from management and will not manage the resources at desirable levels.

Common-pool resources are different from private resources, such as cows. The owner of a cow can exclude others from using the cow and enjoying its products fairly easily. In other terms, private resources are excludable, in contrast to the non-excludability that characterizes common-pool resources. Common-pool resources are also different from public resources, such as the air in the atmosphere. The use of a public resource by one person does not affect the possibilities for immediate use by others. Common-pool resources and public resources are similar, however, in the difficulty of excluding others from their use.

The lack of excludability can cause the overexploitation of the resource in a situation of open **access**. Open access occurs if many entities enjoy use rights to a resource, but none of them have the right to exclude others from using the resource. Under open access, no user has an incentive to manage the resource at desirable levels. They do not have an incentive because, for the lack of excludability, other users would free-ride on any improvement they make to the resource. Although all users together would be better off under a regime of restrained use, each user acts rationally by exploiting the resource at maximum level. Open access to common-pool resources, therefore, tends to lead to resource or **land degradation**, the so-called **tragedy of the commons**.

Three approaches have been proposed for overcoming the free-rider problem in the management of common-pool resources: **private property**, state property, and self-governance. Privatization involves the assignment of **property** rights to real and legal persons, which are presumed to engage in market transactions and direct negotiations to manage the resource at desirable levels. State property regimes rest property rights with the state and rely on the state's administrative capacity to manage the resource at stake. Self-governance, also referred to as common property, is a decentralized approach that gives property rights to collective bodies, such as village communities and user groups, for the coordination of resource management.

Self-governance has gained increasing recognition in resource policy around the world. Governments have recognized the capacity of local communities and user groups to regulate the use of common-pool resources, such as grasslands, irrigation systems, and forests. In addition, many governments have enacted legislation that devolves property rights on common-pool resources to local self-governing bodies. In the process, the local bodies receive the authority to enact rules regulating resource use and protection. The actual outcomes of devolution have been highly varied, however. Legal acts have often not effected corresponding changes in actual property rights and rule-making authority. As for resource conditions, devolution has led to resource improvements in some cases but effected resource degradation or depletion in others.

Grasslands are a type of common-pool resource with important implications for land use. Exclusion is difficult as they are expansive and used for ranging on a seasonal basis. Their low productivity renders exclusion via fencing costly relative to benefits. At the same time, grasslands are subtractable in consumption, as the fodder eaten by one animal is not available to others. Under conditions of open access, herders have incentives to overstock the range. They benefit from the fodder gained by putting on additional animals. The costs of their actions—the degradation of the range and future lack of fodder—are born by all herders together. The actual relevance of this theoretical argument for explaining grassland degradation in the Sahel and Central Asia remains the focus of much debate, however. At the same time, the privatization of grasslands has encountered problems. Rainfall variation motivates significant mobility by herders, requiring flexible resource boundaries not compatible with private property rights. Highly variable biophysical conditions and pre-existing local property systems, therefore, have bolstered calls for expanding the scope of self-governance in managing grasslands. Self-governance presumably affords the flexibility local users require in the face of climatic variability and reduces the cost of exclusion.

Irrigation systems are another type of common-pool resource that has attracted much research. Water is subtractable in consumption, as the consumption of water by upstream farmers forecloses the opportunities of downstream farmers to use that water for irrigation farming. Exclusion is difficult in most irrigation systems because users receive the water from the same access canals. Excluding some water users would require cutting of other water users whose fields are supplied by the same canal. Nonexcludability has hampered the efficiency of water allocation and water **conservation** in irrigation farming. It has also led to conflicts among water users within the same village, between villages, and between upstream and downstream users. Privatization has not become a serious policy option because of the interconnectedness of water users. Instead, governments and international donors have promoted irrigation cooperatives and water user associations as self-governing organizations in irrigation management. Cooperatives and associations typically include all farmers within a small administrative area or supplied by one irrigation canal. They coordinate water allocation among their members, monitor water use, enforce internal use regulations, and resolve conflicts.

Forests are common-pool resources in the sense that forests provide various types of resources that are subtractable in consumption, and in the sense that it is difficult to exclude potential users. Exclusion is difficult because forests are often expansive and located in difficult terrain. Exclusion is also difficult for services rendered by forests to national and international communities, such as the provision of **biodiversity** habitats and absorption of carbon dioxide. The common-pool feature, in combination with open access, has been a primary cause of **deforestation** around the world. Farmers, herders, and loggers have degraded and destroyed forests because they could not reap the benefits of protection or **investment**. They have often rushed to claim their share of forest resources before others could do so. **Forest degradation** and deforestation have occurred in spite of widespread state property on forests. In reaction, governments around the globe are now in a process of devolving authority over forests to local state authorities and transferring property rights on forests to local communities and user groups.

Finally, it is important to draw attention to three issues around the concept of common-pool resources. First, the state of technology mediates the biophysical features of common-pool resources. Most importantly, technology shapes the technical possibilities to exclude potential beneficiaries. This becomes very clear in human-made resource systems. For example, in irrigation systems, the difficulty of excluding potential water users is conditioned by irrigation infrastructure. A shift from larger to smaller canals drastically alters the technical possibilities to exclude some users.

Second, the example of forests indicates that a single resource system may concurrently include goods and services of different biophysical features. Forests provide private, common-pool, and public goods and services in many areas of the world. They supply a range of private goods, such as timber and **non-timber forest products**. They harbor common-pool goods such as game and non-timber forest products. Forests also provide important public goods, for example, habitats for rare species and the capacity to absorb carbon. A single resource system, therefore, may include multiple goods and services characterized by different biophysical features.

Last, but not least, the costs of exclusion depend not only on biophysical features and technology but also on property rights. Exclusion may be difficult in a situation of open access. In contrast, exclusion is easy to achieve where property rights are well established, independent of biophysical features and the state of technology. It is therefore important to keep in mind that the common-pool nature of resources may open up and constrain possibilities of management, but that it is not the only determinant of resource use. The use of common-pool resources is conditioned by property rights, just as it influences property rights. Resource use and property rights are mutually constitutive, as conditioned by the biophysical features of the resource at stake.

See also: **Community Involvement; Degradation Narrative; Desertification; Ecosystem Services; Land Reform; Land Rights.**

Further Reading

Gibson, Clark C., McKean, Margaret A., and Elinor Ostrom, eds., *People and Forests: Communities, Institutions, and Governance*, Cambridge, MA: MIT Press, 2000; Ostrom, Elinor, *Governing the Commons: The Evolution of Institutions for Collective Action*, Cambridge, UK: Cambridge University Press, 1990; Ostrom, Elinor, and Michael McGinnis,

Workshop in Political Theory and Policy Analysis [Online, October 2004], Workshop Web Site, www.indiana.edu/~workshop.

THOMAS SIKOR

Community Involvement. This form of involvement receives increasing recognition of the critical role that it can play in managing land-cover change. Historically, traditional community-based systems of land management by indigenous groups were predominant throughout much of the world. In most countries, the **land rights** of these traditional **institutions** were later derecognized and largely replaced by formal state and private ownership. The "**tragedy of the commons**" expressed the widely held belief that community involvement in land use and management had led to disastrous, free-for-all open **access** situations, and exacerbated problems of land-cover change and **land degradation**. Case studies across the world have now clearly demonstrated that no single type of ownership, whether private, community, or government, is by itself an automatic guarantee of effective management. When community management boundaries are well defined, legitimate, and effectively enforced, however, the social capital generated through community involvement can be very effective for sustainable development and **conservation** over the long term, especially on smaller, local, or regional **scales**.

See also: **Land-Use Policies; Mediating Factor; Participatory Geographic Information System; Participatory Land Management; Property; Reforestation.**

Further Reading

Bray, David Barton, Merino-Pèrez, Letitia, Negreros-Castillo, Patricia, Segura-Warnholtz, Gerardo, Torres-Rojo, Juan-Manuel, and Henricus F. M. Vester, "Mexico's Community-Managed Forests as a Global Model for Sustainable Landscapes," *Conservation Biology* 17 (2003): 672–7; Dietz, Thomas, Ostrom, Elinor, and Paul Stern, "The Struggle to Govern the Commons," *Science* 302 (2003): 1907–12; Gibson, Clark A., McKean, Margaret A., and Elinor Ostrom, *People and Forests: Communities, Institutions, and Governance*, Cambridge, MA: MIT Press, 2003.

HARINI NAGENDRA

Conservation. In the context of land use and land cover, the conservation of soil, water, and nature are important in that they influence land-use decisions, can change land uses and land cover, and their effects can impact land use and land cover over great distances. The relevance of soil and water conservation to land-use and land-cover change lies in two areas: first, the impact of land-use change in increasing erosion rates; and second, in the way that conservation affects land-use decisions and land-cover outcomes.

As natural **vegetation** is converted to rangelands and croplands, the dynamics of **soil erosion** change. Wind and water erosion of soils is in dynamic equilibrium with the environment in areas of natural vegetation: this is known as normal or geologic erosion. When vegetation is disturbed, erosion rates often increase for three reasons: because vegetation canopies are disturbed or destroyed; resistance to erosion in the rooting zone of plants is reduced; and less organic matter is recycled into the soil. This is termed accelerated soil erosion. Normal erosion rates are greatest in mountainous areas because of the effects of steep and—often sparsely vegetated—slopes, and in semi-arid areas due to the synergistic effect of high frequencies of intense precipitation and sparse vegetation cover. Clearly accelerated rates of erosion are high when cultivation and ranching occur in semi-arid and mountainous areas, but they are high also when **forests**—particularly sub-tropical and **tropical humid forests**—are cleared. Paleo-environmental and historical data show that human impact on vegetation has led to increased erosion and **siltation** in lakes and rivers in the past. Stanley Trimble illustrated how the piedmont of the eastern **United States of America** was progressively eroded at the hands of European agricultural settlers from the sixteenth century onward. Well-documented episodes such as the Dust Bowl of the 1930s in the southern **Great Plains** of the United States, and in the steppes of southern **Russia** in the late nineteenth century illustrate the catastrophic decline in soil productivity, and the subsequent dire socio-economic consequences as semi-arid rangelands were converted into croplands by the expansion of settlement. Contemporary rates of accelerated soil erosion are high in most of the world's high **mountain** ranges (e.g., the **Andes** and **Himalayas**); in semi-arid parts of the Americas, southern **Europe**, Africa, Asia, and Australia; where humid tropical forests have been cleared in **Latin America** and the Caribbean, West Africa, southern and Southeast Asia; and in the loess plateau in northern **China**.

Terraces, where trees will be planted to control soil erosion, on a steep hillside in the Palouse wheat country of the Pacific northwest in the United States of America, 1934/35. [Courtesy Library of Congress]

The Dust Bowl in the 1930s led to the creation of the Soil and Water Conservation Service (SWCS) of the U.S. Department of Agriculture. SWCS began scientific research, which has subsequently also been undertaken in many other countries, into erosion and the development of conservation methods that are now applied globally in croplands and rangelands.

Soil and water conservation aims to reduce the effects of accelerated soil erosion and is mainly restricted to productive landscapes, for example, croplands and pastoral systems. However, it can also be applied to commercial **forestry** and in large infrastructure developments, such as major highways.

Reduction in soil loss is usually achieved through either mechanical or agronomic/biological conservation methods. Mechanical conservation methods for soil erosion by water aim to reduce slope lengths and gradients to reduce soil losses; many methods are used, the most common being terracing, contour cultivation, and strip cropping. For wind erosion, such methods aim to reduce wind velocities at the atmosphere-soil interface by establishing windbreaks or deep ploughing to increase surface roughness. Agronomic and biological conservation aims to manipulate crops and cropping practices to reduce raindrop impact or wind speeds. Methods include replacing erosion-promoting crops (e.g., many types of cereals) with erosion-retarding crops (e.g., leafy crops such as beans).

Water conservation aims to reduce runoff from fields into rivers. Surface runoff is a major erosive agent in soil erosion by water, and its control aids the reduction of soil losses from fields. Successful control of soil erosion by water reduces the amounts of topsoil, agrochemicals, and organic matter entering watercourses; thereby reducing siltation and the buildup of chemicals—which causes **eutrophication**—in rivers, lakes, and estuaries downstream in **river basins**.

In considering the impacts of soil conservation on land use and land cover in croplands and rangelands, it is at the fine **scale** in the landscape that the effects are seen, for example, planting linear windbreaks, leaving uncultivated or **fallow** strips of a few meters width in strip-cropping systems, or the adoption of minimum tillage systems. More complex impacts on land-use and land-cover patterns at broader regional and national scales also exist. These are related to the links between the adoption of conservation and the socio-economic characteristics of farms, and the effects of conservation of overall food production levels. For example, the gains in agricultural production that the adoption of soil conservation implies may enable less natural vegetation to be converted to croplands at **agricultural frontiers**, but this effect could be masked by other productivity gains due to agricultural modernization.

Formal efforts to conserve species and habitats (nature conservation) have a history dating back to the nineteenth century in **Europe**. Habitat conservation has had a greater impact on land use and land cover globally than species conservation efforts. There are two main approaches to habitat conservation: the creation of protected areas, and conservation-friendly productive landscapes in existing agricultural areas.

The creation of protected areas initially developed along the lines of the **Yellowstone model**—a model that, when applied to the design of nature reserves, makes strong assumptions about the exclusion of people and restrictions on their **access** to land and other natural resources. Early national parks were established throughout the world using the Yellowstone model in the first part of the twentieth century. However, the strict interpretation of the Yellowstone model has been modified throughout the twentieth century as the rights of indigenous people have been recognized, and as habitat protection was introduced to densely populated areas, for example, **western Europe**. Since 1994, the International Union for Conservation of Nature and Natural Resources (IUCN) has recognized six categories of protected areas (see table). The IUCN reported 102,102 protected areas covering 18.8 million km^2 at the fifth World Parks Congress in 2003, a total area covering 11.5 percent of the earth's surface. Only IUCN categories Ia and Ib have strong elements of exclusion, though access restrictions can be high for II and IV as well. The basis for categories V and VI is that people have and continue to live in these protected areas. However, all protected areas, if well funded and adequately managed, do pose restrictions on people living in and around them. These planning regulations influence the types of **agriculture** and forestry activities that are practiced, and "fossilize" patterns of land use and land cover in some. Protected areas not only impact the parks themselves, but adjacent areas as well. Access regulations for some types

Categories of protected areas	
I	(Ia) Strict nature reserves: for nature protection and scientific enquiry (Ib) Wilderness areas: for wilderness protection
II	National parks: for ecosystem protection and recreation
III	National monuments: for the conservation of natural features such as rock formations
IV	Habitat and species management areas: for wildlife conservation
V	Protected landscapes or seascapes: usually managed for recreation
VI	Managed resource protected area: for the sustainable use of natural resources

Source: International Union for Conservation of Nature and Natural Resources.

of protected areas lead to communities of people who have been evicted from the parks or have had their access restricted to areas around park perimeters. **Agricultural intensification** occurs, and these areas can become degraded, even if (illegal) access of natural resources within the protected area continues.

A strong theme in nature conservation from the 1980s onward, particularly in the developing world, has been the rejection of protectionism based on the Yellowstone model and the use of conservation-with-development. Where conservation-with-development schemes have been implemented, the severe restrictions placed on land use have been relaxed, but land-use **decision-making** is still regulated to some extent.

Simultaneously, in the developed world, there have been moves toward promoting conservation within productive (both agricultural and forested) landscapes. The European Union's adjustments to the Common Agricultural Policy to reduce overproduction of foodstuffs, and set-aside and habitat and species biodiversity action plans in the 1990s in the United Kingdom are examples of such policies. Initiatives based around the policies have promoted buffer strips adjacent to streams to reduce agrochemical flows into watercourses and seed bank strips in fields to augment **biodiversity**. These have similar effects on land use and land cover on a small scale, as wind breaks and strip in soil and water conservation. Such developments have not yet been seen in all developed world croplands, or in the developing world.

See also: **Agrodiversity; Arena of Land Conflict; Biotechnology; Carson, Rachel Louise; Community Involvement; Decentralization; Decision-Making; Degradation Narrative; Desertification; Fire; Hot Spot Identification; Hot Spots of Land-Cover Change; Land Degradation; Land Privatization; Land Reform; Land Rehabilitation; Land-Use History; Land-Use Policies; Marsh, George Perkins; Mediating Factor; Mineral Extraction; Mountains; Pristine Myth; Project on People, Land Management and Ecosystem Conservation; Public Policy; Reforestation; Regrowth; Savannization; Soil Erosion; Swidden Cultivation; Tropical Dry Forest; Urbanization.**

Further Reading

Adams, William M., *Against Extinction: The Story of Conservation*, London: Earthscan, 2004; Toy, Terence J., Foster, George R., and Kenneth G. Renard, *Soil Erosion: Processes, Prediction, Management and Control*, New York: John Wiley and Sons, 2002;

Trimble, Stanley W., *Man-Induced Soil Erosion on the Southern Piedmont, 1700–1970,* Ankeny, Iowa: Soil Conservation Society of America, 1974.

ANDREW MILLINGTON

Consultative Group on International Agricultural Research. An alliance of countries, international and regional organizations, and private foundations supporting fifteen autonomous international agricultural research centers (collectively known as "Future Harvest Centers") that work with national research systems, the private sector, and civil society. The CGIAR mobilizes agricultural science to reduce **poverty**, foster human well-being, promote agricultural growth, and protect the environment. Its mission is to contribute to **food security** and poverty eradication in developing countries through research, partnership, capacity building, and policy support, promoting sustainable agricultural development based on the environmentally sound management of natural resources. The CGIAR formed in 1971 as an outgrowth of earlier initiatives led by the Ford and Rockefeller foundations. The initial emphasis was on major cereals. CGIAR-sponsored research, and that of its partners, has produced more than 500 new varieties of **rice**, **wheat**, and **maize**. Known as the **Green Revolution**, adoption of these varieties significantly raised world food production and **consumption** in developing countries and helped reduce malnutrition for millions of children, particularly in Asia. But the successes of the 1970s and 1980s revealed new challenges, including the need to address poverty reduction and environmental issues directly. In response, the number of CGIAR centers was increased and the scope of activities expanded to include farming systems research; policy and institutional support; livestock; **forestry**; **agroforestry**; and sustainable management of natural resources, including land, water, and biological diversity. The CGIAR holds in trust the world's largest collection of plant genetic resources. A 1998 external review concluded that **investment** in the CGIAR had been the single most effective use of official development assistance, bar none. The **Alternatives to Slash-and-Burn Programme** is one of CGIAR's systemwide programs.

See also: **Agricultural Revolution; Biodiversity; Center for International Forestry Research; Conservation; Food Crops; International Food Policy Research Institute; Public Policy; Sustainable Land Use; World Agroforestry Centre.**

Further Reading

Consultative Group on International Agricultural Research, Nourishing the Future through Scientific Excellence [Online, November 2004], The CGIAR Web Site, www.cgiar.org.

THOMAS P. TOMICH

Consumption. Consumption of resources is a significant underlying factor (or driver) of land-use/cover change. Since the **Industrial Revolution**, when raw materials began to be shipped great distances for manufacturing in **Europe** and North America, the landscapes even in remote parts of the world were transformed by the economic relationships between the core and periphery. Coal, iron, wool, **cotton**, sugarcane, **wheat**, and wood pulp—not to mention countless other commodities—are extracted or produced through the land-based activities of **mineral extraction** (mining), **agriculture**, and **forestry**. Economic **globalization** and growth in consumer demand have only intensified these linkages, and the demand for raw materials continues to grow. Simultaneously, **urbanization** has led to increasing expectations about consumption and a potentially weakened understanding of production-consumption relationships. Economically and politically powerful urban consumers tend to be disconnected from the realities of resource production and largely inattentive to the impacts of their consumption on distant locales. It is estimated that humans appropriate approximately 20–30 percent of global terrestrial net primary productivity.

Although consumption includes the totality of all human end use of materials, some have found it useful to distinguish between the consumption of essentials and the consumption of luxury goods. Gerhard Heilig, for example, has examined the impact of lifestyle-related consumption on global land cover, estimating that more than 214 million hectares are used for the production of lifestyle-related products, such as stimulants, sugar, **tobacco**, oilseeds, and soy (which is used primarily for animal feed). Including the unknown area in illicit drug production, he suggests that up to 20 percent of the world's arable land area may be used for such "luxury" crops.

An identity known as IPAT (environmental impact = population × affluence × technology) has been used to assess the relative contribution of consumption (affluence) to anthropogenic impacts on the environment. Although **population dynamics** and technology have been studied explicitly as drivers of **land-cover change**, consumption has received less attention. This may be due to differing definitions of what is meant by consumption within and between the social and natural sciences, and related difficulties of actually measuring it. Even if one accepts the economist's definition of consumption as total spending on goods and services (as opposed to that portion of income that is saved or invested), data on spending are often difficult to obtain.

See also: **Agricultural Intensification; Coffee; Colonization; Driving Forces; Ecological Footprint; Economic Growth; Economic Restructuring; Systemic Change.**

Further Reading

Heilig, Gerhard K., "Neglected Dimensions of Global Land-Use Change: Reflections and Data," *Population and Development Review* 20 (1994): 831–59; Imhoff, Marc L., Bounoua, Lahouari, Ricketts, Taylor, Loureks, Colby, Harriss, Robert, and William T. Lawrence, "Global Patterns in Human Consumption of Net Primary Production," *Nature* 429 (2004): 870–3.

ALEXANDER DE SHERBININ

Continuous Data. Data that are marked by uninterrupted extension in space or time. **Remote sensing** is an attractive source of land-cover data, as it provides a representation of the earth's surface that is spatially continuous and highly consistent. Examples of continuous data of use to land-cover analyses are **vegetation** indices, for example, the **normalized difference vegetation index**, the tasseled cap analysis, and thermal band data providing surface temperature images, etc. Such data include significant information about the landscape, its variability in space or time, and its surface composition, and as such they are of paramount importance to land-use/cover-change studies. Current reliance on more traditional land-cover classification-type approaches, in much of the **modeling** and fragmentation studies within the land-change arena, may thus be limiting.

See also: **Change Detection; Discrete Data; Probit/Logit Model; Thermal Band Analysis.**

Further Reading

DeFries, Ruth S., Hansen, Matt C., and John R. G. Townshend, "Global Continuous Fields of Vegetation Characteristics: A Linear Mixture Model Applied to Multi-Year 8-Km AVHRR Data," *International Journal of Remote Sensing* 21 (2000): 1389–1414; Lambin, Éric F., "Monitoring Forest Degradation in Tropical Regions by Remote Sensing: Some Methodological Issues," *Global Ecology and Biogeography* 8 (1999): 191–8; Southworth, Jane, Munroe, Darla, and Harini Nagendra, "Land Cover Change and Landscape Fragmentation: Comparing the Utility of Continuous and Discrete Analyses for a Western Honduras Region," *Agriculture, Ecosystems and Environment* 101 (2004): 185–205.

JANE SOUTHWORTH

Contract Farming. A land-use-related element in the **corporate strategies** of large commercial (mainly agricultural) enterprises that seek to gain control over operations at the individual farm level. Contract farming is part of a still-larger strategy in a particular business configuration in which a single and large (usually transnational) corporation coordinates industrial activities through subsidiary companies in the spheres of manufacture and distribution of farm supplies, as well as storage, processing, and distribution of farm commodities and items made from them. In the **tobacco** business, for example, a limited number of multinational tobacco companies act as banks, extending credit to farmers at the beginning of the season (for seed, fertilizer, pesticides, and technical support); in return, farmers agree to sell their crops to these companies at harvest; company inspectors make regular visits to the farms to make sure their guidelines are followed; when the harvest is in and the time comes to pay the farmers, leaf buyers determine the leaf grade and the price; as a common result, they often end up buying producers less than the original loans. Multinational cigarette companies—such as Altria, Japan Tobacco, Philip Morris, British-American Tobacco, and Imperial Tobacco—have made a first step to shifting their manufacturing capacities into low- and middle-income countries where most of today's tobacco is grown and where new and large consumer markets are developing. As these corporations build new factories, enter into joint ventures, and buy formerly state-owned factories, at the same time transnational leaf dealers—such as the three U.S.-based companies Dimon, Standard Commercial, and Universal, which dominate the global world leaf market—have followed; that is, they source tobacco around the world, ensure quality control, and then sell to manufacturers. They work with the major tobacco manufacturers to determine which countries will produce how much tobacco leaf and what kind, regardless of their impacts on local growers and growing environments. Contract farming, as the contractual element in the global **agro-industry**, demonstrates how closely agrarian and **environmental changes** are linked to technological, corporate, and other institutional changes at a truly global **scale**. In both the non-food and agro-food sectors, global change follows the rules of vertical integration, which is a process involving the extent to which successive stages in production and distribution are placed under the control of a large, single firm.

See also: **Decision-Making; Economic Restructuring; Land Rights.**

Further Reading

Little, Peter D., and Michael J. Watts, eds., *Living under Contract: Contract Farming and Agrarian Transformation in Sub-Saharan Africa,* Madison, WI: University of Wisconsin Press, 1994.

HELMUT GEIST

Conversion. Complete replacement of one type of land cover by another. Land-cover conversions are measured by a shift from one land-cover category to another, as is the case in agricultural expansion, **deforestation**, or change in urban extent. In general, land cover is defined by the attributes of the earth's land surface and immediate subsurface, including biota, soil, topography, surface and groundwater, and built-up structures created by humans. Data sets used in **land-cover change** research represent the land surface by a set of spatial units, each associated with attributes that are either a single land-cover category—that is, leading to a discrete representation of land cover—or a set of values for continuous biophysical variables, that is, leading to a continuous representation of land cover. The use of **discrete data** as opposed to **continuous data** has the advantages of concision and clarity, but it has led to an overemphasis on land-cover conversions and a neglect of land-cover **modification**.

See also: **Change Detection; Hot Spots of Land-Cover Change.**

Further Reading

Lambin, Éric F., and Helmut Geist, eds. *Land Use and Land Cover Change: Local Processes, Global Impacts.* Berlin: Springer, 2006.

HELMUT GEIST

Conversion of Land Use and Its Effects (CLUE) Model. Developed at Wageningen University in the Netherlands and first published in 1996, this model is a tool used to better understand the processes that determine changes in the spatial pattern of land-use change and explore possible future changes in land use at the regional scale. The CLUE model consists of two main components. The first supports a spatially explicit methodology

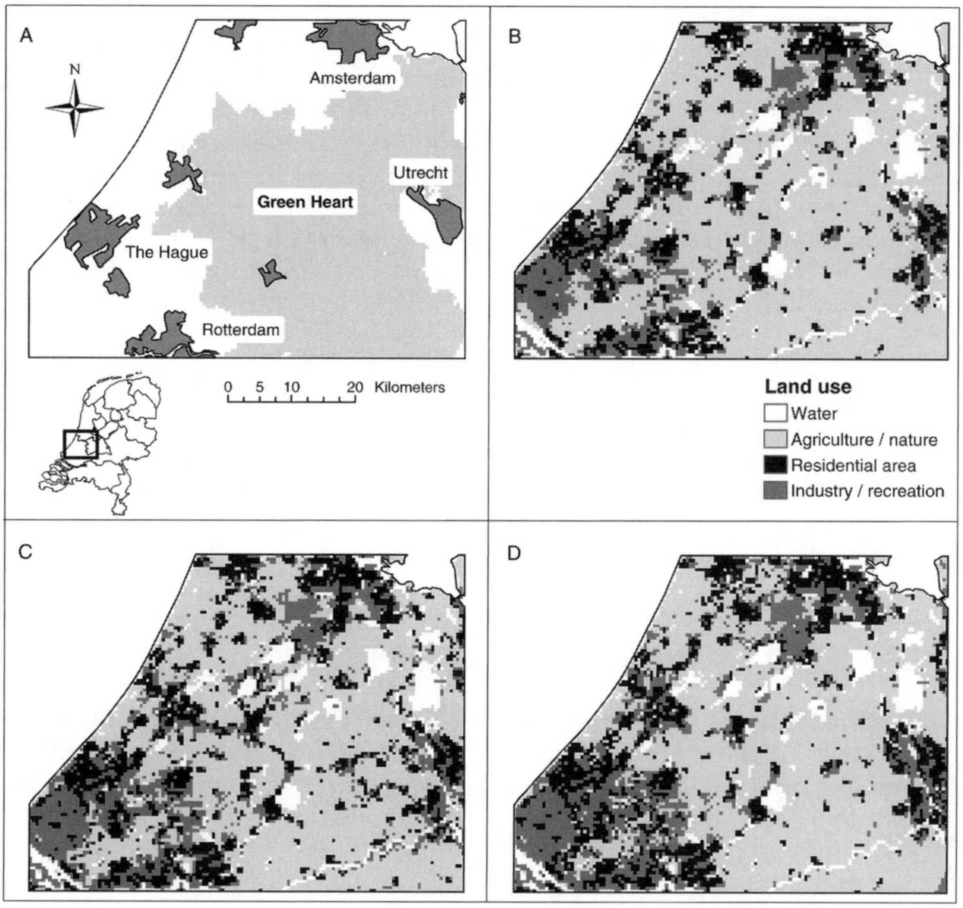

Sample of an application of the CLUE model for the Randstad region within the Netherlands. (A) Location of the Randstad, in light grey the Green Heart area is indicated, (B) Initial land use in 1996, (C) Land use in 2015 for a base scenario, and (D) Land use in 2015 for a scenario with protection of the Green Heart as a green area.

to analyze relations between land-use patterns and their driving factors. Multiscale spatial analysis is used to empirically establish quantified relations between the spatial distribution of land use and spatially distributed factors that represent **proximate causes** and/or underlying **driving forces**. The second component uses the results of this analysis in a dynamic simulation technique to explore changes in land use under scenario conditions. Allocation of future land-use change is based on these relations supplemented with region-specific constraints and land-use type-specific conditions, including time lags, spatial interactions, conversion elasticities, and fixed land-use change trajectories.

The model is widely applicable at resolutions ranging from the subregional (100 km^2) to continental **scales**. At the time of this writing, approximately thirty applications are made, spread over the different regions of the globe and focusing on a wide range of land-use change trajectories including **agricultural intensification**, **deforestation**, **land abandonment**, and **urbanization**.

Most applications result from a collaboration between the core group at Wageningen University and a local or regional university or institute.

The CLUE model is also used within **scenario** studies to facilitate the analysis and visualization of the effects of land-use change. The figure provides a sample of the use of the model for an urban region in the Netherlands. Based on the resulting changes in land-use patterns, the potential effects of land-use change on different sectors can be assessed, for example, landscape fragmentation and **vulnerability** to **land degradation**. The CLUE model can be obtained freely while specific applications and support are provided on a consultancy basis.

See also: **Change Detection; Land-Use System; Modeling.**

Further Reading

Verburg, Peter H., Soepboer, Welmoed, Limpiada, Ramil, Espaldon, Maria V. O., Sharifa, Mastura, and A. (Tom) Veldkamp, "Land Use Change Modeling at the Regional Scale: The CLUE-S Model," *Environmental Management* 30 (2002): 391–405; Wageningen University, Environmental Sciences, CLUE: Conversion of Land Use and Its Effects [Online, February 2005], The CLUE Web Site, www.cluemodel.nl.

PETER H. VERBURG

Coordination of Information on the Environment (CORINE) Database. A pan-European database of biophysical land cover using forty-four classes and available freely from the European Environment Agency. The aim of the CORINE database is to provide qualitative and quantitative data on land cover to those responsible for and interested in European policy on the environment. The data are consistent and comparable across the continent of **Europe**. The database has been made available in a 250 x 250 meter grid format and has been used to create maps of habitats, areas at risk from **urbanization**, land-use/cover change in **coastal zones**, and the location of former agricultural land that has been abandoned.

See also: **Biodiversity; Joint Research Centre (JRC) of the European Commission; Land Abandonment.**

Further Reading

Commission of the European Community, *CORINE Land Cover: Technical Guide*, Luxembourg: Office for Official Publications of the European Community, 1994; European Environment Agency [Online, August 2004], Coordination of Information on the Environment Web Site, www.eea.eu.int; Perdigão, Vanda, and Susan Christensen, eds., *The Lacoast Atlas: Land Cover Changes in European Coastal Zones*, Ispra, Italy: Joint Research Centre of the European Commission, Space Applications Institute, 2000.

KEVIN J. TANSEY

Corporate Strategies. The corporation is a specific form of business organization that emerged in the **United States of America** at the end of the nineteenth century, exemplified by the large, vertically integrated firm. The term describes an entity with legal rights, privileges, and liabilities separate from those of its members, and is often used as a synonym for a large business or firm. Strategy refers both to the direction and scope of an organization over the long term, and to the deliberate actions the organization takes to achieve advantage by configuring its resources to meet market opportunities and fulfill the needs of shareholders. Corporate strategy, then, describes the processes that firms undertake to position themselves advantageously with regard to opportunities, competitors, and resources; it consists of efforts to both determine specific ends and marshal the means to achieve them. Corporate strategy covers a wide range of tasks (marketing, investment, procurement) and perspectives (analysis, planning, evaluation), many of which can have implications for the environment.

A focus on corporate strategy produces an account of the economy in which corporations are the primary actors and in which events can be understood as the outcomes of rational **decision-making**. It assumes that the most productive analytical approach is one that focuses on the differences between corporations, and that these differences can be best explained in terms of strategy. This systematically depresses both structural conditions common to many corporations and the role of non-corporate actors (states, consumers, non-governmental organizations) in contesting and negotiating the terms and content of corporate strategy. Nonetheless, corporate strategy can be a productive entry point for linking economic and **environmental change** in three ways.

First, the size and global reach of corporations make them a critical part of the economy. Since large manufacturing and processing companies control much of the materials transformation within the economy, they represent a significant component of the total anthropogenic flux in global biogeochemical cycles (e.g., sulfur, carbon, nitrogen) and an even greater contribution to the production of persistent organic pollutants. Corporate **agriculture** also exerts extensive control over land use and land cover both directly (through the management of acreage) and indirectly (through purchasing strategies; contract conditions; the development of inputs like fertilizers, seed stocks, and pesticides; and product and production specifications). The direction and content of corporate strategies therefore exert a decisive influence over efforts to achieve industrial transformation and ecological modernization.

Second, strategy needs to be understood comprehensively as the full range of actions by firms, rather than the much more restricted focus on "corporate environmental strategies" that characterizes the business and management literature. Corporate strategy includes critical decisions over where, when, and how to invest; over technology choice; product development; and land acquisition and land-use strategies. While not generally thought of as environmental strategies, each of these can influence the rate and form at which materials are used or the extent to which economic activity modifies the environment.

Third, the notion of "the corporation" as a discrete unit needs to be challenged by emphasizing the relationships between corporations and other components of the economy and society. Corporate strategy, for example, is often closely linked to the needs and perceptions of the investment community: the geographical expansion of mining or oil companies, for example, is driven by a need to control **access** to the limited number of world-class (i.e., large-volume, low-recovery-cost) deposits in order to remain competitive. It is by leveraging this sensitivity to the **perceptions** of the investment community that non-governmental organizations have sought to redirect the scope or direction of corporate strategy in several cases.

See also: **Agro-Industry; Contract Farming; Economic Growth; Economic Restructuring; Investment; Public Policy.**

Further Reading

Johnson, Gerry, and Kevan Scholes, *Exploring Corporate Strategy: Text and Cases,* New York: Prentice Hall, 2001.

GAVIN BRIDGE

Cotton. Typically grown in the semi-arid and humid **grasslands** of the tropics, sub-tropics, and in some areas of the temperate zone under rainfed and irrigated conditions (generally between 37°N and 32°S), cotton is cultivated in more than 100 countries on approximately 2.5 percent of all arable

Picking cotton in some of the poorer land of the cotton zone near Clarksdale, Coahoma County, in the Mississippi Delta, November 1939.
PHOTO: Marion Post Wolcott. [Courtesy Library of Congress]

land. This makes it one of the world's most significant crops in terms of surface area, after food grains and soybeans. The total area globally devoted to cotton has fluctuated, yet shown no tendency to permanently rise since the 1950s (as climbing global production has been more closely related to improving yields). There have been significant reductions in surface area devoted to the crop in some areas of the world, particularly the **United States of America**, northern Brazil, and North Africa, as well as equally dramatic increases in other regions, especially Francophone Africa, Australia, India, Pakistan, and the Middle East. The most commonly cultivated cotton species is *Gossypium hisutum*, or upland cotton.

The textile industry was one of the first manufacturing activities to become organized on a global scale, with mechanized cloth production initially based in **Europe** (acting as a major component of the European **Industrial Revolution**) and cotton grown in the colonies (North America, Asia, and eventually Africa). About 300 years later, cotton cultivation and textile manufacture are still organized globally, albeit with a different spatial distribution.

Over the past three decades, the four leading producers of cotton generally have been **China**, the United States of America, India, and Pakistan. Production gains in these countries have largely been related to increasing yields. Cotton production (but not necessarily surface area) has diminished most dramatically in countries of the former Soviet Union (particularly Uzbekistan), declining from 19 percent of global production in 1980–1981 to 8 percent in 2002–2003.

Generous estimates put human involvement with the production of this crop at one billion people (when family labor; hired farm labor; and workers involved in the transportation, ginning, bailing, and storage of cotton are taken into consideration). There have been heated international debates in recent years concerning the effects of cotton production subsidies in developed countries on the development of some less-developed nations, particularly Francophone Africa.

Other than the expansion and contraction of the surface area it occupies, cotton is associated with at least three environmental issues: pesticide use, **soil degradation**, and problems related to irrigation farming (particularly **salinization** and excessive water diversion).

As improved varieties of cotton are vulnerable to a number of insect pests, particularly foliage feeders and bullworms, the crop frequently is grown with regular pesticide applications. For example, E. Qerke and colleagues estimate that African cotton producers are at risk of losing 85 percent of their cotton crop to pests and diseases, but only sacrifice about 50 percent because of pesticide use. Cotton accounts for more pesticide use than any other crop, about 10 percent of global pesticide consumption when measured by weight of active ingredients.

There are a number of health and environmental concerns related to cotton-associated pesticide use, including human illness and death, loss of plant pollinators, and pesticide resistance. The use of cotton-related pesticides has been a **human health** concern in some developing countries, particularly when safety instructions are not accessible to users (either because of illiteracy or inability to translate a foreign language). Heavy pesticide use has a negative impact on the majority of plant pollinators, especially bees. Bees, for example, not only pollinate cotton, but a number of important **food crops**. Pesticide resistance, the genetic resistance of insect populations to certain pesticides, is an increasing problem in many cotton-growing regions of the world. Globally, over 500 species of insects and mites are resistant to one or more insecticides. Use of cotton-related pesticides may decline with expanded use of genetically engineered cotton varieties (it is estimated that 21 percent of world cotton area was planted in genetically engineered varieties in 2003–2004). However, the use of transgenic cotton may not resolve all of the environmental issues related to pesticide applications. For example, there is concern that the use of Bt cotton will lead to Bt resistance among bullworms.

Cotton is also a very demanding crop in terms of its use of soil nutrients, particularly nitrogen, phosphorus, and potassium. Because of these high nutrient demands, farmers typically apply mineral fertilizers to maintain or bolster yields. However, there are at least three potential problems with the application of mineral fertilizers. First, when mineral fertilizer is applied, increasing yield may come at the expense of simultaneously greater extraction of nutrients, as well as increased losses due to leaching and volatilization. Second, when ammonium fertilizer or urea is applied without regular addition of lime, soils become more acidic (soil acidification). This results in lower yields (because soil acidification makes phosphorus less available), and ultimately in aluminum toxicity. Finally, the runoff of fertilizers into nearby streams, rivers, and lakes may lead to **eutrophication**.

Approximately 55 percent of the world's cotton area is irrigated with water from canals or pumped from underground. The remaining 45 percent is either partially irrigated or entirely dependent on rainwater. With irrigation comes the possibility of salinization, groundwater depletion, and the adverse consequences of excessive diversion of surface water.

One of the most dramatic examples of an ecosystem decimated by cotton-related irrigation is the Aral Sea basin. Soviet planners worked to expand irrigated **agriculture** in Central Asia in the 1950s and 1960s. High on their list of export crops was cotton (made a priority in order to garner hard currency). In 1956, the Kara Kum canal opened, diverting large amounts of water from the Amu Darya (one of two major rivers that flow into the Aral Sea) into the Turkmenistan desert. As the Aral Sea shrank, its salinity increased. Heavy pesticide use in the basin led to increasing concentrations of these chemicals in the water body. As a consequence of both these factors, the formerly large commercial fish catch has shrunk to almost nothing today. The declining water level in the Aral Sea also led to a regional drop in the water table. The regional climate began to change (becoming more continental) as an increasingly large area of the former seabed became exposed. The area now also experiences increasingly severe dust storms.

In addition to studies of cotton cultivation as a form and source of land-use/cover change, a number of investigations have been undertaken to understand the influence of climate change on cotton production. The results of these studies appear to be mixed, with increased warmth and atmospheric carbon dioxide augmenting yields in some settings and depressing them in others.

See also: **Agricultural Intensification; Agricultural Revolution; Biotechnology; Brazilian Cerrado; Cash Crops; Colonization; Consumption; Degradation Narrative; Desertification; Economic Liberalization; Great Plains; Industrialization; Nutrient Cycle.**

Further Reading

Bassett, Thomas J., *The Peasant Cotton Revolution in West Africa: Côte d'Ivoire, 1880–1995,* London: Cambridge University Press, 2001; Bingen, Jim, "Pesticides, Politics and Pest Management: Toward a Political Ecology of Cotton in Sub-Saharan Africa," in *African Environment and Development: Rhetoric, Programs, Realities,* eds. William G. Moseley and B. Ikubolajeh Logan, Burlington, VT: Ashgate Publishing, 2004, pp. 111–26; Clay, Jason, *World Agriculture and the Environment: A Commodity-by-Commodity Guide to Impacts and Practices,* Washington, D.C.: Island Press, 2004; Glantz, Michael H., Rubinstein, Alvin Z., and Igor Zonn, "Tragedy in the Aral Sea Basin: Looking Back to Plan Ahead?" *Global Environmental Change* 3 (1993): 174–98; International Cotton Advisory Committee [Online, September 2004], Cotton Web Site, www.icac.org; Oerke, Erich-Christian, Dehne, Heinz-Wilhelm, Schönbeck, Fritz, and Adolf Weber, *Crop Production and Crop Protection,* New York: Elsevier, 1999.

WILLIAM G. MOSELEY

Cultural Factors. Synonym for socio-political factors that underpin processes of land-use/cover change, or—in more general terms—constitute social processes behind the human relationship to the global environment. From all **driving forces** of land change, cultural factors are most difficult to quantify, because they usually interact in a synergistic manner with other forces such as demography, technology, economy, **institutions**, and policy. They actually shape the creative milieu in which, for example, technological innovation or fertility control happens. The term "culture" usually describes attitudes, value patterns, ideas, norms, and knowledge that shape human behavior. There are also cultural ecosystems services that encompass the nonmaterial benefits that people obtain from ecosystems through spiritual enrichment, cognitive development, reflection, recreation, and esthetic experience; including, for example, knowledge systems, social relations, and esthetic values.

Ronald Inglehart has demonstrated how the phenomenon of cultural change is closely interlinked with social, political, and economic change (synergistically underlying **environmental change**). The question "which factor really explains what is happening?" is, according to Ronald Inglehart, meaningless as long as indicators of change are not analyzed in a comprehensive, all-encompassing way that includes economic, technological, social, political and, above all, cultural determinants. From the perspective of post-materialism as a cultural attribute, it can thus be argued that cultural change has dramatically affected the highly industrialized, advanced societies in **western Europe** in the second half of the twentieth century, a long period of impressive **economic growth** and peace resulting in a shift of major attributes of life satisfaction from socio-economic values (e.g., material growth, social security) toward cultural values such as self-expression and quality of life—that is, a shift from materialistic orientation to one that prefers non-materialistic ambitions. Post-materialists emphasize protecting the environment, they participate actively in environmentalist organizations, and they even earn and accept considerably lower incomes if the quality of life matches their expectations. This cultural shift exerts pressure on policymakers responsible for the locational management of places and the promotion of economic development. While traditional econometric theories and models assert that economic growth originated from companies, labor, capital, technology, or natural resources, there are claims that today's key resource is people; above all, creative people. In western societies, at least, economic growth and the competitiveness of regions and nations are strongly interrelated with the ability of policymakers to attract, cultivate, and mobilize creative people who are environmentally conscious, attach great importance to the quality of life, and esteem a multicultural, tolerant, and inspiring environment for living. Thus, the focus of **public policy** for the promotion of regions and nations has changed. The key dimension for policymakers is now to understand how to attract a substantial mass of creative, entrepreneurial people who are the promoters of post-materialistic societies such as in France, Germany, or the **United States of America**.

In other words, cultural change is influencing both the speed and type of economic development of societies; it shapes the reasons why and how people contribute to economic development; it sets the stage for the level of entrepreneurial activity that people plan to attain; and it is crucial for why people participate in social movements and environmentalist organizations, which kinds of parties they support, and which forms of sociopolitical activities they choose. In a nutshell, cultural factors are crucial for a variety of societal factors or subsystems of society, with potential interactions between the organizational levels of society and between society and the environment. In the study of land change as a forcing function in global environmental change, the U.S. Committee on Global Change Research proposes to make a difference between public values, attitudes, and beliefs, and individual and household behavior.

These categories have been applied to the study of tropical **deforestation**, using a meta-analytical model. Cultural factors of the public domain, as associated with deforestation, have been subdivided into three categories: public non-concern or lack of support for forest protection and **sustainable land use**—such as a frontier mentality, low morale or education, and dominance of other public attitudes like modernization, development, and nation-building; non-concern about the welfare of others and future generations as well as disregard of the "sacredness of nature" or the uniqueness of **ecosystem services**; and beliefs about how environmental conditions affect those things that individuals value. Cultural factors of the individual and household domain associated with deforestation have been subdivided into two categories: first, unconcern by individuals about the forest environment as reflected in increasing levels of demands, aspirations, materials, and energy **consumption** (commonly associated with increased income and commercialization); and second, situation-specific behavior of actors such as rent-seeking, tradition, imitation, and unreflected continuation of inherited **land-use systems**.

Cultural factors relate to several **syndromes** of land change, such as pressures of production on resources through heavy surplus extraction away from land managers. Referring to colonialism, development, and **land degradation** (i.e., the context in which the European mode of civilization spread across the world), **Piers Blaikie** and **Harold Brookfield** provided a classic example of the so-called "culture system" from Java in which taxation and the state as landlord are tied up with typical land-change processes centered around sugar cane, with elements of the system remaining in operation even in modern, independent Indonesia. In this context, Samuel Huntington's concept of culture needs to be mentioned since it links to half a dozen or so contemporary and dominant civilizations, that is, Sinic, Japanese, Hindu, Islamic, Orthodox, Western (further subdivided into European, North American, and Latin American cultures), and African. Most importantly, the fundamental source of conflict in a global, new world will not be primarily ideological or economic; the great divisions among humankind and the dominating sources of conflict will be cultural. Critics argue that cultures or civilizations are not monolithic blocs of the geopolitical scene, but rather constitute living and evolving organisms that need constant exchange, interaction with, and adaptation to their environments. Nonetheless, it was Samuel Huntington who contributed to, or even triggered, the renaissance of culture as a determinant that shapes **globalization** and the world order of the twenty-first century.

See also: **Agroforestry; Decision-Making; Forest Transition; Indigenous Knowledge; Land Rights; Land-Use History; Mediating Factor; Mode of Interaction; Property.**

Further Reading

Audretsch, David B., Grimm, Heike, and Charles Wessner, *Local Heroes in the Global Village: Globalization and New Entrepreneurship Policies*, New York: Springer Science & Media Inc., 2005; Blaikie, Piers M., and Harold Brookfield, *Land Degradation and Society*, London, New York: Routledge, 1987; Boserup, Ester (posthum), "Technological Society and Its Relation to Global Environmental Change," in *Encyclopedia of Global Environmental Change, Volume 5: Social and Economic Dimensions of Global Environmental Change*, eds. Ted Munn and Peter Timmermann, Baffins Lane, Chichester, UK: John Wiley & Sons, 2002, pp. 86–96; Committee on Global Change Research, and National Research Council, *Global Environmental Change: Research Pathways for the Next Decade*, Washington, D.C.: National Academy Press, 1999; Florida, Richard, *The Rise of the Creative Class: And How It's Transforming Work, Leisure, Community and Everyday Life*, New York: Basic Books, 2002; Geist, Helmut J., and Éric F. Lambin, "Proximate Causes and Underlying Driving Forces of Tropical Deforestation," *BioScience* 52 (2002): 143–50; Huntington, Samuel P., *The Clash of Civilizations and the Remaking of World Order*, New York: Simon and Schuster, 1996; Inglehart, Ronald, *Culture Shift in Advanced Industrial Societies*, Princeton, NJ: Princeton University Press, 1990.

HEIKE GRIMM, HELMUT GEIST

Cumulative Change. Aggregate, repeated, and invariably human-driven changes in broad land-cover types worldwide that can affect global biogeochemical cycling and functioning of the earth system, often requiring centuries to millennia to reach a global magnitude. The structure and function of the earth system sustain the biosphere or that portion of the earth system in which life exists. The system's structure includes states or faces of the earth, such as land-cover classes of grassland-savannas, **wetlands**, deserts, tundra, **taiga**, and snow and ice fields. These states are linked to one another and the functioning of the earth system through the biogeochemical cycling.

For example, changes in land-cover states affect the flows of carbon, nitrogen, and water among land, oceans, and atmosphere; these changes, in turn, affect the land-cover states in terms of climate or regional air pollution. For most of human history, societal impacts on the earth system were largely registered through changes in land-cover states: for instance, the clearing of **forest** cover for cultivation, the annual burning of **grasslands** and savannas for grazing, and the impounding of water or depletion of aquifers for irrigation farming. Any individual **land-cover change**, permanent or temporary, need not affect the earth system; but aggregate, repeated individual changes can, as can regionwide changes in land covers that can affect local climate and other biophysical characteristics.

The largest land-cover change generated throughout human history is **deforestation**, primarily for cultivated land cover, both croplands and **pasture**. It is the only cumulative change to have affected biogeochemical states previous to the twentieth century (perhaps thousands of years ago) and has contributed an estimated 20 percent of the current human-derived carbon dioxide in the atmosphere. Grassland expansion from the use of **fire** in hunting and for livestock, in some cases leading to co-evolved human-natural landscapes, is as old as deforestation, but its cumulative contributions to biogeochemical cycling are difficult to determine. Interestingly, tropical deforestation for cultivation, pasture, and timber is the most rapid cumulative land change underway currently, with significant implications for the hydrological cycle. Other cumulative land-use and land-cover changes affecting **systemic changes** include the buildup of irrigated lands, especially wet **rice**, and ruminant livestock, both of which contribute significantly to methane emissions, a more potent **greenhouse gas** than carbon dioxide. The intensification of cultivation worldwide through the use of nitrogen fertilizers contributes significantly to the nitrogen cycle. Land change, especially opening once-forested lands and urban agglomerations, affects the earth's **albedo** and heat balance.

Beyond the consequences for systemic change, cumulative land change has played an important role in global biotic diversity. Ancient human landscape burning apparently assisted in the die-off of Pleistocene megafauna, while current rates of tropical deforestation have led to claims of a new "mass extinction."

See also: **Agricultural Intensification; Atmosphere-Land Interlinkages; Biodiversity; Carbon Cycle; Cattle Ranching; Change Detection; Climate Impacts; Nutrient Cycle; Species Extinction; Water-Land Interlinkages.**

Further Reading

Kasperson, Roger E., Turner, B. L., II, Meyer, William B., Dow, Kirsten, Golding, Dominic, Kasperson, Jeanne X., Mitchell, Robert C., and Samual J. Ratick, "Two Types of Global Environmental Change: Definitional and Spatial-Scale Issues in Their Human Dimensions," *Global Environmental Change: Human and Policy Dimensions* 1 (1990): 14–22; Steffen, Will, Sanderson, Angelina, Tyson, Peter, Jäger, Jill, Matson, Pamela, Moore, Berrien, III, Oldfield, Frank et al., eds., *Global Change and the Earth System: A Planet under Pressure*, Berlin: Springer Verlag, 2004.

B. L. TURNER II

D

Decentralization. National policies aimed at decentralizing the management of natural resources. They are gaining popularity in almost all developing countries, and especially in the tropics where alarming rates of **deforestation** have been witnessed in recent years. The trend toward decentralization has been driven by many factors, prominent among these an increased commitment to more equitable management policies. Highly centralized **forestry** through a strict preservationist approach, once believed to be one of the most effective approaches for **forest** protection, is now regarded by many researchers to have led to increased tenure insecurity and exacerbated land-based conflicts, often contributing to increased forest destruction. Efforts to provide local governments a greater participation in forest management have been initiated internationally, perhaps most visibly in **Latin America** and South Asia. The first set of empirical evaluations coming out in several countries indicates that the impact of these policy changes has been variable. The rhetoric of decentralization has often been louder than actual changes, and some regions are better managed than others. Nevertheless, initial results indicate that democratic decentralization is a powerful force that, if properly managed, can incentivize local governments and communities to provide more effective forest protection and **regrowth**.

See also: **Arena of Land Conflict; Conservation; Land-Use Policies; Participatory Land Management; Public Policy.**

Further Reading

Agrawal, Arun, and Elinor Ostrom, "Collective Action, Property Rights, and Decentralization in Resource Use in India and Nepal," *Politics and Society* 29 (2001): 485–514; Lyes, Ferroukhi, ed., *Municipal Forest Management in Latin America*, Bogor, Indonesia: Center for International Forestry Research, International Development Research Centre, 2003; Ribot, Jesse, and Arun Agrawal, "Accountability in Decentralization," *Journal of Developing Areas* 33 (1999): 473–502.

HARINI NAGENDRA

Decision-Making. The process by which a specific actor or group of actors evaluate possible land-use choices in the context of social and environmental conditions and choose to maintain or alter an existing land use. Research on land-use decision-making has explored how the interaction of factors such as prices, policies, **land tenure**, and climate affect how actors make land management decisions at local, regional, and global **scales** of analysis. Land-use decisions may be made by actors at a variety of different levels, including individual, household, community, company, or government levels. In many cases, the decision-making process is often the product of interaction between agents. A group of individuals may coordinate to preserve a contiguous area of **forest** that spans multiple **landholdings**. Likewise, government organizations may seek input from communities and individuals in the management of publicly owned land. Thus, the decision-making process varies dramatically depending on the actor (or actors) involved and the management context within which the actor is making decisions.

Much of the research on land-use decision-making focuses on specific land-use actions such as timber harvesting, **fallow** periods, or irrigation. But many decisions seemingly disconnected from land use can ultimately have an impact on land-use/cover change. For example, decisions related to fertility and family size, educational opportunities for children, and technological adoption can all ultimately have an impact, albeit over long time periods. Indeed, many theories related to **environmental change** involve macro-level factors such as demographics (Thomas Malthus) or **agricultural intensification** and innovation (**Ester Boserup**). **Driving forces** such as population density, road construction, or technological innovation are linked to subsequent land-use/cover change processes, and the identification of these macro-level drivers is critical for the development and application of policies. However, this macro-level focus does not necessarily provide insight into the decision-making process of individual actors. Behavioral research in economics, psychology, sociology, political science, anthropology, and geography has focused on how individuals and households make land-use decisions at local levels of analysis.

The decision-making framework established within neoclassical economics relies heavily on this definition of actors as self-interested individuals attempting to maximize their benefits (or satisfaction) in the context of an open competitive market. The study of rational choice theory describes human actors (usually individuals) as goal oriented and capable of the comparison and evaluation of complex choices. The concept of utility is one basis in this framework for comparing different decisions and the potential outcomes of those decisions. Utility is the benefit or satisfaction received by an actor as the result of a specific decision or condition. In terms of land-use decision-making, the utility of allocating labor to timber harvesting can be compared to the utility of allocating labor to crop production based on the financial return of these land uses. Actors may receive both pecuniary and non-pecuniary (e.g., esthetic) utility from different land uses. A forest resource may provide a financial return through timber harvesting and a less tangible benefit in terms of recreational opportunities.

A prerequisite for comparing the utility of different land uses is the valuation of environmental resources—a topic of considerable debate in the academic literature. It is relatively straightforward to put an economic value on a hectare of land used for soy production. It is more difficult to place an economic value on a forest resource used for government-sponsored timber harvesting, the extraction of non-timber forest products by surrounding communities, and the species habitat this forest provides. But the process of valuation is one component used by some researchers in exploring the process of land-use decision-making. The process of environmental valuation is further complicated by the consideration of non-pecuniary or non-monetary factors. A landowner whose financial goals are satisfied may have the luxury to consider esthetics in land use decisions, and it is difficult to measure this benefit in the same units as the financial benefit from crop production.

In general, an actor is expected to make the decision with the greatest utility. But a number of issues complicate an actor's ability to effectively evaluate the utility of different actions. Recent research has explored more complex definitions of human actors and challenged the notion of individuals capable of effectively evaluating complex decision-making **scenarios**. Actors are not omniscient with perfect knowledge of the future benefits of decisions in the present. In addition, actors have heterogeneous preferences that lead them to different actions even if posed with the same scenario. Recent research has focused on the issue of complexity in land-use decision-making, such as the interaction between agents or the interaction between climate and prices and how these interactions affect decision-making. Choices such as leasing land versus fallowing, or selective harvesting of timber versus more intensive **agroforestry** involve a large number of social and biophysical factors in the decision-making calculus. Price predictions, climate expectations, and the future availability of labor all present challenges for decision-makers. The ability of all actors to effectively evaluate these choices and arrive at the decision that will deliver the greatest utility can be questioned in the light of these complexities.

A key assumption within many efforts to model land-use decision-making is that of perfect information. Many economic theories rely on the assumption that landowners have perfect and uniform information about their environment, an assumption that has been widely debated. In reality, actors do not have perfect information for evaluating competing land-use decisions and may have considerable uncertainty regarding the information they do have. Researchers in agricultural economics, anthropology, and other fields have explored what sources of information are available to actors and whether all actors have the same **access** to information. For example, new migrants colonizing parcels along the Transamazon Highway in Brazil had less-detailed information about the land suitability within their parcels than long-established migrants. This lack of information affected their ability to select the optimal locations for forest clearing and subsequent agricultural production. Over time the new migrants learned about the pattern of land suitability within their parcels, which ultimately affected their decisions about what areas to use for agricultural production and thereby the regional pattern of land cover. Likewise, landowners in more remote areas may have less access to agricultural extension agents and therefore have slower rates of adoption of new technologies, crop types, and seed varieties.

Risk and uncertainty play a significant role in many land-use decisions. Natural events such as flooding and drought are infrequent and unpredictable but undoubtedly have an impact on land-use decisions. Some landowners in northeast Thailand distribute their landholdings among upland areas and lowland areas so that they are able to have some agricultural production regardless of whether it is a wet or dry year. This system of distributed landholdings helps landowners reduce the risk to years with extreme events. But more generally, some actors can be considered risk averse while others are not, and these different **perceptions** of risk have a significant impact on the decision-making process.

A particular challenge in understanding many land-use decision-making situations is the time horizon associated with the benefits or costs of a particular decision. Myopic actors (those with a very short time horizon) may choose a different set of land uses than might actors who have a very long time horizon. For example, a myopic actor may choose to clear a forested area for the production of a high-value commodity crop while a non-myopic actor may choose to maintain the forest for future revenue from timber harvesting. In order for an actor to evaluate the future potential value of a land-use resource (e.g., forests), it is often necessary to adopt a very long time horizon, but this may conflict with the short-term need for a financial return.

In most land-use/cover-change settings, actors do not make decisions independently of other actors or **institutions**. An institution is a formal or informal rule that governs human behavior. In terms of land-use/cover change, institutions such as zoning restrictions, water rights, and land tenure are all examples of rules that affect what actions or land uses are allowed and how actors evaluate and compare different land-use decisions. In the absence of land tenure rights, actors may not consider the long-term consequences of their land-use actions and may potentially make decisions that result in short-term benefits but also **land degradation** in the long term.

One fundamental question in decision-making frameworks is whether individuals act to maximize their own benefits or whether they also consider the consequences of their actions within a larger pool of individuals (group, region, nation). Garret Hardin's article "**Tragedy of the Commons**" describes individuals utilizing a common resource as trapped in a system of ultimately unsustainable use. Using the definition of individuals as self-interested, he predicted that in the absence of private rights or government intervention, individuals would eventually eradicate a resource through exploitation. Theories of collective action have explored under what conditions actors can coordinate to preserve a resource for long-term use.

Research methods such as experimental economics, game theory, and **agent-based models** are beginning to be applied to some of this complex decision-making within the context of land-use/cover change. The integration of these methodological tools with the rich empirical results that already exist within existing land-use/cover-change studies will likely lead to new insights into the dynamics of decision-making at multiple spatial scales of analysis.

See also: **Arena of Land Conflict; Change Detection; Common-Pool Resources; Conservation; Corporate Strategies; Economic Restructuring; Endogenous Factor; Exogenous Factor; Investments; Land Rights; Land-Use History; Land-Use Policies; Leapfrogging; Modeling; People at Risk; Population Dynamics; Thünen, Johann Heinrich von.**

Further Reading

Boserup, Ester, *The Conditions of Agricultural Growth*, Chicago: Aldine Press, 1965; Boserup, Ester, *Population and Technological Change: A Study of Long-Term Trends*, Chicago: University of Chicago Press, 1981; Gigerenzer, Gerd, Todd, Peter, and the

ABC Research Group, *Simple Heuristics That Make Us Smart*, New York, Oxford, UK: Oxford University Press, 1999; Hardin, Garret, "The Tragedy of the Commons," *Science* 162 (1968): 1243–8; Kahneman, Daniel, Slovic, Paul, and Amos Tversky, *Judgement under Uncertainty: Heuristics and Biases*, Cambridge, UK: Cambridge University Press, 1982; Malthus, Thomas R., *An Essay on the Principle of Population*, reprinted with the variora of 1806, 1807, 1817, and 1826, ed. Patricia James, 2 volumes, Cambridge, UK: Cambridge University Press, 1989 [1803]; Moran, Emilio, *Developing the Amazon*, Bloomington, IN: Indiana University Press, 1981; Van Wey, Leah, Ostrom, Elinor, and Vicky Meretsky, "Theories Underlying the Study of Human-Environment Interactions," in *Seeing the Forest and the Trees: Human-Environment Interactions in Forest Ecosystems*, eds. Emilio F. Moran and Elinor Ostrom, Cambridge, MA: MIT Press, 2005, pp. 23–56; Walker, Robert, Drzyzga, Scott, Li, Yali, Qi, Jiaguo, Caldas, Marcellus, Arima, Eugenio, and Dante Vergara, "A Behavioral Model of Landscape Change in the Amazon Basin: The Colonist Case," *Ecological Applications* 14 (2004): S299–S312.

THOMAS P. EVANS

Deforestation. Change in land cover when **forest** is converted to another land cover or when the tree canopy cover falls below a minimum percentage **threshold**, for example, 10 percent for the **United Nations Food and Agriculture Organization** (FAO). The Global Forest Resources Assessment 2000 of FAO estimated—on the basis of national statistics, inventory reports, estimates by experts, and a pantropical **remote sensing** survey for tropical forests only—that the world's natural forests decreased by 16.1 million hectares per year on average during the 1990s. This equals a loss of 4.2 percent of the natural forest that existed in 1990. However, some natural forests were converted to forest plantations. Gains in forest cover arose from afforestation on land previously under nonforest land use (1.6 million hectares per year globally) and the expansion of natural forests in areas previously under **agriculture**, mostly in **western Europe** and eastern North America (3.6 million hectares per year globally). The net global decrease in forest area was therefore 9.4 million hectares per year from 1990 to 2000. The total net forest change for the temperate regions was positive, but it was negative for the tropical regions. FAO

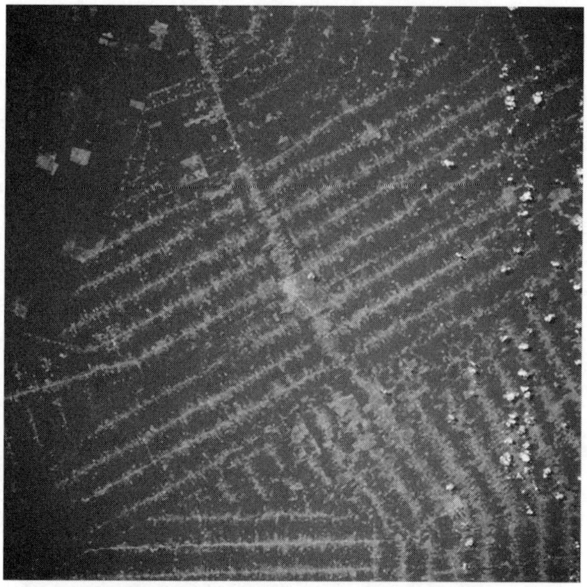

Land clearing in the form of planned agricultural colonization in the southern Amazon River Basin, showing a pattern which typically radiates out from roads or rivers ("herringbone" appearance, "fishbone" pattern of deforestation, with intervals of four to five kilometers between the roads).

PHOTO: Space Shuttle. CREDITS: Image Science and Analysis Laboratory, NASA-Johnson Space Center, 18 Mar. 2005 (Earth from Space—Image Information). http://earth.jsc.nasa.gov//sseop.efs

Deforested land in the Mount Hood National Park, Oregon, in June 1936.
PHOTO: Arthur Rothstein. [Courtesy Library of Congress]

estimated that tropical regions lost 15.2 million hectares of forests per year during the 1990s.

Recent estimates by Frédéric Achard and colleagues from the **Tropical Ecosystem Environment Observations by Satellite (TREES) project** have been provided for only the world's humid tropical forests, based on a sampling strategy of remote sensing data. They revised downward by 23 percent FAO's net rate of change in forest cover for the **tropical humid forests**, which exclude **tropical dry forests**. According to these recent estimates, 5.8 ± 1.4 million hectares of humid tropical forest were lost each year between 1990 and 1997. Forest **regrowth** accounted for 1.0 ± 0.32 million hectares. The annual rate of net cover change in humid tropical forest was 0.43 percent during that period. A further 2.3 ± 0.7 million hectares of forest were visibly degraded. This figure does not include forests affected by selective logging. Southeast Asia has experienced the highest rate of net cover change (0.71 percent per year), whereas Africa and **Latin America** present lower rates (0.36 percent and 0.33 percent, respectively). Latin America, however, lost about the same area of forest as Southeast Asia during the 1990–1997 time period. **Forest degradation** was most extensive in Southeast Asia (0.42 percent per year), lowest in Latin America (0.13 percent per year), and intermediate in Africa (0.21 percent per year). Forest regrowth was more extensive, both in absolute and relative terms, in Southeast Asia than in the other humid tropical regions (0.19 percent for Southeast Asia, 0.04 percent for Latin America, and 0.07 percent for Africa).

These recent assessments of deforestation, as well as another remote sensing survey by Ruth DeFries and colleagues—at a coarser spatial resolution, but covering the entire tropical belt—concur to estimate less deforestation in the 1990s than was observed in the 1980s. Still, it is unlikely that deforestation has significantly slowed down, because differences in methods of assessment and definitions used may account for at least part of the difference. Moreover, deforestation in the dry tropical forests may often be underestimated.

In Latin America, large-scale forest **conversion** and **colonization** for livestock-based agriculture is prevalent, whereas cropland expansion by smallholders dominates in Africa. In Asia, intensified shifting agriculture, including migration into new areas, gradual change of existing areas toward more permanent agriculture, and logging explain most of the deforestation. Within these regions, deforestation is largely confined to a few areas undergoing rapid change, with annual rates of deforestation from 2 to 5 percent. The largest deforestation front is the **arc of deforestation** in **Amazonia**. More scattered areas of forest loss are detected in the Chaco and Atlantic forest areas in **South America**. Central America has significant deforestation fronts in the Yucatán peninsula and at the Nicaraguan border with Honduras and Costa Rica. In Africa, forest-cover change is very rapid in Madagascar, Ivory Coast, and the Congo basin, in small scattered **hot spots of land-cover change**. In Southeast Asia, several deforestation fronts are found around Sumatra, Borneo, Vietnam, Cambodia, and Myanmar.

See also: **Agricultural Frontier; Cash Crops; Cattle Ranching; Driving Forces; Food Crops; Forest Degradation; Forest Transition; Pasture; Pattern to Process; Pristine Myth; Proximate Causes.**

Further Reading

Achard, Frédéric, Eva, Hugh D., Stibig, Hans-Jürgen, Mayaux, Philippe, Gallego, Javier, Richards, Timothy, and Jean-Paul Malingreau, "Determination of Deforestation Rates of the World's Humid Tropical Forests," *Science* 297 (2002): 999–1002; DeFries, Ruth, Houghton, Richard A., Hansen, Mathew C., Field, C. B., Skole, David, and John Townshend, "Carbon Emissions from Tropical Deforestation and Regrowth Based on Satellite Observations for the 1980s and 1990s," *Proceedings of the National Academy of Sciences of the United States of America* 99 (2002): 14,256–61; Geist, Helmut J., and Éric F. Lambin, "Proximate Causes and Underlying Driving Forces of Tropical Deforestation," *BioScience* 52 (2002): 143–50; Lambin, Éric F., and Helmut Geist, eds. *Land Use and Land Cover Change: Local Processes, Global Impacts.* Berlin: Springer, 2006; United Nations Food and Agriculture Organization, *Global Forest Resources Assessment 2000 (FRA 2000): Main Report* (FAO Forestry Paper No. 140), Rome: FAO, 2001.

HELMUT GEIST

Degradation Narrative. A simplification of cause-consequence relationships about the human role in environmental change. In the popular imagination, in development circles, and to a lesser degree in academic literature, degradation narratives (DNs) reflect notions of environmental crisis and predominant views of nature-society relationships that see peasants, pastoralists, and hunters as destroyers of their own environments. To the extent that DNs underscore important environmental issues that require coordinated societal responses, as in the case of tropical **deforestation** or **soil degradation**, they serve a role in drawing attention to these problems, in generating resources to combat them, and are justified for certain times and places. In most cases, however, more nuanced interpretations are required. There is cause for concern when simplistic DNs dominate understanding of land-change dynamics at the expense of uncovering the real complexities and opportunities for **conservation** that exist.

Also called "received wisdom," "environmental orthodoxies," or myths, DNs are usually based on stark images, such as the desertified Sahel, burned rainforest, or eroded hillside, which are recorded by mass media and serve to captivate the public's imagination. While often only snapshots of a particular moment in time, in the context of long-term environmental change, these images of degraded lands create powerful stories: they describe a problem, articulate the main cause of it, and propose a solution. They also include a cast of characters, such as local land managers, who are assigned the role of either agents or victims of **environmental change**. DNs are explicitly programmatic, urging the reader to adopt policy, often in the form of a blueprint or generalized environment and development plan. As such, the narratives provide scripts and justification for development action that frequently ignores local human-environment conditions.

The degree of dominance of a specific DN depends on its correspondence with observed facts and also on the political power of those telling the story. The influence of a narrative is enhanced under the following conditions:

1. Donor experts and user constituents are strongly attached to them.
2. There is political, strategic, or moral pressure on donors to act quickly.
3. There has been little technical or socioeconomic research in, and a weak database in, the recipient nation.
4. The recipient country must rely heavily on expatriate experts for advice.
5. The recipient government is dependent on foreign assistance.
6. The recipient government is weak or authoritarian or both and does not have the institutional capacity to hear and learn from its rural inhabitants.

The pervasive influence of DNs poses a significant challenge for researchers and development workers concerned with land-use and **land-cover change**. In spite of their sometimes weak assumptions—for example, that population growth is axiomatically associated with **land degradation**—or the lack of supporting data, DNs hold sufficient public sway to influence environment and development policies. The narratives persist, in part, because they are compelling stories with clear and simple explanations, but also because they serve the purposes of powerful groups. Their urgent message, which often suggests that if negative trends are to be reversed, then local land management practices need to be transformed and made less destructive, justifies a strong role for non-governmental organizations (NGOs) or the state—**institutions** that seek to regulate local land managers' **access** to, and use of, natural resources.

The roots of the image of the backward and destructive peasantry lie in colonial era thinking, but the view has been perpetuated in contemporary development discourse and policy, providing elites the justification to claim land stewardship responsibilities. For example, concern over tropical deforestation in **Amazonia** has led to the establishment of protected areas from which local inhabitants are excluded; this despite the long occupancy of indigenous peoples in the **forests**, their status as humanized landscapes, concerns over livelihood opportunities, and evidence that forest clearing is linked as much to structural, distal forces of change as to the agency of forest dwellers. There are signs, however, that a new conservation model, inclusive of people, is starting to emerge in Brazil.

DNs exert their greatest influence when they become conventional wisdom, a process that is enhanced as the narratives are translated into institutional arrangements, perhaps via their transferal to recipient aid countries through training, institution building, and **investment** or as they form part of a broader systematic discourse used by the state to claim control over natural resources. When narratives become hegemonic, they obscure a plurality of possible views on cause and effect, thereby thwarting effective environmental management. By ignoring or transforming alternative views, such as those of the common peasant, DNs result in poor policy decisions, cause conflict between peasants and the state and, in the worst cases, may produce real environmental crises. For example, for nearly a century, the predominant view of governments and scientists has held that frequent **fires** set by peasants in the savanna belt of West Africa are progressively causing widespread degradation of woodlands, a conclusion that underpins fire suppression policies. These fears led to burning experiments and the development of a savanna fire ecology model that remains the basis for anti-fire policies. It is increasingly clear, however, that the rural population successfully manages fire through early and patchy burning regimes that reduce the threat of large intense fires and, in some cases, promote woodland expansion. Anti-fire policies are resented by rural populations who rely on proscriptive burning practices to manage livelihood systems, and they cause dissension and conflict between locals and **forestry** agents, which thwarts effective fire management.

Popularized in news media, DNs help to build support among the general public, which holds the potential to influence how donor organizations prioritize environmental problems. For example, during the Sahel droughts of the 1970s and 1980s, an overarching DN, which claimed that the desert was rapidly advancing at the hand of humans, dominated environment and development policy, resulting in innumerable unsuccessful and unwarranted tree-planting projects aimed at halting the desert's advance. More recent **remote sensing** studies demonstrate how **desertification** is more correctly understood as a localized phenomenon.

Given their influence on policy, funding priorities, and the spread of misinformation, scholars of anthropology, environmental history, geography, and historical ecology increasingly target the flaws in DNs. Collectively, these critical works help produce a more balanced understanding of nature-society relationships. First, in contrast to simplistic explanations that focus on proximate land managers rather than social, political, and economic structures, they underscore the notion that there is a web of **driving forces** of land change—debunking myths of crude, linear, causal sequences.

For example, prominent among DNs is the assumption that population growth and **poverty** cause environmental degradation. Rooted in neo-Malthusian causality, this assumption is overly deterministic. Whether or not increasing population is damaging or beneficial depends upon a variety of **mediating factors**, be they institutional, ecological, or technological. While population growth can cause land degradation in the short term, it can also spur agricultural innovation, **agricultural intensification**, and the adoption of conservation techniques. In a widely cited case, high rates of population growth from 1930–1990 in the Machakos region of Kenya correspond with increased agricultural production, improved social welfare, and enhanced environmental services. Similarly, studies in the cottonbelt in Mali demonstrate that increased poverty, which may or may not be linked to population growth, does not always correlate with increased environmental degradation.

Second, critiques of DNs show that they are a function of **scale**, usually based on broad-scale perspectives that emphasize the need to measure, map, and predict environmental change for large areas. As such, they are over-generalizations that ignore the great diversity in local human-environment conditions. In contrast to DNs, in-depth, local-level studies can reveal how land managers successfully work to maintain their environments. The Sahelian desertification narrative is a widely cited example. The narrative still influences policy in the Sahelian countries despite the numerous case studies documenting how peasant farmers and herders adapt to drought and take measures to prevent soil degradation.

Finally, the ecological science underlying many DNs is based on inappropriate models that emphasize equilibrium and homogeneity rather than heterogeneity and **disequilibrium dynamics**. The prevailing scientific view during the colonial era was that once ecosystems reached a climax state, they maintained stability. Stability was viewed as the desired condition, while disturbance was viewed as unnatural and damaging. Indigenous land management systems that involved **slash-and-burn agriculture**, frequent broadcast burning, or intensive grazing of livestock, which caused periodic ecological disturbances over wide areas, were viewed as degrading and often outlawed. These activities were incorporated into grand narratives involving long-term and irreversible trends, such as the **savannization** narrative of West Africa. In stark contrast to the DNs of the past and present, ecologists and biogeographers now recognize the importance of disturbance and the landscape heterogeneity it creates for **biodiversity**.

See also: **Arena of Land Conflict; Colonization; Common-Pool Resources; Forest Degradation; Land-Use History; Pristine Myth; Proximate Causes; Transhumance.**

Further Reading

Forsyth, Tim, *Critical Political Ecology*, London: Routledge, 2003; Hoben, Allan, "Paradigms and Politics: The Cultural Construction of Environmental Policy in Ethiopia," *World Development* 23 (1995): 1007–22; Leach, Melissa, and Robin Mearns, eds., *The Lie of the Land*, Oxford, UK: James Curry, 1996; Roe, Emery M., "Development Narratives, or Making the Best of Blueprint Development," *World Development* 19 (1991): 287–300.

PAUL LARIS, PETER KLEPEIS

Desertification. The process of human-induced **land degradation** in arid, semiarid, and dry subhumid areas of the world, as defined by the **United Nations Environment Programme**. Land degradation is in turn defined as a decline in the productive capacities of land that are irreversible on human time **scales** without significant human effort or **investment**. This two-part definition of desertification is an attempt to add increased rigor to the term, which has historically been plagued with vagueness and confusion, by circumscribing it to longer-term ecological changes brought about by human activities within particular geographical areas of the world. Desertification can occur within the dryland steppe, **grasslands**, and savannas of the world but cannot occur within true deserts or more humid temperate or tropical **forest** environments.

Definition and Stakeholder Perspective

Despite its greater precision compared to past uses, the present definition is not without problems. Continued confusion surrounds the use of the phrases "productive potential" and "irreversible without significant investment" in the definition of land degradation. These phrases are seen to insert economic criteria into a biophysical term. In practice, productive potential does not necessarily refer to strict definitions of biological productivity such as net primary productivity but instead refers to changes in the land's potential for producing wood, fiber, fodder, or crops. The potential for investments in land that reverse or mollify declines in productive potential are as well affected by the land value and the economics of the production on the land. Therefore, when one refers to desertification, one must clarify for whom and for what purpose the productive potential of the land has declined.

During the height of the Dust Bowl years in the Great Plains, dust buried farms and equipment, killed livestock and caused human death and misery, 1935. [Courtesy Library of Congress]

Desertification is an umbrella term for a range of human-initiated biophysical processes that result in changes that are seen as reducing the potential of the land to be used for a particular productive use. These processes most typically result in changes in the productive capacity of dryland soils through changes in their chemistry or structure. Such changes can occur through changes in intact soil profiles through compaction, crusting, **salinization**, and nutrient depletion, or through the burial or removal of soil through erosion. Human-managed activities such as farming, mining, livestock rearing, and timber harvesting are seen as contributing to these soil changes. All of these activities can lead to the erosion and crusting of sensitive soils by reducing vegetative cover, which leads to the exposure of the soil surface to the forces of water and wind. Soil nutrient depletion can occur over time through repetitive removal of nutrients stored in **vegetation** by cropping and livestock grazing. Human-induced salinization is the accumulation of salts in the root zone of soils from excessive evaporation of irrigated water.

Desertification, as defined above, may also result in human-induced changes in vegetation and climate that are not necessarily associated with changes in soils. While much vegetation change in dryland areas can be seen as ephemeral, human disturbance regimes (grazing, farming, logging, setting of fires, etc.) in combination with climatic variability can lead to more persistent shifts in vegetation, resulting in lowered productive potential. The most common examples of such transitions are changes in vegetative structure that result in a relatively persistent change in the importance of woody vegetation compared to grasses and forbs. Bush encroachment is seen as a problem in many parts of the world, since it is often associated with a reduction in the digestibility of vegetative **biomass**, thereby lowering the productive potential of **pastures**. A reduction in tree cover in savannas can be seen as lowering productive potential through a reduction in standing timber, reduced availability of **non-timber forest products**, or a reduction in wildlife habitat. Not only are these transitions often difficult to reverse, but recovery, if possible, is slow.

Desertification, the set of biophysical processes that connect land-use practices to land degradation, is distinguished from desiccation—the drying of the climate, which is seen as an **exogenous factor** in local land-use practices. A complication to this simple distinction comes from climate **modeling** work, suggesting that widespread loss of vegetative cover may contribute to the drying of local and regional climates (e.g., biogeophysical **feedback**). If such work is empirically verified, a shift to a drier climate brought about by local land-use practices could arguably, by lowering productive potential, be classified as a form of desertification.

Desertification's long history of use impedes, by the many meanings that have been assigned to it, recent attempts to clarify the term. The idea that human land use can create desert has a very long history—an idea that had a long history of circulation when adopted by nineteenth-century environmental thinkers such as **George Perkins Marsh**. The level of scientific speculation on this connection increased during the first half of the twentieth century as the British and French confronted semiarid areas of their colonies in South Asia and Africa. Worldwide concern accelerated during the 1930s with the environmental calamity of the Dust Bowl in the **United States of America**. Early uses of the term "desertification," dating back to the 1940s, suggested a singular process by which vegetated areas became barren—the making of the desert. Evidence for desertification was visual—a reduction in vegetated cover. With a few notable exceptions, early observers did not look to understand the mechanisms behind the loss of vegetation cover—drought, human land uses, and the spread of the desert were all seen as part of the process with little understanding of the persistence of the change.

Within wide parts of the desertification literature, the effects of drought and human land use were not distinguished until recently. This, coupled with a recognition that one can only change human land uses in response to the crisis, led to an emphasis on the human causes of desertification and on programs of environmental rehabilitation, with some best described as coercive attempts to drastically change rural people's way of life. For steppe environments of low population density, it is now recognized that much of the change that was diagnosed as anthropogenic and persistent was actually ephemeral and associated with drought conditions.

Causes of Desertification

Comparative analysis of 132 subnational desertification case studies does not support single-factor causation nor irreducible causal complexity. Rather, results reveal a limited set of recurrent core variables that, in varying combinations, drive desertification (see figure). Most prominent root causes or underlying **driving forces** are climatic factors, institutional changes (e.g.,

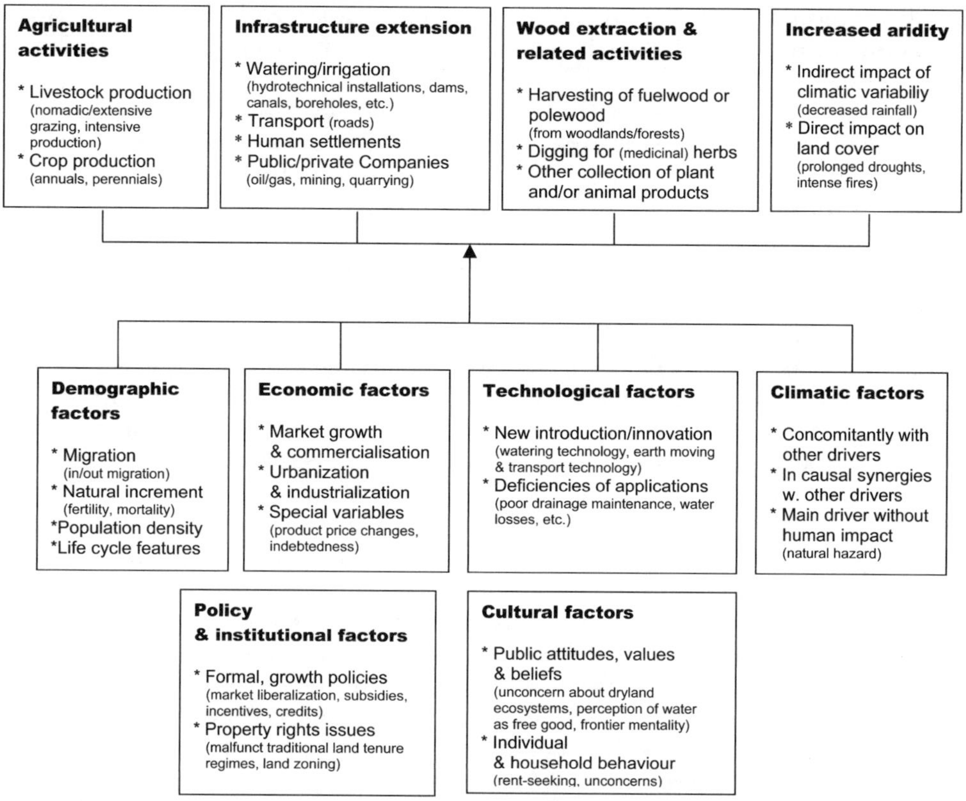

Causes of desertification: six broad clusters of underlying driving forces (fundamental social or biophysical processes) underpin the proximate causes of desertification, which are immediate human or biophysical actions with a direct impact on dryland cover.
[Source: Geist (2005) *The Causes and Progression of Desertification, p. 41*]

in land ownership), national development policies, population growth, and remote economic influences. These in turn can affect **proximate causes** such as local cropland expansion, overgrazing, and infrastructure extension with desertification as a potential but not necessary outcome. Some factors are geographically robust, but most of them are specific to a region and a historical period.

Regional Pathways of Desertification

In Central Asia, two main pathways that have led to desertification are the invasion of grain farming into steppe grazing land, triggering **soil degradation** and overstocking, and the invasion of large-scale hydraulic "cultures" or **agro-industries** into desert ecosystems that historically supported only localized, traditional oasis farming. The most spectacular outcome, notably in low-lying sea region basins (such as the Aral Sea) and in northern **China**, is a widespread increase in desertlike sand cover, which is linked to the exceptionally strong impact of socioeconomic driving forces such as centrally planned frontier **colonization** and (sometimes forced) population movements. But the spread of sand cover, notably in northern China, is also linked to the region's predominantly sandy soils and loess formations and to the geological and climatic predisposition for desert formation of vast basins and plateaus.

In contrast, a typical African pathway of desertification involves the spatial concentration of farmers and pastoralists, very often as a result of national sedentarization policies, around infrastructure nuclei. This local concentration of population results in patches of overgrazing, excessive fuelwood collection, and high cropping intensities, ultimately leading to degraded vegetation and declining soil productivity.

Beefing up of drylands, with little or no involvement of cropping, frequently characterizes the desertification pathways of Australia and of North and **South America**. Historically, these rangeland zones typically shared common patterns of land use, such as the rapid introduction by European settlers of exotic livestock species and commercial pastoralism into ecosystems that had not experienced these uses before. Since the 1950s, however, the price squeeze affecting **agriculture** in industrialized countries has led to different variants of this trajectory. In Australia, the livestock industry and its complex of related infrastructure developed sufficient flexibility to counterbalance droughts and avoid spectacular desertification, and in the U.S. Southwest, principal land uses shifted away from **cattle ranching** to meet urban-driven demand for residential and recreational areas. In contrast, Patagonia (in Chile and Argentina) and northern Mexico suffered from a lack of advanced technologies, alternative land uses, or diversification options to deal with the vagaries of oscillating natural resource productivity. Local farmers find themselves forced to continue raising livestock, often under conditions of impoverishment and deprivation. Consequently, dryland degradation in these areas is not just a historical phenomenon, but continues to advance.

Another common trajectory of dryland change is found in the Mediterranean basin of southern **Europe**. A millennia-old tradition of agropastoral land use has removed nearly all forest cover but has favored a highly resilient *phrygana* (shrub) vegetation, reflecting various stages of soil degradation. The still-valuable agricultural base is at risk only when the mechanization of farming on skeletal soils induces further **soil erosion**, or when grazing on remote **mountain** ranges is followed by devastating **fire**.

Progression of Desertification

A major theme in the desertification literature has been the advancement of the desert along a broad geographical front. The language used to describe this advancement has accorded an expansionary power to the desert with estimated rates of desert advancement circulating widely within governments, international agencies, and the press. While dune mobilization does occur in particular areas, **remote sensing** analysis has not found evidence for the steady advancement of desert edges (typically defined by vegetative cover). Instead, desert boundaries will expand and contract driven largely by local rainfall conditions. The spatial pattern of desertification, as defined above, is not likely to be a broad front along the true desert edge but more patchily distributed around centers of persistently heavy land-use pressure, often in areas that receive more rainfall.

In an effort to synthesize what is known about global, contemporary **hot spots of land-cover change**, that is, areas of rapid dryland change around the world in the past two decades, it was found that Asia had the greatest concentration of desertified land and that existing data did not support the claim of the African Sahel being a desertification hot spot. The study, derived from literature review, field assessments, and regional remote sensing data sets, was done for the **Millennium Ecosystem Assessment** and thus is a direct input into the work of the **United Nations Convention to Combat Desertification**. It estimates that the global extent of desertified or degraded land in all drylands is closer to 10 percent than to 30 percent (or even 70 percent), as claimed in most previous assessments.

Mitigation of Desertification

Early ideas continue to strongly shape our understanding of the social and biophysical processes that contribute to desertification. One legacy of the simple ideas of the past is the paucity of research data, particularly in semi-arid areas of the developing world, concerning the effects of human land use on soil structure and chemistry or that clearly distinguishes the effects of rainfall and human pressure in leading toward more persistent forms of vegetative change. Recent work has shown that the effects of human land-use pressures on soils and vegetation can be strongly conditioned by rainfall—therefore, research directed not only at their independent effects but also their interaction is much needed. Another legacy of these simple ideas is that they persist outside of the scientific community within government and international agencies, non-governmental organizations, and the popular press. As a result, incorrect diagnoses of **environmental change** in semiarid parts of the world continue to shape development and **conservation** programs in the dryland areas of the world. For desertification to remain useful as a term, it is important that the more complex understandings of it within the scientific community be spread more widely across the international development and conservation communities.

See also: **Albedo; Arena of Land Conflict; Biodiversity; Climate Impacts; Cotton; Degradation Narrative; Disequilibrium Dynamics; Exotic Species; Green Wall Project; Pastoral Mobility; Population Dynamics; Savannization; Transhumance.**

Further Reading

Geist, Helmut, *The Causes and Progression of Desertification*, Aldershot, UK, Burlington, VT: Ashgate Publishing Ltd., 2005; Lepers, Erika, Lambin, Éric F., Janetos, Antony C., DeFries, Ruth, Achard, Frédéric, Ramankutty, Navin, and Robert J. Scholes, "A Synthesis of Rapid Land-Cover Change Information for the 1981–2000 Period," *BioScience* 55(2) (2005): 115–24; Middleton, Nick, and David Thomas, eds., *World Atlas of Desertification*, New York: United Nations Environment Programme, 1997; Reynolds, James F., and Mark Stafford Smith, eds., *Global Desertification: Do Humans Cause Deserts?* Berlin: Dahlem University Press, 2002; Swift, Jeremy, "Desertification: Narratives, Winners and Losers," in *The Lie of the Land*, eds. Melissa Leach and Robin Mearns, London: International African Institute, 1996, pp. 73–90; Thomas, David S. G., "Sandstorm in a Teacup? Understanding Desertification," *Geographical Journal* 159 (1993): 318–31; Turner, Matthew, "Environmental Science and Social Causation in the Analysis of Sahelian Pastoralism," in *Political Ecology: An Integrative Approach to Geography and Environment-Development Studies*, eds. Karl S. Zimmerer and Thomas J. Bassett, New York, London: The

Guilford Press, 2003, pp. 159–78; Watts, Michael J., "Social Theory and Environmental Degradation," in *Desert Development: Man and Technology in Sparse Lands*, ed. Yahudi Gradus, Dordrecht, NL: Reidel, 1985, pp. 14–32.

MATTHEW D. TURNER, HELMUT GEIST

Discrete Data. Distinct or unconnected elements. The most common source of land-cover data within studies of land-use/cover change is a classification derived from satellite **remote sensing** images. A discrete classification is generally used for this purpose, to partition a landscape into homogeneous units with discrete boundaries, so that interpreting numbers that refer to the geometry and arrangement of these discrete units is conceptually simple. Land cover, typically categorized into discrete classes, is an obvious example where changes of fundamental interest to environmental and social scientists may not be continuous or observable at all values. **Modeling** land-cover change solely as a discrete process, however, may be more or less appropriate, depending on the relevant research question.

See also: **Change Detection; Continuous Data; Land-Cover Change; Probit/Logit Model.**

Further Reading

Southworth, Jane, Munroe, Darla, and Harini Nagendra, "Land Cover Change and Landscape Fragmentation: Comparing the Utility of Continuous and Discrete Analyses for a Western Honduras Region," *Agriculture, Ecosystems & Environment* 101 (2004): 185–205.

JANE SOUTHWORTH

Disequilibrium Dynamics. Disequilibrium dynamics see tropical arid and semiarid ecosystems as undergoing chaotic shifts among multiple, alternative, relatively stable states, largely driven by non-equilibrium, density-independent, abiotic influences. Disequilibrium thinking comprises an approach and a set of models that challenge the application to these ecosystems of "equilibrium" concepts of carrying capacity, and of linear succession to climax. This view suggests that the more variable rainfall is in space and time, the more variable plant production will be and the greater the potential fluctuations in animal numbers, whether domestic or wild. Savannas may then be poorly represented by equilibrium models, which predict a natural progression along a sequence of successional **vegetation** stages toward a single climax, easily pushed to a state of ecological collapse by the impact of outside influences of grazing pressure or fire. On the contrary, disequilibrium thinking predicts that grazing pressure and **fire** will contribute to the observed state of arid and semiarid rangeland systems at any one time, but are unlikely to be driving any long-term or irreversible trend, and are only two of many influences, the most powerful of which are probably non-anthropogenic factors.

Disequilibrium dynamics models concern the way populations grow, the factors that limit that growth, and the resulting predictions of stability, resilience, persistence or, conversely, degradation of the communities and ecosystems affected by those changing populations. The debate between equilibrium and disequilibrium thinking focuses on the extent to which populations are limited and/or regulated by density-dependent or density-independent factors. The very different interpretations offered by equilibrium as opposed to disequilibrium thinking entail radically different implications for management. Broadly speaking, equilibrium theory understands a growing population as being limited primarily by density-dependent factors. This is illustrated in simplified form in the logistic curve of population growth in the figure titled "Logistic curve of population growth."

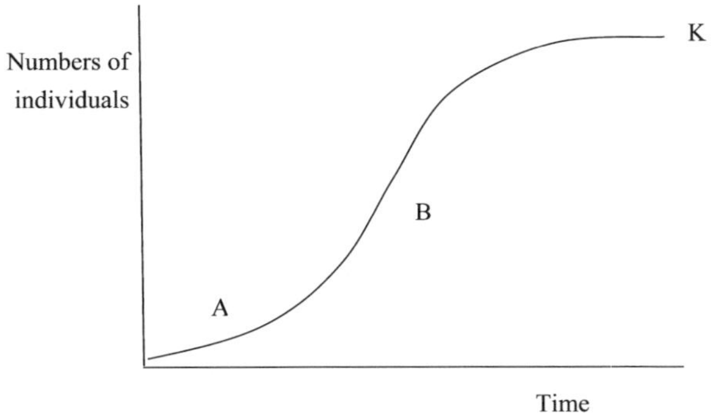

Logistic curve of population growth showing the operation of density-dependent factors.

Equilibrium theory suggests that in a population at low density in a favorable environment, individuals can easily find food and shelter. They grow rapidly, mature early, reproduce successfully, and show low rates of mortality, for example from infectious diseases, which are not easily transmitted at low population densities. This corresponds to point A in the above figure. However, as a population grows, it becomes increasingly hard for individuals to find food, shelter, and suitable breeding sites, and transmission of infectious diseases becomes increasingly problematic. As a result individuals progressively grow more slowly, reach sexual maturity later, and have fewer offspring, who are themselves less likely to survive to sexual maturity or to reproduce successfully. Overall, fertility declines and mortality rises. As a result, population growth rate begins to slow down (point B on the curve in the above figure), and in due course, population numbers stabilize, fluctuating around a plateau that has come to be called K or the carrying capacity, representing a stable and predictable maximum population size given the resources available to exploit in that environment.

Equilibrium thinking sees whole ecosystems as progressing through a linear sequence or Clementsian "succession" of vegetation stages from **colonization** of bare landscapes to the development of a "climax" vegetation type for local conditions of climate and soils. Species with colonizing life history strategies progressively give way to slow-maturing, slow-reproducing specialists with limited dispersal abilities, adapted to the later stages of succession, and locked into intricate symbiotic interrelations with other species. According to this thinking, species populations that overshoot carrying capacity will disrupt climax communities and push the ecosystem back to a lower successional stage, which may be less productive or less diverse. A population crash ensues, and if and when the population eventually stabilizes, this will be at a new, lower value of K. In this equilibrium model, the destructive impact on resources of a temporarily high population effectively resets the carrying capacity at a lower level, perhaps through destruction of particular food species or through **soil erosion** and loss of topsoil. This could happen, for example, where species newly introduced to a favorable environment and released from natural enemies undergo a population eruption, or where species populations are held artificially high. The boom-and-bust patterns characteristic of both vegetation production and livestock numbers in arid and semi-arid lands have tended to be seen in this light (see figure titled "Hypothetical [and much contested] sequence").

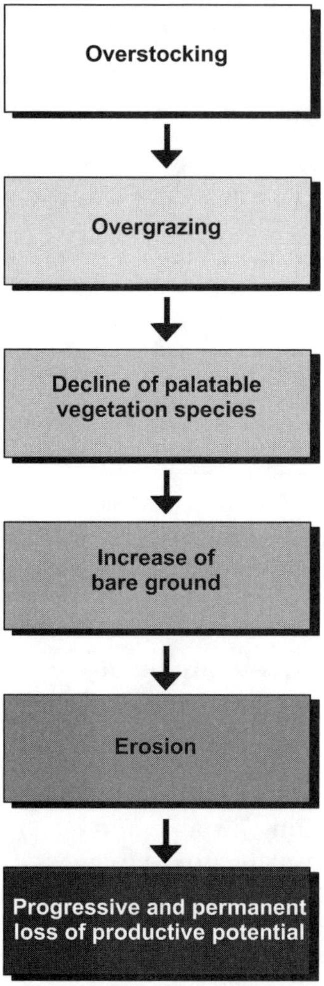

Hypothetical (and much contested) sequence of overstocking, overgrazing, and environmental degradation, as postulated by equilibrium thinking (after Lamprey 1983).

Other authors have gone further and linked this hypothetical sequence of overgrazing and degradation to a postulated positive **feedback** or vicious cycle whereby overgrazing and increasing bare ground could drive climatic deterioration through **albedo** effects. If substantiated, the implication would be that local land use could drive regional climate downturn or, put more bluntly, that overgrazing by pastoralist herds might for example be causing Sahelian droughts, a classic **degradation narrative**. The alternative (more widely accepted) theory is that Sahelian droughts and other extreme climatic events of tropical arid and semiarid Africa correlate closely with El Niño/La Niña events, which correspond to the surface temperature anomalies of the southern oceans. The implications of this alternative theory could not be more different. Rather than causing their own climatic problems by poor land use, overstocking, and overgrazing, Sahelian pastoralists' livestock and livelihoods are in fact jeopardized by global climatic and atmospheric processes that are driven largely by fossil fuel **consumption** in industrialized countries. The two alternative models suggest diametrically opposed management approaches and lead to very different policy decisions with far-reaching implications.

From the 1980s on, major challenges to equilibrium thinking have arisen from both empirical evidence and theoretical developments. The use of equilibrium models and a framework derived from an ecology of relatively more predictable temperate ecosystems is thought to constrain understanding of very variable and unpredictable ecosystems. Disequilibrium thinking focuses on the fact that the lower the mean annual rainfall, the greater the variability and unpredictability of that rainfall in space and time (see figure titled "Rainfall variability as a function of aridity"). Plant-available moisture is the main determinant of forage growth in arid and semi-arid environments, and is closely determined by rainfall. Up to around 1,000 millimeters of annual rainfall, there is a linear increase in plant **biomass** production with increasing rainfall, with the slope of the relationship largely determined by soil nutrient availability (see figure titled "Variation in plant production with rainfall").

Given the great intra- and interannual variability of rainfall in arid systems (see figure titled "Rainfall variability as a function of aridity") and the tight

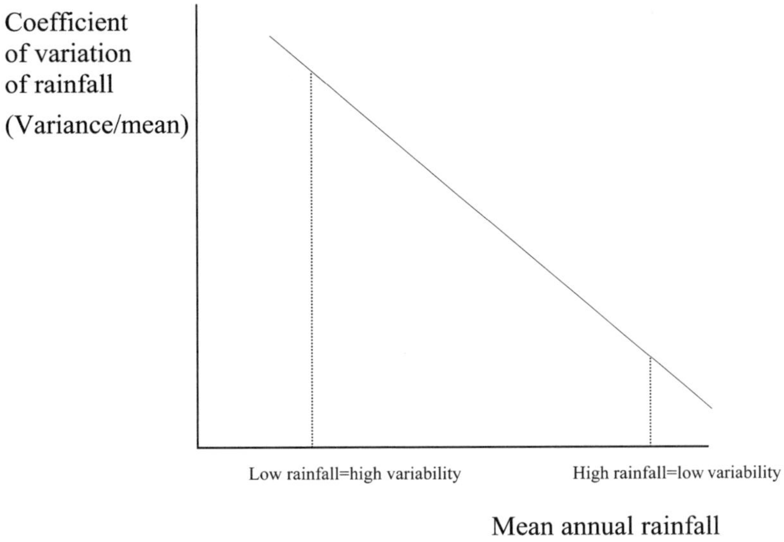

Rainfall variability as a function of aridity.

linkage between rainfall and primary production up to 1,000 millimeters of rainfall per annum (see figure titled "Variation in plant production with rainfall"), it is clear that the "carrying capacity" as outlined in the figure "Logistic curve of population growth" must fluctuate violently within and between years in an arid ecosystem. It then becomes doubtful as to whether the concept is of much use in predicting and managing stocking rates in an arid system. Disequilibrium dynamics models see populations as not primarily limited by density-dependent effects, nor as existing within linear ecological sequences of successional stages. Physical factors in the environment—extreme climatic events, fires, floods, volcanic eruptions—inflict mortality quite independently of the size and density of the population affected. Even a small and intermittently acting density-dependent influence exerts some regulatory effect on a population over time, but density-independent factors can cause potentially massive fluctuations. Disequilibrium dynamics builds on the empirical observation of boom-and-bust arid land systems to predict the following:

1. Disequilibrium ecosystems are dominated by density-independent processes.
2. Massive fluctuations are commonly observed but do not necessarily cause significant (permanent, irreversible) degradation, and changing vegetation states do not represent degraded stages relative to some notional climax successional state.
3. Disequilibrium systems are best understood as undergoing chaotic shifts between multiple alternative stable states, driven primarily by non-anthropogenic factors.

In particular, arid and semiarid tropical systems are characterized by extreme variability in rainfall, plant production, vegetation state, and animal

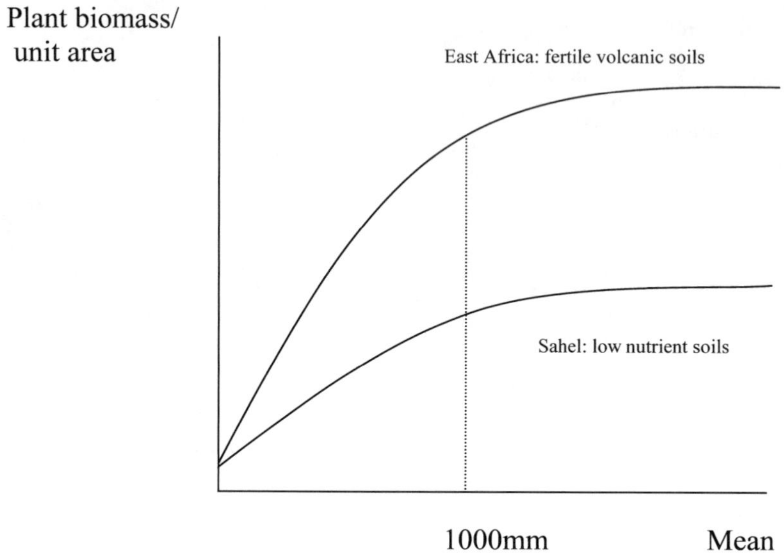

Variation in plant production with rainfall.

biomass, with boom-and-bust fluctuations involving rapid build up, great mobility, and catastrophic die-offs of both wild and domestic herbivores, driven by chaotically interacting random events of rain, fire, flood, drought, and epidemic. The observed vegetation states depend on the past trajectories and current interplay of rainfall, fire, epidemic disease, grazing, and browsing herbivore densities and impacts; and recruitment, regeneration, maturation, and decline of woody species.

Migratory and transhumant systems, both wild and domestic herds, move over great distances to track **pasture** and other resources in these boom-and-bust conditions. Temporary heavy grazing may leave bare landscapes denuded of ground cover in the dry season, returning rapidly to peak production with the next rains. Stephen Sandford predicted considerable variation between and within systems as to the implications of temporary heavy grazing, depending on local ecological factors such as soils and topography on the one hand, and local political and socioeconomic factors on the other. There is still a dearth of data on most systems, though reviews of the changing paradigm outline pathways toward testing the competing hypotheses, and an increasing number of cases are being shown to have no indication of any long-term decline of productive capacity. Detailed work by Matthew Turner and Sian Sullivan shows that individual cases of grazing and/or browsing impacts are complex and differentiated. They neither simply refute nor fully bear out one or other view of rangeland processes. As well as grass growth, woody vegetation cover may vary dramatically among years due to a combination of rainfall, fire, recruitment, browsing, large mammal damage, senescence, and partial, alternative successional trajectories.

Recently there has been a backlash against disequilibrium dynamics models. Some ecologists, often with primarily South African experience, support a conventional view of carrying capacity, overstocking, and **land degradation** in rangelands, on largely theoretical grounds. By contrast,

French-speaking ecologists working in Sahelian rangelands, most East African ecologists, and the great majority of development workers in arid and semi-arid lands find that disequilibrium dynamics offers a framework more consistent with their empirical experience. The two opposed models may be the extremes along a continuum, rather than mutually exclusive theories, and different situations may correspond to different positions along that continuum. The more arid and unpredictable the ecosystem, the more closely it may correspond to a "disequilibrium" system (though disequilibrium dynamics clearly has explanatory power even in relatively less arid systems such as tropical woodland and forest). On the other hand, the overwhelming forces of political economy driving the need to invest in cattle in some rangelands may have led to such high stocking rates that the conventional wisdom may hold some explanatory power in those systems. Also, both models may contribute to complementary insights where either alone could constrain understanding. In a parallel debate, economists and biologists have read the evidence over the determinants and environmental implications of climate change in completely different ways. Economists are interested in the major fluctuations that dominate the system and that call for immediate and major practical interventions. By contrast, biologists have focused on the relatively very small but steady creep of long-term, directional change. These different perspectives have led to major rifts in the **Intergovernmental Panel on Climate Change**. Although both approaches look at the same data series, each focuses on a different aspect of change and emerges with a different understanding of the important features of the system. Each is correct in its own terms, but neither is complete without the other.

However, the changing view of rangeland ecology has fundamental policy implications for sub-Saharan Africa. For example, rather than the destocking of pastoralist systems so commonly advocated, optimal management of disequilibrium systems may mean restocking to accelerate recovery after catastrophic die-offs, and to keep grazing and browsing pressures at a level that avoids excessive spread of woody species, and limits potential eruption of disease vectors. Not surprisingly then, these debates also have major political as well as scientific dimensions (see table below).

Analyses of the environmental discourse and the political ecology dimensions underpinning the debate are now beginning to emerge. In the case of the disequilibrium debate in African rangelands, powerful groups commonly invoke equilibrium theory to support suggestions that local land use drives environmental degradation and to justify attempts at intervention and control. Sian Sullivan and Katherine Homewood state that

> non-equilibrium ideas demote the expert, superior positioning of the scientist by emphasizing unknowability in terms of predicting the behavior of complex systems. They create problems for conservationists wishing to clear . . . landscapes of people and livestock in order to return these spaces to a desired, imagined original undisturbed state of nature. And by emphasizing the significance of local and historical specificities they affirm devolved land use and management as the most appropriate match between people and environment, thus reducing the legitimacy of state-centric, expert-led, top-down policy and planning.

Ecological, production, and tenure correlates of equilibrium and nonequilibrium models (after Oba et al., 2000:37; Sullivan and Homewood, in press)

	Equilibrium	Nonequilibrium
Ecology: plant production	Stable climate Primary production is stable from year to year	Climate varies unpredictably Primary production varies unpredictably (tightly linked to rainfall)
Ecology: animal production	Livestock population is density dependent Change in stocking density produces predictable qualitative and quantitative change in vegetation	Livestock population is density independent Livestock population tracks unpredictable forage production
Herd management and land tenure	Carrying capacity can be predicted (whether ecological or economic, MacNab ([1985]) Stocking density can be tailored to carrying capacity Private tenure of land, water	Carrying capacity calculations not useful guide to forage availability at any given place/time Opportunistic and mobile grazing practices more appropriate Common property resource management systems with fuzzy social and spatial boundaries allow flexible use of range resources

Unfounded allegations by governments and conservation agencies of environmental degradation by, and poor productivity in, mobile pastoralist systems can result in adverse pressures on rural land use and livelihoods. It is important to keep disequilibrium dynamics in view in dealing with any arid land system.

See also: **Climate Impacts; Desertification; Ecological Colonization; Land-Use History; Soil Degradation; Transhumance; Turnover.**

Further Reading

Behnke, Roy, Scoones, I., and Carol Kerven, eds., *Range Ecology at Disequilibrium: New Models of Natural Variability and Pastoral Adaptation in African Savannas*, London: Overseas Development Institute, 1993; Cowling, R. M., "Challenges to the 'New' Rangeland Science," *Trends in Research in Ecology and Evolution* 15 (2000): 303–4; Ellis, James E., and D. M. Swift, "Stability of African Pastoralist Ecosystems: Alternate Paradigms and Implications for Development," *Journal of Range Management* 41 (1988): 450–9; Homewood, Katherine, "Policy, Environment and Development in African Rangelands," *Environmental Science and Policy* 7 (2004): 125–43; MacNab, J., "Carrying Capacity and Related Slippery Shibboleths," *Wildlife Society Bulletin* 13(1985): 403–10; Oba, G., Stenseth, N. C., and W. Lusigi, "New Perspectives on Sustainable Grazing Management in Arid Zones of sub-Saharan Africa," *Bioscience* 50 (2000): 35–51; Sandford, S., "Pastoral Strategies and Desertification: Opportunism and Conservatism in Dry Lands," in *Desertification and Development: Dryland Ecology in Social Perspective*, eds. B. Spooner and H. Mann, London: Academic Press, 1982, pp. 61–80; Sullivan, Sian, and Katherine Homewood, "Nonequilibrium and Nomadology," in *Reclaiming Knowledge for Diversity*, ed. M. Pimbert, London: Earthscan, International Institute for Environment and

Development www.warwick.ac.uk./fac/soc/csgr/research/workingpapers/2003/wp/2203pdf/; Turner, Matthew, "Long Term Effects of Daily Grazing Orbits on Nutrient Availability in Sahelian West Africa," *Journal of Biogeography* 25 (1998): 669–94.

KATHERINE HOMEWOOD

Disintensification. The process of reducing inputs into agricultural land use, sometimes synonymously used for **extensification**. The target of this activity may be the same but less intensively managed land-use type, a transition to another agricultural system or a **fallow** period, or the return to semi-natural **vegetation**. The reasons for and the methods of disintensification are manyfold. Labor shortage in **agriculture** may be one of the globally occurring problems leading either to measures of intensification if sufficient capital and/or technology is available, or in case these options are not available, leading to disintensification, for example by reducing labor-intensive measures like weeding, erosion control, or multiple fruitpicking. Another important reason for the disintensification of land-use practices is that the lack of money causes reduced input of contracted labor or the application of adequate amounts of fertilizers and other agrochemicals. The consequences of disintensifying land use are reduced crop yields or quality of yields (output per harvest) and reduced productivity (output per unit area and time), but may also include increased **soil erosion** or nutrient mining.

While in developing countries disintensification may be more often than not forced upon farmers, in countries of agricultural overproduction economic incentives are provided to encourage farmers to disintensify land use on certain fractions of their terrain. The reasons behind motivating (e.g., farmers in the European Union) to partly reduce the intensity of agricultural practices (eventually down to zero) are twofold. First, in a situation of overproduction and guaranteed prices for many agricultural commodities, it is reasonable to provide incentives to reduce overproduction. Second, land taken out of production is intended to provide habitats for plants and animals, not available in high-intensity agriculture. Additionally, disintensification via reduced fertilizer applications leads to reduced emission of **greenhouse gases** and thus supports the efforts to fulfill the targets of the **Kyoto Protocol**. Both in developing and developed countries, marginal sites, often extensively managed, are given up because agricultural production is no longer economically viable.

Globally, organic or ecofarming plays a marginal but growing role. The decision to ecofarm is usually associated with reducing cropping intensity by reducing the amount of agrochemicals and shifting to more robust, less nutrient- and more labor-demanding, but also lower-yielding, varieties often associated with a different farming system. Thus, in many cases agricultural disintensification will result in changes in land use, while in other cases changes in land cover—for example, from crops to **grassland**, bushland, or even **reforestation**—can evolve from disintensification processes.

See also: **Agricultural Intensification; Driving Forces; Forest Transition; Land Abandonment; LandSHIFT Model.**

JÖRG A. PRIESS

DIVERSITAS. An international program on **biodiversity** science with the mission to understand and predict cycles of interactions in the changes of biological diversity. Cycles are believed to be due to biodiversity's own evolutionary and ecological dynamics but are also a result of deliberate and unintentional human actions. In turn, these cycles affect human societies, which have to adapt to these changes. Therefore an integrative science is promoted, linking biological, ecological, and social disciplines in an effort to produce socially relevant new knowledge and provide the scientific basis for the **conservation** and sustainable use of biodiversity. Together with the **International Geosphere-Biosphere Programme**, the **International Human Dimensions Programme on Global Environmental Change**, and the World Climate Research Programme, DIVERSITAS and these program form the **Earth System Science Partnership** for the integrated study of the earth system.

See also: **Agrodiversity; United Nations Convention on Biological Diversity.**

Further Reading

International Programme on Biodiversity Science [Online, September 2004], The DIVERSITAS Web Site, www.diversitas-international.org.

HELMUT GEIST

Domestication. Process by which plants and animals are selected from wild populations and adapted to special environments created by humans. Domesticated and cultivated plants and animals represent a very small fraction of the total species, but they constitute almost all of the food consumed by humans. Among the roughly 2,500 crop species, just 103 supply over 90 percent of the calories that humans consume, with just three grasses (cereals)—**rice**, **wheat**, and **maize**—supplying over 60 percent of the total. More than fifteen plants are cultivated as sources of fiber, and thousands more as ornamentals or sources of medicine. Of the approximately 50,000 described vertebrate species, thirty to forty species of birds and mammals have been domesticated. Apart from dogs and cats, four species of domestic mammals have a global distribution (except for the Antarctic) and are impressively dominant numerically: cattle (1.3 billion), sheep (1.2 billion), pigs (850 million), and chickens (10 billion). Most of the remaining domesticated animal species are more limited, like those of the Andean camelids (llamas and vicuñas). As in domesticated crops, the degree of intraspecific (genetic) diversity is astonishing. For example, there are about 800 distinct breeds each of cattle and sheep. Although the vast majority of the domesticated animals are used as food, they also provide wool, tallow, bone, manure for fuel and fertilizer, and leather, and they are used widely as draft animals. In the scientific literature, animals have hardly emerged as active agents in shaping the human-environmental conditions so far, except for a strand of work at the interface among archeology, anthropology, and (traditional) cultural geography, as represented by **Carl Ortwin Sauer**, for example. Domesticated plants and animals represent a very small fraction of the total species, but they form an important facet of **biodiversity**. The location where crops and agricultural development originated provides a view of the distribution of this facet of **agrodiversity**. The major crops have their centers of genetic diversity in today's developing countries or regions, largely in the neotropics, Middle East, the Mediterranean and northern Africa region, East Africa, South and Southeast Asia, and **China**. Many countries with higher concentrations of species diversity are located in tropical regions, which also coincide with the location of many of the centers of origin and development of major crops.

See also: **Agricultural Intensification; Agriculture; Cash Crops; Cattle Ranching; Colonization; Exotic Species; Food Crops; Great Plains; Turnover.**

Further Reading

Zohary, Daniel, and Maria Hopf, *Domestication of Plants in the Old World,* 2nd ed., Oxford, UK: Oxford University Press, 1993.

HELMUT GEIST

Driving Forces. Change in land use and land cover can be conceptualized as the result of the operation of pressures or driving forces: an impulse is applied, and a change of state results. The roots of this view lie in the natural sciences, but the approach has been used by both natural and social scientists. Various operational and definitional difficulties have been encountered in applying the approach, but it has proved useful as an analytical device. The notion of linearity implied in the term is further developed by the conventional practice of distinguishing between driving forces acting at different levels, and especially between underlying and **proximate causes**. The latter applies to the immediate agents of change (e.g., individuals practicing forest clearance), whereas the former relates to the general context in which they operate (e.g., political economy, technology). While it is helpful to identify different levels of operation of different factors, it must be acknowledged that simply to postulate two levels is rather simplistic. In practice, tangled webs of factors operate to influence land cover and land use.

One of the conceptual sources of **environmental change** in terms of underlying driving forces is the celebrated expression from the environmental movement in the 1970s that states I = PAT, where *I* stands for environmental impact, *P* is population, *A* is affluence, and *T* is technology. The expression was appealing at a time when there was concern that human populations and their affluence and technology were growing rapidly, and fears about future resource adequacy were widespread. Relatively few validations of the expression by empirical data have been reported, but more important has been the identification of basic or underlying factors that are perceived to drive environmental change, including change in land use and land cover.

The lineage from the origin of IPAT to the adoption of the concept of underlying driving forces by the land-use/cover research community around 1990 is not clear or direct. Nevertheless, B. L. Turner II and William Meyer used the term "human driving forces" and identified "candidate macro-forces" that included population, technological capacity, and economic development. There were, however, significant additions: they added two other factors, namely socio-economic organization and beliefs/attitudes. Since then, the concept has been used especially in relation to tropical **deforestation**, for example by Helmut Geist and Éric Lambin.

While there are variations in the identification and formulation of underlying driving forces, five factors (or groups of factors) commonly appear. They are population, economic development, technology, institutions, and culture. Sometimes attempts are made to isolate the roles of single factors: more commonly combinations or "multi-causal" situations are recognized.

Population

Population (and specifically population growth) has been identified as a driver of change since the days of Thomas Robert Malthus, if not before. A belief that **population dynamics** or change is linked to change in land use is credible and perhaps almost intuitive. The simplicity and elegance of the hypothesis, however, have not always been matched by the clarity and robustness of the conclusions from testing it. Numerous attempts have been made to isolate the influence of population change on land use, and especially on deforestation. Often the result has been that a statistically significant relationship has been established, but that the level of explanation that it offers is very modest.

Much of the attention focused on the demographic factor has been based on an assumption that population growth is detrimental to the environment, and that it results in adverse change in land use or cover. There is, however, an alternative view. Associated with **Ester Boserup**, this view contends that population growth can be a positive impulse leading to **agricultural intensification** and to higher yields per unit area. Casual observation suggests an association between high population densities and intensive land use in some parts of the world, and some verification of a positive relationship also exists. Nevertheless, assumptions of a negative relationship (i.e., more people, adverse change) have been more frequent, perhaps especially since the heydays of neo-Malthusians in the 1960s and 1970s.

The logic of an implied negative relationship is clear: population growth leads to growth in demand for food, which in turn leads to expansion of the cultivated area. **Conversion** of forest to field is the logical outcome of this process. Historically, of course, it is irrefutable that this process has operated. It is also clear that when population growth was reversed, as during the Black Death in fourteenth-century **Europe**, so also was land-use change: the **forest** returned over areas previously cleared for **agriculture**. Nevertheless, it is equally clear that shrinkage of the agricultural area can accompany population growth, as occurred in parts of Europe during the late nineteenth and twentieth centuries.

Perhaps it is not surprising that there is ambivalence or a lack of clarity in the results of many studies of the interrelationship between population and land-use change. One complicating factor is the distribution of population. In many European countries in the nineteenth and twentieth centuries, much of the population growth was in urban areas, far from the **agricultural frontiers**. The land-use effects of this kind of demographic growth are quite different from those of growth in rural areas where the population depends directly on land resources.

A second complicating factor is that of trade and its geographical pattern. Implicit in many analyses is the assumption that a relationship exists between population and land-use change at the local scale. Such an assumption may be valid when a subsistence economy or closed system is under consideration, but not where national or global-scale systems operate. The logical basis for assuming that population growth in a province or municipality drives deforestation at the frontier of cultivation is not clear if much of the growth is concentrated in a town or city.

Third, population has often been viewed as an undifferentiated entity composed simply of a number of individuals. The social organization of the population and its institutions has not always been taken into account. For example, a highly concentrated pattern of land ownership, with little opportunity for the landless population to gain **access** to land, could mean that peasants encroach on the forest in the absence of any other means of survival. Similar population growth in a different situation, in which access to land was easier, could have very different effects, perhaps including Boserupian innovation and intensification of land use. One of the obvious flaws in the original IPAT expression is that there is no space for laws, regulation, or any other institutional response: if data on simple population totals or densities are used in analyzing land-use change, then the same weakness is present.

Population has often been selected as a putative driver of change, notwithstanding the ambivalence of many analyses. The words of William Meyer and B. L. Turner II are apposite and still valid: "We emphasize population because of its prominence in the literature and because it is one of the few candidate driving forces for which there is a simple measure" (1992, p. 52). In other words, we continue to use it because it is easy to use. Perhaps it might be added that it is easily used in regression analysis, one of the most widely deployed tools in work on global change in land use and land cover.

Economic Development or Affluence

If demography has attracted a lot of attention as a postulated driver, so also has **economic growth**, or development. Again, the hypothesized relationship enjoys a logical plausibility: more affluence leads to more demand, more **consumption**, and hence more production. The vast disparities between per capita consumption in rich and poor countries are often cited, and both **poverty** and high levels of affluence have been invoked in attempts to explain change. The gross consumption of the latter can impose its own pressures. In the case of the former, the lack of alternatives can "drive" forest encroachment for **subsistence agriculture** at the individual level; or, at the national level, for products for export. The need to service debt is sometimes quoted as a significant driver in this context.

As in the case of population, simple quantitative measures such as gross domestic product per capita are available. Recent years have seen attempts to relate trends in national agricultural and forest areas to per capita incomes, with suggestions that Kuznets-type curves may describe these trends. In most cases, however, these attempts have failed to progress far beyond the descriptive stage or to demonstrate clear and direct linkages between income levels and land use.

Again, as in the case of demographic factors, it is difficult to deny the existence of an association between level of development and land use. Equally, it seems that no deterministic relationship can be—or at least has been—demonstrated. Nevertheless, certain stages in development are highly significant. For example, when a locality is first integrated into a wider regional, national, or global economy, fundamental changes occur. Whereas previously, local self-sufficiency might require that many products be produced on a small **scale**, the new setting permits the large-scale production of a few products in which local advantage is held. The result can be near-**monoculture** over extensive areas. Numerous examples of such changes could be quoted, including "colonial" plantation crops produced for distant markets. The prerequisite of an efficient transport system is obvious here and is a reminder that economic change and change in technology (including transport technology) are closely interrelated.

Technology

Technology epitomizes the complexity encountered when attempting to analyze land-use change in terms of underlying and proximate factors. It can function at a range of levels, from that of the chainsaw in the forest to the information technology that helps global markets to function. Technology is rather different from population and economic development in that it can be regarded as a passive or facilitative factor rather than a "driver": it provides a set of tools for agents to use but is not an active agent in itself.

Technology also differs from economic development or population in that it is difficult to quantify, and it has been less frequently used in analyses of change. This is not to discount its significance. Without the railroad and steamship in the nineteenth century, for example, neither **wheat** farming in the **Great Plains** nor colonial **coffee** production would have attained their respective scales or significance in terms of **land-cover change**. Similarly, technological change in agriculture, including plant breeding, mechanization, and synthetic fertilizers, has had a profound effect on crop yields and land productivity. In its absence, requirements of land for food production would have been very different.

Thus the relationships between both demography and economy and land-use change have been mediated through technological changes. The low levels of statistical explanation usually established in regression analysis of the relationships between population change and increasing affluence on the one hand and land-use change on the other imply that other factors are involved: technology is one of these. A picture of multicausality emerges: combinations of factors are usually more important than single driving forces. Furthermore, the significance of combinations within technology is as great as that between technology and other driving forces: for example, technological change in agriculture in nineteenth-century Europe occurred alongside the growth of the railway network. Without the latter, the inputs required in the new agriculture would have been difficult to acquire, and the outputs would have been difficult to market.

Institutions

A further group of factors that mediate between demography and economy on the one hand and change in land use on the other are **institutions**, used here as an umbrella concept to cover a wide range of social, political, and legal arrangements that humans have worked out in order to regulate their collective behavior. These include, for example, land laws and **property** rights. A secure and robust system for the recognition and protection of property rights can, arguably, help to discourage exploitative land use and "cut and run" practices. Conversely, systems of **land tenure** that facilitate short-term leases with little need to invest significantly in land may favor exploitative behavior.

The relative merits of private, state, and communal occupation of land have attracted endless debate, and the dominant views have fluctuated over time. Rigorous analysis of their relationship to land-use change, however, is not easily carried out in a way that can lead to general conclusions, and the same applies to the roles of various other institutions. This is not to deny the importance of institutions, but merely to highlight the intellectual challenges encountered in isolating their role. Few would deny the significance of the planning system introduced in Great Britain after World War II, for example, but to measure its effect in quantitative terms is almost impossible. Similarly, at a different scale, colonialism has obviously had a profound effect on global land use and land cover. Its effects, however, are very difficult to measure, not only because they are so profound and pervasive but also because it cannot be known what would have happened if it had not developed.

Culture

Culture is a term that is even broader than institutions: it is used here to describe the attitudes, values, norms, and knowledge of individuals, communities, and societies. The notion of "cultural landscape" is well established, reflecting the idea that the appearance of a piece of territory is shaped by the cultural group who have lived and worked in it. Implicit in the notion is that a link exists between culture and land use. Little success, however, has been achieved in specifying the precise nature of that link. Various reasons can be adduced to explain this: reliance on case studies; problems of quantification; disconnections between attitudes and behavior within culture groups; and (possibly) even political correctness that can deter intercultural comparisons. A further factor is that culture provides the context in which institutions evolve, technology develops and is deployed, economic development proceeds, and population growth occurs. In other words, it operates at a level different from the other underlying factors.

To be unable to measure the significance of culture in relation to land-use change does not mean that it should be ignored or neglected. It has been shown, for example, that groups such as the Mennonites and Amish in North America have distinctive relationships to the land they use. On a different scale, the change in attitudes and values toward forests in parts of the developed world in the late twentieth century is cultural, and it has impacted the location of new forests as well as the ways in which they are designed and used. More generally, the environmental consciousness that developed from the 1960s fed into institutional change, which in turn has had major effects on land use.

Overview

Conceptualizing underlying driving forces is helpful in contributing to an understanding of change in land use and land cover, but only to a certain extent. Several of the difficulties have already been indicated: they include problems of quantification, problems in isolating the role of individual factors, and problems of multi-level operation of some of the candidate factors. They also include multicausality. To some extent, any separation into underlying and proximate factors is arbitrary: in practice a continuum of influences exists, rather than a clearly defined stratification or hierarchy. In addition, the term "driving force" or "driver" can be problematic. While some changes of land use or land cover can clearly be seen to result from certain drivers or impulses, others are more difficult to conceptualize in this way. For example, late nineteenth-century **globalization** produced an impulse for change in areas such as North America and Australia, leading to expanded production of wheat and wool. The availability of these products at low cost reduced the competitiveness of farming in parts of Britain, and hence the **disintensification** or withdrawal of farming in some areas. For such areas, the usefulness of the term "driver" is debatable: arguably globalization was the ultimate driver, harnessed to imperial expansion overseas, but the local influence on land use in Britain could perhaps be more easily regarded as "backwash" than as driver. On the other hand, (physical) climatic change could obviously be a real driver of land-cover change: perhaps the analogy of change in a human climate of culture is not so far-fetched.

The desire to analyze the causes of change with a view to the generation of generalizations is deep-seated. Perhaps, however, we have to acknowledge that the reductionism that has characterized the natural sciences may have limitations when applied to the highly complex human systems with which land-use change is associated. In addition, we may have to acknowledge limitations in the cross-national statistical analyses (and especially regression analyses) that have been so widely used as to almost dominate the field of global land-cover research. Systematic frameworks for the analysis of change at the global level are still youthful, having evolved only since the beginning of the 1990s. Progress has been achieved with them, but a full (or even adequate) understanding of the determinants of change has been elusive.

See also: **Agent-Based Model; Agricultural Revolution; Anthropocene; Biomass; Cattle Ranching; Change Detection; Colonization; Cultural Factors; Decision-Making; Degradation Narrative; Desertification; Economic Livelihood; Economic Restructuring; Exogenous Factor; Exotic Species; Extensification; Forest Anomalies; Hot Spot Identification; Industrial Revolution; Investments; Land Concentration; Land Degradation; Land Rights; Land-Use History; Marsh, George Perkins; Mediating Factor; Metabolism; Pristine Myth; Reforestation; Secondary Vegetation; Technological-Scientific Revolution; Tragedy of Enclosure; Tropical Dry Forest; Urban-Rural Fringe; Vulnerability.**

Further Reading

Boserup, Ester, *The Conditions of Agricultural Growth: The Economics of Agrarian Change under Population Pressure*, London: Earthscan, 1993; Geist, Helmut J., and Éric F. Lambin, "Proximate Causes and Underlying Driving Forces of Tropical Deforestation," *BioScience* 52 (2002): 143–50; Malthus, Thomas R., *An Essay on the Principle of Population, as It Affects the Future Improvement of Society with Remarks on the Speculations of Mr. Godwin, M. Condorcet, and Other Writers*, London: Printed for J. Johnson, 1798; Mather, Alexander S., and Coby L. Needle, "The Relationships of Population and

Forest Trends," *Geographical Journal* 166 (2000): 2–13; Mather, Alexander S., Needle, Coby L., and Jamie Fairbairn, "The Human Drivers of Global Land Cover Change: The Case of Forests," *Hydrological Processes* 12 (1998): 1983–94; Meyer, William B., and B. L. Turner II, eds., *Changes in Land Use and Land Cover: A Global Perspective*, Cambridge, UK: Cambridge University Press, 1994; Meyer, William B., and B. L. Turner II, "Human Population Growth and Global Land-Use/Cover Change," *Annual Review of Ecology and Systematics* 23 (1992): 39–61; O'Neill, Brian C., "IPAT," in *Encyclopedia of Global Change: Environmental Change and Human Society*, eds. Andrew S. Goudie and David J. Cuff, Vol. 1 (A–I), Oxford, UK, New York, NY: Oxford University Press, 2002, pp. 702–6; Turner, B. L. II, and William B. Meyer, "Land Use and Land Cover in Global Environmental Change: Considerations for Study," *International Social Sciences Journal* 43 (1991): 669–79.

ALEXANDER S. MATHER

E

Earth Resources Observation Systems (EROS) Data Center. A data management, systems development, and research field center for the U.S. Geological Survey (USGS), a bureau of the U.S. Department of the Interior. The EROS Data Center (EDC) was established in 1971 to process and distribute data from **Landsat** satellites, as well as **aerial photographs** gathered for the USGS and other agencies. The U.S. Congress designated EDC as the National Satellite **Remote Sensing** Archive in 1992, meaning that all civilian U.S. government remotely sensed data are archived at the EDC. Global coverage from Landsats 1 through 5 and 7 forms the core of the archive. Over the past three decades, more than 630,000 gigabytes of Landsat digital data were placed into the archive. EDC also holds data from other global imaging systems, including data from the **advanced very high resolution radiometer** (AVHRR) carried aboard National Oceanic Atmospheric Administration's polar orbiting weather satellites; declassified intelligence satellite photographs from the CORONA, ARGON, and LANYARD satellite missions; and data from the Shuttle Radar Topography Mission (SRTM). EDC also serves as the Land Processes Distributed Active Archive Center for the NASA Earth Observing System (EOS). EDC handles the higher-level processing and distribution of ASTER data, and the distribution of **moderate resolution imaging spectroradiometer** (MODIS) land products derived from data acquired by the Terra and Aqua satellites. In addition to maintaining earth science data, the EDC scientific staff uses remotely sensed data to study the changing earth. The core of the USGS land-cover research and development capabilities resides at the EDC.

See also: TERRA-ASTER.

Further Reading

United States Geological Survey (USGS) Program, Science for a Changing World [Online, March 2005], The EROS Data Center Web Site, edc.usgs.gov.

THOMAS R. LOVELAND

Earth System Science Partnership. In 2001, the four international global change programs—**DIVERSITAS**, the international program on **biodiversity** science; the **International Geosphere-Biosphere Programme**; the **International Human Dimensions Programme on Global Environmental Change**; and the World Climate Research Programme—developed an alliance called the Earth System Science Partnership (ESS-P). The formal goal of the ESS-P is to facilitate collaborative activities for its integrated study of the earth system, the changes that are occurring to the system, and the implications of these changes for global sustainability. Core ESS-P activities are the following:

1. Joint projects that build on the research of the four programs
2. Integrated regional studies such as the Monsoon Area Integrated Regional Study (MAIRS)
3. Earth system analysis and **modeling**
4. Global change open science conferences
5. START, or Global Change System for Analysis, Research and Training, for capacity building and regional research

The three established joint ESS-P projects are the Global Carbon Project (GCP), launched in 2003, which analyzes the biophysical and human dimensions of the **carbon cycle**; the **Global Environmental Change and Food Systems (GECAFS) project**, launched in 2001 and aiming at determining strategies to cope with the impacts of **global environmental change** on food systems and the environmental and socio-economic consequences of adaptive responses; and the Global Water System Project (GWSP), launched in 2004, which seeks to understand the interconnections between global environmental change and local or regional coupled water-human systems. A fourth joint project on **human health** and global environmental change is currently under development.

Further Reading

Earth System Science Partnership [Online, January 2005], The ESS-P Web Site, www.ess-p.org.

BARBARA GÖBEL

East-Central Europe. East-central Europe (ECE) is part of standard central Europe (CE), which is usually delineated according to its historical evolution. CE is a region that in the past belonged entirely or partly to a rather formal and loose "confederation" of the so-called Holy Roman Empire (AD 962–1806, from the beginning of the sixteenth century called the Holy Roman Empire of the German Nation). It was later ruled by the Habsburg monarchy (1526–1918; from 1867 onward called Austria-Hungary). As opposed to the countries of **west-central Europe** (WCE) such as Austria, Germany, Switzerland, and Liechtenstein, the ECE comprises the region's largest and most populous coastal country of Poland, with 39 million inhabitants; then the inland Czechia (also known as the Czech Republic) and Hungary, each with 10 million; Slovakia, with 5.5 million; and Slovenia, with 2 million; and with a short coastal line by the Adriatic Sea. As a result of the Second World War, the area was divided by the Iron Curtain as a sharp demarcation line separating eastern and **western Europe**: democratic capitalist and communist countries for nearly the entire second half of the twentieth century, creating distinct social and economic systems and thus different land-use changes after 1945. Post-communist **southeast Europe** encompasses the whole Balkan peninsula excluding Greece and has been the least-developed region in Europe, mainly from its heritage under a centuries-long occupation by the Ottoman Empire from the fifteenth until the nineteenth century.

In its physical geography, the ECE extends to two areas. First, there is Hercynian CE, to which the majority of Poland and Czechia belong. The second region is the Alpine-Carpathian one; only Slovenia belongs to the Alpine part, while in Carpathia one finds Slovakia, Hungary, and the eastern quarter of Czechia. Alpine folding during the Tertiary period formed the Alpine-Carpathian sub-region. The Danube River between the cities of Vienna and Bratislava divides the Carpathian part, a bow-shaped mountain range about 1,500 kilometers long. After another rift (the Iron Gate), this part continues to Bulgaria and contains big lowlands such as the Great Hungarian Plain (Alföld) in the Panonian basin. Almost the entire area of Hungary and northern Serbia (Vojvodina) with its southern boundary river Sava is located on this plain (Serbia belongs to southeast Europe). It is the area of the most fertile black soils used for cereal agriculture. The western Carpathians in Slovakia reach heights up to 2,600 meters. Cambisols predominate in their flysh strip. In comparison with the Alps, the climate is more continental, with lower precipitations, which increase with altitude and decrease to the east. Here is a great oscillation of temperature, with lowlands being much warmer. The natural composition of **forests** is disrupted less than that of the Hercynian sub-region. Oak and beech forests predominate, whereas spruce forests prevail in higher altitudes until the upper forest boundary at a height of 1,500 meters in Slovakia. Former steppe **vegetation** was replaced by so-called cultural steppe due to **agricultural intensification**.

According to climate and flora, both areas are transitional zones between the oceanic moderately warm and humid climate of western Europe and the cold and humid boreal climate of continental eastern Europe. The Polish part of Hercynian CE is formed by a plain created by Quaternary continental glaciers (many lakes are there), while the Czech part constitutes highlands of 115–1,600 meters. Cambisols and podzol soils (in Poland also luvisols) predominate; in Czechia one can find also loess black soils. After the glacier's withdrawal about 10,000 BC, mixed forests, which then covered about 90 percent of the total area, emerged. Beginning with the Neolithic revolution (the birth of **agriculture**, ca. 6000–3500 BC), humans began to influence nature at higher **scales**. The amount of forest started to decrease (which was a major trend until the onset of the **Industrial Revolution**), mostly due to large-scale **colonization** from the twelfth to the fourteenth centuries (when the proportion of forests was even lower than today). However, after long wars such as the Thirty Years' War from 1618 to 1648, due to a mass decline of population and settlements, the forest area increased mainly in hilly, less-fertile regions. Today, still-visible remnants of deserted fields and abandoned villages in these new forests indicate landscape features of the Middle Ages. Types of forests changed according to altitude (oak > beech > fir > spruce). This was no longer true at the end of the eighteenth century, when a controlled revitalization of forests set in. This revitalization has led until nowadays to lasting growth of forest areas; however, it has led also to changes in forest composition in favor of spruce and pine **monocultures** (spruce, as a raw material, was the tree of the Industrial Revolution).

East-central Europe covers about 550,000 square kilometers, about 5.3 percent of the total geographic area of **Europe**, and extends up to the Ural **mountain** range, the Caspian Sea, and the Caucasus. Europe has about 770 million inhabitants, out of which ECE's 66 million are 8.5 percent of the population. Average population density reaches 120 per square kilometer and is lower than that of WCE, which is also true for the rate of **urbanization**. East-central Europe has been developing close ties with German-speaking countries. Germs of Polish, Czech, and Hungarian national states were created in the ninth and tenth centuries, whereas Slovaks and Slovenes did not constitute their own statehood until the end of the twentieth century. Hungary and Slovenia, as part of the Habsburg monarchy until the end of the seventeenth century, were a buffer zone against the Ottoman Empire, as was Poland against Russia in the east and together with Czechia against Germany in the west. East-central Europe possessed no colonies, which was one reason why the Industrial Revolution climaxed here forty to fifty years later (e.g., Czechia) than in the United Kingdom. After the collapse of the Austro-Hungarian monarchy in 1918, Hungary lost about two-thirds of its historical territory: Slovakia or Upper Hungary, and Transylvania, Serbian Vojvodina, and Croatia. This economic difference was further deepened after World War II by communism and its rigid centrally planned economy that copied the Soviet Union's style of **industrialization**. The close market alliance (even 70 percent of foreign trade) between ECE and the Soviet Union lasted until the early 1990s. After the fall of communism and consequent transition to a capitalistic market economy, this situation fully changed. Now, the majority of ECE countries' foreign trade is directed to the countries of the European Union (EU), especially Germany. This majority, and especially its standard and quality of living, are expected to increase, because all ECE countries became EU members in 2004. The economies and societies of ECE countries are now undergoing a deep transformation in a democratic environment. The capital cities and their hinterlands, such as the then-historical Polish Upper and Czech Silesia (which constitute the largest conurbation of ECE), northwestern and eastern Czechia, western Slovakia, and Hungary, are highly urbanized and industrialized regions.

East-central Europe's lower level of economic development compared with west-central Europe's is a consequence of its inland location in the heart of Europe and especially of the past. East-central Europe's agricultural sector produced about 3.1 percent of its gross domestic product (GDP) in 2003; the most in Poland was 6 percent. These indices are almost the same as in WCE.

Landscapes of ECE are either lowlands or plains (Hungary, Poland), or hilly and mountainous regions with smaller lowlands (Czechia, Slovakia, Slovenia). They form a rich mosaic of industrial and intensive agricultural areas. The area used for settlement, transportation, and industry covers about 12 percent of the total area, and it continues to grow. This can cause faster outflow of precipitation water. In northwestern Czechia, there are large, open carriers for soft coal mining, with the coal burnt in many big local power plants. Together with adjacent landscapes in eastern Germany (Saxony) and in southwest Poland, the so-called "Black Triangle" formed one of the most-damaged natural environments in Europe until the 1990s. Upper Czechia and Polish Silesia, with their rich hard coal resources, are noted for their deep mines and recently diminishing, formerly huge metallurgy industry. Extensive burning of soft coal with high sulfur content (for electricity production and to provide heating for households) caused—with prevailing east-west direction of winds—soil acidification and serious damage of forests.

Lowland Poland is noted for its agricultural specialization. In Poland, even during the communist period, a private sector, atomized into a few small unproductive farms, was partially preserved there. This situation, which led to a high proportion of employment in agriculture (about 20 percent in contrast to 4 percent in Czechia) and considerably crumbled cultivated land, signifies a problem for the common European market and agricultural policy.

The lowland of Hungary is used for intensive agriculture and has better soil and climatic conditions. Hungarian (but also Czech and Slovak) agriculture, previously typical for its high share of small farms to the extent of twenty hectares, was collectivized and transformed to large cooperatives or state farms during the communist period. After transformation and **land reform** in the 1990s, the cooperatives generally survived and changed their capital forms. Different from countries in southeast Europe in natural conditions as well as historical evolution, socialization of agriculture, and the so-called socialistic industrialization during communism explain why the share of agricultural land in 2002 was highest in Hungary (63 percent of total land), Poland (59 percent), Czechia (54 percent), and Slovakia (50 percent), and only 25 percent in mountainous Slovenia, with a similar trend for arable land. Permanent **grasslands** are most widespread in Slovakia (18 percent) and Slovenia (15 percent). These numbers correlate with the share of forests and woodlands in total land area: Slovenia (55 percent), Slovakia (45 percent), Czechia (34 percent), Poland (30 percent), and Hungary (20 percent).

After 1989, agriculture in ECE was exposed to a much more subsidized EU agriculture and cheap commodities from overseas. Mechanisms of **land rent** started to work (again), and as a consequence, less-fertile soils have been abandoned and transformed to permanent grasslands, the area of which, as

East-Central Europe: land changes 1961–1990–2002 in percent

Land-Cover Category	Share in Total Area			Change in Period		
	1961	1990	2002	1961–1990	1990–2002	1961–2002
Arable land	49.2	45.3	43.2	–7.9	–4.7	–12.3
Permanent cultures	1.8	1.8	1.6	–1.7	–9.4	–10.9
Permanent grassland	14.0	12.9	13.1	–8.0	1.6	–6.5
Agricultural land	65.0	60.0	56.8	–10.9	–3.51	–11.0
Forests-woodland	25.2	28.0	29.6	11.2	5.4	17.2
Other areas	9.8	12.0	12.5	22.5	5.1	28.7
Total	**100.0**	**100.0**	**100.0**	**100.0**	**100.0**	**100.0**

Source: Statistical database of the UN Food and Agricultural Organization (http://www.fao.org).

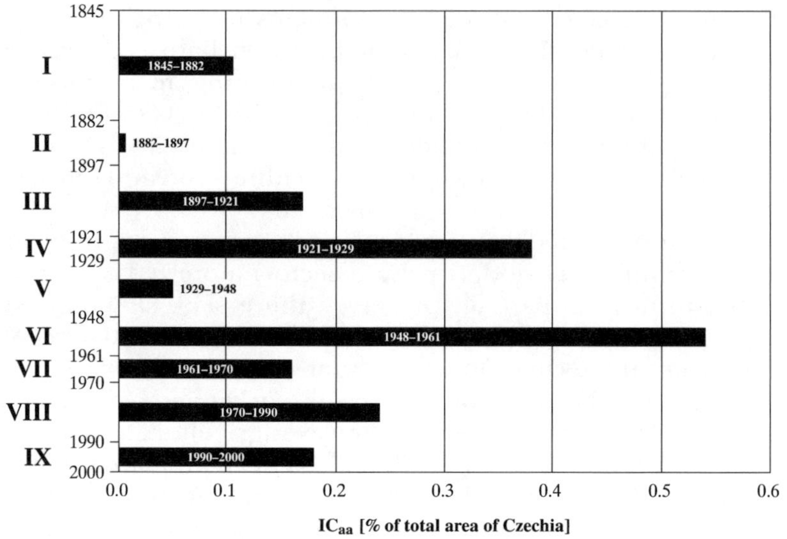

Land-use changes in Czechia 1845–2000 and some of their societal driving forces (IC = average annual index of change in percent); I: Revolution 1848–1849; Abolition of serfdom, land became a free good; climax of Agricultural Revolution; dominant impact of differential rent I; extensive development of agriculture. II: transition to intensification of agriculture; greater impact of differential rent I; competition of cheaper grain from the United States; agrarian drawn-out crisis. III: a boom in agriculture; first phase of technological-scientific revolution in agriculture; the World War I impact. IV: the first land reform; onset of electricity and combustion engine use in agriculture; state monopoly on grain trade; Policy of customary protection of inner agrarian market. V: big economic crisis in 1929–1933; the World War II impact; German occupation of Czechia. VI: the transfer of the Czech Germans; second land reform; extensive development of economy and its nationalization; collectivization of agriculture. VII: economic depression; attempts for intensification of agriculture deepening; completing of collectivization. VIII: stagnation of economy; joining of small cooperatives into larger ones; simplification of rural landscapes by big land tracks; act on agricultural land protection approved. IX: re-introduction of capitalism and market economy; restitution of private property plots; capital transformation of cooperatives and state farms; increase of small farming; competition of more subsidized agricultural products from abroad.

well as forested land, began to increase (again) after about 150 years. These trends for the ECE are demonstrated in the table.

There are long-term cadastral data on land use for wide parts of the ECE that allow one to investigate land-use changes on the macro-regional level, such as in the case of Czechia for nine time periods—see figure titled "Land-use changes in Czechia 1845–2000." These changes are measured by the so-called index of change (IC), representing the area where the change of land use occurred between two time points in the total territory of an area (considering that average annual ICs allow one to observe the basic trends and differences in land-use changes, including a demonstration of selected major societal **driving forces**).

See also: **Agricultural Revolution; Cadastre; Economic Restructuring; Land Privatization; Land Rights; Reforestation; Regrowth; Secondary Vegetation.**

Further Reading

Bičík, Ivan, Chromý, Pavel, Jančák, Vít, Jeleček, Leoš, Štěpánek, Vít, Winklerová, Jana, and Lucie Kupková, "Land Use/Land Cover Changes in Czechia over the Past 150 Years: An Overview," in *Land Use/Cover Change in Selected Regions in the World*, vol. *I*, eds. Himiyama, Yukio, Mather, Alexander, Bičík, Ivan, and Elena V. Milanova, Asahikawa, Japan: International Geographical Union Study Group on Land Use/Cover Change, Hokkaido University of Education, 2001, pp. 29–39; Bičík, Ivan, Jeleček, Leoš, and Vít Štěpánek, "Land-Use Changes and Their Social Driving Forces in Czechia in the 19th and 20th Centuries," *Land Use Policy* 18 (2001): 65–73; Feranec, Ján, Šúri, Marcel, Ot'ahel, Ján, Cebecauer, Tomáš, Pravda, Ján, Kolář Jan, Soukup, Tomáš, et al., "Landscape Change Detection, Analysis and Assessment in the Phare Countries: The Czech Republic, Hungary, Romania and Slovak Republic," in *Land Use/Cover Change in Selected Regions in the World*, vol. I, eds. Himiyama, Yukio, Mather, Alexander, Bičík, Ivan, and Elena V. Milanova, Asahikawa, Japan: International Geographical Union Study Group on Land Use/Cover Change, Hokkaido University of Education, 2001, pp. 53–60; Gabrovec, Matej, Kladnik, Drago, and Franci Petek, "Land Use Changes in the 20th Century in Slovenia," in *Land Use/Cover Change in Selected Regions in the World, Vol. I*, eds. Himiyama, Yukio, Mather, Alexander, Bičík, Ivan, and Elena V. Milanova, Asahikawa, Japan: International Geographical Union Study Group on Land Use/Cover Change, Hokkaido University of Education, 2001, pp. 41–52; Kostrowicki, Jerzy, *Land Utilisation in East-Central Europe: Case Studies*, Warsaw, Poland: Geographica Polonica, 5, 1965; Magocsi, Paul R., *Historical Atlas of East Central Europe*, Seattle, WA, London: University of Washington Press, 1993; United Nations Food and Agricultural Organization, Statistical Database [Online, December 2004], The FAO Web Site, www.fao.org; U.S. Central Intelligence Agency, The World Factbook [Online, December 2004], The CIA Publication Web Site, www.cia.gov/cia/publications; Zagoroff, Slavcho, and Stanley P. King, eds., *The Agrarian Economy of the Danubian Countries, 1933–1945*, Stanford, CA: Stanford University Press, 1955.

LEOŠ JELEČEK

Ecological Colonization. Entry and spread of a species (or a gene) into an area, habitat, or population from which it was previously absent. The first species to invade a new territory are called pioneers. Over time a range of species arrives and the habitat matures. In this process, called succession, early colonizers are often replaced and the species composition of the habitat changes. Factors influencing ecological colonization are facilitation (species positively affecting other species, facilitating colonization), inhibition (species negatively affecting colonizing species, hampering colonization) and competition (species competing for limited resources with mutual negative effects). In ecology, the opposite of colonization is extinction. According to the ecological notion of **colonization**, humans seem to be one of the most successful species, as they are able to invade almost all types of habitats. At least with respect to terrestrial environments, this process is today almost completed, because humans inhabit almost all parts of the earth. In the context of land-use/cover change, the ecological significance of colonization can be assessed by analyzing changes in natural systems induced by colonization. Indicators can assess, for example, ecosystem changes in terms of energy availability, **biomass** stocks (standing crop), biomass **turnover**; or changes in other ecosystem processes, in land cover, in species composition, or distortions of food webs.

See also: **Biodiversity; Disequilibrium Dynamics; Exotic Species; Human Appropriation of Net Primary Production; Species Extinction.**

Further Reading

Crosby, Alfred W., *Ecological Imperialism: The Biological Expansion of Europe, 900–1900,* Cambridge, UK: Cambridge University Press, 1986; Elton, Charles Sutherland, *The Ecology of Invasions by Animals and Plants,* London: Methuen, 1958; Fischer-Kowalski, Marina, and Helmut Haberl, "Sustainable Development: Socio-Economic Metabolism and Colonization of Nature," *International Social Science Journal* 158 (1998): 573–87; Gallagher, Richard, and Betsy Carpenter, "Human-Dominated Ecosystems," *Science* 277 (1997): 485.

KARLHEINZ ERB

Ecological Footprint. The total area required for sustaining socio-economic processes. The concept of the ecological footprint (EF) was originally developed in the early 1990s by William Rees and Mathis Wackernagel. The basic methodology has been subjected to a variety of adaptations, extensions, and modifications by several authors since then.

Two basically distinct approaches exist for calculating EFs in the strict sense. The more popular systems approach calculates footprints for socio-economic units, such as nations, cities, regions, and households. The second approach calculates footprints for (industrial) processes, products, and services, analogously to life cycle assessments.

The EF serves as a comprehensive indicator of ecological sustainability. In general, footprint assessments are based on the notion that the availability of bioproductive area is a major limiting factor for a sustainable economy. With this assumption, the ecological footprint ultimately adopts the concept of carrying capacity, which is a population-ecological concept expressing the maximum sustainable population density of a given species on a certain area—and applies it to human society.

In the systems approach, the EF assesses how much biologically productive area is needed to sustain the socio-economic **metabolism** of a country or other politically defined entities. In a second step, this overall demand is compared with biocapacity, that is, the amount of bioproductive area available (on the global or regional level). If the EF (as area demand) exceeds the available biocapacity, this is interpreted as ecological deficit (or "overshoot"), being a situation in which human **consumption** exceeds ecological limits and hence is not sustainable. The EF is a measure of strong sustainability, assessing the utilization of the environment in physical terms and comparing the extent of resource use with the availability of resources.

In practice, footprint calculations include only a limited range of environmental concerns that are directly related to land use and consumption (conservative approach). Area demand—typically consisting of land-use classes such as arable land, grassland, **forests**, fishing ground area, built-up area, and (hypothetical) CO_2-absorption land—is calculated from the quantity of domestic production plus imports minus exports; hence, the footprint focuses on what is usually called the apparent consumption of resources.

In its original version, EF assessments rely on globally average productivities; that is, EF assessments appraise the area that would be required to sustainably produce the goods and services consumed in a country on a global average. Also, the biocapacity is calculated as global average area equivalent. This global average approach is applied in cross-country comparisons, as it makes different consumption levels of different countries comparable, regardless of local or regional differences in production conditions like climate, soil characteristics, or technology.

Ecological footprinting has attracted criticism and praise in equal measures. Its advantages are its comprehensiveness, its strength as a communication tool, and its ability to make the ecological significance of the resource consumption in different countries comparable. Its ability to demonstrate the extent to which humanity is exceeding the ecological limits, and the share of regions in this, has attracted considerable attention in both science and the general public. Moreover, the EF methodology contributed to studies of the global distribution of **access** to natural resources and of the consequences of trade in re-allocating environmental pressures between world regions. On the other hand, the EF has been criticized as an oversimplification of the sustainability question (e.g., pointing out that human welfare is a multidimensional issue), for its restriction to a few single environmental concerns, for how it addresses issues of **greenhouse gas** emissions, for not discerning sustainable and non-sustainable land use, and for ignoring the possibility of a multifunctional use of ecosystems. The approach applying world average yields (or productivity) has been criticized because it results in global average areas that cannot directly be related to actual land use and because it is problematic when applied in time series analysis.

Despite its rather specific definition, the term "ecological footprint" is used synonymously with other indicators of human impacts on terrestrial ecosystems, above all the **human appropriation of net primary production**.

See also: **Change Detection; Ecosystem Services.**

Further Reading

Global Footprint Network, Advancing the Science of Sustainability [Online, October 2004], Ecological Footprint Web Site, www.footprintnetwork.org; Haberl, Helmut, Wackernagel, Mathis, and Thomas Wrbka, eds., *Land Use and Sustainability Indicators: Land Use Policy* 21 (2004): 193–320; Wackernagel, Mathis, and William E. Rees, *Our Ecological Footprint: Reducing Human Impact on the Earth*, Gabriola Island, BC, Philadelphia: New Society Publishers, 1996; Wackernagel, Mathis, Schulz, Niels B., Deumling, Diana, Linares, Alejandro C., Jenkins, Martin, Kapos, Valerie, Monfreda, Chad, et al., "Tracking the Ecological Overshoot of the Human Economy," *Proceedings of the National Academy of Sciences* 99 (2002): 9266–71.

KARLHEINZ ERB

Ecology and Society. An electronic, half-yearly, multidisciplinary journal devoted to the rapid dissemination—especially in the form of special issues—of current research into the natural sciences, social sciences, and the humanities. Formerly called *Conservation Ecology*, the journal is concerned with the relationship between society and life-supporting ecosystems (on which human well-being ultimately depends). The journal addresses the foundations for sustainable social-ecological systems; that is, the management, stewardship, and sustainable use of ecological systems, resources, and biological diversity at all levels; the role of natural systems in social and political systems and the effects of social, economic, and political **institutions** on ecological systems and services; and the means by which desired ecological, social, and political states can be gained and sustained.

See also: **Agrodiversity; Biodiversity; Conservation; Ecosystem Services; Indigenous Knowledge; Metabolism; Resilience; Sustainable Land Use.**

Further Reading

Ecology and Society: A Journal of Integrative Science for Resilience and Sustainability [Online, February 2005], The E&S Web Site, www.ecologyandsociety.org.

HELMUT GEIST

Economic Growth. The question of the relationship among economic growth, land use, and cover change specifically, and environmental quality generally, is often framed within the context of the environmental Kuznets curve (EKC) hypothesis, so-named after Simon Kuznets for his findings in the 1950s of an inverted-U-shaped relationship between income inequality and per capita income for different countries. The EKC hypothesis states that at early stages of economic development and associated low levels of national income, most of a society is in **subsistence agriculture** with little negative impact on the environment. Subsequently, as a society further develops, with associated increases in population, it requires more agricultural products. Agricultural land use increases at both the extensive and intensive margins, resource extraction increases, and the structure of economic production becomes more industrial. As a result, waste production increases and the resource base is degraded, decreasing environmental quality. As a society develops further, the composition of the structure of economic output changes as production moves into cleaner service industries and improved technologies are used for production, leading to improvements in environmental quality. In addition, as incomes increase with further economic development, there is an increase in the demand for a clean environment, which leads to new regulations on resource use and pollution production. As a result, along the society's development path, the relationship between economic growth and environmental quality is U-shaped.

It has been argued that economic growth can be sustainable in the long run if the proceeds from converting natural capital, such as the **conversion** of forested land, are invested in physical capital, but the question remains whether this occurs "too late" for a developing country, that is, if natural capital or ecological **resilience** is compromised by the time a country attains enough income to demand higher levels of environmental quality. If the EKC is true, then the argument has been made that free trade will lead to environmental improvements in developing countries as their incomes increase.

Most of the empirical research that attempts to test the EKC hypothesis has focused on industrial pollution, such as different air and water pollutants. The empirical evidence on the existence of the EKC is mixed, depending upon the time frame, pollutants, and methodology used to estimate the relationship; but in many cases studies do find a U-shaped relationship between the concentration of these pollutants and per capita income, with a turning point between $5,000 USD and $10,000 USD.

However, one of the criticisms of the EKC is that these models are missing environmentally destructive activities, such as excessive logging and pollution-intensive industrial production, which are simply moving from higher-income countries to lower-income countries. This area of research is known as the "pollution haven" hypothesis, which has focused on the role of **institutions** and government policies of developing countries in attracting polluting industries.

Research that has focused on the relationship between economic growth and land use has focused on either **forest** or **agriculture**. There are a number of ways that economic growth can affect forest cover: increases in the demand for forest products, such as logwood for construction and fuelwood for energy use; increased use of land for agricultural production and improvements in agricultural technologies; the greater demand for forests as a source of environmental goods such as recreation, **carbon sequestration**, and **biodiversity**; and how relative labor costs change with economic growth. For example, in the New England states of the **United States of America**, over one half of the forest was cut for agricultural use after **colonization**, but by 1990, forested area surpassed that which existed in 1790, all the while incomes were increasing in the region. It has been argued that this change in land use was due to a decrease in the returns to agriculture in the region, as more productive western lands became available, leading to large-scale abandonment of agricultural land with the resulting return of the forest.

The relationship between population and forest cover will depend upon the level of development in a country, such as the available agricultural technology and trade patterns. For example, the increased demand for agricultural products could be met through **agricultural intensification** instead of **extensification**, and through the use of high-yielding seeds and chemical fertilizer, thereby reducing the need to convert further forested land into agricultural production. Empirical results on this relationship have also been mixed, but with some studies showing a slowdown of **deforestation** in developing countries with increases in economic growth. In a study using cross-country time-series data for developing countries for 1980–1995, countries were divided into "open economies" and "closed economies," defined by the amount of imports and exports as part of the national economy. For closed economies, that is, those that do not participate to a great extent in international trade, the demand for wood products must be met domestically, while for open economies, this demand could be met through imports. Empirical results demonstrated no clear relationship between economic growth and changes in forest cover for open economies, while for closed economies there was a clear significant positive relationship between changes in incomes and forest cover, suggesting that as demand for forest products in these countries increases with increases in income, domestic supply also expands, leading to the growth in forest cover. Therefore, macroeconomic policies beyond population levels and growth rates are important considerations as well.

The question of the relationship between population growth and agricultural land use has been studied since at least Thomas Robert Malthus in the eighteenth century. Malthus predicted that population would grow much faster than agricultural production, necessarily leading to famine. While this has not been the case as of yet, as improvements in technology have so far outpaced increases in population, it is a potential issue as population continues to grow with associated pressure on the amount of remaining available arable land.

The relationship between economic growth and agricultural land use appears to be closely related to levels of economic development. An EKC for agricultural land would predict that the percentage of land in agricultural use would also follow a U-shaped curve as income levels rose. Important issues to consider besides income levels are population density and international trade, as well as **property** rights. While greater income will increase the demand for agricultural products, demand increases at a slower rate at higher income levels as basic food requirements are met. All else being equal, population density will increase the demand for agricultural land, although the ability to import agricultural products can substitute for domestic production. Finally, as agricultural production requires **investments**, secure property rights will facilitate agricultural land use.

Empirical results from a study of 121 countries during 1965–1987 does show some support for an EKC of the relationship between the percentage of land in agricultural use in a country and income levels. At very low levels of income, there is rapid growth in agricultural land use, until income reaches $1,500 USD per capita, after which there is a gradual reduction in agricultural land use. Although the results show that at very high income levels there is an increase again in agricultural land use, this is perhaps driven by government policies in the European Union and the United States to keep land in agricultural use. The result for population density is very significant and is found to be the principal cause for determining the amount of land in agricultural use, while the results for secure property rights as well as trade relationships were inconclusive in this study.

While there are many paths through which economic growth can cause land-use change, the empirical results discussed above are mixed. It can be difficult to tease out the impact of individual variables on land-use change. For example, population may be endogenous to land conversion, due to unobserved government policies that encourage development of targeted areas, or that population may be collinear with government policies. If the former is the case, then including population as an **exogenous factor** of land-use change will produce a biased estimate and lead to misleading policy conclusions. If the latter is the case, then the estimates will be unbiased but inefficient, leading to a potential false interpretation of the significance of variables in explaining land-use change. To address these issues, care must be taken in the empirical specification of a model, such as using a temporally lagged value of population in the analysis.

See also: **Consumption; Driving Forces; Economic Liberalization; Economic Restructuring; Endogenous Factor; Forest Transition; Industrial Revolution; Industrialization; Land Abandonment; Mineral Extraction; Population Dynamics; Public Policy; Systemic Change; Tradeoffs; Urbanization; Urban-Rural Fringe.**

Further Reading

Foster, Andrew D., and Mark R. Rosenzweig, "Economic Growth and the Rise of Forests," *The Quarterly Journal of Economics* 118 (2003): 601–37; James, Alexander N., "Agricultural Land Use and Economic Growth: Environmental Implications of the Kuznets Curve," *International Journal of Sustainable Development* 2 (1999): 530–53; Kuznets, Simon, "Economic Growth and Income Inequality," *American Economics Review* 45 (1955): 1–28; Malthus, Thomas R., *An Essay on the Principle of Population, as It Affects the Future Improvement of Society with Remarks on the Speculations of Mr. Godwin, M. Condorcet, and Other Writers,* London: Printed for J. Johnson, 1798; Panayotou, Theodore, *Economic Growth and the Environment (CID Working Paper No. 56)*, Boston: Harvard University, Center for International Development, 2000; Panayotou, Theodore, "Economic Growth and the Environment," in *Handbook of Environmental Economics,* eds. Karl-Goran Maler and J. Vincent, Elsevier Science, 2003.

JACQUELINE M. GEOGHEGAN

Economic Liberalization. Actions of national, regional, and municipal governments to expand the scope of the market and the logic of competitiveness in the organization of economic and social life. Frequently stylized as an epochal transition from state to market, national policies of economic liberalization—and the processes of **economic restructuring** that they entail—have largely defined international political economy for the last two decades. Although they show historic and geographical variation, national policies of economic liberalization proceed along a number of common axes: the reduction or elimination of tariff barriers, which increases the penetration of national markets by foreign goods; the liberalization of interest and exchange rates, allowing values to float and exposing them to market influences; the rollback of restrictions on foreign **investment**, asset ownership, and the repatriation of profits; and the reduction of state controls on entry and exit into the economy. Other core themes of liberalization include fiscal discipline, the assigning of individual or **private property** rights (privatization), and the creation of market or quasi-market mechanisms (marketization) for their exchange.

The roots of liberalization lie in two factors. The first is liberal theories of the economic and social benefits of the "self-regulating market," particularly the bodies of thought that developed around the so-called Vienna school in the 1930s (centered on the work of Friedrich von Hayek) and at the University of Chicago in the 1960s (centered on the work of Milton Friedman), and which subsequently were incorporated into the "neo-liberal" political movements that brought to power Margaret Thatcher in the United Kingdom and Ronald Reagan in the United States. The second factor is the international "debt-cum-banking crisis" of the early 1980s, and the call of many Third World countries for a new international economic order based on an expansion of development assistance, technology transfer, and commodity price support. The "structural adjustment" policies prescribed by multilateral lending **institutions** in response to the debt crisis, together with international agreements on trade and investment such as the General Agreement on Tariffs and Trade (GATT) and the World Trade Organization (WTO)—the shelved Multilateral Agreement on Investment, and regional agreements such as the proposed thirty-four-country Free Trade Area of the Americas—have served to "roll out" neo-liberal policies that "roll back" the state across the developing world. The technocratic approach to international development policy that emerged in the 1990s and became known as the Washington Consensus has economic liberalization at its core.

The environmental implications of economic liberalization are contested. The impacts of trade policies, foreign direct investment, and marketization on the ecologies of **forests**, fisheries, **grasslands**, and urban-industrial environments are a focal point of popular challenges to **globalization**. The multiple components of economic liberalization and the diversity of its institutional forms make it impossible to draw simple assessments of whether its implications are "good" or "bad." Proponents of the environmental benefits of economic liberalization highlight its potential role in rationalizing the allocation of land uses according to Ricardian comparative advantage; undercutting perverse subsidies and price supports that sustain costly and environmentally destructive forms of land use (government payments to **cotton** and sugar growers in the United States, for example); the creation of markets as a way to internalize negative externalities (such as regional emissions trading in the United States for sulfur dioxide, or global carbon credits under the **Kyoto Protocol**); and the adoption of "environmental" accounting practices (environmental cost-benefit analysis, national environmental accounts, etc.) that allow for an evaluation of the impacts of development on stocks of both social and natural capital. Proponents also argue for a convergence between the market logics of liberalization and the environmental imperatives of sustainability, and find some support for this position in ecological modernization theory; argue for a "democratization" of environmental regulation via an increased role of civic or "informal" modes of regulation that complement the formal administrative mechanisms of the state; and, in the final analysis, hinge their arguments on the assertion that growth (and/or the elimination of **poverty**) are prerequisites for environmental improvement.

Detractors, on the other hand, highlight the expansionary dynamics unleashed by trade and investment policies; the marketization (and subsequent concentration) of resource rights that can undermine the economic security of rural people; the potential for a competitive "race to the bottom" in environmental standards; and the incommensurability of market exchange and the energy and material exchanges that comprise ecosystems (and the simplification of ecologies that occurs when ecosystems are commodified and marketized). Detractors also take issue with liberalization's elevation of the business agenda and its privileging of economy as the primary dimension of social life. The North American Free Trade Agreement (NAFTA), for example, demonstrates how policies of economic liberalization can enfranchise corporations, limit the sovereign powers of states, and circumscribe the rights of citizens to a clean environment: Chapter 11 of NAFTA, for example, enables private corporations to sue states when regulatory decisions (including regulations established to protect the environment and human health) negatively affect the values of their assets in a way that is "tantamount to expropriation."

See also: **Arena of Land Conflict; Economic Livelihood; Land Privatization; Land Reforms; Land Rights; Mineral Extraction.**

Further Reading

Bridge, Gavin, "Grounding Globalization: The Prospects and Perils of Linking Globalization to Environmental Outcomes," *Economic Geography* 78 (2002): 361–86; McCarthy, James, and Scott Prudham, "Neoliberal Nature and the Nature of Neoliberalism," *Geoforum* 35 (2004): 275–84; Peck, Jamie, and Adam Tickell, "Neoliberalizing Space," *Antipode* 34 (2002) 380–404; Williamson, John, *What Should the Bank Think about the Washington Consensus*, 1999 [Online, December 2004], Institute for International Economics Web Site, www.iie.com/publications/papers/williamson0799.htm; Yergin, Daniel, and Joseph Stanislaw, *The Commanding Heights: The Battle for the World Economy*, Free Press, 2002.

GAVIN BRIDGE

Economic Livelihood. A livelihood that comprises the capabilities, assets (including both material and social resources), and activities required for a means of living. Economic livelihoods refer to the productive assets (i.e., labor, land, and financial capital) and material endowments that individuals and households employ to make their livings. In rural economies, household economic livelihoods establish a direct relationship with land-use change, since the latter is shaped by the quality and capacity of economic resources that households have at their disposal to make their living and accumulate economic wealth. While wealthier households have larger capacities to mobilize economic resources (i.e., hire labor force, buy inputs for agricultural production, and borrow financial credit), poorer rural households have limited capacities to invest productive resources in their lots.

Several studies have explored the implications that economic livelihoods have in land-use change, specifically on forest **conversion** to **agriculture**. There are two contrasting approaches about the implications that economic livelihoods have on **deforestation**. The first, labeled the "impoverishment school," argues that **poverty** is the main **driving force** of forest conversion to agriculture. In contrast, the "neoclassical approach" states that individuals and households react to economic incentives to use **forest**, or to convert it to agricultural land uses, seeking to enlarge their economic benefit.

The impoverishment approach suggests that smallholders are the principal agents of deforestation, although this argument does not link that to the assets that poor people have at their disposal. Their clearing activity is often linked to population growth in a gradual push process of rural labor surplus, resulting in progressive impoverishment, **land degradation**, and deterministic labor expulsion to the **agricultural frontier**. This process can be exacerbated by external shocks, such as competing land demands from urban entrepreneurs, which endanger the traditional balances among peoples' multiple forest use. In this view, poor households self-perpetuate themselves in poverty, leading to deforestation and land degradation, because these are the limited options available to smallholders to secure their survival. This perspective considers that higher farm prices and **agricultural intensification** strategies would induce peasants to improve their economic livelihoods and to deforest less because alleviated poverty reduces the push toward forest conversion and degradation.

The neoclassical approach, on the other hand, argues that landholders are not uniformly endowed with productive resources, and each will make use of their economic livelihoods to look for their self-interest, which is to enlarge their economic profit from using their available resources (i.e., family or hired labor, land, and some financial assets). In this view, smallholders' primary goal is not mere survival but looking for basic capital accumulation. Farming systems are viewed as more flexible in absorbing shocks, and they may also generate local technological changes that facilitate labor absorption. Labor is sectorally mobile so that alternative incomes exist. In opposition to the impoverishment framework, increasing agricultural prices and agricultural intensification will tend to induce higher production incentives and more deforestation, as off-farm labor is deliberately reduced. Among the conditions that will prevent forest loss in excess are avoiding market distortions that impede to get prices right, and uncertainties that prevent secure land ownership.

Although the two foregoing frameworks are contradictory, both may coexist in the real world. Indeed, different behaviors and explanations can coexist within the same region. Even more, rural households' behaviors may not respond strictly to any of the two frameworks but to a combination of the two, as well as to external conditions shaping households' **decision-making**.

See also: **Access; Degradation Narrative; Economic Liberalization; Human Immunodeficiency Virus (HIV)/Acquired Immunodeficiency Syndrome (AIDS); Investments; People at Risk; Population Dynamics; Transhumance.**

Further Reading

Carney, Diana, *Sustainable Rural Livelihoods: What Contribution Can We Make?* London: Department for International Development, 1998; Ellis, Frank, *Rural Livelihoods and Diversity in Developing Countries*, New York: Oxford University Press, 2000; Kaimowitz, David, and Arild Angelsen, *Economic Models of Tropical Deforestation: A Review*, Bogor, Indonesia: Center for International Forestry Research, 1998; Wunder, Sven, *The Economics of Deforestation: The Example of Ecuador*, New York: St. Martin's Press, 2000.

PABLO PACHECO

Economic Restructuring. A general concept from political economy that refers to processes of structural and institutional transition within local, national, and global economies. Classically applied to such transitions as Fordism to flexible specialization, industrial to post/de-industrial, or international to global economy, the term expresses a commitment to analyzing change as an expression of the internal contradictions of capitalism, rather than an effect of **exogenous factors**; it is, however, often used as a general descriptor of institutional change. Used in this way, economic restructuring encompasses shifts in corporate organization, technological conditions, energy regimes, and the geography of production, and captures the restless process of "creative destruction" that characterizes capitalist development. At the global **scale**, for example, restructuring can describe the constellation of technological, financial, and institutional changes in the second half of the nineteenth century that underpinned the geographical separation of bulk commodity production and **consumption** (e.g., **wheat**, wool, coal, copper, iron) and which gave rise to the massive land-use changes of the "old international division of labor": coal-based **industrialization** and **urbanization** in **western Europe** and the northeastern **United States of America**, and an extensive, global periphery specializing in primary commodity production. Restructuring also refers to the emergence of a "new international division of labor" in the second half of the twentieth century (exemplified by a rapid expansion of manufacturing in a handful of former developing countries, the de-industrialization of former industrial cores, and the associated **environmental changes** such as the mobilization of heavy metals and the creation of urban "brownfields." More recently, economic restructuring has been used to refer to the institutional transitions from state to market undertaken in eastern Europe and the former Soviet Union, and to the programs of structural adjustment visited on many of the world's developing countries.

Although restructuring is unspecific with regard to geographic and historical context, the value of the term lies in its insistence that processes of corporate, technological, and institutional change be considered together as a totality, rather than as separate variables. This makes restructuring an effective way of approaching the multiple mechanisms through which processes of economic restructuring can drive local and regional environmental change. Consider, for example, the restructuring of developing economies following the adoption of **economic liberalization** policies. Schematically, one can identify six mechanisms by which liberalization produces environmental change.

First, there is a "scale effect" whereby restructuring promotes expansion and **economic growth**, increasing the throughput of natural resources and the scale and rate of environmental change.

Second, there is a "technique effect" in which restructuring alters the technologies of production via either innovation or inward **spatial diffusion** of best-practice technologies and management practices that reduce pollution per unit output and improve the efficiency of materials transformation, or the off-shoring to the developing world of standard production technology that no longer meets environmental requirements of industrial economies.

Third, there is a "composition effect" in which falling barriers to international trade and investment promote competitive specialization and the sectoral restructuring of national economies. The "pollution haven" hypothesis, for example, asserts that competition for **investments** will promote a "race to the bottom" in which developing countries specialize in resource- and pollution-intensive industries; while work on the restructuring of global **agriculture** and the rise of "new agricultural economies" highlights land-use change as a result of specialization in the production of high-value food exports.

Fourth, there is a "property-regime effect" whereby the **institutions** of **property** through which land and resources are managed are progressively privatized and the social and environmental commons enclosed. The creation of private, exchangeable rights to resources such as land, water, or plant and animal life embeds resources in a distinctive management regime, one that prioritizes land-use practices that maximize the production of exchange values rather than use value.

Fifth, there is a "governance effect" whereby the restructuring of the state's role in resource management leads to an expanding influence of non-traditional institutions and actors on land-use decisions. The erosion of the insulating effects of national borders, for example, can increase the exposure of firms to large, affluent markets that demand products (and occasionally production conditions) that meet certain environmental standards, while non-governmental organizations exert an increasing influence over the environmental practices of corporations.

Sixth, there is a "metric effect" in which restructuring introduces new criteria and decision tools for administering the environment. The current round of economic **globalization**, for example, has increased the scope for market-based regulatory instruments and introduced new rationalities and metrics for the management of environment and resources, such as cost-benefit analysis.

See also: **Agricultural Revolution; Biotechnology; Corporate Strategies; Decision-Making; East-Central Europe; Industrial Revolution; Land Privatization; Land Reform; Mineral Extraction; Private Property; Technological-Scientific Revolution; Urban-Rural Fringe.**

Further Reading

Chisholm, Michael, "The Increasing Separation of Production and Consumption," in Turner, B. L. II, Clark, William C., Kates, Robert W., Richards, John F., Mathews, Jessica T., and William B. Meyer, eds., *The Earth as Transformed by Human Action: Global and Regional Changes in the Biosphere over the Past 300 Years*, Cambridge, UK: Cambridge University Press, 1990, pp. 87–102; Dicken, Peter, *Global Shift: Reshaping the Global Economic Map in the 21st Century*, 4th ed., New York: Guilford Press, 2003; Frederickson, Per, ed., *Trade, Global Policy, and the Environment*, Washington, D.C.: World Bank, 1999.

GAVIN BRIDGE

Ecosystem Services. Flows of goods and services from ecosystems that provide life-sustaining benefits to humans. Examples include air and water purification; maintenance of carbon dioxide/oxygen balance in the atmosphere; temperature and climate stabilization; provision of genetic, species, and ecosystem-level **biodiversity**; **nutrient cycling**, pollination, and seed dispersal; groundwater recharge; detoxification and decomposition of wastes; photosynthesis; **carbon sequestration**; soil **conservation** and fertility maintenance; and provision of esthetic and cultural benefits. The flow of these services is potentially compromised by the process of land-use and **land-cover change**. For example, **conversion** of **forest** and **wetland** ecosystems into urban land can result in loss of water filtration, groundwater recharge, and soil preservation services, as natural areas and **vegetation** are replaced by cultivated landscapes and impermeable surfaces.

Rudolph de Groot and Gretchen Daily each categorize ecosystem services according to the type of ecosystem functions that generate the services. Rudolph de Groot's categories include regulation functions, habitat functions, production functions, and information functions, while Gretchen Daily's similar characterization includes production of goods, regeneration processes, stabilizing processes, and life-fulfilling functions. In general, these functions provide a set of services that are essential to maintenance of human life, and, as demonstrated by Robert Costanza and colleagues, cannot be replicated through human technologies at any reasonable economic cost. Thus, the ecosystems generating these services can be classified as critical natural capital—natural resources whose depletion would compromise a sustainable future for the human race. The function of such natural capital may be compromised when depletion reaches a critical **threshold**, resulting in non-linear declines in productivity as capital levels are linearly depleted. As well, since many ecosystem services can be classified as public goods and are not directly traded in the marketplace, economic theory predicts that the natural capital is likely to be maintained at lower than socially optimal levels. These arguments imply that land housing natural ecosystems may be undervalued in the global marketplace, leading to too much conversion into uses that do not generate ecosystem services.

See also: **Carbon Cycle; Ecological Footprint; Millennium Ecosystem Assessment; Nutrient Cycle; Public Policy; Tradeoffs; Water-Land Interlinkages; Wetlands.**

Further Reading

Constanza, Robert, d'Arge, R., de Groot, Rudolph, Farber, S., Grasso, M., Hannon, B., Limburg, K., et al., "The Value of the World's Ecosystem Services and Natural Capital," *Nature* 387 (1997): 253–9; Daily, Gretchen, ed., *Nature's Services: Societal Dependence on Natural Ecosystems,* Covelo, CA: Island Press, 1997; de Groot, Rudolf S., Wilson, M. A., and R. M. J. Boumans, "A Typology for the Classification, Description and Valuation of Ecosystem Functions, Goods and Services," *Ecological Economics* 41 (2002): 393–408; Stanford University, Center for Conservation Biology, Ecosystem Services [Online, August 2004], The CCB Web Site, www.stanford.edu/group/CCB/ecosvcs.

DAWN CASSANDRA PARKER

Endogenous Factor. Factor upon which local land managers exert the highest possible control in order to ensure **sustainable land use** in their surroundings. At first glance, only some local-scale factors such as resource scarcity, increased **vulnerability**, and changes in social organization are endogenous to **decision-making** and thus uncontrollable by these communities, even though they may be influenced by exogenous factors as well. Other **syndromes** of land change, such as changing market opportunities and policy intervention, but also biophysical perturbations such as earthquake and flooding, are mostly exogenous factors. A notable case of the operation of endogenous factors is the **tragedy of enclosure** in which the coupled effects of corporate, institutional, and technological change generate social as well as environmental change. **Mediating factors** are crucial for the quality and power of responses by local land managers. Also, the exact combination of endogenous versus **exogenous factors** in the web of **driving forces** depends upon the regional and historical context. Moreover, as the temporal **scale** of analysis expands, all causes of land-use change—from demographic changes to technological innovations—become endogenous to the human-environment system and are affected in some degree by land-use change. Actually, the changes in **ecosystem services** that result from land-use change lead to important **feedbacks** on the drivers of land-use change.

See also: **Agent-Based Model; Community Involvement; Desertification; Economic Growth; Land-Use History.**

Further Reading

Lambin, Éric F., Geist, Helmut J., and Erika Lepers, "Dynamics of Land-Use and Land-Cover Change in Tropical Regions," *Annual Review of Environment and Resources* 28 (2003): 205–41; Lambin, Éric F., and Helmut J. Geist, "The Land Managers Who Have Lost Control of Their Land Use: Implications for Sustainability," *Tropical Ecology* 44 (2003): 15–24.

HELMUT GEIST

Environment. A usually monthly, peer-reviewed journal of the Washington, DC-based Helen Dwight Reid Educational Foundation that seeks to fulfill an educational and charitable mission in environmental science and policy issues through publication of this as well as other journals and magazines.

See also: **Environmental Change; Public Policy.**

Further Reading

Heldref Publications, Environment: Where Science and Policy Meet [Online, December 2004], The Environment Web Site, www.heldref.org.

HELMUT GEIST

Environmental Change. A continuous process of alternations in environmental conditions on all temporal and spatial **scales** (see figure). When we look back beyond the period of instrumental measurements and documentary records, the history of environmental change has to be reconstructed from natural archives such as ice cores, tree rings, corals, marine and lake sediment cores, peat stratigraphy, loess, and stalagmites. Converting the signals contained in these archives into proxy records of environmental change requires that they be calibrated by linking their present-day variability in time or space to variations in the properties they are designed to reconstruct.

On the timescale of the last half million years, **global environmental change** has been dominated by successive alternations between long glacial periods and shorter interglacial intervals. The **Holocene** period, within which we live, is the latest of these interglacials. The sequence of alternations between glacial and interglacial conditions was, in broad terms, synchronous over the whole earth. The rhythm of alternation is cyclic, with interglacial conditions recurring roughly every 100,000 years. The underlying reason for this rather regular periodicity lies in the pattern or orbital changes

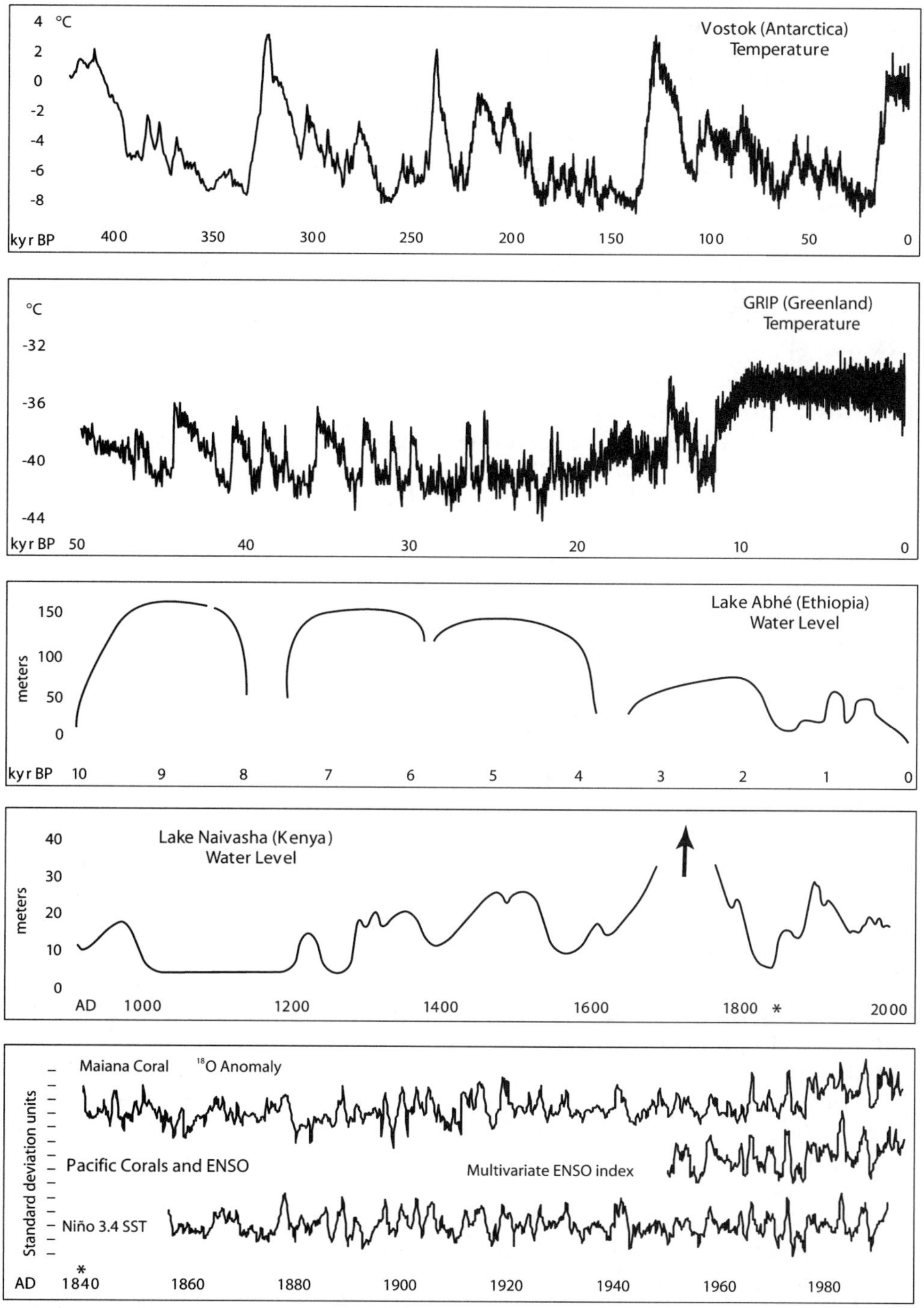

Indicators of environmental changes.

undergone by the earth. These modulate changes in insolation received at the earth's surface and, in so doing, provide the pacing mechanism for broad-scale global change. Whereas the pattern of orbital changes is continuously smooth and regular, at each transition from glacial to interglacial conditions, the response of the earth system to these smooth changes is remarkably abrupt. Additional processes serve to amplify the orbital changes and translate their effects into responses that are out of all proportion to the initial changes. These are periods of rapid global warming that precede human impacts on the earth system, and as such, they are of outstanding interest.

Of the various processes that may serve to amplify the effects of orbital changes, two are generally acknowledged to be of crucial importance. Changing ice extent as the major continental glaciers of the Northern Hemisphere wax and wane brings with it a range of important **feedbacks**, for example, through changes in surface reflectivity (**albedo**), sea ice, ocean salinity, and circulation. All these serve to amplify the effects of the smooth orbital changes. Each shift from glacial to interglacial conditions is also marked by a steep rise in atmospheric **greenhouse gas** concentrations, notably carbon dioxide (CO_2) and methane. The effect of these increases on surface air temperature also makes a major contribution to amplifying the orbital signals. In view of the likely role of greenhouse gases in current and future global change, the role of these gases during glacial/interglacial transitions has received a great deal of attention. One key question is the phasing of the increases in temperature, greenhouse gas concentrations, and ice melting. Most of the evidence so far suggests that a modest temperature response to orbital forcing preceded greenhouse gas and ice volume changes. It also seems likely that greenhouse gas concentrations began to increase before the onset of widespread Northern Hemisphere deglaciation.

There is, as yet, no full consensus as to the mechanisms controlling the sequestration and release of greenhouse gases. In the case of methane, most of the evidence points to the importance of the terrestrial biosphere in tropical and, to a lesser extent, high northern latitudes as the major sources and sinks, though there remains the possibility that the release of methane trapped in liquid form (clathrates) beneath marine sediments on continental margins and beneath permafrost may have contributed significantly to the rise in concentrations during deglaciation. In the case of CO_2, there is general agreement that the dominant source/sink on glacial/interglacial timescales must be the ocean. Less clear are the mechanisms responsible for the rapid shifts in exchange between ocean and atmosphere. Changes in alkalinity, solubility, marine productivity, and circulation may all be involved. One hypothesis invokes a link between dust deposition, marine productivity, and atmospheric greenhouse gas concentrations. According to this hypothesis, the well-documented high dust concentrations during glacial times fertilize areas of the southern ocean where productivity is limited by iron. The increased productivity leads to a drawdown of carbon as dead organisms sink toward the seabed. This leads to reduced atmospheric carbon dioxide, as the "biological pump" serves to sequester carbon in the deep ocean. The cessation of dust deposition at the end of each glaciation reverses the processes. Present-day experiments confirm the credibility of this mechanism as a contributor to changing atmospheric carbon dioxide concentrations without either quantifying it or establishing it as a feasible approach to significantly reducing atmospheric CO_2 in the future.

Once we examine the record of environmental change on timescales shorter than glacial/interglacial cycles, the pattern of broadly synchronous and parallel global changes in both hemispheres breaks down. During the long glacial periods that dominate the record, evidence is accumulating, mostly from the last glacial stages, that changes at each pole were often in antiphase: that is, strong warming in the Northern Hemisphere coincided with cooling in Antarctica and vice versa. This probably resulted from changes in circulation in the Atlantic Ocean. Under normal conditions, deep water is formed in the North Atlantic as strongly saline water sinks. The North Atlantic Deep Water (NADW) flows south, and a reverse surface current brings warm water to high latitudes in the Northern Hemisphere. This Atlantic meridional overturning circulation (MOC) forms part of the thermohaline circulation that links the Atlantic to the Indian and Pacific oceans. During the last glaciation, the formation of NADW was interrupted or significantly reduced during some twenty-two episodes. The most severe disruptions brought icebergs and ice-rafted detritus preserved in marine sediments (known as Heinrich layers) far south in the North Atlantic. During each of these episodes, the Antarctic became warmer. There are growing indications that the antiphase pattern was also typical of the less-severe oscillations (the so-called Dansgaard/Oeschger events). There are also signs that they were typical of earlier glacial periods. The most likely explanation of these changes involves a bi-polar "seesaw" mechanism whereby a reduction or total cessation in NADW formation disrupts both the deep water and surface limbs of the MOC, warming the southern ocean and cooling the North Atlantic. Several hypotheses have been proposed as the trigger for instability in the MOC. All agree that freshening of the surface waters of the North Atlantic is an essential part of the process, whether through ice melt or drainage from large glacial lakes. Similar freshening of the North Atlantic is occurring at the present day, and the possibility that it may lead to major changes in climate at some point in the future cannot be ruled out. During glacial times, the effects of the bi-polar seesaw were translated into major changes in atmospheric circulation that affected climate over much of the earth.

The pattern of antiphase changes dominated the climate record until the opening of the Holocene. Indeed, there was at least one similar episode during the early Holocene around 8,200 years ago, linked to the sudden release of fresh water into the North Atlantic from a large glacial lake. This episode serves as a reminder that MOC instability can occur under the mean conditions prevailing during interglacial as well as glacial times.

The broad pattern of change described so far has focused on temperature changes and on the record from high latitudes in each hemisphere. The evidence from the full range of environmental archives suggests that glacial times were not only colder, but generally much drier than interglacials. The long-term pattern of hydrological change in low latitudes is, to some degree, out of phase with high-latitude changes in temperature. Tropical monsoon systems seem to have been modulated not so much by the 100,000-year pacing to which glacial/interglacial changes are linked, but to a shorter orbital period of around 23,000 years. One consequence of this is that major hydrological changes at low latitudes took place alongside the relative invariance of the temperature record at high latitudes during the Holocene. From a human standpoint, these low-latitude hydrological changes are of great significance. Effective moisture is more important than temperature changes for most human populations; moreover, the largest concentrations of people are to be found in lower latitudes where monsoon regimes dominate the climate.

Over much of Asia and Africa, the early-mid Holocene appears to have been much more humid than the later Holocene. The shift from more- to less-moist climate seems to have taken place 4,000–6,000 years ago and to have been, depending on the archive and region, sudden, stepped, or, in some cases, quite irregular. One of the most dramatic and best-documented changes was in the region of the present-day Sahara. Prior to about 5,500 years ago, the region was well supplied with water, and there were numerous lakes, lush **vegetation**, a rich aquatic fauna, and widespread human settlements. Models seeking to simulate the climate of the period show that changes in insolation, sea surface temperatures and feedback from water bodies, continuous vegetation, and more humid soils were all important factors in the development of a "green" Sahara at this time. The demise of these feedback processes, superimposed on gradual, orbitally driven insolation changes, appears to have led to a rapid transition to desert conditions. The extent to which human activities contributed to this shift remains an open question.

The paleo-record from glacial and early-mid-Holocene times is of vital importance for our understanding of contemporary and future climate changes, for it provides the empirical evidence against which to test the performance of climate models under conditions of forcing and feedback that differ from those prevailing in the more recent past. Only if models can reproduce, within reasonable error margins, past climates associated with different boundary conditions can they be expected to form a realistic basis for projecting future climates.

As we move into the later part of the Holocene the focus changes, for this is the period during which the boundary conditions of the earth system most closely approximated those prevailing prior to **industrialization**. It is also the period during which the growing impact of human activities becomes increasingly evident in the paleo-record.

Annually resolved records of pre-industrial, late-Holocene climate variability reveal several features of considerable future significance. In many parts of the world, the duration of the most severe droughts during the early part of the last millennium greatly exceeded that of the worst droughts during the period of instrumental records—a reminder that even without anthropogenically driven climate change, natural climate variability includes extremes beyond the range documented by direct measurements. Further important inferences may be drawn from century- to millennial-scale reconstructions of modes of climate variability, such as the El Niño/Southern Oscillation. Far from being a relatively regular phenomenon every three to five years, its strength, frequency, and spatial influence have varied continuously and unpredictably, making the task of predicting its future behavior and likely interaction with any anthropogenic forcing extremely difficult.

The record of pre-industrial climate variability plays a vital role in developing future climate projections. In the first place, it is the pattern upon which future changes will be superimposed and with which they will interact. It is also the background "noise" against which anthropogenic forcing must be detected. One of the key bases for projecting future climate rests on the skill with which climate models can capture the pattern and sequence of climate changes over the last thousand years and identify the forcing and feedback mechanisms, both natural and anthropogenic, responsible for the changes recorded. The combination of empirically based and model-simulated reconstructions now allow confident ascription of pre-industrial global climate changes over the last thousand years mainly to solar variability and volcanic influences. Climate change from 1970 onward, however, is only explicable by invoking the growing influence of surface warming by greenhouse gases. Many uncertainties remain regarding, for example, the sensitivity of surface temperatures to both greenhouse gases and sulphate aerosols, the importance of changes in the terrestrial biosphere through activities such as **deforestation**, and the extent to which pre-industrial human activities, for example, long-term forest clearance for **agriculture**, **biomass** burning, and **rice** paddy cultivation, generated a sufficiently large increase in atmospheric greenhouse gas concentrations to have led to global warming before AD 1700. Despite these uncertainties, the vast majority of studies now confirm the reality of ongoing surface warming as a result of increased atmospheric greenhouse gas concentrations. Along with this conclusion come rather confident projections of rising sea levels and a likely increase in the occurrence and severity of many climate extremes.

Environmental change is more than climate change. The impact on the environment as a result of human activities over the last three hundred years has greatly exceeded the impact of climate change, despite the fact that the amplitude of climate change over that period, as much of the world emerged from the Little Ice Age, has been large relative to the total range of late Holocene variability. Among the human impacts additional to the release of greenhouse gases and sulphate aerosols (which have opposing effects on the radiative balance of the atmosphere) are the following: domination of the global nitrogen or **nutrient cycle** by anthropogenic processes; consequent **eutrophication** of lakes and coastal waters, leading to zones of severe oxygen depletion; long-term reduction of stratospheric ozone through a relatively brief period of enhanced chlorofluorocarbon release; exhaustion of soil resources over many parts of the world through **salinization**, **soil erosion**, and **land degradation** through nutrient depletion; acidification of surface waters and poorly buffered soils mainly through sulfur deposition from fossil fuel combustion; the addition of a wide variety of contaminants and pollutants to ecosystems; the spread of invasive species through both conscious and unintended introductions; and loss of **biodiversity** with important implications for the functional **resilience** of ecosystems. In many cases, these effects reinforce the impacts of climate change, as for example, in the case of the world's coral reefs. Several of these processes—deforestation and **soil degradation**, for example—have a long history stretching back well into human prehistory. Others have become globally significant only over the last

200–300 years; and in most cases, there has been a steep rise in their impacts over the last fifty years.

Putting together both the long-term and short-term records of environmental change, several important generic characteristics become apparent. Few if any components of the earth system respond in a smooth, linear fashion to the main forcing mechanisms both natural and anthropogenic: non-linear responses, delays in response followed by dramatic changes as **thresholds** are crossed, hysteresis (whereby the trajectories that systems follow as they are perturbed from an initial state differ from those followed as they recover from the perturbation), and strong elements of emergent behavior or contingency. One implication of emergent behavior is that many features of the earth system are contingent upon their histories. This adds greatly to the difficulties involved in developing quantitative predictive statements about many aspects of environmental change on the basis of a limited number of studies.

See also: **Atmosphere-Land Interlinkages; Boreal Zone; Carbon Sequestration; Century Ecosystem Model; Change Detection; Decision-Making; Degradation Narrative; Desertification; Exotic Species; Human Impact on Terrestrial Ecosystems (HITE) Initiative; Integrated Model to Assess the Global Environment; Land-Use History; Mineral Extraction; Paleo Perspectives; Past Global Changes (PAGES) Project; Water-Land Interlinkages.**

Further Reading

Alverson, Keith, Bradley, Raymond S., and Thomas F. Pedersen, eds., *Palaeoclimate, Global Change and the Future,* Berlin: Springer Verlag, 2003; Oldfield, Frank, *Environmental Change: Key Issues and Alternative Perspectives,* Cambridge, UK: Cambridge University Press, 2005; Steffen, Will, Sanderson, Angelina, Tyson, Peter D., Jäger, Jill, Matson, Pamela, Moore III, Berrien, Oldfield, et al., *Global Change and the Earth System: A Planet under Pressure,* Berlin: Springer Verlag, 2004.

FRANK OLDFIELD

ENVISAT ASAR. The advanced synthetic aperture radar (ASAR) was built upon the experience gained with the **European remote sensing (ERS-1/-2) satellites**' synthetic aperture radar (SAR) instruments to extend earth observation with C-band radar. ASAR was launched on the European Space Agency's ENVISAT satellite in 2002. Flexible swath positioning (15–45 degrees incidence angle) and swath coverage (56–450 km), dual polarization (horizontal HH, vertical VV, or cross-polarization HV or VH), and multiple resolution modes (30-meter image products, 150-meter-wide swath products, and 1,000-meter global monitoring products) allow average revisit times of seven days at the equator and nearly every two days at 70° latitude. ASAR data users are **remote sensing** scientists who develop data processing and analysis techniques; the earth science community who use it for **vegetation**, sea ice, glacier, and snow monitoring; scientists who use it for oceanography and **coastal zone** processes; and users of commercial applications such as ship routing through sea ice, oil slick monitoring, nowcasting of ocean fronts, **monitoring** of **agriculture** and **forestry**, water management, and flood mapping.

See also: **Along Track Scanning Radiometers (ATSR) World Fire Atlas; GLOBCOVER; GLOBICE; Medium Resolution Imaging Spectrometer.**

Further Reading

European Space Agency, ENVISAT Data Products [Online, December 2004], The ASAR Web Site, envisat.esa.int/dataproducts/asar.

CHRISTIANE SCHMULLIUS

Europe. With a combination of high levels of development and relatively dense populations, human impact on land cover in Europe is probably greater than in any other part of the world. The severity of this impact varies, however, as do the natural environment and **vegetation**. The land area of the continent is divided among some forty countries, with widely varying histories and political complexions. They range in size from the Russian Federation to San Marino, and in climate from Iceland to Greece.

Land Cover

Most of Europe was once clothed in **forest**: coniferous in the north, deciduous in central latitudes, and Mediterranean in the south. Much of the remainder was under tundra and bog, with steppe **grassland** in the extreme southeast. The great latitudinal belts of climate and vegetation were complicated both by maritime influences and mountain chains. Centuries of human activity have modified but not completely erased this pattern. Three features in particular stand out in the map of European land cover. One is the still-extensive tract of boreal forest, stretching from Sweden through Finland to northern **Russia**. Another is the great belt of cropland from southeast England and the Paris basin across the North European Plain to the Black Sea and the Russian steppes. The brown forest soils underlying much of the temperate deciduous zone have proved attractive for **agriculture**, while the podsols of much of the **boreal zone** have been much less rewarding. The third feature is the distinctive signature of the Alps and Carpathians across the central latitudinal band of Europe, picked out from the surrounding plains by the relative extent of the remaining forest and other seminatural habitats.

The Clearing of The Forest

The advance of agriculture across much of Europe was accompanied by the retreat of the forest. This began in prehistoric times, but probably accelerated during the Dark Ages and the medieval period. Detailed charting of the forest retreat is impossible, although heroic attempts have been made to depict the process in both statistical and cartographic forms. For example, Michael Williams contends that 497,000 km^2 were cleared, of which about 200,000 km^2 were cleared before 1650; while others prepared maps that purported to show the extents of forest and woodland in the central part of Europe in 900 and 1900, as reproduced in the work of Henry Darby.

Legacies of the former forest linger on in place names: the theme of forest clearance was as prominent in Saxon England in the Dark Ages as it was in **east-central Europe** under Germanic **colonization** in later centuries. The process was severely interrupted by the Black Death: in the fourteenth century, the catastrophic decline in population was followed by a return of the forest across the lands of depleted or deserted villages. Dramatic though it was, this episode was only a temporary interruption to the process that was to persist until the nineteenth century. The dominant historical theme is simply that of the retreat of the forest across the plains and up the mountain valleys of Europe, in the face of the rising human population. Today, the retreat has largely halted, and the process has gone into reverse over much of the continent.

If the dominant theme is forest clearance, a significant sub-theme is drainage of **wetlands**. Especially in early modern and modern times, wetlands have been the focus of ambitious projects to increase arable extent and food production. In seventeenth-century England, for example, the fens were the scene of such large-scale activity, and in the Netherlands the scale was even grander. With drainage, low-lying alluvial areas became fertile farmland, as wetland gave way to **wheat** field. Now the drive has all but ceased: with food surpluses a greater problem than food shortages, there is little incentive to carry out major drainage projects. Indeed, with wetland habitats being increasingly valued in environmental terms, there is some incentive to reverse the drainage process.

The stories of European forests and, to a lesser extent, wetlands are more than mere postscripts to narratives of progressively increasingly human impact on land cover through time. Forests, clearly, and wetlands more tentatively, demonstrate that the relationship between land cover and humans is not linear, but that reversals or inversions can and do occur.

The Structure of Land Cover and Land Use

The boundaries of the continent do not coincide with the boundaries of individual countries, and therefore the compilation of continental totals from national figures is not straightforward. The Russian Federation (RF) is listed under the heading "Europe" in statistics of the **United Nations Food and Agriculture Organization** (FAO) and some other international statistics, although the greater part of its land area is in Asia. If RF statistics from sources such as FAO are included with European data, they have a dominating effect. If excluded, a large part of the continent is omitted. Similar problems occur on a much smaller scale in the cases of Turkey and Kazakhstan. These countries are usually listed as Asian, although parts of them are in Europe. Some of the apparent continental totals as listed by FAO and also in the assessments done by David Stanner and Philippe Bourdeau and by the European Environment Agency (EEA) therefore need to be viewed with caution. While data from the **Coordination of Information on the Environment (CORINE) database** are available for various unit areas such as biogeographical provinces and **watersheds** as well as for countries, coverage is not comprehensive.

The two tables below summarize the structure of land use/cover. Both suffer from limitations, not least in terms of definitions of categories, but they provide an indication of the area breakdown of land use/cover in the continent.

Although the two tables refer to different geographical areas, there is close agreement on the relative extent of forests, which account for around one-third of the land area. On the other hand, the relative arable extent differs, but in each of the tables it is relatively high in terms of intercontinental comparisons. Even the lower of the two figures is 50 percent higher than the corresponding figure for Asia, the continent with the next-highest arable extent. Perhaps this reflects both the relative extent of the continent that is climatically and edaphically suitable for crop growth, and the high population densities—and demand for food—that have characterized much of the continent.

Another distinctive feature of European land use/cover might be expected to be urban extent. The table titled "Major land-use types," however, suggests that the overall continental extent remains low, although some countries such as the Netherlands (13.5 percent) and Belgium (17.6 percent) are heavily urbanized in land area as well as demography (EEA figures relate to 1995).

The apparent difference in the figures given for permanent crops in the two tables is large, but the significance of "concentrations of" should be noted in the table "Major land-use types." In other words, the figure in this table relates to complexes in which permanent crops are characteristic but not sole features. These crops include vineyards and olive groves, and they tend to be spatially concentrated in parts of central and southern Europe respectively. In the latter, recent decades have seen a radical change in farming systems and landscapes.

Structure of land use in Europe 2000: major categories (excluding the Russian Federation)

	ha (×1000)	Percent
Land area	571,101	100
Agricultural area	271,834	47.6
Arable	165,061	28.9
Permanent crops	15,047	2.6
Permanent pasture	91,726	16.1
Forest area	187,859	32.9
Other wooded land	29,484	5.1

Source: compiled from data in FAOSTAT.

Major land-use types: continent of Europe 1994

	Percentage land cover
Forest	33
Arable	24
Concentrations of permanent crops	16
Grassland	2
Urban	1
Mountain and tundra	24

Source: based on Stanners and Bourdieu 1995 (Chapter 3).

Trends

Behind the static pattern depicted in the tables lies a dynamic process of land-use change. A persistent theme in the narrative of European land use is its dynamism. Changing technologies and political economies have been powerful influences. In detail, the land-use trends and matrices of change are complex, but three major trajectories can be identified. One is the sustained expansion of the forest area. A second is a partly complementary shrinkage of the agricultural area. Third, the urban area continues to expand, though both the total population and urban population have been relatively static in recent times.

Forests have expanded for nearly two centuries in France and some other countries, but in other cases the increase began more recently. This expansion has followed the **forest transition**, and has been dramatic in some countries. In Portugal, for instance, the relative forest extent increased from around 5 percent in the 1870s to 40 percent today, but in most parts of the continent the growth has been less spectacular at the national level though equally impressive within individual parts of countries. For example, the Massif Central saw major expansion, and it and Les Landes in southwest France now stand out in continental land-cover maps because of their forest extent.

While forests have expanded, the agricultural area has decreased. Agricultural land has been transferred both to forest and to urban use, but typically, different qualities and grades of land are involved in these transfers. Land that is urbanized is often fertile and highly productive: that which is afforested is usually more marginal and more remote from centers of population. It follows that the implications of the loss of agricultural land (in terms of food production) differ widely. The urban area continues to expand, despite a relative stagnation of the population. Concern continues to be expressed about this expansion. In the past this concern focused on food production: at present, it is geared more to environmental issues such as "soil sealing" and the loss of **biodiversity**.

Land Policy and Planning

Modification of land cover and change in land use are now regulated across most of the continent, and in theory individual decisions on the use of land are subject to public control. In practice, both regulatory mechanisms and broad **public policies** vary from country to country, although measures of environmental protection in some form are in place across most of the continent. If the environment is a relatively recent concern, farming and farmers have a much longer history of policy intervention.

Policy issues have affected European agriculture for more than a hundred years. The opening up of the New World, combined with the advent of steamships and railroads, made available large quantities of cheap food in the latter part of the nineteenth century. The response from France and several of the other large mainland countries was protectionism: tariff barriers were erected to protect domestic farmers from this new competition. Britain, on the other hand, adopted a *laissez-faire* approach, exposing its farmers to intense competition. A hundred years on, the legacy remains in the form of vast contrasts in average farm sizes. Other countries again, such as Denmark and the Netherlands, took a third way, encouraging diversification into products in which there was less competition from the New World (e.g., dairy and horticultural products).

State intervention has had a major influence on European agriculture. The disruption to food supplies resulting from world wars brought the state into agriculture even in Britain. Protection from cheap imports became firmly established both in the member countries of the European Union and its predecessors and others such as Switzerland and Norway. Without state aid in one form or another, agriculture would have contracted far more rapidly. Furthermore, the incentives to expand production through **agricultural intensification**, understandable in a post-World War II Europe that was short of food, were slow to be withdrawn, and led to the loss of semi-natural habitats through drainage and through **conversion** of heath or permanent **pasture** to intensively cultivated land. The combination of technological change and state support led to dramatic increases in yields and production, but these were achieved at some environmental cost.

In the former Soviet bloc countries, state intervention took more direct forms, with the imposition of state and collective farms over extensive areas. This process was not monolithic, for example, being much less prominent in Poland than in Ukraine, and contrasts in terms of patterns of land use and land cover remain. More recently, restitution in much of the former Soviet bloc has restored land to small landowners, although the full consequences of this process, in terms of land use/cover, have still to be worked out.

Land Use/Cover: The European Pattern

The intensity of human activities and their impact on land cover vary greatly across the continent and do so at a variety of geographical **scales**, ranging from the local to the continental. Locally, around each city there is a shadow zone of high human impact, but there is also a continental-scale pattern. At this scale, Olof Jonasson highlighted a pattern of high-intensity agriculture in a core area around the Low Countries in the 1920s, with intensity decreasing radially outward across the continent in a pattern reminiscent of that described by **Johann Heinrich von Thünen**. This pattern is still evident in a range of manifestations such as fertilizer application and value-added-per-unit agricultural area. The core is characterized by intensive horticulture and agriculture (as well as by extensive **urbanization**), while the periphery has been less thoroughly modified. Indeed, David Stanners and Philippe Bourdeau's assessment indicated that "natural" conditions in Europe were largely confined to the extremities of the continent, in the form of islands, peninsulas, and **mountains**. The pattern, however, is asymmetric: extents of modification are less in the far north and east than in the south and west. The pattern is also complicated by the occurrence of mountains and plains. In particular, the mountains that stretch across central Europe destroy the simplicity of the pattern, while the plains that lie between the mountains (e.g., the Hungarian Plain and the Po Valley) serve as secondary centers of high intensity.

This pattern serves as a reminder that the determinants of land cover and land use are both varied and complex. At one level, the physical environment of climate and relief exerts a significant influence that may have grown rather than diminished over time, being translated into terms of relative economic advantage and disadvantage. Population distribution is a further determinant, as is the set of **land-use policies** (in the broadest sense) that have been implemented over the years. These factors are not independent of each other: disentangling and isolating their relative influences on the European pattern of land cover and land use is, to say the least, a major challenge.

See also: **Cadastre; Coastal Zone; Land Abandonment; Land Reform; Reforestation; Regrowth; Transhumance; West-Central Europe; Western Europe.**

Further Reading

Darby, Henry C., "The Clearing of the Woodland in Europe," in *Man's Role in Changing the Face of the Earth,* ed. William L. Thomas, Chicago: Chicago University Press, 1956, pp. 183–216; European Environment Agency, *Environmental Signals* (annual), Copenhagen: European Environmental Agency; Eurostat, European Union in Figures [Online, January 2005], The Eurostat Database Web Site, epp.eurostat.cec.eu.int; Jonasson, Olof, "Agricultural Regions of Europe," *Economic Geography* 1 (1925): 277–94 and 2 (1926): 19–48; Stanners, David, and Philippe Bourdeau, eds., *Europe's Environment: The Dobříš Assessment,* Copenhagen: European Environmental Agency, 1995; United Nations Economic Commission for Europe and Food and Agriculture Organization, Forest Resources of Europe, CIS, North America, Australia, Japan and New Zealand, New York: United Nations, 2000; United Nations Food and Agriculture Organization, *FAO Production Yearbooks* (annual), Rome: FAO; Williams, Michael, "Forests," in *The Earth as Transformed by Human Action,* eds. B. L. Turner II, William C. Clark, Robert W. Kates, John F. Richards, Jessica T. Mathews, and William B. Meyer, Cambridge, UK: Cambridge University Press, 1990, pp. 179–202.

ALEXANDER S. MATHER

European Remote Sensing (ERS-1/-2) Satellites. Identical satellites launched by the European Space Agency in 1991 and 1995. Both satellites carry synthetic aperture radar (SAR) instruments to image the earth's surface with VV-polarized 5.6 centimeter wavelengths (C-band), a repeat-pass coverage of thirty-five days, and an average spatial resolution of twenty-five meters. Applications of SAR data are on the oceans (oil spills, ship detection, mapping of wind and current fronts, eddies and internal waves, and bottom topography in shallow waters), on the land (mapping and **monitoring** of **forestry** and **agriculture**, especially in the tropics and boreal **biomes** due to cloud penetration capabilities), and of geological features (flood mapping and, at high latitudes, regional ice monitoring). In 1996, ERS-1 and -2 were flown in tandem mission, delivering the first-ever operational interferometric (InSAR) products, that is, digital elevation models and detection of small surface movements caused by earthquakes, landslides, or glacier advancement, from space.

See also: **Along Track Scanning Radiometers (ATSR) World Fire Atlas; Boreal Zone; Change Detection; Coastal Zones; ENVISAT ASAR; Global Land Cover Map of the Year 2000; Global Terrestrial Observing System; GLOBCARBON; GLOBICE; Remote Sensing; Water-Land Interlinkages.**

Further Reading

European Space Agency, Earth Observation [Online, December 2004], The European Remote Sensing Satellites Web Site, earth.esa.int/ers.

CHRISTIANE SCHMULLIUS

Eutrophication. A condition in an aquatic ecosystem where increases in the anthropogenic supply of nutrients and organic matter change the composition of primary producers, particularly stimulating algal blooms.

Eutrophication is a widespread environmental problem in rivers, lakes, **coastal zones**, and terrestrial ecosystems. It is arguably the biggest pollution problem facing estuaries all over the world. Such changes in producers then lead to other changes that propagate up through estuarine food webs.

Nutrients can overstimulate the growth of macro and microalgae, which may have intensive consequences contrasting only in the breadth of effects. Macroalgal blooms lack direct chemical toxicity, have a broader range of ecological effects, and last longer. Microalgal blooms are usually referred as "harmful algal blooms" because episodic increases in abundance of these organisms have direct chemical-based toxic effects on animal or **human health**, and can cause **biomass** decomposition and anoxia. Such conditions interfere with the recreational use of lakes and estuaries, and the health and diversity of indigenous fish, plant, and animal populations. Productivity and species composition are altered, with a few opportunistic species proliferating.

Human activity has profoundly altered the global biogeochemical cycle of nitrogen, duplicating its input rate into the terrestrial **nutrient cycle**. The external supplies of nitrogen and phosphorus to aquatic ecosystems are derived from a wide variety of sources, including groundwater, pluvial, and atmospheric inputs. Agricultural runoff, fertilizer industries, urban runoff, sewage discharges, leaking septic systems, eroded stream banks, and similar sources can increase the flow of nutrients and organic substances into aquatic systems.

Although eutrophication is a natural process in the aging of lakes and estuaries, human activities are greatly accelerating this phenomenon by increasing the rate at which nutrients and organic substances enter aquatic ecosystems from their surrounding **watersheds**. Although there is a fundamental understanding of the processes that cause eutrophication, one still needs to develop new methods and improve others to detect incipient eutrophication in ecosystems, providing information before major changes in species composition and abundance take place.

See also: **Biodiversity; Cotton; Water-Land Interlinkages; Wetland.**

Further Reading

Nixon, Scott W., "Coastal Eutrophication: A Definition, Social Causes, and Future Concerns," *Ophelia* 41 (1995): 199–220; Smith, Val H., Tilman, David G., and Jeffrey C. Nekola, "Eutrophication: Impacts of Excess Nutrient Inputs on Freshwater, Marine, and Terrestrial Ecosystems," *Environmental Pollution* 100 (1999): 179–96; Vitousek, Peter M., Aber, J., Howarth, R. W., Likens, G. E., Matson, Pamela A., Schindler, D. W., Schlesinger, W. H., and David G. Tilman, "Human Alteration of the Global Nitrogen Cycle: Causes and Consequences," *Ecological Applications* 7 (1997): 737–50.

HELENA FREITAS

Exogenous Factor. External force or factor outside the control of local land managers. A considerable part of the social and biophysical underlying **driving forces** of land change are often exogenous to the local communities managing land, and are thus uncontrollable by these communities (there are only some **endogenous factors** to decision-makers at the local **scale**).

High-level causes of land change (such as changing market opportunities and policy intervention) as well as biophysical perturbations (such as earthquake, flooding, and drought) are mostly exogenous. However, **mediating factors** are crucial for the response of land managers to these external forces, that is, responses are strongly mediated by local factors such as gender, education, and access to land. **Land privatization**, for example, initiated by the state or international donors, often exerts impacts on agricultural production that often do not outweigh the high costs of land registration programs and have several negative consequences on equity, since women and other marginalized groups often become excluded from **access** to land.

The exact combination of exogenous versus endogenous factors in the web of driving forces depends on the regional and historical context. **Land-use history** helps to reveal external forces that often operate not only in distal space, but also in distal time. In particular, non-linear situations and processes affected by stochastic events represent external shocks to local land managers, and even larger regional **land-use systems**. However, in order to ensure **sustainable land use**, local land managers need to exert the highest possible control on the driving forces of land change in their surroundings.

See also: **Decision-Making; Desertification; Feedback.**

Further Reading

Lambin, Éric F., and Helmut J. Geist, "The Land Managers Who Have Lost Control of Their Land Use: Implications for Sustainability," *Tropical Ecology* 44 (2003): 15–24.

HELMUT GEIST

Exotic Species. Any organism (microorganism, fungus, plant, animal) that is found in a habitat or ecosystem in which it did not evolve. Also called alien species, foreign species, non-native species, or invaders, these organisms are introduced through either accidental or intentional means and often disrupt the natural balance of the ecosystem into which they are introduced, with further implications for life support functions and the livelihood of people. Virtually unchecked, the population of the new species can grow because there are no natural predators. The growing population, in turn, may then outcompete native species for fundamental ecosystem functions such as food, water, and living space. When this occurs, the native species often abandons its habitat or begins to die off, which can possibly lead to **species extinction**. The invasions of exotic species have been an integral part of the evolution of life on earth. However, land-use activities by humans have significantly altered the process and greatly accelerated the rate of invasions. Different from pressures such as climate change, exotic species introductions can be related directly to land-use/cover change, in addition to habitat **conversion/modification** and pollinator losses, all impacting **biodiversity**. For example, crop plants such as New World **maize** and **tobacco**, as well as commercial lumber trees, have been introduced intentionally on a global **scale**. As for lumber trees, especially in developing nations, indigenous **forests** have been cleared to plant fast-growing eucalyptus or pine trees. These plantations are still actively promoted by international development agencies, although they have often proven uneconomical and usually far worse than indigenous species in terms of **ecosystem services** (such as erosion control and wildlife habitat). In some respects, invaders are a product of **globalization** (or "green" or "ecological" imperialism), notably due to increased trade and travel among **Europe**, Australia, the Americas, and Africa, especially **South Africa**. The current development of trade links with Asian countries may result in further invaders. It remains difficult to estimate the global scale, for example, of plant invasions. However, limited data exist for specific regions. Dov Sax, for example, quotes a rate in excess of 1,000 vascular plants for each of California, Hawaii, and New Zealand.

See also: **Biodiversity Novelties; Domestication; Driving Forces; Ecological Colonization; Proximate Causes.**

Further Reading

Cosby, Alfred W., *Biological Imperialism*, Cambridge, UK: Cambridge University Press, 1986; Grove, Richard, *Green Imperialism*, Cambridge, UK: Cambridge University Press, 1995; Mooney, Hal A., and Richard J. Hobbs, eds., *Invasive Species in a Changing World*, Covelo, CA: Island Press, 2000; Sax, Dov F., "Latitudinal Gradients and Geographic Ranges of Exotic Species: Implications for Biogeography," *Journal of Biogeography* 28 (2001): 139–50; Van Driesche, Jason, and Roy Van Driesche, *Nature out of Place: Biological Invasions in the Global Age*, Washington, DC: Island Press, 2000.

HELMUT GEIST

Extensification. Land extensification in its classical sense is the expansion of the **agricultural frontier** by clearing new land for **slash-and-burn agriculture**, for example, with subsistence in food as the main **driving force**. Mining, dam building, industrial development, and **urbanization** may pass as its modern form, albeit in non-agricultural settings (e.g., drains replaced rivers in great urban areas such as metropolitan Tokyo, the largest concreted area on earth). Major historical forms of extensification include **conversion** and (under certain circumstances) consolidation. The former involves expansion of arable land through annexation of grazing **pastures**, **deforestation**, and drainage of **wetlands**, and so on; the latter denotes the post-conversion processes of clearing one's own land or appropriating a neighbor's land for more active use. From the historical perspective of population-landscape interactions, extensification constitutes only one step in the multiphased societal response to land constraints undertaken either after means of demographic adjustments (e.g., postponing marriage and reducing fertility) were exhausted, or alongside with them. Together with **agricultural intensification** (involving more inputs per unit of land to raise productivity), extensification is classified primarily as a physical type of social response to land constraints, often applied in sequence or in combination with organizational, managerial, tenurial, distributive, and technological types of response. It is the large socio-environmental context wherein a community copes with land scarcity that determines the timing, sequence, and relative importance of these different types of response, leading to differential environmental consequences.

For much of human existence, extensification has been a primary process involved in land-cover change; as such, it is a more reliable proximate indicator of land-use change observable from remote sensors and land transaction records. This is largely the case when land appeared limitless while population densities rose too high for comfort, or the natural resource base declined. In historical context, extensification has overwhelmingly been a response to fast-rising populations. Various data show that the sixty years from 1860 to 1920 saw an area larger than India brought under cultivation, more than half of which took place in the temperate lands of North America and in the region that became the Soviet Union. Extensification of similar scale took place in the subsequent sixty years when most of the potentially productive temperate lands of the Northern Hemisphere, including **Europe** and East Asia, became occupied—a process that did not run out of steam until the rate of population growth began to slacken.

Nowadays, the new "frontier lands" under population stress are in Africa, South Asia, and **South America**. However, the linear view of predicting the rate of extensification solely on rising population densities needs to be qualified. First, this view is largely based on the assumption of constant technology, the critique of which led **Ester Boserup** to postulate that land tended to be used more intensively as it became scarcer in relation to increasing population. Indeed, past increases in agricultural production through land **modification** occurred as a result of both extensification and intensification. Another major qualification lies in the variant relationship between households' structures, their changing capacity to migrate over the course of a life cycle, and migration patterns; variation in these may explain which option between extensification and intensification will be taken as the dominant coping strategy, and for that matter, whether temporary or permanent migration occurs; different choices will have different landscape impacts. Relevant examples may be found in **Amazonia**, where relatively thinly spread populations of settlers have cleared huge areas of **forest** for pasture—something violating the linear view of causation. Trade, transportation, and economic differentiation are relevant **mediating factors** as well, because where other economic activities allow, purchasing food from elsewhere can reduce the pressure for extensification, as evidenced in the historical decline of New England **agriculture** and subsequent **reforestation** in the region. Changing values and needs of the population must also be considered (in addition to rising population density) if we are to consider the modern manifestation of extensification: that it occurs in the "urban frontier" and is driven by insatiable human needs for space, as what we see behind the phenomenon of **urban sprawl** in the **United States of America** and more recently in **China**, for example. In its traditional sense, given that most of the potentially available land on earth is presently under tropical forests, further extensification is undesirable on both ecological and economic grounds, for high costs would be incurred to provide the necessary infrastructure. This does not suggest that extensification should be cast in absolutely negative light, because thinly populated landscapes can also suffer ecological degradation (e.g., oilfields of western Siberia) whereas extensification in the form of well-planned urbanization may well be a boon to the environment in such areas (e.g., Western China). Still an option as a coping strategy, its applicability must be evaluated in light of precautionary principles and sustainability.

See also: **Agricultural Revolution; Cultural Factors; Disintensification; Hot Spots of Land-Cover Change; Industrialization; Land Degradation; Proximate Causes; Savannization; Technological-Scientific Revolution.**

Further Reading

Boserup, Ester, *The Conditions of Agricultural Growth: The Economies of Agrarian Change under Population Pressure*, London: Allen and Unwin, 1965; Boserup, Ester, *Population and Technology*, Oxford, UK: Basil Blackwell, 1981; Boserup, Ester (posthum), "Technological Society and Its Relation to Global Environmental Change," in *Encyclopedia of Global Environmental Change, Volume 5: Social and Economic Dimensions of Global Environmental Change*, eds. Ted Munn and Peter Timmermann, Baffins Lane, Chichester, UK: John Wiley & Sons, 2002, pp. 86–96.

KE CHEN

F

Fallow. Cultivated land that is allowed to lie idle during the growing season. In **swidden cultivation**, fields are cultivated and harvested for one or more years, and then left to lie fallow for varying periods to allow secondary forest to regrow. **Harold Brookfield** and Christine Padoch argue that in swidden cultivation, the fallow is a managed resource, from which a great deal of food and other materials are obtained. A deeply embedded concept of "abandoned" field sites is slowly being displaced by research data on successions through field crops into managed forests, whose plant contents are in a large measure the result of conscious planning. Even where the forest that succeeds farming is not closely managed, it is used. Swidden cultivation and its associated stages of successional **vegetation** in the fallows are not captured well in traditional land-cover classes of **agriculture**, plantation, **forest**, etc.

Wheat-fallow farming in Nez Perce County, Idaho, in July 1941 (darker fields are summer fallow).
PHOTO: Russell Lee. [Courtesy Library of Congress]

For example, as much as 26 percent of all land in Southeast Asia falls into an "other" category where "other" is defined as shrub, brush, **pasture**, waste, and other land-cover categories, many of which are actually successional vegetation on fallowed land. Failure to understand the swidden agricultural system and forest regeneration on fallowed lands has led scientists to overestimate the amount of **deforestation** that has occurred.

See also: **Agricultural Intensification; Agricultural Revolution; Agrodiversity; Change Detection; Conservation; Decision-Making; Industrial Revolution; Land Abandonment; Reforestation; Regrowth; Secondary Vegetation; Slash-and-Burn Agriculture.**

Further Reading

Brookfield, Harold, and Christine Padoch, "Appreciating Agrodiversity: A Look at the Dynamism and Diversity of Indigenous Farming Practices," *Environment* 36 (1994): 7–11; Brown, Sandra, and Ariel E. Lugo, "Tropical Secondary Forests," *Journal of Tropical Ecology* 6 (1990): 1–32; Fox, Jefferson, "How Blaming 'Slash and Burn' Farmers is Deforesting Mainland Southeast Asia," *Asia Pacific Issues* 47 (2000): 1–9.

JEFFERSON FOX

Feedback. Social and/or ecological response to **land-cover change** that affects the rate of land change. Feedbacks usually accelerate or attenuate processes of land-use change. "Positive" feedback means an amplifying impact that leads to an acceleration of change, often in the context of **poverty** and **land degradation**. In such cases, the positive feedback loop leads to a rapid degradation of ecosystems and the impoverishment of societies using these ecosystems. "Negative" feedback means an attenuating mechanism or dampening impact upon land use, and henceforth of land use on the natural environment, often in the context of innovation and a transition toward **sustainable land use**. Institutional and technological innovations, in particular, may lead to negative feedback loops, causing a decrease in the rate of change or even a reversal of the land-cover change trend. Together with the environmental and **land-use history** of each region (which defines the **initial conditions** for each subsequent round of land-use change), and the particular combination of underlying **driving forces** and **proximate causes** triggering and driving land change, feedbacks are one of the three sets of factors leading to variations in the pathways or trajectories of land change. It is mainly the feedback structure that explains why land use is never static, but constantly changing in response to the dynamic interaction between drivers and feedback from land-use change to these drivers. In other words, human-environmental systems are complex adaptive systems in which properties (such as land use) emerge from the interactions among the various components of the entire system, which themselves feed back to influence the subsequent development of these interactions. Thus, the exact future of the behavior of coupled human-environment systems is often unpredictable because it is emergent rather than predetermined.

See also: **Agent-Based Model; Atmosphere-Land Interlinkages; Disequilibrium Dynamics; Institutions; Integrated Model; Land-Use System; Mode of Interaction; Scale; Syndrome.**

Further Reading

Lambin, Éric F., Geist, Helmut J., and Erika Lepers, "Dynamics of Land Use and Cover Change in Tropical and Subtropical Regions," *Annual Review of Environment and Resources* 28 (2003): 205–41; Lambin, Éric F., and Helmut J. Geist, "Regional Differences in Tropical Deforestation," *Environment* 45 (2003): 22–36.

HELMUT GEIST

Fire. One of the most important factors in the dynamics of many terrestrial ecosystems that contain vegetation fuel, weather conditions, and ignition sources to start and spread it. Whereas **vegetation** and weather are a result of the climatic and ecological characteristics of a given site, in many areas ignition is largely associated with human activity. If ignition occurs, a fire propagates only if sufficient fine, dead fuels are present and dry. A fire burning through vegetation in a landscape is sometimes called a wildfire.

Fuel type depends on the characteristics of the vegetation. Fuel moisture content depends on current and antecedent conditions of rain, ambient relative humidity, and temperature, and also on the structure of the fuel type present. Fine fuels, such as grass, dry rapidly and can sustain fire shortly after rainfall, while heavier fuels require larger time lags for moisture to be driven off. Climatic factors determine the structure and availability of fuels. The most important factor controlling aboveground primary production (and thus **biomass** accumulation and fuel) is the water balance, followed by soil fertility. Arid areas cannot support enough vegetation for burning. On the other hand, **forests** with heavy biomass loads occur primarily where high fuel moisture

Fires in the San Bernandino Mountains of Southern California on October 26, 2003.

PHOTO: International Space Station. CREDITS: Image Science and Analysis Laboratory, NASA-Johnson Space Center, 18 Mar. 2005 (Earth from Space—Image Information). http://earth.jsc.nasa.gov/sseop.efs

A once virgin stand of fir, spruce, and cedar along the Yachats River, Orgeon, has been devasted by the combined effects of fire and logging in June 1936.
PHOTO: Arthur Rothstein. [Courtesy Library of Congress]

hampers burning, conditions that obviate the propagation of fires even with high fuel availability. Thus, fires are more frequent in ecosystems where adequate fuel loading and seasonally dry periods coincide. Fires are very frequent in savannas of the seasonally dry tropics and in Mediterranean-type shrub ecosystems, where dry, hot summers facilitate burning of the biomass accumulated during a wet and mild winter/spring season.

Fires exert direct and indirect effects on individual organisms and on assemblages of species. The response of a species to a particular fire depends on its life history and is affected by the fire event itself, by processes occurring after the fire, and by the history of fire occurrence on the site. Changes in species composition following a fire, or through a sequence of fires, will be determined by the fates of individual organisms and by the vital attributes of the component species. Over time, the occurrence of fires can influence the selection of species, and specific fire intervals may be required to maintain a given community structure. When a fire occurs, a very fast process of decomposition takes place. Fires promote fast organic matter recycling, inducing nutrient release from various reservoirs and opening new spaces for **ecological colonization**. The occurrence of fires thus opens new opportunities for the establishment of organisms, and it promotes the dynamic process of sequential species replacements, or succession. The classical view of succession describes it as a directed process moving toward equilibrium. First, grasses establish, and then longer-lived shrub species dominate, followed by several types of trees, culminating in an equilibrium community dominated by shade-tolerant species. It is now well accepted that ecological systems do not necessarily behave in orderly sequential patterns and that the interaction of natural disturbances with physical features such as topography can significantly influence, or even determine, the structure and function of ecological systems. Succession is seen as a manifestation of species' life history, and for a given set of environmental conditions the sequence of establishment after a fire is a function of initial floristic composition and differential dispersal, growth, and survival. Fires can shape a landscape and influence ecosystem dynamics by creating a spatial and temporal heterogeneous patch mosaic that generates a variety of regeneration environments. This mosaic can drive the abundance and distribution of organisms by determining areas suitable for **colonization**, by conditioning the movement of organisms among the burned-area mosaic, and by promoting the mix of different vegetation types in adjacent patches. Ecosystems that depend on the spatial and temporal dynamics of fires are called fire-driven ecosystems. In some environments, fires are so dominant that strong adaptations to fire have developed. For example, resprouting and seed recruitment after fire are traits prevalent in vegetation species of fire-driven ecosystems.

A fire is a physical process that begins with endothermic reactions and ends with exothermic reactions. It consists of several phases: pre-heating or thermal dehydration; volatilization of waxes, oils, and other compounds followed by pyrolysis—the thermal degradation of fuels; and then flaming and glowing combustion. Ignition is the transition between pre-heating and combustion. Combustion in wildfire is a process that requires fuels (plant material consisting mainly of cellulose and hemi-cellulose), oxygen, and heat and produces carbon dioxide, water, and heat. During combustion in a wildfire, large quantities of aerosols and gases are released into the atmosphere. A spreading fire can be seen as a series of ignitions, with the heat of the fire successively raising fuels ahead to ignition temperature via a contagion process. The three basic heat transfer mechanisms in a spreading fire are conduction, radiation, and convection. In surface understory fires, radiation is the primary heat transfer mechanism; whereas in crown fires, convection plays a major role. The rate of spread (meters/minute) is one of the characteristics that determine fire behavior, and can be defined as the ratio between the heat flux received from the source (e.g., the burning flaming front) and the heat required for ignition of the available fuel. Other characteristics of fire behavior are its intensity (heat release per unit time—kJ/m^2 per minute) and flame length (meters).

The three basic types of fire are surface fires, which spread by flaming combustion through fuels at the surface (grass, shrubs, downed woody material, litter, etc.); crown fires, which burn through the tree crowns; and ground fires, which burn organic fuels underneath the surface, for example, in forest bogs. Fire behavior is sometimes also classified according to the final fire size. Large fires are usually associated with severe weather conditions and thus, in many instances, correspond to high intensity, fast-spreading fires. Three interacting factors control the behavior of a particular fire event: fuels, weather, and topography. Topography, although variable in space, is essentially static. Fuels are moderately variable in space and in time, and weather changes very rapidly both in space and time. Wind speed, and direction in particular, exert a very strong control over the shape, size, and dominant spread directions of a given fire event.

The many individual fire events over a given landscape, with their typical behavior, constitute the patterns of a fire regime. Distributions of fire size and frequency through time (including seasonality), fire severity, fire intensity, and spatial pattern of burns define the fire regime of a given site. A fire regime characterizes the features of historic, natural fires that have been typical for a given ecosystem. Fire frequency is the reciprocal of the fire-return interval, which is the average time required to burn an accumulated area equal to the size of the whole landscape once. Fire regime descriptions can go from infrequent light surface fires to short return interval crown fires. However, full characterization of a fire regime requires studying historical data that can document or serve as a basis for fitting statistical distributions. One important landscape attribute that can be derived with knowledge of fire history is the instantaneous risk of fire of a given location, which is a function of the number of years that have elapsed since it last burned. The instantaneous risk of fire can be drawn from fuel hazard functions. Some authors suggest the use of theoretical statistical distributions as hazard functions for given ecosystems where not enough data is available; a controversy is now open on whether hazards tend to decline or increase with time since the last event. Fire regimes have a direct effect on landscape properties. Fires generate patterns that can interact with landscape-level phenomena through a **feedback** effect: patterns in landscapes can be the results of past fires and in turn can influence the spread of future fires. Recently burned areas usually function as an obstacle to the spread of new fires. The degree to which a landscape is connected with respect to fire depends on fuel distribution and on meteorology, with a controversy currently open on whether the former or the latter exerts the strongest control.

Fire regimes are strongly conditioned by ignition frequency and spatial distribution. Lightning is the dominant form of natural ignition; however, lightning is rare during the dry months in most ecosystems and is known to account for only 1–10 percent of all ignitions. Thus, "natural" fire regimes are constrained to those areas of the globe where humans do not act as an ignition source. Most fire regimes are the result of the interaction between human activity and nature, and in many regions humans have altered the natural fire regime either by dispersing ignition (in time and space) or through removal of fire by changing the vegetation cover.

Fire has been used since early times: humans have used fire to hunt and harvest natural products, to ward off predators, to maintain or change habitats against succession trends, and to open space for **agriculture**. Man has learned to build on the fire dynamics of a site to better exploit potential outcomes. Some important ecosystems of the world have been maintained or directed through this interaction. For example, the woody savannas of northern Australia and **southern Africa** are ecosystems strongly controlled by anthropogenic and natural fire. Even though the primary determinants of the composition and structure of savannas are moisture and nutrients, frequent fires help maintain the grass coverage and keep woody vegetation at desirable levels for agriculture and herding. Humans set fires in synchrony with interannual and seasonal weather variations, thus promoting early-season burns that prevent late-season severe fires, induce nutrient release, and clear land. These fire regimes are associated with ancient practices that have existed since pre-historic times.

Fire is the most ancient tool used by man and is one of the strongest means by which man can manipulate and shape landscapes, drive **land-cover change**, and condition land use. However, land-use/cover changes also determine fire occurrence. Land-cover **conversion** has many different motivations and is strongly dictated by socio-economic needs. Converting a forest to a conducted vineyard will most certainly eliminate fires, leaving a forest without conduction will promote fuel build up and fires, and setting land aside for **conservation** will induce the establishment of a new fire regime. Land-cover **modification**, such as that due to selective logging in rain forests (which increases light penetration and dryness of surface fuels) can alter the ecological characteristics and the fire dynamics of a site. Controlling fire for specific management objectives requires control over both fuels and ignitions. Fuel manipulation, through promotion of both vertical and spatial discontinuities, is one of the techniques of fire management. Priorities of fire management vary with social, economic, institutional, and circumstantial factors. For instance, in conservation areas, fire management usually aims to maintain the full range of natural communities, minimize wildfire risk, and protect infrastructures.

The understanding that emissions of **greenhouse gases** (mainly carbon dioxide) and aerosols from local or regional anthropogenic fires can produce environmental impacts over spatially remote regions through atmospheric circulation processes has been the main driver of a new research area called biomass burning. Through their atmospheric input, fire regimes can affect global climate. Conversely, global change can drive major future shifts in current fire regimes through changes in climate. Thus the study of fire dynamics has evolved from a focus on localized effects on plants and communities or concerns about pollution, smoke, and land-use practices; to its role in ecosystem and landscape processes; and to its impact on **biodiversity**, **biome** sustainability, and global anthropogenic activities.

See also: **Advanced Very High Resolution Radiometer; Along Track Scanning Radiometers (ATSR) World Fire Atlas; Atmosphere-Land Interlinkages; Biodiversity; Climate Impacts; Cumulative Change; Degradation Narrative; Disequilibrium Dynamics; Global Burnt Area 2000; GLOBCARBON; Grassland; Great Plains; Hot Spots of Land-Cover Change; Pristine Myth; Savannization; Secondary Vegetation; Slash-and-Burn Agriculture; Tropical Humid Forest; Turnover.**

Further Reading

Cochrane, Mark A., "Synergistic Interactions between Habitat Fragmentation and Fire in Evergreen Tropical Forests," *Conservation Biology* 15 (2001): 1515–21; Kasischke, E. S., Williams, D., and D. Barry, "Analysis of the Patterns of Large Fires in the Boreal Forest Region of Alaska," *International Journal of Wildland Fire* 11 (2002): 131–44; Page, S. E., Siegert, F., Rieley, J. O., Boehm, H. D. V., Jayal, A., and S. Limin, "The Amount of Carbon Released from Peat and Forest Fires in Indonesia during 1997," *Nature* 420 (2002): 61–5; Pereira José M. C., Pereira, B. S., Barbosa, P., Stroppiana, D., de Vasconcelos, Maria J. P., and Jean-Marie Grégoire, "Satellite Monitoring of Fire in the EXPRESSO Study Area during the 1996 Dry Season Experiment: Active Fires, Burnt Area, and Atmospheric Emissions," *Journal of Geophysical Research–Atmosphere* 104 (1999): 30, 701–12; Siegert, F., Ruecker, G., Hinrichs, A., and A. A. Hoffmann,

"Increased Damage from Fires in Logged Forests during Droughts Caused by El Niño," *Nature* 414 (2001): 437–40.

MARIA JOSÉ P. DE VASCONCELOS

Food Crops. Refers to the agricultural production of staple crops for human nutrition. Food crops are either used for a household's home **consumption** or sold on domestic or export markets. In many developing countries, food crop production is targeted mainly for home consumption (**subsistence agriculture**), but surplus production is increasingly often marketed domestically or on export markets to earn cash income. **Rice** is a prime example as the main subsistence food crop in Southeast Asia, but at the same time also a major export earner for countries like Thailand and Vietnam. **Maize** is a major subsistence crop in large parts of **Latin America**, while it turned into one of the most important agricultural cash crops, for example, in the **United States of America** and Argentina.

Cultivation and Production

Examples of major food crops include the basic staples, which are cereals (**wheat**, rice, maize, barley, sorghum, millet, oats, rye), pulses (legumes like beans, peas, chickpeas, cow peas), and roots and tubers (potatoes, **cassava**, sweet potatoes, yams), as well as soybeans and groundnuts. According to statistics of the **United Nations Food and Agriculture Organization**, the harvested areas of all cereal crops increased from 6.48 to 6.74 million km^2 between 1961 and 2003, pulses are planted on 0.71 million km^2 in 2003 (compared to 0.63 million km^2 in 1961), and roots and tubers cover 0.53 million km^2 of the earth's surface in 2003, up from 0.48 million km^2 in 1961.

Cereals as members of the grass family (*Gramineae*) are by far the most important food crops in terms of area harvested. The three major cereals

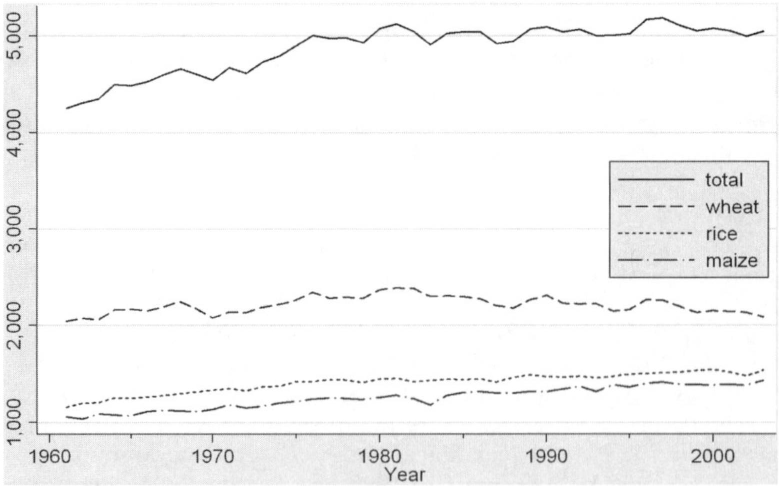

Harvested areas of major food crops (1,000 sqkm).
SOURCE: FAOSTAT data, 2004

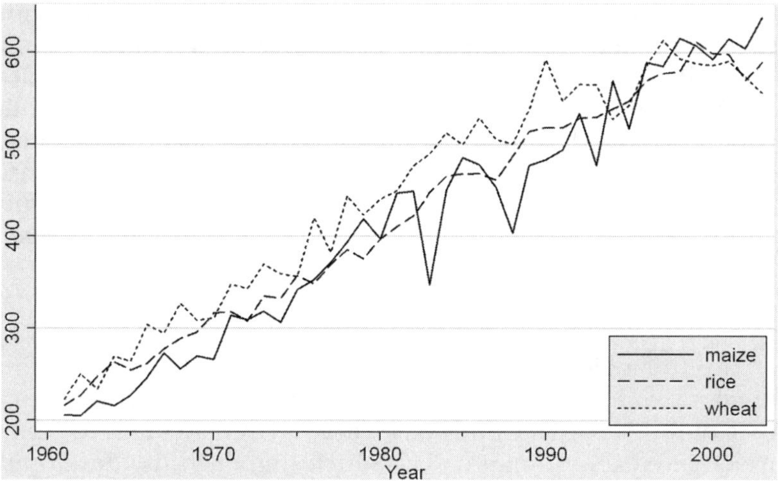

Food crop production (Mio. tons).
SOURCE: FAOSTAT data, 2004

wheat, maize, and rice constitute the most important food staples for the majority of the world's population in both developed and developing countries. The area devoted to these three food crops increased from 4.25 million km^2 to 5.05 million km^2 between 1961 and 2003 (see figure "Harvested areas of major food crops") while production almost tripled from 634 million tons to 1,784 tons (see figure "Food crop production"). The three major food crops occupy 30 percent of the world's total 15.34 million km^2 of cropland.

Production of the three major food crops increased during this period by 277 percent, while the amount of land allocated rose by 19 percent. The increases in output are primarily due to increases in crop productivity due to the introduction of high-yielding varieties, mineral fertilization, improved pest control, and increasing area under irrigation. Little of the increase in production is due to the expansion of agricultural areas.

Land-Use Change

In developing countries, food crops are the main subsistence crops, as traditional smallholders attempt to secure the household's food consumption requirements first before diversifying production activities into the cultivation of commercially marketed **cash crops**. If the combination of natural, physical, social, and economic capital endowments as well as market **access** allow, traditional smallholders intensify existing food crop production to sell surpluses and/or alter their production orientation into the cultivation of non-food crops. Higher yields and emerging market possibilities permit farmers to purchase the required food with cash earnings from marketed crop production. The need for more employment and the desire for increasing cash incomes is often the motivation for this increasing market orientation and commercialization of production. In many tropical areas, this evolution is accompanied by abandonment of more marginal upland areas and more intensive farming in lowland valleys or upland areas with suitable natural and geophysical conditions.

The expansion of permanent agricultural cultivation for both home consumption and market sales is identified as one of the main causes of **deforestation**. Especially in upland tropical areas, the expansion of areas devoted to food and cash crop production is often associated with forest cover decrease. Under conditions of continued population growth, farmers practicing land-intensive farming systems such as the various forms of shifting cultivation have to expand agricultural areas in order to satisfy consumption requirements in the absence of possibilities to intensify production on their existing land resources.

Nutrition

Cereals are the major food crops in terms of the role they play for human diets. Cereals provide over 48 percent of worldwide human calorie intake. The importance of food crops for diets varies across countries and income groups. The share of cereals in human diets is much higher in the developing countries, with 54 percent compared to 31 percent in developed countries. The three major cereals **wheat**, maize, and rice constitute the most important food staples for the majority of the world's population in both developed and developing countries. Rice is by far the most important food crop in developing countries, while it has less importance in terms of area and calorie intake in the developed world. Other regionally important food crops include cassava (or manioc) as one of the major food crops in sub-Saharan Africa and Oceania, and maize and beans in large parts of Latin America.

See also: **Agricultural Intensification; Agriculture; Cascades; Domestication; Food Security; Global Environmental Change and Food Systems (GECAFS) Project; Land Abandonment; Landholding; Land-Use System; Proximate Causes; Swidden Cultivation.**

Further Reading

Barbier, Edward B., *Cash Crops, Food Crops and Agricultural Sustainability*, London: International Institute for Environment and Development, 1987; Clay, Jason, *World Agriculture and the Environment: A Commodity-by-Commodity Guide to Impacts and Practices*, Washington, D.C.: Island Press, 2004; Ruthenberg, Hans, *Farming Systems in the Tropics*, 3rd ed., Oxford, UK: Clarendon Press, 1980; United Nations Food and Agriculture Organization, FAOSTAT Data [Online, August 2004], The FAO Web Site, www.faostat.fao.org; United Nations Food and Agriculture Organization, *Summary of Food and Agricultural Statistics 2003*, Rome: Food and Agriculture Organization of the United Nations, 2003.

DANIEL MÜLLER

Food Security. Defined by the World Bank as "**access** by all people at all times to enough food for an active, healthy life." It is a lack of food security, and often related **poverty**, that frequently has been posited as a potential cause and/or consequence of land-use and **land-cover change**. Food security may be conceived of at a variety of spatial and social **scales**, from the national to the local and from the community to the individual. Historically, many national governments and relief agencies have operationalized food security at the national level via an accounting procedure known as the "food balance sheet approach." This approach is used to establish whether there is adequate food supply to meet demand; it calculates national food needs (population times per capita grain needs) and compares them to the sum of agricultural production, stocks, and net imports (imports minus exports). Amartya Sen's work on exchange entitlements, and the unexpected Sahelian famine of the mid-1980s, demonstrated that the food balance sheet approach was seriously flawed in equating food supply with food access. Sen's concept of entitlement refers to a person's legitimate claims to available food. While food supply or availability might be sufficient at the national level, it may be inaccessible to certain segments of society due to high prices or insufficient income (and, as such, there has been a growing attention to poverty in the food security literature).

Many agencies increasingly conceptualize food security at the household level. Of course, power imbalances between men and women, as well as adults and children, mean that food resources are not always distributed according to need within the household. At the household level, food insecurity and **vulnerability** are closely related concepts, wherein the former is the result of the latter in the event of a perturbation, such as a drought or insect infestation. Vulnerability in the food security/hunger literature has at least four major components: the risk or probability that a production or income-generation strategy will fail in a given geographic context, the exposure of a household's livelihood to particular risks, the capacity to cope with potential shortfalls through reserves and alternative strategies in the short term, and the ability to recover over the longer term.

A decline in surface **biomass** is thought to have repercussions for livelihood activities that are most directly linked to this cover, such as wild food collection, hunting, firewood harvesting, some artisanal activities (e.g., beekeeping, shea butter production), and browsing and grazing of livestock. Change in surface **biomass** may also have a number of less-perceptible indirect effects (such as water quality that is impacted by land-use and land-cover change). Households respond to the effects of land-cover change on livelihood through adaptation, that is, by changing the mix and intensity of economic activities or by undertaking management responses to declining land cover. Adaptation may lead to an increase or decrease in surface biomass.

See also: **Agricultural Intensification; Cassava; Economic Livelihood; Human Immunodeficiency Virus (HIV)/Acquired Immunodeficiency Syndrome (AIDS); Land Degradation; Land-Use Policies; People at Risk; Regrowth.**

Further Reading

Sen, Amartya, *Poverty and Famines,* Oxford, UK: Clarendon, 1981; Watts, Michael, and Hans-Georg Bohle, "The Space of Vulnerability: The Causal Structure of Hunger and Famine," *Progress in Human Geography* 17 (1993): 43–67; World Bank, *Poverty and Hunger: Issues and Options for Food Security in Developing Countries.* Washington, DC: World Bank, 1986.

WILLIAM G. MOSELEY

Forest. A type of **biome** dominated by trees and occupying nearly one-third of the earth's surface. Forests are resources that supply humans with a wide range of products and **ecosystem services**. They are usually rich reserves of biodiversity, and they play key roles in hydrological and climatic systems. Symbolically, they are often seen as representing environmental purity; and damage to them, through pollution or deforestation, is perceived as an environmental ill. Indeed, forests have sometimes become emblematic environmental issues.

Definitions

Forests are usually easy to recognize in general terms but hard to delineate at the margins, and therefore difficult to define in detail. Problems of definition have long beset attempts to chart trends in forest extent. Not until 2000 was a standard definition agreed for global application. Logically, it is problematic in that it combines elements of land cover and land use, and actual and potential land cover. The definition used by the **United Nations Food and Agriculture Organization** (FAO) (2001, p. 363) is as follows:

> Forest includes natural forests and forest plantations. It is used to refer to land with a tree canopy cover of more than 10 percent and area of more than 0.5 ha. Forests are determined both by the presence of trees and the absence of other predominant land uses. The trees should be able to reach a minimum height of 5 meters. Young stands that have not yet but are expected to reach a crown density of 10 percent and tree height of 5 m are included under forest, as are temporarily unstocked areas. The term includes forests used for purposes of production, protection, multiple-use or **conservation** (i.e., forest in national parks, nature reserves and other protected areas), as well as forest stands on agricultural lands (e.g., windbreaks and shelterbelts of trees with a width of more than 20 m), and rubberwood plantations and cork oak stands. The term specifically excludes stands of trees established primarily for agricultural production, for example fruit tree plantations. It also excludes trees planted in **agroforestry** systems.

Furthermore, there are problems in defining natural forests and forest plantations. According to the FAO, a natural forest is a forest composed of indigenous trees and not classified as forest plantation. Forest plantation, on the other hand, is a forest established by planting and/or seeding in the process of afforestation or **reforestation**. It consists of introduced or **exotic species** or, in some cases, indigenous species.

Such definitions are pragmatic, and their acceptance is helpful in achieving a global standard. It is important, however, that they be borne in mind in using recent FAO data on forest area and trends. When comparing recent statistics with earlier data, it is crucial that definitions be checked: it is all too easy to reach false conclusions arising from the tacit assumption of constancy in definition.

Forest Extent and Distribution

On the basis of these definitions, the FAO estimated the global forest area in 2000 to extend to 3,869 million hectares, or 30 percent of the world's land area. This total is made up of 3,682 million hectares of "natural forest" and 187 million hectares of "forest plantation." Distribution is uneven. A primary determinant is climate. Much of the earth's surface is too cold or too dry to support forests, and therefore high latitudes and deserts are mostly treeless. Climate is of course a major factor influencing the nature of the forest as well as its extent, and evidence of a latitudinal zonation of forest is clear. Equatorial forests, along with **tropical humid forests** and **tropical dry forests**, give way poleward to savanna and eventually to desert. On the other hand, the **taiga** or great northern coniferous forest belt of the **boreal zone**, stretching across North America and Eurasia, gives way equatorward to deciduous forest of the temperate zone and then to steppe.

In detail, the forest distributional pattern is far more complex than this broad latitudinal zonation would suggest. One of the main complicating factors is the existence of **mountain** chains, and especially those trending in north-south directions. A further complication is that both the extent and distribution of the forest have varied greatly in recent geological times, and especially since the end of the last glacial period. This era is also that in which the relationship between humans and forests has been worked out. The notion that humans have impacted upon a constant, "natural" forest is simplistic. In some parts of the world, "natural" and anthropic change have been concurrent. Nevertheless, there is no doubt that humans have had a significant effect on the distribution and to some extent on the composition of the forest.

Forests and Humans

The effects of humans have varied spatially. To date, they have been greater in the temperate forest zone than in the boreal and equatorial latitudes. Much of the forest that once clothed **Europe**, for example, has been removed, and much of what remains has been modified. **Agriculture** has been a primary factor in this process: as a generalization, the soil and climate of this zone have been favorable for permanent cultivation, and the forest has given way to farmland. This has been true to a much lesser degree in both the boreal and equatorial zones. In the former, harsh climate and infertile soil have restricted agriculture, while in the latter—at least until recently—clearance for agriculture has been relatively limited. This is not to say that these forests have been completely free from human impact. There has, for example, been the widespread practice of shifting agriculture or **swidden cultivation**. Nevertheless, greater relative extents of forest have survived in high and low latitudes than in mid-latitudes.

Forests as Resources

Throughout history, forests have supplied many of the physical needs and wants of humans, including food, fodder, medicines, fuel, tools, and building materials. In recent times, wood—as an energy source, construction material, or raw material—has usually been the dominant forest product in terms of value and volume. Total production (i.e., removals) in 2002 amounted to 3,384 million m^3, equivalent on average to around 0.87 m^3 per hectare of forest. Of this volume, just under half (47 percent) was in the form of industrial wood, with the remainder fuelwood. There are, however, great variations around that global proportion. In Africa, for example, only 12 percent of the total production in 2002 was industrial wood, whereas in Europe the corresponding proportion was 82 percent.

Increasingly, an emphasis has been placed on the "full" value of forests, including the values of non-market forest functions such as **carbon sequestration** and maintenance of **biodiversity**. While problems exist over the quantification of these values, there is little doubt that they are getting increasing attention in forest management, which in modern times has focused largely on timber production.

Forest Trends

One of the ways in which human use of the forest resource is manifested is in forest trends and the dynamics of forest areas. In general terms, forests in the tropics have been shrinking in recent decades, and those in non-tropical areas have been expanding. From 1990 to 2000, for example, the net "total forest" area in the tropics contracted by an average of 12.3 million hectares per year, while that in the non-tropics expanded by 2.9 million hectares. The latter part of the twentieth century saw rapid tropical **deforestation**, and a growth of concern about the consequences of that deforestation. To a more limited degree, there has been some concern about the expansion of forests in some parts of the non-tropical world, and notably in the alpine mountains of Europe. Both of these trends are linked to some extent to trends in agriculture, and to expansion of agricultural areas in the tropics and contraction in Europe and in some other parts of the developed world. From Neolithic times onward, agriculture has probably been the main determinant of trends in forest areas.

Knowledge of how to create "artificial" forests in the form of plantations has long existed, but until recently it was usually deployed only in very small and localized areas. In recent decades, however, the rate of expansion of forest plantations has accelerated spectacularly. According to FAO data, the global forest plantation area is estimated to have increased from 17.8 million hectares in 1980 to 43.6 million hectares in 1990 and 187 million hectares in 2000. Accompanying this increase has been a geographic dispersal. Previously, plantations were largely restricted to the developed world, but more recently they have begun to increase rapidly in Africa, Asia, and **Latin America**, and especially in Oceania. They are of growing and disproportionate importance in timber production: amounting to around 5 percent of the global forest area, they are estimated to account for 35 percent of global roundwood production. It is sometimes argued that they can help to avert pressures that would otherwise have been concentrated on natural forests. On the other hand, a substantial area of plantations, amounting to just over half that in the tropics, has resulted from **conversion** of natural forests, with a loss of biodiversity. Perhaps this process epitomizes the application of management to forests in recent times, and in turn a desire to achieve greater forest production and productivity.

See also: **Anthropocene; Arc of Deforestation; Carbon Cycle; Decision-Making; Driving Forces; Economic Growth; Extensification; Forest Anomalies; Forest Transition; Forestry; Hot Spots of Land-Cover Change; Land Rights; Land-Use History; Non-Timber**

Forest Products; Pristine Myth; Regrowth; Secondary Vegetation; Tropical Ecosystem Environment Observations by Satellite (TREES) Project; Turnover.

Further Reading

Perlin, John, *A Forest Journey: The Role of Wood in the Development of Civilization,* New York: W.W. Norton, 1989; United Nations Food and Agriculture Organization, *Global Forest Resources Assessment 2000: Main Report* (FAO Forestry Paper No. 140), Rome: FAO, 2001; United Nations Food and Agriculture Organization, *State of the World's Forests 2003,* Rome: FAO, 2003; Westoby, Jack C., *Introduction to World Forestry: People and Their Trees,* Oxford, UK: Blackwell, 1989.

ALEXANDER S. MATHER

Forest Anomalies. Areas of stable **forest** or **regrowth** that are found in landscapes susceptible to **deforestation** pressures. **Remote sensing** analysis of time-series data provides a particularly effective approach to locating forest anomalies. Since these anomalies are located in regions where there are strong **proximate causes** and **driving forces** of deforestation (such as roads or markets), they indicate the likely locations of effective forest management **institutions** that are working to mitigate the overall trend toward deforestation in these regions. Thus, searching for forest anomalies can provide a powerful approach for quickly zooming in on effective institutions of forest management in a landscape. A range of institutional mechanisms can be associated with forest anomalies through private, community, or government ownership.

See also: **Center for the Study of Institutions, Population and Environmental Change; Property.**

Further Reading

Schweik, Charles, Nagendra, Harini, and Deb Ranjan Sinha, "Using Satellites to Search for Forest Management Innovations in Nepal," *Ambio* 32 (2003): 312–9.

HARINI NAGENDRA

Forest Degradation. A synonym for forest disturbance, damage, deterioration, impoverishment, decline, or cryptic deforestation that means a slight to severe **modification** of forest cover, mainly due to **proximate causes** of land-use/cover change and/or "natural" disturbances such as **fires**, insect pests, diseases, storms, wind damages, and air pollution. The **United Nations Food and Agriculture Organization** defines forest degradation as a process leading to a temporary or permanent deterioration in the density or structure of **vegetation** cover or its species composition, thus leading to a lower productive capacity of the **forest**. As with most processes of land-cover modification, few quantitative data exist at the global scale. Recent estimates by researchers from the **Tropical Ecosystem Environment Observations by Satellite (TREES) project** have been provided for only the world's humid tropical forests. According to these estimates, forest degradation during 1990–1997 was most extensive in Southeast Asia (0.42 percent per year), lowest in **Latin America** (0.13 percent per year), and intermediate in Africa (0.21 percent per year).

Forest degradation due to the combined effects of logging and burning along the County K Highway in Forest County, Wisconsin, in May 1937.
PHOTO: Russell Lee. [Courtesy Library of Congress]

Though lowest in Latin America, local-scale evidence suggests a prominent role of forest degradation in all forest cover changes there. In **Amazonia**, for example, every year forest impoverishment caused by selective logging and fires affects an area at least as large as the area affected by the **conversion** of old-growth forest. Even though harvest intensities in the Amazon region range from one to about nine trees per hectare, selective logging can lead to substantial damage to the residual stands. Gregory Asner and colleagues estimate that logged areas were equivalent to 60 to 123 percent of the reported deforestation area in the Brazilian Amazon between 1999 and 2002. Most logging in the area is conducted by poorly trained workers with minimal planning. Waste and high levels of collateral damage are common. The construction of logging infrastructure such as logging roads and decks is also an important source of mortality, damage, and ground and canopy disturbance. The underlying **driving forces** of forest degradation in the tropics more or less conform with those of **deforestation**, but synergistic interactions among fire, selective logging, and habitat fragmentation as well as between demographic factors and fuelwood harvesting potentially play a much stronger role. On the latter, it has been found that small consuming units (or nuclear households) in the **Sudan-Sahel** cause relatively more forest degradation through fuelwood collection than large, extended families, which seems especially true for peri-urban environments of developing countries. Another recent study by Erika Lepers and colleagues—synthesizing information on rapid **land-cover change** for the period 1981–2000—found that forest degradation in Eurasia has been growing in recent years, mostly related to unsustainable logging activities or increase in fire frequency. Especially in Siberia, the frequency of fires—actually a natural disturbance factor in the

boreal zone—has increased in recent years. Over 7.5 million hectares per year were burned over a six-year period in the late 1990s in **Russia** alone. Further, there has been concern about "forest dieback" (*Waldsterben*) in temperate forests, mainly in the conifers of central Europe, as evidenced by widespread needle yellowing and loss of species such as *Picea abies.* Acid precipitation and its effect on soil pH, soil structure, and aluminum mobilization were implicated as the primary physiological mechanism. However, more recent studies have documented substantial forest growth in the very same regions, tentatively attributed to a CO_2 fertilization signal (although changes in atmospheric pollutant levels and altered forest management practices such as curtailing of litter raking may also play important roles).

See also: **Himalayas; Hot Spots of Land-Cover Change; Tradeoffs; Tropical Dry Forest; Tropical Humid Forest.**

Further Reading

Asner, Gregory P., Knapp, David E., Broadbent, Eben N., Oliveira, Paulo J. C., Keller, Michael and José N. Silva, "Selective Logging in the Brazilian Amazon, *Science* 310 (2005): 480–82; Lepers, Erika, Lambin, Éric F., Janetos, Anthony C., DeFries, Ruth, Achard, Frédéric, Ramankutty, Navin, and Robert J. Scholes, "A Synthesis of Information on Rapid Land-Cover Change for the Period 1981–2000," *BioScience* 55 (2005): 115–24; Nepstad, Daniel C., Veríssimo, Adalberto, Alencar, Ane, Nobre, Carlos, Lima, Eirivelthon, Lefèbvre, Paul, Schlesinger, Peter, Potter, Christopher, Cochrane, Mark, and Vanessa Brooks, "Large-Scale Impoverishment of Amazonian Forests by Logging and Fire," *Nature* 398 (1999): 505–8; Skelly J. M., and J. L. Innes, "Waldsterben in the Forests of Central Europe and Eastern North America: Fantasy or Reality?" *Plant Disease* 78 (1994): 1021–32.

HELMUT GEIST

Forest Transition. The passage, in modern times, from net **deforestation** to net **reforestation** in a defined area, usually a country. From around 1800 onward, this transition has been effected in most of the developed world. From the date of transition, reforestation is normally sustained and has resulted in significant forest expansion. Episodes of reforestation following deforestation have occurred in the past. Following the Black Death in medieval Europe, for example, a drastic decline in population and curtailment of **agriculture** resulted in the reforestation of abandoned land. Similarly, the decline of some ancient civilizations (such as the Maya in Central America) was followed by regeneration of the forest. Modern forest expansion in the developed world, however, has been achieved without such catastrophic events. It has proceeded alongside continuing population growth and economic development.

Form and Timing

The forest transition is easy to recognize but is difficult to chart in reliable detail. It is comparatively easy to track the expansionist phase of the curve of forest area, although the varying proportions of natural regeneration and planting are a complication. It is usually more difficult to plot the trajectory of past deforestation. In addition to complications arising from the changing definitions of **forest** are those stemming from the use of non-standard units of measurement, and in particular from the lack of systematic inventory. In most countries, rigorous survey has occurred only in the last two centuries (at the most), and hence the reconstruction of most of the forest-area curve cannot be an exact science. It follows that it is difficult to identify the exact date of the forest transition. Nevertheless, it seems that this transition was achieved in Denmark in the early part of the nineteenth century, in France around the mid-nineteenth century, and in Portugal around the 1870s. In these and in other European countries, this followed centuries or millennia when deforestation was the prevailing trend, interrupted only by events such as the Black Death. In parts of the New World, a similar transition was achieved over a shorter timescale: deforestation in New England, for example, was followed by reforestation by the early part of the twentieth century, as agricultural land there was abandoned after the opening up of more fertile areas to the west. An earlier forest transition probably occurred in **Japan** in the eighteenth century, but the forest-area curve is even more difficult to reconstruct there than in **Europe**. There is speculation that the forest transition may also take place in the developing world: there is increasing evidence, for example from parts of Southeast Asia and central and **South America** that it is currently happening.

The Causes of the Forest Transition

Multiple causes are probably responsible in most cases, operating in different combinations in different parts of the world. These causes can include technical, economic, and demographic factors as well as policy-related influences. Many of them may relate to "modernization" or "development," both in the fullest sense. If this is so, it may be unrewarding to isolate individual factors.

A Theoretical Basis

An almost imperceptible process may underlie the passage to forest expansion: on its own, it may have little evident significance, but its effect may be magnified by other factors. This process is that of progressive adjustment of the locational distribution of arable land to that of fertile land. Initially, there may be a random element in the distribution of human settlement and agriculture over a territory. With a process of learning behavior over time, however, humans may adjust their pattern of arable land more closely to the more fertile parts of the territory. This could mean that the same amount of food is produced on a smaller area, and that more marginal areas are abandoned and are available for forest regeneration. In other words, the forest area may expand. In practice, this process on its own would operate only if demand for food were static (or growing more slowly than the increase in food production made possible by gravitation to the more productive areas). On the other hand, the effect of the process could be magnified if crop yields were to increase, and it would be facilitated if a transport network were developed that would connect fertile and marginal areas. The need for local self-sufficiency would weaken in the latter, and agricultural land could be abandoned.

Agricultural Intensification

Some national forest transitions in Europe appear to be closely linked to changes in agriculture. The changing agricultural technology and rising yields that took place during the nineteenth century were important factors. Increasing food production per unit area lessened the need to expand the area of production, and on the contrary led to a reduction in that area. The rewards of **agricultural intensification** were probably greatest in the most fertile areas, so that again the poorer areas were increasingly marginalized. This could happen at the national **scale** (e.g., that of France) but also more locally. For example, the introduction of sown grasses (usually on better land in the valleys) could reduce the necessity to graze livestock on mountain **pastures**. A reduction in grazing and browsing would facilitate the regeneration of scrub and forest.

The **Green Revolution** in recent decades may resemble that of the **Agricultural Revolution** in nineteenth-century Europe, and may be seen by some as a precursor to a forest transition in some parts of the developing world. The effect of agricultural changes, however, may be contingent on the complication of trade. Policies favorable to free trade could increase food imports: the effect of these could be similar to those of increases in yields. Conversely, food exports could be favored, encouraging in turn the clearance of more forest to allow expansion of the arable area.

Demographic Change

Population growth has often been seen as a "driver" of deforestation, implying that population decline might be favorable to forest expansion. There is little evidence, however, to suggest that forest transition is directly linked to demographic transition. In most European countries experiencing the forest transition in the nineteenth century, there was an acceleration of population growth.

Nevertheless, this is not to deny that demographic factors may be significant. In most European countries, the nineteenth century witnessed a radical redistribution of population. **Industrialization** and **urbanization** sucked people from the rural land, and in particular from the remoter and more marginal parts of the land. In France, for example, areas such as the Massif Central experienced a major "rural exodus," which in turn was followed by reforestation on abandoned land. In parts of the developing world, the demographic pressures that resulted in the **conversion** of forest into field, in order to produce food for otherwise landless people, have weakened: the next generation may head off to the city, leaving the fields to be recovered by the forest. This causal factor is sometimes called "scarcity of labor" or "labor shortage": a reduction in labor input to agriculture is followed by **disintensification** and withdrawal of agriculture. The labor is withdrawn because it can achieve better rewards elsewhere: there is a retrenchment of agriculture and a freeing of land for reforestation. Whether a term such as labor shortage in this context is helpful to understanding is debatable.

Energy and Transport

Both agricultural intensification and the "rural exodus" may be accompanied in time by significant developments in energy and transport. Prior to the nineteenth century, most European countries depended strongly on woodfuel. During the nineteenth century, fossil fuels replaced that dependence in whole or in part, both for domestic and industrial purposes. This process was facilitated by the spread of the railway, which permitted the bringing in of coal and, later, of oil. This change, of course, lessened pressures on the forest. The use of the forest as a fuel source was not necessarily unsustainable: nevertheless, a weakening of that use could reduce the likelihood of continued deforestation. In other words, a switch of dependence to the "underground forest," representing fossil fuels, could have contributed to the forest transition, especially in conjunction with the factors outlined above.

Demand for Forest Products

It can be hypothesized that an increase in demand for forest products could stimulate an expansion of the forest area. While this hypothesis is credible, there is little evidence that it explains most developed-world forest transitions. Imports can be brought in from elsewhere in the world, and there may be complications stemming from increases in demand for food, alongside those for wood. In addition, it fails to explain expansion of the forest area through natural regeneration on abandoned farmland.

Silviculture and Sustainability

At around the same time dramatic changes were taking place in agriculture, rural demography, energy, and transport, significant changes were occurring in attitudes to the forest and in approaches to its management. In the context of concerns about timber shortages in Germany around 1800 (and especially shortages of woodfuel), paradigm shifts occurred in silviculture. The concept of the "sustained-yield forest" emerged, in which timber production could be maintained indefinitely at a constant level through the harvesting and replanting of (for example) one compartment in a fifty-compartment forest over a fifty-year period. The forest became a timber farm: whereas traditionally it had provided a range of products for the local community, timber production now became the primary function. Grazing of livestock was excluded: agriculture and forest became functionally and spatially segregated. At the same time, forest codes and other laws might be introduced in an effort to prevent further deforestation, and in the context of developing statehood in nineteenth-century Europe, might be more effectively enforced than in the past.

Forest Policy and the Role of the State

In several European countries, a crisis, real or perceived, triggered state intervention in forest-related matters in a new way or on a new scale. In countries such as France and Switzerland, floods in alpine areas attracted state attention and were attributed to excessive deforestation in the headwaters of catchments. The eventual result was a clampdown on further deforestation and the beginnings of active reforestation. In the case of Britain, a very different type of crisis had a similar result. During World War I, the difficulty of importing timber (especially for the pit props essential in coal mining) led to the establishing of a state forest service: the Forestry Commission. This new body was given the task of expanding the forest area within the country, which had fallen to less than 5 percent of the land area. Once begun, this process continued, and the forest area is now more than twice its extent at the beginning of the twentieth century.

The Contingency of Causes

Previous efforts had been made by the state, both in Britain and elsewhere, to halt deforestation and encourage reforestation. The classic example is that of Jean Colbert in France in the late seventeenth century, when attempts were made to improve the French forest resource on which the French Navy then depended. Until the modern period, however, most of these attempts were ineffective. This is probably partly because of weakness of enforcement, and in turn because of the weakness of the state prior to its nineteenth-century development. If this is so, it may point to the multifaceted nature of "development" as the underlying causal factor of the forest transition. Development, in this context, probably means far more than simply **economic growth**: it will extend also to state **institutions** and infrastructure, including not only powers of enforcement but also, for example, provision for silvicultural training and education. It may also extend to civil society: in several European societies, interest groups or non-governmental organizations played important roles in promoting state intervention and in highlighting the relationship, as perceived, between floods and deforestation.

The example of Jean Colbert may point to the limits of state capability to intervene effectively. Effective intervention in the nineteenth century occurred alongside a set of agricultural, demographic, and technical trends that were favorable to reforestation. If these favorable trends had not been present, then the fate of intervention might have been similar to that of earlier centuries.

Environmental Kuznets Curves

The curve of forest area against time, or more precisely the curve of rates of deforestation against time, is reminiscent of the environmental Kuznets curve. This does not necessarily mean, however, that simple economic-development variables such as gross domestic product (GDP) per capita are closely related to reforestation in a direct sense, although they may have a role as proxies for wider aspects of development. Nor does it follow that a rapid drive for economic development would necessarily lead to forest expansion. There are, however, cases where deforestation has proceeded rapidly during early phases of development, as attempts have been made to convert natural capital in the form of forests into human capital by rapid exploitation of the forest (e.g., for conversion to plywood for export markets). It should also be borne in mind that some countries have maintained high or expanding forest proportions while importing much of their requirement of forest products and while being implicated in the continuing diminution of forest resources in other countries.

Forest Futures

Perceptions of current deforestation in the tropics may be influenced by the concept of the forest transition. It can be argued that a "turn round" will ultimately happen in the tropics, as it happened in Europe. This argument is, however, simplistic: it does not necessarily follow that what happened in Europe in the past will happen in the tropics in the future. This is especially so when understanding of the forest transition is still limited. Also, it can be argued that recent forest expansion does not compensate for the loss of natural forest, especially if much of the expansion is in the form of plantations. Typically, these have lower levels of **biodiversity** than the earlier forests. Again, however, there is a counterargument: given time, secondary forest, established on abandoned agricultural land, may eventually develop high levels of biodiversity.

The forest transition is an interesting phenomenon, which suggests that negative environmental trends (such as that of deforestation) can be halted and reversed while economic development proceeds. It is clearly one case where the well-known but poorly grounded expression I = PAT does not apply (*I* representing environmental impact, *P* population, *A* affluence, and *T* technology). A balanced view, both of its causal factors and its significance in influencing our **perception** of current deforestation, is desirable, as is a fuller understanding. The concept has been documented only since the early 1990s: the phrase "forest-transition theory" began to appear in the literature in the early 2000s. That theory, however, is still in its infancy, and much work is required to develop it more fully.

See also: **Cultural Factors; Driving Forces; Forest Anomalies; Human Appropriation of Net Primary Production; Land Abandonment; Land-Use Policies; Land-Use Transition; Mountains; Proximate Causes; Regrowth; Secondary Vegetation.**

Further Reading

Mather, Alexander S., "The Forest Transition," *Area* 24 (1992): 227–37; Mather, Alexander S., Fairbairn, Jamie, and Coby L. Needle, "The Course and Drivers of the Forest Transition: The Case of France," *Journal of Rural Studies* 15 (1999): 65–90; Perz, Stephen G., and David L. Skole, "Secondary Forest Expansion in the Brazilian Amazon and the Refinement of Forest Transition Theory," *Society and Natural Resources* 16 (2003): 277–94; Rudel, Thomas K., "Is There a Forest Transition? Deforestation, Reforestation, and Development," *Rural Sociology* 63 (1998): 533–52; Rudel, Thomas K., Bates Diane, and Rafael Machinguiashi, "A Tropical Forest Transition? Agricultural Change, Out-Migration, and Secondary Forests in the Ecuadorian Amazon," *Annals of the Association of American Geographers* 92 (2002): 87–102; Walker, Robert, "Deforestation and Economic Development," *Canadian Journal of Regional Science* 16 (1993): 481–97.

ALEXANDER S. MATHER

Forestry. The science of managing **forests**, encompassing silviculture—the science of tending and growing trees. Local and **indigenous knowledge** of trees and forests has existed for as long as humans have dwelled in or near the forest. "Scientific" forestry is usually regarded as a phenomenon of the modern period, and especially of the post-1800 period. Since then, it has been applied throughout the developed world and increasingly also in the developing world. While it is usually applied in a quest for "improvement," in terms of goals such as better timber increments, its application has not been without problems, and clashes between scientific and local knowledge have formed one dimension of power struggles between competing interests and value systems.

The application of forestry is sometimes likened to the advent of **agriculture**: it can be seen as a shift from the equivalent of hunting and gathering to that of farming. As in agriculture, it can lead to significant increases in yield. It is also often associated, however, with major reductions in **biodiversity**, as "weed" species are removed in the hope of channeling ecosystem productivity into a few desired species. As in agriculture and other applied sciences, evolution has occurred through time, not only in terms of technical knowledge but also in the values underlying the deployment of that knowledge.

"Modern" Forestry

The evolution of forestry in modern times can be seen as a response to scarcity: concerns about forest resources prompted attempts to manage forests in better, more productive ways. One of the milestones was the publication in 1664 of John Evelyn's *Silva*. Concern about the adequacy of accessible English forest resources, in the context both of destruction during the Civil War and maritime competition with the Dutch, led to the involvement of the Royal Society and in turn to John Evelyn's tome, subtitled *A Discourse of Forest-Trees and the Propagation of Timber in His Majesty's Dominions*. The book had much impact and is claimed to have led to the planting of millions of trees. Perhaps it codified and disseminated existing knowledge at least as much as it produced new knowledge, and perhaps it is true that it focused more on trees than on forests. It did not solve the problem of forest resource management in England: this problem was to reappear in heightened form at the time of the Napoleonic Wars.

Perhaps the most formative stage in the evolution of modern forestry was that of Germany in the late eighteenth century. The production and processing of minerals in the Harz Mountains required large amounts of timber, but the forest resource was shrinking as more timber was being harvested than was being added through growth. Concern about this state of affairs led to what could be regarded as a new regime or paradigm, in the form of "sustained yield" forestry. At the heart of this paradigm was an acknowledgment that timber should not be removed at a rate faster than growth. This principle was translated into practice by dividing a forest area up into a number of divisions equivalent to the number of years in the proposed rotation. If this rotation were, for example, fifty years, then the forest would have fifty divisions, one of which would be harvested and replanted each year. In this way, the timber yield of the forest would be sustained in perpetuity (assuming, of course, that soil deterioration or other ecological problems were avoided). Also, the species to be planted were those that would offer the maximum timber yield: in the German forests, spruce was widely used. Clearly, there were significant landscape effects arising from this checkerboard pattern, with clear-felled areas integral to it. There were also other consequences. One was that this "scientific" approach depended heavily on quantification of trees and areas. Ever since, quantification, positivism, and reductionism (e.g., to the goal of timber production) have been closely associated with this emergence of "scientific" or "rational" forestry, which itself could be seen as a product and manifestation of the Age of Enlightenment. The forest became a tree farm, or a "machine" for producing timber: other forest values were subordinated to this primary goal.

While some have suggested that the roots of recent concepts of sustainability can be traced back to the emergence of sustained-yield forestry, this relationship should be viewed with caution. The only "sustained" element in sustained-yield forestry was the level of timber production. At a time when the forest resource was declining, this was probably seen as a major advance. But the desirability of a sustained level of timber production through periods of fluctuating demand is questionable. Constant supply along with declining demand would mean decreases in price, and therefore potential financial problems for the forest managers. In other words, "sustained yield" in this context is a much narrower concept of sustainability than (for example) in the definition put forward in the so-called Brundtland report. The fact that the "sustained yield" forest might be a near-**monoculture** of species selected for timber production casts further doubt, in the minds of some, on the permanent "sustainability" of management.

Nevertheless, the influence of German sustained-yield forestry has been profound. The individual most closely associated with it, Georg Ludwig Hartig (1764–1837), went on to have a major influence on German education in forestry, and the students emerging from that education proceeded to have an international influence (e.g., Dietrich Brandis [1824–1907] in India and Carl Schenck [1868–1955] in the **United States of America**). It was to Dietrich Brandis and the German forestry tradition that Britain looked when it became concerned about the state of Indian forests. When it sought to improve forests and forestry in the homeland, it again looked to the German tradition, this time to William Schlich.

Colonial Forestry

The "scientific" forestry paradigm was exported around the world by the colonial powers, and especially by the British. Initial concerns, for example in India, were focused on naval timber, but even when the procurement of naval timber became less vital (after the introduction of iron and steel ships), the emphasis on timber persisted. The legacy of colonial forestry has been significant, and not always benign. In detail, the history is complex, especially in India, but a number of common themes are evident. The practice of forestry was usually informed by the German tradition and geared to the needs and worldview of the colonial power. Extensive areas were appropriated by the British as forest reserves, and indigenous peoples excluded. In the context of the preoccupation with timber, **non-timber forest products** were often discounted as "minor forest products," though these were sometimes of vital importance for local peoples. The clash among cultures, traditions, and worldviews did not occur on a level playing field: one side was backed by power not available to the other. On the one hand, there could be an alienation of the local people and the development of antipathy toward the forest and its management. On the other hand, local knowledge of the forest and its trees and range of resources might be discounted by the colonial masters. Problems often persisted after independence: forestland, having been appropriated by the colonial power, was passed to the independent state as its successor. Thereafter management, to the extent that it could be provided, was likely to be based on the "scientific" tradition and practiced by national foresters schooled in that tradition either in the territory in question or in the imperial homeland.

A Changing Paradigm?

While the dominance of the German model of "scientific" forest was persistent and pervasive, it was not unchallenged. Even during the nineteenth century, opposition with a more Romantic view resisted it, and in the latter part of the twentieth century challenges have been effective. In many parts of the world, there has been a recent "greening" of forestry. Differing terminology is applied, but a common theme is a greater emphasis on multiple forest products (as opposed simply to timber production) and in particular on environmental values related to the forests' **ecosystem services**. Greater **community involvement** in forest management has been sought, with a weakening of an explicit "top down" approach. In the United States, this shift is associated with the emergence of "forest ecosystem management." While definitions vary, there is agreement that this approach is more holistic and less reductionist than traditional "scientific" forestry, and that it seeks to manage forests in ways that mimic ecological processes. In short, "sustainable" forest management, seeking to achieve sustainability concurrently in economic, environmental, and social dimensions, has emerged to challenge in an effective manner the legacy of sustained-yield forestry in the form of "scientific" forestry. Forestry evolved later than agriculture, but its goals are comparable. As in agriculture, the intensity of management is variable, ranging from the creation of artificial plantations of **exotic species** to slight **modification** of otherwise natural forests through selective felling or enrichment planting. The model of "scientific" forestry that evolved in response to timber scarcity in Germany at the end of the eighteenth century has spread around the world. In recent years, however, it has faced a challenge from models that are less pre-occupied with timber production and that take a more benign approach to the management of forest ecosystems.

See also: **Access; Arena of Land Conflict; Biomass; Colonization; Conservation; Cultural Factors; Decentralization; Ecological Colonization; Land Rights; Metabolism; Mineral Extraction; Reforestation; Regrowth; Secondary Vegetation; Sustainable Land Use; Tragedy of Enclosure; Turnover; Yellowstone Model.**

Further Reading

Heske, Franz, *German Forestry*, New Haven, CT: Yale University Press, 1938; McManus, Phil, "Histories of Forestry," *Environment and History* 5 (1999): 185–208; Schlich, William, Fisher, William R., Hess, Richard, and Karl Gayer, *Schlich's Manual of Forestry*, London: Bradbury-Agnew, 1906–11; Westoby, Jack C., *Introduction to World Forestry: People and Their Trees*, Oxford, UK: Blackwell, 1989.

ALEXANDER S. MATHER

Frontiers in Ecology and the Environment. The new members' journal of the Ecological Society of America (ESA), which covers all aspects of ecology, the environment, and related disciplines. International in scope and interdisciplinary in approach, the journal focuses on current ecological issues and environmental challenges and addresses professional ecologists as well as scientists working in related disciplines.

Further Reading

Ecological Society of America [Online, January 2005], The Frontiers in Ecology and the Environment Web Site, www.frontiersinecology.org.

HELMUT GEIST

G

Geographical Information System. A computer-based system for analyzing spatially referenced data. From the earliest experiences of these systems in the 1960s, such as the Canada Geographical Information System (CGIS) and the Land Use and Natural Resources Inventory of New York State (LUNR), the development of the geographical information system (GIS) discipline has largely focused on the management of land use. The fact that the storing and analysis of voluminous geographical data is more easily undertaken by computer probably explains the continuously more widespread efforts of automated map production. The recent dramatic progress of the information technology (IT) industry and the significant reductions in the price of GIS hardware and software over the last thirty years accounted for the success of such digitally based approaches. Currently, this technology and discipline are inherent to most land-use studies. Advances in Internet technology bring spatial data to the individual user. In this way, geographical knowledge is shared and, it is to be hoped, awareness of our precious environment is intensified.

The approach of GIS to spatial data has been embraced by numerous disciplines and divergent views that benefit from a standardized toolbox to process their data flows. Since the essence of GIS is its heterogenous and multidisciplinary character, many different definitions, quite similar methods, and a wide variety of applications co-exist. The numerous definitions illustrate the debate about the focus of the GIS field from the GIS technology (hardware and software), the GIS database (geographical and related data), and the GIS infrastructure (staff, facilities, and supporting elements). As a literal interpretation, a GIS is an information system about geography that describes and interprets the physical location of resources and objects in the real world. The term "spatial" or "geospatial" is often synonymous with "geographic." The GIS is a system in its way of setting different elements together and organizing them as a whole. By the term "information," we can refer to the observations, collection of data, investigations, interpretation, **modeling**, and finally the increase of the understanding that is the goal of the system.

The interest of GIS is the ability to organize and integrate apparently disparate data sets together by specific techniques such as spatial searching and overlay. Data acquisition is the most important but also the most expensive and time-consuming part of implementing GIS technology. Digital geospatial data that are used in a GIS come from a variety of sources such as user input; maps; and tabular data such as census lists, **remote sensing** imagery, and other products. Collected data must then be converted into a suitable, digital format. Data sets used in GIS methods combine graphic and nongraphic data (attributes, topological relationships, and map annotation). Details of graphic features are stored in vector points, lines, and polygons or raster grid cell formats, called **pixels**. In GIS databases, the geographic data are mapped into separate files, layers, or themes, georeferenced in through a common coordinate system. In GIS functions, a distinction can be made between analyses (spatial or attribute query) and more trivial processing tasks like updating and simple error removal. The real power of GIS lies in the ability to consider different attributes together, mainly as a decision support tool, but also in support of science.

The people involved in GIS are, on the one hand, the developers, and on the other, the so-called end users. The latter are people who will benefit from the numerous applications such as cartography, geography, environmental science, social science, business, and everyday life. An example of a GIS application in cartography is National Geographic's MapMachine. This tool provides user-personalized maps with online functionalities. GIS applications vary from very large geographic **scales** (utility management) to small geographic scales (global systems). Automatic mapping/facility management (AM/FM) applications use networks and are generated toward management of public utilities (e.g., water, electricity). Land information system (LIS) applications are developed by local governments and public utilities to manage **land tenure**, land use (both rural and urban), infrastructure, and services (**fire** combat, emergency services) on the basis of cadastral data. GIS-based **land evaluation** studies at individual, regional, and global levels such as the World Soil Resources Map produced by the **United Nations Food and Agriculture Organization** belong to the older applications of GIS in which data on soil, terrain height, and water availability are combined to retrieve information on the suitability of growing a particular crop or applying a specific land use.

See also: **Cadastre; Georeferencing; Hot Spot Identification; Integrated Model; Land Cover Classification System; Participatory Geographic Information System; Pixelizing the Social; Socializing the Pixel.**

Further Reading

Burrough, Peter A., and Rachael A. McDonnell, *Principles of Geographical Information Systems*, Oxford, UK: Oxford University Press, 1998; Van Der Wel, Frans J. M., "Geographical Information Systems," in *Encyclopedia of Life Support Systems*, developed under the auspices of UNESCO, Eolss Publishers, Oxford, UK [Online, September 2004], The EOLSS Web Site, www.eolss.net.

NATHALIE STEPHENNE

Georeferencing. An abbreviation of geographical referencing, which is the process of linking spatial data to the appropriate real-world coordinates using a spatial reference or coordinate system. A large number of potential coordinate systems exist at the level of countries, regions, hemispheres, and for the globe. Geographic data is said to be georeferenced (or spatially referenced) if it is associated with a known position in a coordinate system. Georeferenced data can be viewed, overlaid, queried, and analyzed together with other geographic data. The most commonly used coordinate systems include the geographic coordinate system (lines of latitudes and longitudes expressed in degrees) and projected or planar coordinate systems such as Universal Transverse Mercator; and state plane coordinates, expressed in easting and northing coordinate pairs in a metric system. One important application is the georeferencing of **aerial photographs** and satellite images to derive and validate land-use and **land-cover change**. The transformation to a map coordinate system is commonly achieved by means of points of known position, that is, ground control points. These can be collected by ground-truthing using the **global positioning system** or by registering the image to an already-referenced data source such as a topographic base map.

See also: **Change Detection; Geographical Information System; Pixelizing the Social; Remote Sensing; Socializing the Pixel; Validation.**

Further Reading

Lillesand, Thomas M., Kiefer, Ralph W., and Jonathan W. Chipman, *Remote Sensing and Image Interpretation*, 5th ed., New York: John Wiley & Sons, 2004.

DANIEL MÜLLER

George Perkins Marsh Institute. An interdepartmental unit of Clark University in Worcester, Massachusetts, named after the namesake prescient environmentalist whose seminal book *Man and Nature* (1864), together with the estate study of **Johann Heinrich von Thünen** (1826), pioneered the study of land-use change. Institutional research centers around the same fundamental question that concerned **George Perkins Marsh**: what is and ought to be our relationship with nature? The more balanced view of this relationship that set him apart from later preservationists has left a mark on the tradition of the institute: combining basic and applied research often undertaken in interdisciplinary, team-based mode on human dimensions of **global environmental change**, best reputed in the fields of environmental hazards, land-use/cover change, and international development. The institute is internationally renowned for assisting the United Nations Institute for Training and Research in developing analytical capacities to address **environmental changes**, developing IDRISI **geographical information system** software, and for its leading role in developing the **Land-Use/Cover Change (LUCC) project**.

Further Reading

Clark University [Online, December, 2004], The George Perkins Marsh Institute Web Site, www.clarku.edu/departments/marsh.

KE CHEN

Global Burnt Area 2000. A global scale inventory of the vegetated areas burnt in the year 2000 derived from **SPOT Vegetation** satellite data. The Global Burnt Area 2000 (GBA 2000) data set provided the first independent estimates of the area of **vegetation** burnt at the global **scale**. Biomass burning occurs in almost every vegetated ecosystem on earth. **Fire** is used in agricultural practices (such as in southern **Russia**), land management practices (such as in African national parks), and for **forest** clearance (such as in **Amazonia**). The implication of **biomass** burning on changes in the climate system, global warming, and the **carbon cycle** is not well understood. Vegetation fires act as a direct disturbance to the terrestrial ecosystems, which play a role as sources or sinks of carbon at local, regional, and global scales. Emissions from biomass burning are known to contribute significantly to the injection of pollutants into the atmosphere. **Greenhouse gases** and carbonaceous aerosols emitted by the vegetation fires impact the radiation balance at the surface, on the acidification of precipitation, on air quality, and on **human health**. Biomass burning is thought to contribute up to 40 percent, 16 percent, and 43 percent of the total emissions of anthropogenic origin for carbon dioxide, methane, and carbon monoxide, respectively. The methods required in GBA 2000 to detect a burnt area differed from one ecosystem or climatic zone to another (e.g., boreal forest, tropical forest, **grasslands**). The GBA 2000 strategy used a network approach for the development and testing of a series of regional algorithms for processing the daily SPOT Vegetation imagery into burnt area maps. Seven algorithms were selected for processing the entire year 2000 data set and integrated into a global processing chain. Monthly and annual maps at a minimum mapping resolution of one km^2 are available to the research community at no cost. In addition, statistical data sets, such as the area burnt per type of vegetation cover and sets for geographical entities of particular interests (e.g., countries, protected areas), and data sets at 0.5° and 1° grids at global level, suitable for the climate change **modeling** community, have been developed. The **GLOBCARBON** project of the European Space Agency is utilizing the algorithms developed in GBA 2000 to produce a multiannual inventory of the years 1999–2003.

See also: **Atmosphere-Land Interlinkages; Climate Impacts; Global Land Cover Map of the Year 2000; GLOBSCAR.**

Further Reading

Joint Research Center of the European Union [Online, August 2004], The Global Burnt Area 2000 Project Web Site, www.gvm.jrc.it/fire/gba2000; Tansey, Kevin, Grégoire, Jean-Marie, Stroppiana, Daniela, Sousa, Adelia, Silva, João M. N., Pereira, José M. C., Boschetti, Luigi, et al., "Vegetation Burning in the Year 2000: Global Burned Area Estimates from SPOT VEGETATION Data," *Journal of Geophysical Research* 109 (2004): D14S03, doi:10.1029/2003JD003598.

KEVIN J. TANSEY

Global Environmental Change. A wide range of changes in the physical conditions of the earth's land, oceans, and atmosphere that result from the interactions of humans and natural processes. Global environmental change (GEC) affects the quality of human life and sustainable development on a worldwide scale. GEC covers a range of issues from such global-scale problems as climate change, **biodiversity**, and **desertification** through problems experienced at the local level, such as waste management, water pollution, and air pollution. GEC is of global concern, but it has local implications. While it has consequences for ecosystems, humans, and human infrastructures at multiple spatial, temporal, and sociopolitical **scales**, GEC results from the individual and collective actions of people.

The earth's physical and biological systems have always been in a dynamic state of change. **Environmental change** is more characteristic of the earth system than constancy is, but GEC can be rapid and threatening. As a consequence, GEC has become a political issue of considerable importance, forcing itself before the world and its citizens. At the Earth Summit in Rio de Janeiro in 1992, the international community affirmed that current patterns of economic and social development were not sustainable. The **Intergovernmental Panel on Climate Change** (IPCC) was established in 1988 to address the impacts, adaptation, and mitigation of climate change, a GEC issue of widespread concern.

Human-induced GEC is often moderated through land-use/cover change. Land change is central to the GEC debate as both a driver and a consequence of GEC. Important land-use/cover changes that have contributed to GEC include **deforestation**, **urbanization**, **industrialization**, and the expansion of **agriculture**. These changes are often the cause of environmental issues such as **soil erosion**, changes in regional water cycles and **nutrient cycles**, freshwater usage and quality, the **eutrophication** of water bodies, and **greenhouse gas** (GHG) emissions. Because of this, management of land-use/cover change has the potential to mitigate some of the negative aspects of GEC. For example, **reforestation** can contribute to **carbon sequestration**, de-intensifying agricultural production can improve water quality and abate GHG emissions, and urban planning can be used to minimize atmospheric pollution.

Land-use/cover change is, however, also a consequence of GEC. **Population dynamics** and **economic growth** are important drivers of both GEC and land-use/cover change. A growing and increasingly affluent population has in the past increased the need for food and fiber production commensurate with agricultural expansion and deforestation. Providing housing for this population has led to rapid urbanization. Climate change will have widespread implications at global and local scales for human land-use activities. This includes the effects of climate change on the potential distribution of agricultural crops and their productivity, **forestry** growth rates, the location of leisure activities and tourism, and other processes that affect the provision of ecosystem goods and **ecosystem services**.

There will always be a need for people to use natural resources and to develop. Some human activities, such as urbanization and the development of transport networks, agricultural production, and **mineral extraction**, however, are prone to causing pollution and contributing to GEC. Whenever these activities occur, there is an increase in the amount of waste material and mineral emissions. This is an unavoidable consequence of human activities, but society should seek to minimize the rate of production of waste, avoid degradation of the environment, and learn to manage the global ecosystem in a sustainable way.

See also: **Atmosphere-Land Interlinkages; Driving Forces; Earth System Science Partnership; Proximate Causes; Public Policy; Sustainable Land Use; Water-Land Interlinkages.**

Further Reading

Steffen, Will, Sanderson, Angelina, Tyson, Peter D., Jäger, Jill, Matson, Pamela A., Moore, Berrien III, Oldfield, Frank, et al., *Global Change and the Earth System: A Planet under Pressure* (The IGBP Series), Berlin: Springer, 2004.

MARK ROUNSEVELL

Global Environmental Change. An international, peer-reviewed journal established in 1990 to address the human, ecological, and **public policy** dimensions of **environmental change** at the global scale. It deals, in brief, with the environmental processes threatening the sustainability of life on earth. The journal interprets **global environmental change** as the outcome of processes that manifest in localities but have consequences on multiple spatial, temporal, and socio-political **scales**. The journal publishes on, but is not limited to, **deforestation**, **desertification**, **soil degradation**, **species extinction**, sea level rise, acid precipitation, destruction of the ozone layer, atmospheric warming/cooling, nuclear winter, the emergence of new technological hazards, and the worsening effects of natural disasters. Part A ("Human and Policy Dimensions") spans the social and natural sciences and publishes articles across areas including **biodiversity** and **ecosystem services**, water resources, climate change, international agreements, North-South relations, **institutions**, governance, and **land change as a forcing function in global environmental change**. Part B ("Environmental Hazards") addresses the human and policy dimensions of hazards. It spans the full range of hazard events from extreme geological, hydrological, atmospheric, and biological events (such as earthquakes, floods, storms, and epidemics) to technological failures and malfunctions such as industrial explosions, **fires**, and toxic material releases, thus highlighting issues of human exposure, **vulnerability**, awareness, response, and risk.

See also: **People at Risk.**

Further Reading

Elsevier Science [Online, September 2004], The Global Environmental Change Journal Web Site, www.elsevier.com.

HELMUT GEIST

Global Environmental Change and Food Systems (GECAFS) Project. A comprehensive research program of the **Earth System Science Partnership** that focuses on understanding the links between **food security** and **global environmental change**. Launched in 2001, GECAFS aims at an improved understanding of the relationship between food systems and the earth system, and wants to deliver tools designed both to help reduce **vulnerability** to change and to analyze the socioeconomic and environmental consequences of potential management and policy adaptation options.

See also: **Biomass; Food Crops.**

Further Reading

Global Environmental Change and Food Systems [Online, January 2005], The GECAFS Web Site, www.gecafs.org.

HELMUT GEIST

Global Land Cover Map of the Year 2000. A new land-cover database for the year 2000 that was produced by an international partnership of thirty research groups coordinated by the **Joint Research Centre (JRC) of the European Commission**. The database contains land-cover maps with detailed, regionally relevant map legends and a global product that combines all regional classes into one consistent legend. Each continent is dealt with as a separate region. The database is designed to serve users from science programs, policymaking bodies, environmental convention secretariats, international and non-governmental organizations, and development-aid projects.

Although 70 percent of our planet's surface is covered by water, it has long been known that land cover affects our climate by influencing energy, water, and gas exchanges with the atmosphere, and through acting as a source and sink in biogeochemical cycles. Accurate land surface parameterization is therefore needed to improve our understanding of the state of the global climate system and its variability, and also to monitor the forcing of the climate system. These requirements call for a land-cover classification where divisions among cover types are determined for use in biogeochemical models.

Humans have mapped their immediate environments throughout history. The advent of suitable earth observing (EO) satellites in the early 1970s provided a means of uniform observation, but globally consistent maps treating all parts of the world equally and representing actual land cover over a fixed, contiguous period of time only became available in the 1990s. The first of these was begun in 1990 to meet the scientific needs of the **International Geosphere-Biosphere Programme** (IGBP). The IGBP's land-cover mapping activities were based on data from the **advanced very high resolution radiometer** (AVHRR) sensor with a nominal 1.1 kilometer resolution, collected between 1992 and 1993; the validated data set was published toward the end of the decade. The final legend of the **IGBP-DIS global 1-km land cover data set** mapped global land cover into seventeen classes and continues to be used for global land-cover maps destined for the **modeling** communities today. There are now several instantaneous assessments of land cover based on data collected by EO satellites, such as the Global Land Cover Map of the Year 2000 (GLC-2000).

GLC-2000 is derived from the analysis of one year of daily satellite imagery from the VEGETATION sensor onboard SPOT 4 satellite. In addition, data from other sensors—such as the **along track scanning radiometers** (ATSR) sensor onboard the **European remote sensing (ERS-1/-2) satellites**, radar sensors onboard JERS and ERS satellites, and DMSP sensors—have been used to solve specific problems, in particular in regions with persistent cloud cover, especially in equatorial regions and for identification of urban areas. Detailed legend definition, image classification, and map-quality assurance are carried out region-by-region. A statistically valid accuracy assessment has been performed on the final global product.

The regional maps have been published individually, and the global data set, which includes twenty-four land-cover classes, is being published by the JRC in association with the **United Nations Food and Agriculture Organization** and the **United Nations Environment Programme**. The database has been adopted by **Millennium Ecosystem Assessment** as the baseline for 2000 and is being used by the World Conservation Monitoring Centre in their report on the state of the world's protected areas. The GLC-2000 database and the more detailed regional products are freely available to science and policy users.

The GLC product is accompanied by a "sister" product on **Global Burnt Area 2000**, which documents the areas burnt in the year 2000 from the same **SPOT Vegetation** data set.

Such exercise of global land-cover mapping can now be repeated regularly in the future. There are already programs in **Europe** and in the **United States of America** to develop new land-cover databases for 2005 and beyond with an improved detail (at 300-meter resolution).

See also: **Conservation; Global Land Cover Network; GLOBCOVER; JERS Radar Mosaics; Remote Sensing.**

Further Reading

Joint Research Centre of the European Commission, Institute for Environment and Sustainability, Global Vegetation Monitoring Unit, Terrestrial Ecosystem Monitoring Unit [Online, September 2005], The Global Land Cover Map of the Year 2000 Web Site, www-gvm.jrc.it/glc2000.

ETIENNE BARTHOLOMÉ, ALAN BELWARD,
HUGH EVA, JEAN-MARIE GRÉGOIRE, PHILIPPE
MAYAUX, HANS-JÜRGEN STIBIG

Global Land Cover Network. Initiated in 2004 by the **United Nations Food and Agriculture Organization** (FAO) and the **United Nations Environment Programme** (UNEP), with the financial support of the government of Italy, to answer the needs of both national and international communities for standardized land-cover data. The Global Land Cover Network (GLCN) is based on the recommendations of **Agenda 21** for coordinated, systematic, and harmonized collection and assessment of data on land cover and environmental conditions, especially for **monitoring** of the environment. It answers the specific declaration and request by the participants of the Strategies for Global Land Cover Mapping and Monitoring conference held in Artiminio, Italy, on May 6–8, 2002, for the development of an international harmonized land-cover mapping effort. The objectives of the project are to improve the availability of reliable and standardized information on land cover and **land-cover change** at the national, regional, and global levels, especially for the user community in developing countries that have difficulty accessing reliable, consistent, and updated information. GLCN also undertakes national capacity building in land cover data production and management, and develops effective communication links among relevant organizations through regional networks. Project activities are supported by "topic centers," which have a number of functions, including providing assistance to all land-cover mapping and monitoring projects; advising on land-cover mapping methodologies; acting as neutral international clearinghouses; developing regional products; and disseminating information on products and services, databases, and meta-databases. GLCN sub-programs already include the **AFRICOVER** and ASIACOVER initiatives, and networks are currently being developed for Central and Latin America as well as for the Middle East. During this process, GLCN assists and integrates with other initiatives to ensure that a broad and diverse community is fully engaged in the development, application, and use of the outputs. A number of applications were developed and are used in previous FAO and UNEP mapping initiatives. These include the **Land Cover Classification System** (LCCS), a comprehensive method for the description, characterization, classification, and comparison of land cover anywhere in the world, at any **scale**. These applications further include image processing, interpretation and mapping systems, and geographical vector interpretation systems (GeoVIS) aimed at enhancing both the use and the updating of products, and the dissemination of data and information through specially designed, interactive, Web-based databases.

See also: **Geographical Information System; Global Land Cover Map of the Year 2000; Global Terrestrial Observing System.**

Further Reading

The Global Land Cover Network [Online, January 2005], The GLCN Web Site, www.GLCN.org.

FREDERIK PISCHKE, REUBEN SESSA

Global Land Project. A scientific project, jointly sponsored by the **International Geosphere-Biosphere Programme** and the **International Human Dimensions Programme on Global Environmental Change**, on **land change as a forcing function in global environmental change**. Human transformations of ecosystems and landscapes are the largest source of change on earth, affecting the ability of the biosphere to sustain life. As a species, mankind has become ever more capable of appropriating and altering the earth's resources for human needs. **Agricultural intensification** and diversification of land use and advances in technology have led to rapid changes in biogeochemical cycles, hydrologic processes, and landscape dynamics. Changes in land use and management affect the states, properties, and functions of ecosystems. In turn, these consequences affect human well-being. There is a need for improved understanding of how human actions affect natural processes of the terrestrial biosphere, and an even greater need to evaluate the consequences of these changes. The Global Land Project (GLP) is at a planning stage and will be implemented in late 2005. Its overarching research goal is to measure, model, and understand the coupled human-environmental system ("land system") as part of broader efforts to address changes in earth processes and subsequent human consequences in the following areas:

1. To identify the agents, structures, and nature of change in coupled human-environment systems on land and quantify their effects on the coupled systems
2. To assess how the provision of **ecosystem services** is affected by the changes in these couplings
3. To identify the character and dynamics of vulnerable human-environment land systems to interacting perturbations, including climate change

Three thematic areas emerge from the research framework: the causes and nature of land-system change, the consequences of land-system change, and integrating analysis and **modeling** for land sustainability. Central to the GLP implementation strategy are the networks and research activities from the Global Change and Terrestrial Ecosystems (GCTE) project and the **Land-Use/Cover Change (LUCC) project**. Activities inherited from GCTE and LUCC research have been incorporated into the GLP science plan, since both projects have been discontinued. In addition, joint activities will be developed with other projects of the **Earth System Science Partnership**.

Working groups and networks of researchers will serve a fundamental role in the implementation of the GLP science plan. The GLP's guiding principle is that studies will be place-based, carried out by interdisciplinary teams, will be well aware of the need to be able to scale both up and down and across disciplines, and will define the relationship of the research to the broader coupled human-environmental framework. Research activities will include the following:

1. Case studies, manipulative studies, and comparative studies
2. Networks of experimental and case studies across gradients of land systems
3. Long-term observations and experiments
4. Process models (e.g., **vegetation**/ecosystem, agro-ecosystem, **agent-based models**)
5. Land-use **meta-analyses**
6. Integrated analytical tools, including models and practical advances in field techniques
7. **Decision-making** models
8. Integrated regional studies
9. Interdisciplinary database development and archival systems services
10. Evaluation of the coupled human-environmental land system's **vulnerability** to **global environmental changes**

See also: **Integrated Land Ecosystem-Atmosphere Processes Study; Land-Ocean Interactions in the Coastal Zone (LOICZ) Project.**

Further Reading

Global Land Project, Science Plan and Implementation Strategy, IGBP Report No. 53/IHDP Report No. 19, Stockholm, IGBP Secretariat, 2005, Steffen, Will, Sanderson, Angelina, Tyson, Peter D., Jäger, Jill, Matson, Pamela A., Moore III, Berrien, Oldfield, Frank, et al., *Global Change and the Earth System: A Planet under Pressure*, Berlin: Springer Verlag, 2004.

JOÃO M. F. MORAIS

Global Observation of Forest and Land Cover Dynamics. A panel of the **Global Terrestrial Observing System** (GTOS). Its objective is to improve the quality and availability of observations of **forests** at regional and global **scales** and to produce useful, timely, and validated information products from these data for a wide variety of users. Global Observation of Forest and Land Cover Dynamics (GOFC-GOLD) grew out of an active community of researchers studying forest-cover changes, particularly in the tropics, during the 1990s. A variety of novel forest-cover maps and statistics, most based on satellite observations, became available, but were often in disagreement with one another and with official national statistics, and could not provide a reliable global picture of the state of the world's forests. There was a clear need to rationalize the effort, standardize the methods and definitions, and ensure repeatability so that valid cover-change observations could be made. At the same time it became obvious that a simple view of **deforestation** was untenable. Partial clearing of trees can also occur, and land cycles between forest and other cover forms. As a result, a rigorous approach to forest cover change needs to make observations outside the forest margin of other land uses as well. This prompted the change in emphasis from "Global Observations of Forest Cover" to "Global Observations of Land Cover Dynamics." It was originally developed as a pilot project by the Committee on Earth Observation Satellites (CEOS) as part of their Integrated Global Observing Strategy (IGOS).

GOFC-GOLD has three themes: land-cover characteristics and change, fire **monitoring** and mapping, and biophysical processes.

The land-cover theme focuses on products related to land-cover classification (both forest and non-forest); and on measurements of forest attributes, such as leaf type and longevity, canopy cover, and height. It proposes a systematic global observation program at moderate spatial resolutions (250–1,000 meters), repeated every five years, combined with periodic mapping of key areas at fine resolution (about twenty-five meters). These products find users in a variety of fields, including global change scientists, resource managers, **biodiversity** conservators, and climate change policymakers.

The fire theme focuses on developing, testing, and implementing methods for fire monitoring. **Fire** is one of the main mechanisms modifying forest cover. It also has a significant impact on atmospheric chemistry and the global carbon cycle. The global increase in fires following the 1997–1998 El Niño event emphasized the need for more complete, accurate, and timely information regarding fire location, extent, and intensity. A variety of existing satellite-based sensors can be used for near-real-time active fire detection as well as post-fire scar mapping. GOFC-GOLD coordinates a number of regional networks for validation of the products. The products are finding use in national and regional disaster management systems, as well as in research programs such as the Global Carbon Project.

The biophysical processes theme is building an operational framework to receive and continue observations of leaf area index (LAI), fraction of absorbed photosynthetic radiation (FAPR), and net primary productivity, that are currently in the research domain. Potential users include policymakers in the UN conventions on climate change and combating **desertification** and resource managers at national and sub-national scales.

See also: **Global Land Cover Network GLOBCARBON; GLOBCOVER; Remote Sensing; United Nations Convention to Combat Desertification.**

Further Reading

United Nations Food and Agriculture Organization [Online, November 2004], Global Observation of Forest and Land Cover Dynamics Web Site, www.fao.org/gtos/gofc-gold.

ROBERT J. SCHOLES

Global Positioning System. A radio-navigation network comprising a constellation of at least twenty-four satellites (in six orbital planes) first launched in 1978, and their groundstations. Global positioning systems (GPS) work on the premise of trilateration: they determine a location on the earth by using the positions of the satellites in orbit. Each GPS satellite is equipped with a highly accurate atomic clock and a unique identifying code called a pseudorandom code (PRN). The time difference from the radio signal being transmitted by the satellite to the receiver on earth is used to determine a distance, or range. A minimum of four range measurements is required to determine an XYZ (latitude, longitude, elevation) location on the earth. In order to accurately determine the range from a GPS satellite to a GPS receiver, the location of the satellite in its orbit must be precisely known. Several GPS groundstations across the earth track the positions of the satellites and upload their orbit positional data (ephemerides) to the orbiting satellite, which are then received by a GPS receiver. The GPS system is used by both civilians and the U.S. military (C/A, or coarse acquisition mode), though more accurate positional data are available to the military (P, or precise code). On May 1, 2000, President Clinton ordered the intentional scrambling (selective availability, or SA) of the signal turned off, making accurate GPS readings more realizable by recreational GPS users at the 10–20 meter horizontal error compared to over 100 meters of random, horizontal error.

The soon-to-be-developed European Galileo system and the Russian GloNass system are compatible with GPS, providing even greater coverage. Commercial subscription services such as Omnistar and WAAS (wide area augmentation system) now allow for real-time differential correction to provide greater accuracy of location. However, for survey-grade accuracy, post-collection differential correction is still required. GPS equipment is also used onboard airborne platforms in conjunction with IMU (inertial measurement unit), in conjunction with the collection of **airborne light detection and ranging** (laser), optical, infrared, microwave, and videography data. Recreational-level GPS units are now routinely used in fishing, camping, mountaineering, and geocaching.

See also: **Georeferencing Remote Sensing.**

Further Reading

Spencer, John, Frizzelle, Brian G., Page, Philip H., and John B. Vogler, *Global Positioning System: A Field Guide for the Social Sciences*, Malden, MA: Blackwell Publishing, 2003; Trimble Navigation Limited [Online, October 2004], The Trimble Global Positioning System Tutorial Web Site, www.trimble.com.

KELLEY A. CREWS-MEYER, AMY L. NEUENSCHWANDER

Global Terrestrial Observing System. One of three collaborative earth-observing systems established by the international community following the first Earth Summit, dealing with the land surface, freshwater systems, and the cryosphere (ice caps, glaciers, and sea ice). Coastal systems are covered in coordination with the Global Ocean Observing System, and the near-surface climate by a joint panel with the Global Climate Observing System.

The Global Terrestrial Observing System (GTOS) was established in 1996 with the mission of providing policymakers, resource managers, and researchers with access to the data they need to detect, quantify, locate, understand, and warn of changes (especially reductions) in the capacity of terrestrial ecosystems to support sustainable development. It has five sponsors: the **United Nations Food and Agriculture Organization** (FAO); **United Nations Environment Programme** (UNEP); United Nations Education, Science and Culture Organization (UNESCO); World Meteorological Organization (WMO); and the International Council of Science (ICSU). GTOS has a director and small secretariat hosted by FAO in Rome.

The GTOS implementation plan covers five issues: changes in **land quality**, availability of freshwater resources, loss of **biodiversity**, climate change, and pollution and toxic substances. The implementation of modules covering these topics is phased, and achieved through the work of specialist panels. Those to do with climate (the Terrestrial Observing Panel for Climate, TOPC) and land cover (**Global Observation of Forest and Land Cover Dynamics**, GOFC/GOLD) are fully operational, as is the Terrestrial Ecosystem Monitoring Sites (TEMS) database. The Terrestrial Carbon Observations (TCO) began implementation in 2004, in partnership with the Global Carbon Project research program. At that time, a global **biodiversity** observation system was being designed in collaboration with the **DIVERSITAS** research program.

GTOS is not a primary source of data, nor is it a research program. It is an integration mechanism; data and site directory; and promoter of consistent, shared Earth observation products. The observations are made by sites, networks of sites, and remote-sensing agencies around the world. The organizing framework is the Global Hierarchical Observing System, Terrestrial (GHOST). This approach recognizes that some relatively simple observations are taken periodically at very many locations, whereas other complex sets of measurements are usually made on a continuous basis, but in only a limited number of places. In between is a continuum of observational intensities, which GHOST organizes into five "tiers." By combining the spatial coverage provided by tier 5 (which includes, for instance, space-based **remote sensing**) with the accuracy and process understanding provided by tiers 1–4, reliable, ongoing global coverage can be obtained in a cost-effective way.

Further Reading

United Nations Food and Agriculture Organization [Online, November 2004], The Global Terrestrial Observing System Web Site, www.fao.org/GTOS.

ROBERT J. SCHOLES

Globalization. Contemporary **global environmental change** is intimately linked to processes of globalization such as **population dynamics**, the growth of international trade and financial flows, increasing transport and energy **consumption**, and accelerated **urbanization**, which are transforming the earth system in many, often irreversible ways. Globalization is the growing and accelerated interconnectedness of the world in an economic, political, social, and cultural sense. Due to almost unrestrained flows of goods, capital, information, and people, the world is becoming more and more a single place ("a global village" or "a global ecumene"), in which distant peoples share knowledge and lifestyles and different **institutions** function as parts of one complex system. Although globalization processes are not new phenomena in human history, the **scale**, quality, and speed of the globalization of the last decades are without precedence.

Globalization implies a time-space compression. People increasingly perceive local situations and events as being influenced by external, often unpredictable factors that they can only control in very limited ways. At the same time, there is a growing recognition that local events or actions taken in one place may have direct and immediate consequences in other locales worldwide, with often unforeseeable results and diffuse **feedbacks**. The everyday experience of an increasing degree of interconnectedness among local, national, regional, and global dimensions that cannot be grasped by the individual in its whole complexity and dynamic is restructuring basic relations among humans, subjects and objects, and humans and their environment. Globalization is modifying in many places the "traditional" configurations of agency and structure. It also redefines the roles and the relations between centers and peripheries of power and production. As the world turns faster, today's periphery may be tomorrow's center; today's satellite may be tomorrow's metropolis. Globalization entails a shift from basically two-dimensional spaces with sharp boundaries and clearly defined centers and peripheries, to a multidimensional global space with unbounded, often discontinuous and interpenetrating sub-spaces.

Globalization contains the dialectic between expansion and limitation. While on the one hand the dynamic of economic globalization pushes for the removal of all boundaries associated with cultural, social, and political spaces; on the other hand, the environmental limits of the earth stand out more clearly than ever before.

Another particularity of globalization is that it reduces diversity. Similar cultural, social, political, and economic patterns and convergent practices are observable throughout the globe (e.g., similar means and modes of communication and networking, symbolic resources, consumption patterns, institutional settings, production processes, commercial practices). A synonym for this is "the Cocacolonization of the world." The dominant image of the world is that of an increasingly homogenous space. However, many studies show that globalization can also trigger the redefinition of "difference" and the creation of new diversity. The experience of a growing harmonization is, for example, currently contested by the emergence of strong ethnic, fundamentalist, regional, and national movements.

See also: **Colonization; Cultural Factors; Driving Forces; Economic Liberalization; Economic Restructuring; Exogenous Factor; Industrialization; Investments; Public Policy.**

Further Reading

Appadurai, A., *Modernity at Large: Cultural Dimensions of Globalization*, Minneapolis, MN: University of Minnesota Press, 1996; Harvey, David, *The Conditions of Postmodernity*, London: Blackwell, 1989; Sachs, Wolfgang, *Planet Dialectics: Explorations in Environment and Development*, London: Zed Books, 1999.

BARBARA GÖBEL

GLOBCARBON. An initiative of the European Space Agency (ESA) to develop fully calibrated global estimates of land products from earth observation data. The key objective is the observation of terrestrial carbon dynamics to yield an understanding of the spatial and temporal patterns of the **carbon cycle**. Four product themes were developed for the period 1997–2003:

1. **Fire** location, timing, and area affected
2. Reflectance
3. Fraction of absorbed photosynthetically active radiation (FAPAR) and leaf area index (LAI)
4. Growth cycle, timing, duration, and spatial and temporal variability of **vegetation**

The project will utilize satellite image data from a range of sources including **advanced very high resolution radiometers**, **SPOT Vegetation**, **European remote sensing (ERS-1/-2) satellites**, along track scanning radiometers, and **moderate resolution imaging spectroradiometers**. Expected customers are likely to be the Global Carbon Cycle joint project of the **International**

Geosphere-Biosphere Programme, **International Human Dimensions Programme on Global Environmental Change**, World Climate Research Programme (WCRP), and the **Intergovernmental Panel on Climate Change**, all to determine the spatial and temporal distribution of global terrestrial carbon.

See also: **Along Track Scanning Radiometers (ATSR) World Fire Atlas; Atmosphere-Land Interlinkages; Climate Impacts; Global Observation of Forest and Land Cover Dynamics; GLOBSCAR; Remote Sensing.**

KEVIN J. TANSEY

GLOBCOVER. An initiative of the European Space Agency to develop a global land-cover map for the year 2005 using satellite data from the **medium resolution imaging spectrometer** onboard the ENVISAT satellite. The products will be developed at a spatial resolution of approximately 300 meters, one of the highest-resolution global data sets to be made available to the research community. The initiative has evolved into an international project with the **United Nations Food and Agricultural Organization**, the **United Nations Environment Programme**, the **Joint Research Centre (JRC) of the European Commission**, the **International Geosphere-Biosphere Programme**, and **Global Observation of Forest Cover and Land Dynamics**. The product is intended to complement previous activities such as the **Global Land Cover Map of the Year 2000** project.

See also: **Land-Cover Change; Remote Sensing.**

KEVIN J. TANSEY

GLOBICE. An initiative of the European Space Agency to derive information on Arctic sea ice properties. This information will improve our understanding of the role of the Arctic in global climate functioning. Information will be provided on a range of processes that affect climate over the Arctic at three- to six-day intervals, including mass balance, heat transfer, and momentum transfer between the Arctic Ocean and the atmosphere. The project will utilize a satellite archive of thirteen years of **European Remote Sensing (ERS-1/-2) satellites** and ENVISAT data. Products will include seasonal and yearly variations of sea ice as important components for climate **modeling** and will provide a better understanding of the impact of Arctic ice-motion on the climate.

See also: **Atmosphere-Land Interlinkages; Climate Impacts; Remote Sensing.**

KEVIN J. TANSEY

GLOBSCAR. An initiative of the European Space Agency to map the global distribution of burnt area using the along-track scanning radiometers

(ATSR-2 instrument) onboard the ERS-2 satellite. The ATSR World Burned Surface Atlas has been developed at a spatial resolution of one kilometer at monthly time intervals. The GLOBSCAR and **Global Burnt Area 2000** products are complementary data sets, as they both estimate global burnt area for the year 2000. A simple, dual algorithm approach was adopted. The results are available in a variety of formats including maps and geographical coordinates of the burnt areas and available at no cost to the research community through the Internet. The methodology will be used in the **GLOBCARBON** project to yield multiannual products.

See also: **Along Track Scanning Radiometers (ATSR) World Fire Atlas; Carbon Cycle; Climate Impacts; European Remote Sensing (ERS-1/-2) Satellites; Fire; Remote Sensing.**

Further Reading

European Space Agency [Online, August 2004], The GLOBSCAR: ATSR World Burned Surface Atlas Web Site, shark1.esrin.esa.it/ionia/FIRE/BS/ATSR; Simon, Muriel, Plummer, Stephen, Fierens, Freddy, Hoelzemann, Judith, and Olivier Arino, "Burnt Area Detection at Global Scale Using ATSR-2: The GLOBSCAR Products and their Qualification," *Journal of Geophysical Research* 109 (2004): D14S02, doi:10.1029/2003JD003622.

KEVIN J. TANSEY

GLOBWETLANDS. An initiative of the European Space Agency to develop **remote sensing** methods for a range of different wetland types focusing on North and **South America**, Africa, Asia, and **Europe**. Satellite imagery, inventory maps, and digital elevation models of approximately fifty **wetlands** and surrounding catchments will be analyzed. The project will support the 1971 **RAMSAR Convention on Wetlands**, establishing a framework for the stewardship and preservation of wetlands. GLOBWETLANDS will provide information about difficult and inaccessible terrain, describe the local topography, map the types of wetland **vegetation**, and monitor land-use/cover change.

See also: **Land-Cover Change; Monitoring; Peatlands; Watershed.**

KEVIN J. TANSEY

Grassland. Large territory of grasses, flowers, and herbs. Albeit dominated by grasses and grasslike plants, grassland **biomes** have a number of herbaceous flowering plants and a variety of scattered trees and bushes. Latitude, soil, and local climates for the most part determine what kinds of plants grow. Grassland biomes can be found in the middle latitudes, in the interiors of continents. They can have either moist continental climates or dry subtropical climates, and can include ecosystems from a large number of regions and environments. Grasslands can be divided in two main groups: tropical grasslands, called savannas, and temperate grasslands. Temperate grasslands can be further subdivided into prairies, which are grasslands with tall grasses; and steppes, which are grasslands with short grasses. Tropical

Extensive, semiarid grasslands with smoke plumes due to burning in Mongolia along Herlen River and Hulun Lake in China and two smaller lakes in Russia, May 1997 (object along the bottom is an array of solar panels that are part of the Russian MIR spacecraft).

PHOTO: Shuttle, Mir Space Station. CREDITS: Image Science and Analysis Laboratory, NASA-Johnson Space Center, 18 Mar. 2005 (Earth from Space—Image Information). http://earth.jsc.nasa.gov/sseop.efs

grasslands grow where temperatures are fairly high all year round and rainfall occurs seasonally. In tropical and subtropical grasslands, the length of the growing season is determined by how long the rainy season lasts; whereas in the temperate grasslands, the length of the growing season is determined by temperature. Temperate grasslands are found where the summers are hot, the winters cold, and rainfall is low throughout the year. The amount of rainfall is very important in determining which areas are grasslands, because it is hard for trees to compete with grasses in places where the uppers layers of soil are moist during part of the year but where deeper layers of soil are always dry.

Savanna

Savannas are grasslands that may be associated with open formations of shrubs and trees. The word *savanna* is a Spanish variation of a Native American word for "plains." Savannas cover almost half the surface of Africa (central Africa) and large areas of Australia, **South America**, and India. Climate is the most important factor in establishing a savanna. Savannas are always found in warm or hot climates where the annual rainfall is about 50–130 centimeters per year. It is critical that the rainfall be concentrated in six to eight months of the year, followed by a long period of drought when fires can occur. If the rain were well distributed throughout the year, many such areas would become tropical forest.

The soil of the savanna drains water rapidly. It has only a thin layer of humus (the organic portion of the soil after partial decomposition of plant or animal matter), which provides vegetation with nutrients. The savanna's principal **vegetation** consists of grasses and forbs. Different savannas support diverse grasses depending on rainfall and soil conditions. Because the savanna supports such a large number of species competing for living space, usually only one or a few kinds of grass are more successful than the others in a specific area. For example, in drier savannas such as those on the Serengeti plains, the dominant grasses on well-drained soils are Rhodes grass and red oat grass; throughout the East African savannas, star grasses are dominant; the lemon grasses are common in many Ugandan savannas. Deciduous trees and shrubs are spread across the open landscape.

Fire can be a relevant component of these ecosystems, as recurrent fires may impair the establishment of woody species, thus preventing the evolution of the ecological succession toward formations of shrubs and forests. Grasses can survive fires because they grow from the bottom instead of the top. Their stems can grow again after being burned off. Although the dry stems and leaves of grasses are consumed by fire, the grasses' deep roots remain undamaged. These roots, with all their starch reserves, are ready to send up new growth when the soil becomes moist. The scattered shrubs can also subsist on food reserves in their roots while they wait for the time to grow above the soil again. Unlike grasses and shrubs, trees survive a fire by retaining some moisture in all their aboveground parts throughout the dry season. Sometimes they have a corky bark or trunks covered with resinous bark, both being fire resistant. Many grasslands in Australia, Africa, and South America are maintained by fires.

The world's greatest diversity of ungulates is found on the savannas of Africa, including buffalo, giraffes, impalas, gazelles, wildebeest, plains zebras, rhinoceroses, elephants, and warthogs. The vast herds of herbivores support carnivores that include lions, leopards, cheetahs, jackals, and hyenas. Other animals characteristic of this biome (which do not all occur in the same savanna) include buffaloes, kangaroos, mice, moles, gophers, ground squirrels, snakes, worms, termites, and beetles.

Temperate Grasslands

Temperate grasslands are characterized as having grasses as the dominant vegetation. Trees and large shrubs are absent. Temperatures vary more from summer to winter, and the amount of rainfall is less in temperate grasslands than in savannas. Precipitation in the temperate grasslands usually occurs in the late spring and early summer. The annual average is about 50.8–88.9 centimeters (20–35 inches). The amount of annual rainfall influences the height of grassland vegetation, with taller grasses in wetter regions.

The temperature range is very large over the course of the year. Temperate grasslands have hot summers and cold winters: summer temperatures can be well over 38°C, while winter temperatures can be as low as −40°C.

The major expressions are the *veldts* of South Africa, the *puszta* of Hungary, the *pampas* of Argentina and Uruguay, the steppes of the former Soviet Union (a large area of grassland that extends from the Ukraine and **Russia** all the way to Siberia, with very cold and dry climate because there is no nearby ocean to get moisture from, and with winds from the arctic not blocked by any **mountains** either), and the plains and prairies of central North America. Examples from temperate prairies may be found in North America, such as the prairies of the central lowlands and high plains of the **United States of America** and **Canada**.

As in the savanna, seasonal drought and occasional fires are very important to **biodiversity**. However, their effects are not as dramatic in temperate grasslands as they are in savannas. The soil of the temperate grasslands is deep and dark, with fertile upper layers. It is nutrient-rich from the growth and decay of deep, many-branched grass roots. The rotted roots grasp the soil together and provide a food source for living plants. Each different species of grass grows best in a particular grassland environment. The seasonal drought, occasional fires, and grazing by large mammals all prevent woody shrubs and trees from invading and becoming established. However, a few trees, such as cottonwoods, oaks, and willows grow in river valleys; and some non-woody plants, specifically a few hundred species of flowers, grow among the grasses. The various species of grasses include purple needlegrass, blue grama, buffalo grass, and galleta. Flowers include asters, blazing stars, coneflowers, goldenrods, sunflowers, clovers, and wild indigos.

The fauna (which do not all occur in the same temperate grassland) include gazelles, zebras, rhinoceroses, wild horses, lions, wolves, prairie dogs, jack rabbits, deer, mice, coyotes, foxes, skunks, badgers, blackbirds, grouses, meadowlarks, quails, sparrows, hawks, owls, snakes, grasshoppers, leafhoppers, and spiders. The temperate grassland fauna are very low in diversity, especially compared to the tropical grasslands or savannas of Africa. In North America, the dominant herbivores are the bison and the pronghorn (the latter being sole member of the Nearctic endemic family, *Antilocapridae*). Rodent herbivores include the pocket gopher (another Nearctic endemic), ground squirrels, and the prairie dog. Carnivores include the coyote (actually an omnivore), badger, and the endangered black-footed ferret, the last two being members of the weasel family. The importance of the herbivores is common to all of them, but especially in the savanna where there is a great diversity of ungulate herbivores. Compared to tropical savannas, the temperate systems have much less faunal diversity.

Steppes

Steppes are dry areas of grassland with hot summers and cold winters. They receive 25.4–50.8 centimeters (10–20 inches) of rainfall a year. Steppes occur in the interiors of North America and **Europe**. Plants growing in steppes include blue grama and buffalo grass, cacti, sagebrush, and small relatives of the sunflower. Steppe fauna include badgers, hawks, owls, and snakes. Today, people use steppes to graze livestock and to grow wheat and other crops.

Overgrazing, plowing, and excess salts left behind by irrigation have damaged some steppes. On the Russian steppes the extinct fauna include wisent (*Bison bonasus*), tarpan or wild horse, and saiga antelope, among others. Polecats and other members of the weasel family are among the larger, extant carnivores.

Conservation

The world's natural grasslands are fragile and dynamic ecosystems that are habitat to many uniquely adapted animals and plants. But most of the world's natural grasslands, including huge areas of the Russian steppes and American prairies, have been converted to **wheat** or corn fields, and the East African savannas are facing the same threat. On the other hand, grassland wildlife must compete with domestic livestock, which are too often kept in such high numbers that overgrazing and **soil erosion** occur. In fact, livestock grazing represents a major human alteration of natural disturbance regimes in grasslands throughout the world, having severe impacts on plant communities.

Apart from the outright loss of grasslands to **agriculture** and urban development, grasslands also face major threats from the introduction of **exotic species**, changes in disturbance regimes, and habitat fragmentation.

In the World Wildlife Fund's "Global 200" ranking—the world's most important ecoregions—a number of grassland ecosystems are listed. They include five areas of temperate grasslands such as the Mongolian steppes, North American tallgrass prairie, and Patagonian grasslands, and eleven tropical grassland ecosystems in South America, Africa, Asia, and Australasia.

See also: **Agricultural Revolution; Brazilian Cerrado; Common-Pool Resources; Cotton; Desertification; Great Plains; Pasture; Savannization; Transhumance; Turnover.**

Further Reading

Axelrod, D. I., "Rise of the Grassland Biome, Central North America," *Botanical Review* 51 (1985): 163–201; McNaughton, Sam J., "Ecology of a Grazing System: The Serengeti," *Ecological Monographs* 55 (1985): 291–320; Solbrig, Otto, "Ecological Constraints to Savanna Land Use," in *The World's Savannas: Economic Driving Forces, Ecological Constraints and Policy Options for Sustainable Land Use,* eds. M. D. Young and Otto T. Solbrig, Man and the Biosphere Series 12, Paris: UNESCO, Parthenon Publishing Group, 1993, 21–44.

HELENA FREITAS

Great Plains. The U.S. Great Plains extend from the Midwest savanna west to the front range of the Rocky Mountains, and from **Canada** to Texas. Dry winters and wetter summers characterize a climate shaped by three air masses, each lending its own seasonal dynamics. A Pacific air mass, dominant in winter, brings air dried by crossing several **mountain** ranges before reaching the plains. A polar air mass also shapes winter weather, creating a strong north-south gradient in air temperature and snow cover. In summer, the

Field preparation for cotton growing in Childress County, Texas Panhandle, in June 1938.

PHOTO: Dorothea Lange. [Courtesy Library of Congress]

westerly flow weakens and polar air retreats north, allowing moist subtropical fronts from the Atlantic Ocean to bring occasional rain.

Two major climatic gradients control the distribution of natural ecosystems and land use: an east-to-west gradient of decreasing precipitation and a north-to-south gradient of increasing temperature. Mean annual precipitation ranges from 1,200 millimeters per year to less than 300 millimeters, although rainfall varies widely around the mean. Average temperature varies from 0°C to greater than 20°C in the south. Plant species composition shifts from tallgrass prairie to shortgrass steppe, with decreasing precipitation. These gradients also affect net primary production and soil organic carbon.

When people first entered the Great Plains, it was not yet a **grassland**. At the end of the Ice Age 12,000 years ago, glaciers reached to the northern third of the region, across Canada and into Montana and the Dakotas. South of the ice front were 300 kilometers of arctic tundra. In the central plains was a transitional ecosystem shifting from wet spruce and pine forest to grassland. Open forest was interspersed with stretches of wet, boggy grasslands. This forest and parkland environment reached as far south as northern Texas.

Between 12,000 and 9,000 years ago, the climate warmed and dried, glaciers retreated, **forest** gave way to grass, new animal species migrated in (including bison), and people arrived. As the ice cap melted northward, the tundra zone retreated with it, and behind that the coniferous forest gave way to larger grass openings. Scattered patches of forest persisted in northern New Mexico as late as 6,000 years ago, and relict stands of ponderosa pine can still be found in canyons there. By about 6,000 years ago, the Ice Age plains habitat had changed to the grassland we now consider to be the region's natural vegetation.

The earliest nomadic big game hunters had little impact on land cover and **vegetation**, although their hunting practices dramatically changed the faunal composition of the Great Plains. Later broad-spectrum hunters and gatherers likewise had few far-reaching ecological impacts. The introduction of **agriculture**, in the first centuries AD, initiated human impacts on land cover. Plains woodland cultures occupied the eastern plains between about 250 BC and AD 1000, establishing villages along the Kansas, Missouri, and other plains rivers. They continued to hunt and to gather wild foods, but the introduction of agriculture was significant. Still, the villages were small, no more than about four hectares in size, with only a few dwellings. Farming occurred exclusively in river valleys, not on the wide upland plains. Woodland Indians could mobilize only human labor to turn the soil. They did not have any domesticated animals, such as horses or oxen, that might have done some of the work for them. It is a key difference from the Euro-American farm system, and one that restricted Indians to very small valley plots.

Between AD 900 and 1700, plains woodland horticultural societies shifted to full-fledged agricultural settlements. Along all the permanently flowing plains rivers, Indians planted more corn, beans, and squash, and came to rely on farming for much more of their annual subsistence. Population density increased, as did the number of villages along the Missouri, Platte, Republican, Solomon, Arkansas, and other plains rivers. Human labor continued to provide all farm power so that while Indian populations expanded, farming continued to be restricted to river valleys. On the upland plains, where prairie sod was too tough for wood and bone tools, Indians supplemented their crops with small game, wild plants, and occasional bison. On the northern plains, Indians pushed the environmental limits of agriculture, taking advantage of warmer periods to expand corn agriculture up the Missouri River and its tributaries into central South and North Dakota. On the central plains of Kansas, Nebraska, and Colorado, Indian farmers pushed the limits of soil moisture, enjoying a relatively wetter climate between AD 900 and 1250, then retreating eastward for some 200 years of drought, and moving west again after 1450, when rainfall increased.

Indians managed their landscapes with complexity and sophistication. Larger populations provided enough human labor to thoroughly control fire regimes. **Fire** was the Indians' most powerful tool to manage land cover. By careful fire use, Indians expanded the size of the grassland, created more habitat conducive to bison, and thus raised bison populations. Before Europeans brought their broadaxes to North America, Indians had already cleared the continent of over a million hectares of forest and replaced it with grassland. Only on the driest shortgrass prairies of the western plains can grassland be considered the natural dominant vegetation. Eastward, on the mixed-grass prairies, intentional burning prevented brush and small trees from overgrowing grass. All of the eastern tallgrass prairie, running through the eastern Dakotas, Nebraska, eastern Kansas, Oklahoma, and Texas, was Indian-created habitat. Without Indian fire the tallgrass prairie would have been forest. Indians expanded the size of the grassland considerably after about AD 1000. It was about that time that bison first crossed the Mississippi River into newly created prairie peninsulas that extended through Iowa into forested Illinois. In the deciduous forests of the East, Indians opened swaths of grassland by burning. Some were small, others as vast as the Shenandoah Valley. Bison first appeared in the southeast in the 1500s, then in Pennsylvania, Massachusetts, and even into Maine in the 1600s.

On the northern Great Plains, agricultural Indians developed a spring and fall burning cycle designed to manage bison. In the autumn they burned land far from the village, which removed vegetation and encouraged bison to graze closer to the settlement during winter. In spring they burned outlying areas to lure bison away from the village. In these ways, hectare by hectare, over several hundred years, Indians reshaped plains land cover. They converted an enormous area from forest to prairie, increasing bison populations to all-time highs, about 28 million by 1700. It was this cultural and natural legacy—a vast grassland rich in bison and burned on average every three to five years—that plains village Indians passed on to the final episode of Indian occupation, the invasive plains horse Indians who came and went in a blaze of glory as Euro-Americans began to contest ownership of the land.

Beginning around 1690, invasive Indian bands flooded the Great Plains from the Rocky Mountains, Great Lakes, and elsewhere. Tribes such as the Sioux, Cheyenne, and Comanche contested **access** to bison with pre-existing agricultural villages. By 1750, horses were dispersed throughout the Great Plains where they wreaked havoc among Indian cultures. Horses functioned as technological innovations often do: they gave new power over nature to their possessors. The horseback chase was deadly for bison herds. Almost as soon as they acquired horses and moved into the plains—at least by 1790—Indians began overhunting bison, killing more than annual reproduction replaced. Subsistence required six bison per year, on average, for every woman, man, and child. The 10,000 Indians on the southern plains killed 60,000 bison per year, roughly 18,000 animals more than the herd naturally generated. Southern plains tribes killed an additional 20,000 bison per year for trade, all of which went unreplaced. Between 1780 and 1840, Indians overhunted the southern bison herd by 2 million animals, reducing it by 25 percent.

Horses also competed with bison for grass. Plains Indians were hunters, but also horse pastoralists, moving large equine herds across the grassland. Indians maintained about six horses per person. The 2,400 Arapahos along the upper Arkansas in 1855 maintained 15,000 horses. In a normal year the Arapaho horses would have needed over forty hectares of grazing range per day to remain healthy. Horse culture meant that Indians had to move their camps frequently in search of new grass. A stay of only a few weeks in one place could consume all the grass within grazing range of a village. During an especially dry year, the Arapaho horses would need over 250 hectares of grazing per day. In 1855 some 11,500 Comanches, Kiowas, Apaches, Cheyennes, and Arapahos owned about 70,000 horses. On the west-central plains, Indians maintained between 100,000 and 150,000 horses.

By the 1840s, Indians across the plains reported fewer bison, more hungry people, and starving horses.

The transition from Native American to Euro-American land management in the Great Plains happened quickly, between 1850 and 1880, and was accompanied by two key ecological changes. First, domesticated cattle replaced wild bison. Pressured by grazing competition and overhunting, bison populations collapsed in the 1870s. Euro-American hunters alone killed over 4 million bison in five years; by 1900 only a thousand remained. For a decade the enormous grazing niche was open on the plains, and grasses grew luxuriantly. Soon Euro-American ranchers began moving cattle onto the plains. By the middle 1880s, there were 7 million cattle in the region, one-fourth the previous bison population. Only around 1900 did Euro-Americans, with some 13 million cattle, fill the niche vacated by bison. A second important change was the shift from Indian fire management based on controlled burns to Euro-American fire management devoted to fire suppression. Especially as farmers entered the region, prairie fires that had supported grass and destroyed trees became less frequent.

Euro-Americans transformed the Great Plains from a managed grassland supporting wild animals to an agricultural system based on domestic animals and farm crops. Most of the region was poorly suited to crop agriculture. The western plains were too dry for reliable crop production; other areas were too sandy or steeply sloped for plowing. Nearly two-thirds of the Great Plains were used as **pasture** for cattle, sheep, and horses. Ranchers fenced pastures, developed windmills to water livestock, and grew some feed crops and hay to support them through winter. Ranchers planted non-native grasses in some pastures, and some invasive plant species entered the plains. Most of the region remains in native vegetation, used for grazing but never plowed. Cattle mimic the ecological impact of bison. Light to moderate grazing usually increases plant diversity and percent of ground cover compared to ungrazed grassland. Heavy overgrazing can damage pastures, but highly variable rainfall controls the vigor of grassland plants much more than does livestock management. Especially dry years, not unusual in the plains, can cause plant cover to decline dramatically, regardless of grazing pressure, while wet years see rapid plant recovery. The greatest ecological impact of cattle is in riparian zones along watercourses, where trampling seriously degrades habitat.

The greatest ecological impact of Euro-American settlement came in those areas well-suited to crops. Arable land makes up one-third of the Great Plains, mostly along the eastern edge where average rainfall is higher. Plowing the region's farmable land—46 million hectares—was an enormous job. Farmers started the task in the 1860s, but most of the plow-up happened between 1900 and 1935. Despite its large size, only ten crops occupy significant area in the region. In the eastern, wettest parts of the plains, farmers grow **maize** for livestock feed. They tried growing maize farther west, but quickly discovered that it failed in dry years; and replaced it with sorghum, an African grain that served the same function but was more drought-hardy. Some 23 million hectares of land were devoted to **wheat** production during most of the twentieth century. Farmers grow winter wheat in the central plains of Oklahoma, Kansas, Nebraska, and Colorado. In North and South Dakota, spring wheat is dominant. In the driest areas, farmers must rotate wheat with **fallow** to accumulate two years' precipitation for one crop. Hay is also important on the northern plains, where cold winters require ranchers to feed animals for several months each year. Most of the 8 million hectares of hay come from domesticated grasses and legumes. Oats, barley, rye, and **cotton** make up only a small percentage of total area devoted to crops, but are regionally important in some places. Soybeans entered into rotation with corn on the eastern plains after 1950.

Euro-American settlement and land-use patterns required fifty years to establish, but have been fairly stable since 1940. Settlement is characterized by small towns of a few hundred to a few thousand people, with farmsteads disbursed through the countryside. Roads run straight north-south or east-west, following an intense grid spaced at one-mile intervals. Fire suppression by Euro-Americans meant the **regrowth** of trees throughout the eastern plains. Today the former tallgrass prairie, wherever not converted to cropland, grows dense forest: along rivers, on hills, in towns, in groves around farmsteads, in parks and preserves, and anywhere farmers do not actively suppress trees.

The greatest change in Great Plains land use and land cover accompanied the cultural transition from Native American to Euro-American occupation, a transformation completed between 1875 and 1925. Since then, farmers have maintained about 46 million hectares of cropland and 104 million hectares of pastureland in the U.S. portion of the region. Most of the cropland is distributed on the eastern half of the plains; most of the pasture on the western half. Within the arable third of the **biome**, farmers plant wheat, hay, maize, sorghum, oats, barley, rye, cotton, and soy, in declining proportion of total area. Basic land use—the division between cropland and pasture and the emphasis on particular crops—has changed little since 1940 and appears stable.

See also: **Agricultural Revolution; Cattle Ranching; Century Ecosystem Model; Colonization; Conservation; Conversion; Disequilibrium Dynamics; Domestication; Exotic Species; Hot Spots of Land-Cover Change; Pastoral Mobility; Spatial Diffusion; Species Extinction; United States of America.**

Further Reading

Cunfer, Geoff, *On the Great Plains: Agriculture and* Environment, College Station, TX: Texas A&M University Press, 2005; Flores, Dan, *The Natural West: Environmental History in the Great Plains and Rocky Mountains,* Oklahoma City: University of Oklahoma Press, 2001; Parton, William J., Gutmann, Myron P., and William R. Travis, "Sustainability and Historical Land-Use Change in the Great Plains: The Case of Eastern Colorado," *Great Plains Research* 13 (2003): 97–125; West, Elliott, *The Contested Plains: Indians, Goldseekers, and the Rush to Colorado,* Kansas City: University Press of Kansas, 1998.

GEOFF CUNFER

Green Revolution. Refers to substantial increases in cereal yields in many developing countries in the 1960s and 1970s that were achieved by improved varieties, use of synthetic fertilizers and pesticides, increased irrigation, and new machinery.

Population continuously increased after World War II in many developing countries, which drastically increased the demand for food. Since crop yields were low and stagnant for centuries in many of these countries and extension of agricultural area was limited, increasing food demand rapidly outran agricultural production. As a result, famine, undernutrition, and malnutrition became a constant threat to the majority of people in developing countries.

In response to this situation, a joint effort by many organizations, officials, thousands of scientists, and millions of farmers was taken in the 1940s and 1950s to improve the productivity of important crops. Norman Borlaug has been a central figure in this process and received the Nobel Peace Prize for this work in 1970.

Drastic increases in yields were realized over large areas within a relatively short time period as had never been obtained before in the history of **agriculture**. Yield increases were first realized for **wheat** in Mexico and in turn for **maize** and **rice** and significantly improved food supply in many developing countries. Progress was achieved by the development of new varieties, for example, dwarf wheat and rice varieties, and through improved production technologies that included the use of synthetic fertilizers, pesticides, and other agrochemicals, improved water supply through irrigation, and advances in mechanization.

In the 1970s, achievements of the Green Revolution were critically discussed in the wake of the energy crisis and the growing awareness of the long-term environmental consequences. Sustainability of agricultural production became increasingly important; and problems such as **soil erosion**, compaction and decline of soil fertility, ground water contamination, and rural development issues received growing attention.

However, population in many developing countries has continued to grow throughout the 1980s and 1990s, and global population has been projected to reach 8 billion people by 2030. Required increases in food production will have to be realized by further increases in crop yields.

This has ushered the debate for another, "Second Green Revolution." The scope for yield improvement is seen in further intensifying agricultural production, mainly in developing countries together with advances in **biotechnology** to develop new varieties with improved characteristics for resource use and resistance against pests, diseases, and weeds. The impacts of such technologies on the environment are largely unknown and are critically discussed. The limited investment capital of smallholders in developing countries may also hamper the introduction of new technologies in these countries.

See also: **Agricultural Intensification; Agricultural Revolution; Cash Crops; Food Crops; Technological-Scientific Revolution.**

Further Reading

Borlaug, Norman E., "Ending World Hunger: The Promise of Biotechnology and the Threat of Antiscience Zealotry," *Plant Physiology* 124 (2000): 487–90; Borlaug, Norman E., "The Green Revolution, Peace, and Humanity" [Online, February 2005], The 1970 Nobel Lecture Web Site, www.nobelprize.org; Calderini, Daniel F., and Gustavo A. Slafer, "Changes in Yield and Yield Stability in Wheat during the Twentieth Century,"

Field Crops Research 57 (1998): 335–47; Evans, Lloyd T., *Feeding the Ten Billion*, Cambridge, UK: Cambridge University Press, 1998.

FRANK EWERT

Green Wall Project. A grand ecological project being executed in the north of **China**; also called the "Green Great Wall Project." It is an action to plant trees in 551 counties of thirteen provinces in north China. The area involved is nearly 406.9×10^4 km^2, which is almost equal to half the whole area of China. Therefore, the Green Wall Project is possibly the largest ecological project in human history. The afforestation program was launched in 1978 to restore **vegetation**, stop the encroachment of desert sands, and improve peoples' living conditions. Early in its history, north China had been a beautiful place with rich **forests** and grass. However, due to population growth, wars, and construction activities, forestland and **grassland** disappeared and deserts were created, advancing step by step. From Xinjiang Autonomous Region to Heilongjiang Province, eight large deserts and four big sandy lands have gradually formed connected **desertification** areas. The total area of desertification land is up to 149×10^4 km^2. The unsound environmental conditions have impoverished the people living in the regions, and the development of the economy and society has been blocked. Therefore it was considered to be urgent and necessary to build a network of forest belts to halt the advancing desert and amend the quality of water and soil into a productive and sustainable state. According to the construction planning of the project, the whole project can be divided into three phases and eight periods. It is anticipated to require seventy-three years to finish. The area of tree planting will reach $3{,}560 \times 10^4$ km^2. Based on the existing forests, the main measures include the following: planting trees by human labor, raising trees by sealing **mountains** and sandy land passes, and using aerial seeding. Simultaneously, the project not only focuses on the structural combination of arbor, shrub, and grass, but also on the establishment of a network of a variety of trees; and finally makes a large windbreak system with reasonable structure and stable function.

See also: **Land Rehabilitation; Maathai, Wangari M.; Poverty; Reforestation; Regrowth.**

QI LU

Greenhouse Gases. Gases so-called because of their role in enhancing the greenhouse effect that leads to global warming and climate change. There are several potential greenhouse gases (GHGs) including carbon dioxide (CO_2), methane (CH_4), nitrous oxide (N_2O), nitrogen oxides (NO_x), carbon monoxide (CO), non-methane volatile organic compounds (NMVOCs), sulfur dioxide (SO_2), chlorofluorocarbons (CFCs) and hydrochlorofluorocarbons (HCFCs), hydrofluorocarbons (HFCs), perfluorocarbons (PFCs), and sulfur hexafluoride (SF_6); although CO_2 is the dominant GHG in terms of radiative forcing. Among the other gases, CH_4 and N_2O

are considered to be the most important. Atmospheric concentrations of GHGs depend on the net balance of gaseous fluxes between various global sources (emissions) and sinks (absorption and sequestration). Emissions of greenhouse gases (notably CO_2) derive primarily from human industrial activities based on the use of fossil fuels, and from the gaseous fluxes between various pools within the terrestrial biosphere. Industrial emissions are affected by **driving forces** such as **population dynamics**, **economic growth**, and technological development; and are mediated though land-use change, as this influences the extent and location of human activities such as **urbanization** and **industrialization**.

The terrestrial biosphere is central to the regulation of carbon and nitrogen cycles, so that changes in land use and land cover can have profound effects on GHG fluxes between different terrestrial pools. According to the **Intergovernmental Panel on Climate Change** (IPCC), the most important categories of land-use emissions are CO_2 from net **deforestation**, CH_4 from **rice** cultivation, CH_4 from enteric fermentation of cattle, and N_2O from fertilizer application. These sources account for almost all of the land-use emissions of CO_2, over half of the land-use emissions of CH_4, and about 80 percent of the land-use emissions of N_2O. Land-use emissions make up only a small percentage of global CO_2 emissions (23 percent) because of the overriding effect of industrial emissions in developed countries. They make, however, a very large contribution to the total global emissions of CH_4 (74 percent) and N_2O. Furthermore, land-use emissions are an important part of CO_2 emissions from developing countries (45 percent) and an even larger percentage of their total CH_4 emissions (78 percent) and N_2O emissions (76 percent). Thus, the contribution of land-use emissions to total emissions of GHGs is important, and the estimation of future trends in GHG fluxes from land use are relevant to the assessment of climate change and its mitigation.

See also: **Agricultural Intensification; Atmosphere-Land Interlinkages; Carbon Cycle; Carbon Sequestration; Cattle Ranching; Environmental Change; Fire; Kyoto Protocol; Nutrient Cycle; Systemic Change.**

Further Reading

Nakićenović, Nebojsa, and Rob Swart, eds., *Emissions Scenarios: Special Report of the Intergovernmental Panel on Climate Change*, Cambridge, UK: Cambridge University Press, 2000; Watson, Robert T., Noble, Ian R., Bolin, Bert, Ravindranath, N. H., Verardo, David J., and David J. Dokken, eds., *Land Use, Land-Use Change, and Forestry: Special Report of the Intergovernmental Panel on Climate Change*, Cambridge, UK: Cambridge University Press, 2000.

MARK ROUNSEVELL

H

Himalayas. The earth's highest **mountain** region, containing nine of the world's ten highest peaks. This mountain range extends for nearly 2,500 kilometers along the northern fringes of the Indian subcontinent, creating a barrier between the Tibetan plateau to the north and the Gangetic alluvial plains of the Indian subcontinent to the south. These mountains range from the Indus River in Pakistan across northern India, **China**, Nepal, and Bhutan to the Brahmaputra River in Tibet and the Myanmar border.

The Himalayas are among the world's youngest mountains, and continue to exhibit tectonic activity. This mountain region consists of two nearly parallel ranges, with a general increase in elevation from south to north and from west to east. The northern range is called the Trans-Himalayas, and the southern range further comprises three parallel zones: the Great Himalayas (with an average elevation of 6,100 meters and where the highest peaks are found), the Middle or Lesser Himalayas (with an elevation range of 2,130–4,570 meters), and the southernmost sub-Himalayas or Outer Himalayas (with an elevation range of 610–1,520 meters). The northern slopes of the Himalayas receive little rainfall, while the snow-covered and glaciated southern peaks receive most of the northwest monsoon's rainfall, giving rise to several major Asian rivers, including the Indus, Sutlej, Ganges, and Brahmaputra.

The Great Himalayas are one of the few remaining inaccessible regions of the world today, though mountain tourism presents increasing environmental pressure on these isolated landscapes. Below the timber line, these mountains are covered with birch, spruce, juniper, fir, and cypress. At higher elevations, alpine meadows dominate the landscape just below the snow line, with wildflowers, mosses, lichens, rhododendrons, and shrubs. These meadows are subject to grazing pressure from nomadic pastoral communities that inhabit this region. The valleys contain isolated clustered communities, which practice a limited amount of agriculture in the cold mountains, growing one crop a year.

The Middle Himalayas present a complex landscape with forest-covered slopes dominated by pine and oak, intermingled with terraced hill and valley agriculture. Population density is highly variable, with a few densely populated valley towns and market centers that are separated by intervening mountain ranges that present obstacles to easy accessibility. As domestic and national tourism expands in the region, road and air transport networks have grown, linking the densely populated cities with the rest of the moderately populated area.

The sub-Himalayan ranges narrow in width from west to east, gradually disappearing as they reach northeastern India and Bhutan. This zone includes

Snowcapped Himalayas (left) and the Tibetan Plateau (right), the "rooftop of the world" (lower foothills on the southwest side of the Himalaya and the Indo-Gangetic Plain of northern India in the southwest).

PHOTO: Space Shuttle. CREDITS: Image Science and Analysis Laboratory, NASA-Johnson Space Center, 18 Mar. 2005 (Earth from Space—Image Information). http://earth.jsc.nasa.gov/sseop.efs

the Siwalik mountain range; the long, spindle-shaped Dun valleys; and the Terai and Duar regions at the foothills, with highly fertile soils. These regions were heavily malarial and densely covered by tropical moist deciduous forests until the 1950s, but since then, the lowered incidence of **malaria** and extensive migration from the middle hills have converted this fertile region into one of the most heavily populated and intensively farmed parts of the Himalayas.

The Himalayan landscape represents a complex interface between people and nature, and there has been considerable discussion about the trajectory of **land-cover change** in the region. Scholarly discussions of the direction and causes of land-cover change in the region have varied considerably over the last four decades. In the mid-1970s, what was subsequently termed the "theory of Himalayan degradation" received wide publicity, with the hypothesis that deforestation in the Himalayas was rapidly reaching catastrophic proportions because of population increase. The literature on this supposed crisis contained alarming descriptions of rapidly exploding mountain populations that were unable to meet their subsistence needs from the nearby forests and agriculture lands, thus denuding nearby forests and extending

Mount Everest, highest point on the planet, in late November 2003.
PHOTO: International Space Station. CREDITS: Image Science and Analysis Laboratory, NASA-Johnson Space Center, 18 Mar. 2005 (Earth from Space—Image Information). http://earth.jsc.nasa.gov/sseop.efs

agriculture onto previously forested slopes, leading to landslides, surface runoff, and erosion. This led to a further decrease in soil productivity, creating a vicious cycle in which more land had to be cleared for planting more food to meet the needs of an ever-increasing population. Further, the increased surface runoff on the denuded slopes was predicted to lead to increased flooding and **siltation** on the plains of the two major rivers in the Indian subcontinent: the Ganges and Brahmaputra. Soon, accounts of monsoon-related floods in Bangladesh blamed the deforestation in the Himalayas as the cause. Predictions were made that all accessible forests in the hills of Nepal would disappear by the year 1995, and in the Terai by the year 2000. The envisaged denudation of all **vegetation** on the mountain slopes was foretold to have serious effects, with rapid erosion and water runoff giving rise to downstream effects as serious as large-scale flooding, in areas as far off as Bangladesh.

This model is now believed to be overly simplistic and exaggerated. The rich social, ecological, physical, and cultural heterogeneity in the region makes it difficult, if not impossible, to generalize trends across the entire Himalayan range and identify a single unidirectional trend of land-cover change. Expansion of road networks and an increase in tourism have contributed to a large extent to the recent **deforestation** in the region. However, even within specific regions of the Himalayas, there has been much recent evidence to support the claims that the deforestation crisis in this region has been exaggerated.

We now know through extensive analyses of **aerial photography** and satellite images, corroborated through oral history accounts and archival research, that the process of large-scale forest clearing had begun in much of Nepal and colonial India as far back as the eighteenth century. This contradicts previously held beliefs that most deforestation in the Himalayas had occurred in recent decades, after the 1950s. Commercial clearing of mountain timber was a widespread practice in the Indian Garhwal and Kumaon areas until the 1970s, when it was restricted by a popular mass uprising against timber harvesting—the Chipko movement. In the Nepalese middle hills, the Gorkha rulers encouraged the **conversion** of **forest** land to **agriculture** in the late eighteenth and early nineteenth centuries, which provided them with an opportunity to increase their tax base.

Thus in the Pakistani, Nepalese, and Indian Himalayan regions where the population densities are the highest, there has been little reduction in forest cover since the early to mid-twentieth century, although there has been considerable deterioration in the quality of forest habitat and wildlife populations. Major losses in forest cover have occurred in recent decades, but in the lowlands of the Terai and Duars, not in the hills. This distinction is important to make, since it was environmental or **land degradation** in the hills that was believed to be responsible for landslides, soil erosion, and flooding in Bangladesh.

Thus, a simple model of unidirectional land-cover change from forest to agriculture, caused by population increase, fails to adequately describe the complexity of land-cover change in this highly varied and socioculturally and biophysically complex mountain region. It also fails to take into account the diverse array of institutional responses that mountain communities are capable of when faced with such change. Several studies in the Indian central and western Himalayas, and in Nepal, where the worst crises were predicted, have demonstrated that there has actually been a net regrowth of forests in several regions in the hills, following the response of local communities to **forest degradation**. The Kumaon region in India has some extremely stable forest boundaries maintained by local communities for hundreds of years. Nepal is now internationally recognized as one of the most progressive proponents of community forestry. Much of this regrowth is scale-dependent however, and mostly confined to smaller patches of **reforestation** occurring in substantial parts of the landscape.

In contrast, extensive degradation of alpine **pastures** has been reported in the high mountains due to overgrazing, leading to a loss in forest and **grasslands** on the upper slopes. Thus, there have been significant internal exchanges between different land-cover types. With grazing becoming more difficult for the high mountain nomadic pastoralists, shifting cultivation and grazing lands in the high mountains have been abandoned in several regions. Recent studies show that the impact of grazing has diminished in recent years across the Hindu Kush Himalayas, with an increase in stall feeding of sheep, goats, and cattle.

The eastern Himalayas present a somewhat different picture of land-cover change, and have not been as extensively studied as the western and central Himalayas. Thus, our knowledge of the extent of land-cover change and the **driving forces** in this region remains somewhat limited. To the east of Nepal, the forests in the Sikkim and Darjeeling hills have been cut down to a large extent to make way for tea plantations and other **cash crops**, leading to extensive deforestation, to the extent that some of the protected areas represent the few remaining regions of original cover in these parts. In contrast, the relatively small mountain country of Bhutan has managed to maintain a significant percentage of its forest cover through a strict policy of restricting outside influences. Further to the east, constant conflict and violent insurgent movements have had their impact on forest cover in several of the northeast Indian states. Although China has had an extensive history of forest clearing in the Himalayan mountains, the Chinese government has now banned logging in many of its forests, particularly those on mountain slopes, to protect downstream areas from **soil erosion** and flooding. This has resulted in substantial **regrowth**, but with the result that there is a growing demand for forest products that have to be imported from other parts of the world, leading to a net deforestation elsewhere.

See also: **Access; Change Detection; Conservation; Degradation Narrative; Feedback; Institutions; Land-Use History; Perception.**

Further Reading

Awasthi, Anjali, Uniyal, Sanjay Kumar, Rawat, Gopal S., and Asha Rajvanshi, "Forest Resource Availability and Its Use by the Migratory Villages of Uttarkashi, Garhwal Himalaya (India)," *Forest Ecology and Management* 174 (2003): 13–24; Buch-Hansen, Mogens, "Environment: A Liability and an Asset for Economic Development: Some Views on Environmental Protection with Economic Development in Bhutan," *International Journal of Sustainable Development and World Ecology* 4 (1997): 17–27; Gamble, Douglas W., and Vernon Meentemeyer, "The Role of Scale in Research on the Himalaya-Ganges-Brahmaputra Interaction," *Mountain Research and Development* 16 (1996): 149–55; Gautam, Chinta Mani, and Teiji Watanabe, "Reliability of Land Use/ Land Cover Assessment in Montane Nepal: A Case Study in the Kanchenjunga Conservation Area (KCA)," *Mountain Research and Development* 24 (2004): 35–43; Ives, Jack D., *Himalayan Perceptions: Environmental Change and the Well-Being of Mountain Peoples*, London and New York: Routledge and Taylor and Francis Group, 2004; Ives, Jack D., and Bruno Messerli, *The Himalayan Dilemma*, London and New York: Routledge and United Nations University Press, 1989; Nusser, Marcus, "Understanding Cultural Landscape Transformation: A Re-Photographic Survey in Chitral, Eastern Hindukush, Pakistan," *Landscape and Urban Planning* 57 (2001): 241–55; Schweik, Charles M., Nagendra, Harini, and Deb Ranjan Sinha, "Using Satellite Imagery to Locate Innovative Forest Management Practices in Nepal," *Ambio* 32 (2003): 312–9.

HARINI NAGENDRA

History Database of the Global Environment. In order to create a consistent estimate of historical land-use/cover change for the past three centuries, on a spatially explicit basis and with a satisfactory resolution for global climate change modelers, the History Database of the Global Environment (HYDE) was developed at the **National Institute of Public Health and the Environment. Driving forces** for **global environmental change** occur at several spatial and temporal **scales** and dimensions, and differ often among regions. Therefore a geographically explicit **modeling** approach is required. To improve the accuracy and credibility of projections made with these models, it is crucial that they be tested against reliable data sets. HYDE presents both general topics such as land use/cover, population, livestock, gross domestic product, and value addition of industry and/or services, as well as specific data categories concerning energy/economy, atmosphere/ocean, and the terrestrial environment. Most data have been organized at the country level for the period 1890–1990, and where possible for the period 1700–1990. Population and land use are available with geographic detail on a 0.5 × 0.5 longitude/latitude grid.

See also: **BIOME 300 Project; Scenario.**

Further Reading

Klein Goldewijk, Kees, "Estimating Global Land Use Change over the Past 300 Years: The HYDE Database," *Global Biogeochemical Cycles* 15 (2001): 417–33; Klein Goldewijk, Kees, and J. J. Battjes, *A Hundred Year (1890–1990) Database for Integrated Environmental Assessments* (HYDE Version 1.1), Bilthoven, NL: National Institute of Public Health and the Environment, 1997.

KEES KLEIN GOLDEWIJK

Holocene. The interglacial interval period within which we now live and which began some 11,600 years ago with a period of rapid, dramatic warming over much of the Northern Hemisphere and beyond. During the first half of the Holocene, many of the changes set in motion from the last glacial retreat continued. Ongoing ice melt led to the pulsed release of freshwater that disrupted ocean circulation at least once, around 8,200 years ago. Global sea level only reached something close to its present level around 6,000 years ago. **Vegetation** spread over previously glaciated lands in response to increasing warmth and maturing soils, and the terrestrial biosphere as a whole both adapted and contributed to the changed environmental conditions.

The increase in Northern Hemisphere summer insolation that appears to have triggered initial deglaciation peaked at the beginning of the Holocene and slowly declined thereafter, leading to higher summer temperatures during the first half of the Holocene. Insolation changes also strengthened monsoon systems during the first half of the Holocene. Many records of Holocene climate change identify a "climate optimum," usually reflecting higher temperatures, at some stage during the early- to mid-Holocene.

The amplitude of temperature changes in the temperate zone or high latitudes was less than that during glacial times, but it was still significant in human terms. The variations in effective precipitation and hydrological regimes that took place in lower latitudes were of immense human significance. Many of the most dramatic changes in climate, especially at lower latitudes during the mid- to late-Holocene, appear to be linked to solar variability on sub-millennial timescales. There is growing evidence from both the Old World and New World that prolonged droughts during prehistory were instrumental in the decline of civilizations, though the key to understanding these episodes lies in better documenting the interactions among the biophysical processes, cultural changes, and human impacts on the environment.

The climate of the later parts of the Holocene is often discussed in terms of modes of variability, the chief of which is the El Niño Southern Oscillation (ENSO). Other modes including the linked North Atlantic (NAO) and Arctic Oscillations (AO) and the Pacific Decadal Oscillation (PDO) modulate short-term climate variability over a range of timescales. Each of these modes varies in strength, periodicity, and spatial influence through time. One of the challenges in the detection and attribution of a significant portion of current climate change to increasing anthropogenic **greenhouse gases** concentrations lies in disentangling all the indicators of change from the modes of variability. Already, there are indications that recent circum-Arctic warming lies outside the range of variability associated with natural modes of variability. A further challenge arises from disentangling the most recent trends from the changes in global climate over the Medieval Warm Period (prior to AD 1300) and the subsequent Little Ice Age that drew to a close during the nineteenth century. Despite continuing uncertainties about the degree to which late-nineteenth-century and early-twentieth-century warming can be ascribed to human activities, there is a consensus around the ascription of post-1970 changes to anthropogenic greenhouse gases.

See also: **Environmental Change; Paleo Perspectives; Past Global Changes (PAGES) Project; Pristine Myth.**

Further Reading

Mackay Anson, Battarbee, Richard W., Birks, H. John B., and Frank Oldfield, eds., *Global Change in the Holocene*, London: Arnold, 2003.

FRANK OLDFIELD

Hot Spot Identification. The notion of hot spots has emerged over the last several decades, most notably as a useful tool for both researchers and policymakers alike. **Land-cover change** is occurring on a broad scale, but these changes are not uniform across space and are often concentrated in small areas. Certain areas are changing much more quickly and dramatically than others. These **hot spots of land-cover change** are of great concern. Their identification allows researchers and policy makers to monitor these areas and target interventions that facilitate beneficial land-cover changes and prevent adverse effects like **deforestation** in **biodiversity**-rich areas.

Different definitions of a "land-cover hot spot" prevail in the literature:

1. Hot spots can be designated as areas with a rate of land-cover change above a certain threshold (e.g., a yearly deforestation rate above 3 percent).
2. Hot spots can be seen as areas under threat of future land-cover changes. This predictive approach is used to estimate likely future changes of land cover.
3. A location might qualify as a hot spot of land-cover change if recent changes have had a severe impact in terms of, for example, socioeconomic **vulnerability** of local populations, adverse impacts in biodiversity-rich areas, or severe losses in the potential to sequester carbon.

Biodiversity hotspots is a prominent term in the ecological sciences used to indicate areas with exceptional concentrations of endemic areas that are experiencing rapid loss of habitat, or areas with a concentration of rare species. The term "hotspot" is useful in that it can be based on quantifiable indicators that have scientific validity and credibility, and can also be easily understood by policymakers and the general public, facilitating the creation of real, comparable policy priorities.

Hot Spot Research

Éric Lambin and Daniele Ehrlich propose a methodology to detect tropical deforestation "hot spots" at regional to global spatial **scales**. They use a definition of areas that experienced high deforestation rates in the present and/or recent past. Due to data limitations, particularly for socioeconomic indicators, hierarchical decision rules are applied to first identify hot spots at broader spatial scales. At these scales, they use available remote sensing indicators to limit the area of interest. The inclusion of more variables reduces the size of the identified hot spots and therefore further refines data requirements. The resulting hot spot maps are validated using additional data. However, this approach still limits the detection of hot spots to available data sets. Currently, data that explicitly connect local socioeconomic processes to deforestation are not widely available.

The **Tropical Ecosystem Environment Observations by Satellite (TREES) project** of Frédéric Achard and colleagues developed a methodology adapted from the hierarchical approach employed by Lambin and Ehrlich to identify tropical deforestation hot spots based on the use of spatial indicators. TREES established a global database containing indicators derived from remote sensing with available socioeconomic indicators. This database is used in a spatial model that identifies deforestation hot spot areas. After the hot spots are identified and measured, a set of indictors is defined by expert consultation. These indicators are used in spatial modeling, which results in a "deforestation risk map."

Chandra Giri and colleagues identify areas experiencing major land-cover changes by overlaying **advanced very high resolution radiometer** satellite data from different years and different seasons for Southeast Asia. These areas are termed "hot spots" and are investigated in detail with medium-resolution satellite data (**Landsat** and **SPOT Vegetation**) and field surveys to determine the major **driving forces** of land-cover changes.

Usefulness of Hot Spot Identification

Hot spot identification can aid in myriad applications of interest to land-use/cover change researchers. First and foremost, hot spot identification can be a useful first step for locating particular areas that warrant further investigation. Spatial data, particularly land-cover data derived from satellite remote sensing, are increasingly available; and it may be easier and more cost effective to conduct a larger-scale scan of land change through remote methods and then determine which areas warrant more costly and detailed field study. Hot spots can also be derived from a **modeling** context. Once a spatial model of land-cover change has been developed, it may be possible to generate a surface representing the probability of future land-cover changes. This surface can then be overlaid on the land-cover map corresponding to the current situation to highlight, at a fine spatial scale, areas at risk of land-cover change. Locations characterized by a high probability possess those spatial attributes that, in the recent past, have been associated with a high frequency of a specific land-cover change. Such analyses can then aid researchers in speculating on the nature and location of future land-cover changes at a local or landscape scale.

Most importantly, hot spot identification can help set priorities for researchers and policymakers to prioritize **conservation** efforts, land management interventions, and other related projects. Discussion of hot spots can be used to raise public awareness about the need to monitor and manage critical natural resources quickly and effectively. Particularly if the methodology involves widely accepted, quantifiable scientific criteria for rapid, significant change, the label "hot spot" implies an urgency that cannot be easily ignored.

Challenges and Future Directions

Though the concept of hot spots and hot spot identification is a useful tool for land-change researchers, there are limitations to this methodology that warrant careful consideration. First, data limitations—ubiquitous to land-use/cover change research—confound hot spot identification and international comparisons. Data vary considerably from region to region in terms of level of aggregation, quality, and frequency of collection. Frequently, socioeconomic indicators are only available for relatively large administrative units that do not correspond to spatially available **geographical information system** and **remote sensing** variables. Moreover, in many countries socioeconomic data are aggregated to protect confidentiality.

A second problematic factor concerns the possible incongruence of managing biophysical versus *human* systems. An implicit assumption for hot spot identification in land-change research is that, ideally, one would strive for locally environmentally and economically **sustainable land use**. The potential exists, however, for fragile ecosystems to be impacted to such an extent even from minimally invasive land use to the point where the biophysical and human systems are in conflict. It may simply not be feasible to protect both. Therefore, land-change researchers must take care to define what the ultimate policy goal is to best manage such conflicts, thus considering **tradeoffs**.

Last, hot spot identification is necessarily based on past change—that is, targeting those areas that have recently experienced rapid change. However, this methodology then may or may not allow predicting or forecasting future change in other areas. One can attempt to characterize the human and biophysical factors that created the "hot spot" (i.e., socioeconomic and/or biological **vulnerability**), but these factors may not necessarily be important predictors of change in other areas. Also, it may be equally important to monitor the parts of the world that are not undergoing rapid change, in order to determine which factors might be preventing change in those regions. Therefore, hot spot identification alone, while a powerful tool, should not be used in isolation, and would be best seen in the larger context of studying land change.

See also: **Arena of Land Conflict; Change Detection; Monitoring; Probit/Logit Model.**

Further Reading

Achard, Frédéric, Eva, Hugh, Glinni, A., Mayaux, Philippe, Richards, Timothy, and Hans-Jürgen Stibig, *Identification of Deforestation Hot Spot Areas in the Humid Tropics*, TREES Series B, N°4, EUR 18079 EN, Luxembourg: European Commission, 1998; Giri, Chandra, Defourny, Pierre, and Surendra Shrestha, "Land Cover Characterization and Mapping of Continental Southeast Asia Using Multi-Resolution Satellite Sensor Data," *International Journal of Remote Sensing*, 24 (2003): 4181–96; Lambin, Éric F., and Daniele Ehrlich, "The Identification of Tropical Deforestation Fronts at Broad Spatial Scales," *International Journal of Remote Sensing* 18 (1997): 3551–68; Myers, Norman, Mittermeier, Russell A., Mittermeier, Christina G., da Fonseca, Gustavo A. B., and Jennifer Kent, "Biodiversity Hotspots for Conservation Priorities," *Nature* 403 (2000): 853–58; Ramankutty, Navin, Graumlich, Lisa, Achard, Frédéric, Alves, Diogenes, Chhabra, Abha, DeFries, Ruth, Foley, Jonathan, Geist, Helmut, Houghton, Richard, Klein Goldewijk, Kees, Lambin, et al., "Global Land Cover Change: Recent Progress, Remaining Challenges," in *Land Use and Land Cover Change: Local Processes, Global Impacts*, eds. Éric F. Lambin and Helmut Geist, Berlin: Springer-Verlag, 2006.

DANIEL MÜLLER, DARLA K. MUNROE

Hot Spots of Land-Cover Change. Applying it to tropical forest cover, the concept of "hot spots" was first studied by Norman Myers in 1993. Fourteen "deforestation hot spots" were identified and described around the tropical belt. The Conservation International Institute has further developed this concept of "hot spots": they are defined as areas with a high **biodiversity** (the region must support 1,500 endemic plant species) and having lost more than 70 percent of their original habitat. The twenty-five identified "biodiversity hot spots" are concentrating 44 percent of the existing plant species on only 1.4 percent of earth's land area.

Although the **conservation** or biodiversity aspects are usually deemed more important, particularly when considering locations that are experiencing a severe impact on **ecosystem services** or where ecosystem services are particularly vulnerable to future change, the current dynamics of all land-cover types are now considered for an enlarged definition of "hot spots" of change.

Even if information is not complete globally, several products are now available that depict snapshots of many important regions with substantial **land-cover change**: European **Russia**, **South America** and Africa, parts of East Asia and Southeast Asia, and the continental **United States of America** and **Canada**, for example. There has been a profusion of information and studies based on satellite **remote sensing** data and other data sources, with multiple examples of studies and resultant databases of rapid land-cover change and ecosystem disturbances in important regions of the world: **deforestation** in the pan-tropical forest belt; **fire** frequency globally and regionally in South America, **Southern Africa**, and parts of Russia; and the influence of **urbanization** in selected cities around the world.

The types of change that can be included in the "hot spot" concept are deforestation and **forest degradation**, afforestation and **reforestation**, land degradation in the drylands and hyper-arid zones of the world (referred to below as **desertification**), expansion of cropland and **pasture** and **land abandonment**, and urban settlements.

A "deforestation hot spot area" is an area where major changes of land cover were or are occurring, or are expected to take place in subsequent years. Deforestation is the most-measured process of land-cover change at a regional **scale**. Deforestation and forest degradation are much more frequent in the tropics than in the other parts of the world. In particular, the Amazon basin and Southeast Asia concentrate many of the deforestation hot spots. However, more data sets cover the **tropical humid forests** than the **boreal zone** or **taiga**. Moreover, the frequency of **fires**, which are a natural disturbance factor in boreal ecosystems, have been increasing in recent years, having reached over 7 million hectares per year in the late 1990s. Even though deforestation is one of the best-studied processes of land-cover change, regional gaps in spatially explicit data persist.

The Asian continent includes most of the main areas of degraded land. In this case also, not all the drylands and hyper-arid zones of the world are well covered by desertification studies. Major gaps occur around the Mediterranean basin, in eastern Africa, in parts of South America (north of Argentina, Paraguay, Bolivia, Peru, and Ecuador) and in the United States. If all the continents were evenly covered by desertification data, the global distribution of the most degraded land could be different, but the patterns observed in Asia would likely remain the same.

The main areas of recent cropland increase are spread across all continents. They are principally located in the Far East, in Bangladesh, in Southeast Asia, in the Great Lakes region of eastern Africa, along the southern border of the Amazon basin in **Latin America**, and in the **Great Plains** region of the United States. Some regions of cropland increase are also seen in the Middle East, along the Indus Valley, and in Central Asia. North America accounts for the majority of the main areas of cropland decrease (lowlands of southeastern United States), followed by Asia (eastern part of China) and South America (Brazil and Argentina). Some areas of decrease in cropland extent are located in the other continents, except for Africa, where no decrease in cropland is identified.

The changes in urbanization represent usually the most rapidly changing cities with more than 750,000 inhabitants. The most populated areas are located mainly in the Indo-Gangetic plain of northern India, on the plains and northern plateaus of **China**, and on the Indonesian island of Java. The most populated cities of more than 750,000 inhabitants are located mainly on the eastern coast of the United States, in **western Europe** (EU-15), in India, and in East Asia, whereas the most-changing cities are located all over the tropical belt.

See also: **Arc of Deforestation; Land-Use/Cover Change (LUCC) Project; Millennium Ecosystem Assessment; Southeast Asia, Mainland; Tropical Dry Forest.**

Further Reading

Achard, Frédéric, Eva, Hugh, Glinni, Ariella, Mayaux, Philippe, Richards, Timothy, and Hans-Jürgen Stibig, *Identification of Deforestation Hot Spot Areas in the Humid Tropics,* Luxembourg: European Commission, 1998; Lepers, Erika, Lambin, Éric F., Janetos,

Antony C., DeFries, Ruth, Achard, Frédéric, Ramankutty, Navin, and Robert J. Scholes, "A Synthesis of Rapid Land-Cover Change Information for the 1981–2000 Period," *BioScience* 55 (2005): 115–24; Mittermeier, Russell A., Myers, Norman, Gil, Pablo Robles, and Cristina G. Mittermeier, eds., *Hotspots: Earth's Biologically Richest and Most Endangered Terrestrial Ecoregions,* Mexico City, Mexico: CEMEX, 1999; Myers, Norman, "Tropical Forests: The Main Deforestation Fronts," *Environmental Conservation* 20 (1993): 9–16.

FRÉDÉRIC ACHARD, HUGH EVA,
PHILIPPE MAYAUX, HANS-JÜRGEN STIBIG

Human Appropriation of Net Primary Production. Humans depend on the primary production of ecosystems at least for their food supply. The amount of net primary production (NPP) used for human purposes or foregone due to land use is denoted as "human appropriation of net primary production," abbreviated HANPP. Primary production is the process in which solar energy (or another inorganic energy source) is converted into trophic energy, that is, energy stored chemically within organisms. Photosynthesis of green plants is the most important primary production process, and the only one discussed here. Gross primary production (GPP) is the total amount of energy fixed by plants. NPP is defined as GPP minus plant respiration.

Humans appropriate NPP through two interrelated processes: changes in the productivity (NPP per m^2 and year) of ecosystems caused by land use and present or past **land degradation**; and harvest of NPP, either directly or through livestock grazing. HANPP can be defined as the difference between the NPP of potential vegetation and the fraction remaining in ecosystems of the NPP of actual (currently prevailing) **vegetation** (see figure).

HANPP is a measure of the extent to which humans dominate ecosystems. The best available estimate of current global terrestrial HANPP is 20–40 percent of potential NPP, but uncertainty is large. Global HANPP time series are lacking. Because NPP serves as energy input to all heterotroph food

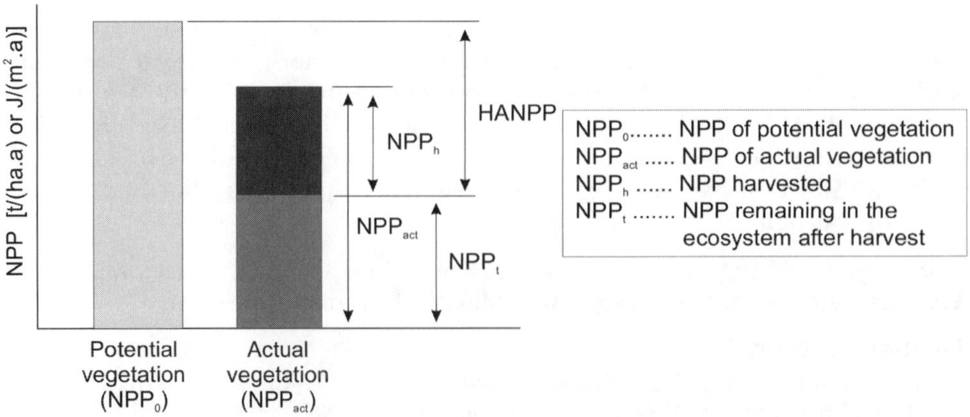

Definition of human appropriation of net primary production (HANPP)

chains, all heterotroph species besides humans and their domesticated animals must dwell on the remaining 60–80 percent. It is thus not only plausible that HANPP could be a major pressure on **biodiversity**, but there is also empirical evidence to support this notion. Since the part of NPP not consumed by heterotrophs accumulates in the ecosystem, which thus absorbs carbon from the atmosphere, HANPP is also relevant for human-induced changes in the ecosystem's carbon balance. There may exist long time delays between changes in HANPP and vegetation-related carbon flows, however.

HANPP can also be seen as an indicator of the intensity of socio-economic **colonization** of terrestrial ecosystems, that is, the process in which humans intentionally alter the functioning of these systems according to their needs and wants. One important function of colonization is the provision of agricultural and forest **biomass** for socio-economic **metabolism**. Both hunter-gatherers and agricultural society depend almost exclusively on NPP for their supply of primary energy, but the way in which they use ecosystems is very different. While hunter-gatherers put little, if any, effort into altering ecosystems and consequently can use only one ten-thousandth or less of the NPP of the regions they inhabit, HANPP of agricultural societies is often large: they actively change terrestrial ecosystems in order to promote the growth of plants useful for humans and are thus able to use a much larger proportion of NPP.

Case studies conducted thus far have, however, shown that agricultural societies seem seldom to be able to utilize the production potential of their territory. As a consequence, NPP_{act} of agriculturalists is mostly much lower than NPP_0, and this discrepancy is the principal cause for their large HANPP. Fridolin Krausmann demonstrated for Austria from 1830–1990 that **industrialization** was related to an increase in NPP_{act} that was larger than the considerable (+70 percent) increase in harvest occurring in this period. As a result, aboveground HANPP declined during industrialization, demonstrating that there are considerable potentials to increase agricultural area efficiency. The finding that HANPP and biomass harvest can be decoupled to a considerable extent has discredited earlier claims that HANPP was a simple, straightforward measure of ecological limits. Note, however, that this increase in area efficiency was possible only because **agriculture** and **forestry** ceased to be the almost-sole source of societal energy. Only because new area-independent sources of energy became available were industrial societies able to energetically "subsidize" agriculture, sometimes even to the extent that the energy return on **investment** of many agricultural products fell below 1. Moreover, the improvement of area efficiency had considerable other environmental effects, such as surging CO_2 emissions, breakup of **nutrient cycles**, **soil erosion**, and pesticide use.

HANPP is defined with reference to a particular land area, implying that HANPP can, in principle, be evaluated on any spatial level. Using **remote sensing** and statistical data, it is possible to determine HANPP in great spatial detail, thus revealing patterns in land-use intensity. It is important to recognize that a national HANPP calculation does not consider demands on other territories caused by the **consumption** of imported biomass—contrary to other measures such as the **ecological footprint**. Demands on a nation's territory by people from outside (e.g., biomass export) are also not separated in HANPP calculations (in contrast to ecological footprint calculations).

HANPP is measured in the same units usually used to evaluate NPP, that is, either as substance flow (kilograms of carbon per year), as material flow (kilograms of dry matter biomass per year), or as energy flow (joules per year). Results derived in these different units are similar but not identical, because calorific values and carbon content of dry matter biomass can vary to some extent.

See also: **Agricultural Intensification; Carbon Cycle; Cascades; Pattern Metrics.**

Further Reading

Haberl, Helmut, Erb, Karlheinz, Krausmann, Fridolin, Loibl, Wolfgang, Schulz, Niels B., and Helga Weisz, "Changes in Ecosystem Processes Induced by Land Use: Human Appropriation of Net Primary Production and Its Influence on Standing Crop in Austria," *Global Biogeochemical Cycles* 15 (2001): 929–42; Haberl, Helmut, Schulz, Niels B., Plutzar, Christoph, Erb, Karlheinz, Krausmann, Fridolin, Loibl, Wolfgang, Moser, Dietmar et al., "Human Appropriation of Net Primary Production and Species Diversity in Agricultural Landscapes," *Agriculture, Ecosystems & Environment* 102 (2004): 213–8; Haberl, Helmut, Wackernagel, Mathis, Krausmann, Fridolin, Erb, Karlheinz, and Chad Monfreda, "Ecological Footprints and Human Appropriation of Net Primary Production: A Comparison," *Land Use Policy* 21 (2004): 279–88; Imhoff, Marc L., Bounoua, Lahouari, Ricketts, Taylor, Loucks, Colby, Hariss, Robert, and William T. Lawrence, "Global Patterns in Human Consumption of Net Primary Production," *Nature* 429 (2004): 870–3; Krausmann, Fridolin, "Land Use and Industrial Modernisation: An Empirical Analysis of Human Influence on the Functioning of Ecosystems in Austria 1830–1995," *Land Use Policy* 18 (2001): 17–26; Vitousek, Peter M., Ehrlich, Paul R., Ehrlich, Anne H., and Pamela A. Matson, "Human Appropriation of the Products of Photosynthesis," *BioScience* 36 (1986): 363–73; Wright, David H., "Human Impacts on the Energy Flow through Natural Ecosystems, and Implications for Species Endangerment," *Ambio* 19 (1990): 189–94.

HELMUT HABERL

Human Health. Usually a synonym for an absence of disease, connoting strength, feeling good, and having a good functional capacity, which can be impacted by land-use-driven changes of ecosystem goods and **ecosystem services**. With a primary focus on infectious diseases (not considering, for example, local and transboundary air pollution from **fire** and dust), Jonathan Patz and Douglas Norris have documented disease outbreaks and new emergence events that follow four broad types of land-use/cover changes, including urbanization and urban sprawl, agricultural development such as irrigation and intensified production, road construction and forest removal in the tropical zone, and regeneration of temperate forests. Habitat and/or climatic change substantially affect intermediate invertebrate hosts involved in many prevalent diseases such as **African trypanosomiasis**. Therefore, vector-borne disease biology is considered to be a foundation for analyzing the effects of land-use/cover change (e.g., a **rice** paddy may provide up to ten times more mosquito habitat than a riverine system with equal surface area). In some cases, **tradeoffs** need to be considered between the provision of food and the promotion of deadly diseases, or between human health and ecosystem sustainability. In the latter situation, draining swamps may reduce vector-borne disease hazards but also may destroy important ecosystem properties of **wetlands** and their inherent services, for example.

Urbanization is associated with a range of health problems, including vector-borne diseases such as dengue and malaria as well as diarrheal and respiratory diseases. Air and water pollution together with overcrowding and inadequate infrastructure can trigger these conditions. Water sanitation and hygiene interventions can greatly reduce water-related diseases. **Urban sprawl** poses health challenges stemming from heat waves exacerbated by the "urban heat island" effect (e.g., over 20,000 people died during the 2003 summer heat wave in **Europe**). It also has the potential to worsen ozone air pollution (a potent lung irritant) via the urban heat island effect. In addition, impervious roadways and concrete surfaces such as parking lots exacerbate urban stormwater runoff and thus increase the risk of water contamination because combined sewer systems contain untreated human and industrial waste, which carries solids, oxygen-demanding substances, ammonia, other potential toxics, and pathogenic microorganisms (associated with human disease and fecal pollution) to the receiving waters.

Dam construction, irrigation farming, and **agricultural intensification** have long been associated with diseases such as schistosomiasis and filariasis. Especially in the tropics, the construction of reservoirs, irrigation canals, retention ponds, and dams can lead to a shift in vector populations (such as snails and mosquitoes), their larvae, and their parasites. Crop irrigation increases breeding sites for disease vectors. For example, the Aswan High Dam in Egypt caused soil moisture to increase from the associated irrigation schemes in the Southern Nile delta, thus triggering proliferation of the *Culex p. pipiens* mosquito and subsequent increases in *Bancroftian filariasis.* In addition, large-scale use of pesticides has had deleterious effects on farm workers, including hormone disruption and immune suppression. The use of pesticides in growing **cotton**, a preferred irrigation crop, has been a human health concern in some developing countries, particularly when safety instructions are not accessible to users (either because of illiteracy or inability to translate a foreign language). Better land and water management methods—such as increasing water flow and depth in retention ponds—are needed to address the tradeoffs between expanded food production and altered habitats promoting deadly diseases.

Road construction and **deforestation** can increase the nature and number of breeding sites for vector-born diseases such as onchocerciasis and **malaria**. New roads into old-growth forests not only facilitate logging, ranching, crop farming, mining, and tourism, but also provide **access** for new human, livestock, vector, and parasite populations. Across the tropical zone, deforestation with subsequent changes in land use and human settlement patterns has coincided with an upsurge of malaria and/or its vectors. For example, cleared lands and culverts that collect rainwater are in some areas far more suitable than **forest** for malaria-transmitting anopheline breeding.

In forests of the temperate zone, tick populations have changed due to **regrowth** on former croplands and because **secondary vegetation** has become more prevalent. In the **United States of America**, for example, human infection rates for Lyme disease have risen over time with the increase in contact between humans and ticks. Habitat fragmentation due to urban sprawl and road building usually exacerbates the loss of species diversity (which could decrease the risk of disease through a "dilution effect"), thus increasing the risk of Lyme disease.

See also: **Carson, Rachel Louise; Eutrophication; Human Immunodeficiency Virus (HIV)/Acquired Immunodeficiency Syndrome (AIDS); Indigenous Knowledge; Integrated Assessment; Land Degradation; Reforestation; Tobacco; Water-Land Interlinkages.**

Further Reading

McMichael, Tony, *Human Frontiers, Environments and Disease: Past Patterns, Uncertain Futures.* Cambridge, UK: Cambridge University Press, 2001 (reprinted in 2003); Patz, Jonathan A., and Douglas Norris, "Land-Use Change and Human Health," in *Ecosystems and Land Use Change* (Geophysical Monograph No. 153), eds. Ruth DeFries, Gregory Asner, and Richard Houghton, Washington, DC: American Geophysical Union, 2004, pp. 159–67.

HELMUT GEIST

Human Immunodeficiency Virus (HIV)/Acquired Immunodeficiency Syndrome (AIDS). Human immunodeficiency virus (HIV) and its associated disease, acquired immunodeficiency syndrome (AIDS), are not directly related to land-use change in the same way as vector-borne diseases such as **malaria** or **African trypanosomiasis**, which are driven by the habitat of the disease vector. In the case of HIV/AIDS, a virus transmitted predominantly through sexual activities, blood transfusion, intravenous drug use, and mother to fetus/infant transmission, the associations of the virus and disease with land-use/cover change are much more subtle and relate instead to socioeconomic processes. HIV is believed to have emerged in humans sometime between the 1940s and the 1960s after crossing the species barrier from similar infections (simian immunodeficiency viruses) in African monkeys. This species jump has been associated with land-use changes. HIV etiology, however, remains in debate.

Current HIV/AIDS impacts land use predominantly through close ties between agricultural and **land-use systems**, and livelihood strategies in developing counties. For example, AIDS losses of the most economically valuable members of communities—young adults—contribute to economic and social upheavals that can affect agricultural activities, land use, and **land tenure**. Families limited to grandparents and young children, for example, will not be able to manage cultivation and livestock to the same extent as families with fully active adults. Moreover, medical expenses during illness contribute to increased **poverty** and social and economic **vulnerability**. Tony Barnett and **Piers Blaikie**, who explore the potential impact of HIV/AIDS on farming systems in Africa, suggest that HIV/AIDS could have significant impacts on rural food and income resources associated with loss of labor and resulting agricultural coping strategies.

Alternatively, land use and land cover can have reciprocal impacts on rates of HIV/AIDS. Changes to land use through environmental, resource, or **land degradation**, for example, may contribute to social and economic changes within affected communities. Such impacts may include modifications in the risk factors and risk activities for HIV transmission. Scott Drimie notes several potential examples of land-use change processes that may contribute to increasing HIV/AIDS rates through increased vulnerability, inability to implement control and prevention systems, and absence of **decision-making** power regarding risk activities. These include land ownership inequality, urban migration, marginal agriculture, and environmental degradation. Ann Danaiya Usher, in her discussion of AIDS and ecological crisis in Thailand, provides an example of the processes linking land-use change and HIV. She describes a process of increasing **deforestation** in a community in Thailand, where she observed limited community **land rights** and decreasing availability of forested areas for **subsistence agriculture**. An increasingly market-oriented economy associated with the marginalization of rural people traditionally dependent on **forest** resources resulted in increased poverty and migration of people into cities and industrial zones, argues Usher. In the case of young females, many entered the urban sex industry. This dramatic shift in rural and sexual culture in Thailand resulting from deforestation policies is associated with the emergence and explosion of HIV rates in the country. In this case, the removal of forested land cover, and the threat to livelihood security of traditional forestland use by rural communities, may have contributed to the social and cultural changes of the sexual culture and industry in Thailand, and its associated HIV/AIDS epidemic. Godfrey Linge and Doug Porter also extensively explore this link between development and HIV rates in Asia and the Pacific. As noted by Charles Hunt, HIV/AIDS can exhibit urban-rural dichotomies, such as that observed in much of Africa, where AIDS predominates in urban areas; transmission pathways are generally associated with urban processes.

These examples indicate a strong, though potentially very complex, reciprocal relationship between HIV/AIDS and land-use/cover patterns, where associations are observed primarily within socio-economic processes and vulnerabilities. For HIV/AIDS, it may be particularly pertinent to note the widespread potential for positive **feedback** processes resulting in negative impacts on both natural resources and HIV/AIDS in marginalized countries. Degradation in land-cover resources may exacerbate HIV rates, and increasing HIV rates may, in turn, further exacerbate pressure on natural land-cover resources.

See also: **Human Health; People at Risk.**

Further Reading

Barnett, Tony, and Piers Blaikie, *AIDS in Africa: Its Present and Future Impact,* New York: The Guilford Press, 1992; Corless, Inge B., "HIV/AIDS," in *Emerging Infectious Disease: Trends and Issues,* eds. Felica R. Lashley and Jerry D. Durham, New York: Springer Verlag, 2002, pp. 161–70; Drimie, Scott E., "HIV/AIDS and Land: Case Studies from Kenya, Lesotho and South Africa," *Development Southern Africa* 20 (2003): 647–58; Hunt, Charles W., "Migrant Labor and Sexually Transmitted Disease: AIDS in Africa," in *AIDS: Readings on a Global Crisis,* ed. Elizabeth R. Bethel, Boston: Allyn and Bacon,

1995, pp. 137–56; Linge, Godfrey, and Doug Porter, *No Place for Borders: The HIV/AIDS Epidemic and Development in Asia and the Pacific*, New York: St. Martin's Press, 1997; Southern African Regional Poverty Network (SARPN) [Online, November 2004], The HIV/AIDS Web Site, www.sarpn.org.za/rpp/hiv.php; Usher, Ann Danaiya, "After the Forest: AIDS as Ecological Collapse in Thailand," in *Close to Home: Women Reconnect Ecology, Health and Development Worldwide*, ed. Vandana Shiva, Philadelphia: New Society Publishers, 1994, pp. 10–42.

LEA BERRANG FORD

Human Impact on Terrestrial Ecosystems (HITE) Initiative. Part of Focus 5 of the range of activities subsumed within the **Past Global Changes (PAGES) project** of the **International Geosphere-Biosphere Programme**. HITE aims to promote and integrate case studies that reconstruct the long-term human-environment dynamics that underlie present ecosystems. It seeks to do this by furthering the use of paleo records to document and understand interactions between human activity and ecosystem response through time. The key questions addressed by HITE include the following: What are the major human impacts that have influenced the ecosystems that we see in the present day? How can information about past conditions influence suitable land use and management strategies? What insights do paleodata sets give on the frequency and magnitude of ecological changes? Among the past human activities that HITE seeks to reconstruct are **deforestation**/afforestation/**reforestation**, cultivation/grazing, the use of **fire**, the alteration of nutrient sediment and water fluxes, habitat destruction/transformation, and the history and impact of environmental pollution in terrestrial ecosystems. Underlying the HITE initiative is a commitment to use paleorecords to deepen understanding of ecosystem processes during the whole period over which human activities have had a significant effect on terrestrial ecosystems. Many such processes are part of trajectories of change much longer than the short span of direct measurements or observations.

See also: **Environmental Change; Nutrient Cycle; Paleo Perspectives.**

Further Reading

Past Global Changes (PAGES) Project, Focus 5: Past Ecosystem Processes and Human-Environment Interactions [Online, January 2005], The HITE Web Site, www.liv.ac.uk/geography/hite.

FRANK OLDFIELD

I

IGBP-DIS Global 1-Km Land Cover Data Set. Researchers from the U.S. Geological Survey, University of Nebraska–Lincoln, and the **Joint Research Centre (JRC) of the European Commission** in Ispra, Italy, completed development of a one-kilometer-resolution global land-cover characteristics database in 1998 for use in a wide range of continental- to global-scale environmental studies. The project was carried out as an **International Geosphere-Biosphere Programme** initiative with support from the U.S. Environmental Protection Agency, NASA, and the **United Nations Environment Programme**. This database provides a unique view of the patterns of biogeographical and ecoclimatic diversity of earth's global land surface, and presents a detailed interpretation of the extent of human development. The one-kilometer global land-cover database was developed through a continent-by-continent unsupervised classification of one-kilometer monthly **advanced very high resolution radiometer** and **normalized difference vegetation index** composites covering 1992–1993. Extensive post-classification stratification was necessary to resolve spectral/temporal confusion among disparate land-cover types. The complete global database consists of seven separate land-cover data sets used for global environmental **modeling** and assessment. The data sets include IGBP-DISCover, the U.S. Geological Survey Anderson System, Simple Biosphere Model, Simple Biosphere Model 2, Biosphere-Atmosphere Transfer Scheme, Olson Ecosystems, Running Global Remote Sensing Land Cover, and seasonal land-cover regions. The IGBP-DISCover global land-cover product is particularly significant. A formal accuracy assessment of the DISCover data layer was completed in 1999, making DISCover the only validated global land-cover map produced to date.

See also: **BIOME 300 Project; Global Land Cover Map of the Year 2000; GLOBCOVER; Integrated Assessment; Land-Cover Change; Remote Sensing; United States Geological Survey (USGS) Program.**

Further Reading

Loveland, Thomas R., Reed, B. C., Brown, J. F., Ohlen, D. O., Zhu, J., Yang, L., and J. W. Merchant, "Development of a Global Land Cover Characteristics Database and IGBP DISCover from 1-Km AVHRR Data," *International Journal of Remote Sensing* 21 (2000): 1303–30; Loveland, Thomas R., Zhu, Z., Ohlen, D. O., Brown, J. F., Reed, B. C., and L. Yang, "An Analysis of the IGBP Global Land-Cover Characterization Process," *Photogrammetric Engineering and Remote Sensing* 65 (1999): 1021–32.

THOMAS R. LOVELAND

IKONOS. Commercial sector satellite of Space Imaging (USA) that provides very high spatial resolution **remote sensing** information, that is, the

pixels in the panchromatic band of IKONOS data cover a 1 × 1 meter area on the ground (multispectral information: four-meter resolution). The satellite is flown at an altitude of 681 kilometers and has a revisit period of four days. Two independent sensors scan the earth's surface in a strip of 11 km × (up to) 1,000 km. The quality of images comes close to that of **aerial photography** products, like the one-meter resolution digitized photographs that the Russian SPIN-2 has been producing since 1998 (with both satellites for the first time providing high-resolution data for non-military purposes). From declassified data, generated by the satellite since it was launched in 1999, there is indication that these data will have great utility for **vegetation** classification and **land-cover change** analysis. For example, IKONOS—as well as other satellites such as **Landsat** Thematic Mapper, **advanced very high resolution radiometer**, **SPOT Vegetation**, and **medium resolution imaging spectrometer**—have sensors that are needed to acquire data at red and near-infrared wavelengths (and at a range of temporal and spatial resolutions) so that the **normalized difference vegetation index** is easy to compute. On the other hand, high-resolution satellites such as IKONOS cannot really be used for general large-area environmental **monitoring**, because of the small size of their images and the lack of global coverage. Nonetheless, IKONOS imagery may be useful for studying particular land-change events such as earthquake damage or localized flooding, or they can be used in place of air photos for cartographic work. It has been emphasized that IKONOS images might have the potential to revolutionize **geographical information systems** (GIS) methodologies.

See also: **Change Detection; Global Terrestrial Observing System.**

Further Reading

Cracknell, A. P., Satellite Remote Sensing [Online, March 2005], The Encyclopedia of Life Support Systems Web Site, www.eolss.net; Space Imaging, Ikonos [Online, March 2005], The Ikonos Web Site, www.spaceimaging.com/products/ikonos.

HELMUT GEIST

Indigenous Knowledge. The dynamic experience of a particular community, gained through successive generations, to develop a particular **land-use system** that is well adapted to the potentials and constraints of a specific region.

There are currently a variety of terms with somewhat similar meanings: "indigenous knowledge," "local knowledge," "traditional knowledge," and "traditional ecological knowledge." Fikret Berkes defines traditional ecological knowledge as a cumulative body of knowledge, practices, and belief, evolving by adaptive processes and handed down through generations by cultural transmission, about the relationship of living beings (including humans) with one another and their environment. Indigenous knowledge results from a continuous process of experimentation, innovation, and adaptation, that is, "adaptive indigenous knowledge." This knowledge serves as the basis for practical, local-level **decision-making** in such diverse domains as animal health and **human health**, **agriculture** and food production, and natural resource management and **conservation**.

From an historical point of view, we can distinguish three positions with regard to local knowledge. In a first period, indigenous knowledge, particularly in the African context, has long been ignored and maligned by outsiders. The role of nineteenth-century colonialism in this attitude has been well documented. Secondly, indigenous knowledge has been compared to "scientific," "western," "international," or "modern" knowledge generated within universities, international organizations, and private firms using a formal scientific approach. In this vision, scientific knowledge was assumed to be superior to all other forms of knowledge. For Paul Richards, this agricultural science approach implied ethnocentric bias, elitism, ignorance or disregard of local ecological and sociological conditions, inappropriateness of temperate-region techniques, and inadequate field research. Local knowledge should not be in opposition to science, since it is not at all a pseudoscience or an anti-science. Finally, since the last decade of the twentieth century, studies that depicted local communities and their knowledge as primitive, simple, and static have been logically countered by authors who describe the adaptive capabilities of many indigenous natural resource management systems. The recent but large interest in the contribution of indigenous knowledge was catalyzed by the United Nations Conference on Environment and Development in 1992.

Three basic models have progressively replaced each other: technology transfer, farmer-first, and the participatory approach. The first model involves top-down technology transfer from researchers to farmers. The second emphasizes the important role that farmers have to play in research and agriculture from the bottom up. In 1993, John Thompson brought attention to the fact that populist strategies encountered the same sorts of problems as interventionist programs. The concept of powerful outsiders helping powerless insiders is always present: knowledge, whether "indigenous" or "scientific," is inclusive in the sense that it is the result of many great decisions and selective assimilations of previous beliefs, but at the same time exclusive of other possible frames of conceptualization and understanding. The third model is a participatory approach that integrates and extends the first two models. The participatory approach relies on the involvement of researchers and farmers, as well as other stakeholders. In the last two approaches, the peasant's ecological sense and innovative talents are seen as two of the most significant rural resources. Gloria Emeagwali details numerous successful projects and case studies illustrating the use of local and indigenous knowledge in the development of cost-effective and sustainable survival strategies.

In the past ten years, interest in the role of indigenous knowledge in sustainable development has increased in national governments as well as in inter- and non-governmental organizations. Nevertheless, indigenous knowledge continues to be occasionally disregarded—in development planning, for example. Moreover, this knowledge, often linked to oral tradition, is at risk of becoming extinct. There is a need to protect the knowledge generated and perpetuated by local communities, through awareness-raising, training programs, international **property** rights arrangements, and **validation** procedures. One cost-effective way in which indigenous knowledge is now systematically recorded, stored, and shared is through the global network of indigenous knowledge resource centers.

See also: **Agrodiversity; Colonization; Community Involvement; Cultural Factors Endogenous Factor; Forestry; Land-Use History; Land-Use Planning; Participatory Geographic Information System; Participatory Land Management; People at Risk; Project on People, Land Management and Ecosystem Conservation; Resilience; Sustainable Land Use.**

Further Reading

Berkes, Fikret, Traditional Ecological Knowledge, Indigenous Knowledge and Development Monitor [Online, February 2005], The Nuffic Web Site, www.nuffic.nl/ciran/ikdm; Emeagwali, Gloria, African Indigenous Knowledge Systems (AIK): Implications for the Curriculum [Online, February 2005], The Africa History Web Site, www.africahistory.net; Richards, Paul, *Indigenous Agricultural Revolution: Ecology and Food Production in West Africa,* London: Hutchison University Library, 1985; Thompson, John, Beyond Farmer First: Research Programme and Workshop, Indigenous Knowledge and Development Monitor [Online, February 2005], The Nuffic Web Site, www.nuffic.nl/ciran/ikdm.

NATHALIE STEPHENNE

Industrial Revolution. The transformation of production forces, especially manufactory and handicraft production, into large-scale factory production, using machines and new chemical technological processes. The most important element was the "universal engine" of the Industrial Revolution, such as the double-action steam engine, and then later the machine production of machines. The period in which systematic use of new machines or chemical technological processes in industry was applied is considered to be the beginning of the Industrial Revolution. This was completed when factory production in the main (e.g., textile, machinery, mining) industries predominated the older production forms, which had been manufacture and handicraft. The Industrial Revolution (IR) has been developed in four phases: the initial stage, and then the stages of development, expansion, and culmination. Formerly, the term "Industrial Revolution" was applied to characterize such changes in Great Britain, or related contexts. In other contexts, it is named the "First Industrial Revolution," and the following **technological-scientific revolution** was given the label of Second Industrial Revolution. Still others use the term "Third Industrial Revolution," which began in the second half of the twentieth century and was based on the use of high technology, computers, nuclear energy, space research, and massive application of **biotechnology** science, especially of biology (genetics) and physics (new sources of energy). In theoretical terms, such revolutionary processes could be called a scientific-technological revolution, or stages of growth. The notion of "revolution" here does not mean acceleration of some part of the historical process (or time) as, for example, political upheaval, but expresses a slower and especially fundamental change of content or quality.

Historical and geographical roots and consequences made the IR spread over **Europe** from west to east. In the United Kingdom, it begun approximately in the 1780s and climaxed in the 1820s. In France and Germany, it culminated in the 1860s; in whole Austria-Hungary in the late 1880s; in the **United States of America**, Argentina, and **Japan** in the 1870s; and in **Russia** in the 1890s, in some regions only.

The IR had societal, economic, social, geographic, and cultural aspects. It contributed to the fall of feudalism (and the abolition of slavery in the United States) and to the rapid development in capitalism of free competition with its greed for free labor forces. It contributed to the processes of **industrialization** and **urbanization**, to a revolution in transport (railways, steamboats, telegraph, etc.), to the rapid spreading of international trade, and to the birth of a worker class and its organization, among other developments.

Of particular importance were geographic aspects. The steam engine, for example, had removed the former localized dependency of industrial production on the position of especially hydro-energy resources (based on stream declivity in mountainous regions), as well as the dependency on raw materials (ores) and fuel (coal) deposits. This facilitated the concentration of industry into regions with a better position to transport industrial products across national and international markets. Subsequently, growing demand for food stimulated the progress of the **Agricultural Revolution**, which caused fundamental land use and **land-cover changes**, including a reversal of their historical trends.

Agricultural and industrial revolutions had direct or indirect relations established through a system of **feedback** mechanisms. The decline of **fallow**, for example, meant an increase in the area of arable land. In **western** and **west-central Europe**, the Agricultural Revolution caused a decline of sheep breeding, especially in fertile lowlands, and triggered an orientation of the wool industry to imported wool. The production of root crops (e.g., sugarbeets and potatoes) stimulated the development of the sugar and distillery industries. The industrial waste from these and other branches of the food industry improved livestock feeding. The use of synthetic fertilizers in turn stimulated appropriate branches of the chemical industry. The greed for new and more effective iron machines gave birth to engineering that produced new and sophisticated agricultural machines. In this way, **agriculture** linked with industry and had to adopt its rational principles of the organization of production. Wood was replaced by coal and became a valuable raw and construction material in the paper and pulp industry and for the construction of railways, for example.

See also: **Agro-Industry; Cotton.**

Further Reading

Ashton, Thomas S., *The Industrial Revolution, 1760–1830,* New York, London: University of Oxford Press, 1968; Hobsbawn, Eric J., *The Age of Revolution: Europe 1789–1848,* New York: Vintage Books, 1996; Olson, James S., *Encyclopedia of the Industrial Revolution in America,* Westport, CT: Greenwood Press, 2002; Purš, Jaroslav, "Complex Revolution of the Modern Age and Industrial Revolution," *Historica* XIX (1980): 135–70; Rostow, Walt W., *The Stages of Economic Growth: A Non-Communist Manifesto,* 3rd ed., Cambridge, UK, New York: Cambridge University Press, 1990; Toynbee, Arnold, *The Industrial Revolution,* Boston: Beacon Press, 1956.

LEOŠ JELEČEK

Industrialization. The transition from agricultural society to an economy based on large-scale, machine-assisted production of goods by a concentrated, usually urban labor force. This transition was made possible by large-scale technological changes. In a first phase that began in the middle of the eighteenth century, industrialization relied above all on coal and steam power. Electrification and the wide use of internal combustion engines characterize the second phase of industrialization, beginning in the 1920s. In economic terms, industrialization is characterized by an increasing share of the industrial sector compared to agriculture and handicraft. The share of industry in gross domestic product (GDP) and in total labor force is growing, and both total and per capita GDP increased rapidly during industrialization. This is accompanied by a surge in labor productivity in both industry and agriculture. Since its beginnings roughly 250 years ago, and especially after World War II, industrialization has affected practically every region of the world. The degree of industrialization, however, differs widely across countries: economists and political scientists distinguish between fully industrialized countries (e.g., **Europe**, **United States of America**, **Japan**), newly industrialized countries (e.g., Mexico, India, **South Africa**), and the least-developed countries (most African countries). Hence, industrialization is a global and still-ongoing process. Besides its effects on the overall

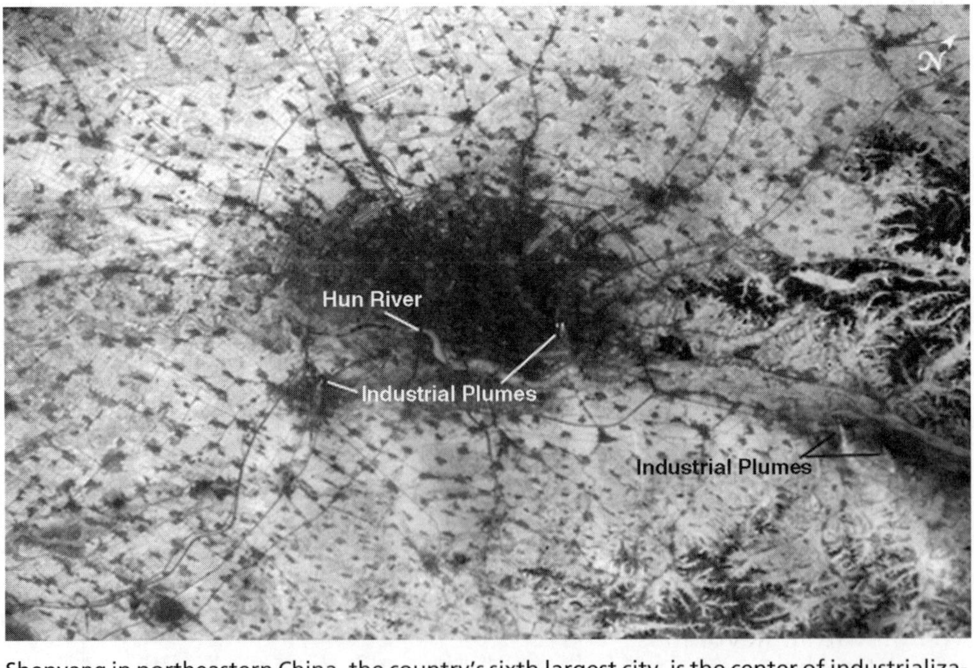

Shenyang in northeastern China, the country's sixth largest city, is the center of industrialization (coal mining, metal smelting, petrochemical processing) in what is historically known as Manchuria (snow cover in January 2005 provides a contrast between urban-industrial zones and the rural regions, with plumes from industrial facilities visible in the image).
PHOTO: International Space Station. CREDITS: Image Science and Analysis Laboratory, NASA-Johnson Space center, 18 Mar. 2005 (Earth from Space - Image Information). http://earth.jsc.nasa.gov/sseop.efs

economic and social structure, industrialization is also related to major biophysical transformation processes, fundamentally altering nature-society relations, including land use. From the perspective of social **metabolism**, industrialization results in the transformation of the socio-economic energy system, in the substitution of fossil fuels (more recently, nuclear power and other kinds of energy also) for **biomass**, resulting in an overall increase in socio-economic material and energy flows. This process has major implications for the spatial patterns of social systems and land use, including large-scale spatial specialization and concentration processes of population and production.

The Pre-industrial State: Solar Energy-based Agrarian Society

Before the **Industrial Revolution**, the energetic basis of almost all socio-economic activities was solar energy, mostly various types of biomass. Under this socio-ecological regime, **agriculture** and **forestry** were the central element of the socio-economic energy system and provided primary energy for human and animal work, space heating, and process heat. The almost complete reliance on biomass, an area-based source of energy, imposed a decisive bottleneck for socio-**economic growth** under pre-industrial conditions.

Until the invention of the steam engine in the eighteenth century, it was practically impossible to convert heat energy from combustion into mechanical work: apart from the direct use of hydropower or windpower, the only means to produce mechanical energy was human or animal work. Energy provision was therefore by and large related to specific types of land use: heat generation was based on wood and thus **forests**, human work on cropland, and animal power on **grassland**. Far-distance transportation of bulk materials was hampered by high energy costs, especially in the case of land transport. As a consequence, the specific local requirement for a mix of energy services was reflected in the local combination of cropland, grassland, and woodland. The long-term functioning of the solar energy system was intimately related to the issue of maintaining soil fertility. Due to extremely limited possibilities of external replacement of plant nutrients, it was essential to keep **nutrient cycles** largely closed at a local level. In combination with the transport limitations, this represented limits for spatial concentration and **urbanization**, as cities acted as large-scale sinks for biomass and essential plant nutrients.

Pre-industrial societies are characterized by comparatively low population densities, high proportion of rural population (over 75 percent), locally integrated **land-use systems** that combine different types of land use, and land-use intensity according to local needs. Under temperate climate conditions, this typically was a mix of cropland, **pasture**, and woodland on the local level, allowing for maximum average population densities of 30–50 persons/km^2 in the late nineteenth century.

The Industrial Revolution: Coal and the Transformation of the Energy System

The initial phase of industrialization is called the Industrial Revolution and started in the United Kingdom in the middle of the eighteenth century. During the first half of the nineteenth century, the Industrial Revolution spread over a number of northern and central European countries (e.g.,

Germany, France, and the Netherlands) and reached the United States, Japan, and Argentina in the late nineteenth century. The Industrial Revolution was characterized by a rapid substitution of coal for firewood and the increasing importance of the iron industry in combination with widespread application of the steam engine in industrial processes and railroad transportation; it thus set off the transformation of the socio-economic energy system. It resulted in at least a partial emancipation of energy provision from land use and allowed societies to overcome some of the essential pre-industrial limits to growth.

As early as the 1860s, British coal consumption equaled an amount of energy the provision of which would have required a forest of the size of Great Britain itself. The use of coal triggered unprecedented growth of production and population. Between 1830 and 1910, population density in Great Britain doubled to 150 persons/km^2, urbanization increased up to 80 percent, and per capita coal consumption and GDP grew seven- and sixfold, respectively. The expansion of the railroad system and the use of steamships resulted in an increasing national and global division of labor and large-scale transfers of industrial raw materials, biomass, and people over large distances. This established increasingly tight relations between land use and **consumption** patterns in distant regions of the globe: with the railroad-driven spread of grain production in the U.S. Midwest, grain exports from the United States to Europe increased. In the late nineteenth century, about 60 percent of Great Britain's food supply was imported from the United States. Food imports (and yield increases) allowed a reduction of British cropland by 30 percent from 1850 to 1910, despite a twofold increase in food demand in Britain during this period.

The increasing use of coal resulted in decreasing pressure on forest resources. Particularly forests around industrial centers and urban agglomerations were required less and less for energy production, and forested area began to grow in **western Europe**. However, agriculture as the most significant human activity related to land use was not yet fundamentally transformed during the first phase of industrialization: although new crops and more effective nutrient management increased area productivity in many European countries throughout the nineteenth century, agriculture was based almost solely on animate power, and had to rely on natural processes to maintain soil fertility. European population and its food demand more than tripled during the nineteenth century, resulting among other things in increasing **poverty** of urban masses. Economists such as Thomas Robert Malthus anticipated sustainability problems stemming from these undampened growth processes, the concentrated demand of urban populations, and large-scale transfers of biomass and plant nutrients.

The Second Phase of Industrialization: Electrification, Motorization, and the Industrialization of Agriculture

Beginning in the 1920s in the United States and Europe, a second phase of industrialization took off and then rapidly changed the global economy after World War II. Within a few decades the transformation of energy systems from biomass to fossil and other fuels in industrialized countries was more or less complete, and thus also the decoupling of energy supply from land use in many regions of the world. Technically, this phase was and is driven by electric power and the internal combustion engine, both closely related to the large-scale exploitation of oil and natural gas deposits. It is accompanied by further population growth and urbanization as well as rapidly increasing per capita energy and material consumption on a global level.

The effects of this second phase of industrialization on the land were even more drastic than the transformation processes of the nineteenth century. A crucial element with respect to land use was the industrialization of agriculture, often termed the **Green Revolution**, which boosted agricultural labor productivity and yields through mechanization, use of new plant varieties, and a host of agrochemicals. All of this required considerably more energy inputs, most of them (directly or indirectly) fossil energy. Energy inputs grew more quickly than yields, so that the energy efficiency of agriculture fell. In industrialized countries, yield (harvest per unit area) increased by impressive factors—between three and ten over pre-industrial levels. Increased land and labor productivity and falling transport costs triggered a process of spatial concentration of agriculture: the best soils were used with increasing intensity, whereas marginal land was taken out of production and reforested. This resulted in the disintegration of previously locally integrated and optimized land-use systems based on a specific local combination of different intensities and types of land use. Livestock farming was increasingly decoupled from crop production, and large regions were dedicated to the production of a very limited number of crops. While the industrialization of agriculture allowed for dramatic increases in global food production, it was also accompanied by a number of negative impacts on land-use systems such as **soil erosion** and compaction, groundwater depletion and pollution, loss of **biodiversity**, and traditional cultural landscapes.

However, while land-use intensity in many newly industrialized countries and least-developed countries is still increasing and cropland is expanding at the expense of woodlands and grasslands, in fully industrialized countries afflicted by agricultural overproduction, the reverse trend is observable: since the 1980s, fertilizer and pesticide inputs have been reduced, organic farming has gained importance, and the area of land in **conservation** programs has been increasing. Land demand for urban dwelling and transport infrastructure, however, is still increasing quickly. In European industrialized countries, between five and ten square meters per capita are converted into built-up land annually. With respect to the industrial sector, many highly industrialized countries are characterized by deindustrialization processes resulting in the reduction of heavy industries and mining. To a certain extent this reflects the externalization of energy- and labor-intensive production processes to newly developing and least-developed countries. Globally, industrialization and its increasing demand for natural resources are major driving forces for **deforestation**, **soil degradation**, and groundwater depletion.

See also: **Agro-Industry; Economic Restructuring; Forest Transition; Globalization; Mineral Extraction; Reforestation.**

Further Reading

Cipolla, Carlo, *The Fontana Economic History of Europe*, Brighton, UK: Harvester Press, 1976; McNeill, John R., *Something New under the Sun: An Environmental History of the*

Twentieth Century, London: Allen Lane, 2000; Pomeranz, Kenneth, *The Great Divergence: China, Europe, and the Making of the Modern World Economy*, Princeton, NJ: Princeton University Press, 2000; Sieferle, Rolf P., *The Subterranean Forest: Energy Systems and the Industrial Revolution*, Cambridge, UK: The White Horse Press, 2001; Wrigley, Edward A., *Continuity, Chance and Change: The Character of the Industrial Revolution in England*, Cambridge, UK: Cambridge University Press, 1988.

FRIDOLIN KRAUSMANN

Initial Conditions. A set of environmental and socio-economic factors that leads to strong regional variations of the processes and patterns of land-use/cover change. Initial conditions relate to the environmental history as well as to the **land-use history** of each location, region, or otherwise defined case of land change. Together, the histories of biophysical land, surface attributes and the socio-economic setting define the initial conditions for each subsequent round of land-use change. Initial conditions constitute one set of factors for the analysis of trajectories of land change, together with a particular combination of causes or underlying **driving forces** (triggering and driving land-use change) and the structure of **feedbacks** (social and ecological responses to changes of land cover affecting the rates of change). Based on a detailed description of the environmental and land-use history of land change per major geographical area—with information taken from hundreds of tropical forest and dryland case studies—Helmut Geist and Éric Lambin included initial conditions in a consideration of detailed pathways of tropical **deforestation** and **desertification** worldwide. Not surprisingly, these pathways are nearly as diverse as the histories, cultures, and ecosystems of the major dryland and forest areas in the world.

See also: **Agent-Based Model; Land Quality; Land-Use System; Pattern to Process; Proximate Causes.**

Further Reading

Geist, Helmut, *The Causes and Progression of Desertification* (Ashgate Studies in Environmental Policy and Practice), Aldershot, UK, Burlington, VT: Ashgate Publishing, 2005; Lambin, Éric F., and Helmut J. Geist, "Regional Differences in Tropical Deforestation," *Environment* 45 (2003): 22–36.

HELMUT GEIST

Institutions. Defined most clearly by John Commons in 1934 as "collective action in control of individual action," institutions are those norms through which groups self-organize for collective gain. Distinct from organizations, the term "institution" refers to habits, rules, and laws governing behavior. Mundane examples include four-way stop signs to avoid wrecks and the practice of queuing at the post office. More profound and complex examples take the form of property law and redistributive systems, from potlatch to farm subsidies.

Revitalized more recently by Elinor Ostrom and other scholars, theories of institutions serve to explain why predicted "prisoner's dilemmas," those failures of collective interest resulting from the rational pursuit of individual gain, do not generally occur. This body of research grew from the empirical observation that social groups, both simple and complex, tend to form systems of cooperation that offset the apparently rational tendency of individuals to seek their own good at collective expense.

Emerging from debates over the merit and persistence of **common-pool resources** in particular, including community- and group-controlled **pasture**, **forest**, and groundwater resources, investigation of institutions has recorded a staggering range of **land tenure** systems, access rights, and group decision arrangements around the world, which together often work to sustain renewable resources, though they are neither held privately nor exclusively by state authorities. These ethnographic findings bolster the notions that people commonly organize without free-riding, shirking, or acting opportunistically; and the outcomes they attain are often more optimal than those they might achieve individually, therefore demonstrating that cooperation is rational and does not require altruism *per se.*

For understanding land-use/cover change, an appreciation of institutions is essential. An individual, for example, may or may not convert a parcel of land from forest to farm based on exclusive profit or preference, but such decisions are enmeshed in collective systems of rules and norms—some formally written, others only informally understood—that both restrict some behaviors and encourage others. Since many land-cover-relevant policy measures take the form of spatially explicit institutions (e.g., conservation zones or enclosures), serious, sophisticated, and robust explanation of **land-cover change** usually demands some consideration of institutions. Research has shown that, controlling for environmental and demographic conditions, variations in institutional form governing **access** and use of natural resources produces significantly different land-cover outcomes.

In the case of land-use/cover change, such institutional forms can vary in a number of ways. They may be formal or informal, local or exotic, strong or weak, with varying and complex implications for land cover. Formal institutions (e.g., state land-use laws) are in no way exclusively or consistently stronger than informal ones (e.g., local rules and restrictions), for example. In a study comparing formal and informal land management institutions in semiarid pasture and forest lands in India, for example, it has been shown that local and state institutions produce divergent responses in local producer behavior, including cutting, collecting, and grazing. But whereas local, informal, non-state **fallow** pasture management was proven to coincide with greater and denser cover of desirable perennial grass species than formal state pasturage, state **forestry** was demonstrated to preserve greater tree cover than traditional taboos. These somewhat complex results suggest that neither state nor local institutions are "superior" in **conservation**. It reveals, moreover, that land-use and land-cover outcomes are directed by the differing ways decision-makers respond to sources of power and forms of control in specific contexts. In this case, where state authorities may be able to enforce forestry rules, state control over pasturage, viewed locally as illegitimate authority, results in institutional weakness and failure, leading to overgrazing and a notable shift in the profile of grass and tree cover in village lands.

Elsewhere, research in central Nepal using time series analysis of satellite images between 1976 and 2000 has shown similar complexity of the institutional effect. Here, semi-government forests, those legally under state control but with local de facto community claims of ownership, have been shown to reforest faster than exclusively local or fully state-controlled areas. Again, the role of institutions is profound, but subtle, and reflects the regionally determined legitimacy of specific forms of mixed **property**. Methodologically, consideration of institutions like these in land-use and land-cover analysis presents several challenges. Most commonly, the inclusion of institutions in research takes the form of incorporating property layers into analysis. Ownership and security of tenure are recorded and relatively easy to map. They are also of obvious immediate importance, since insecure tenure may lead to changes in **decision-making** behavior either through overexploitation, where managers reap short-term benefits in anticipation of loss of future control; or underdevelopment, where decision-makers choose not to invest in land improvements, again anticipating dispossession. Such property data are commonly available from state land management authorities, moreover, sometimes in digital map formats.

While some institutions represent formal property codes, however, many are simply perceived standards or collective habits. Discerning meaningful norms, rather than simple and legally formal property claims, typically requires intensive face-to-face interviews, long-term participant observation, and detailed behavioral observation. This is especially important where de facto rules tend to overshadow written and legal rules, or where multiple and contradictory formal systems vie against one another. In post-independence Tanzania, for example, competing systems for settling land disputes, including traditional village councils and ward tribunals, together created tenure insecurity and the domination of traditional norms. Determining how such local systems work requires innovative land resource research and higher levels of local participation in **land-use planning** than occurs traditionally.

A second problem emerges from the spatial coherence of institutional forms. While some institutions, like property lines, are "hard" and easily mapped and incorporated into **geographical information systems**, many institutions have overlapping, inchoate borders. The edges of a sacred grove, for example, in which strong restrictions are observed against tree cutting, may taper off with distance from a shrine, rather than being explicitly marked. Such conditions may require fuzzy boundaries, probability functions, or other tools to properly reflect their "soft" edges.

Institutional boundaries, even where they are coherent, may not be perennially persistent and may change over time, moreover. Long-fallow land, for example, may be managed as community pasturage with rules against cutting or collecting during dry season, but turn into exclusive property during the rainy season.

These realities have complicated, though by no means thwarted, the incorporation of institutional form and variation into land-use/cover change **modeling**. Large-scale forest-cover change, for example, can to a degree be attributed simply to topography and human settlement history through regression modeling, but examination of residuals and unexplained variations shows that at higher resolution, rules of use and access are crucial to predicting change. Such initial successes show the need for more finely resolved spatial social data and a better understanding of the often-intangible social, legal, and cultural influences on individual decision-making.

See also: **Arena of Land Conflict; Community Involvement; Driving Forces; Economic Restructuring; Forest Anomalies; Forest Transition; Investments; Land Rights; Land-Use Policies; Mediating Factor; Public Policy; Tradeoffs; Tragedy of the Commons; Transhumance; Vulnerability.**

Further Reading

Commons, John R., *Institutional Economics,* New York: Macmillan, 1934; Dalal, Clayton, Dent, Barry, and David Dent, *Knowledge of the Land: Land Resources Information and Its Use in Rural Development,* Oxford, UK: Oxford University Press, 2001; Gautam, A. P., Shivakoti, G. P., and E. L. Webb, "Forest Cover Change, Physiography, Local Economy, and Institutions in a Mountain Watershed in Nepal," *Environmental Management* 33 (2004): 48–61; McConnell, William J., Sweeney, S. P., and B. Mulley, "Physical and Social Access to Land: Spatio-Temporal Patterns of Agricultural Expansion in Madagascar," *Agriculture, Ecosystems & Environment* 101 (2004): 171–84; Ostrom, Elinor, *Governing the Commons: The Evolution of Institutions for Collective Action,* Cambridge, UK: Cambridge University Press, 1990.

PAUL ROBBINS

Integrated Assessment. A structured process of dealing with complex issues, using knowledge from various scientific disciplines and/or stakeholders, such that integrated insights are made available to decision-makers. Multiple diverse approaches are needed and have been developed to address the complex and value-laden process of integrated assessment (IA) that operates at a variety of levels and needs to inform a variety of stakeholders. Land-use and cover change are typical examples of complex issues that cannot be treated either by a single discipline or at a single scale, and land-use models have been referred to as a type of integrated assessment model (IAM).

There are many more IA definitions, but in any of those, "integrated" conveys a message of multidisciplinarity and "assessment" a message of policy relevance. The terms *IA* and *IAM* seem to originate in scientific and **public policy** efforts to understand and control acid deposition in the 1970s in **Europe** and North America. The establishment of the **Intergovernmental Panel on Climate Change** marked the turning point toward defining IA as a multidisciplinary process. The first generation of IAMs was rooted in systems analysis and global **modeling**. The second generation focused on the global-scale climate change issue. The third and current generation—addressing more meso-scale issues—deals with sustainable development, including many non-environmental issues like **human health**, land use, and water.

The issue of **scale**—in time, space, and quantity—is of fundamental importance. By definition, IA has to deal with complex issues that operate at multiple scales. One of the aims of IA is therefore to address this observed need and to develop heuristics, procedures, and tools to address spatial and temporal scale issues. Transition and the management of transitions is a second pivotal issue to IA. A transition can be described as a gradual continuous process of change where the structural character of society transforms. The perspective of transition management is used as a new approach to the management of complex policy problems.

Two types of IA methods can be distinguished: analytical methods (including **scenario** development and models) and participatory methods that aim to involve non-scientists in the process. IAMs are heuristic tools that amplify our insights into the present and future **driving forces** behind the complex structures in our society. The ultimate goal of these models is to serve as tools in integrative planning. Scenario development covers a broad range of methods and issues including participatory techniques, multiscale scenarios, surprises, use of historical analogies in scenarios, and internal consistency and plausibility of scenarios. IA has an explicit participatory dimension, both through participation by scientists from various disciplines as well as through participation of stakeholders and societal actors in the assessment process. Participatory methods is an umbrella term for describing interactive approaches that actively involve a range of stakeholders, varying from decision-makers to laypeople.

See also: **Decision-Making; Integrated Model to Assess the Global Environment; Land-Use Planning; Millennium Ecosystem Assessment.**

Further Reading

Rotmans, Jan, "Methods for IA: The Challenges and Opportunities Ahead," *Environmental Modeling and Assessment* 3 (1998): 155–79; Rotmans, Jan, and Dale Rothman, eds., *Scaling in Integrated Assessment*, Linne, NL: Swets & Zeitlinger Publishers, 2002; Tol, Richard, and Pier Vellinga, "The European Forum on Integrated Environmental Assessment," *Environmental Modeling and Assessment* 3 (1998): 181–91; Van Asselt, Marjolein, and Nicole Rijkens-Klomp, "A Look in the Mirror: Reflection on Participation in Integrated Assessment from a Methodological Perspective," *Global Environmental Change* 12 (2002): 167–84.

KASPER KOK

Integrated Land Ecosystem-Atmosphere Processes Study. A project on land-atmosphere interactions of the **International Geosphere-Biosphere Programme** that addresses the interconnections among biospheric processes, atmospheric composition, land cover, and energy and water exchange at the land surface for the functioning of the climate system. The multitude of interactions between the various compounds and forms of energy that interact between terrestrial surfaces and the atmosphere on all temporal and spatial scales present a major challenge for quantifying the role of each compound in the land-atmosphere system. Consequently, the objective of the Integrated Land Ecosystem-Atmosphere Processes Study (iLEAPS) can be summarized as providing understanding of how interacting physical, chemical, and biological processes transport and transform energy and matter through the land-atmosphere interface. Research under the auspices of iLEAPS addresses the following four questions: How do interacting physical, chemical, and biological processes transport and transform energy, momentum, and materials through the land-atmosphere system? What are the implications for the dynamics of the earth system? How did the terrestrial-ecosystem/atmosphere system function under pre-industrial conditions, and how are human activities influencing it? And, to what extent does the **vegetation** determine the physical and chemical environment on various temporal and spatial scales?

iLEAPS aims to facilitate state-of-the art research to improve the understanding of **atmosphere-land interlinkages**, and thus the earth system. Research addresses scientific issues of regional and global significance by considering multiple **scales** and their interactions and **feedbacks**. Process-based understanding of the land-atmosphere system on a broad variety of scales requires incorporating measurements together with **modeling** and integration from the onset and throughout research activities.

See also: **Climate Impacts; Greenhouse Gases; Land-Ocean Interactions in the Coastal Zone (LOICZ) Project; Land-Use/Cover Change (LUCC) Project; Global Land Project.**

Further Reading

iLEAPS [Online, April 2004], The Integrated Land Ecosystem-Atmosphere Process Study Web Site, www.atm.helsinki.fi/ILEAPS.

ALMUT ARNETH

Integrated Model. Combination of one or more separate models across space, discipline, or functional hierarchy in order to address relevant questions related to the causes and consequences of land-use and **land-cover change**. Integrated models are often necessary for **modeling** land change due to the spatial and temporal complexity of interactions between social and biophysical systems inherent in land-use and land-cover change. While a broader range of integrated models are discussed by Helen Briassoulis, this review focuses on those models that contain a spatially explicit land-use/land-cover change module. This module may be constructed using a variety of techniques, including statistical models; computable general equilibrium models; optimization techniques such as linear, integer, and dynamic programming; Markov and **cellular automaton** models; and agent/individual-based models. Integrated models may be either positive (focusing on what might happen) or normative (focusing on what should happen).

The simplest type of integrated land-change model may combine two or more of the above techniques in order to construct a model that projects land-use/cover change over space and time. Models may combine techniques to build a model at a single spatial **scale**. For example, statistical techniques may be used to identify landowner types, cropping system characteristics, or the land-price effects of proximity to an undesirable land use, and these estimates may be used to parameterize a cellular automaton or **agent-based model**. Models may also use different modeling techniques at various spatial scales. For example, an aggregate model of supply and demand for products for which land is a production input may be used to estimate the total amount of land in each land-use category required in a region. Then, these quantities may be allocated over space using algorithms based on transition probabilities derived from statistical regression or cellular automata. Additional information related to the ease of **land-use transitions** and feasible land-use strategies may be used to constrain the land-use change allocation algorithm. Alternatively, a simulated land market can be combined with a linear programming-based model of landowner **decision-making** to simultaneously determine the quantity and location of land-use change. These examples can be classified as positive in their focus. A normative application may use linear programming to determine the optimal proportion of land in each use in order to meet a particular objective, such as meeting an employment objective or habitat preservation goal, and then use suitability indices based upon **geographical information systems** to allocate these land uses across space. Alternatively, integer-based mathematical programming can also be used to jointly determine the quantity, pattern, and location of preserved habitat. Because normative models generally focus on socially optimal land-use allocation, they are more likely to be equilibrium-based, use formal optimization techniques, and be fairly broad in their spatial extent.

More complex integrated models may combine socioeconomic and biophysical models in order to explore **feedbacks** between the socioeconomic and biophysical systems. Such models may be used to investigate linkages between the socioeconomic **driving forces** of land-use change and such environmental impacts as **soil degradation**, groundwater depletion, habitat fragmentation, **biodiversity** loss, and global climate change. Integrated models have been constructed that include biophysical models of crop regrowth, soil fertility, **carbon sequestration**, animal species migration, hydrologic networks, spread of invasive species, and air quality. A variety of techniques, including statistical, cellular-automaton, and agent-based models, have been used for the land-use/cover change component of these models. These models are often used for projection, policy, and **scenario** analysis, potentially for the purpose of **integrated assessment**. Development of integrated land-change models can present many challenges. Many are common to development of all land change models; however, these challenges are often magnified for integrated models. Peter Verburg and colleagues provide an excellent review of many of these modeling challenges. Models focused only on projection of land-use/cover change may face challenges in identifying the appropriate scale at which to model each driver, identifying cross-scale linkages and feedbacks, and building these cross-scale processes into the model. Models that integrate components across disciplines face an even greater set of challenges. Often, disciplinary models ideally operate at differing and non-nested spatial and temporal scales. Many integrated models have been developed by linking or adapting a series of previously developed disciplinary models. Guy Engelen and colleagues provide an excellent discussion of the challenges of developing the Modulus model. As model complexity increases, understanding model behavior and communicating model structure can become exponentially more challenging. Researchers interested in sensitivity and scenario analysis may face challenges in constructing appropriate confidence intervals for their projections, as the variance of their final projection may be a function of the variance of multiple stochastic components that feed into the final projection. Finally, researchers may face challenges in communicating the results of complex models to policymakers.

See also: **Conversion of Land Use and its Effects (CLUE) Model; Integrated Model to Assess the Global Environment; LandSHIFT Model; Land-Use System; Probit/Logit Model; Transition Matrix; Validation.**

Further Reading

Briassoulis, Helen, Analysis of Land Use Change: Theoretical and Modeling Approaches [Online, August 2004], The Web Book of Regional Science Web Site, www.rri.wvu.edu/regscweb; Engelen, Guy, "Development of a Decision Support System for the Integrated Assessment of Policies Related to Desertification and Land Degradation in the Mediterranean," in *Climate Change in the Mediterranean: Socio-Economic Perspectives of Impacts, Vulnerability, and Adaptation*, eds. Carlo Giupponi and M. Shechter, Cheltenham, UK: Edward Elgar, 2003, pp. 159–95; Parker, Dawn Cassandra, Land-Use Modeling Techniques and Applications [Online, February 2005], Dawn Parker's homepage, mason.gmu.edu/~dparker3; Verburg, Peter H., Schot, Paul, Dijst, Martin, and A. (Tom) Veldkamp, "Land-Use Change Modeling: Current Practice and Research Priorities," *GeoJournal* (forthcoming); White, Roger, Straatman, Baas, and Guy Engelen, Planning Scenario Visualization and Assessment: A Cellular Automata Based Integrated Spatial Decision Support System [Online, February 2005], The Center for Spatially Integrated Social Science Web Site, www.csiss.org/best-practices/siss/21.

DAWN CASSANDRA PARKER

Integrated Model to Assess the Global Environment. The IMAGE 1.0 was an early climate model that simulated the physical carbon flows on a global aggregated scale. In the 1990s, the IMAGE model was extended with a more detailed terrestrial **carbon cycle** model, a natural **vegetation** model, and a land-cover model using a grid scale of 0.5 × 0.5 degrees. Further extensions and linkages with energy, demographic, and macroeconomic models led to the IMAGE 2.1 model and, subsequently, the IMAGE 2.2 model. To enhance its use for interactive policy support and education, the model interface has been made more open and transparent.

The IMAGE 2.1 model has been used for two important and quite different applications: the Delft Dialogue workshops and the **Intergovernmental Panel on Climate Change** *Special Report on Emission Scenarios* (SRES) analysis. The Delft workshops were constructed to improve the dialog between the scientific world and the policymakers. These workshops resulted in the construction of the Safe Landing Analysis to determine the short-term range of emissions that would still lead to desirable stabilization levels of carbon dioxide (CO_2) concentrations in the long term. The SRES analysis resulted in a set of **greenhouse gas** emission scenarios from now until 2100, based on the different ways the world might evolve in the coming years.

The IMAGE 2.2 model has played an important role in the **United Nations Environment Programme**'s Global Environmental Outlook 3 and the **Millennium Ecosystem Assessment**, in quantifying **scenarios** to explore the future of food provision, climate change, and ecosystems. In more specific contexts, the model has been used to assess the climate change consequences of energy strategies in **China**, to evaluate cost-efficient post-Kyoto strategies for **Europe** toward specified climate parameter targets, and to assist in the construction of four land-use scenarios for Europe in order to identify the consequences of **economic liberalization** for people, profit, and the planet.

In recent years the IMAGE model has been expanding into a **modeling** framework used for broader analyses than climate change only. The model is extended with water use, biodiversity impact, and food supply modules; and the linkages with demographic, macroeconomic, and energy models are strengthened. The various modules are integrated and include many **feedbacks**, such as the effects of climate change on vegetation and of the CO_2 concentration on crop productivity. The socio-economic modules supply data on population size and age distribution, on economic activity levels, and on energy demand and supply for seventeen world regions. These, too, are linked in several ways, for instance in the land-use/cover-based estimate of the regional potentials of **biomass**-derived fuels. Data from many different sources are used to calibrate the energy, climate, and land-use variables over a period from 1970 to 1995. IMAGE simulations cover the 1995–2100 period.

The overall objective of the IMAGE modeling framework is to explore the long-term dynamics of **global environmental change**, in particular, **land change as a forcing function in global environmental change**, and to be able to evaluate specific issues of sustainable development in the broader context of global socio-ecological system evolution.

See also: **Integrated Model; Kyoto Protocol; National Institute of Public Health and the Environment.**

Further Reading

Alcamo, Joseph, Leemans, Rik, and E. Kreileman, eds., *Global Change Scenarios of the 21st Century: Results from the IMAGE 2.1 Model*, London: Pergamon & Elseviers Science, 1998; De Vries, Bert, Bollen, J., Bouwman, L., den Elzen, M., Janssen, M., and E. Kreileman, "Greenhouse Gas Emissions in an Equity-, Environment- and Service-Oriented World: An IMAGE-Based Scenario for the 21st Century," *Technological Forecasting and Social Change* 63 (2000): 137–74; IMAGE Team, *The IMAGE 2.2 Implementation of the SRES Scenarios: A Comprehensive Analysis of Emissions, Climate Change and Impacts in the 21st Century*, RIVM CD-ROM publication 481508018, Bilthoven, NL: National Institute for Public Health and the Environment, 2001; IMAGE Team of the National Institute for Public Health and the Environment [Online, October 2004], The IMAGE 2.2 Model Web Site, www.rivm.nl/image.

BERT J. M. DE VRIES, BAS EICKHOUT

Intergovernmental Panel on Climate Change. Established in 1988 by the World Meteorological Organization and the **United Nations Environment Programme**, the Intergovernmental Panel on Climate Change (IPCC) aims to assess the scientific, technical, and socio-economic information relevant for the understanding of climate change, its potential impacts, and options for adaptation and mitigation. The IPCC does not carry out research nor does it monitor climate-related data or other relevant parameters.

Instead, it undertakes comprehensive assessments based mainly on peer-reviewed and published scientific and technical literature.

The IPCC has three working groups that deal with the scientific assessments and a task force that is responsible for the National Greenhouse Gas Inventories Programme. Working group I assesses the scientific basis for climate change. Working group II assesses the impacts, adaptation, and **vulnerability** of socio-economic and natural systems to climate change. Working group III assesses options for limiting **greenhouse gas** emissions in order to mitigate climate change. The working groups have published three assessment reports (1990, 1995, and 2001) and are currently working on a further report for publication in 2007. The IPCC also publishes special reports and technical papers and supports the UN Framework Convention on Climate Change (UNFCCC). The work of the IPCC was instrumental in the development of the UNFCCC in 1992, and the **Kyoto Protocol** in 1997.

A number of IPCC publications are especially relevant to the issues of land-use/cover change. The assessment reports have examined the impacts of climate change on many land-based sectors, including **agriculture**, **forestry**, human settlements and **coastal zones**, and **mountains**. The assessments have influenced land-use/cover change research by highlighting the role of climate as a determinant of land-use patterns and the ways in which people make decisions when confronted with climatic uncertainties and the interactions between climate and other drivers of **environmental change**.

The *Special Report on Land Use, Land-Use Change and Forestry* examines the scientific and technical implications of **carbon sequestration** and the global **carbon cycle** with respect to human land-use activities. The flows of carbon between different pools and how carbon stocks change in response to afforestation, **reforestation**, **deforestation**, and other land-use activities are major factors in the exchange of carbon between the atmosphere and the biosphere. The report also looks at the potential role of land use and **land-cover change** in mitigating global warming, and carbon accounting strategies linked to the Kyoto Protocol. The *Special Report on Emissions Scenarios* (SRES) produced a new range of future climate change projections based on a set of internally consistent, narrative storylines. SRES covered a wide range of **driving forces** of future greenhouse gas emissions, from demographic to technological and economic developments, and was explicit about the role of land-use change within future emissions. Although global in extent, SRES has provided a consistent basis on which many subsequent climate change impact and adaptation studies have been based as well as providing a framework for the interpretation of alternative future land-use change **scenarios** at a range of geographic **scales**.

Through its various activities, the IPCC has stimulated discussion and analysis of the role of climate, climate change, and variability in the world around us. It has also highlighted the role of land-use change as a potential consequence of climate change, but also as a mediator of changes in the climate system.

See also: **Atmosphere-Land Interlinkages; Climate Impacts; Decision-Making; Disequilibrium Dynamics; Integrated Assessment; Integrated Model to Assess the Global Environment; Public Policy.**

Further Reading

Houghton, J. T., Ding, Y., Griggs, D. J., Noguer, M., van der Linden, P. J., and D. Xiaosu, eds., *Climate Change 2001: The Scientific Basis*, contribution of working group I to the third assessment report of the IPCC, Cambridge, UK: Cambridge University Press, 2001; McCarthy, James J., Canziani, Osvaldo F., Leary, Neil A., Dokken, David J., and Kasey S. White, eds., *Climate Change 2001: Impacts, Adaptation & Vulnerability*, contribution of working group II to the third assessment report of the IPCC, Cambridge, UK: Cambridge University Press, 2001; Metz, Bert, Davidson, Ogunlade, Swart, Rob, and Jiahua Pan, eds., *Climate Change 2001: Mitigation*, contribution of working group III to the third assessment report of the IPCC, Cambridge, UK: Cambridge University Press, 2001.

MARK ROUNSEVELL

International Food Policy Research Institute. One of the fifteen food and environmental research organizations located around the world, and known as the Future Harvest centers, that conduct research in partnership with farmers, scientists, and policymakers to help alleviate **poverty** and increase **food security** while protecting the natural resource base. Associated with the **Consultative Group on International Agricultural Research**, the mission of the International Food Policy Research Institute (IFPRI) is to provide policy solutions that cut hunger and malnutrition based upon two key premises: sound and appropriate local, national, and international **public policies** are essential to achieving sustainable food security and nutritional improvement; and, research and the dissemination of its results are critical inputs into the process of raising the quality of debate and formulating sound and appropriate food policies. Through its policy analyses, which are of broad relevance to the developing world, IFPRI directly supports developing-country policymakers, non-governmental organizations, and civil society in their efforts to help the poor.

See also: **Agriculture; Biotechnology; Cassava; Food Crops; Institutions; People at Risk; Vulnerability.**

Further Reading

International Food Policy Research Institute, Sustainable Solutions for Ending Hunger and Poverty [Online, February 2005] The IFPRI Web Site, www.ipfri.org.

HELMUT GEIST

International Geographical Union (IGU) Commission on Land-Use and Land-Cover Change. Established in 1996, the major aim of IGU-LUCC is the enhancement of research on land-use/cover change in general, and in particular to contribute to the **Land-Use/Cover Change (LUCC) project**. The membership is open to anyone interested in its activities, and there are some 300 members from about fifty countries/regions. IGU-LUCC entered its third four-year term in August 2004. It intends to continue its mission under the new framework of the **Global Land Project**

after the LUCC project is finished in 2005. IGU-LUCC enjoys wide support from the world geographic community, which has enabled it to organize or sponsor a number of meetings, to publish books and atlases, to promote projects related with land change in many countries, and to disseminate through its network information on the meetings and other activities organized by the LUCC project or other related organizations as well as by itself. The Commission's meetings and publications are closely related with each other, and they together represent its effort to encourage comprehensive regional land-use/cover change studies with strong human dimensional, comparative, and multiple-scale views. **Urbanization**, rural sustainability, technology transfer, core-periphery dichotomy, and land reclamation are among their focal issues. The China-Japan Comparative Study of Land Use/Cover Changes Project (CJLUC, 2001–2005), which IGU-LUCC supports, is an example of such a bi-national regional study. The Commission is also involved in some regional land change community initiatives, such as the "International Colloquium on Land Use/Cover Change and Environmental Problems in Asia," which has been successively held in **Japan**, India, Indonesia, and **China** in order to tackle regional environmental problems. The Commission's concern on land-use policy is shown in its commitment to the ***Land Use Policy*** journal.

The Commission puts high priority on historical information bases for land-use/cover change, which is demonstrated in some of its publications, especially that of the atlas series titled *Land Use/Cover Changes in Selected Regions in the World.* Over twenty-five countries/regions will have participated in this scheme by early 2005, each of which represents substantial research work related to historical information bases. In addition to maps of land use/cover and its change, the following items are included in each contribution as basic elements: data sources and preparation; land-use/cover classification; study methods; important facts, processes, **driving forces**, and mechanisms of land-use/cover changes; and regional differences, similarities, linkages, and patterns.

Further Reading

Bičík, Ivan, Chromý, Pavel, Jančák, Vit, and Helena Janů, eds., *Land Use/Land Cover Changes in the Period of Globalization*, Prague, CZ: Charles University Faculty of Science, International Geographical Union Commission on Land-Use and Land-Cover Change, 2002; Himiyama, Yukio, Hwang, Manik, and Toshiaki Ichinose, eds., *Land-Use Changes in Comparative Perspective*, Enfield, NH: Science Publishers, 2002; Himiyama, Yukio, Mather, Alexander, Bičík, Ivan, and Elena V. Milanova, eds., *Land Use/Cover Changes in Selected Regions in the World, Volumes I–IV*, Asahikawa, JP: International Geographical Union Commission on Land-Use and Land-Cover Change, 2001, 2002, 2005; Hwang, Manik, and Yukio Himiyama, eds., *Land Use and Land Cover Change—The Contribution of Geography*, Seoul, Korea: College of Education, Seoul National University, International Geographical Union Commission on Land-Use and Land-Cover Change, 2000; International Geographical Union Commission on Land-Use and Land-Cover Change [Online, October 2004], IGU-LUCC Web Site, www.abdn.ac.uk/~geo028/IGULUCC; Mather, Alexander, ed., *Land Use and Rural Sustainability*, Proceedings of Conference on Land Use and Rural Sustainability, Aberdeen, Scotland 10–14, August 2004, Aberdeen: University of Aberdeen, 2005; Milanova, Elena, Himiyama, Yukio, and Ivan Bičík, eds., *Understanding Land-Use and Land-Cover Change in Global and Regional Context*, Enfield, NH: Science Publishers, 2005; Singh, R. B., Fox, Jefferson, and Yukio Himiyama, eds., *Land Use and Cover Change*, Enfield, NH: Science Publishers, 2001.

YUKIO HIMIYAMA

International Geosphere-Biosphere Programme. One of the four international research programs in global change science. The Programme, or IGBP, was established in 1986 by the International Council for Science to address the need for improved scientific understanding of the interactions between humans and the earth's environment. An international and interdisciplinary scientific research effort was required because acquiring this scientific understanding was well beyond the scope of individual countries and regions as well as beyond individual disciplines, and beyond the traditional divide between the natural and the social sciences. The objectives of IGBP are to describe and understand the interactive physical, chemical, and biological processes that regulate the earth system, the unique environment it provides for life, the changes that are occurring in this system, and the manner in which they are influenced by human actions.

IGBP has its science currently organized around six projects, reflecting the major compartments of the earth system (land, oceans, atmosphere) and their interfaces (land-atmosphere, land-ocean, and atmosphere-ocean). Additionally, two framework integration activities deal with paleo-aspects of the earth system and with global, integrative **modeling**. Together, these activities aim to do the following:

1. Develop common international frameworks for collaborative research based on agreed agendas
2. Form research networks to tackle focused scientific questions
3. Promote standardized methodologies
4. Guide and facilitate the construction of global databases
5. Undertake model intercomparisons and comparisons with data
6. Facilitate efficient patterns of resource allocation
7. Undertake analysis, synthesis, and integration activities on broad earth system themes

Each IGBP project is guided by a scientific steering committee that undertakes the detailed planning and implementation. The projects are supported by International Project Offices (IPO), whose staff coordinate and facilitate each project's work. The network of IPOs and the IGBP secretariat together form the basic infrastructure that supports IGBP's research activities.

Together with **DIVERSITAS**, the **International Human Dimensions Programme on Global Environmental Change**, and the World Climate Research Programme, IGBP and these programs form the **Earth System Science Partnership**(ESS-P) for the integrated study of the earth system. Several activities are promoted under ESS-P, namely earth system analysis and modeling; joint projects on issues of global sustainability (**carbon cycle** and energy systems, food systems, water resources and **water-land interlinkages**, and **human**

health); regional activities, including capacity building, networking, and integrated regional studies; and Global Change Open Science Conferences, the first of which was held in Amsterdam in 2001.

See also: **Global Land Cover Map of the Year 2000; Global Land Project; GLOBCARBON; GLOBCOVER; Integrated Land Ecosystem-Atmosphere Processes Study; Land-Ocean Interactions in the Coastal Zone (LOICZ) Project; Land-Use/Cover Change (LUCC) Project; Past Global Changes (PAGES) Project.**

Further Reading

International Geosphere-Biosphere Programme Secretariat [Online, April 2004], The International Geosphere-Biosphere Programme (IGBP) Web Site, www.igbp.kva.se; Steffen, Will, Jäger, Jill, Carson, David C., and Clare Bradshaw, eds. *Challenges of a Changing Earth,* Proceedings of the Global Change Open Science Conference, Amsterdam, NL, 10–13 July 2001. Berlin: Springer Verlag, 2002; Steffen, Will, Sanderson, Angelina, Tyson, Peter D., Jäger, Jill, Matson, Pamela A., Moore III, Berrien, Oldfield, Frank et al., *Global Change and the Earth System: A Planet under Pressure,* Berlin: Springer Verlag, 2004.

ALMUT ARNETH

International Human Dimensions Programme on Global Environmental Change. An international, interdisciplinary, non-governmental science organization dedicated to promoting and coordinating research, capacity building, and networking on the human dimensions of global environmental change. Initially launched in 1990 by the International Social Science Council (ISSC), the Human Dimensions of Global Environmental Change Programme (HDP) was designed as the social science partner to the other global change programs (**DIVERSITAS**, **International Geosphere-Biosphere Programme** [IGBP] and World Climate Research Programme). In 1996, the International Council for Science (ICSU) joined ISSC as a co-sponsor, and the name changed to International Human Dimensions Programme on Global Environmental Change (IHDP).

IHDP's mission is to generate scientific, policy-relevant knowledge on coupled human-environmental systems and achieve comprehensive understanding of global environmental change processes and their consequences for sustainable development. It makes contributions to explore the anthropogenic **driving forces** of global environmental change, the impact of such change on human welfare, and societal responses to mitigate and adapt to **global environmental change**.

IHDP has four established core research projects fostering research from diverse disciplinary and regional perspectives and leading to a broad range of products:

1. Global Environmental Change and Human Security, launched in 1999, has its main thematic areas on **vulnerability**, and environmental conflict and cooperation.
2. Industrial Transformation, launched in 2000, analyzes the transformation of socio-economic systems (in particular **thresholds** and transitions) in

order to explore pathways to decouple **economic growth** from the related environmental degradation.

3. Institutional Dimensions of Global Environmental Change, launched in 1999, studies the role of **institutions** in causing and responding to global environmental change, the effectiveness of institutional innovations, and the prospects of (re)designing institutions to confront environmental challenges.
4. The **Land-Use/Cover Change (LUCC) project**, co-sponsored with IGBP and launched in 1995, focuses on the analysis of land-use and land-cover dynamics (comparison of case studies) and the development of regional and global integrated models.

In addition, since 2004, IHDP has been the scientific co-sponsor (together with IGBP) of the "new" **Land-Ocean Interactions in the Coastal Zone (LOICZ) project**. It provides an integrated framework to address the sustainable human management of coastal systems (with particular emphasis on vulnerability of coasts and risks for human uses). At the beginning of 2005, two new core research projects will be launched: the Global Environmental Change and **Urbanization** Project and the **Global Land Project** (the latter co-sponsored with IGBP).

IHDP's core research projects are linked by four crosscutting questions that crystallize key aspects of human dimensions research. These are vulnerability/**resilience**/adaptation, thresholds/transitions, governance, and knowledge/social learning. Besides coordinating research, IHDP is actively engaged in capacity building (e.g., the biennial International Human Dimensions Workshops with different thematic foci). IHDP is also co-organizer of the biennial Open Meetings of the Human Dimensions of Global Environmental Change Research Community, which have been of key relevance for networking and the development of integrative global change research agendas. IHDP is an active partner of the alliance of the four global change programs called the **Earth System Science Partnership**.

See also: **Global Environmental Change and Food Systems (GECAFS) Project; Population-Environment Research Network.**

Further Reading

International Human Dimensions Programme on Global Environmental Change [Online, January 2005], The IHDP Web Site, www.ihdp.org.

BARBARA GÖBEL

International Institute for Applied Systems Analysis. A nongovernmental research organization located in Laxenburg, Austria, near Vienna, which is sponsored by sixteen national member organizations in countries from Africa, Asia, **Europe**, and North America. The research of the International Institute for Applied Systems Analysis (IIASA) covers international, interdisciplinary, and policy-related science on environmental, economic, technological, and social issues in the context of human dimensions of global change. The scientific activities relate to interdisciplinary topics of global concern with respect to a changing world (environment, energy, water, terrestrial biota, demography, land-use/**land-cover change**, theory of systems analysis, etc.). The work carried out is based on systems analysis approaches and advanced analytical approaches. The methods and tools generated are useful for both decision-makers and the scientific community. Founded in 1972, IIASA has become a reputed center for innovative research, international interdisciplinary collaboration, conferences and workshops, and networking among scientists around the world.

See also: **Agricultural Intensification.**

Further Reading

The International Institute for Applied Systems Analysis [Online, December 2004], The IIASA Web Site, www.iiasa.ac.at.

ANATOLY SHVIDENKO

Investments. Capital resources committed with the expectation that their market value will increase over time. Investment is one of the primary categories for analyzing the circulation of capital within an economy (along with production, trade, and distribution) and can take several forms, including foreign direct investment (FDI), portfolio investment (equities), and bank credits. The growth and geographical spread of cross-border investment flows are central to the historical development of the global economy. Capital export from northern **Europe** during the nineteenth century incorporated the peripheries of southern Europe, **Latin America**, Asia, and coastal Africa into the global economy as commodity producers, and drove extensive land use and **land-cover change** in the form of plantations, ranches, mines, and infrastructure (railroads and ports). FDI and portfolio investment flows surged in the 1990s following the adoption of **economic liberalization** policies in developing countries and the de-regulation of financial markets.

It is widely recognized that the temporal volatility and geographical mobility of capital exert a significant influence on land use. There are, however, few systematic studies of investment's role as a **driving force** of environmental change. During the 1980s researchers began to consider the role of international debt (i.e., banking investments) and structural adjustment in the developing world in driving landscape **conversion** for the production of export crops. The Asian financial crisis of the late 1990s demonstrated how the spatial switching of portfolio investment could have serious socio-economic and environmental consequences, as people (and governments) turned to the exploitation of their environment to stave off the economic effects of falling currencies. It is the multinational corporation that has received the most attention in connection with the environmental effects of investment: recent work centers on the potential for leveraging the interests of shareholders—and, more broadly, stakeholders—to better align corporate and environmental performance (see, for example, work on corporate social responsibility). By contrast, there has been comparatively little work on global institutional initiatives to channel investment into environmentally responsible projects: implemented by the World Bank, the **United Nations Environment Programme**, and United Nations Development Programme, the Global Environmental Facility funds the incremental costs of implementing national development projects (such as power generation) in ways that achieve global environmental benefits (by covering the difference in costs between conventional fossil fuel generation and less polluting, renewable power options).

A more comprehensive approach to investment would address two objectives: address both investment and disinvestment (abandoned farms and contaminated industrial sites, such as New York's infamous Love Canal, attest to **environmental changes** that can occur when capital is withdrawn); and focus on how capital flows condition land use across space and over time. Investment exerts an exogenous influence on the dynamics of local and regional land-cover change and structures place-specific ecologies (e.g., tropical **forests** and **mountain** environments) according to the interests and objectives of investors located elsewhere. To realize an increase in the value of an investment, particular patterns of land use (and, more generally, whole ecologies) must be maintained over time, often against natural processes. In **agriculture**, for example, the need to demonstrate a return (to shareholders, or to repay bank loans) bounds decisions about what crop to grow, the extent to which soil fertility is mined rather than stewarded, the application of inputs (pesticides, seed stock, herbicides, etc.), and even the topography of the landscape (e.g., re-contouring, irrigation, or drainage).

See also: **Agent-Based Model; Agro-Industry; Cattle Ranching; Coffee; Corporate Strategies; Decision-Making; Degradation Narrative; Economic Liberalization; Economic Restructuring; Globalization; Land Abandonment; Land Privatization; Land Rehabilitation; Land Rights; Land Tenure; Mediating Factor; Mineral Extraction; Multiplier Effect; Urbanization.**

Further Reading

Cronon, William, *Nature's Metropolis: Chicago and the Great West*, New York: W.W. Norton, 1991; Dicken, Peter, *Global Shift: Reshaping the Global Economic Map in the 21st Century*, 4th ed., New York: Guilford Press, 2003; Tucker, Richard, *Insatiable Appetite: The United States and the Ecological Destruction of the Tropical World*, Berkeley, CA: University of California Press, 2000.

GAVIN BRIDGE

J–K

Japan. A mountainous island country located in the northern middle latitude to the east of the Eurasian continent facing the Pacific Ocean, with about two-thirds of its land on forest-covered highlands, while its narrow lowlands are used intensively for urban, agricultural, and other uses. The geographical setting of Japan offers a wealth of water, heat energy, oceanic, and volcanic resources, but the very conditions that support the people sometimes cause disasters. Land-use/cover change in Japan has been under the influence of these conditions, and driven by the changing population, people's values, international as well as domestic socio-economic conditions, and government policy.

Before the Meiji Restoration (–1867)

The beginning of the *Tokugawa* (*Edo*) period (1603–1867) marked the end of the century-long *Sengoku* period, or period of war. During the *Sengoku* period, feudal lords made efforts to reclaim land for **agriculture** in order to consolidate the economic bases of their territories. Land reclamation during this period was mainly located in the warmer western Japan. Toward the end of the *Sengoku* period, they also became increasingly committed to afforestation and forest **conservation** in order to rehabilitate their war-worn territories.

During much of the *Tokugawa* period, from 1641 to 1854, the government took a strict policy of national seclusion. The majority of people lived from agriculture, which was then more or less sustainable, and the carrying capacity of the country was determined on the basis of local renewable resources and pre-industrial technology. In the early *Tokugawa* period, until about 1720, land reclamation for cultivation and other agricultural development gained momentum. This period is characterized by large-scale irrigation works and reclamation of bay-head areas, ponds, lakes, and estuaries, which were backed by the capital provided by the emerging merchant class. Agricultural development in the cooler Tohoku, or northeast, district is also a notable event of this period. Most of the alluvial plains in the country except on the island of Hokkaido came to be devoted to paddy field, and the area of paddy field increased substantially during this period.

During the remaining *Tokugawa* period after 1720, places that had been considered of limited agricultural use, such as lower slope lands around volcanoes and diluvial terraces, were turned more vigorously into paddy fields. However, large-scale projects decreased on the whole, partly because of rising costs and natural disasters, and partly because of the lack of potential sites for land reclamation.

The middle of the nineteenth century is considered to be the period when the long-lasting *Tokugawa* period and the policy of national seclusion neared their ends; that is, the period represents the last days of the closed Japan. The locations of the major urban centers at that time are almost identical to those of today, except Hokkaido, suggesting that the inter-urban system of Japan had been established by then. The table titled "Land use structure in Japan" shows that **forest** occupied 68.6 percent of the country; while agricultural land, rough land, and settlement occupied 14.4 percent, 11.8 percent, and 1.7 percent, respectively.

After the Meiji Restoration (1868–)

The *Meiji Restoration*, which was the result of growing pressure from the great powers and the change of the domestic power structure, converted the whole structure of the country, and the population and industrial activities started to grow rapidly. The present population of Japan is 127,590,000 (2004); this is four times that of the late *Tokugawa* period. These changes influenced land use enormously, as shown below.

Urban Land Use

Urbanization, or expansion of settlement, was one of the most conspicuous land-use/cover changes in Japan during the twentieth century, and it was particularly eminent in the major plains, especially in the Tokyo, Kei-Han-Shin, and Nagoya metropolitan areas. Urban expansion is not only an extremely rapid land change itself, but also a major driver of other types of land-use change. According to the table, the increase in the rate of settlement area for 1850–1900 was only 4.8 percent, while it rose to 51.4 percent for 1900–1950 and even 108 percent for 1950–1985, showing an increasing speed of settlement growth. The population began to concentrate in large urban centers in the latter half of the 1950s, when the *Kodo Keizai Seicho*, or period of high **economic growth**, set in. The figure below shows that conversion of

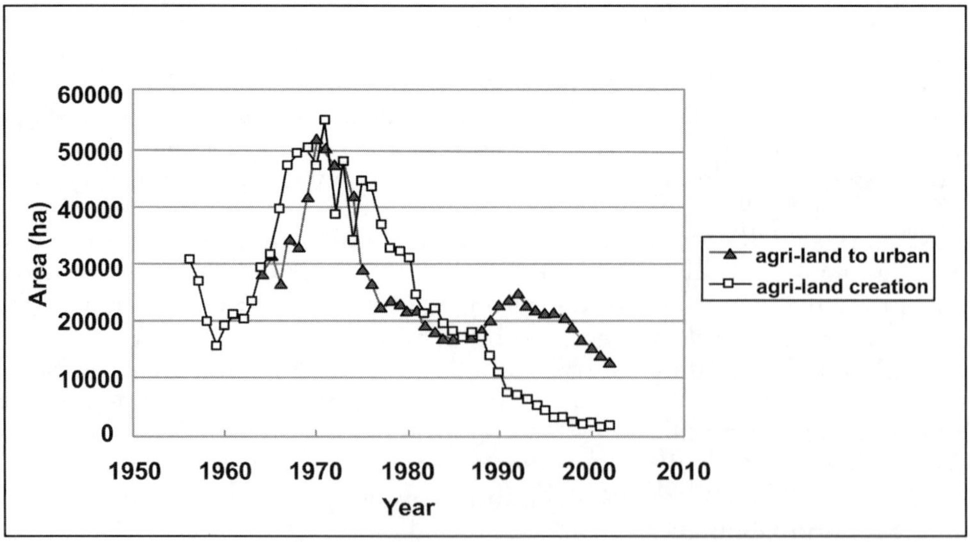

Agricultural land creation and conversion to urban use in Japan (after the Ministry of Agriculture, Forestry, and Fisheries, Japan)

agricultural land to urban land was extremely high from the late 1960s to the early 1970s, when the *Retto-Kaizo*, or rebuilding Japan, boom prevailed. This is the period when ordinary people started to use private cars and many moved to the suburbs, where land prices and rent were lower and the environment was better than that of inner-city areas. This out-migration extended urban areas enormously, but was somewhat restrained after the oil shock of 1973. The "bubble economy" around 1990 activated urban development, which invaded agricultural land; but after its collapse in late 1991, urban expansion was in constant decline until today.

Urban expansion in Japan was made possible at the sacrifice of not only agricultural land, but also other land uses, including inland water surface, **wetland**, brushwood and rough land, and shallow coastal waters, some of which are regarded as ecologically important. Agricultural land adjacent to urban areas fell prey to **urban sprawl**, but the effect of urban expansion on other land uses was largely indirect during the 1960s to 1980s, that is, it was intermediated by agricultural land, which invaded other uses when being invaded by urban expansion, as the figure describes.

Agricultural Land Use

The percentage of agricultural land in Japan increased from 14.4 percent around 1850 to 16.7 percent around 1900, but has been relatively stable since then at about 17 percent, thanks to the government's determination to maintain the figure by land reclamation using huge amount of public money.

The increase in agricultural land before 1900 is attributed mainly to that of dry fields in Hokkaido. The development of new agricultural land in Hokkaido continued well into the 1980s and was supported by strong government incentives, filling the loss of existing agricultural land in other parts of the country. The Kanto Plain, which is the largest plain in Japan and has Tokyo and many other cities in it, lost large areas of agricultural land for the sake of urban expansion, but the new development of agricultural land in other parts of the plain more or less compensated the loss.

Rice has long been the staple food in Japan, and it has been self-sufficient until today except for some extreme years. The area of paddy field reached its maximum in the 1960s at about 35,000 km^2 thanks to the government's agricultural development policy. It started to decline in the early 1970s because of a shift of the policy to ease overproduction of rice, and dropped to as low as 2,660 km^2 in 2000. However, paddy-field area during the twentieth century was relatively stable at around 30,000 km^2.

The increase of orchard after the Second World War was tremendous. It increased by about 500 percent between about 1950 and 1985. The increase was partly due to growing fruit **consumption** and partly due to the government's policy. Mulberry gardens spread in the late nineteenth and early twentieth centuries, and receded rapidly afterward. Silk was one of the most important export goods in Japan before the war; but the war, which stopped Japanese silk exports to the **United States of America**, and the invention of nylon and other synthetic fibers destroyed the industry.

Forest and Wetland

The total area of forest in Japan has been fairly stable at around 66 percent since about 1850, as shown in the table. Some forests were converted to agricultural land or settlement, but there were also efforts at afforestation. Both the government and the landowners in mountainous areas took action in forest rehabilitation after World War II. Many farmer-landowners planted needle-leaved trees mainly because of the sharp increase of timber prices due to the vast shortage of timber during the post-war period, and because of the governmental subsidies as well. The boom of afforestation continued into the *Kodo Keizai Seicho* period, because the progress of economic activities created high demand for timber, and the price of timber rose up further. Thus, the areas deforested before or during the war had been converted to forest by the 1960s. Thanks to such forest rehabilitation, the vast decrease of forest in Hokkaido because of agricultural development during the twentieth century was more or less offset.

Land use structure in Japan ca. 1850, 1900, 1950, and 1985 (in Percent)

Land Use	ca. 1850	ca. 1900	ca. 1950	ca. 1985
Urban	2.86	4.16	6.10	10.11
Settlement	1.65	1.73	2.62	5.45
Road	1.22	2.32	3.23	4.41
Railway	0.0	0.11	0.25	0.24
Agricultural	14.38	16.72	16.94	17.21
Paddy field	9.66	9.30	9.74	9.59
Dry field	4.71	6.24	5.98	5.85
Mulberry garden	0.0	0.92	0.95	0.39
Tea garden	0.0	0.04	0.04	0.13
Orchard	0.0	0.11	0.20	1.22
Other tree crops	0.0	0.10	0.04	0.04
Forest	68.59	65.44	66.88	66.58
Broad leaved forest	30.07	26.49	25.49	14.25
Coniferous forest	12.30	11.85	11.76	10.34
Mixed forest	25.36	26.27	28.91	40.61
Bamboo	0.87	0.83	0.72	1.37
Other	14.16	13.68	10.08	6.10
Rough land	11.84	11.23	7.84	3.68
Marshy land	0.43	0.41	0.27	0.20
Water	1.89	1.94	1.95	2.04
Golf course	0.0	0.0	0.0	0.18
Other	0.0	0.10	0.02	0.0

(The data for ca. 1900, ca. 1950, and ca. 1985 have been provided by the author mainly based on 1:50,000 topographic maps, while those for ca. 1850 by S. Arizono. See *Atlas—Environmental Changes in Modern Japan* (1995) Y. Himiyama et al., eds. Asakura Shoten, Tokyo, and *Land Use and Cover Change* (2001) R.B. Singh, J. Fox, Y. Himiyama eds., Science Publishers.)

The government decided in 1963 to start importing cheap timber from overseas. It lowered the price of domestic timber in the 1970s and 1980s, and many forest owners in **mountain** villages abandoned the management of their planted trees, which needed to be cared for. The management of the newly planted areas is a very important issue not only for the maintenance of forest resources, but also for the protection of the mountain environments. It is also important to note that Japan has been one of the greatest importers of timber and timber products for decades, and it has been responsible at least partly for the extensive **deforestation** of tropical forest in Asia, while its huge domestic timber resource remains dormant. Widespread development of leisure facilities such as golf courses and ski grounds in the 1980s and the 1990s has also added new environmental and economic dimensions to the use of forest in Japan.

Most of the wetland in Japan, excepting Hokkaido, was reclaimed mainly for agricultural purposes before the Meiji Restoration, and only about 1,520 km^2 remained at the beginning of the twentieth century. The area is now 730 km^2, over 90 percent of which is in Hokkaido. The Kushiro Mire, the largest wetland in the country, was registered with the **RAMSAR convention on wetlands** in 1980 and became the first wetland national park in Japan in 1987.

See also: **Urban-Rural Fringe.**

Further Reading

Himiyama, Yukio, Hwang, Manik and Toshiaki Ichinose, eds., *Land-Use Changes in Comparative Perspective*, Enfield, NH: Science Publishers, 2002; Kornhauser, David, *Japan: Geographical Background to Urban-Industrial Development*, 2nd ed., Harlow, UK: Longman Scientific & Technical, 1982; Singh, R. B., Fox, Jefferson, and Yukio Himiyama, eds., *Land Use and Cover Change*, Enfield, NH: Science Publishers, 2001; Totman, Conrad, *The Green Archipelago: Forestry in Preindustrial Japan*, Berkeley, Los Angeles: University of California Press, 1989.

YUKIO HIMIYAMA

JERS Radar Mosaics. A series of satellite image mosaic products developed in collaboration with the newly named Japanese Aerospace Exploration Agency (JAXA), focusing on the mapping of **forests** at the continental **scale**. The regions that were mapped from 1995 to 1997 were Central and **South America**, western and central Africa, and Southeast Asia. The boreal forests of North America were mapped between 1997 and 1998. The mosaics have been constructed from radar images acquired at L-band (a wavelength of twenty-three centimeters) by the JERS satellite. Radar data are sensitive to forest parameters (species and ground cover) and surface moisture levels. The mosaics are available, at no cost, through the Internet at a mapping resolution of approximately 100 meters. The products have been used extensively to map the land use and land cover of tropical rain forest and hydrological applications by proving information on **deforestation**, quantities of **biomass**, and flooding.

See also: **Global Land Cover Map of the Year 2000; Remote Sensing.**

Further Reading

Japanese Aerospace Exploration Agency [Online, September 2004], The JERS-1 SAR Global Forest Mapping Program Web Site, www.eorc.nasda.go.jp/JERS-1/GFMPl.

KEVIN J. TANSEY

Joint Research Centre (JRC) of the European Commission. A directorate-general of the European Commission (EC) with the mission to provide customer-driven scientific and technical support for the conception, development, implementation, and **monitoring** of European Union (EU) policies to tackle transnational and global problems. Activities of the Joint Research Centre (JRC) include development of common databases for harmonized exchange of information across **Europe**, **validation** of methods in areas of concern to citizens (such as environmental protection), and development of reference measurements systems using highly accurate facilities.

The JRC functions as a reference center of science and technology for the EU. Close to the policymaking process, it serves the common interest of its member states, while being independent of special interests, whether private or national.

The seven JRC institutes are located on five separate sites and include the Institute for Environment and Sustainability (IES) in Italy. Among its key competencies, the IES focuses on providing accurate information on changes in world **vegetation** cover and the effect on the regional and global environment; and on research related to spatial characterization of European territory, land resources, and **biodiversity**; natural hazards; and protection of soil resources.

See also: **Coordination of Information on the Environment (CORINE) Database; GLOBCOVER; Global Land Cover Map of the Year 2000; Tropical Ecosystem Environment Observations by Satellite (TREES) Project.**

Further Reading

Joint Research Centre of the European Commission [Online, October 2005], The JRC Web Site, www.jrc.cec.eu.int.

FRÉDÉRIC ACHARD

Journal of Arid Environments. A monthly international journal that publishes original scientific and technical research articles and reviews on climate, geomorphology, geology, geography, botany, zoology, anthropology, human ecology, sociology, and technical development in arid, semiarid, and desert environments. Specific research areas include **soil erosion**, deposition of sediments, dust storms, dunes, climatology, **vegetation** studies, the impact of grazing, wildlife **conservation**, desert peoples, oasis settlement, land classification, land improvement, water control, **remote sensing**, and **desertification**.

See also: **Climate Impacts; Land Degradation; Land Rehabilitation; Regrowth.**

Further Reading

Elsevier [Online, December 2004], The Journal of Arid Environments Web Site, www.elsevier.com.

HELMUT GEIST

Journal of Land Use Science. A new journal for 2006 that promotes "land use science" as an integrated approach that couples natural and human systems, thus fostering interdisciplinary collaboration of land-change scientists (i.e., social, economic, behavioral, ecological, biological, and atmospheric scientists).

See also: **Global Land Project.**

Further Reading

Taylor and Francis [Online, October 2005], The *Journal of Land Use Science* Web site, www.tandf.co.uk/Journals.

HELMUT GEIST

Kyoto Protocol. A legally binding commitment on February 16, 2005, for developed countries ("Annex I") as part of the 128 parties of the convention, agreed by consensus in 1997, to reduce their collective emissions of six **greenhouse gases** by at least 5.2 percent by 2008–2012 compared to 1990 levels. The reports published by the **Intergovernmental Panel on Climate**

Change (IPCC) have shown that the biosphere is a strong determinant of the chemical composition of the atmosphere. The IPCC includes human activities that change the way land is used in national calculations of greenhouse gas emissions required for compliance with the United Nations Framework Convention on Climate Change (UNFCCC) under its article 4. The Kyoto Protocol makes provision for the use of biological sources and sinks to meet commitments. Annex-I parties to the Protocol are thus required to monitor and measure changes in carbon stocks and in greenhouse gas emissions resulting from human-induced land use, land-use change, and **forestry** (LULUCF) project activities under Article 3, Article 6 ("Joint Implementation"), and Article 12 ("Clean Development Mechanism"). Moreover, countries can transfer the issued carbon credits.

Under Article 3.3 of the Protocol, the parties decided that greenhouse gas removals and emissions through certain activities—namely, afforestation and **reforestation**, since 1990—were accounted for in meeting the Protocol's emissions targets. Conversely, activities that deplete forests, namely **deforestation**, will be subtracted from the amount of emissions that an Annex I party may emit over its commitment period.

The Kyoto Protocol also establishes "flexible mechanisms," including an emissions trading regime and the so-called Joint Implementation (JI) under Article 6 and Clean Development Mechanism (CDM) under Article 12. Developed countries may reach their greenhouse gas reductions commitments not only by decreasing national emissions but also by implementing projects through these flexible mechanisms, including ones that increase sinks of carbon. The JI and CDM are mechanisms under which the cooperative production and transfer of carbon credits between Annex-I parties or with non–Annex I parties are encouraged. Annex I parties will be credited for carbon sequestered in approved projects.

Verification and review of project-based LULUCF activities under the Kyoto Protocol and in particular its Clean Development Mechanism can be achieved through the use of satellite **remote sensing** monitoring systems. In particular, technical capabilities exist now to ascertain with a high degree of certainty (forest versus non-forest), at a satisfactory spatial resolution (20–30 meters) and at a low cost (most of the data are freely accessible from the Web), the land use for the beginning of the year 1990 for virtually any place on the face of the earth and any following land-use/cover change.

See also: **Atmosphere-Land Interlinkages; Carbon Cycle; Carbon Sequestration; Climate Impacts; Monitoring.**

Further Reading

Mollicone, Danilo, Achard, Frédéric, Eva, Hugh, Belward, Alan, Federici, Sandro, Lumicisi, Antonio, Rizzo, Valeria, Stibig, Hans-Jürgen, and Riccardo Valentini, *Land Use Change Monitoring in the Framework of the UNFCCC and Its Kyoto Protocol: Report on Current Capabilities of Satellite Remote Sensing Technology*, Luxembourg: European Commission, 2003 [Online, October 2005], www.gvm.jrc.it/tem/PDF_publis/2003/Mollicone_COP9_2003.pdf; United Nations Framework Convention on Climate Change [Online, November 2004], The UNFCCC Web Site, http://unfccc.int.

FRÉDÉRIC ACHARD, DANILO MOLLICONE

Our Earth's Changing Land

International Geographical Union
COMMISSION ON LAND USE/COVER CHANGE

This work was supported by the International Geographical Union's Commission on Land Use/Cover Change (IGU-LUCC)

The Commission involves a network of geographers with research interests in land use/cover change. Its objectives are:

- To promote geographical research on land use/cover changes, at scales ranging from the local to the global
- To stimulate the production and the use of land-use information bases of both the present and the past
- To coordinate the comparative study and the model study of land use/cover changes and their driving forces in different regions.

Contents

List of Entries

Guide to Related Topics

Analysis of Land Change

Aerial Photography
Auto-Correlation
Cadastre
Change Detection
Conversion
Cumulative Change
Decision-Making
Disequilibrium Dynamics
Driving Forces
Feedback
Georeferencing
Hot Spot Identification
Human Appropriation of Net Primary Production
Initial Conditions
Land Cover Classification System
Land Evaluation
Land-Use History
Land-Use System
Land-Use Transition
Mediating Factor
Meta-Analysis
Metabolism
Mode of Interaction
Multiplier Effect
Paleo Perspectives
Pattern Metrics
Pattern to Process
Pixelizing the Social
Proximate Causes
Remote Sensing
Scale
Socializing the Pixel
Spatial Diffusion
Syndrome
Systemic Change
Thermal Band Analysis
Tradeoffs
Transition Matrix
Validation

Biomass Burning

Advanced Very High Resolution Radiometer
Along Track Scanning Radiometers World Fire Atlas
Alternatives to Slash-and-Burn (ASB) Programme
Amazonia
Atmosphere-Land Interlinkages
Biodiversity
Biomass
Boreal Zone
Brazilian Cerrado
Canada
Carbon Cycle
Climate Impacts
Cumulative Change
Degradation Narrative
Disequilibrium Dynamics
Ecological Colonization
Fire
Global Burnt Area 2000
Global Observation of Forest and Land Cover Dynamics
GLOBCARBON
Grassland
Great Plains
Hot Spots of Land-Cover Change
Human Impact on Terrestrial Ecosystems (HITE) Initiative
LANDFIRE Project
Modification
Mountains
Nutrient Cycle
Paleo Perspectives
Pasture
Pristine Myth
Remote Sensing
Russia
Savannization
Secondary Vegetation
Slash-and-Burn Agriculture
Southern Africa
Southern Cone Region
Taiga
Threshold

Tropical Humid Forest
Turnover
United States of America
Vegetation

Case Studies

Amazonia
Boreal Zone
Brazilian Cerrado
Canada
China
Coastal Zone
East-Central Europe
Europe
Great Plains
Himalayas
Japan
Latin America
Meta-Analysis
Miombo
Mountains
River Basin
Russia
South Africa
South America
Southeast Asia, Mainland
Southeast Europe
Southern Africa
Southern Cone Region
Sudan-Sahel
Taiga
United States of America
West-Central Europe
Western Europe

Causes

Access
Agricultural Intensification
Agriculture
Agro-Industry
Arena of Land Conflict
Cattle Ranching
Colonization
Consumption
Contract Farming
Conversion
Cultural Factors
Cumulative Change
Decision-Making
Deforestation
Degradation Narrative
Desertification
Driving Forces
Economic Growth
Economic Liberalization
Economic Livelihood
Economic Restructuring
Endogenous Factor
Exogenous Factor
Extensification
Fire
Forest Degradation
Forest Transition
Global Environmental Change
Globalization
Green Revolution
Industrialization
Institutions
Investments
Land Abandonment
Land Concentration
Land Degradation
Land Fragmentation
Land Rehabilitation
Land Rent
Land Rights
Land Tenure
Land-Use Policies
Land-Use Transition
Leapfrogging
Mineral Extraction
Mode of Interaction
Modification
Pasture
Population Dynamics
Poverty
Property
Proximate Causes
Reforestation
Regrowth
Remittances Landscape
Slash-and-Burn Agriculture
Soil Degradation
Soil Erosion
Subsistence Agriculture
Suburbanization
Swidden Cultivation
Syndrome
Systemic Change
Technological-Scientific Revolution
Tragedy of Enclosure
Tragedy of the Commons
Transhumance
Urbanization
Urban Sprawl

Conventions and Environmental Agreements

Agenda 21
Conservation
Cropland Change
Forest
Institutions
Kyoto Protocol
Public Policy
Land Evaluation
Land-Use Planning
Land-Use Policies
Millennium Ecosystem Assessment
Ramsar Convention on Wetlands
United Nations Convention on Biological Diversity
United Nations Convention to Combat Desertification
United Nations Environment Programme
United Nations Food and Agriculture Organization

Cropland Change

Agricultural Intensification
Agricultural Revolution
Agriculture
Agrodiversity
Agro-Industry
BIOME 300 Project
Biotechnology
Cash Crops
Cassava
Coffee
Colonization
Consultative Group on International Agricultural Research
Consumption
Contract Farming
Conversion
Cotton
Cumulative Change
Decision-Making
Desertification
Disintensification
Domestication
Economic Livelihood
Eutrophication
Extensification
Fallow
Food Crops
Food Security
Globalization
Green Revolution
Hot Spots of Land-Cover Change
Human Appropriation of Net Primary Production
Initial Conditions
International Food Policy Research Institute
Land Abandonment
Land Concentration
Land Degradation
Land Rent
Land Rights
Land Tenure
Landholding
Land-Use History
Land-Use Legacy
Land-Use Policies
Land-Use System
Maize
Modification
Monoculture
Nutrient Cycle
Property
Proximate Causes
Remittances Landscape
Rice
Slash-and-Burn Agriculture
Soil Degradation
Soil Erosion
Spatial Diffusion
Species Extinction
Subsistence Agriculture
Swidden Cultivation
Tobacco
Urban-Rural Fringe
Wheat

Data

Aerial Photography
BIOME 300 Project
Cadastre
Center for International Earth Science Information Network
Center for Sustainability and the Global Environment
Change Detection
Continuous Data
Coordination of Information on the Environment (CORINE) Database
Discrete Data
Earth Resources Observation Systems (EROS) Data Center
Forest
Geographical Information System

Dryland Change

Forest Change

Vegetation
World Agroforestry Centre

Impacts

African Trypanosomiasis
Agrodiversity
Albedo
Atmosphere-Land Interlinkages
Biodiversity
Biodiversity Novelties
Biomass
Carbon Cycle
Carbon Sequestration
Cascades
Climate Impacts
Cumulative Change
Ecological Footprint
Ecosystem Services
Environmental Change
Eutrophication
Exotic Species
Feedback
Fire
Global Environmental Change
GLOBCARBON
Greenhouse Gases
Holocene
Human Appropriation of Net Primary Production
Human Health
Human Immunodeficiency Virus (HIV)/ Acquired Immunodeficiency Syndrome (AIDS)
Land Fragmentation
Land Quality
Malaria
Metabolism
Nutrient Cycle
People at Risk
Salinization
Savannization
Species Extinction
Systemic Change
Threshold
Tradeoffs
Water-Land Interlinkages
Watershed

Journals

Agriculture, Ecosystems & Environment
Ambio, A Journal of the Human Environment
BioScience
Ecology and Society
Environment
Frontiers in Ecology and the Environment
Global Environmental Change
Journal of Arid Environments
Journal of Land Use Science
Land Degradation & Development
Land Use Policy

Land-Cover Conversion and Modification

Agricultural Frontier
Agricultural Intensification
Anthropocene
Arc of Deforestation
BIOME 300 Project
Change Detection
Conversion
Cumulative Change
Deforestation
Desertification
Ecological Colonization
Eutrophication
Extensification
Fallow
Fire
Forest Degradation
Forest Transition
Holocene
Hot Spot Identification
Hot Spots of Land-Cover Change
Human Appropriation of Net Primary Production
Land Abandonment
Land Degradation
Land Rehabilitation
Land-Use Transition
Land-Use/Cover Change (LUCC) Project
Leapfrogging
Mineral Extraction
Modification
Monoculture
NASA Land-Cover/Land-Use Change (LCLUC) Program
Pasture
Peatlands
Proximate Causes
Reforestation
Regrowth
Savannization
Secondary Vegetation
Siltation
Soil Degradation
Soil Erosion

Suburbanization
Sustainable Land Use
Urban Sprawl
Urbanization
Vegetation
Watershed
Wetlands

Land-Cover Products

AFRICOVER
Along Track Scanning Radiometers (ATSR) World Fire Atlas
BIOME 300 Project
Coordination of Information on the Environment (CORINE) Database
Earth Resources Observation Systems (EROS) Data Center
European Remote Sensing (ERS-1/-2) Satellites
Global Burnt Area 2000
Global Land Cover Map of the Year 2000
Global Land Cover Network
Global Terrestrial Observing System
GLOBCARBON
GLOBCOVER
GLOBICE
GLOBSCAR
GLOBWETLANDS
History Database of the Global Environment
Hot Spots of Land-Cover Change
IGBP-DIS Global 1-Km Land Cover Data Set
Sahel Land Cover
Tropical Ecosystem Environment Observations by Satellite (TREES) Project
United States Geological Survey (USGS) Program

Land-Use Dynamics

Access
Agricultural Frontier
Agricultural Intensification
Agricultural Revolution
Agrodiversity
Agroforestry
Agro-Industry
Arena of Land Conflict
Biomass
BIOME 300 Project
Biotechnology
Cattle Ranching
Colonization
Common-Pool Resources
Community Involvement
Conservation
Consumption
Contract Farming
Conversion
Corporate Strategies
Cotton
Cultural Factors
Cumulative Change
Decentralization
Decision-Making
Deforestation
Degradation Narrative
Desertification
Disequilibrium Dynamics
Disintensification
Domestication
Driving Forces
Ecological Colonization
Economic Growth
Economic Liberalization
Economic Livelihood
Economic Restructuring
Exotic Species
Extensification
Fallow
Feedback
Fire
Food Crops
Food Security
Forest Anomalies
Forest Degradation
Forest Transition
Forestry
Globalization
Green Revolution
Green Wall Project
Human Appropriation of Net Primary Production
Human Impact on Terrestrial Ecosystems (HITE) Initiative
Indigenous Knowledge
Industrial Revolution
Industrialization
Initial Conditions
Institutions
Investments
Land Abandonment
Land Concentration
Land Privatization
Land Reform

Land Rent
Land Rights
Land Stability
Land Tenure
Landholding
Land-Use History
Land-Use Legacy
Land-Use Planning
Land-Use Policies
Land-Use System
Land-Use Transition
Leapfrogging
Mineral Extraction
Modification
Monoculture
Multiplier Effect
Non-Timber Forest Products
Participatory Land Management
Pastoral Mobility
Pasture
People at Risk
Perception
Population Dynamics
Poverty
Private Property
Property
Proximate Causes
Resilience
Slash-and-Burn Agriculture
Socializing the Pixel
Spatial Diffusion
Subsistence Agriculture
Suburbanization
Sustainable Land Use
Swidden Cultivation
Syndrome
Technological-Scientific Revolution
Tragedy of Enclosure
Tragedy of the Commons
Transhumance
Urban Sprawl
Urbanization

Models and Scenarios

Agent-Based Model
Cellular Automaton
Century Ecosystem Model
Conversion of Land Use and its Effects (CLUE) Model
Decision-Making
Disequilibrium Dynamics
Integrated Assessment
Integrated Model
Integrated Model to Assess the Global Environment
Intergovernmental Panel on Climate Change
LandSHIFT Model
Land-Use/Cover Change (LUCC) Project
Modeling
Participatory Geographic Information System
Probit/Logit Model
Sahelian Land Use (SALU) Model
Scale
Scenario
Socializing the Pixel
Spatial Diffusion
Tradeoffs
Tragedy of Enclosure
Tragedy of the Commons
Validation
Vulnerability
Yellowstone Model

People

Blaikie, Piers
Boserup, Ester
Brookfield, Harold Chilingworth
Carson, Rachel Louise
Maathai, Wangari M.
Marsh, George Perkins
Sauer, Carl Ortwin
Thünen, Johann Heinrich von

Programs, Projects, and Organizations

Agenda 21
Alternatives to Slash-and-Burn (ASB) Programme
Anthropological Center for Training and Research on Global Environmental Change
BIOME 300 Project
Center for International Earth Science Information Network
Center for International Forestry Research
Center for Sustainability and the Global Environment
Center for the Study of Institutions, Population and Environmental Change
Consultative Group on International Agricultural Research
DIVERSITAS
Earth Resources Observation Systems (EROS) Data Center

Earth System Science Partnership
George Perkins Marsh Institute
Global Environmental Change and Food Systems (GECAFS) Project
Global Land Cover Network
Global Land Project
Global Observation of Forest and Land Cover Dynamics
Global Terrestrial Observing System
Green Wall Project
Human Impact on Terrestrial Ecosystems (HITE) Initiative
Integrated Land Ecosystem-Atmosphere Processes Study
Intergovernmental Panel on Climate Change
International Food Policy Research Institute
International Geographical Union (IGU) Commission on Land-Use and Land-Cover Change
International Geosphere-Biosphere Programme
International Human Dimensions Programme on Global Environmental Change
International Institute for Applied Systems Analysis
Joint Research Centre (JRC) of the European Commission
Land Degradation Assessment in Drylands (LADA) Project
LANDFIRE Project
Land-Ocean Interactions in the Coastal Zone (LOICZ) Project
Land-Use/Cover Change (LUCC) Project
Large-Scale Biosphere-Atmosphere Experiment in Amazonia
Millennium Ecosystem Assessment
NASA Land-Cover/Land-Use Change (LCLUC) Program
National Institute of Public Health and the Environment
Past Global Changes (PAGES) Project
Population-Environment Research Network
Project on People, Land Management and Ecosystem Conservation
Tropical Ecosystem Environment Observations by Satellite (TREES) Project
United Nations Environment Programme
United Nations Food and Agriculture Organization
United States Geological Survey (USGS) Program
World Agroforestry Centre

Rangeland Change

Access
Agricultural Frontier
Amazonia
Arc of Deforestation
Cattle Ranching
Common-Pool Resources
Conversion
Desertification
Disequilibrium Dynamics
Grassland
Great Plains
Hot Spots of Land-Cover Change
Land Concentration
Land Degradation Assessment in Drylands (LADA) Project
Large-Scale Biosphere-Atmosphere Experiment in Amazonia
Modification
Pastoral Mobility
Pasture
Resilience
Savannization
South America
Southern Africa
Sudan-Sahel
Tragedy of the Commons
Transhumance
United Nations Convention to Combat Desertification

Satellite Imagery

Advanced Very High Resolution Radiometer
AFRICOVER
Airborne Light Detection and Ranging
Albedo
Along Track Scanning Radiometers (ATSR) World Fire Atlas
Center for International Earth Science Information Network
Center for Sustainability and the Global Environment
Center for the Study of Institutions, Population and Environmental Change
Change Detection
Continuous Data
Coordination of Information on the Environment (CORINE) Database

Discrete Data
Earth Resources Observation Systems (EROS) Data Center
ENVISAT ASAR
European Remote Sensing (ERS-1/-2) Satellites
Georeferencing
Global Environmental Change and Food Systems (GECAFS) Project
Global Land Cover Network
Global Observation of Forest and Land Cover Dynamics
Global Positioning System
Global Terrestrial Observing System
Hot Spot Identification
IKONOS
JERS Radar Mosaics
Joint Research Centre (JRC) of the European Commission
Land Cover Classification System
Landsat
Medium Resolution Imaging Spectrometer
Moderate Resolution Imaging Spectroradiometer
NASA Land-Cover/Land-Use Change (LCLUC) Program
Normalized Difference Vegetation Index
Pattern Metrics
Pattern to Process
Pixel
Pixelizing the Social
Remote Sensing
Socializing the Pixel
SPOT Vegetation
TERRA-ASTER
Thermal Band Analysis
Transition Matrix
Tropical Dry Forest
Tropical Ecosystem Environment Observations by Satellite (TREES) Project
United States Geological Survey (USGS) Program
Validation

Urban-Industrial Change

Anthropocene
Atmosphere-Land Interlinkages
Carbon Cycle
Cascades
Consumption
Corporate Strategies
Economic Growth
Economic Restructuring
Forest Transition
Globalization
Greenhouse Gases
Industrial Revolution
Industrialization
Institutions
Japan
Kyoto Protocol
Land-Use Transition
Metabolism
Mineral Extraction
Population Dynamics
Remittances Landscape
Suburbanization
Syndrome
Systemic Change
Technological-Scientific Revolution
Urbanization
Urban Sprawl
Urban-Rural Fringe
Western Europe

L

Land Abandonment. The withdrawal of land from production, which often follows marginalization processes in agriculture and rural areas. Land abandonment leads to changes in **vegetation** and landscapes (most of the land would turn into **forest** if left unmanaged). It may provide environmental benefits, in particular if pollution by agricultural chemicals were reduced. However, abandonment may also entail losses of **biodiversity** in cases where much of the land was **grassland** with high nature value for botanical interest, as habitats for birds or traditional landscapes (mainly semi-natural grasslands and other areas important for birds).

Agricultural marginalization is a process that is driven by a combination of social, economic, political, and environmental factors, by which in certain areas **agriculture** ceases to be viable under an existing **land-use system**. Although marginalization of agricultural land is not new, its extent and speed of growth are evident most clearly in remote areas under harsh climatic and soil conditions, but also observed more locally. **Driving forces** behind marginalization of land can be found in a range of factors at various spatial **scales**. The consequences operate on the local and regional scales, making their appearance in a change of the socio-economic position of the rural population in areas in question. However, **vulnerability** to marginalization depends on the local or regional, social, economic, political, and environmental conditions.

Competitive advantages of production may therefore result from a range of socio-economic factors like **access** to markets and the availability of proper **institutions**, skilled labor, credits available to farmers, infrastructure, international politics, **cultural factors**, and human choices. A major driving force behind vulnerability of rural areas (or marginalization of agriculture) therefore is economic marginality. In areas marginal for agriculture, agricultural productivity is often low because of climatic constraints, poor soils, and poor accessibility of agricultural lands or traditional low-input agriculture. Declining trends in agricultural production could result from farming becoming less viable, getting marginalized, and land eventually becoming abandoned. Marginalization has sometimes far-reaching consequences—social, economic, and ecological. Marginalization and abandonment of agricultural land have become a serious threat to **biodiversity** and landscape features as well. The effects of marginalization and abandonment on flora and fauna can vary considerably; the ecological consequences can be complex, site-specific, and far from uniform, even in a relatively small geographical region. Marginalization does not necessarily lead to progressively more extensive farming systems followed by abandonment. There may be a combination of **agricultural intensification** and **extensification** within the same farm or region, or the restructuring of **landholdings** or new land uses such as **forestry**. Both ecological and socio-economic aspects of marginalization could be solved through the provision of payments in support of biodiversity **conservation** and as rewards for income loss due to application of specific land management practices.

See also: **African Trypanosomiasis; Agricultural Frontier; Biodiversity; Coffee; Economic Growth; Food Crops; Forest Transition; Hot Spots of Land-Cover Change; Investments; Mountains; Pasture; Pristine Myth; Reforestation; Remittances Landscape; Secondary Vegetation.**

Further Reading

Baldock, D., Beaufoy, G., Brouwer, Floor, and F. Godeschalk, *Farming at the Margins: Abandonment or Redeployment of Agricultural Land in Europe*, London, UK, the Hague, NL: Institute for European Environmental Policy, Agricultural Economics Research Institute, 1996.

FLOOR BROUWER

Land Concentration. Amassing of large amounts of land in the hands of a few individuals, corporations, or **institutions**. It involves processes that result in a particular structure of bimodal land distribution in which a large number have very little land and a very few have a great deal of land. Land concentration involves two processes: one that leads to the acquisition of land by some, and another that leads to its loss by others. The dynamics of land concentration are largely mediated by economic, social, and political factors. Land concentration can take place in several contexts. In some cases, it occurs in situations in which customary rights are negated to local people and land usurpation takes place by national or local elites through land encroachment. In other situations, particularly in those in which market economies are dominant, land transactions may tend to favor the concentration of land in few hands. Finally, land concentration is a common phenomenon in situations in which there are still open **access** areas, particularly in frontiers for agricultural expansion.

The three different situations often have taken place in tropical contexts where there are strong pressures for forest **conversion** to agricultural uses. In tropical areas, particularly of **Latin America**, land concentration of public land has occurred in different historical contexts, mainly in those where there were increasing economic interests for land as a means of production, or to support speculative purposes. This process, rather than being stopped, was stimulated by political patronage mechanisms in support of national or local elites who concentrated the lands, and who often held the power to legitimate or enforce land ownership structures by legal or semi-legal means. Land concentration has also taken place in areas already occupied by indigenous communities, or small farmers. Land encroachment has been the most-used mechanisms for ranchers and loggers to usurp communal lands, in contexts in which the lack of **property** rights and the absence of legal mechanisms to protect such rights both constituted the norm rather than the exception.

In areas dominated by small-scale farmers, land concentration tends to take place through land transactions. While the reasons that large landholders have to buy land, or to amend new areas to existing rural establishments, are strongly linked to economic expectations, the motivations that land sellers have to sell their land are somewhat ambiguous. On one hand, large landholders (i.e., large-scale agricultural producers or ranchers) buy the land expecting to increase their economic benefit from using such land on productive activities (i.e., raising cattle or producing grains), or prompted by some sort of speculative interest by which they expect either to increase their profits by selling the land at some time in the future, or to obtain some institutional rent from holding, such as **land rights** (i.e., cheap credit). On the other hand, small farmers tend to sell their lands in two different circumstances. First, they sell due to their inability to sustain productive activities in precarious conditions. Second, they sell in cases in which the resources they are offered for the land exceed the estimated benefits they will obtain from working on the land. The two dynamics lead to higher land concentration.

See also: **Agricultural Frontier; Cattle Ranching; Land Fragmentation; Landholdings; Land Privatization; Land Rent; Land Tenure; Mediating Factor; Tragedy of Enclosure.**

PABLO PACHECO

Land Cover Classification System. Land-cover assessment and **monitoring** of its dynamics are essential requirements for the sustainable management of natural resources and for environmental protection. Land-cover data are also essential for environmental, **food security**, and humanitarian programs of many United Nations, international, and national institutions. At present there is no internationally accepted land-cover classification system. Some are more universally applicable than others, but none have been accepted as the international standard. The increasing number of regional and global initiatives involving land-cover mapping and monitoring requires a close interaction among them, in order to ensure their benefits. Since land cover has also become a basic data layer in **geographical information systems**, the synergy among land-cover mapping projects has a high priority.

To solve this problem, the land cover classification system (LCCS) was developed and implemented by the **United Nations Food and Agriculture Organization** (FAO) and the **United Nations Environment Programme** (UNEP) with the financial support of the government of Italy. LCCS was developed during the **AFRICOVER** project in 1994, to allow the harmonization of different nomenclatures, legends, and classifications of twelve African countries. It is a comprehensive methodology for the description, characterization, classification, and comparison of land-cover data, identified anywhere in the world, at any **scale**. LCCS allows the description of different land-cover features in a standardized way.

Classification Methodology

Classification systems exist in hierarchical and non-hierarchical formats. Most systems are hierarchically structured. They start with broad-level classes, which allow further systematic subdivision into more detailed subclasses. The classification process can be done *a posteriori* or *a priori. A posteriori* classification is based upon the definition of classes after the field samples have been collected. Although it is flexible and adaptable, it is often developed for a specific area being described; the classes therefore often need to be adapted for each new location. For such a classification, it is therefore difficult to define standardized classes. In an *a priori* classification system, all the classes are defined before the data is collected. The main advantage is that a standard set of classes and methodology is created, which is independent of the geographic area. The disadvantage is that an enormous number of predefined classes have to be created to allow an accurate description of all the possible land-cover types occurring in the world.

The classification system used in LCCS is an *a priori* classification system, but LCCS is unique as it provides a scale-independent method of classifying land cover. The approach supports all types of land-cover monitoring and enables a comparison of land-cover classes regardless of data source, sector, or country. This is achieved by the use of a set of universally valid classification criteria that uniquely identify the land-cover classes worldwide. The system works on two levels. On the first level, eight major land-cover types are defined. At the second level, land-cover classes are created by the combination of sets of predefined classifiers. These classifiers have been carefully defined to describe different land-cover variations present inside each major land-cover type and also to avoid inappropriate combinations of classifiers. The number of classifiers used determines the level at which the land cover is classified. The system is highly flexible but allows land-cover classes to be clearly characterized, thus providing internal consistency. The system is truly hierarchical and applicable to a variety of mapping scales and in any geographic location. It can be used as a reference standard system because its diagnostic criteria allow correlation with existing classifications and legends.

Software

A user-friendly interface has been built around the set of databases holding the numerous classifiers and attributes. The software facilitates the user and prevents errors in creating the required land-cover classes. The LCCS software contains three main modules. The classification module allows the user to define the land-cover class using two main phases. In the initial dichotomous phase, the major land-cover type is defined by selecting between the two main land-cover types, "primarily vegetated" and "primarily non-vegetated," and then selecting the more detailed classification fields. After having defined the main land-cover type, the user can set up a generic class using only the results of this initial phase; or can proceed to the modular-hierarchical phase, where more detailed land-cover classes can be created by combining different sets of predefined classifiers. The classifiers available depend on the main land-cover type and the classifiers previously selected (see figure).

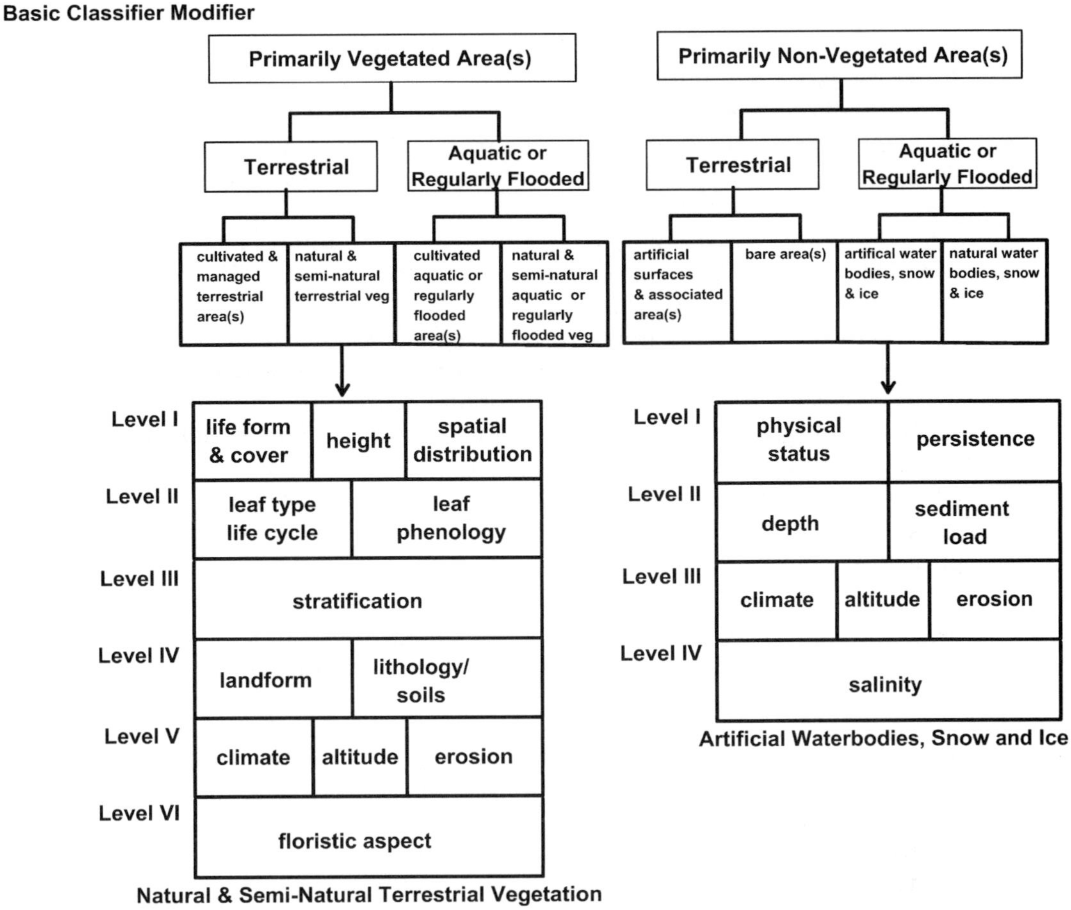

The LCCS Classification Module—the initial dichotomous phase and the modular-hierarchical phase.

The main purpose of the legend module is to take the land-cover classes identified and store them in a hierarchical structure that groups the classes according to the main land-cover type. The legend usually contains only a subset of the classification, namely those classes that are applicable in the area (to be) interpreted or mapped. In addition to providing a hierarchical structure, the legend also offers the ability to display, edit, and add user-defined attributes to a land-cover class. Standard descriptions are provided for the classes identified and the classifiers used, and all this information can be exported in various formats. In the legend section, mixed mapping units can be used, which comprise two, three, or more classes.

The translator module allows the translation of existing classifications and legends into LCCS. The reference base that is created can be used to compare classes of translated classifications and legends and their attributes, including the direct comparison of the individual land-cover classes.

Use of LCCS

The LCCS concept is rapidly being adopted by both national and international initiatives. Organizations that have already adopted LCCS include

FAO, the **Global Land Cover Map of the Year 2000** initiative of the **Joint Research Centre (JRC) of the European Commission**, the Global Land Use Cover Classification (GLUCC) working group of the **International Geosphere-Biosphere Programme**, and UNEP. Through the AFRICOVER project, a number of countries have also adopted LCCS, including Burundi, the Democratic Republic of the Congo, Egypt, Eritrea, Kenya, Rwanda, Somalia, Sudan, Tanzania, and Uganda. Numerous other countries from other regions of the world are now also adopting LCCS to implement their national mapping programs.

In addition, LCCS is one of the main tools of the **Global Land Cover Network** (GLCN), which has been developed by FAO and UNEP to increase the reliability of land-cover information for a large user community. GLCN provides direction, focus, and guidance for harmonization of land-cover mapping based on LCCS. AFRICOVER and ASIACOVER are the first regional outreach programs that are supported by GLCN, but additional networks are now being developed in the Caribbean, Central America, Central Asia, **east-central Europe**, **Latin America**, and the Middle East.

Future Developments

At the beginning of 2005, a new version of LCCS (version 2.0) was released with many developments and improvements, including a new user interface, modified class names and class descriptions, increased availability of classifiers and environmental attributes, and increased flexibility in class creation. The software and user manual of this new version of LCCS will be available in a number of languages including Arabic, English, French, and Spanish.

An LCCS Web Site has now been created with free information, documentation, and software. In addition, online assistance is provided to facilitate users in the development of their own LCCS initiatives and projects. A forum on LCCS has also been created to promote discussions among LCCS users. To further assist users, regional and national training workshops in LCCS and other mapping and GIS tools have, and are, being organized by GLCN with the support of UNEP and the governments of Italy and the Netherlands. In addition, self-tuition tools will soon be available to allow individuals to acquire the skills to use the LCCS software in their own initiatives and programs.

LCCS has the potential to become accepted as the international land-cover classification standard because of its inherent flexibility, applicability in all climatic zones and environmental conditions, and compatibility with the existing classification systems. LCCS has been submitted for approval to become a standard of the International Organization for Standardization (ISO). ISO is the world's largest developer of standards, with a network of 147 national standards institutes with a central secretariat in Geneva, Switzerland.

See also: **Conservation; Remote Sensing.**

Further Reading

Di Gregorio, Antonio, and Louisa J. M. Jansen, *Land Cover Classification System*, Rome: United Nations Food and Agriculture Organization, 2001; Di Gregorio, Antonio, and Louisa J. M. Jansen, *Land Cover Classification System: Classification Concepts and User Manual* [Online, October 2004], The United Nations Food and Agriculture Organization Web Site, www.fao.org/WAICENT/FAOINFO/SUSTDEV/EIdirect/EIre0062.htm; International Organization for Standardization [Online, November

2004], The ISO Web Site, www.iso.org; Global Land Cover Network; The Land Cover Classification System [Online, December 2004], The LCCS Web Site, www.glcn-lccs.org.

SOPHIA LINKE, REUBEN SESSA

Land Degradation. Loss of life-supporting functions due to land use that is coupled with natural processes, which will remain an important global issue for the twenty-first century because of its adverse impact on agronomic productivity, the environment, and its effect on **food security** and the quality of life. The livelihoods of more than 900 million people in some 100 countries are now directly and adversely affected by land degradation due to complex biophysical and socio-economic interactions and regionally varied mechanisms. Unless the current rate of land degradation is slowed and reversed, the food security of humanity will be threatened—and the number of people adversely affected is expected to double to 1.8 billion until 2025.

Concept and Definition

There are biophysical (e.g., land management), socioeconomic (e.g., income and **land tenure**), and political (e.g., incentives and political stability) forces that determine causes and extent of land degradation. Land degradation has no single attribute, but comprises a bunch of processes. Processes such as **soil erosion**, loss of **biodiversity**, or **eutrophication** take place on different time **scales**. Often the degradation effects caused by human activities are delayed, and therefore people are not aware of the potential impacts on land. Apart from degraded ecosystems and deteriorated natural resources (e.g., soil, wood, and species), land degradation has socio-economic impacts (e.g., productivity decline and loss of income) on people.

The main function of land is to support human life, that is, the production function of land is one **perception** of defining land degradation. Apart from the production function there are other important functions of land, however, that sustain human life, such as the recreation function, cultural function, habitat function, and regulation function. An anthropocentric definition of land degradation that integrates the different perceptions of land (use) and keeps a multifunctional perspective on land in mind is the following: land degradation is the loss of a sustained economic, cultural, or ecological function due to human activity in combination with natural processes.

Land degradation results from a "mismatch" between functional capability of land and its utilization by people. Land degradation is therefore closely related to **resilience**, sensitivity, and carrying capacity of land, as well as **vulnerability** of people living on and from these lands.

Causes

The most frequently expressed main causes of land degradation are overgrazing of rangelands, overcultivation of cropland, waterlogging and **salinization** of irrigated land, **deforestation** and overlogging, and pollution and industrial causes. These **proximate causes** include factors such as **agricultural intensification**, cropland expansion, livestock extension, fuelwood extraction, infrastructure extension, settlement extension, non-adapted technologies, habitat fragmentation, and introduction of invasive species.

The underlying **driving forces** comprise elements such as population growth and density, migration patterns, **land-use policies**, tenure rights, institutional capability, education, gender policy, conflicts, and **poverty**. Without a comprehensive and regional specific analysis of proximate causes and underlying **driving forces**, it will be not possible to deal adequately with sustained solutions of land-degradation problems.

Main Processes

The main mechanisms that initiate land degradation include physical, chemical, and biological processes. Important among physical processes are unsustainable use of natural resources, removal of functional landscape elements, and a decline in soil structure, all leading to crusting, compaction, erosion, desertification, anaerobism, and environmental pollution.

Significant chemical processes include acidification, leaching, salinization, decrease in cation retention capacity, and fertility depletion. Biological processes include reduction in carbon stocks in **vegetation** and soils as well as a decline in species diversity. Many of these processes are related to eutrophication of surface water, contamination of groundwater, and emissions of trace gases (CO_2, CH_4, N_2O, NO_x) from terrestrial/aquatic ecosystems to the atmosphere.

The Dynamic of Land Degradation

Land degradation is a phenomenon with a history of centuries or even millennia (e.g., medieval soil erosion and **forest degradation** in Central Europe or historical salinization of arable land in Mesopotamia). Given a long history of land degradation, the threat and the dynamic of land degradation have considerably increased during the last sixty years.

Global **wetlands** have decreased about 50 percent since 1900. The majority of the remaining wetland is degraded or under threat of degradation. The intensity of these problems is closely related to the intensity of human activity in and around the wetlands. Further pressure to drain land for **agriculture** is highest in tropical and subtropical areas of Asia, Africa, and **South America**.

More than 39 percent or about 553 million hectares of primary **forests** are threatened with degradation or extinction by human actions such as logging, mining, clearing for agriculture, and overhunting. In 1992, the World Resources Institute stated that since 1950, about 580 million hectares of fertile land have been lost due to soil erosion caused by deforestation. Overgrazing caused damage on 20 percent or 685 million hectares of the world's **pasture** and rangelands in the last decades. Recent losses have been most severe in Africa and Asia.

According to the **United Nations Environment Programme**, about 19.5 million hectares of different land-cover types are assessed to be degraded due to **urbanization** and industrial processes. Often highly productive agricultural land is lost.

Soil degradation was assessed to have increased by about 2 billion hectares since 1945, which accounts for nearly 18 percent of the earth's vegetated land. As a consequence, productivity has been significantly reduced on roughly 60 percent of these 2 billion hectares. The current annual impact of land degradation on agricultural land is assessed to be about 21 million hectares of land that became uneconomic for crop production, and an additional 6 million hectares of land that are irreversibly lost for production.

Soil erosion is one of the major causes of reduced potential for food production in both developed and developing countries. For example, the **United States of America** has lost approximately one-third of its topsoil since farming began less than 300 years ago, and it continues to lose twelve tons per hectare per year. However, in certain regions of Africa, India, and Central America, soil erosion reaches up to 100 tons per hectare per year, with severe impacts on food security and the regional economy.

On average, the impact of soil degradation on agricultural productivity in recent decades was assessed to be about 0.5 percent per year, which to a large degree foils yield increases as a result of technical development.

The dry areas of the world are the origin of a large number of globally important cereals and food legumes, such as barley, **wheat**, fava beans, and lentils. In these dry areas, biodiversity is seriously eroding through the degradation of natural habitats, the intensification and expansion of cultivation, and overgrazing in natural rangelands.

A Global Overview

Available statistics show a large variation in the extent and rate of land degradation because of different definitions and terminology. A study, for example, by Harold E. Dregne and Nan-Ting Chou focused on land degradation (soil degradation and degradation of vegetation) in drylands, which is known as **desertification**, while the report "Global Assessment of Human Induced Soil Degradation" (GLASOD) by L. Roel Oldeman and colleagues emphasizes different aspects of soil degradation. The tables "Estimates of all degraded lands in dry areas" and "Assessment of soil degradation" give an overview of both studies, while the table "Regional soil degradation of main land-cover types" highlights regional soil degradation of main land-cover types according to the GLASOD assessment. Both assessments, however, are qualitative assessments carried out by regional experts and do not refer to a quantitative and reliable database. Until now, such a global database of harmonized quantitative data focusing on soil and land degradation has been lacking.

According to calculations performed by Harold Dregne and Nan-Ting Chou, about 3.6 billion hectares of dryland areas are degraded, an area that comprises about 70 percent of all global drylands (see table "Estimates of all degraded lands in dry areas"). Drylands cover a large part of the developed and the developing world, but the countries of the developing world are most seriously affected because 1.4 billion people live on or from degraded land, of which most are less-favored areas, including marginal agricultural areas and woodland

Estimates of all degraded lands (in million ha) in dry areas (Dregne and Chou, 1994)

Continent	Total Area	Degraded Area†	Percent Degraded
Africa	1,433	1,046	73
Asia	1,881	1,342	71
Australasia	701	376	54
Europe	146	94	65
North America	578	429	74
South America	421	306	73
Total	**5,160**	**3,592**	**70**

† Comprises land and vegetation.

Assessment of soil degradation (in million ha) (Oldeman et al., 1992)

Continent	Total Area†	Degraded Area	Percent Degraded	Share of Seriously* Degraded Area
Africa	2,234	494	22	65
Asia	3,771	748	20	61
Australasia	787	103	13	6
Europe	949	219	23	72
North America	1,810	95	5	83
South America	1,993	306	15	65
Total	**11,544**	**1,965**	**17**	**62**

† Without wastelands (e.g. deserts, bare land).

* Seriously degraded area comprises degradation classes from moderate to extreme (Oldeman et al., 1992).

areas. The marginal drylands of the world experience enormous pressure on their environments, caused by human utilization and recurrent droughts. The resilience of dryland resources (i.e., soil and vegetation) is usually low, and this is why drylands are particularly susceptible to degradation. The considerably higher amount of degraded area calculated in the study by Harold Dregne and Nan-Ting Chou compared to the GLASOD assessment results from consideration of vegetation degradation.

In the GLASOD study, 2 billion hectares are assessed to be prone to soil degradation. From these 2 billion hectares, around 562 million hectares belong to arable land, 685 million hectares are permanent **pastures**, and 719 million hectares are forests or woodlands (see table "Regional soil degradation of main land-cover types"). As can be seen in the table "Assessment of soil degradation," land degradation is not evenly distributed. **Europe** and Africa show the highest amount of degraded area, while in North America and **South America**, soil degradation is limited to 5 percent and 15 percent of the total land area, respectively. However, these numbers reflect only one aspect of the picture of soil degradation and could be misleading. Looking at the soil degradation of arable land (see table "Regional soil degradation of main land cover types") reveals that both Africa and South America face considerable degraded area. Apart from the extent of soil degradation, the GLASOD assessment describes the severity of soil degradation, which is related to adverse effects on site productivity. Rosamund Naylor, based on the GLASOD assessment, stated that about 38 percent of the degraded land is estimated to be

Regional soil degradation of main land-cover types (in million ha) according to Oldeman et al. (1992)

	Arable Land			Permanent Pasture			Forests/Woodlands		
Continent	Total	Degraded	%	Total	Degraded	%	Total	Degraded	%
Africa	187	121	65	793	243	31	683	130	19
Asia	536	206	38	978	197	20	1,273	344	27
Australasia	49	8	16	439	84	19	156	12	8
Europe	287	72	25	156	54	35	353	92	26
North America	236	63	26	274	29	11	621	4	1
South America	180	92	51	572	78	14	962	137	14
Total	**1,475**	**562**	**38**	**3,212**	**685**	**21**	**4,048**	**719**	**18**

lightly degraded (leading to a loss in agricultural productivity of up to 10 percent, yet retaining potential for full recovery), 46 percent moderately degraded (leading to a 10–25 percent reduction in agricultural productivity and capable of being restored only through considerable financial and technical investment), 15 percent severely degraded (exhibiting a 25–50 percent loss in agricultural productivity and reclaimable only with substantial international assistance), and 0.5 percent extremely degraded (incapable of supporting agriculture and irreclaimable). The last column in the table "Assessment of soil degradation" shows the regional share of seriously degraded soils (moderately to extremely degraded soils) with an agricultural productivity decline of at least 10 percent.

A Regional View

Africa

In Africa, **conversion** of marginal land, forests, and wetlands due to expansion of agriculture is a major driver of land degradation. Land degradation is linked to rural poverty and stems to a large degree from colonial imbalances in land distribution, lack of incentives for **conservation**, insecure land tenure, and the failure to provide for diversified rural production systems. Livelihoods of pastoralists and wildlife as well as habitats and biodiversity are threatened by draining wetlands for agriculture. Loss of natural habitats has reduced vegetation cover and exposed soils to wind and water erosion. Soil erosion reduces the productivity of land, requiring farmers to apply more and more fertilizers, which causes additional economic pressure. Soil erosion also causes **siltation** of reservoirs and rivers, and increased risk of flooding.

Asia

In Asia, land degradation problems are directly related to land-use practices, particularly agricultural expansion and intensification. Biological degradation, soil fertility depletion, soil erosion, and soil acidification are of particular concern. Large-scale clearance of forestland has caused a decline in soil structure and substantial loss in biodiversity. Due to a large extent of steep land, severe water erosion occurs in the **Himalayas**, Central Asia, and **China**. Agricultural mismanagement—for example, the application of acid fertilizers—has led to substantial soil degradation in Cambodia, Malaysia, Thailand, and Vietnam. Irrigation farming is responsible for salinization in parts of northern India and Bangladesh. Asia comprises large areas of drylands in Central Asia, south Asia, and northeast Asia, which are seriously affected by wind erosion. In south and southeast Asia, soil contamination from lead and arsenic is a prevalent problem.

Australia

Land degradation is one of the most important environmental issues in Australia. More than half of Australia's farmland requires treatment for land degradation. Large-scale mining causes the loss of vegetation cover and the lowering of the groundwater water table. Half the major forests and about 35 percent of Australia's woodlands have been cleared or severely modified.

Various combinations of soil erosion, salinity, acidification, soil structure decline, and waterlogging affect much of Australia's cropping land. Apart from losses in agricultural productivity, the impacts of soil degradation include damage to water resources, biodiversity, pipelines, houses, and roads.

Europe

In Europe, **urban sprawl**, an increasing transportation network, and highly intensive agriculture are major reasons for land degradation. The increase in sealed surfaces together with a decrease in forest cover has increased the frequency and size of storm runoff, causing flooding, mudflows, and landslides. Soil erosion in Europe is mainly caused by water and is largely a result of unsustainable agricultural practices, clearcutting of forests, and overgrazing. Soil erosion causes particular problems in the Mediterranean region of Europe. Soil contamination occurs throughout Europe, and is particularly severe in urban areas due to industrial activities and inadequate waste disposal, as well as in areas with a long tradition of heavy industry, mining and military activities, and accidents. In eastern Europe, large irrigation and hydroelectric projects together with poor water management have resulted in salinization and waterlogging of large areas, especially in Azerbaijan, Belarus, the Russian Federation, and Ukraine.

North America

In North America, land degradation is associated with agricultural expansion, intensification, and **industrialization**. One of the key issues of agricultural land use is the use of chemical pesticides, which have contributed to increased food production but have also had important environmental and **human health** effects. North America accounts for 36 percent of world pesticide use. In **Canada**, the land area treated with chemical pesticides increased 3.5 times between 1970 and 1995. Some of the direct pressures leading to degradation due to erosion as well as chemical and physical degradation have been agricultural expansion, intensification, and overgrazing in arid lands. However, conservation measures have led to significant declines in erosion over the past thirty years. The expansion of farmland and urban areas, especially into wetlands, destroyed habitats and contributed to a substantial loss of biodiversity. North America's forests are becoming increasingly fragmented and biologically deteriorated due to large-scale logging by the timber industry.

South America

In South America, the agricultural expansion and the intensified use of natural resources have triggered land degradation. Soil erosion is the main cause of land degradation, while nutrient depletion is also a serious issue, largely driven by **agricultural intensification**. Further nutrient depletion is closely linked to rural poverty, which, in turn, has contributed to greater environmental degradation and land deterioration. Large-scale farming leads to erosion and compaction due to mechanization, as well as salinization because of improper irrigation and chemical pollution. Expansion of permanent pastures into previously forested areas is still the main source of deforestation in **Amazonia**. However, smallholder-farming systems increase deforestation as well, and lead to erosion and loss of soil fertility because they are used intensively without allowing for adequate **fallow** periods.

Institutional Framework and Policy Response

Important global conventions related to land degradation are the **United Nations Convention to Combat Desertification**, the **United Nations Convention on Biological Diversity**, and the United Nations Framework Convention on Climatic Change. To bring these conventions into action, an infrastructure is needed to cover funding mechanisms, research and development centers, and coordinated international programs. On the international level, organizations such as the **United Nations Food and Agriculture Organization** (FAO), the **United Nations Environment Programme** (UNEP), and the United Nations Development Programme are the main institutions coping with land degradation. These organizations have gained significant expertise and experience with land degradation on both the regional and global scales. One example of a joint international effort on land degradation is the **Land Degradation Assessment in Drylands (LADA) Project**. This project is carried out in partnership by international organizations such as FAO and UNEP to develop and test an effective assessment methodology for land degradation in drylands.

International funding mechanisms such as the Global Environment Facility (GEF) and organizations can support national governments, but land degradation needs to be tackled at the local and regional scales, with direct intervention in the affected areas taking into account competition among concurrent uses of land and soil (e.g., food production, living space, infrastructure and industrial production, and conflicts between private and public use of land resources). Particularly in developing countries, land degradation is about rural people and their **institutions**. To avoid further land degradation, these people need both sectoral and macroeconomic policies that positively affect farm incomes as well as independent local institutions that take care of future land management.

See also: **Access; Arena of Land Conflict; Blaikie, Piers; Brookfield, Harold Chilingworth; Carbon Cycle; Common-Pool Resources; Decision-Making; Degradation Narrative; Disequilibrium Dynamics; Exotic Species; Extensification; Hot Spots of Land-Cover Change; Land Rights; Marsh, George Perkins; Nutrient Cycle; Private Property.**

Further Reading

Barrow, Christopher J., *Land Degradation: Development and Breakdown of Terrestrial Environments*, Cambridge, UK: Cambridge University Press, 1991; Blaikie, Piers, and Harold Brookfield, *Land Degradation and Society*, London: Methuen, 1987; Carman, Bill, ed., Challenges and Opportunities for Integrated Land Management [Online, May 2003], The International Development Centre Web Site, web.idrc.ca; Daily, Gretchen C., "Restoring Value to the World's Degraded Lands," *Science* 269 (1995): 350–4; Dregne, Harold E., and Nan-Ting Chou, "Global Desertification Dimensions and Costs," in *Degradation and Restoration of Arid Lands*, eds. Harold E. Dregne, Lubbock, TX: Texas Technical University, 1994, pp. 249–81; Middleton, Nick, and David Thomas, eds., *World Atlas of Desertification*, 2nd ed., London: Arnold, 1997; Naylor, Rosamund N., "Energy and Resource Constraints on Intensive Agricultural Production," *Annual Review of Energy and Environment* 21 (1996): 99–123; Oldeman, L. Roel, "The Global Extent of Land Degradation," in *Land Resilience and Sustainable*

Land Use, eds. Dennis J. Greenland and Ivan Szabolcs, Wallingford, UK: CAB International, 1994, pp. 99–118; Pagiola, Stefan, *The Global Environmental Benefits of Land Degradation Control on Agricultural Land*, Washington, DC: The World Bank, 1997; Stuip, Mishka A. M., Baker, James C., and Willem Oosterberg, The Socio-Economics of Wetlands [Online, April 2004], The Wetlands International Web Site, www.wetlands.org; United Nations Environment Programme, GEO: Global Environment Outlook 3: Past, Present and Future Perspectives [Online, July 2002], The UNEP Web Site, www.unep.org; Watson, Robert T., Dixon, John A., Hamburg, Steven P., Janetos, Anthony C., and Richard H. Moss, Protecting Our Planet, Securing Our Future: Linkages among Global Environmental Issues and Human Needs [Online, November 1998], The World Bank Group Web Site, www-esd.worldbank.org/planet; World Resources Institute, *World Resources 2000–2001: People and Ecosystems: The Fraying Web of Life*, Washington, DC: WRI, 2000.

GERALD BUSCH

Land Degradation & Development. An international journal that seeks to promote recognition, **monitoring**, control, and rehabilitation of degradation in terrestrial environments. The journal focuses on the character, causes, impacts, and scale of **land degradation**. It addresses the history, current status, and future trends of land degradation (including the potential for **sustainable land use** or management); and discusses avoidance, mitigation, and control of land degradation, as well as remedial actions to rehabilitate or restore degraded land.

See also: **Arena of Land Conflict; Deforestation; Desertification; Driving Forces; Land Rehabilitation; Mineral Extraction; Perception; Proximate Causes; Soil Erosion; Species Extinction; Urban Sprawl.**

Further Reading

Wiley Publishers, Land Degradation & Development Journal [Online, December 2004], The Publishers Web Site, www.wileyeurope.com.

HELMUT GEIST

Land Degradation Assessment in Drylands (LADA) Project. One of the most recent attempts, funded by the Global Environmental Facility, to assess the magnitude of **desertification** and to provide a baseline for **monitoring**. In earlier attempts associated with the **United Nations Convention to Combat Desertification**, among others, case studies were collected from national governments, focusing on randomly selected sites rather than applying a well-conceived framework for analysis. This effort has been improved for pilot analysis and case studies related to the Land Degradation Assessment in Drylands (LADA) project, with national case studies stemming from Argentina, Egypt, Mexico, Uzbekistan, Senegal, Kenya, **South Africa**, and Malaysia. Different from the analysis of **hot spots of land-cover change** associated with the **Millennium Ecosystem Assessment**, LADA is a programmatic rather than a scientific effort of the **United Nations Food and Agriculture Organization** and other partners. It responds to the need to strengthen support to combat **land degradation**. LADA aims to generate up-to-date ecological, social, and economic and technical information, including a combination of traditional or **indigenous knowledge** and modern science, to guide integrated and cross-sectoral **land-use planning** and land management in drylands. LADA's main objective is foremost to develop and implement strategies, tools, and methods to assess and quantify the nature, extent, severity, and impacts of dryland degradation on ecosystems, **watersheds**, and **river basins**, and carbon storage in drylands at a range of spatial and temporal **scales**. The project will also build national, regional, and global assessment capacities to enable the design and planning of interventions to mitigate land degradation and establish **sustainable land use** and land management practices.

See also: **Carbon Cycle; United Nations Environment Programme.**

Further Reading

United Nations Food and Agriculture Organization, Land and Water Development Division, Land Degradation Assessment in Drylands [Online, February 2005], The LADA Web Site, www.fao.org/ag/agl/agll/lada.

HELMUT GEIST

Land Evaluation. An assessment of the present and potential suitability of land for different land uses (**agriculture, forestry**, or other) in order to promote specific land-use patterns. Land evaluation is a necessary and initial step in the process of **land-use planning**. This understanding of the **land-use system** is also called analysis or diagnosis. Since the 1950s, the evolution of the agricultural system approach and the development of **geographical information systems** (GIS) as well as of dynamic simulation models allowed a more extensive coverage of spatial and temporal variations in land typologies. Non-soil, economic considerations and **indigenous knowledge** are also progressively taken into account. For over thirty years, the **United Nations Food and Agriculture Organization** (FAO) has been applying a land-planning method that is quite largely used. It is not only a soil survey interpretation, but also an evaluation of climatic, landform, and socio-economic conditions. The FAO land evaluation framework is a semi-quantitative method that compares the performance of different land uses. The method involves a consultation of indigenous people to formulate land-utilization types, that is, the relevant land uses and their ecological requirements. The next steps of the method are compilation or cartographic design of a land-use database; matching each land-mapping unit in terms of a suitability assessment (taking into account the productivity potentials under different uses, the environmental and economic impacts, and a social analysis); and development planning. Since 1976, computerized databases and **modeling** programs have been interfaced with GIS and transform the qualitative matching in a quantitative and automated one, which is called agro-ecological zoning.

See also: **Integrated Assessment; Land Quality; Land Rehabilitation.**

Further Reading

Verheye, Willy, Parviz Koohafkan, and Freddy Nachtergaele, The FAO Guidelines for Land Evaluation, in *Encyclopedia of Life Support Systems*, developed under the auspices of the UNESCO, Eolss Publishers, Oxford, UK [Online, September 2004], The EOLSS Web Site, www.eolss.net.

NATHALIE STEPHENNE

Land Fragmentation. Land tenure structures in developing rural economies are often bimodal, which means that most of the land is owned or retained by a few people, and large populations have only small plots of land to make their livings. In this context, land fragmentation of individual **landholdings** takes place mainly through two main mechanisms. First, it occurs when the landholding **property** is divided among different family members according to specific inheritance rules for **land rights** transference that varies among regions; and second, when a portion of the land is sold in the land market. The two processes mentioned above constitute forces contributing to land fragmentation.

The first trend is somewhat straightforward due to which land rights are inherited after the person who held the property rights passed away, or decides to transfer his/her land rights. In this case, intensity of land fragmentation depends on the density of family members with potential rights over such pieces of land, as well as the inheritance rules' specificities. The second trend responds to broader economic determinants. In this case, large-scale landholders tend to fragment their landholdings in a situation in which they may want to achieve a medium-size scale that could meet some economies of scale that would fit better their intended activity (e.g., semi-intensive beef cattle production, intensive milk production, grain production, etc.). In other cases, landowners divide their establishments in the absence of any other valuable assets to sell to recover from economic crisis, or to invest the obtained capital in non-agricultural activities whose profits are higher than those that originated in **agriculture**. These latter cases often take place in rural economies in which land rights are relatively secure and there are no legal constraints to transfer them.

Land fragmentation is not commonly seen in tropical frontier areas. It takes place only in the so-labeled consolidated frontiers in which land tenure rights are guaranteed legally. The latter is in part the outcome of the **land tenure** evolution in the agricultural frontiers by which, after the initial rush driven by last speculation interests, original landholders begin to value the occupied land, either by legal or illegal means, through more developed land markets. The evolution of **agricultural frontiers** leads to growing investment interest from outsiders to buy a piece of land, and this in turn tends to increase the land prices. Larger **investments** in the frontiers along with trends to landholdings fragmentation tend to accelerate the economic activities in the frontiers, which leads to higher **deforestation**. The latter, in the Latin American context, often is an outcome of the transition from large-scale extensive **cattle ranching** toward semi-intensive systems of cattle ranching.

Finally, there is a third dynamic not mentioned above of induced land fragmentation. This results from the implementation of redistributive agrarian reforms. Though in the past, administrative agrarian reforms were applied through expropriation of the land rights mechanism to medium- and large-scale landholdings, currently there are sponsored mechanisms—with strong involvement of the World Bank—toward negotiated agrarian reforms assisted by the land markets. The underlying notion is to pulverize large landholdings by allocating land property rights and helping small landholders to negotiate and buy the land from large landholders through land credits.

See also: **Amazonia; Arena of Land Conflict; Biodiversity; Land Concentration; Land Reform; Pattern Metrics; South America.**

PABLO PACHECO

Land Privatization. A shift in **property** from state or other collective entities toward private actors. This shift can occur by way of expanding the types of **land rights** accorded to land users, or by granting them **private property** (ownership) or a bundle of rights close to private property. Land privatization can also take other forms. Reducing the control rights held by the state or other collective entities can enhance the protection of individual land users from interference into their land rights. Privatization can also expand the duration of land rights or strengthen them by way of their recognition and enforcement through the state. Land privatization programs have covered croplands and land used for urban and industrial purposes around the world. Their scope has been limited for **forests** and **grasslands**. Privatization has often been motivated by the presumed effects of private property on the efficiency of land management. Other primary motivations have been political and social, such as the use of land titles as signifiers of political participation and social equity.

Africa has witnessed a vivid debate about the benefits of land privatization. The debate was first stirred up by land registration programs in the 1980s and is now being revived in the wake of heightened attention to **institutions** and governance in development. Until today, rural people across much of Africa held use rights to land that were overlapping and not registered with the state. African land rights have been perceived as constraints on agricultural performance and causes of unsustainable land management. In reaction, the national governments, with support by the World Bank and other international donors, set out in the 1980s to demarcate, register, and certify land held so far under customary law. Land titling was conceived as a panacea to overcome agricultural stagnation and rural **poverty**. Proponents of land registration expected it to create secure land rights by clarifying local property relations. Secure rights would motivate landowners to invest in agriculture and perform land **conservation** practices. Land titles would also serve as collateral for landowners to obtain bank credit. Furthermore, land registration would create the basis for the development of land markets to facilitate the efficient allocation of land, in particular **land concentration**, in the hands of those with **access** to capital and modern technology.

Land registration has not delivered the expected rise in agricultural production and improvement in rural living standards, however. First, its effects on the security of land rights have been limited because insecurity originates from many other sources than a lack of clarity. For example, land users may have insecure land rights because of abuse of power by local elites and encroachment by outsiders. Land registration may even decrease the security of land rights by adding another layer of social control legitimating claims on land. Second, the lack of agricultural **investment** and conservation efforts is much more often due to inappropriate technology than insufficient security. Even in the absence of registration, local people adjust **land tenure** to emerging conservation requirements, granting investors long-term land rights. Third, land registration often does not increase the use of credit because that is constrained by factors independent of land rights. Farmers are reluctant to use land as collateral, as they are afraid to lose their most important productive resource in case of repayment problems. Banks, in turn, are reluctant to accept land titles as collateral because banks often cannot foreclose on land for political reasons. In addition, credit supply to agriculture tends to be very limited, as investments in other sectors yield higher returns. Finally, land registration rarely contributes to the development of land markets, as customary land tenure already includes mechanisms for land transfers despite the absence of ownership, and recognizes land sales when **agriculture** becomes more commercialized. In addition, the effects of land markets on agricultural productivity are dubious, as smaller **landholdings** are often more productive than large ones. Overall, the debate indicates that the benefits of land privatization are specific to political and economic contexts. Its effects on agricultural production often do not outweigh the high costs of land registration programs and negative consequences on equity, as women and other marginalized groups become excluded from access to land. More broadly, experience with land registration demonstrates that property is not an exogenous variable to be defined by states and international donors. The effects of land privatization on actual property relationships depend on the broader social dynamics of resource use.

Land privatization came to the fore again when previously socialist countries around the world embarked on massive property reforms after 1989. Land privatization programs in post-socialist countries transfer land rights to private entities by way of distribution and restitution. Distribution refers to policies that grant agricultural workers or the rural population equitable shares in the available land. Restitution implies the return of land to the historical owners (or their heirs), typically using as a reference point the period between the **land reforms** conducted by the new socialist governments after World War II and agricultural collectivization. There is great variation in the land privatization programs instituted by post-socialist governments. This variation reflects the highly political nature of land legislation, as many kinds of political actors have sought to influence the new land laws. The actors have asserted different goals for land privatization, often emphasizing notions of historical justice, political participation, and social equity more than more-narrow objectives of land management.

Land privatization has effected a massive shift toward private land rights in **west-central**, **east-central**, and **southeast Europe**. Today, most agricultural and urban land is no longer under the management of state units or collective entities but under private management. At the same time, closer looks at actual property relationships reveal high levels of heterogeneity and complexity. Local implementation of land privatization has witnessed intense political negotiations among villagers, urban residents, agricultural managers, ethnic groups, and emerging entrepreneurs. The political struggles have resulted in distributions and kinds of land rights that are specific to local political economic conditions and are often quite different from national legislation. For example, agricultural associations and large tenants control land markets and agricultural production in some regions, limiting the property rights held by the new landowners in practice; or, landowners often do not comply with the obligations tied to their land rights, such as the requirements to cultivate the land and protect valuable **biodiversity**. In addition, serious constraints limit the exercise of land rights, such as the lack of machinery, agricultural credit, inputs, and competitive product markets. As for land use, land privatization is associated with radical changes in land use across central and eastern Europe. Agricultural output dropped drastically during the first half of the 1990s. Cropping structures have changed in response to changes in farm structures and agricultural markets. There is also some evidence suggesting a link between land privatization and overgrazing, the destruction of irrigation systems, **soil erosion**, **deforestation**, and **urban sprawl**.

Land privatization has taken a different form in East Asia and the former Soviet Union. The shift toward private property has not been as complete as in central and eastern Europe. States tend to retain formal ownership to land, granting use rights only to private entities. Individual property rights include very limited rights of alienation only. As for policy, the emphasis has been on distributing land to agricultural workers previously employed in state and collective farms. In many places, collective entities have retained control rights over land, exercising their authority by way of redistributions of land and restrictions on allowable land use. Land privatization has contributed to strong increases in agricultural output in **China** and Vietnam, in stark contrast to the general output decline observed in central and eastern Europe. Cropping structures in East Asia have changed from **rice** toward other crops, such as vegetable and fruit trees.

See also: **Cadastre; Economic Restructuring; Exogenous Factor; Land-Use Policies; Public Policy; Transhumance.**

Further Reading

Bruce, John W., and Shem E. Migot-Adholla, eds., *Searching for Land Tenure Security in Africa*, Dubuque, IA: Kendall/Hunt Publishing House, 1993; University of Wisconsin–Madison, Land Tenure Center [Online, October 2004], Land Tenure Center Web Site, www.wisc.edu/ltc/; Verdery, Katherine. *The Vanishing Hectare: Property and Value in Postsocialist Transylvania*, Ithaca, NY: Cornell University Press, 2003.

THOMAS SIKOR

Land Quality. The condition of land relative to the requirements of land use, that is, **agriculture**, **forestry**, **conservation**, and environmental management. Indicators of land quality (LQIs) are needed to raise awareness of agricultural and environmental issues, and as instruments for **monitoring** progress toward or away from **sustainable land use**. The World Bank has been leading an international coalition to develop a program on land quality that is probably the most advanced effort to address the integration of socioeconomic (land management) data with biophysical information in the definition and development of appropriate indicators, and how to scale and aggregate indicators from local to global **scales**. In this regard, one of the requirements for basic or core LQIs is that land quality is to be assessed for specific agro-ecological zones (AEZs) in a country, allowing for a further breakdown by specific types of land use and land management. This requires the development and application of spatially located and georeferenced temporal databases of variables that are still under construction.

See also: **Agrodiversity; Georeferencing; Global Terrestrial Observing System; Land Evaluation; Mediating Factor; Proximate Causes.**

Further Reading

Dumanski, Julian, and Christian Pieri, Land Quality Indicators (LQI): Monitoring and Evaluation, in *Encyclopedia of Life Support Systems*, developed under the auspices of UNESCO, Eolss Publishers, Oxford, UK [Online, October 2003], The EOLSS Web Site, www.eolss.net.

HELMUT GEIST

Land Reform. A politically, socially, and economically desirable or urgent need to change ways of **land tenure** or its ownership, and/or the redistribution of agricultural and **forest** lands owned by landlords and other big owners among farmers and landless persons. In communist countries, land reforms (LRs) involved the forcible joining of farmers' land into big cooperatives or state estates. A less-frequent definition concerns such innovations as improvement of farms (by reallocations of plots via exchange, sale, or lease), large-scale operations, mechanization, and so on, in order to change the field patterns optimal for economic efficiency.

The history of land reforms began in antiquity and had some common features with historical development during feudalism, capitalism, and communism. LRs have been connected particularly to the transition from feudalism to capitalism in **Europe** and America for the last three centuries. The process climaxed in the period of bourgeois revolutions; first in the Netherlands and England in the seventeenth century, in France at the end of the eighteenth century, in almost the whole rest of Europe in the years 1848–1849, and in the **United States of America** during the Civil War in 1861–1865.

The major goals of LRs have been to solve social tensions in the rural areas and to decrease relative overpopulation through parceling of large **landholdings**, in order to at least partially satisfy the common greed of small farmers for land. In contrast, the so-called enclosures in England from the late fifteenth to the nineteenth centuries brought an expropriation of peasant land and common fields, and then conversion of small farmers into wage-earning labor forces in agriculture, or predominantly enabled them to move to the cities. This has led to a rise of very big estates enclosing big tracts of **pastures** or arable land.

Land reforms have been proclaimed usually by the state or by some interest groups, mostly under political pressure and with the aim to remove survivals of the feudal system, which became a brake in the development of capitalism in **agriculture**, realized through the payments as monetary fixed tributes by serfs—hereditary tenants to feudal noblemen, or state or church landlords; tithes to clergy; or manorial labor for big landowners (the amount of time for such labor differed according to social and economic situations in the country or territory). LRs have introduced payments based on real land **property** of all, that is, payment of land taxes to the state, or tributes for leased land in order to farm larger homesteads or estates.

Land reforms have greatly influenced land use and land cover. In most cases, they have brought about an increase of agricultural land (except permanent **grasslands**) and primarily arable land. Small farmers who gained permanent grasslands usually ploughed them up for better securing of their self-sufficiency or for the market. Former landless peasants (workers in big farms or in industry) became part-time farmers for self-sufficiency. This situation was typical until the 1950s, mostly in mountainous regions in **west-central**, **east-central**, and **southeast Europe**. This has led to a higher share of arable land in some of these regions, which, as a consequence of **urbanization** and the **technological-scientific revolution** in agriculture, resulted in afforestation again in the second half of the twentieth century. These processes have enforced the disappearing of small farms and their joining with bigger farms and ranches. Consequently, the permanent decrease in the number of farms and ranches triggered a concentration of landownership or land tenure. It lasted until the present as a general feature of agricultural development, not only in Europe and North America.

Capitalism has eliminated feudal survivals that tied peasants to land, for example, through the abolishment of serfdom in the Austrian monarchy in 1848, and of slavery in the United States in 1863; and made land goods on a free market and enabled its free inheriting. Already in 1841, the so-called "federal land fund" started selling federal land or free lands to farmers at very cheap prices on the basis of laws in the United States. The first "Homestead Act," from 1862, was one of the most important from those laws. A colonist could gain about 195 hectares (480 acres) for a mere $200. The fund was abolished to ensure nature **conservation** on about 24 percent of U.S. territory (minus Alaska) as late as 1935. Thus, it led to extreme use of agricultural land, such as in the **Great Plains** in the United States, as well as to considerable wind erosion there (the so-called Dust Bowl in the dry first half of the 1930s). Water erosion prevailed in humid regions, where the layer of mold has been reduced, for example, by nearly one half in Illinois. A process comparable to the Dust Bowl event took place in the Soviet Union after tilling up of prairies in the 1970s in former Soviet Central Asia (Kazakhstan).

The following ways and goals of land reforms give evidence of the application of various so-called "ways of development of capitalism in agriculture." The "American way" has specifically created economically efficient large farms oriented toward market production. After the French Revolution in 1789, the "French way" realized political and social goals by abolishing serfdom, when the land of clergy and emigrated nobility was expropriated by state and sold to farmers. This led to an increase of arable land area, and above all to the crumbling of agricultural land among small farms, which currently burdens the agricultural policy of the European Union (EU).

The so-called "Prussian way" of land reform occurred on the borders of the former Prussia in 1939, and also in the Austrian monarchy. After the revolution of 1848–1849, big estates of nobility were not parceled or expropriated but transferred to capitalistic ways of farming based on the use of free labor as waged workers and particularly using the elements of the **Agricultural Revolution** and the technological-scientific revolution in agriculture. The capital obtained by noblemen from deliberated farmers markedly helped them to realize such innovations. In order to become owners of land cultivated by farmers in serfdom, these farmers had to pay to their former landlords a part of the value of land, which they had only in hereditary tenure or lease before revolution. Land reform in **Russia** was similar to Prussian and Austrian reform; but serfdom was abolished later, in 1861. Only homesteads became hereditary property of farmers; the land they had in tenure became common land, and peasants had to pay redemption to village authorities to receive it in ownership. The state compensated about 80 percent of the allotment land-market value to landlords.

After the First World War, further land reforms aimed to economically strengthen small farmers toward large estates, accomplished in some succeeding states of the abolished Habsburg monarchy and in Romania, for example. The government redeemed a part of large-estate land and sold it to peasants who mostly tilled it. Similar land reforms were realized in many countries in North Africa and the Middle East, usually following independence or political revolution.

Land reforms in communist countries had a very specific character. The aims were particularly political and only then economic. Individual character of farming didn't satisfy collectivism. Large estates and small estates were expropriated without any compensation. Small peasants were forced by political and sometimes even police pressure to put their land into cooperatives, become its members, and farm collectively. Thus, large farms were created, the areas of which became a comparative advantage through the development of mechanization and modernization. They farmed big track-fields that brought about simplification of the rural landscape and **soil erosion**. Another desirable result of such reforms was the release of many peasants as new labor forces for the extensive growing industry.

The fall of communism has been followed by restitution of land property "back" into the state before so-called socialist collectivization. Former members of cooperatives or their descendants received back their land (if they asked for it). After capital transformation of most cooperatives, they usually farm the same area of land, since their size enables greater effectiveness of production. They still farm the land of "old-new" owners, paying them very small tribute. These owners are usually not interested in doing business in agriculture because their farms are very small; or because they live in the cities or have jobs in other sectors of the economy, and thus have lost their affinity for farming.

See also: **Agro-Industry; Cadastre; Common-Pool Resources; Industrialization; Land Privatization; Land Rent; Land Rights; Private Property; South Africa.**

Further Reading

Ginter, Donald E., *A Measure of Wealth: The English Land Tax in Historical Analysis*, Montreal, Kingston, JM, London: McGill-Queen's University Press and Hambledon Press, 1992; Stanis, Vladimir E., *Socialist Transformation of Agriculture: Theory and Practice*, Moscow: Progress Publishers, 1976; Tuma, Elias H., *Twenty-Six Centuries of Agrarian*

Reform: A Comparative Analysis, Berkeley, CA: University of California Press, 1965; Turnock, David, ed., *Privatisation in Rural Eastern Europe: The Process of Restitution and Restructuring*, Cheltenham, UK, Northampton, MA, USA: Edward Elgar Publishing, Inc., 1998.

LEOŠ JELEČEK

Land Rehabilitation. Amelioration of land, by engineering or biological restoration measures, that has been destroyed by mining or quarrying, landfills, occupation, and pollution in the course of production and construction; or demolished by natural disasters such as flood, coastal invasion, mud-flow, and sandstorm to a state where land can be used for farming or other purposes. It is often executed after good land has been demolished through industrial activities like mining, tile making, or coal-based electricity generation. In a much broader sense, land rehabilitation also includes readjusting and rearranging land-use structures by engineering or biological measures to increase the area of arable land; improve the quality and productivity of land; and finally amend the general conditions of production, living, and the environment. Land rehabilitation needs to follow the principles of pursuing integrated benefits in economy, society, and ecology and boosting harmony among socio-economic development, rational resources use, and **conservation**. Its operation should be in concordance with **land-use planning** and rely upon the advancement of technical methods. Moreover, the operation should also take into consideration market mechanisms. The main characteristics of land rehabilitation can be summarized as large **investment** and long-term return. Therefore, land rehabilitation is a large and complex project involving many factors of economy, society, and technology. Above all, it is an important way to complement farmland or offer construction land for countries with large populations and limited land resources, such as **China** with its limited cultivated land resources. Land rehabilitation is an effective means to keep the balance between population and land resources through resolving the conflict between them. In China, land rehabilitation intends to increase the area of farmland, to improve the quality of farmland, to optimize structures of land use, to ameliorate the conditions of production and living in rural regions, to protect the environment, and to likewise promote **agricultural intensification**. Land rehabilitation includes four courses of action. First, level off land, consolidate scattered plots, build terraces or fix breeding surfaces, and merge villages by engineering or biological measures. Second, build agricultural engineering works such as roads, wells, canals, slopes management, and forest shelter. Third, control sandy land, **salinization** and polluted land, meliorate soils, and renew **vegetation**. Fourth, ensure secure **land rights** as well as proper land use and make regular land-change investigations and registrations. In China, the policy of land rehabilitation follows the rule that who destroys land has to find answers for land rehabilitation. In the process of practice, land rehabilitation planning is scientifically worked out and functions as the head of all operations. Land and resources bodies are in charge of confirming the land-use purpose after rehabilitation. In 2001, a project about land consolidation and rehabilitation and **sustainable land use**, supported by the **United Nations Development Programme**

(UNDP), was implemented in China. The main items of this project include policy research, ability improvement, demonstration works, and information systems construction.

See also: **Arena of Land Conflict; Forestry; Green Wall Project; Land Evaluation; Land Reform; Reforestation; Technological-Scientific Revolution.**

QI LU

Land Rent. Payment for a plot of land. The price for a plot of agricultural or urban land derives from its real economic output and the average rate of deposit interest, with usually no land transfers happening if interest rates are higher than economic output. Land rent is a basic economic category that expresses and influences land-use/cover change, valid for both capitalist and communist systems. Differences in land characteristics, such as location and productivity, and/or differences in social power around the ownership of land, usually determine the precise level of rent on any specific tract of land, thus producing what is called differential (land) rent.

Two forms of differential rent can be distinguished. The extensive differential rent, or differential rent I (DR I), represents an extra profit that arises because all plots of land have equal amounts of capital **investment** but are different in terms of natural fertility and in their position to the market of agricultural products. DR I relates to the extensive growth of **agriculture** based on the extension of cultivated, especially arable land. Though losing its basic impact and importance, historically, advancing **industrialization** and **urbanization** created new location conditions, including growing demand for food. The intensive differential rent, or differential rent II (DR II), represents an extra profit that is created by unequal capital investments into plots of land with same degrees of natural fertility. Thus, DR II relates to **agricultural intensification**, that is, more effective use of capital (artificial fertilizer, machinery, genetic engineering) creating more fertile land, but also to **land abandonment** in areas with low prospects for intensification, such as those at high or remote locations. In both cases, the differential rent is determined by the difference in cost between any productive plot of land and the so-called marginal plot. The marginal plot is land so unproductive that the cost, for example, of producing a crop there is equal to its market price, thereby leaving no room for rent. There is an equivalent to the land rent model developed by **Johann Heinrich von Thünen**. While calculated in a similar manner, differential rent is interpreted differently in neoclassical and Marxist land economics. In the neoclassical interpretation of Thünen, for example, it is portrayed as an index of natural scarcity whether of locations or differentially fertile lands, while in the Marxist account it is an expression of the monopoly power of capital as a whole.

See also: **Agricultural Revolution; Land Reform; Land Rights; Technological-Scientific Revolution.**

Further Reading

Marx, Charles, *Capital: A Critique of Political Economy*, vol. III/2, Section IV, New York: International Publishers, 1967; Thünen, Johann Heinrich von, *Von Thünen's*

Isolated State: An English Translation of Der Isolierte Staat, Oxford, New York: Pergamon, 1966.

LEOŠ JELEČEK

Land Rights. Enforceable claims on land. The holders of land rights are able to make decisions about land (legal rights) and receive a share of the benefits derived from land (economic rights). Land rights and obligations together constitute **land tenure**, or **property** in land. Land rights involve complex social relationships among actors, including many different types. The rights pertaining to one parcel of land are typically many and may not be bundled in the hands of one single actor, instead belonging to several different actors. As land rights are embedded in broader social relations, they display a tremendous diversity around the world. Simple classifications of land rights as private, state, or communal do not capture the diversity. This diversity also makes clear statements about the relations between land rights and land use impossible. Land rights and land use are generally thought to influence each other. The nature of their relationship is highly dependent on contextual factors, however.

Like property rights in general, land rights are of many types. A simple distinction is between use rights (usufruct), such as rights of cultivation, passage, grazing, and collection; and control rights, such as the rights to exclude other claims, transfer land, make major alterations, and sell land. Land rights are often specific to particular resources associated with land. For example, the right to cultivate land does not automatically imply a right to the trees standing on the land. Another distinction is between private ownership (or freehold) and leasehold. Land ownership, or **private property** in land, accords landholders far-ranging rights to land, including rights of alienation. In contrast, the rights pertaining to land held under leasehold are more restricted and are often tied to more extensive obligations. Common forms of leasehold are land rentals, under which tenants pay landlords a fixed rent; and sharecropping, under which tenants pay landlords a fixed share of the harvest. Nevertheless, there are many mixed forms combining elements of ownership and leasehold. For example, **contract farmers** typically retain ownership rights to land, yet their contracts with agribusiness severely limit the exercise of the rights.

Land rights can be further distinguished by the social actors (legal entity) holding them. Land rights are public if they belong to organized political collectivities at the international, national, or local level. In most cases, states exercise public land rights on behalf of the political entities they represent. Land rights are private if they belong to natural or legal persons, such as individuals, families, corporations, and partnerships. Communal land rights are more complex. They combine control rights held by an organized collectivity with individual use rights allocated to their members. The members' rights are use rights, possibly including inheritance rights but not a right to sell the land. The degree of collective control may vary among different types of land. For example, communities may allocate arable land to individual members but keep **grasslands** open to all members for joint use. Again, there are also many mixed forms of property combining public, private, and communal elements.

Land rights differ from each other by the nature of authority backing up and enforcing claims. Land rights specified in state laws and regulations are called statutory. In contrast, customary land rights derive their legitimacy from social conventions and norms. In many settings, statutory and customary laws coexist side by side, emanating from states, religious bodies, international agreements, local customs, etc. Common is the juxtaposition of statutory rights based on European systems of law and customary rights, having their origin in attempts by colonial regimes to maintain indigenous legal systems. Land rights in such situations of legal pluralism are especially complex, as the rights legitimated by different legal systems rarely match. The rights are often hybrids, deriving their legitimacy from multiple legal orderings. Examples of such hybrid land rights are *Adat* laws in Indonesia and Indian reservations in the **United States of America**.

Land rights are key elements of social organization. As with property, land rights find attention in the social and cultural norms of an organized social collectivity, are the subject of laws and regulations, take their expression in actual relationships, and resonate with social practices. Land rights are embedded in broader social dynamics and serve the maintenance of social control. For example, land rights used to play an important role in the social organization of the Lozi in southern Africa, as described by Max Gluckman in his seminal work on property. Land rights combined use rights (estates of production) belonging to individual Lozi with control rights (estates of administration) held by the king, village headmen, and household heads. Rights to land consisted of a series of allocations from the king to the ultimate user, entitling every involved actor to a share in the benefits derived from land. The relation of each actor to property was a function of their sociopolitical status and depended on the fulfillment of associated obligations. In colonial India, land rights were a key element in the strategies of local power holders to maintain social control. The Indian estate holders used their control over land to expand their control over people, hence political power and wealth. To them, control over land generated economic benefits and political gains. These two examples illustrate the general observation that land rights have many functions. They influence not only the efficiency of resource use but also the patterns of political participation, processes of social continuity, and measures of social protection, among others. Land rights also include nested layers of social control in many instances, granting several units of social aggregation overlapping control rights.

The embeddedness in broader social relations implies that gender relations and land rights influence each other. In many countries, land rights belong to households. Intrahousehold relations, as shaped by gender norms and relations, shape the distribution of discretionary powers and obligations among household members. Male household heads often concentrate significant **decision-making** powers, while female household members assume a large share of labor duties. Development projects have often contributed to the gendered division of discretionary powers and labor duties in the name of modernization. For example, male farmers in Gambia have developed fruit orchards with the assistance of development projects, outcrowding females' vegetable gardens previously cultivated on the same land. Also in Gambia, the spread of commodity production in the uplands has undermined women's ability to market part of lowland **rice** production, as households consume an increasing share of rice outputs. In addition, small-scale pump-irrigated rice projects have enabled men to establish individual control over new land and draw on female labor obligations to cultivate those. Another influence of gendered notions of property is that women often gain and lose land rights through marriage, as households pass on land rights to male offspring only. Sons are considered direct successors in the family line, in contrast to female offspring, who are thought to leave the households upon marriage. These insights demonstrate that actual land rights are often quite different from those written into land laws and regulations, a central insight in the study of land rights and, more generally, property.

Land rights in advanced industrialized societies are characterized by a stark discrepancy between cultural norms on the one hand and laws and regulations on the other. Notions of private property take a prominent role at the level of social and cultural norms. Landholders are thought to enjoy private ownership of land, including extensive use and control rights, especially rights of alienation. States are accorded minimal administrative rights, such as the authority to issue title, tax land, regulate the type of use, monitor actual use, regulate transfers, and arbitrate disputes. Land rights are thought to be private and exclusive, in the sense that all rights pertaining to a particular parcel of land are bundled in the hands of one private actor. The reality is different. Land rights in the European Union are highly regulated by common European law, national laws and regulations, and local conventions. For example, codes of good agricultural practice and standards on the level of pollutants in water and air limit the range of land management practices available to farmers. British laws grant the public a right of trespassing agricultural land and decide about the **conversion** of agricultural land to other uses, including the associated financial benefits. The sale of farmland in Germany is subject to numerous written and unwritten rules, for example, requiring the seller to offer the land to fellow villagers before approaching outsiders. In the United States, 42 percent of all land is public, including federal, state, and local government holdings. On private land, the public holds the rights of eminent domain, taxation, zoning, and regulating land use. In addition to private and public lands, there is a large diversity of mixed property forms, such as common property estates, Indian estates, and public utility estates.

Notions of private, exclusive landownership have also informed the design of land legislation in post-socialist societies. **Land privatization** programs have restituted and distributed agricultural land to private actors in an unprecedented effort of land reform. Statistical data on **west-central**, **east-central**, and **southeast Europe** show a radical move from state and collective to private ownership. Nevertheless, ethnographic studies report persisting differences between actual property relations and the private, exclusive land rights promoted in land laws. Katherine Verdery has coined the term "fuzziness" to describe the ambiguity of land rights in post-socialist Europe. This fuzziness originates from multiple sources: unresolved conflicts over the ownership of single plots; overlapping claims to plots, resulting from legally pluralistic settings; contextual constraints on the exercise of the newly accorded land rights; the blurring of the boundary between private and public, related to private actors' efforts to push debts and liabilities onto the public shoulder; different meanings attributed to land, land rights being often seen as signifiers of citizenship, personhood, and social justice; different visions informing claims on land; and the absence of generally recognized rules and patterns of behavior around private property in the context of post-socialism. Fuzziness is also a key characteristic of emerging land rights in central and East Asia. The difference to post-socialist Europe is that land laws in central and East Asia have not fully adopted the notions of exclusive, private land rights. Many states grant landholders long-term usufruct rights to agricultural land only, excluding the right of permanent alienation. Land remains under the formal ownership of "the people." Intermediate layers of authority retain significant degree of control over land, allowing local and state authorities to maintain strong leverage on land use by designating land-use regulations, and allowing local communities to re-allocate use rights to agricultural land every few years. This is, in part, a legacy of socialist land rights, which granted control rights in land to multiple layers of the state.

Ambiguity and conflict have also been central themes in research on land rights in Africa. Land rights are the focus of complex material and symbolic struggles, resulting in overlapping and highly variable rights. Underlying the struggles is the overlap of customary law and European systems of property law introduced by colonial regimes. Customary law has, with some modifications, survived into the postcolonial period. Customary legal systems typically recognize several intermediate layers of authority between the ultimate sovereign (the state nowadays) and land users, resting with customary chiefs, village headmen, and family heads. The intermediate layers of authority retain residual rights on land and connect the right to land with obligations for land users, such as political allegiance, labor services, and product shares. At the same time, African land rights are rife with contestations over single resources and the specific rights associated with them. Just as in post-socialist societies, material and symbolic struggles blur the boundary between private and public, are as much about the meaning attributed to resources as they are about the resources themselves, and are informed by conflicting visions of a desirable future. States' views of rural resources as commodities contributing to foreign currency earnings are in stark contrast to local people's visions and the meanings they attribute to land.

Land rights have been the targets of repeated attempts at **land reform**. They have taken the form of redistribution, seeking a change in agrarian structure either through state-mandated land redistributions or market-assisted allocation. Redistributive land reforms have sought to achieve different goals. Some have promoted agricultural commercialization and accumulation through a landed elite, commercial farmers, or peasants. Others attempted to create shifts in the dominant rural class from capitalist landed elites to smallholders or peasants. Post-socialist redistributions have moved control over land from state and collective to private actors, being a form of land privatization. Another type of land reform does not seek to change the agrarian structure but wants to modify land rights. This type of tenurial reform has aimed at the abolition of absentee landlordism, eradication of sharecropping, replacement of sharecropping with fixed money rent, regulation of that rent, and land consolidation to eliminate fragmented holdings. Sharecropping has been a primary target of reform because of its perceived negative effects on the efficiency of land use and long-term **investments**. In Africa, land reform has often taken the form of land registration. In the 1980s, land registration programs sought to eliminate the perceived lack of secure land rights through formal titling. The outcomes of the programs in terms of their effects on agricultural performance have been largely disappointing, however. Moreover, the focus on private, exclusive land rights has weakened the land rights belonging to some actors, such as women. Overall, the outcomes of land reforms have mostly failed their goals. Legal and regulatory reforms have exerted only limited effects on actual land rights. Even where landholders have expanded their legal rights on land, they have often not been able to translate those into substantive economic benefits in the absence of accompanying measures such as credit programs.

Land rights and land use mutually influence each other, although it is difficult to come up with generalizable statements about the nature of their relationship. Programs strengthening private property rights have been shown to encourage agricultural investments and more efficient land use in some cases. The absence of enforceable land rights is often a primary driver of **deforestation**. Nationalization of land rights, in turn, has affected the management of **forests** in positive and negative ways. Current devolution programs in **forestry** and grassland management assume that assigning land rights to local communities will improve local land use. Overall, the effects of land rights on land use are highly varied and context-dependent. Vice versa, the influences of land use on the nature of land rights are equally complex. The emergence of private land rights has been hypothesized to result from increasing land scarcity due to rising population density, technical change, and expansion of trade. At the same time, there are many examples where land scarcity has not led to private land rights. Similarly, the use of **common-pool resources** such as grasslands and forests is thought to be prone to communal land rights. The world's grasslands and forests are under a large diversity of land rights, however. These insights indicate that land rights are complex social relationships, defying simple relations with land use. Land rights carry not only economic values but also social meanings, and they are embedded in broader social relations.

See also: **Access; Agro-Industry; Arena of Land Conflict; Cadastre; Community Involvement; Cultural Factors; Driving Forces; Institutions; Land Rent; Land Tenure; Mediating Factor; Transhumance.**

Further Reading

Agarwal, Bina, *A Field of One's Own: Gender and Land Rights in South Asia*, Cambridge, UK: Cambridge University Press, 1994; Bassett, Thomas, and Donald E. Crummey, eds., *Land in African Agrarian Systems*, Madison, WI: University of Wisconsin Press, 1993; Geisler, Charles, and Gail Danecker, eds., *Property and Values: Alternatives to Public and Private Ownership*, Washington, D.C.: Island Press, 2000; Gluckman, Max, *The Ideas in Barotse Jurisprudence*. Manchester, UK: Manchester University Press, 1972; Sturgeon,

Janet, and Thomas Sikor, eds., "Postsocialist Property Relations: Ethnographies of Agrarian Change in Eastern Europe and East Asia," *Conservation and Society* 2 (2004), special issue; University of Wisconsin–Madison, Land Tenure Center [Online, October 2004], Land Tenure Center Web Site, www.wisc.edu/ltc/; Verdery, Katherine, *The Vanishing Hectare: Property and Value in Postsocialist Transylvania*, Ithaca, NY: Cornell University Press, 2003.

THOMAS SIKOR

Land Stability. The scale-dependent notion of "no change" in land-cover attributes, which often contradicts prevailing scientific views or underlying assumptions, especially neo-Malthusian views of land change in **degradation narratives**. Areas of "no change" show up as diagonal elements in table formats derived from the analysis of land-cover images in post-classification "from-to" **change detection**. The Tri-Academy Panel on Population and Land Use provides examples that are actually case studies from the world's most populous countries, that is, India (Kerala, Haryana), **China** (Jitai basin region, Perl River delta), and the **United States of America** (south Florida, Chicago). Stability in land use and land cover is referred to in two respects: first, contrary to common **perceptions**, forest areas seem to be stable (or even increase) in study regions that have high population density; and second, total agricultural land in most study regions did not undergo major changes in the 1970–1995 period, even though the populations in these regions did grow significantly. Local-scale studies confirm these trends, but also stress the scale dependency of observations. For example, **agricultural intensification** and growing agricultural workforces are dominant processes associated with land change in Southeast Asia during the recent decades, but areas of **disintensification** as well as areas of production stability can be found as well. The process of land stability has involved swidden cultivators in the highlands but many more lowlanders. Summarizing results from eight sites where changes in land cover in Southeast Asian upland areas over the last fifty years were documented, Jefferson Fox and John Vogler found that, at the level of secondary forest **regrowth** after swidden cultivation, land cover remained stable across the region throughout this fifty-year period. However, the level of abstraction hid major changes that occurred within the sub-categories of **secondary vegetation** as well as in the size and number of land-cover fragments (actually, intensification of **swidden cultivation** and introduction of **cash crops**). In the lowlands, few farmers in Thailand, for example, rely solely on **agriculture** to meet their needs, and rural households are dividing their time between farm and non-farm activities instead. Thus, they construct **economic livelihoods** that are increasingly hybrid, both spatially and sectorally, which might explain the observation of land stability there. Areas of stable **forest** that are found in landscapes susceptible to **deforestation** pressures are called **forest anomalies**. Since these anomalies are located in regions where roads and/or markets primarily drive forest-cover changes, Charles Schweik and colleagues assume that they indicate the likely locations of effective forest management **institutions** that are working to mitigate the overall trend toward deforestation in these regions. Several studies in the Indian central and western **Himalayas**, and in Nepal (where the worst crisis conditions were predicted), have demonstrated that there has actually been a net regrowth of forests in several regions in the hills, following local responses of forest communities to degradation; and some extremely stable forest boundaries even maintained by local communities for hundreds of years, such as in the Kumaon region in India. Obviously, the identification of land stability can change with varying **scales** of analysis. On the other hand, it cannot be excluded that an understanding—especially by ecologists—of changing land-cover attributes due to human activities might be based on inappropriate models that emphasize equilibrium and homogeneity rather than heterogeneity and **disequilibrium dynamics**. For example, the prevailing scientific view during the colonial era was that once ecosystems reached a climax state, they maintained stability. Stability was viewed as the desired condition, while disturbance was viewed as unnatural and damaging. In this regard, indigenous land management systems that involved **slash-and-burn agriculture** in humid forest ecosystems or intensive grazing of livestock by pastoralists in drylands (both practices potentially causing periodic ecological disturbances over wide areas) were often misperceived as degrading and often outlawed.

Further Reading

Fox, Jefferson, and John Vogler, "Land-Use and Land-Cover Change in Montane Mainland Southeast Asia," *Environmental Management* 36 (2005): 394–403; Schweik, Charles M., Nagendra, Harini, and Deb Ranjan Sinha, "Using Satellite Imagery to Locate Innovative Forest Management Practices in Nepal," *Ambio* 32 (2003): 312–9; Tri-Academy Panel, "Population and Land Use in India, China, and the United States: Context, Observations, and Findings," in *Growing Populations, Changing Landscapes: Studies from India, China, and the United States*, eds. Indian National Science Academy, Chinese Academy of Sciences, and U.S. National Academy of Sciences, Washington, D.C.: National Academy Press, 2001, pp. 9–72.

HELMUT GEIST

Land Tenure. The mode by which land is held or owned. It may vary from place to place within the same locality, or between an urban area and its surroundings (hence, the distinction between "urban" and "rural" tenure). The word "tenure" the (from the French *tenir*, meaning "to hold") has come to mean primarily holding of rights to own, occupy, use, control, and transfer land. The meaning of ownership is emphasized in the Anglo-American legalist approach to tenure studies, which deems landed rights as "**property**" vested in an individual or group and applicable separately to land or development on it. This approach places the focus on rights, not people or land, hence downplaying the role of the stakeholder, geographies, and technologies in shaping tenurial arrangements.

Several complications merit brief explication. First, land tenure is a multidimensional concept. It comprises the complete "bundle" of landed rights made up of many separate "sticks" that are often assigned to a multitude of landholders (quantity dimension). It can be classified into general tenure types, the "public," "communal" (or **common-pool resources**), and "private" being the primary division (quality dimension). These dimensions may be used separately or together to identify principal tenure forms: for example, ownership and tenancy are two broad classes defined in terms of quantity of interest vested in land, while the latter may be further classified into freehold and leasehold types according to duration of such interests. The spatial and hierarchical arrangements of tenure add further complications: for example, not only may several forms of tenure co-exist on the same plot, but tenure boundaries may not be coterminous with operating units. Remnants of the feudal tenure system may provide for a hierarchy of successive groups according to rank from the landowner to the farm worker. Second, it might be more appropriate to think of tenure status as a continuum (e.g., using quantity/quality of vested interests in land as a common denominator), rather than in terms of mutually exclusive classes. The continuum may run from occupancy with no legal interest (as represented by squatters) to unqualified public interest in land control (as manifest in the rights of public domain, policing, taxation, and spending vested in the U.S. government), while "sharecropping" and "hired man" may fall in between. Third, it is necessary to distinguish *de jure* from *de facto* (and formal from informal) tenure status with regard to tenure security. Although tenure security is widely accepted as a precondition for land stakeholders to invest in any improvement, occupancy alone sometimes provides security of tenure, a fact well known to squatters. Notably, different legally acceptable systems may operate in adjacent locations, so that legal tenants migrating from one location may be considered to be behaving illegally in their new locations.

We may speak of different land tenure systems (the "statutory," "customary," "Islamic," etc.) in terms of their origins, basic principles, and structures. Most actions that affect the tenure systems occur externally, where breaks with the past were first developed in practice and later formalized in statutory enactments. They often embody divergent principles: for example, many customary tenure systems regard land as sacred, and man's role as one of stewardship to protect the interests of future generations, where land is distributed by community leaders according to needs, rather than through payment. In the **United States of America**, on another hand, the overarching tenure principle has long been freedom of action without personal responsibility, where the economist's free market theory and the policymaker's practice of *laissez faire* complement the legalists' championing of freedom of contract (as seen in the "fee simple" estate). In terms of tenure structure, socialist countries vest most landed rights in the state, whereas capitalist countries restrict public ownership to a narrow range. In a larger picture, the manner by which a society distributes and assigns rights in land is an important indicator of that society, since they can be held to reflect rights in other areas of social life. Therefore, **land reform** essentially means tenure reform.

Land tenure is critical to understanding land-use/cover change processes, for it provides for an orderly way of land use, a means of conflict prevention in land transactions, and more importantly, a frame for land-use **decision-making** (if we deem it a key component in the "structure-agency" formulation with regard to structure and **institutions**). It helps establish the size of feasible farming operations, and influences urban land value and options for **access** by the urban poor. However, tenure studies remain underdeveloped in land-use/cover-change research.

See also: **Agricultural Revolution; Arena of Land Conflict; Cadastre; Human Immunodeficiency Virus (HIV)/Acquired Immunodeficiency Syndrome (AIDS); Investments; Land Privatization; Land Rights; Mineral Extraction; Population Dynamics; Private Property; Transhumance.**

KE CHEN

Land Use Policy. Launched in 1984, this interdisciplinary journal is currently produced in four issues per year by Elsevier Science Ltd. It publishes reports of original research and critical reviews relating to any aspect of land-use policy, broadly defined. Submissions are subjected to independent external review before acceptance for publication. Special thematic issues are published from time to time. Book reviews are also carried. Papers on urban and transport-related issues are accepted, but the majority of articles relate to rural themes. Typical topics include tropical **deforestation**, the European Union's Common Agricultural Policy, and matters of **land-use planning** in the **urban-rural fringe**. A key feature is an emphasis on implications for land-use policy. The journal is international in its readership and author origins. It seeks to provide a stimulating forum for researchers and practitioners involved in the broad field of land-use policy.

See also: **Land-Use Policies.**

Further Reading

Elsevier Science [Online, December 2004], The Land Use Policy Journal Web Site, www.elsevier.com.

ALEXANDER S. MATHER

LANDFIRE Project. An initiative started in 2003 to provide detailed, consistent, and comprehensive landscape-scale wildfire-related vegetation and other ecological variables for the **United States of America**. LANDFIRE data sets support **fire** management planning, prioritization of fuel treatments, collaboration, community and firefighter protection, and effective resource allocation. LANDFIRE includes detailed natural vegetation characteristics maps that correspond to fuels. Mapping fire fuels requires data and information about **vegetation** types, structure, and green **biomass**. Combined with extensive field-based vegetation descriptions, **Landsat** 7 ETM+ imagery are being used as the primary source of the vegetation characteristics data, a general land-cover classification, a **forest** -type group classification, and a tree canopy density data layer. The LANDFIRE initiative is a collaborative effort among several federal agencies, including the U.S. Forest Service, U.S. Geological Survey, and the Department of the Interior.

See also: **United States Geological Survey (USGS) Program.**

Further Reading

United States Geological Survey, Landfire [Online, November 2004], The LANDFIRE Web Site, www.landfire.gov.

THOMAS R. LOVELAND

Landholding. The economic attributes of rural producers are strongly associated with the landholding size, but also to the way in which the production process is organized, and the levels of economic or **agricultural intensification**. In this light, looking at the size of landholdings provides the main distinction among rural agricultural producers that differentiates small- and large-scale farmers, or some other types. Hence, landholding size by itself is a powerful variable that allows for separating the attributes of two quite different groups of rural producers. It is well known that land-use decisions made at the farm level, and their implications on forest **conversion** to agricultural uses, tend to vary greatly in small- versus large-scale landholdings. Landholding size, therefore, can be seen as a **mediating factor** that contributes to explain **deforestation** in the tropical landscapes.

Land-use change processes and patterns are different for small- versus large-scale landholders, though their actions in some cases are interlinked. In the tropical regions, deforestation taking place in large-scale landholding is often done through clearing of larger chunks of **forests**, taking the shape of large continuous and symmetric patches. The trajectories of forest conversion are relatively simple, since forest is in most of the cases definitely converted to agricultural uses, mainly to **pasture**, and it is alternatively devoted to grains production, though some proportion may be abandoned.

Land-use change trajectories taking place in small-scale landholdings, and their imprints on the landscape, are both quite more complex. There is no single dominant land-use trajectory in small-sized lots. Forest cleared for **agriculture** may remain for a couple of years under temporal crops, or it may be converted to pasture after the first year to avoid forest succession. In other cases, forest **regrowth** is allowed once the soil fertility is exhausted with temporal crops. Some other smallholders favor permanent crops intermingled with temporal crops and forest succession areas. The diversity of situations is due to the fact that small-scale farmers often choose diversified systems of production as a way to secure their livelihoods, which leads to diversified land uses whose final imprints on the landscape will consist of multiple and dispersed small, asymmetric patches of cleared areas in combination with forest regrowth areas in different stages of succession.

Large-scale landholders' main motivation is to achieve the largest economic benefits. They are often better connected to markets, use industrial inputs, and hire labor forces. Hence, in these groups of producers, land-use change decisions, and particularly deforestation, are contingent upon the profit they will obtain from expanding arable land for commercial crops, or pasture for beef cattle production. The medium- and large-scale agriculturalists or ranchers, therefore, will deforest more in cases in which the prices of commodities are higher than their production costs. The latter is likely to happen when they benefit from prices of support, subsidized and cheap credit, preferential export taxes, favorable exchange rates for export commodities, and low prices of inputs. The medium- and large-scale establishments have in diverse contexts benefited from policies that have for different reasons tended to provide them additional institutional rents (cheap credit, access to roads, etc.).

Small-scale farmers have lower **access** to assets. These types of farmers are highly diversified because they may have a high discount rate, specializing in low-return and low-risk activities, and combining subsistence with market behaviors. While wealthier households may prefer to make **investments** in higher-return activities, such as expanding their **cash crops**, the poorest may choose to spend their economic assets in producing subsistence crops. Their difference with respect to large-scale landholders is that while the latter are able to mobilize resources to expand their scale of operations or to intensify their productive system, smallholders are constrained by their **perception** of risk, higher cost of access to financial resources, and scarce factors of production to allocate them among competing on-farm activities. Furthermore, policies have negatively biased small farmers, and limited them to expand their access to assets.

In short, the differences in the holding of land and other assets can largely influence local natural resources use by affecting landholders' resource-allocation decisions. Large-scale landholders with better access to assets are less constrained by financial resources, are better suited to interact in the markets, and adopt new technologies more readily, but this often leads to more simple land-use decisions that favor permanent forest conversion to agricultural uses. In contrast, smallholder farmers are more constrained in their access to factors of production and financial markets, and food production is at the center of their economic strategies. These factors lead smallholders to adopt more complex land-use strategies, resulting in less-homogenous landscapes in which temporal and permanent crop areas coexist intermingled with secondary forest.

See also: **Agricultural Frontier; Agrodiversity; Corporate Strategies; Decision-Making; Economic Livelihood; Food Crops; Land Abandonment; Land Concentration; Land Fragmentation; Land Privatization; Land Reform; Land Rent; Land-Use Policies; Property; Public Policy; Subsistence Agriculture.**

PABLO PACHECO

Land-Ocean Interactions in the Coastal Zone (LOICZ) Project. A joint core project of the **International Geosphere-Biosphere Programme** and the **International Human Dimensions Programme on Global Environmental Change** since 2004. The project supports research aimed at informing scientists, policymakers, managers, and stakeholders on the relevance of **global environmental change** in the **coastal zone**.

See also: **Water-Land Interlinkages.**

Further Reading

Land-Ocean Interactions in the Coastal Zone [Online, February 2005], The LOICZ Web Site, www.loicz.org.

HELMUT GEIST

Landsat. A series of satellites designed to gather data on the earth's resources, that is, to characterize, monitor, manage, explore, and observe the earth's land surface. Landsat is an abbreviation for Landsat thematic mapper. It provides a widely used type of long-term satellite imagery—along with the **advanced very high resolution radiometer** as a second source of remote sensing information—covering the majority of the earth's surface at thirty-meter resolution. The system records the intensity of radiation reflected from the earth's atmosphere-surface system in five to seven spectral bands in the visible, near-infrared, and thermal infrared parts of the spectrum (with the information used to extract surface images). A Landsat scene is approximately 185 kilometers on a side, and each location on earth is revisited at about sixteen-to-eighteen-day intervals. The Landsat data record started in 1972, when Landsat-1 was developed and launched by the U.S. National Aeronautics and Space Administration, or NASA. The multispectral scanner on-board Landsat-1 was so valuable that a version of it was flown on each of the first five missions. Since 1982, when Landsat-4 was launched, the primary instrument has been the thematic mapper. Landsat-6 failed to reach orbit in 1993, but Landsat-5 continued to provide coverage well beyond its design lifetime. In 1999, Landsat-7 was launched to provide continuity and enhanced coverage. The Landsat program passed the Land Remote Sensing Policy Act and was then approved by Congress in 1992. The act allows a continuous collection and utilization of terrestrial **remote sensing** data from space in order to study and understand human impacts on the global environment and to better manage the earth's natural resources.

See also: **Change Detection; Earth Resources Observation Systems (EROS) Data Center; Global Terrestrial Observing System; LANDFIRE Project; NASA Land-Cover/Land-Use Change (LCLUC) Program; Normalized Difference Vegetation Index; Pixel; Sahel Land Cover; United States Geological Survey (USGS) Program.**

Further Reading

National Aeronautics and Space Administration, Landsat 7 [Online, March 2005], The NASA Web Site, landsat.gsfc.nasa.gov.

HELMUT GEIST

LandSHIFT Model. A global land-use change model currently under development at the Centre for Environmental Systems Research in Kassel, Germany. The abbreviation means "Land Simulation to Harmonize and Integrate Freshwater Availability and the Terrestrial Environment." The principal goal of the model is to translate regional demand trends for land-intensive commodities (such as field crops) into area requirements for land-use types, and to allocate these on a grid of five arc minutes resolution. Allocation assumes an idealized behavior of land users in that they seek for optimal conditions, constrained by the competition among different land-use types and regional socio-economic settings. The procedure accounts for intrinsic **feedbacks** by considering the consequences of land use and land-use changes on **land quality**, land scarcity, and water availability. LandSHIFT is expected to provide insights into the dynamic behavior of regional **land-use systems** within a globally consistent framework, driven by **scenarios** of climate change as well as socio-economic and technological development.

The set of land-use types involves **agriculture**, or agricultural land use, which is characterized by major crops and a set of management variables such as irrigation, fertilization rate, crop rotations, and/or cropping intensities. Remaining land-use types include **forestry**, urban/built-up land, and a set of **biomes** representing unused or natural land.

To allocate demands, a rule setting is applied that first translates a spatial inventory of suitability factors into a spatial measure of preference for a specific land-use type. These factors include, for example, terrain slopes, potential crop yield under specified management conditions, and **access** to market centers as a derivative of proxy data such as distance to infrastructure and distance to urban centers. Second, allocation rules account for the competitive capacity of different land-use types as well as for the **resilience** of land-use types against **conversion**. A set of spatial indicators will support the decision of how scenario assumptions about technological development and management will be spatially realized. Such indicators include measures for the abundance of suitable land, as well as maps quantifying the sensitivity of crop yields on specific management activities such as irrigation.

In terms of implementation, LandSHIFT will be coupled to two external models. First, potential crop yield will be dynamically simulated by applying the process model DayCent, which was developed by William Parton and colleagues. The adapted version of this agro-ecosystem model is able to simulate yields for major crops under specified management conditions, as well as related nitrogen, carbon, and water fluxes in the plant-soil system. DayCent allows for calculating crop water requirements as well as the impact of irrigation on yield levels, and accounts for the feedback of changing nutrient levels on crop growth. Second, the global water cycle model WaterGAP, developed by Joseph Alcamo and colleagues, is applied to compute water withdrawals and water availability on watershed levels. Water users include municipalities, industries, power plants, and agriculture. Based on daily water balances, river flows are routed through a global flow routing scheme. Both DayCent and WaterGAP are able to account for the potential impact of changes in the climate system on crop yields and water balances.

See also: **Agent-Based Model; Agricultural Intensification; Century Ecosystem Model; Disintensification; Driving Forces; Land Degradation; Modeling; Proximate Causes; Urbanization; Water-Land Interlinkages.**

Further Reading

Alcamo, Joseph, Doell, Petra, Henrichs, Thomas, Kaspar, Frank, Lehner, Bernhard, Roesch, Thomas, and Stefan Siebert, "Development and Testing of the WaterGAP 2 Global Model of Water Use and Availability," *Hydrological Sciences Journal* 48 (2003): 317–37; Parton, William J., Holland, Elizabeth A., Del Grosso, Steve J., Hartman,

Melanie D., Martin, Robin E., Mosier, Arvin R., Ojima, Dennis S., and David S. Schimel, "Generalized Model for NOx and N_2O Emissions from Soils," *Journal of Geophysical Research-Atmospheres* 106/D15 (2001): 17,403–19; Center for Environmental Systems Research [Online, August 2004], The Center for Environmental Systems Web Site, www.usf.uni-kassel.de/usf.

MAIK HEISTERMANN, JÖRG A. PRIESS

Land-Use History. Exploration of land change in both the distant past and over the last 300 years, and the period of the most fundamental land-cover change with many consequences for today's landscape configurations. Aside from long-term climatic shifts, changes in land cover are primarily linked to land use. Landscapes are therefore understood fully only by knowing their land-use history. This historical perspective directly informs understanding of contemporary land conditions by assessing types and rates of land change and identifying biophysical and social change processes.

The bulk of land-use history is produced by practitioners in the fields of ecology, history, and geography. Historical ecologists and physical geographers consider the interactions among the land, human activity, and natural disturbance to better understand ecological processes and changes in the structure and function of ecosystems (the magnitude and character of land change). Toward these ends, they draw on a range of tools and data sources, including pollen and macrobotanical remains, soil chemistry, climate reconstruction, radio carbon dating, cadastral information, vegetation inventories, and remotely sensed imagery.

Research in the humanities and social sciences shifts the focus of land-use history to the impact of changes in socio-economic and political organization on land use and land cover. While addressed in diverse ways, some of which rely on the same data sources as historical ecology but that also include oral histories and the written record, environmental historians and human geographers tend to rely on a narrative to weave together the role of **institutions**, **cultural factors**, and political economic characteristics in either driving land change or the reverse ways in which land conditions shape cultural norms.

In addition to addressing the questions pertinent to the practitioner's core interests, collectively, land-use history research has produced a range of important outputs. Three general contributions are emphasized here. First, land-use history empirically documents the spatial **scale**, pattern, and pace of historical land change, which helps uncover the ecological impacts of different land uses, and the role of biophysical **feedbacks** on use systems, such as rates of soil and **vegetation** recovery. The provision of historical data also contributes to understanding of biogeochemical cycles (carbon, in particular) and regional climate change.

Efforts to distinguish between natural and social causes of land change are often inconclusive. Descriptive information on historical land use, when combined with an understanding of the relationships among climate, natural perturbations (e.g., wind, **fire**, and disease), and vegetation, proves invaluable when trying to determine the relative influence of anthropogenic and biophysical forces in causing land change. In the northeastern **United States of**

America, for example, changes in **forest** composition and structure over the past 300 years are linked to both environmental variation and land use, but with climate and geology exerting a greater influence at broader scales of analysis. In another example, a recurring debate surrounds the degree to which **grassland** degradation (e.g., **soil erosion**, shrub encroachment, and falling productivity) in regions around the world is linked to climate shifts or overgrazing. In West Africa, what once were thought to be forest remnants of larger forests vastly diminished by local inhabitants are now known to be forest islands, created by land managers in what had been open grasslands. Here, land-use history has served to highlight the human role in landscape change, but it also challenges the conventional wisdom of **degradation narratives** and reinforces new ecological models that emphasize disturbance and non-equilibrium systems, which can only be understood over the long term.

Second, the reconstruction of prehistoric or indigenous landscapes and the ways in which humans have modified and transformed the land over the long term are fundamental to debunking the myth of pristine nature. Most areas on earth, from Antarctica to the Amazon basin, show a human imprint. Demonstrating the social construction of nature from the distant past to the present corrects misperceptions that human impact on the earth is recent and underscores human agency in altering biophysical environments. Such studies also demonstrate the role of humans in maintaining certain "desirable" landscapes, such as forest cover.

Recognition of the historical impact of humans on nature is relevant today. In the context of debates about sustainable development, for example, trends in the past may point to alternatives for the future. The embrace of cultural landscapes may be a necessary part of a transition to more sustainable human-environment conditions. Awareness of the outcomes of past land-use decisions, therefore, may influence choices about ongoing and future land use. Which human-environment conditions lead to more resilient or more vulnerable use systems? For example, plant and animal species (thus **biodiversity**) adapt to, and evolve with, a given disturbance regime, which in some areas is a function of long-term land use. Future alterations of past anthropogenic disturbance regimes may have implications for biodiversity. In the face of increasing dominance by human beings over land resources, which technologies and management regimes have been the most sustainable?

Land-use histories can also identify issues of path dependency, or the degree to which past land uses affect the range of use options available to present-day land managers. In the southern Yucatán peninsula, for example, past logging of hardwoods has eliminated this economic activity as a contemporary option in many locations, and long-term **slash-and-burn agriculture** seems to contribute to the rise of bracken fern (*Pteridium aquilinum*), a weed that impedes processes of **secondary vegetation** or succession and that effectively takes land out of agricultural production for an extended period of time.

Third, land-use histories inform understanding of the **driving forces** of land-use and **land-cover change** as well as efforts to model land-change dynamics. Explanations for land change are clustered, minimally, into two broad categories. The first explains land use as a product of political, economic, and social structures; and the second focuses on the immediate agents of change, usually local land managers. Land-use histories examine the relative roles of these drivers of land change as they cross distinct human-environment episodes. Within any given episode, for example, household demographics and economic explanations may be identified as the critical factors; whereas, in other periods, key land transformations may be precipitated by state policy.

Comparing regional land-use histories supports the rising consensus in nature-society studies that a web of driving forces of land change exists, the exact combination of which depends on regional and historical contexts. Land-use histories often identify the degree to which change trends are nonlinear and affected by stochastic events. These events may be represented by external shocks to regional **land-use systems** (forces emanating from outside of the region and beyond the control of local inhabitants), such as federal development policy or natural perturbations, and they are forces often operating not only in distal space but in distal time. In the southern Yucatán peninsula, the most important land-use changes in the last 100 years—depleted hardwood reserves, large areas of permanently cleared forest, the establishment of **conservation** and extractive reserve areas—are all linked to particular historical episodes in which there were government policy decisions about **colonization**, federal land grants to the timber industry, agricultural development policy, and international pressure to conserve the region's tropical forests.

In **modeling** efforts, whether backcasting (to use models to "predict" the recent past, thus testing the strength of the models and their parameters) or projecting future land use, understanding the role of these external shocks is critical. Is the recent historical episode typical of land-use change in broader temporal scales, and therefore an appropriate period on which to base assumptions about future trends? By defining the **initial conditions** upon which current land-use patterns were founded, land-use history is fundamental (or should be fundamental) in setting the parameters of models and their suitability for the region in question. Models that do not include the effect of federal policy, for example, are apt to be ineffective in projecting land use beyond the short term.

See also: **BIOME 300 Project; Boserup, Ester; Cadastre; Decision-Making; Disequilibrium Dynamics; Environmental Change; Exogenous Factor; Mediating Factors; Pattern to Process; Pristine Myth; Remote Sensing.**

Further Reading

Fairhead, James, and Leach, Melissa, *Misreading the African Landscape*, Cambridge, UK: Cambridge University Press, 1996; Foster, David R., "Insights from Historical Geography to Ecology and Conservation Lessons from the New England Landscape," *Journal of Biogeography* 29 (2002): 1269–590; Klepeis, Peter, and B. L. Turner II, "Integrated Land History and Global Change Science: The Example of the Southern Yucatán Peninsular Region Project," *Land Use Policy* 18 (2001): 27–39; Lambin, Éric F., and Helmut J. Geist, "Regional Differences in Tropical Deforestation," *Environment* 45 (2003): 22–36.

PETER KLEPEIS, PAUL LARIS

Land-Use Legacy. The notion among ecologists and conservationists that land-use activities continue to influence ecosystem structure and function for decades or centuries—or even longer—after those activities have ceased. The recognition that legacies of land use can be remarkably persistent contradicts the pervasive assumption or **pristine myth** that society's **modification** of nature is a relatively recent phenomenon, and that remote or uninhabited lands are untouched or pristine wilderness areas. Indeed, there is growing recognition that most "natural" areas actually have more cultural history than assumed. For example, the **Intergovernmental Panel on Climate Change** estimates that roughly one-fifth of the total anthropogenic emissions of **greenhouse gases** during the 1990s were from changes in land use. Likewise, it has been estimated that the human impact on the terrestrial water cycle during the last fifty years has likely exceeded natural forcings of continental aquatic systems in many parts of the world. In the study of land-use/cover change, land-use legacies are best understood through a careful consideration of the environmental and **land-use history** of a given site or area, that is, that which constitutes the **initial conditions** of any land change. These conditions, together with an exploration of the drivers of change and the structure of feedbacks (or responses) to these drivers, do best account for typical pathways or trajectories of land change.

See also: **Degradation Narrative; Ecological Footprint; Ecosystem Services.**

Further Reading

Foster, David, Swanson, Fredrick, Aber, John, Burke, Ingrid, Brokaw, Nicholas, Tilman, David, and Alan Knapp, "The Importance of Land-Use Legacies to Ecology and Conservation," *BioScience* 53 (2003): 77–88; Turner, B. L. II, and Susannah McCandless, "How Humankind Came to Rival Nature: A Brief History of the Human-Environment Condition and the Lessons Learned," in *Earth System Analysis for Sustainability*, eds. Clark, William C., Crutzen, Paul, and Schellnhuber, H. John, Cambridge, MA: MIT Press, 2005, pp. 227–43.

HELMUT GEIST

Land-Use Planning. Choosing, strategically in time and space, the uses of land in a sustainable perspective. As the land is a functional interface between the ground surface with all the physical, chemical, and biological features that influence the use of resources, land planning means the rational and optimal integration of all the dimensions of the dynamic **land-use system**. The **United Nations Environment Programme**, in collaboration with the **United Nations Food and Agriculture Organization** (FAO) and the United Nations Education, Science and Culture Organization (UNESCO), defined sustainable land use as the management of land to ensure the continuous satisfaction of human needs for present and future generations. **Sustainable land use** then automatically involves the optimization of the land-use potential and the planning of current and future uses. The implementation of land-use planning is a real challenge for land-use planners at global, regional, and national scales, but mainly for land users themselves that have in the past sometimes been excluded from the planning process. The evolution of land-planning theories, the definition of the land-planning process, and the current applications of these methodologies at different scales will be briefly described here.

Over time, land-use planning has developed into more of a science than a tool for development. Theory and methodological frameworks emerged from different scientific disciplines as landscape planning, ecosystem management, and environmental planning in the nineteenth and twentieth centuries. To better understand the reality of the land-use system, scientists built models that represented the use of resources that satisfied human needs. They evolved from models with one quantitative variable (population or climate) to more complex but qualitative land-use systems (e.g., the perspective of agro-ecosystems or **agrodiversity** that includes **indigenous knowledge**), and to quantitative **integrated models** of land-use change. Experiences of **land degradation** are progressively explained by the **vulnerability** of the population rather than by a simple misuse of the land. In the twentieth century, soil mapping widely developed and became a major method of land-resource assessment, which was based not only on field surveys but also on **remote sensing** techniques. **Decision-making** on sustainable land use requires an integrated and multidisciplinary approach of understanding land use in the long term. In its guidelines for integrated land planning, FAO proposes a step-by-step method for physical **land evaluation** that involves a consultation to formulate the relevant land uses. The **International Institute for Applied Systems Analysis**, in collaboration with FAO, developed a decision-making software tool that has been applied in Kenya. Other initiatives are currently tested, for example on the Loess Plateau in **China**. An adequate institutional context and the effective participation of all stakeholders in the negotiations is a necessity in the implementation of such participative land-use planning experiences.

The concept of planning has been used in various disciplines to describe the four steps in the implementation of plans in one organization. The first element of planning is the articulation of well-defined objectives. Land-use planning deals with multiple objectives and multi-agents with reference to specific spatial and time **scales**. The optimization of objectives requires weights and priorities in the short and long terms. Compromise goals result from **tradeoffs** between two somewhat contradictory dimensions of land use: ecological **conservation** and socio-economic subsistence. The objectives should be expressed as explicitly and quantitatively as possible, for example in terms of performance indicators. Second, the understanding of the organization and its resources is called analysis, diagnosis, or evaluation. Understanding the land-use system and its environment involves the classification of resources, production functions, populations, and influences. This analysis should define the strengths and weaknesses of all parts of the system and identify the opportunities of change. In this phase, the building of a detailed **geographical information system** (GIS) is an interesting approach. The third element of planning is the definition of possible options to achieve the objectives. In this definition, the decision support system managing a GIS proposes **scenarios** and simulations to the land user itself. These simulations consider the optimum strategies in land-use allocation and land-use changes. Development of such systems at local or regional scales is an extraordinary challenge of this century. Fourth, design and implementation of plans are achieved by a collaboration of several teams of people assembling different skills. Implementing development programs, plans, and activities should be done by governments, communities, and individuals, respectively. This fourth step in strategic planning deals with the identification and negotiation of major development priorities. This involves using a set of practical tools (land evaluation, farming systems analysis, etc.) to integrate existing information, analyzing **land-use policies** and proposals, and negotiating conflicts and competing interests.

Most of the world's countries have set up national land policies. Most developed countries have put in place physical plans or land zoning. In the United States, the first zoning ordinance was passed in New York City in 1916, and by the 1930s, most states had adopted zoning laws. From 1969 onward, a federal law required every city and county in the state to have a comprehensive land-use plan. Within the European Union (EU), many guidelines are formulated into EU directives at the supra-national level. Many developing countries have a national environmental action plan (NEAP) as well, often established with the assistance of the World Bank. Unfortunately, the strategies have been developed, but their implementation is often not yet operational. Planning at the local level (community or village) involves the identification of concrete actions within the strategic plan. Indigenous communities usually have more experience of local conditions than outsiders when managing natural resources like **pastures**, **forests**, water, and wildlife. Technical **institutions** and development organizations often have experience and knowledge of the outside world (e.g., market information and outsider demand for resources use) that is unavailable to those at the local level. Bringing these different sources of information and knowledge together opens more perspectives for sustainable development.

See also: **Community Involvement; Decentralization; Integrated Assessment; Land Rehabilitation; Modeling; Participatory Land Management; Public Policy; River Basin; Urban Sprawl; Watershed.**

Further Reading

Beinat, Euro, and Peter Nijkamp, "Land Use Planning," in *Encyclopedia of Global Change: Environmental Change and Human Society*, vol. 2 (J–Z), eds. Andrew S. Goudie and David J. Cuff, Oxford, UK, New York: Oxford University Press, 2002, pp. 27–33; Botequilha, Leitão A., and Jack Ahern, "Applying Landscape Ecological Concepts and Metrics in Sustainable Landscape Planning," *Landscape and Urban Planning* 59 (2002): 65–93; De Wit, Paul, and Willy Verheye, Land Use Planning for Sustainable Development, in *Encyclopedia of Life Support Systems*, developed under the auspices of UNESCO, Eolss Publishers, Oxford, UK [Online, September 2004], The EOLSS Web Site, www.eolss.net.

NATHALIE STEPHENNE

Land-Use Policies. Land use and changes to land cover are increasingly affected by policies of various kinds. A policy usually consists of one or more defined goals and the means of achieving them: policy objectives are distinguished from policy instruments. Usually the term "policy" is used in relation to government bodies, but social groups such as environmental NGOs (non-governmental organizations) may have policies that they pursue on land they control and with which they seek to influence government

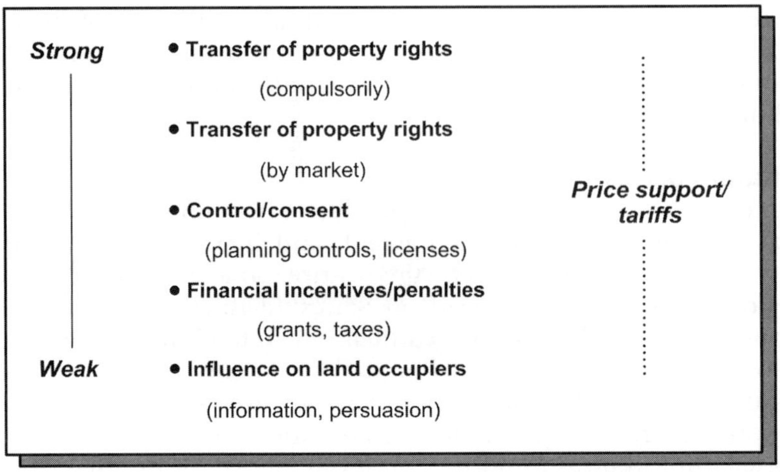

Policy instruments (the "strength" of price support is variable, depending on the level of support provided).

Paralleling agricultural policies are those relating to sectors such as **forestry**, **conservation**, and housing. All imply that an aim has been identified and that the state or other body has resolved to attempt to achieve it.

Policy Instruments

The range of potential instruments is wide: they have different characteristics, especially in terms of strength, and their selection usually involves issues of political judgment and acceptability as well as potential effectiveness. It is sometimes asserted that the number of instruments in a policy should be equal to the number of objectives, but this "law" has not always been observed in land-use policies.

Both the carrot and the stick have been widely employed (see figure titled "Policy instruments"). The latter can involve the compulsory acquisition of land and the banning of various practices or processes by legal means (e.g., removing trees or building houses without the appropriate permission). Land-use policies (*sensu stricto*) can use devices such as zoning to identify the types of land use considered appropriate in particular areas: often the implications in terms of land values are enormous. On the other hand, various incentives can be offered to encourage actions deemed desirable. For example, grants may be offered for tree-planting if forest expansion is a policy objective. They can be offered in undifferentiated form over a whole country, or targeted in particular areas.

Policy Trends

As a sweeping generalization, it can be said that through time, policies have become more numerous and complex, and policy instruments more diverse and sophisticated. More types of land use have been brought under the influence of policy, and a widening range of tools has been employed. Environmental conservation has been a particular growth area: there has been extensive use of the "protected area" or "designated area" as a basic tool, but in recent decades a wider and more imaginative range of instruments has been deployed.

The rise in significance of land-use policies parallels, in large degree, the rise of the state in the modern era, and it continues despite the alleged weakening of the state in recent decades. Some policies exist on paper but have little observable effect: to be effective, they must be backed by a capability of enforcement and implementation that states in weak or embryonic form may not possess. Nevertheless, recent decades have witnessed a trend away from traditional "top-down" approaches in which policies are simply devised and imposed by government bodies, and toward greater participation in policy formulation by the people directly affected by policies.

One of the dimensions of complexity is vertical or hierarchical. In theory, the policies pursued by local government bodies should "nest" within those of the national government. Increasingly, supra-national and global bodies are also becoming involved, through measures such as the Common Agricultural Policy and the **United Nations Convention on Biological Diversity**. Policies, in both the strict and the broad sense, have become major elements of the context in which land is used. Issues such as their evolution and the nature of **feedback** between change in land use and change of policy will increasingly attract the attention of researchers in the field of land use/cover change.

See also: **Access; Agrodiversity; Arena of Land Conflict; Community Involvement; Decentralization; Decision-Making; Green Wall Project; Land Concentration; Land Degradation; Landholding; Land-Use Planning; Participatory Land Management; Perception; Public Policy; Reforestation; Tradeoffs.**

Further Reading

European Commission, CAP Reform: A Long-Term Perspective for Sustainable Agriculture [Online, December 2004], The EC Web Site, europa.eu.int; Organisation for Economic Cooperation and Development, *Agricultural Policies in OECD Countries: At a Glance*, Paris: OECD, 2004; Platt, Rutherford H., *Land Use and Society: Geography, Law, and Public Policy*, Washington, DC: Island Press, 2004; Reid, Robin S., Tomich, Thomas P., Xu, Jianchu, Geist, Helmut, DeFries, Ruth S., Liu, Jiangho, Alves, Diogenes, Agbola, Babatunde, Lambin, Éric F., Chabbra, Abha, et al., "Linking Land-Use/Cover Change Science and Policy: Current Lessons and Future Integration," in *Land Use and Land Cover Change: Local Processes, Global Impacts*, eds. Éric F. Lambin and Helmut Geist, Berlin: Springer-Verlag, 2006; Rydin, Yvonne, *Urban and Environmental Planning in the UK*, London: Palgrave MacMillan, 2003; World Bank, Sustaining Forests: A Development Strategy [Online, December 2004], The World Bank Web Site, www.worldbank.org.

ALEXANDER S. MATHER

Land-Use System. Systemic or semi-systemic representation and characterization of the intricate web of relationships, factors, processes, and structures that impinge on land use (change) and—to the extent that the dynamics of land-use is addressed—**land-cover change**. The simplest form of a land-use system (LUS) consists of one land-use type applied to one land unit, assuming uniform fields. Multiple land uses can be conceptualized as aggregations of such simple systems: for example, intercropping can be considered as concurrent single use sharing the same land unit, whereas rotations are their sequences. Its plural form speaks to the fact that relationships and processes that are manifest in different land-use contexts may vary widely across places; hence, we have myriad land-use systems—for example, **urbanization**, **agriculture**, **forestry**, and **mineral extraction**—coexisting over heterogeneous landscapes.

Diversity of Approaches

There is a wide diversity of LUS constructs in applications, ranging from those conceived in the context of **land evaluation** to those proposed in global change research by the **Land-Use/Cover Change (LUCC) project**, for example (see figure title "Framework"). The former is chiefly concerned with spatial organization of such land-use elements as soil series, land units, and land utilization types; and their interrelationships in maintaining the functioning of the production system are defined mostly in economic and biophysical terms. Although it captures many operational details of land use at the field/parcel level, it often does not account for land-use change, much less their broad-scale **driving forces**. The latter attempts to identify and explain cross-scale, interactive mechanisms of change in both land use/land cover by enlisting a different set of analytical elements (drivers, agents of change, **proximate causes**, mitigators, etc.), which are woven into a broader analytical framework to capture multiple dimensions of LUS.

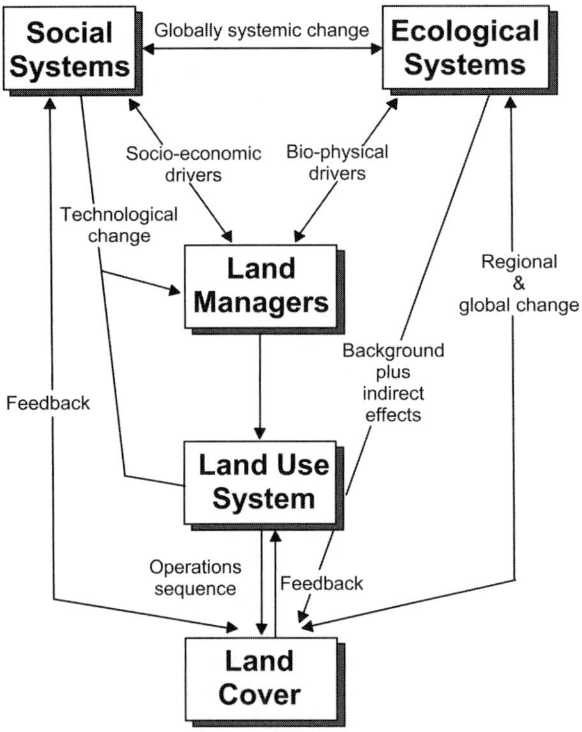

Framework for understanding land-use/cover situations.

SOURCE: Science/Research plan of the Land-Use/Cover Change (LUCC) project, 1995, p. 35.

This diversity may be attributed to differences in purpose of study, level of aggregation, dynamics considered, and functional specification, etc.; but more significantly, it is owing to the lack of theoretically informed conceptualizations of LUS that are widely accepted—those that are able to address the "what" of land-use change and the "why" of this change. In other words, such conceptualization should facilitate researchers not only to describe the patterns of land-use/land-cover changes and their impacts, but also to explain mechanisms of these changes in spatially explicit terms, in that both changes and impacts are treated as phenomena occurring on an area of the earth's surface with particular, spatially variable properties. As noted by Helen Briassoulis, many existing land-use theories refer only to either the "what" or the "why" questions, but rarely both; they attempt to describe but rarely explain. When they do address the "what," they tend not to deal directly with land-use change but with changes in its determinants, while the "why" part is often "black boxed," leaving the user to figure out the underlying mechanisms. In addition, they tend to be procrustean in choosing drivers to be considered, or fail to offer exact conceptual and operational definitions of these drivers, or not treat land-use change and their drivers at the same level of definitional detail. Moreover, many of them are not spatially explicit (in the sense that they tend to reduce land to a point in space, or land as independent variable is totally absent from analysis); and if they do, they do not cover the full array of existing and potential uses of land, their changing patterns, and drivers. Many of these inevitably filter into conceptualizations of LUS in reality. Given the lack of theoretical integration to ground LUS conceptualizations, instrumentalist use of theoretical assumptions becomes a prevailing approach. Systems analysis is something up for grabs for such use, due to substantive and practical imperatives working from both within the land-change community and without. From within, strong need exists to accord a measure of coherence to an intellectual terrain that has become increasingly fragmented by growing diversity of disciplinary inputs, crying for more integrative theoretical schemata. Now that the feasibility and desirability of seeking a general theory of land-use change remains questionable, systems analysis offers a convenient instrument for ostensible if not truly transdisciplinary integration. From without (and in a more practical sense) land-use researchers well-steeped in social science tradition have long envied (and feel inferior to) natural science researchers in earth systems science and climate change study in terms of prestige and funding; for them, systems approaches provide a partial bridge to the language barrier that has long hindered them from communicating and competing with natural science programs operating in the broad field of global change research, such as the Global Change and Terrestrial Ecosystem (GCTE) project of the **International Geosphere-Biosphere Programme**.

Elements of the Systems Approach

The very notion of LUS implies applying systems approaches (not necessarily systems theory per se) to the study of land use. Although most of the systems concepts reviewed below found their way into land-change research mainly through the "back door" of (systems) ecology, their semantic and theoretical lineages can be traced back to classical systems literature of natural sciences (notably cybernetics, informatics, dissipate structure, and synergetics). They have inspired subsequent research in hierarchy, complexity, adaptive behaviors, and nonlinear and spatial dynamics. Systems concepts recently enlisted to describe LUS as complex, dynamic systems include complexity, openness, connectivity, hierarchy, stability/resilience, non-linearity, and entropy.

With regard to complexity, lying at the interface of multiple, cross-scale social and ecological systems, LUS is complex in that it comprises biophysical, socio-economic, and ecological components (subsystems), with attributes of openness, spatiotemporal dynamics, adaptability, and multiple objectives; it represents a myriad of **feedback** mechanisms that intermediate among land uses, their driving forces, and land covers, either strengthening or attenuating their interrelationships or rendering them non-linear. To better understand the behavior of LUS as complex systems, two interrelated questions that pertain to two key dimensions of complexity need to be addressed: How is LUS functionally structured and embedded with other socio-ecological systems? (according to Kasper Kok, this is a question of continuity as well as one of "structural complexity"); and, How do changes in these embedded systems (disturbances and chaos in the shorter term, transformation in the longer term) affect the attributes of LUS and vice versa? The latter is a question of change as well as one of "functional complexity."

With regard to openness, land-use systems bear features of **metabolism**, that is, they are in constant exchange of materials, energy, and information with the outside world (e.g., through trade, tourism, pollution, and species invasion).

Connectivity refers to the fact that spatially distant locations influence each other through multiple ways: either directly (e.g., **soil erosion** leading to sedimentation) or indirectly (e.g., **land degradation** at one location triggering migration, thus leading to **deforestation** at another location). The neighborhood effect in spatial analysis is also a relevant example.

Hierarchy describes both the scale of observation as pertains to LUS, and also explains how the interplay of the two key dimensions of complexity ("structural" and "functional") drive and shape the evolution of LUS. LUS is influenced by a great number of factors operating at different scales, while the observed patterns and processes often differ with regard to the **scale** of observation. Across the hierarchically nested levels, variables and processes operating at higher levels often steer and constrain lower-level variables and processes (e.g., dam building at basin level triggering irrigation and inundation in tributaries), but higher-level phenomena might emerge from lower-level dynamics—this is the property of emergence (e.g., massive overdraft by individual households leading to aquifer degradation, thus triggering landscape change).

Concerning stability and **resilience**, the hierarchy of LUS is not static; it evolves. The evolving hierarchy may reach a point where it becomes vulnerable to change, and a combination of internal and external forces will flip LUS either into a new qualitatively different phase of stability (e.g., agricultural adaptation in the U.S. **Great Plains** after the Dust Bowl), or into decline or collapse. In this context, resilience refers to the ability of LUS to absorb the impact of changes and continue to function.

Nonlinearity refers to the scale-dependence of observations (e.g., rainfall-runoff relationships), an intrinsic property of any dynamic system. It is popularly interpreted as "the sum of the parts do not equal the whole." Non-linear relationships tend to become more linear at coarser scales of observation, as evidenced by those among PAT variables (population, affluence, and technology). As Kasper Kok notes, "There are two fundamental problems related to structural complexity when describing the land-use system. First, processes that are established at a certain scale cannot be linearly translated and used at higher scales because of the aggregation effect. Second, processes that are established at a certain level cannot be translated to other levels because of differences in the hierarchical structure" (2001, pp. 7–8).

Entropy is a measure of disorder that tends to increase while LUS evolves (e.g., accumulative land degradation and fragmentation of landscapes). Entropy created within LUS plus entropy added by natural disasters is offset by negative entropy generated through sound land management and technology progress, and so on, for LUS to maintain its normal functioning.

Needless to say, these concepts do not do full justice to what the rich systems theory has to offer. The influence of each term on the study of land use (and the way LUS is delineated) is not even, depending on the level of abstraction it means to describe and the context of real world situation it can successfully approximate. While risking crude oversimplification, two main streams of research that conceptualize LUS in (semi-)systems terms (explicitly or implicitly) may be loosely identified: one is "observe, describe, and integrate to understanding," and another is "model to understand." A third, hybrid, "act to understand" stream may be identified, mainly in the context of **land-use planning**.

Observation, Description, and Integration

Examples of LUS conceptualized in the first stream are found mainly in traditional land-use literature. This stream contributes to the better understanding of LUS mainly through taking stock of existing research findings, clarifying theoretical concepts, and suggesting new research questions and alternative explanatory frameworks, often in combination with in-depth (not necessarily rigorous) case studies. In some cases, vivid land-change stories are told, through which empirical and interpretative baselines are established to help assess the validity of systems propositions and identify random events that significantly impinge on land-use/cover changes. Its method of inquiry is mainly qualitative, allowing a combination of different reasoning modes (from inductive to dialectic). Illustrative of this stream are the following characteristic statements:

1. The role of biophysical drivers in effecting land-use change is intermediated by socio-economic drivers, mainly through influencing the response of land owners/managers to land-cover change (e.g., no farming on marginal lands).
2. Land-cover changes resulting from land-use changes may feed back on **decision-making**, causing perhaps new rounds of land-use change and a vicious cycle of land degradation (e.g., shifting cultivators in **Latin America**).
3. Special attention should be paid to **cumulative change** and cumulative impacts, those that, though not physically connected through a globally operating system, can reach a global scale and status when their occurrence in many places adds up (e.g., **wetland** drainage and grassland degradation, **biodiversity** loss, and species transfer).
4. Certain regional impacts of land-use change may take a long time to show up in that they play out as chains of events responding to slow environmental alterations, often resulting in the delayed and sudden occurrence of harms (e.g., chemical soil pollution).

Evidently, these exemplary statements variously capture ramifications of some of the above systems concepts, albeit loosely and on abstract levels. The broad scoping of LUS, though not clearly defined, is implicit in the analytical and explanatory frameworks they propose. Because these conceptualizations are cast in abstract terms, however, operationalization in rigorous research design is often difficult. In terms of theoretical position, they are eclectic at best, which come out as both an advantage and a disadvantage.

Modeling

The second stream seeks to use quantitative models as tools to describe LUS (in the narrow sense of expressing operationally the relationships wherein), and explain (statistically) or mimic parts of its systemic character. It is overwhelmingly deductive and reductive, which is both its strength and limitation. Among many types of land-use models, the **conversion of land use and its effects (CLUE) model** or modeling framework is used as an illustrative example, because it represents the recent advance in developing systems-based toolboxes for land-use analysis.

CLUE and its supplementary modeling tools represent a new genre of land-use **modeling** effort that attempts to address, all at once, several critical land-change needs identified above: integration, linking with natural science programs, spatial explicitness, and scaling (as well as decision support). It is systemically structured, at least in a mechanical sense, as evidenced by its various modules that pertain to the main systems components of LUS as it is conceptualized (and operationalized) in CLUE. It is rigor, at least in the mathematical sense but not necessarily in the logical sense. It does manage to describe operationally many aspects of LUS as complex dynamic systems; but due to hoary data problems, the systems view embodied in CLUE has to be heavily truncated in terms of factors, relationships, and processes to be examined. Most importantly, it lacks substantive explanatory power with regard to the mechanisms of changes, as with most symbolic models. This forces us to critically reconsider the modeling fervor in land-use research and its potential influence on sound LUS conceptualization. To qualify as "substantive" in this regard, the explanation of change should tell the land-use change story and reason about the processes of changes at the same level drivers of change operate; the explanation should be theoretically informed and get into the causality of the relationship under scrutiny and map out the causal structure to the finest possible scale as the availability of (not necessarily digital) data allow.

Clearly, statistical association does not pass for causal explanation in this sense, which is what the search for causal processes of land-use change is all about. Unqualified employment of quantitative procedures raises other concerns as well: for example, used alone, statistical analysis cannot capture many qualitative aspects of LUS critical to the understanding of land-use change, such as cultural, political, and institutional factors (i.e., systems of **land tenure**) that may work as both constraints and drivers. In some cases, the rich reality is often chopped to fit the peculiar specifications of a model, hence making a model self-referential and not easily falsifiable. For another instance, statistical prediction that is the main purpose of many projective models is based on relationships existing in the past—an equilibrium assumption that may not easily hold for LUS in a rapidly changing world, if the specified time scale does not match the time horizon on which the modeled LUS operates. This is both a technical and a theoretical concern, of course, which requires a corresponding theory of change that seems to be lacking

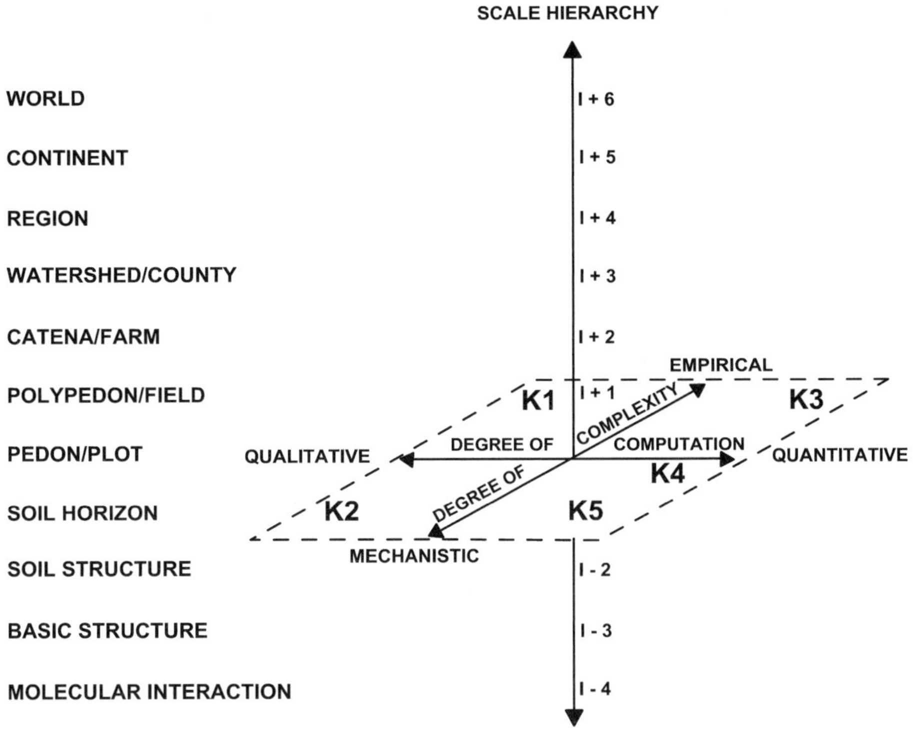

in the majority of mathematical models. Further problems arise from the general deprivation of theoretical underpinnings in operational models (and lack of interest in this by modelers): because theoretical assumptions underlie all models built, predicative value as well as functional specification of a given model have to be evaluated ultimately in light of the assumptions (implicitly embodied or explicitly articulated) and logical integrity, not by map validation, which unfortunately has become a peculiar niche business for a handful of researchers interested in technically fixing quantitative inaccuracies of models. Arguably, perfectly validated maps will never become a substitute for theoretically informed and methodologically sound analyses. In the worst cases, statistical **validation** may be used in the same way a drunkard employs a lamppost: not for illumination but for expelling the fear of darkness (in this case, the fear of lack of theoretical direction). In an overall assessment, different types of operational models conceptualize LUS differently, with some being more realistic, less abstract, and more spatially explicit than others; but their potential tends to be compromised by two common problems: the paucity of relevant data and the anemia of theories. The former is delimiting, but the latter is crippling.

The foregoing suggests that, in the context of applying systems analysis to the multidisciplinary study of land use/change, the observation by **Piers Blaikie** and **Harold Brookfield** that there is a "theoretical and methodological failure to combine social and natural science" is rightly put. Until we can find a way to move beyond the watershed dividing the two intellectual domains, no LUS construct will serve the dual goals of "**socializing the pixel**" and "**pixelizing the social**" equally well. The merits of different LUS constructs have to be evaluated ultimately in view of the value to scientific research and to management decision and policymaking. However, the criteria required in these two contexts do not always conform to each other: in the former case the criteria rest mainly on their conceptual soundness and potential in aiding hypothesis testing and theoretical development, while in the latter they are based primarily on its capabilities of helping identify **hot spots of land-cover change** and options/constraints, and on closing the inevitable gap between science and realities. The importance of scales of observation and levels of analysis is a crosscutting theme for both streams of research reviewed above, as well as for theories and models; hence they must be seriously tackled. The meaningful and useful analysis of LUS to support policymaking also hinges upon specification of land units, land-utilization types, and LUS at the right scales for effective management/policy intervention to be applied. To address these combined needs, it has been argued that LUS and its elements have to be defined differently at different spatial scales using a methodological typology that links scale specification of LUS to different level decision-making needs: the operational, the tactical, and the strategic, as illustrated in the second figure.

See also: **Agent-Based Model; Agricultural Intensification; Agrodiversity; Agroforestry; Cattle Ranching; Economic Livelihood; Exogenous Factor; Human Immunodeficiency Virus (HIV)/Acquired Immunodeficiency Syndrome (AIDS); Mediating Factor; Property; Slash-and-Burn Agriculture; Subsistence Agriculture; Swidden Cultivation; Thünen, Johann Heinrich von.**

Further Reading

Blaikie, Piers, and Harold Brookfield, *Land Degradation and Society*, London: Routledge, 1987; Briassoulis, Helen, Analysis of Land Use Change: Theoretical and Modeling Approaches [Online, August 2004], The Web Book of Regional Science Web Site, www.rri.wvu.edu/regscweb; Driessen, P. M., "Biophysical Sustainability of Land Use Systems," *ITC Journal* 3/4 (1997): 243–7; Kok, Kasper, Scaling the Land Use System: A Modeling Approach with Case Studies for Central America, PhD dissertation, Wageningen University, NL, 2001; Turner, B. L. II, and William B. Meyer, "Global Land Use and Land Cover Change: An Overview," in *Changes in Land Use and Land Cover: A Global Perspective*, eds. William B. Meyer and B. L. Turner II, Cambridge, UK: Cambridge University Press, 1994, pp. 3–10.

KE CHEN

Land-Use Transition. A possible development path or trajectory of a **land-use system** where the direction, size, and rate of change can be influenced through policy and some specific circumstances. A transition in land use is mainly but not necessarily a process of societal change in which the structural character of society (or a complex subsystem of society) and the functional character of land use (including biophysical attributes) transform, usually in a series of transitions. Broad transformations—such as those from shifting cultivation or rotational farming to sedentary, permanent, and high-intensity farming—usually result from a set of connected changes that reinforce each other but take place in several different components of the system. Multiple causality and co-evolution of different sectors (e.g., policy, meteorology, economy, hydrology, technology, and demography), caused by interacting developments, are central to the concept of transitions in land use. Thus, land-use transitions must be viewed as multiple and reversible dynamics: they do not follow a fixed pattern, and they are not deterministic. A transition in land use is not set in advance, and there is large variability in specific passages or pathways. Also, there is a strong notion of instability and indeterminacy. The theory of land-use transition has been applied in land-use change studies at different spatial and temporal **scales**. Though still in its infancy, the concept of **forest transition** is most advanced.

See also: **Boserup, Ester; Economic Restructuring; Metabolism; Public Policy.**

Further Reading

Lambin, Éric F., Geist, Helmut J., and Erika Lepers, "Dynamics of Land Use and Cover Change in Tropical and Subtropical Regions," *Annual Review of Environment and Resources* 28 (2003): 205–41.

HELMUT GEIST

Land-Use/Cover Change (LUCC) Project. A worldwide, interdisciplinary joint core project on land change of the **International Geosphere-Biosphere Programme** (IGBP) and the **International Human Dimensions Programme on Global Environmental Change**, formalized in 1995 and phased out in October 2005. The three missions of the LUCC project had been to build a compendium of information about local land-use and cover dynamics, to identify a small number of robust principles that could better knit together local insights into a predictive science, and to foster the development of common models that might then become widely available to scientists and stakeholders. In order to implement the project's science or research plan, centered around six major research themes, three interlocking strategies were encapsulated in three research foci. These had been the development of case studies that analyzed and modeled the processes of land-use change and land management in a range of generalized global situations (focus 1: land-use dynamics). After half a decade of operation, this strategy was complemented by the use of **meta-analysis** generating a general understanding, for example, of the causes, rates, and pathways of land change while preserving the descriptive richness of local studies. The second major research theme was the development of empirical, diagnostic models of **land-cover change** through direct observations and measurements of the explanatory factors (focus 2: land-cover dynamics). The third theme was the development of **integrated models** as well as of prognostic regional and global models (focus 3: integrated **modeling**). To meet its mission goals, the LUCC project reached out for a large scientific community to generate a wealth of results on its fundamental science questions. Two particular concerns had been the facilitation of interdisciplinary research work between the social and natural sciences, and to globalize research on land-change processes—such as **deforestation** and **desertification**—by contrasting results obtained from a variety of regions and geographic situations. Historically, three major periods of operation can be distinguished. From 1995 to 1998, major efforts were undertaken to establish the foundations of the project through the establishment of basic support infrastructure (i.e., scientific steering committee as core group, international project office, and regional networks); to develop and catalyze land-change science (e.g., through associated groups such as the **International Geographical Union (IGU) Commission on Land-Use and Land-Cover Change** and the **NASA Land-Cover/Land-Use Change (LCLUC) Program**); and to move toward full implementation. Following completion of the implementation strategy in 1999, major headway was made toward synthesizing land-change results and methodologies up to about 2002, while support infrastructure and dialog with funding agencies in the Americas, **Europe**, and Asia continued to be maintained. During 2003–2005, full implementation of LUCC's mission goals was achieved through final synthesis work in the form of workshops, conferences, and publications, with implications drawn for a transition toward **sustainable land use**, among other actions. Since its formalization in 1995, the LUCC project has welcomed applications by research groups, institutions, and agencies worldwide to join the community through formal status. The number of endorsed projects—such as the **Tropical Ecosystem Environment Observations by Satellite (TREES) project** and the **Large-Scale Biosphere-Atmosphere Experiment in Amazonia**—increased from two in 1997 to more than fifty, and about a dozen regional research networks were established in addition. In 2005, the scientific and institutional memory of the LUCC project was given to the **Global Land Project**, which also integrates the scientific and institutional memory of IGBP's Global Change and Terrestrial Ecosystems (GCTE) project.

See also: **BIOME 300 Project; Change Detection; Hot Spots of Land-Cover Change; Land-Use System; Socializing the Pixel; Syndrome.**

Further Reading

de Sherbinin, Alex, A Guide to Land-Use and Land-Cover Change (LUCC) Research: A CIESIN Thematic Guide [Online, March 2005], The CIESIN Web Site, sedac.ciesin.columbia.edu/tg; Geist, Helmut, The IGBP-IHDP Joint Core Project on Land-Use and Land-Cover Change (LUCC) [Online, March 2005], The Encyclopedia of Life Support Systems Web Site, www.eolss.net; Lambin, Éric, and Helmut Geist, eds., *Land Use and Land Cover Change: Local Processes, Global Impacts*, Berlin: Springer, 2006; Land-Use/Cover Change Project [Online, March 2005], The LUCC Web Site, www.geo.ucl.ac.be/LUCC.

HELMUT GEIST

Large-Scale Biosphere-Atmosphere Experiment in Amazonia. An integrated, interdisciplinary study led by Brazil (and perhaps the largest research initiative or regional global change study) to explore the climatological, ecological, and hydrological functioning of the Amazon, the impact of land-use/cover change on these functions, and the interactions between **Amazonia** and the earth system. Planning for the Large-Scale Biosphere-Atmosphere Experiment in Amazonia (LBA) began in 1992, and field research programs began in 1998. Apart from having a substantial education and training component, LBA focuses on two key questions, as laid down in the science plan: How does Amazonia currently function as a regional entity? and, How will changes in land use and climate affect the biological, chemical, and physical functions of Amazonia, including the sustainability of development in the region and the influence of Amazonia on global climate? The LBA's core research focuses on the effects of tropical forest **conversion**, **regrowth**, **fire**, and selective logging on the **carbon cycle**, **nutrient cycle**, trace gas fluxes, and the prospects for **sustainable land use** in the region, with as much quantification of conditions and processes done as possible. Field research is carried out along two transects spanning gradients from wet, aseasonal **forests** to relatively dry and seasonal forest or savannas. LBA is part of a set of continental-scale terrestrial transects of the **International Geosphere-Biosphere Programme**, and it has been endorsed by the **Land-Use/Cover Change (LUCC) project**. Moving beyond the initial stages of LBA, and initiating a first major synthesis effort, the need has been identified to better integrate the human dimensions of **environmental** change, by abandoning, for example, the focus on spatially delimited transects and exploring typical pathways of land-use/cover change instead (such as **land degradation** and/or **agricultural intensification**). Several new research areas were considered essential to fostering the understanding of major processes behind large-scale changes in Amazonia. These are land-use/cover changes in relation to human **population dynamics** (mainly mobility) and **urbanization**, logistics (or infrastructure such as roads) and regional development, agrarian and **land tenure** structures, agricultural and economic systems, and environmental valuation.

See also: **Anthropological Center for Training and Research on Global Environmental Change; Arc of Deforestation; Brazilian Cerrado; Cattle Ranching; Deforestation; Forest Degradation; Secondary Vegetation; South America.**

Further Reading

Alves, Diogenes S., "An Analysis of the Geographical Patterns of Deforestation in Brazilian Amazonia in the 1991–1996 period," in *Land Use and Deforestation in the Amazon*, eds. Charles H. Wood and R. Porro, Gainesville: University Press of Florida, 2002, pp. 95–106; Keller, Michael, Silva-Dias, Maria Assunção, Nepstad, Daniel C., and Meinrat O. Andreae, "The Large-Scale Biosphere-Atmosphere Experiment in Amazonia: Analyzing Regional Land Use Change Effects," in *Ecosystems and Land Use Change*, eds. Ruth S. DeFries, Gregory P. Asner, and Richard A. Houghton (Geophysical Monograph 153), Washington, DC: American Geophysical Union, 2004, pp. 321–34; Large-Scale Biosphere-Atmosphere Experiment in Amazonia Science Plan [Online, March 2005], The LBA Web Site, lba.inpa.gov.br.

HELMUT GEIST

Latin America. Countries in the Americas that were originally colonized by Spain and Portugal. They are located from Mexico and Central America in the Northern Hemisphere, to certain islands of the Caribbean, and throughout most of the South American continent. Land cover is strikingly variable over this vast region, ranging from the world's largest tropical rainforest areas and the arid vegetation of large deserts, to the heterogeneity of natural environments in long and high mountain chains. Land use can be characterized in terms of strong asymmetries that characterize both the physical and human geographies of the region. Not only can examples of virtually all the world's climates and resulting climatic contrasts be found, but there are deep structural imbalances in power, land, wealth, and **access** to basic human needs that divide peoples and affect land use and changes in land use. The earlier exploitive economic systems set up by the colonizing powers of Spain and Portugal in the fifteenth and sixteenth centuries were replaced by more complicated systems that perpetuated inequalities between and among the landless and landowners, the poor and rich, and ethnic and racial minorities and majorities. In addition to these historical legacies, current **land-use systems** are rapidly changing due to the growing demands and power of cities, the **colonization** of new agricultural frontiers, and increasingly intertwined socioeconomic fates precipitated by migration and **globalization**.

Due to their locations, Mexico, Central America, and the Caribbean have been most affected in recent decades by national and regional geopolitics and economics as influenced by decisions of and opportunities within the **United States of America**. This influence is especially notable in land-use changes visible near transnational borders, for example between the United States and Mexico, where industrialized or urban landscapes have replaced rural agricultural systems, and where limited supplies of water for **agriculture** and growing cities are topics of daily concern. However, habitat alterations for natural resource extraction, **urbanization**, and the effects of growing populations are also to be found in many other landscapes of the region. For example, changes in distant landscapes are now tied together through globalization and trade ties among North and South American countries and with the rest of the world.

Originally, the humid areas of Mexico, the Caribbean islands, and Central America were subtropical and tropical environments dominated by **forests** and shrublands. Northern Mexico is subtropical and affected by occasional cold air masses intruding from the North America continent from October to February. This climatic regime limits the types of tropical agriculture that can be practiced: it controls how far to the north citrics and **coffee** can be successfully raised. The **mountains** also create many elevational zones and microclimates that are home to oak and pine forests on hillsides and higher elevations, with shrublands and agricultural fields covering lower areas and valley bottoms. There are also extensive desert environments, for example along the northwestern coast of Mexico, in intermontane valleys, and on leeward sides of Caribbean mountains. These are often utilized as rangelands, with herders moving flocks of cattle, sheep, and goats. In some places, land management includes techniques such as terraces and rock walls that store erratic rain as soil moisture, permitting the growth of crops in lands otherwise marginal for agriculture.

The tropical latitudes extend from about 23°N to 23°S, and here the climate systems are regulated by the convergence of the tropical trade winds in the Intertropical Convergence Zone (ITCZ), which moves seasonally, often bringing rains and a wet season when it is overhead, and clear skies and a dry season when it moves away. These latitudes include Latin American countries from Mexico southward to Peru and northern Argentina. Especially in Central America and the Caribbean, the trade winds also carry in atmospheric disturbances from July to September that can create very damaging hurricanes and tropical storms. The humid lowlands were originally covered by extensive tropical rain forests, occasionally opened up by hurricanes or treefalls. Beginning two millennia ago, these areas were utilized by large human populations in some areas, particularly in the Mayan-dominated areas of Mexico and Central America. The Caribbean islands were also home to large indigenous populations, with resulting **conversion** of forested landscapes for agriculture. Many other people lived in the highlands of these areas, interspersing agricultural fields among forests and shrublands. The arrival of European colonists brought grazing animals and the need for more pastoral land-use systems, steel axes that facilitated tree cutting, and permanent towns and cities that prospered with trade.

Tropical **South America** has very extensive humid lowlands because of the large width of South America at those latitudes. There are tropical rain forests in the Chocó of Colombia, in the Orinoco River basin of Venezuela and Colombia, and across the immense Amazon region, including large parts of Brazil, Ecuador, Peru, and portions of other countries. There are also sizable areas of tropical savannas and herbaceous **wetlands**, especially in sites with poor drainage or with enough seasonality in rainfall that the grasses burn in the dry season; these are located around the northern, eastern, and southern limits of the tropical forests of the Amazon, characterizing, for example much of Amazonian drainages of lowland Bolivia. Coastal margins from Brazil to northernmost Peru often have mangrove forests.

The most conspicuous **land-cover changes** in tropical Latin America have been associated with the expansion of the **agricultural frontier** in the South American lowlands and equivalent areas made accessible by road or rivers in Central America and Mexico. Beginning in the 1960s and continuing to the present, this landscape alteration has often resulted from organized or spontaneous colonization, and has included the clearing of forests, increased burning for the establishment and maintenance of **pastures**, cutting of mangroves to establish shrimp-raising ponds, and the growing of crops such as **maize** and **cassava** for subsistence and others for sale. A recent phenomenon is the additional conversion of tropical savannas to agriculture, especially for soybeans, and particularly in central Brazil. The most notable sites of **deforestation** and the most rapid expansion in area of agricultural lands are in tropical lowlands that have a pronounced dry season, which facilitates tasks such as slash burning and promoting the **regrowth** of grasses. The tropical environments that are perpetually humid are generally still dominated by wet forests, except along roads and rivers. In some places, indigenous groups continue to use these areas in manners that hardly alter forest structure and cover, by extracting fish from rivers, game from the forests, and crops from small agricultural fields that approximate the size of areas disturbed naturally by treefalls.

Coastal western South America includes some of the driest places on earth, due to the limitation of atmospheric humidity and uplift caused by cold-water currents of the eastern Pacific Ocean from about 4°S to 35°S, plus a rain shadow effect created by the Andes. These drylands are in coastal Peru and northern Chile. The land cover is **mesquite** woodlands or **tropical dry**

forests in places that receive seasonal rainfall, but much of the area is too arid for most **vegetation** types to be present and instead is barren or has a light cover of seasonal or ephemeral herbs. Land use consists of the opportunistic grazing of goats or irrigated agriculture in places where water from the Andes can be tapped into. The northern part of Peru and the southern part of Ecuador are the places most affected by the ENSO (El Niño Southern Oscillation), which is triggered by shifts in pools of cool water in the Pacific Ocean, causing a low-pressure atmospheric system to form every three to eight years above these generally arid areas and creating destructive precipitation events. An interesting phenomenon is that dryland areas receive ample humidity in an El Niño year and become covered by herbs or by the seedlings of dry forest trees, "greening up" in satellite imagery.

The tropical highlands and uplands of the **Andes** Mountains (Venezuela south to Bolivia) are places of continuity in terms of land use, given that they have been utilized by people for one to ten millennia, depending on the location. These are home to extensively terraced landscapes, often with intricate systems of irrigation farming, maintained by hand tools and physical labor. Land cover includes fields and pastures, in addition to shrublands; high-elevation **grasslands**; rocky slopes; glaciers; and, in places isolated from roads and settlements, Andean forests of evergreen broadleaf trees. The biggest recent land-cover changes have been driven by deforestation along penetration roads connecting the highlands to the humid lowlands, alterations in land use due to changes in types of agriculture and sizes of local populations, and by the growth of highland cities. Other upland areas in South America were not home to large pre-Hispanic populations, but have been converted by colonists in the past several centuries. For example, the subtropical forests that once covered the coastal mountains of southeastern Brazil are now almost completely gone, cut for agriculture and pastureland.

It is likely that the tropical highlands will be among the places in the world most altered by the effects of future climate change. Ice caps have become smaller along the entire length of the Andes. There is other evidence that precipitation and cloud formation have been altered, in some places perhaps due to deforestation and changed evapotranspiration processes, but in others due to the lifting of cloud banks caused by warmer temperatures. The alteration of physical environmental conditions will have many important consequences for land use, and for nature reserves and national parks established to protect the **biodiversity** of particular locations. A partial solution being pursued throughout Latin America is to attempt to connect nature reserves together using **conservation** corridors, so that no one protected area becomes completely isolated. In addition, the uplands serve as sources of water for cities and agriculture. Changed hydrological regimes are already altering plans for supplying water to growing urban areas and for the production of hydroelectric power.

The temperate zones of South America, to the south of about 35°S latitude, have seasonally cold climates, with winter months from April to September. The westerlies bring in humid air masses from the Pacific Ocean, so the wettest environments are on the coast and in the western Andes. These are covered by wetlands and bogs or by humid forests. The exceptions are the areas that have been converted from cover by native Andean forests to plantations of exotic tree species. Large areas of Argentina are covered by humid to dry grasslands, characterized by strong winds, flat to rolling topography, and gigantic ranches for the raising of cattle and sheep.

Throughout Latin America the growth of cities has been the land-use change with the most profound implications for altering land covers and shifting socioeconomic processes. Many people have moved from rural areas to growing cities, beginning in the 1940s, but increasingly so in the 1980s and continuing to the present. Concurrent population growth in the rural areas has kept population size relatively constant, despite this out-migration. The consequences for land use include the switching of agriculture to **cash crops** and other items bought by urban populations, more raising of livestock for meat, more tree plantations for pulp and timber needs, and the continual search for more drinking water for cities. Livelihood decisions made within households on the basis of their needs, goals, and resources available (i.e., social and natural capital) affect land use on lands they control. Thus, **land tenure** and other constraints affect the rates and locations of land-cover **modifications**. The last two decades have also been years of increased democratization in Latin America, with large urban populations increasingly controlling national natural resource decisions, and increased globalization resulting in some needed products being purchased from other countries rather than bought from nearby rural populations.

The environmental shifts and dilemmas created by these global, regional, and national changes are most profound in the island nations and in other countries with limited land or resources. In addition, the places with the largest disparities in wealth, especially in terms of the distribution of lands and land tenure, also have inherent social instabilities that can cause economic crises or social injustices. These twenty-first century preoccupations also rest uncomfortably atop legacies of extractive economic systems that characterized Spain's and Portugal's colonial heritages. The borders of the Latin American countries have often been the source of conflicts in the past, and tend to be foci of rapid land-cover change today.

The dynamics of land cover reflect these broadly social processes, just as other seasonal or episodic changes in land cover are due to altering physical environmental controls of precipitation, temperature, and natural disturbances. These changes can now be tracked using **remote sensing** and analyzed using computations and models powered through **geographical information systems** and computerized simulations. Such methods have helped elucidate the current **hot spots of land-cover change**, some of which coincide spatially with Latin America's hot spots of biological diversity. These are obviously places of special global concern, and include not only the tropical forests and grasslands, but the unique upland, dryland, and wetland environments to be found in the region. Mitigation efforts, such as **carbon sequestration** by establishing tree plantations, nature reserves, extractive reserves, and environmental restoration are all taking place; all of these endeavors can alter land cover. Predicting the location and direction of land-cover change due to roads, migration, and urban expansion is greatly facilitated with these techniques. **Land-use planning** for more appropriate land use and for the sustainable extraction of aquatic and forest products now routinely uses these inherently geographical approaches. These projects are being carried out both by governmental agencies and nongovernmental organizations in many countries in Latin America.

See also: **Amazonia; Arc of Deforestation; Arena of Land Conflict; Brazilian Cerrado; Cattle Ranching; Coastal Zone; Decentralization; Decision-Making; Economic Livelihood; Exotic Species; Food Crops; Investments; Land Concentration; Poverty; Remittances Landscape; Slash-and-Burn Agriculture; Southern Cone Region; Subsistence Agriculture; Tropical Humid Forest.**

Further Reading

Achard, Frédéric, Eva, Hugh D., Stibig, Hans-Jürgen, Mayaux, Philippe, Gallego, Javier, Richards, Timothy, and Jean-Paul Malingreau, "Determination of Deforestation Rates of the World's Humid Tropical Forests," *Science* 297 (2002): 999–1002; Blouet, Brian W., and Olwyn M. Blouet, eds., *Latin America and the Caribbean: A Systematic and Regional Survey*, 4th ed., New York: John Wiley, 2002; Bulmer-Thomas, Victor, *The Economic History of Latin America Since Independence*, 2nd ed., Cambridge, UK: Cambridge University Press, 2003; Denevan, William M., *Cultivated Landscapes of Native Amazonia and the Andes*, Oxford, UK: Oxford University Press, 2001; Doolittle, William E., *Cultivated Landscapes of Native North America*, Oxford, UK: Oxford University Press, 2001; Economic Commission for Latin America and the Caribbean [Online, August 2004], The CEPAL Web Site, www.eclac.cl; Gwynne, Robert N., and Cristóbal Kay, eds., *Latin America Transformed: Globalization and Modernity*, 2nd ed., London: Arnold, 2004; Kellman, Martin, and Rosanne Tackaberry, *Tropical Environments: The Functioning and Management of Tropical Ecosystems*, London: Routledge, 1997; Kelly, Philip, *Checkerboards and Shatterbelts: The Geopolitics of South America*, Austin, TX: University of Texas Press, 1997; Knapp, Gregory, ed., *Latin America in the 21st Century: Challenges and Solutions*, Austin, TX: University of Texas Press, 2002; Latin American Network Information Center [Online, August 2004], The LANIC Web Site, lanic.utexas.edu; Thorp, Rosemary, *Progress, Poverty and Exclusion: An Economic History of Latin America in the 20th Century*, Baltimore, MD: Johns Hopkins University Press, 1998; Whitmore, Thomas M., and B. L. Turner II, *Cultivated Landscapes of Middle America on the Eve of Conquest*, Oxford, UK: Oxford University Press, 2001.

KENNETH R. YOUNG

Leapfrogging. A spatial process of urban development in which land closer to urban areas is skipped over while land further away is developed. The result is a discontinuous pattern of urban and rural land rather than a contiguous urban area. "Leapfrog," "scattered," and "fragmented development" are often used interchangeably to mean a pattern of discontinuous development, but leapfrog development implies such patterns with respect to an urban gradient, whereas scattered and fragmented development do not. Because the intervening rural land may be developed at a later date, these patterns can be transient. While some studies have provided evidence of extensive discontinuous development patterns in specific **urban-rural fringe** areas of the **United States of America**, it is unclear how pervasive these patterns are in general.

Explanations of leapfrog development range from exogenously driven processes of land speculation and land heterogeneity to endogenous spatial externalities. The traditional urban economics model explains leapfrogging as a temporary phenomenon resulting from optimal intertemporal **decision-making** in which developers generate higher returns by delaying the development of land that is closer to the city. Other theories appeal to the existence of dispersive forces such as congestion externalities and open-space amenities. Dawn Parker and Vicky Meretsky demonstrate that fragmented urban patterns can emerge from such effects at the urban fringe using an **agent-based model** of land use. Elena Irwin and Nancy Bockstael find empirical evidence of such externalities using data from the Washington, DC region and demonstrate that their magnitude is sufficiently large to generate systematic patterns of discontinuous residential development.

See also: **Endogenous Factor; Exogenous Factor; Land Fragmentation; Modeling; Suburbanization; Urbanization; Urban Sprawl.**

Further Reading

Irwin, Elena G., and Nancy E. Bockstael, "Interacting Agents, Spatial Externalities and the Endogenous Evolution of Land Use Patterns," *Journal of Economic Geography* 2 (2002): 31–54; Parker, Dawn C., and Vicky Meretsky, "Measuring Pattern Outcomes in an Agent-Based Model of Edge-Effect Externalities Using Spatial Metrics," *Agriculture, Ecosystems, and Environment* 101 (2004): 233–50.

ELENA G. IRWIN

M

Maathai, Wangari M. (1940–). Environmental and political activist, first woman in eastern and central Africa to earn a doctorate degree, honorary professor of biology (veterinary anatomy), assistant minister for Environment, Natural Resources and Wildlife in Kenya's ninth parliament since 2002, and winner of the 2004 Nobel Peace Prize. With the recognition by the Norwegian Nobel Committee that "peace on earth depends on our ability to secure our living environment," the "can-do person" Wangari Maathai was awarded for her holistic or integrative approach to sustainable development, which interlinked women's rights, democracy, ecological restoration, and grassroots activism in favor of sustainable development. She founded the Green Belt Movement, which for thirty years has mobilized women to plant about 30 million trees in Kenya and other African countries. More than two decades ago, she became a member and then the chair of the Environment Liaison Centre International (ELCI) in Nairobi, Kenya. The ELCI emerged from the first Earth Summit in Stockholm in 1972 and was located in Nairobi in 1974 to track international environmental processes (with the **United Nations Environment Programme** there), and to support effective techniques for networking of non-governmental organizations (NGOs) in developing countries. Locally and in Africa, ELCI campaigns relate to **forests**, **sustainable land use**, and air pollution. ELCI facilitates NGO involvement at the world summits for sustainable development and organizes the NGO community on the **United Nations Convention on Biological Diversity**. In the most recent past, Maathai was involved in campaigns against land grabbing and unsustainable forest management practices.

See also: **Arena of Land Conflict; Community Involvement; Conservation; Economic Livelihood; Forestry; Green Wall Project; Institutions; Reforestation.**

Further Reading

Environment Liaison Centre International [Online, February 2005], The ELCI Web Site, www.elci.org; Greenbelt Movement, Biographies [Online, February 2005], The GBM Web Site, www.greenbeltmovement.org; Maathai, Wangari [Online, February 2005], The Official Web Site of Prof. Wangari Maathai, www.wangarimaathai.or.ke.

HELMUT GEIST

Maize. After **wheat** and **rice**, the most important cereal worldwide. Maize (or corn) is one of the most widely distributed cereals grown in the tropics and subtropics and in the temperate zones, between 58° N (**Canada**, **Russia**) and 42° S (New Zealand, **South America**) and from below sea level (Caspian Sea plain) up to 3,600 meters high (Peruvian **Andes**).

Soil erosion in an agricultural field that has been cleared of timber and put into corn (maize) in Hamilton County, Illinois, in February 1937 (note stumps of trees and beginning of erosion gullies).
[Courtesy Library of Congress]

Maize is used as a staple food particularly in the tropics; as feed for livestock and poultry, particularly in the industrialized countries of the temperate zones; and as raw materials for many industrial products. Maize is a starch plant, and the ripe grain is prepared for human **consumption** in multiple ways. In 2000, about 19 percent of the total maize harvested in the world was used for human consumption. Most maize (about 65 percent in 2000), however, is used to feed livestock. Silage maize has become one of the most important feed crops in industrialized western countries.

Maize originated in Central America (Mexico). **Domestication** of maize started about 8,000 years ago. Cultivation of maize and its importance as a **food crop** are closely related to the rise of pre-Columbian Central and South American civilizations. In the fifteenth and sixteenth centuries, maize was introduced into **Europe** and Southeast Asia. Due to its adaptability and productivity, maize spread rapidly around the globe. It is the most domesticated crop, but its genetic origin and evolution are not well understood. Maize is a tall annual cereal grass bearing kernels on large ears. It has a single predominant stem and, different from other cereals, the formation of basal branches (tillers), which are limited and only of importance in low-density stands. Maize is a C4 plant with typically higher rates of photosynthesis compared to non-C4 plants, and requires warm temperatures and adequate soil water supply. It ideally grows in tropical and subtropical conditions and is less suitable for semiarid and equatorial climates. Minimum temperature for germinations is about 10°C, and average daily temperature during growth should be at least

20°C. The period from planting to harvesting varies considerably depending on cultivar and climatic conditions. Plants may be harvested between 70 and 200 days for very early or very late cultivars, respectively. The large number of cultivars of different maturity types allows cultivation of maize across a wide range of temperature conditions.

Maize is a highly variable species. Cultivars are divided into groups according to the structure and shape of the grain. Important cultivar groups are dent, flint, pod, pop, soft, sweet, and waxy maize, or corn.

Maize evolved into a highly productive crop within a relatively short time period. Productivity of maize was significantly increased in the early twentieth century by the discovery and development of hybrid maize, created by crossing of two or three inbred lines. Cultivation of hybrid maize requires adequate facilities for seed production and is commonly used in high-input farming with high fertilizer use. The larger area of maize (about 66 percent of the total global maize area) is, however, in developing countries characterized by low-input smallholder farming. In these areas, local, unimproved cultivars are commonly grown. Resources for production of hybrid maize seeds and intensification of maize production are generally not available.

The advantages of maize compared to other cereals with respect to its cultivation and processing and its enormous genetic variability have favored its use, geographical distribution, and production. Maize is presently grown in more countries than any other cereal, and in recent years maize production has even exceeded the production of paddy rice and wheat.

The area used for maize production has steadily increased over the last centuries due to increasing demand for food, mainly from population growth. Different from other important cereals, maize's harvested area has continued to increase up until present days. About 145 million hectares are presently cultivated with maize, which is almost 10 percent of the total arable land, and represent an increase of maize-harvested area of about 40 percent between 1961 and 2004. However, increases in maize area are mainly observed in developing countries, with more than 60 percent between 1961 and 2004. Maize-harvested area in developed countries increased only marginally, but productivity in these countries is high. About 57 percent of the global total maize production in 2004 came from developed countries.

Most maize is produced in the **United States of America**, which contributes about 42 percent (2004) to the global production. Other important producers are **China**, Brazil, Mexico, and France with 19 percent, 6 percent, 3 percent, and 2 percent of respective global maize production.

Global average maize yields have more than doubled over the last four decades and have reached almost five tons per hectare. Yields, however, differ considerably among countries, ranging between one and nine tons per hectare. Yields are more than twice as high in developed than in developing countries, which is largely due to technological, economic, and organizational reasons. In many developing countries of the tropics and subtropics, maize is grown in small farms with low inputs such as fertilizer, improved seed, water, and labor for crop weeding.

Demand for maize will further increase with population growth and economic wealth. Future production of maize will depend on biophysical and socio-economic factors. Climate change has been projected to affect the productivity of maize and its regional distribution. Increasing temperatures may drastically reduce yields in some regions, for example, in Mexico and in Mediterranean countries, or increase yields in other regions, for example, **western Europe** and North America. A northward shift of the maize-growing area has been projected for Europe and Canada. Increasing atmospheric CO_2 concentration may ameliorate some of the negative effects of climate change. However, since maize is a C4 plant, effects of increasing CO_2 on growth are less pronounced than for cereals with a C3 pathway of carbon fixation such as wheat and rice, and climate change effects may be more evident for maize than for C3 cereals. Impacts of climate change will also depend on the **vulnerability** of the agricultural systems, including their ability to adapt to changes in climatic conditions. Effectiveness of adaptation options, such as change of cultivar and/or crop, irrigation, and fertilizer will depend on the economic and social conditions of the specific agricultural system. These production conditions are less favorable in most developing countries and pose severe risks on future maize production in these countries due to climate change. The yield potential of maize is higher compared to that of wheat and rice. It will depend on the socio-economic development of the developing world to determine whether what extent this potential can be utilized in the future and the negative environmental effects associated with its intensive production can be minimized.

See also: **Agricultural Intensification; Biodiversity Novelties; Cash Crops; Exotic Species; Monoculture; Population Dynamics; Turnover.**

Further Reading

Koopmans, A., ten Have, H, and Subandi, "Zea Mays L," in *Plant Resources of South-East Asia Handbook 10: Cereals,* eds. Gerard J. H. Grubben and Soetjipto Partohardjono, Leiden, NL: Backhuys Publishers, 1996, pp. 143–49; Parry, Martin L., Rosenzweig, Cynthia, Iglesias, Ana, Livermore, Matthew, and Günther Fischer, "Effects of Climate Change on Global Food Production under SRES Emissions and Socio-Economic Scenarios," *Global Environmental Change* 14 (2004): 53–67; Ramankutty, Navin, and Jonathan A. Foley, "Estimating Historical Changes in Global Land Cover: Croplands from 1700 to 1992," *Global Biogeochemical Cycles* 13 (1999): 997–1027; United Nations Food and Agriculture Organization [Online, February 2005], The FAO Web Site, www.fao.org.

FRANK EWERT

Malaria. A human disease caused by the vector-borne, protozoal parasite *Plasmodium spp*, affecting people in the Caribbean, **South America**, Southeast Asia, India, and sub-Saharan Africa. Malaria, together with tuberculosis and **human immunodeficiency virus (HIV)/acquired immunodeficiency syndrome (AIDS)**, is considered to be one of the most significant public health threats to development in tropical countries of the developing world. According to estimates of the World Health Organization (WHO) of the United Nations, at least a million malaria deaths occur each year, 90 percent of them in sub-Saharan Africa. Mortality is particularly high in infants and young children. Transmitted by the *Anopheles* mosquito, the plasmodium parasites affect only humans and have no other significant animal host or reservoir. In recent years, malaria has experienced a resurgence, which has been primarily associated with increased drug resistance, increased insecticide resistance, global climate change, and local land-use changes.

The links between land use and related land-cover change and malaria are predominantly associated with the impact of land-use/cover change on the habitat and survival of the *Anopheles* mosquito vector. In the case of malaria, temperature, humidity, sun exposure, and water availability can affect reproduction and breeding site availability for the mosquito vectors, though different vector and parasite species have unique temperature and environmental preferences and associated distributions. Many mosquito species breed well in stagnant water with high sunlight, such as irrigation ditches and pools. Malaria mosquitoes respond to rapid changes in their habitat and are particularly susceptible to temperature and moisture changes. As noted by Jonathan Patz and Nathan Wolfe, land-cover **modifications** resulting in changes to temperature and moisture can cause dramatic increases in vector populations and malarial transmission rates.

The name malaria derives from associations with land cover; it comes from *mal aria*, which is Italian for "bad air." This association is due to an historical association of the disease with fetid swamps and marshes. Up until the eighteenth and nineteenth centuries, malaria occurred in the **United States of America**, some parts of **Canada**, and most northern European countries. Paul Reiter argues that a dramatic decline in malaria rates in England and northwestern **Europe** during the late nineteenth century was due, in a large part, to land-use changes, including improved drainage and extensive land reclamation, increased cultivation of root crops, and increased livestock-keeping. The latter may have diverted mosquitoes from humans to animals, reducing potential transmission of the parasite within the human population, since malaria has no significant animal host. More recently, however, the contraction of malaria's extent has reversed, and the disease has extended into areas of Central America, northern South America, tropical and subtropical Asia, and parts of the Mediterranean and former Soviet Union. This reemergence has been associated with human **population dynamics** (such as changes in population distribution and density), **urbanization**, **deforestation**, drug resistance, sociopolitical instability, international travel, agricultural change, and climate change. Of these causal factors, a significant number are strongly correlated with patterns of land-use/cover change.

Land-use factors such as agricultural development, population increase, and urbanization have been associated both with declines and expansions of malarial distributions. Historical expansion of malaria in West Africa was associated with **conversion** of rain **forest** into **agriculture**; increased mosquito habitat and population growth provided breeding habitat in cleared areas with sunlight-exposed pools, as well as availability of human hosts. More recent expansion of malaria in frontier zones such as Southeast Asia and **Amazonia** is similarly associated with deforestation and conversion to agricultural cropping. By contrast, nineteenth-century disease decline in Europe has been linked to agricultural change as the resulted of improved livestock husbandry, therefore diverting mosquito feeding from humans to animals.

Decline of disease in Africa following World War II was predominantly the result of wide-scale spraying of dichloro-diphenyl trichloroethane (DDT), a pesticide lauded for its efficacy at drastically reducing mosquito densities. This campaign, however, encountered the emergence of DDT resistance in mosquitoes in conjunction with detection of persistent environmental impacts. Fifty mosquito species have now been found to be resistant to one or more insecticides, including DDT. In addition to DDT, other insecticide and larvacides are used to reduce mosquito populations and malarial rates, but associated drug resistances and environmental impacts have been noted. Use of chemicals for purposes other than malarial control has also contributed to mosquito drug resistances: the use of some pesticides and fertilizers for increased crop production has been implicated in increased mosquito resistance to DDT.

In the case of population density changes, malaria requires a minimum density of humans in order to support a continued reservoir for transmission of the parasite. Generally, malaria does not occur in high-density, established urban areas (though this has been observed in some cities of South Asia). Warm, developing cities with semi-urban slums, however, can provide reservoirs of disease surrounding urban areas, resulting in high urban infection rates. As noted by Irwin Sherman, up to 10–20 percent of incidence in some countries is attributable to urban and peri-urban malaria, due primarily to the presence of slums.

Land-development changes, including creation of dams, irrigation schemes, commercial tree cropping, deforestation, and agricultural development, have been associated with increases in mosquito vector habitat as well as increased contact between mosquitoes and humans. David Molyneux and Irwin Sherman summarize a range of examples of increased abundance of mosquito vectors following forest disturbance. Deforestation in West Africa, for example, resulted in the creation of stagnant pools ideal for larval habitat, while expansion of the species *Anopheles darlingi* is associated with deforestation in Amazonia. In Southeast Asia, adaptation of several species of mosquito to new plantations has been found. In some cases, forest conversion or **land-cover change** may decrease the habitat availability of one mosquito species while increasing available habitat for an alternate mosquito species; whether the former or the latter species is a better vector of malarial parasites will determine whether there will be an associated decrease or increase in malarial risk. Research by Kim Lindblade and colleagues found that replacement of natural swamps with agricultural cropping in Uganda led to increased temperatures and potential for increased malarial transmission. This is supported by Katrin Kuhn and colleagues, who note that mosquito densities are often associated with irrigated cropland (such as **rice** paddies) and marshland. Creation of micro-habitats such as these provides adequate sunlight and moisture appropriate for breeding in a range of mosquito species.

The types of **vegetation** fragmentation noted in the above examples, and their associated impacts on increased malarial transmission potentials, are particularly prominent in frontier areas such as the Amazon basin and in **agricultural frontiers** of Southeast Asia. Wannapa Suwonkerd and colleagues, for example, found that increased landscape diversity and forest **fragmentation** in Chiang Mai province of northern Thailand were associated with decreases in mosquito densities over twenty-three years. Irwin Sherman notes that temporary migrant workers traveling into malarial zones often become infected with malaria, particularly in association with deforestation projects in frontier forest areas.

Dam construction is another common example of land development associated with malarial incidence change. Tedros Ghebreyesus and colleagues, in a community-based study of the effect of microdam construction on malaria incidence in Ethiopia, found that the projects resulted in significantly increased transmission in local communities. Where there are unmanaged zones of vegetative growth around reservoir edges, dam development has been associated with the creation of excellent mosquito habitat. More recently, potential impacts of climate change on infectious disease rates and distributions has drawn attention to possible changes in the distribution and magnitude of global malaria risk. It is notable that climate change models predict significant increases in *P. falciparum* malaria under almost all scenarios. Jonathan Patz and colleagues provide further discussion on the potential impacts of climate change on malaria.

Reciprocally, malaria can have impacts on land use, particularly in developing countries, where disease burdens may undermine health and welfare, debilitate the active members of communities, and strain resources; these impacts affect **people at risk** in **southern Africa**.

See also: **Climate Impacts; Human Health.**

Further Reading

Ghebreyesus, Tedros A., Haile, Mitiku, Witten, Karen H., Getachew, Asefaw, Yohannes, Ambachew M., Yohannes, Mekonnen, Teklehaimanot, Hailay D., Lindsay, Steven W., and Peter Byass, "Incidence of Malaria among Children Living near Dams in Northern Ethiopia: Community Based Incidence Survey," *British Medical Journal* 319 (1999): 663–6; Kuhn, Katrin G., Campbell-Lendrum, Diarmid H., and Clive R. Davies, "A Continental Risk Map for Malaria Mosquito (*Diptera: Culicidae*) Vectors in Europe," *Journal of Medical Entomology* 39 (2002): 621–30; Lindblade, Kim A., Walker, Edward D., Onapa, Ambrose W., Katungu, Justus, and Mark L. Wilson, "Land Use Change Alters Malaria Transmission Parameters by Modifying Temperature in a Highland Area of Uganda," *Tropical Medicine and International Health* 5 (2000): 263–74; McMichael, Anthony J., *Human Frontiers, Environment and Disease: Past Patterns, Uncertain Futures*, Cambridge, UK: Cambridge University Press, 2001; Molyneux, David H., "Vector-Borne Infections and Health Related to Landscape Changes," in *Conservation Medicine: Ecology and Health in Practice*, eds. A. Alonso Aguirre, Richard S. Ostfeld, Gary M. Tabor, Carol House, and Mary C. Pearl, New York: Oxford University Press, 2002, pp. 194–206; Oaks, Stanley C., Mitchell, Violaine S., Pearson, Greg W., and Charles C. J. Carpenter, eds., *Malaria: Obstacles and Opportunities*, Washington, DC: National Academy Press, 1991; Patz, Jonathan A., and Nathan D. Wolfe, "Global Ecological Change and Human Health," in *Conservation Medicine: Ecology and Health in Practice*, eds. A. Alonso Aguirre, Richard S. Ostfeld, Gary M. Tabor, Carol House, and Mary C. Pearl, New York: Oxford University Press, 2002, pp. 167–81; Reiter, Paul, "From Shakespeare to Defoe: Malaria in England in the Little Ice Age," *Emerging Infectious Disease* 6 (2000): 1–11; Sherman, Irwin W., ed., *Malaria: Parasite Biology, Pathogenesis, and Protection*, Washington, DC: ASM Press, 1998; Suwonkerd, Wannapa, Overgaard, Hans J., Tsuda, Yoshio, Prajakwong, Somsak, and Masahiro Takagi, "Malaria Vector Densities in Transmission and Non-Transmission Areas During 23 Years and Land Use in Chiang Mai Province, Northern Thailand," *Basic and Applied Ecology* 3

(2002): 197–207; World Health Organization of the United Nations, Roll Back Malaria [Online, October 2004], The RBM Web Site, www.rbm.who.int.

LEA BERRANG FORD

Marsh, George Perkins (1802–1882). A polymath, multilingual American conservationist from Vermont who combined careers in law, farming, business, politics (U.S. congressman from 1843–1849), diplomacy (U.S. envoy to Turkey from 1849–1853 and Italy from 1861–1882), and scholarly writing. He is best known for his insights into the human imprint upon ecosystems and environmental **conservation**. David Lowenthal states that up until 1864, when Marsh wrote *Man and Nature*, most human impacts on the globe were regarded as benign or, where deleterious, trivial and self-correcting. However, Marsh demonstrated that intended as well as unintended consequences of human interventions—such as settlement or **colonization**, **deforestation**, **desertification**, and **species extinction**—deranged the normal stability of, for example, **soil erosion** and restorative processes, which otherwise altered the planet at a slow (geological) pace only. Based on extensive studies and travel, he portrayed the earth as a principally harmonious world torn apart by human greed and ignorance. Nonetheless, he never doubted that continued mastery and management of land resources was essential to civilization. Celebrated in a 1955 Princeton University symposium and again at Clark University in 1989, *Man and Nature* was republished and remains in print to this day. Although the essential message was not entirely novel—writers such as the German geographer Alexander von Humboldt, and others in France, Italy, and India, had expressed similar concerns—George Perkins Marsh was the one who synthesized in an accessible format a wide array of existing knowledge and experience (he could read in twenty languages). His approach, together with the rational utility work of **Johann Heinrich von Thünen**, has pioneered contemporary scientific studies dealing with land-use/cover changes worldwide.

George Perkins Marsh (scholar, whig congressman from Vermont, and U.S. Minister to Turkey and Italy), ca. 1850.
PHOTO: Mathew B. Brady. [Courtesy Library of Congress]

See also: **Driving Forces; Environmental Change; George Perkins Marsh Institute; Land Degradation.**

Further Reading

Lowenthal, David, "Marsh, George Perkins," in *Encyclopedia of Global Change: Environmental Change and Human Society*, vol. 2, eds. Andrew S. Goudie and David J. Cuff, Oxford, New York: Oxford University Press, pp. 60–61; Marsh, George P., *Man and Nature, or Physical Geography as Modified by Human Action*, Cambridge, MA: Harvard University Press, 1864 (reprint; ed. David Lowenthal, 1965).

HELMUT GEIST

Mediating Factor. A causative factor of land change that shapes, modifies, or intervenes in the interplay between underlying **driving forces** and **proximate causes** and that, if related to **feedbacks**, can affect the rate of land change. Often-cited examples of mediating factors are gender, ethnic affiliation, class, and wealth status (and thus power relations), but also biophysical properties such as land suitability (e.g., degree of slopes, soil moisture, etc.). For example, the **conversion** of tropical forests into agricultural uses is often found to be mediated by the unequal relations between well-endowed large-scale farmers or corporate agricultural enterprises and smallholders eking out a living, thus creating "entrepreneurial" versus "populist" **agricultural frontiers** with rather distinct land uses. Likewise, all categories of **agrodiversity** in settled agricultural zones—that is, biophysical diversity, management diversity, agro-biodiversity, and organizational diversity—are shaped by factors that play out differently at various time and spatial **scales**. For example, crop choice and types of conservation practices often differ between poor and rich farmers, thus affecting the pattern of management diversity, and feeding back to enlarge differences in natural **land quality**. In a broader sense, mediating factors that operate through demographic drivers can shape the trajectory of land change toward degradation or restoration. Whether or not increasing population is damaging or beneficial depends upon a variety of institutional, ecological, and technological factors: population growth can cause **land degradation** in the short term, but it can also spur innovation and **agricultural intensification** as well as the adoption of conservation techniques. Mediating factors are crucial also for the response of land managers to external forces, that is, feedbacks are strongly mediated by local factors such as **access** to land, gender, education, and institutional arrangements. In particular, **institutions** need to be considered at various scales to identify those local mediating factors that—together with peoples' adaptive strategies or responses to changing market opportunities—drive land-use change. Juanita Sundberg, for example, illustrates how local participation in natural resource **conservation** is strongly mediated by a community's interaction with non-local actors such as national governments, transnational corporations, and international non-governmental organizations. Seen together with other examples, these "conservation encounters" can shape landscapes and livelihoods in rather contradictory ways. In the Mayan zone, for example, local evidence of high **deforestation** can be found close to locations where exceptionally low rates of deforestation have occurred, with intervening institutional factors making the difference. Another example relates to institutional settings that allow or impede the creation of purchasing power: especially if linked to opportunities of trade and transportation of food, purchasing power can reduce the pressure for **extensification**, as evidenced in the historical decline of New England **agriculture** and subsequent **reforestation** in the region. Mediating factors can form a **multiplier effect**. For example, the environmental effects of an external capital injection—be they foreign direct **investments** or incomes creating **remittances landscapes**—are not limited to the injection itself, but are mediated or magnified by the ways in which capital then circulates through the institutional structure of the economy.

See also: **Arena of Land Conflict; Community Involvement; Exogenous Factor; Land Concentration; Landholding; Land-Use History; Tradeoffs.**

Further Reading

Agrawal, Arun, and Gautam N. Yadama, "How Do Local Institutions Mediate Market and Population Pressures on Resources? Forest *Panchayats* in Kumaon, India," *Development and Change* 28 (1997): 435–65; Bray, David Barton, Ellis, Edward A., Armijo-Canto, Natalia, and Christopher T. Beck, "The Institutional Drivers of Sustainable Landscapes: A Case Study of the 'Mayan Zone' in Quintana Roo, Mexico," *Land Use Policy* 21 (2004): 333–46; Sundberg, Juanita, "Strategies for Authenticity and Space in the Maya Biosphere Reserve, Petén, Guatemala," in *Political Ecology: An Integrative Approach to Geography and Environment-Development Studies*, eds. Karl S. Zimmerer and Thomas J. Bassett, New York, London: The Guilford Press, 2003, pp. 50–69.

HELMUT GEIST

Medium Resolution Imaging Spectrometer. Launched in 2002 onboard the European Space Agency (ESA) ENVISAT satellite, the medium resolution imaging spectrometer (MERIS) has fifteen spectral bands between 0.39 and 1.04 μm (visible to near-infrared wavelengths). Data are acquired at a spatial resolution of 300 meters at nadir. The primary mission of MERIS is the measurement of sea color. It also has great potential to contribute to the mapping and **monitoring** of **land-cover change**. The **normalized difference vegetation index** can be computed. Good temporal resolution is achieved with a repeat-pass interval of only three days, thus ensuring frequent observations of the earth's surface.

See also: **Change Detection; GLOBCOVER; Remote Sensing.**

Further Reading

European Space Agency [Online, August 2004], The Medium Resolution Imaging Spectrometer Web Site, envisat.esa.int/instruments/meris.

KEVIN J. TANSEY

Meta-Analysis. Comparison of disparate case studies as a means of pattern recognition or providing insights about land-change dynamics at the meso- and macroscales. The purpose of meta-analytical research is to combine findings from separate but largely similar local-scale studies that allow for the application of a variety of analytical techniques such as structured literature review or formal statistical approaches, generating a general understanding, for example, of the causes, rates, and pathways of land change while preserving the descriptive richness of these studies. The potential errors produced by data biases and other methodological problems associated with the comparative method have early been raised and dealt with by Charles Ragin, Benedetto Matarazzo and Peter Nijkamp, and Thomas Cook and colleagues, among others. Given the many limitations for original research following standardized data protocols to arrive at a general understanding of land change dynamics beyond the local to subnational level, meta-analytical frameworks for analyzing case studies of **global environmental change** issues have been increasingly recognized as important analytical tools, i.e., to draw inferences on common issues with different but allied empirical backgrounds. For example, Camille Parmesan and Gary Yohe evaluated results from thousands of cases linking climate change and population ecology to better understand the assumed relevance of an **ecological footprint** of climate change upon ecosystems, while Peter Nijkamp and colleagues ran a meta-analytical comparison of empirical results related to the same notion. Likewise, Helmut Geist and Éric Lambin used a meta-analytical framework to determine whether **proximate causes** and underlying **driving forces** of tropical **deforestation** and **desertification** fall into any patterns, also identifying **mediating factors**, **feedbacks**, cross-scalar dynamics, and typical pathways.

See also: **Global Land Project; Land-Use/Cover Change (LUCC) Project; Non-Timber Forest Products; Population Dynamics; Scale.**

Further Reading

Cook, Thomas D., Cooper, H., Cordray, D., Hartmann, H., Hedges, L., Light, R., Louis, T., and F. Mosteller, *Meta-Analysis for Explanation: A Casebook*, New York: Russell Sage Foundation, 1992; Geist, Helmut J., and Éric F. Lambin, "Proximate Causes and Underlying Driving Forces of Tropical Deforestation," *BioScience* 52 (2002): 143–50; Geist, Helmut J., and Éric F. Lambin, "Dynamic Causal Patterns of Desertification," *BioScience* 54 (2004): 817–29; Matarazzo, Benedetto, and Peter Nijkamp, "Meta-Analysis for Comparative Environmental Case Studies: Methodological Issues," *International Journal of Social Economics* 24 (1997): 799–811; Nijkamp, Peter, Rossi, Emilia, and Gabriela Vindigni, "Ecological Footprints in Plural: A Meta-Analytic Comparison of Empirical Results," *Regional Studies* 38 (2004): 747–65. Parmesan, Camille, and Gary Yohe, "A Globally Coherent Fingerprint of Climate Change Impacts across Natural Systems," *Nature* 421 (2003): 37–42. Ragin, Charles, *The Comparative Method: Moving Beyond Qualitative and Quantitative Strategies*, Berkeley: University of California Press, 1987.

HELMUT GEIST

Metabolism. A biological notion that denotes the processes that supply organisms with energy and chemical building blocks needed for growth and maintenance of their structure and functions. The notion is also used to describe energy, material, and substance flows associated with the functioning of socio-economic systems. The importance of socio-economic metabolism for land use and vice versa is increasingly recognized.

The original biological notion was coined by the biologist Jakob Moleschott (1822–1893), who described metabolism as the exchange of matter between an organism and its environment. Nowadays, biological metabolism is primarily researched by biochemists, and the notion has changed accordingly. Modern biology textbooks describe the metabolism of an organism as the sum of biological reactions that convert raw materials obtained from the environment into the macromolecules that make up the organism or are needed for its energy supply. Two main metabolic pathways are usually distinguished: catabolism denotes all biological processes by which food is broken down to simple molecules and energy is produced, while the synthesis of new molecules from simple building blocks is termed anabolism. Regulatory mechanisms that govern these complex processes are usually also included in the biological notion of metabolism. Important examples of metabolic processes are photosynthesis; the breakdown of food proteins to amino acids; energy-generating reactions such as the citric acid cycle; and the synthesis of proteins, DNA, or other important macromolecules. However, even though most ecologists acknowledge that energy flows and **nutrient cycles** in ecosystems are to some extent self-organized, biologists generally do not extend the metabolism notion to the supra-organismic level (i.e., to populations, communities, or ecosystems).

The use of the metabolism analogy to describe exchanges of energy, materials, or even single substances such as metals, carbon, nitrogen, etc., between societies and their environment can be traced back to Karl Marx (1818–1883) and Friedrich Engels (1820–1895), who borrowed the notion from Moleschott and used it in the book *Das Kapital* (*Capital*) to describe the appropriation of natural substances to human requirements in the labor process. Issues of material flows and the possible threat of related resource exhaustion played a considerable role in the 1955 conference "Man's Role in Changing the Face of the Earth" in Princeton, New Jersey.

Contemporary socio-economic metabolism research is largely divided into two strands: methods to account for inputs required (energy, materials, etc.) or outputs generated (wastes, emissions) in the production chain of products or services, an approach called life cycle analysis (LCA); and methods to analyze the energy, materials, or substance flows through geographically defined socio-economic systems (villages or cities, nation-states, supranational entities, or humanity as a whole) or parts thereof (e.g., economic sectors). The two approaches are clearly complementary and should, in principle, also be compatible; in practice, however, the two are researched by only loosely related communities. The first approach is pursued mostly by engineers and others who seek to optimize products; accordingly, analyses of substance flows (e.g., toxic or other hazardous substances) play an important role here, although accounts of the amount of land area required for a production process have also been performed. Within the second approach, most work in the last decades has gone into developing tools to account for nationwide material flows (material flow analysis, or MFA), but other spatial **scales** (e.g., cities, local communities) have also received attention. Analogous methods to account for energy flows have been proposed, and the concept can also be extended to substances such as carbon. These issues are currently researched by a community dominated by biologists, economists, and social scientists who are interested in the ability of MFA to provide accounts of national material (energy, carbon, etc.) flows in good accordance with monetary national accounts (i.e., the System of National

Accounts), thus allowing analyses of interrelations between **economic growth** and a nation's physical throughput. The metabolism approach has recently also been used in environmental history to explore changes in society-nature interaction over long periods of time.

Many aspects of socio-economic metabolism are relevant for land-use change: extraction of resources requires land area, as does their transport, processing, **consumption**, and disposal. The provision of **biomass** usually requires the largest proportion of this area. Several approaches have been used to relate systemwide analyses of socio-economic metabolism and land use. **Ecological footprint** studies basically ask how much area is needed for the provision of biomass and other resources, to host infrastructure, and to absorb carbon from fossil fuel combustion of a defined population, mostly that of a nation-state. Analyses of the **human appropriation of net primary production** demonstrate how much of the biological productivity of a defined area is harvested to support socio-economic metabolism, or foregone due to land **conversions** or land management. Historical analyses of the long-term development of land use and socio-economic metabolism have revealed that changes in socio-economic metabolism, above all those occurring during transitions from agricultural to industrial society, fundamentally alter the role of **agriculture** and **forestry** for the society's energy system and are thus a major **driving force** behind many important changes in land use.

See also: **Analysis of Land Change; Carbon Cycle; Economic Restructuring; Industrialization; Mineral Extraction; Pattern Metrics.**

Further Reading

Ayres, Robert U., and Udo E. Simonis, eds., *Industrial Metabolism: Restructuring for Sustainable Development*, Tokyo: United Nations University Press, 1994; Fischer-Kowalski, Marina, "Society's Metabolism: The Intellectual History of Materials Flow Analysis, Part I, 1860–1970," *Journal of Industrial Ecology* 2 (1998): 61–78; Haberl, Helmut, Wackernagel, Mathis, and Thomas Wrbka, eds., "Land Use and Sustainability Indicators," *Land Use Policy* 21 (2004): 193–320; Krausmann, Fridolin, Haberl, Helmut, Schulz, Niels B., Erb, Karlheinz, Darge, Ekkehard, and Veronik Gaube, "Land-Use Change and Socioeconomic Metabolism in Austria, Part I: Driving Forces of Land-Use Change 1950–1995," *Land Use Policy* 20 (2003): 1–20; Martinez-Alier, Joan, *Ecological Economics: Energy, Environment and Society*, Oxford, UK: Basil Blackwell, 1987; Purves, William K., Sadava, David, Orians, Gordon H., and H. Craig Heller, *Life: The Science of Biology*, 7th ed., Sunderland, MA: Sinauer, 2003.

HELMUT HABERL

Millennium Ecosystem Assessment. An international work program that was designed to meet the needs of the public and decision-makers for scientific information concerning the consequences of ecosystem change for human well-being, including options for responding to those changes. The Millennium Ecosystem Assessment (MA) was launched in 2001 and will help, among others, to meet assessment needs of the **United Nations Convention on Biological Diversity**, the **United Nations Convention to Combat Desertification**, and the **Ramsar Convention on Wetlands**. The MA synthesizes information from the scientific literature and data sets, and makes use of knowledge held by the private sector, practitioners, local communities, and indigenous peoples. Designed as a multiscale assessment, the MA consists of interlinked assessments undertaken at local, **watershed**, national, regional, and global **scales**. It established mechanisms to incorporate information and knowledge from non-peer-reviewed sources including local and **indigenous knowledge**. More than 500 authors are involved in four expert working groups preparing the global assessment, and hundreds more are undertaking more than a dozen subglobal assessments. Subglobal assessments approved or linked to the MA include, among others, the tropical forest sites of the **Alternatives to Slash-and-Burn Programme** of the **Consultative Group on International Agricultural Research**. All of the MA findings undergo rigorous peer review. If the MA proves to be useful to its stakeholders, it is anticipated that an assessment process modeled on the MA will be repeated every five to ten years and that ecosystem assessments will be regularly conducted at national or sub-national scales.

See also: **Decision-Making; Desertification; Global Land Cover Map of the Year 2000; Hot Spots of Land-Cover Change; Integrated Assessment; Integrated Model to Assess the Global Environment; United Nations Environment Programme.**

Further Reading

Millennium Ecosystem Assessment, Strengthening Capacity to Manage Ecosystems Sustainably for Human Well-Being [Online, January 2005], The MA Web Site, www.millenniumassessment.org.

HELMUT GEIST

Mineral Extraction. Economic, technological, political, and social practices through which non-renewable resources are defined, located, segregated, and removed from the immediate environment. Like **agriculture**, the extraction of minerals occurs at the interface of biophysical and socioeconomic processes and is a "pressure point" through which anthropogenic activity can drive land-use change. Unlike agriculture, however, mineral extraction accounts for a very small proportion (less than 1 percent) of the terrestrial land surface, although the degree of land transformation is often extreme. The significance of mineral extraction as a driver of global land-use change, therefore, is not its cumulative areal extent but rather the location of mineral development in areas of high **biodiversity** and/or wilderness value; as well as the secondary and third-order impacts on physical ecosystems as minerals mobilized by extraction and processing move off-site (i.e., acid mine drainage, tailings disposal, or smelter emissions); and the indirect effects of mineral investment on the value of land, the structure of land ownership, and the demand of mineral extraction and processing for other environmental inputs such as wood and water.

Mineral extraction—and the role of mineral extraction in land **conversion**—surged during the nineteenth and twentieth centuries, underscoring the "prodigal and peculiar" nature of environmental transformation in the modern period. **Industrialization**, **urbanization**, and resource-intensive forms of **economic growth** expanded the demand for, and range of, minerals wrested from the earth: aluminum, nickel, oil, and natural gas, for example,

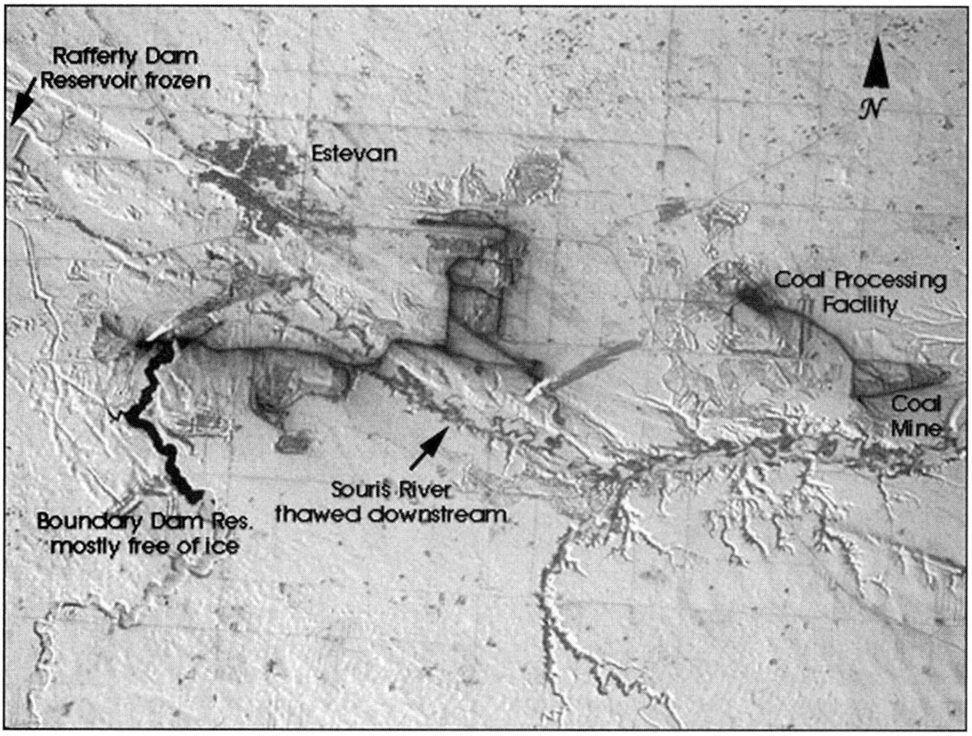

Estevan coalfield along the Souris River in Saskatchewan, Canada, where coal is mined from a trench at the surface, processed, and then used to fire two power plants, on February 17, 2001 (areas of active surface mining and processing appear black; roads used for transporting the coal are also dusted black).

PHOTO: Space Shuttle. CREDITS: Image Science and Analysis Laboratory, NASA-Johnson Space Center, 18 Mar. 2005 (Earth from Space—Image Information). http://earth.jsc.nasa.gov/sseop.efs

had no significant commercial applications 150 years ago, yet today account for a significant portion of the flow of minerals into the economy. By tapping the "extra-somatic" energy of fossil fuels, it has been possible to extract and process progressively lower-quality minerals; to develop mineral resources at great distance from centers of industry; and to produce minerals at lower unit cost, driving down their price and expanding demand still further. Demand has been met, in the main, by increased extraction of (primary) materials rather than the recycling of (secondary) materials already in the economy, as reflected in steeply upward trending resource-production curves for most minerals. The figure above provides examples of four leading nonferrous metals for the period 1850–2000.

Mineral Resources Defined

Minerals are non-living, scarce materials found in the earth's crust that have social utility and/or economic value. They include both organic and inorganic substances and occur in solid (coal, nickel, silica), liquid (oil, brine), and gaseous states (methane, helium). Minerals are conventionally classified as fuels (oil, coal, gas), industrials (potash, salt, sulfur), metals (copper, zinc, bauxite), construction materials (sand, gravel, building stone)

Gold Coin Mine in Victor, Colorado, ca. 1900.
PHOTO: William Henry Jackson. [Courtesy Library of Congress]

and gemstones (ruby, jade), although some materials fall into more than one category. Diamonds, for example, are both industrial and gemstones; oil is a fuel mineral and an industrial feedstock; and clays and sands are used in both construction and in industrial applications. Some minerals are consumed (fuels, solvents) or dissipated in use (applications of metals like zinc and lead in paints) so that re-aggregation and recovery are economically unfeasible. Other minerals can be recycled: in the **United States of America**, for example, recycling represents 55 percent of the apparent supply of iron and steel, 30 percent of copper, 65 percent of lead, and 35 percent of aluminum. Despite an increase in the recycling of metals, resource extraction remains the primary means of meeting new metal demand, the vast majority of construction minerals, and the sole means of supplying demand for fuels and industrial minerals.

As defined here, minerals refer to a broad class of materials drawn from the earth. This definition has greater utility for research on land-use/land-cover change than the more specific definitions found in the chemical and mineralogical literature, which focus on a mineral's physical properties (atomic structure, degree of homogeneity) or mode of formation. The more inclusive definition emphasizes characteristics of minerals that are economically and environmentally significant, and includes the following:

1. Minerals are non-renewable resources and, as such, the contribution of mineral extraction to sustainable development centers on the successful

Results of deforestation during the early mining days in San Juan County, Colorado, September 1940.

PHOTO: Russell Lee. [Courtesy Library of Congress]

conversion of exhaustible natural assets into renewable forms of natural or human capital.

2. Minerals are scarce rather than ubiquitous (their occurrence in extractable quantities is limited to particular locations), and so frequently there is competition among extra-local interests for **access** to and control of mineral resources.
3. As a consequence, the direction and rate of land-use change in extractive economies is strongly influenced by relations of power between extra-local interests and local land users.
4. Whether or not a particular mineral (uranium or coal-bed methane, for example) is valued as a resource depends on geological knowledge and the economic feasibility of extraction, both of which are historically and geographically variable.
5. Thus, mineral extraction is an inherently dynamic form of land use, the "boom and bust" histories of which translate into relatively localized, yet intense forms of land-use change. For example, land use can shift to mineral extraction as new mines open, or from mineral extraction as existing operations close in response to the accumulation of geological knowledge (the development of diamond mining in northern **Canada** since the 1990s), changes in the price of minerals or the costs of extraction (the low costs of leaching technology made it profitable to develop low-grade copper and gold projects worldwide beginning in the late 1970s), a shift in the

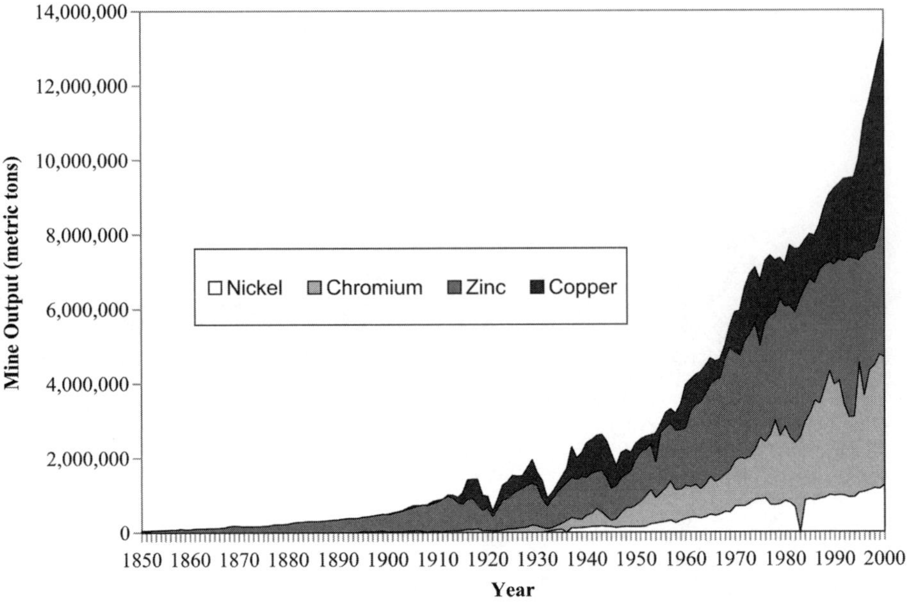

World mine output of four leading nonferrous metals, 1850–2000.

perceived risk/reward ratio of mineral **investments** relative to other investment options, or a change in the perceived risk/reward ratio of one mineral region over another (the expansion of mineral investment in the Peruvian **Andes** following the adoption of economic liberalization policies in Peru in the early 1990s).

Coupling Extraction and Land-Use/Cover Change

Conventionally, the concept of extraction applies to the engineering processes of removing minerals from the earth. From this perspective, the land-use changes arising from mineral extraction are best understood by referring to the technologies of extraction and via a focus on **environmental changes** at the site of extraction. Efforts to mitigate the land-use changes of extraction, for example, would center on the introduction of cleaner technology and/or best-practice environmental management techniques. This traditional view of discrete, localized impacts arising from the choice of technology or management practice has been challenged from a number of perspectives. These critiques provide the framework for a more comprehensive approach to extraction that recognizes its ecological, geographical, and institutional components.

First, research in environmental science has shown how the initial impacts of mineral extraction can be magnified or transmitted over a broad area via ecosystem processes. This work includes studies of the effect of emissions and/or solid wastes on the biological productivity of receiving ecosystems; the contribution of mineral extraction and processing to biogeochemical cycles that are regional and global in scope (mining and smelting account for 67 percent and 55 percent of the total anthropogenic release of copper and zinc, for example); and the role of mineral extraction in driving habitat fragmentation. Recent geographical shifts in the location of mining investment in the global South have intensified longstanding concerns about the impact of mining on global biodiversity and critical ecosystems. The World Resources Institute reports that 75 percent of active mines and exploration areas overlap with areas of high **conservation** value and areas of **watershed** stress, and that nearly one-third of all active mines and exploration sites are located within intact ecosystems of high conservation value.

Second, the longstanding engagement of ecological economics with thermodynamic constraints on resource production has shown that mineral extraction rests on an exchange (and geographical transfer) of organizational states: highly ordered, economically useful forms of energy are expended to add order to materials that are initially disorganized. Access to low-entropy forms of energy at low cost has underpinned the scaling up of "earth moving" capacities during the twentieth century: for example, the introduction of open pit mining for metals, or the shift from underground methods to "mountaintop removal" in Appalachian coal mining. Other work has examined how the energy exchanges of extraction produce disorder in the surrounding environment and how, as a result, extractive economies are at the center of a fundamentally uneven geographical exchange: mineral extraction requires the import and expenditure of low-entropy energy to produce an ordered material form (the useful mineral) that is subsequently exported, leaving disorder (ecosystem disruption) as a by-product. Energy analysis reveals, therefore, how the accumulation of low-entropy states in the core (the land-use changes of urbanization and industrialization that pile up sand, gravel, cement, iron, and copper, for example) and the production of high-entropy states in the periphery (the land-use changes of extraction that dismantle tropical forest cover and introduce contaminants to surface and groundwater, for example) are two sides of the same coin (although this account is not informed by energy analysis).

Third, extraction is a political and economic process through which value is created, appropriated, and transferred. The social histories of iconic mineral discoveries—the silver resources of the Comstock Lode, Nevada, in 1859, and oil at Spindletop, Texas, in 1901—vividly demonstrate how mineral production drives changes in the value of land and other critical inputs that can extend well beyond the site of extraction. Rising values for wood and water in a mine's hinterland, for example, can lead to the commodification and exchange of these resources (trees are felled for mine timbering or for the production of charcoal fuel; water is diverted for use in mining), producing extensive **land-cover change**: development of the Comstock Lode, for example, drew so extensively on lumber from the Lake Tahoe region that its mines have been called the "Tomb of the Sierras." Changes in resource ownership can also introduce new land management priorities and rules for resource access/use that affect land use and land cover. In addition, the attractiveness of mineral resources to national governments seeking to meet the requirements of multilateral lending **institutions** or deliver on economic goals means that mineral extraction is frequently the vanguard of economy-wide initiatives to commodify land and resources, create markets in resource rights, and "enclose the commons." Mineral law reform, for example, has been a central element of **economic liberalization** in many developing countries, while the prospect of investment in mineral extraction has influenced the priority and pace of land titling programs and the direction of reforms in water resource law (to facilitate water transfers away from traditional agricultural uses).

Fourth, mineral extraction may be organized in several different ways, with different implications for land-use change. Although the bulk of mineral output is produced by large, capital-intensive corporations operating a relatively small number of very large mines, some minerals (particularly gold and gemstones from placer deposits) are also extracted using labor-intensive methods. This informal mining sector is characterized by low capital intensity, relatively small volumes of output per operation, and insecurity of **land tenure**. Although the extent of informal sector mining can wax and wane in response to the same mineral price trends that govern corporate mining, the dynamics of the informal sector are quite different. Informal sector mining, for example, is often a response to **poverty** and a lack of other economic opportunities, and so it can expand even when mineral prices are relatively low. Addressing the environmental effects of mineral extraction, therefore, requires an approach that recognizes these organizational differences and that can assess their significance for addressing land-use change.

See also: **Amazonia; Arena of Land Conflict; Boreal Zone; Consumption; Economic Restructuring; Ecosystem Services; Land Rights; Multiplier Effect; South Africa; South America; Taiga.**

Further Reading

Brechin, Gray, *Imperial San Francisco: Urban Power, Earthly Ruin*, Berkeley, CA: University of California Press, 2001; Bridge, Gavin, "Contested Terrain: Mining and the Environment," *Annual Review of Energy and Resources* 29 (2004): 205–59; Bunker, Stephen, *Underdeveloping the Amazon: Extraction, Unequal Exchange, and the Failure of the Modern State*, Urbana, IL: University of Illinois Press, 1985; McNeill, J. R., *Something New under the Sun: An Environmental History of the Twentieth-Century World*, New York: Norton, 2000; Moore, Johnnie, and Samuel Luoma, "Hazardous Wastes from Large-Scale Metal Extraction," *Environmental Science and Technology* 24 (1990): 1278–85; Nriagu, Jerome, "History of Global Metal Pollution," *Science* 272 (1996): 223–4; World Resources Institute, *Mining and Critical Ecosystems: Managing the Risks*, Washington DC: World Resources Institute, 2004.

GAVIN BRIDGE

Miombo. An abbreviation for miombo woodlands, which form the world's largest area of **tropical dry forest** in southeastern and central Africa, that is, 5.5 out of a total of 7.7 million square kilometers. *Miombo* is a vernacular word that has been adopted by ecologists to describe woodlands dominated by trees in the genera *Brachystegia*, *Julbernardia*, and *Isoberlinia*. The ecosystem extends from Tanzania and Congo in the north; through Zambia, Malawi, and eastern Angola; to Zimbabwe and Mozambique in the south. Miombo features a hot, seasonally wet climatic zone the soils of which—mainly derived from acid crystalline bedrock—are predominantly infertile. The woodlands have contracted to their present pattern of distribution in the past 1,000 years. Measured against degrees of "human disturbance," based on data from the 1970s and 1980s, it has been estimated that about 62 percent of the woodland cover has been undisturbed. However, all factors so far contributing to the **conservation** of the woodlands are seen to have drastically changed during the most recent past, and many formations are considered to be heavily modified to mostly **secondary vegetation**. The most important underlying **driving forces** of the present dynamics of miombo transformation are considered to be "macro-level phenomena," mainly national policies related to **agriculture**, **forestry**, energy, and population. Rooted in colonial interventions, these policies have grown in strength and extent since the past decades. At the proximate level, most of the **land-cover changes** in the miombo woodlands relate to agricultural economies based on export and colonial settler crops such as tea, **coffee**, **cotton**, and **tobacco**. These cash-cropping systems, practiced on commercial as well as smallholder farms, have been increasingly drawn into the world market, and they went through various stages of both cropland extension and **agricultural intensification**.

See also: **Cash Crops; Colonization; Economic Liberalization.**

Further Reading

Campbell, Bruce M., ed., *The Miombo in Transition: Woodlands and Welfare in Africa*, Bogor, Indonesia: Center for International Forestry Research, 1996; Geist, Helmut, "Soil Mining and Societal Responses: The Case of Tobacco in Eastern Miombo Highlands," in *Coping with Changing Environments: Social Dimensions of Endangered Ecosystems in the Developing World*, eds. Beate Lohnert and Helmut Geist, Aldershot, UK, Brookfield, VA: Ashgate, 1999, pp. 119–48.

HELMUT GEIST

Mode of Interaction. The interaction between the various causes or processes of land-use/cover change show different patterns that can be approached by using meta-analytical frameworks of analysis (see figure). First, one cause may completely dominate the other causes, assuming that land use in a given locality is influenced by whatever factor exerts the greatest constraint. The proportion of cases, however, in which dominant, single, or key factors operate at either the proximate or underlying level (or connecting the two levels) has been found to be low in cases of tropical **deforestation** and **desertification** (around 10 percent of the cases). Second, factors driving land change can be connected as causal chains, that is, interconnected in such a way that one or several variables drive one or several other variables. Again, the proportion of cases in which pure chain-logical causation dominates has been found to be low (less than 10 percent of the cases). Third, different factors can intervene in concomitant occurrence or concurrently, that is, independent but synchronous operation of factors leading to land change. Concomitant occurrence of causes is much more widespread than single-factor or chain-logical causation (around 25 percent of the cases). Fourth, factors and their interactions as detailed above may also play together in synergetic factor combinations, that is, several mutually interacting variables driving land-use/cover change and producing an enhanced or increased effect due to reciprocal action and **feedbacks** between causes. The pattern

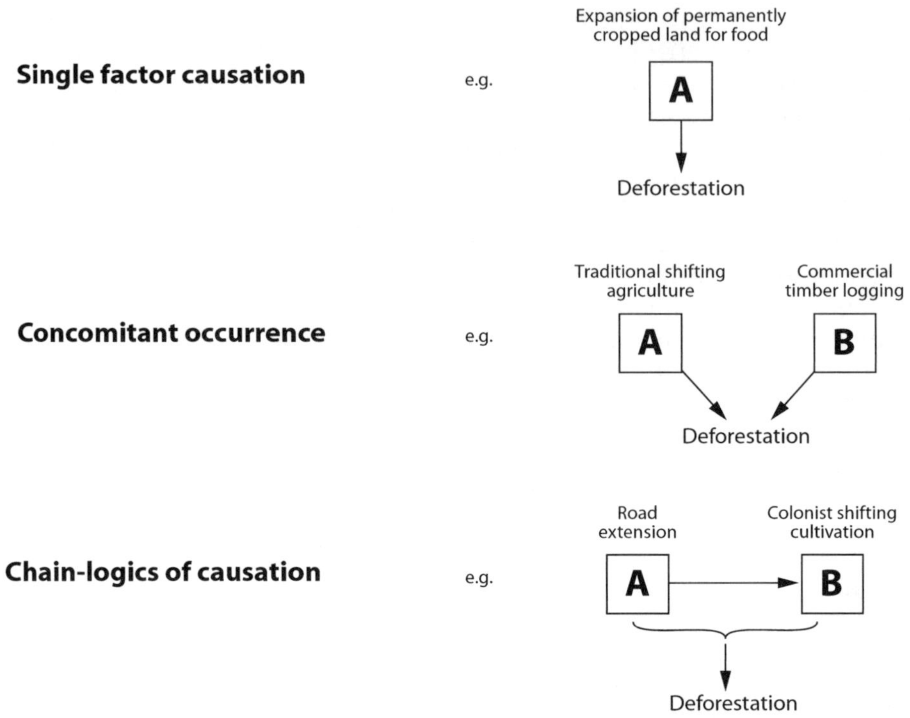

Three possible, but not widespread, modes of interaction.

of causal synergies has actually been identified as the most common type of interaction (inherent to 70–90 percent of the deforestation and desertification cases under study). Éric Lambin and colleagues propose to apply the term "land use = *f* (pressures, opportunities, policies, vulnerability, and social organization)," with the function *f* having forms that account for strong interactions among the causes of land change. In detail, these are pressures = *f* (population of resource users, labor availability, quantity of resources, and sensitivity of resources); opportunities = *f* (market prices, production costs, transportation costs, and technology); policies = *f* (subsidies, taxes, **property** rights, infrastructure, and governance); **vulnerability** = *f* (exposure to external perturbation, sensitivity, and coping capacity); and social organization = *f* (resource **access**, income distribution, household features, and urban-rural interactions). Though still in its infancy, the study of **forest transitions** points to such strong interaction of multiple causative factors in land-change processes, with land change as an emergent property of complex adaptive systems.

See also: **Cultural Factors; Driving Forces; Economic Growth; Land-Use Policies; Land-Use Transition; Mediating Factor; Meta-Analysis; Population Dynamics; Proximate Causes; Resilience; Scale; Syndrome.**

Further Reading

Lambin, Éric F., Geist, Helmut J., and Erika Lepers, "Dynamics of Land Use and Cover Change in Tropical and Subtropical Regions," *Annual Review of Environment and Resources* 28 (2003): 205–41.

HELMUT GEIST

Modeling. Land-use/cover-change models provide simplified representations of the spatial and/or temporal dynamics of land use/cover. Since real-life experiments in land-use change are difficult to perform, computer models can be seen as a computational laboratory in which hypotheses about the processes of land-use change can be tested. Furthermore, models provide a structured way of analyzing complex interactions; scientists can make assumptions on the most important mechanisms of land-use change and then test these hypotheses through sensitivity analysis. Once a functioning and validated model of the **land-use system** is constructed, projections of future developments can be made. The user can explore system functioning through "what if" **scenarios** and explore the response of land use to these changes. Scenario simulations can provide insight in alternative futures or allow the evaluation of the effects and **tradeoffs** of different scenarios. Possible scenario simulations include the evaluation of the effect of changes in the agricultural sector (e.g., due to changes in market conditions or agricultural policies) on land-use patterns. Finally, land-use change models allow the translation of scenario conditions to visualizations of land-use patterns that can be used as a communication and learning environment for stakeholders involved in land-use **decision-making**. Projections can be used as early warning systems for the effects of future land-use changes and indicate **hot spots of land-cover change** that are priority areas for in-depth analysis and policy intervention. **Johann Heinrich von Thünen** provided one of the first spatially explicit land-use models based on economic theory. His work is based on the concept of **land rents**, which are closely related to the potential profit a farmer can make from growing a crop. As this profit is related not just to the value of a crop at the market but also to the cost to transport the products to that market, rent for any particular crop falls off with distance from the market. When farmers have a choice of crops to grow, they will, obviously, choose the most profitable one. Spatially, this will result in a series of concentric rings around the market, with crops with the highest transport costs relative to the market price grown nearest to the city.

Since the acknowledgment of the importance of land-use/cover change for **global environmental change** and following the research agenda set by the international **Land-Use/Cover Change (LUCC) project**, a large variety of modeling approaches have been developed to address different aspects of land-use change, based on different scientific disciplines.

Typology of Model Approaches

Different authors have provided reviews of land-use models using classification systems, often based on the dominant technique used in the model or the underlying disciplinary theory. For **deforestation** models, an overview is provided by Éric Lambin as well as by David Kaimowitz and Arild Angelsen, while Eric Miller and colleagues present a review of integrated urban models. Éric Lambin and colleagues review models for **agricultural intensification**, while Nancy Bockstael and Elena Irwin review a number of land-use models in terms of economic theory foundations. Chetan Agarwal and colleagues review nineteen models based on their spatial, temporal, and human-choice complexity, while Helen Briassoulis and Peter Verburg give overviews of all types

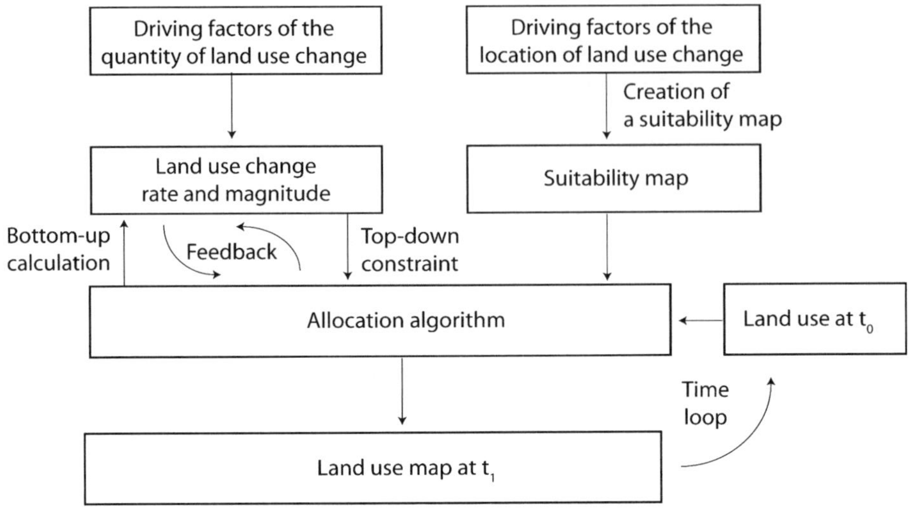

Main components of spatially explicit land-use/cover change models.

explicit land-use models. One of the important parts of the model structure relates to the distinction between the calculation of the magnitude of change and the allocation of change. Both calculations are based on a set of driving factors, some steering the magnitude of change and some others only steering the location of change. Sometimes the same driving factor can influence the quantity and location of change. Demographic drivers are often seen as important factors that can (partly) explain the rate and quantity of land-use change. However, the spatial distributions of population and migration patterns are important determinants for the location of land-use change as well.

Based on the interpretation of one or more driving factors that are supposed to be determinants of the location of land-use change, a so-called suitability or preference map is created that indicates the suitability of a location for a specific land-use type relative to the suitability of other locations. The translation of the driving factors into a suitability map can vary between the direct use of potential productivities of agricultural crops to complex empirical or rule-based systems that combine a larger set of location factors. The selection of the driving factors used in the model and its translation to a suitability map are two of the main components of a land-use model. A wide variety of approaches exist, including the following three, described below.

Rule-based systems based on expert knowledge: In this approach, a set of selected variables are assumed to best represent the decision-making concerning the location of land-use change. If it is assumed, for example, that **agriculture** is preferably located at locations where a high production can be attained, the suitability map may represent the potential or water-limited production of agricultural crops at each location. An example of this approach is the land-allocation module of the **integrated model to assess the global environment** (IMAGE).

Rule-based systems based on theory: In this approach, assumptions about decision-making are made strictly following theory. Most often, use is made of economic reasoning. In these applications, the suitability map often consists of (relative) land rent. Given the absence of data on land rent and the large number of factors determining land prices, empirical techniques are often used to estimate land rent from a number of determining factors.

Suitability maps based on empirical analysis: This is probably the most widely used approach. Because the interaction between land use and its driving factors is mostly complex and region-dependent, it is not easy to define a rule-based system for deriving a suitability map. Therefore, modelers often make use of empirical methods to explore and quantify the relation between observed (historic) land-use patterns and the spatial variation in a set of supposed driving factors. Common techniques include multiple regression, logistic or multinomial regression, and the use of neural nets. Markov models are also based on empirical analysis. They use observed transitions in land use to estimate the probability of future transitions based on the current land cover at a location.

Neighborhood interactions and ***cellular automata***: Land-use patterns most often show a strong degree of spatial correlation, due to spatial clustering in driving factors and spatial interactions among locations of land use. In many models, especially those focusing on **urbanization**, special attention is given to spatial interactions that determine the suitability of a location. Cellular automata update the suitability or transition probability each time step, based on an analysis of neighboring land uses and a rule-based system that translates the neighborhood characteristics into suitabilities or transition probabilities. Examples of models making use of neighborhood interactions include those of Claudia de Almeida and colleagues, Keith Clarke and Leonard Gaydos, and Fulong Wu.

The construction of a suitability map does not directly lead to the determination of changes in land-use patterns. The actual pattern of land-use change is also determined by the requirements for the different land-use types and competition among land uses. Therefore, in most models, a rule-based system allocates the actual land-use changes based on the suitability map. These rules range from using a simple cut-off value to select the locations with the highest suitability from the suitability map, to dynamic modeling of competition between land uses based on land-use type-specific characteristics. These rule sets are incorporated in an allocation algorithm.

A wide variety of approaches are used to calculate the claims of different land uses for space. In a number of models, a bottom-up approach is chosen, in which the spatial dynamics and allocation rules determine the aggregated quantity of land-use change. However, often a top-down approach is chosen, in which a set of rules or a modeling approach is used to calculate the quantity of change based on a set of driving factors. This quantity is used as a constraint in the actual allocation procedure. In a few cases a hybrid approach is chosen in which the land-use requirements are dynamically adapted through a **feedback** between the allocation and the calculation of land-use requirements to account for land availability, for example.

Which Model Should be Chosen?

It is not possible to indicate what model approach or specific model is most suitable for land-use/cover change modeling. The choice of model is largely dependent on the research or policy questions that need to be answered. Not all model approaches are capable of answering all questions. Furthermore, the research questions might pose restrictions on the applicability and suitability of a particular question, in terms of **scale**, that is, by its spatial and temporal resolution and extent and dominant land-use change process. For example, a spatially explicit cellular automata model might be well suited to explore urban growth dynamics but be incapable of exploring the driving factors of agricultural transitions.

A number of recent models combine different theories and techniques to better account for the different processes and make best use of the specific strengths of different approaches. Such a hybrid model is often capable of simulating multiple land-use types at the same time using different techniques to specify the suitability map and allocation rules depending on the specific land-use type rather than having a similar procedure for all land-use types.

When models are used to improve and test our understanding of the driving factors of land-use change, the combined use of inductive and deductive modeling approaches might lead to additional insights. Inductive techniques might suggest possible driving factors that can be tested on causality in deductive approaches, leading to a more direct linkage between process- and pattern-based techniques in land-use analysis. In case models are used to assess the impacts of different storylines in scenario studies, other constraints will be imposed on the choice of model. Most scenarios assume changes in attitude and decision-making in line with changes in worldview. Such scenarios cannot be simulated by land-use models in which the transition rules are solely determined by (empirical) analysis of past trends. Similar considerations apply to all studies with temporal extent longer than one or two decades.

Decision Support and Uncertainty

The application of land-use models in policy decision support systems is limited. This is partly due to the relatively recent development of full-fledged modeling approaches in this field and the explorative state of most model approaches. Diversity of approaches is another drawback for users, indicating the absence of a model that might perfectly suit their needs. Furthermore, most **validation** exercises have indicated that uncertainty in land-use simulations is high. These limitations are common for models of complex integrated systems; and, although progress is made, some of these constraints are inherent to the land-use system. Therefore, visualization issues and adequate presentation of the results are most important to communicate land-use modeling results to policymakers and other stakeholders.

See also: **Century Ecosystem Model; Change Detection; Continuous Data; Discrete Data; History Database of the Global Environment; Hot Spot Identification; Integrated Assessment; Integrated Model; LandSHIFT Model; Land-Use History; Land-Use Planning; Land Use Transition; Pattern to Process; Probit/Logit Model; Socializing the Pixel.**

Further Reading

Agarwal, Chetan, Green, Glen M., Grove, J. Morgan, Evans, Tom P., and Charles M. Schweik, "*A Review and Assessment of Land Use Change Models: Dynamics of Space, Time, and Human Choice*," Bloomington: Indiana University, Center for the Study of Institutions, Population, and Environmental Change, and USDA Forest Service, 2001; Angelsen, Arild, "Agricultural Expansion and Deforestation: Modelling the Impact of Population, Market forces and Property Rights," *Journal of Development Economics* 58 (1999): 185–218; Bockstael, Nancy E., and Elena G. Irwin, "Economics and the Land Use-Environment Link," in *The International Yearbook of Environmental and Resource Economics 1999/2000*, eds. Henk Folmer and Thomas Tietenberg, Cheltenham, UK: Edward Elgar Publishing, 2000, pp. 1–54; Bousquet, François, and Christophe Le Page, "Multi-Agent Simulations and Ecosystem Management: A Review," *Ecological Modelling* 176 (2004): 313–32; Briassoulis, Helen, Analysis of Land Use Change: Theoretical and Modeling Approaches [Online, August 2004], The Web Book of Regional Science Web Site, www.rri.wvu.edu/regscweb; Brown, Daniel G., Pijanowski, Bryan C., and Jiunn-Der Duh, "Modeling the Relationships between Land Use and Land Cover on Private Lands in the Upper Midwest, USA," *Journal of Environmental Management* 59 (2000): 247–63; Chomitz, Kenneth M., and David A. Gray, "Roads, Land Use, and Deforestation: A Spatial Model Applied to Belize," *The World Bank Economic Review* 10 (1996): 487–512; Clarke, Keith C., and Leonard J. Gaydos, "Loose-Coupling a Cellular Automaton Model and GIS: Long-Term Urban Growth Prediction for San Francisco and Washington/Baltimore," *International Journal of Geographical Information Science* 12 (1998): 699–714; de Almeida, Claudia M., Batty, Michael, Monteiro, Antonio M. V., Camara, Gilberto, Soares-Filho, Britaldo S., Cerqueira, Gustavo C., and Cerqueira L. Paennachin, "Stochastic Cellular Automata Modeling of Urban Land Use Dynamics: Empirical Development and Estimation," *Computers, Environment and Urban Systems* 27 (2003): 481–509; Kaimowitz, David, and Arild Angelsen, *Economic Models of Tropical Deforestation: A Review*, Bogor, Indonesia: Center for International Forestry Research, 1998; Lambin, Éric F., Rounsevell, Mark D. A., and Helmut J. Geist, "Are Agricultural Land-Use Models Able to Predict Changes in Land-Use Intensity?" *Agriculture, Ecosystems and Environment* 82 (2000): 321–31; Lambin, Éric F., "Modelling and Monitoring Land-Cover Change Processes in Tropical Regions," *Progress in Physical Geography* 21 (1997): 375–93; Miller, Eric J., Kriger, David S., and John D. Hunt, *TCRP Web Document 9: Integrated Urban Models for Simulation of Transit and Land-Use Policies: Final Report*, Toronto: University of Toronto Joint Program in Transportation and DELCAN Corporation, 1999; Overmars, Koen P., and Peter H. Verburg, "Analysis of Land Use Drivers at the Watershed and Household Level: Linking Two Paradigms at the Philippine Forest Fringe," *International Journal of Geographical Information Science* 19, no. 2 (2005): 125–52; Parker, Dawn C., Manson, Steven M., Janssen, Marco A., Hoffman, Matthew, and Peter Deadman, "Multi-Agent Systems for the Simulation of Land-Use and Land-Cover Change: A Review," *Annals of the Association of American Geographers* 93 (2003): 314–37; Pontius, Robert G., Cornell, Joseph D., and Charles A. S. Hall, "Modeling the Spatial Pattern of Land-Use Change with GEOMOD2: Application and Validation for Costa Rica," *Agriculture, Ecosystems and Environment* 85 (2001): 191–203; Veldkamp, A. (Tom), and Éric F. Lambin, "Editorial: Predicting Land-Use Change," *Agriculture, Ecosystems and Environment* 85 (2001): 1–6; Veldkamp, A. (Tom), and Peter H. Verburg, "Modelling Land Use Change and Environmental Impact: Introduction to the Special Issue," *Journal of Environmental Management* 72 (2004): 1–3; Verburg, Peter H., and A. (Tom) Veldkamp, "Editorial: Spatial Modeling to Explore Land Use Dynamics," *International Journal of Geographical Information Science* 19, no. 2 (2005): 99–102; Verburg, Peter H., Kok, Kasper, Pontius, Robert Gilmore, Jr., and A. (Tom) Veldkamp, "Modelling Land Use and Land Cover Change," in *Land Use and Land Cover Change: Local Processes, Global Impacts*, eds. Éric Lambin and Helmut Geist, Berlin: Springer, 2006;

What to monitor in a global context? Human-induced transformations of ecosystems and landscapes are the largest sources of change on earth, affecting the ability of the biosphere to sustain life. Acknowledging both mankind's influence and its dependence on the environment, the first joint global earth monitoring endeavor, Global Earth Observation System of Systems (GEOSS), is designed around nine "societal benefit areas." The GEOSS ten-year implementation plan lists observational requirements, monitoring architecture, temporal targets, and performance indicators for each area in a comprehensive tabular format (observable counts given for reference): disasters, 32 observables; health, 18; energy, 31; climate, 44; water, 54; weather, 41; ecosystems, 28; **agriculture**, 32; and **biodiversity**, 12 observables.

Good Practice Procedures

Monitoring efforts range from local, to regional, to continental and global endeavors. In general, monitoring procedures should be clearly described in an implementation phase (i.e., database of strategies, implementation plan, project implementation, implementation assessment, and implementation report) and an effectiveness phase (i.e., list of desired outcomes, indicators, effectiveness assessment, and effectiveness report).

Local/regional monitoring programs should incorporate the following steps. First, plan a successful monitoring/indicators program: be specific about goals; select the most important variables or the most practical ones (and be diverse) for the specific site or region monitored; be sure to have indicators that change over a gradient of human influences; build upon existing monitoring data and accepted standards and prioritize what needs to be done next; promote agreement by the scientific and management communities on the critical questions; embrace principles of ecologically sustainable development; specify how information collected will be used in the decision-making process; and analyze cost effectiveness. Second, carry out a successful monitoring/indicators program: sampling should be regionwide, for many years, on a frequent time scale, and collaboration among agencies is important (particularly in times of limited funds) both for sampling and sharing data; utilize different monitoring groups (e.g., citizen volunteers, academics, state agencies, and federal agencies) and make data accessible; couple monitoring with experiments to determine whether the assumptions were correct; be open to new methods, techniques, and equipment; and use a long monitoring time frame (longer than a political time frame). Third, work up the data from a successful monitoring/indicators program: use a sufficient level of data resolution to assure accuracy and precision; provide synthesis to turn data into useful information; put results into a larger perspective; track cumulative impacts; gain new insights by repeated assessments and observing responses to management actions; be sure to use reliable data in order to maintain credibility; communicate regularly to the public and to decision-makers; and establish monitoring data as an integral management tool.

The framework for a national environmental monitoring and assessment program (EMAP) has been published by the U.S. Environmental Protection Agency. The EMAP approach has exemplary character for national to continental monitoring programs. It elaborates on the scientific barriers with respect to the chosen indicators and a nationally comprehensive monitoring framework and implementation efficacy with respect to capacity-building of governmental and public bodies. EMAP is shaped by its statistical design (resource sampling: sampling site selection, classification for strata), ecological indicators (indicator development, measurement, responsiveness, and variability), reference conditions, and key research issues. Special attention is given to landscape research and landscape indicators. **Remote sensing** techniques have been applied to create a national land-cover database from **Landsat** Thematic Mapper images taken from 1991 to 1993. Work has been undertaken on a revised national land-cover database for the year 2000. The landscape indicators depend on the quality of the remotely sensed landscape data from which they are built. Problems to be investigated are related to the scale of the remote sensing data—and thus of the indicators—and the number of attributes (e.g., land-cover classes).

Comparable efforts on strengthening national environmental monitoring and information systems in **Europe** are led by the European Environment Agency (EEA) with its European Environment Information and Observation Network (EIONET) and its **Coordination of Information on the Environment (CORINE) database**. Environmental assessment was undertaken in 1995 and 1998 and focused on the following twelve prominent European environmental problems: climate change, stratospheric ozone depletion, acidification, tropospheric ozone, chemicals, waste, biodiversity, inland waters, marine and coastal environment, **soil degradation**, urban environment, and technological hazards.

Responsible for global environmental monitoring activities is the **United Nations Environment Programme** (UNEP). In 1972, UNEP established the Global Environmental Monitoring System (GEMS). GEMS includes the assessment of a list of priority pollutants and monitoring of natural resources. Monitoring is recognized as a global problem with respect to institutional, spatial, and temporal scales and under consideration of the need for agricultural, **forestry**, and freshwater sustainability. In 2003, the UN Framework Convention on Climate Change (UNFCCC) published the *Good Practice Guidance for Land Use, Land-Use Change and Forestry* (GPG-LULUCF), which provides methods for estimating, measuring, monitoring, and reporting on carbon stock changes and **greenhouse gas** emissions from LULUCF activities for the **Kyoto Protocol**—that is, choice of estimation method, quality assurance procedures, documentation, assessment of inventory, and quantification of uncertainties.

Monitoring Programs

Two major, politically driven programs were launched at the beginning of this millennium, both reflecting the necessity and urgency to monitor the state of the earth's environment: the European GMES and the global Group on Earth Observation (GEO) with its GEOSS. The overall aim of GMES is to support Europe's goals regarding sustainable development and global governance, by facilitating the provision of quality data, information, and knowledge. GMES forums bring together the main stakeholders in Europe involved in the use of observations from space and *in situ* facilities for better management of the environment and enhanced security of Europe's citizens.

The draft GEOSS ten-year implementation plan states (October 15, 2004) that the purpose of GEOSS, as stated in the declaration of the first earth observation summit, is to achieve comprehensive, coordinated, and sustained earth observations for the benefit of humankind. GEOSS implementation will enable improved monitoring of the state of the earth, increased understanding of dynamic earth processes, enhanced prediction of the earth system, and further implementation of international environmental treaty obligations. GEOSS aspires to encompass all areas of the world and to cover *in situ* observations as well as airborne and space-based observations. The focus of GEOSS is on observations relevant to large parts of the world and issues that require global information to be addressed optimally.

There are international efforts at regional to global scales, partly preparing or influencing GMES and GEOSS, which are programs that often address specific monitoring gaps about earth surface dynamics to improve the availability and dissemination of baseline data, monitoring capabilities and resources, and institutional stability.

The three aims of the European Commission's CORINE program are to compile information on the state of the environment, to coordinate the compilation of data and the organization of information within the European member states or at the international level, and to ensure that information is consistent and that data are compatible. The working scale is 1:100,000; and the land-cover nomenclature is organized into three levels with the first level having five headings, the second level having fifteen headings, and the third level having forty-four headings. A further objective of CORINE is to obtain information about the way the environment is changing. The basis has been accomplished with the two Europe-wide CORINE coverages for the years 1990 and 2000.

The aim of Monitoring of Land-Use/Cover Dynamics (MOLAND) is to provide a spatial planning tool that can be used for assessing, monitoring, and **modeling** the development of urban and regional environments in Europe. The project was initiated in 1998 under the name Monitoring Urban Dynamics (MURBANDY). Particular effort is dedicated to the development of easy handling indicators, since potential users of the results include city managers, local administrators, and planners, and even up to international bodies. The spatial dimension is used to develop sets of indicators having a degree of standardization that allows for comparison within Europe. The three specific aims are to produce quantitative information on the evolution of land use and transport networks, from 1950 onward, in study areas subject to infrastructural changes (e.g., **urbanization**, construction of transport links); to develop methods for performing a harmonized analysis of historical trends, including socio-economic aspects, impact of legislation, landscape fragmentation, etc.; and to develop models for the harmonized simulation of future Europe-wide **scenarios** at local and regional scales. Since 2004, MOLAND is contributing to the evaluation and analysis of impact of extreme weather events, in the frame of research on adaptation strategies to cope with climate change. To date, the MOLAND methodology has been applied to an extensive network of cities and regions for an approximate total coverage in Europe of 70,000 km^2.

Agro-environmental indicators have been monitored in Europe since 2003 through the Land Use/Cover Area Frame Statistical Survey (LUCAS) collaboration among EEA, the Agriculture and Environment Directorate-Generals, Eurostat, and the **Joint Research Centre (JRC) of the European Commission**. LUCAS uses land-cover information, administrative data, and geo-referenced statistical surveys that are important for providing information on fluxes, stocks, and pressure indicators; and it explores data sets for the European Union. It observed that the integration of different data sets was possible and stresses the importance of data being available at periodic intervals.

The Multi-Resolution Land Characteristics (MRLC) consortium is a group of U.S. federal agencies that joined in order to develop a land-cover data set called the National Land Cover Dataset (NLCD). In 1999, a second-generation MRLC consortium was formed to coordinate the production of a comprehensive land-cover database for the United States of America called NLCD 2001, using Landsat 7 ETM+ data. The key component of this land-cover mapping effort is a database approach, which provides flexibility in developing and applying suites of independent data layers. These independent standardized data layers or themes will be useful not only within the land-cover classification but as data components for other applications. NLCD is a twenty-one-class land-cover classification scheme applied consistently over the United States. The spatial resolution of the data is thirty meters. The classification system used is modified from the Anderson land-use and land-cover classification system. No attempt was made to derive classes that were extremely difficult or "impractical" to obtain using Landsat TM data, such as the level III urban classes. In addition, some Anderson level II classes were consolidated into a single NLCD class.

The **United Nations Food and Agriculture Organization** and UNEP established the **Global Land Cover Network** (GLCN) cooperative program as an integral part of the **Global Terrestrial Observation Strategy** (GTOS). The objective of GLCN is to provide direction, focus, and guidance for harmonization of land-cover mapping and monitoring at national, regional, and global levels. The involvement of reference institutions ensures adequate monitoring standards and effectiveness of monitoring modalities. Close collaboration and linkage with ongoing regional and global geospatial information initiatives as well as environmental conventions will be pursued: for example, UNFCCC, the **Millennium Ecosystem Assessment**, and research projects that focus on measurement of regional **land-cover change** (e.g., the **Land-Use/Cover Change [LUCC] project** and the **Large Scale Biosphere-Atmosphere Experiment in Amazonia**).

Terrestrial Ecosystem Monitoring Sites (TEMS) is an international directory of sites (named T sites) and networks that carry out long-term terrestrial monitoring and research activities. The database provides information on the who, what, and where in long-term terrestrial monitoring that can be useful to both the scientific community and policymakers.

The LULUCF report of the **Intergovernmental Panel on Climate Change** addressed the implications of differing definitions of land use, land-use change, and forestry activities: definitions of these terms as used in the Climate Convention and the Kyoto Protocol, definitions of accounting and reporting, methodological definitions, and sustainable development considerations. The technical requirements to measure, monitor, and verify

Kyoto Protocol requirements will be significantly affected by these definitions. The good practice monitoring procedures as described earlier are mirrored in this report as a list including accounting objectives and inventory techniques, that is, transparency, consistency, comparability, completeness and precision, uncertainty and accuracy, verifiability, costs, and efficiency.

In September 2004, a new theme was prepared for the Integrated Global Observing Strategy (IGOS), that is, Integrated Global Observations of Land (IGOL), with the focus on human influence on the environment and vice versa. Technical deficiencies in monitoring land cover are addressed, such as spatial and temporal resolution requirements for earth observation data. Implementation strategies are suggested, for example, to obtain annual estimates of **deforestation** and land-use change. The importance of the socioeconomic component is incorporated, considering management practices, **food security**, human settlements, and infrastructure. The IGOL strategy builds on experience and requirements from the space agencies, GTOS, UN organizations, the **International Geosphere-Biosphere Programme** and the **International Human Dimension Programme on Global Environmental Change**, GEOSS, and **Global Observation of Forest and Land Cover Dynamics** (GOFC-GOLD) as the first international program since the 1990s with a comprehensive global land-cover change observation strategy. GOFC-GOLD is also responsible for coordinating harmonization and validation of global land cover and land-cover change products. GOFC-GOLD's harmonization tool is FAO's **Land Cover Classification System**, which can be utilized as a legend translation system. Global **validation** relies strongly on GLCN.

Global Monitoring Products

In 1978, global earth observation data—that is, the **advanced very high resolution radiometer** (AVHRR) with low geometric resolution and high temporal repetition frequency—became available. Global cloud-free image mosaics and higher-level products such as land-cover maps were generated. New sensors such as the **moderate resolution imaging spectrometer** (MODIS) allow the retrieval of global biophysical products, for example, net primary production, based on operational data supply and complex processing schemes. The following global advanced map products exist from earth observation: **IGBP-DIS global 1-km land cover data set** and University of Maryland AVHRR products (1.1 kilometer resolution, years 1992–1993); MODIS products, for example, of land cover (1 kilometer since 2001, future 500-meter resolution) and **vegetation** continuous fields (500 meters, since 2001); the **Global Land Cover Map of the Year 2000** (GLC 2000) (1 kilometer, **SPOT Vegetation**, year 2000); and—in preparation—**GLOBCOVER** 2005 (300 meters, **medium resolution imaging spectrometer** [MERIS], year 2005).

To improve global monitoring research with respect to temporal extent as well as information content, a combined usage of the above-listed products would be of advantage. Cross-comparisons between products however illustrate difficulties. Chandra Giri and colleagues, for example, compared GLC 2000 and the corresponding MODIS land-cover products. The authors found general agreement at the class aggregate level except for savannas/shrublands and **wetlands**, increasing disagreement when comparing detailed land-cover classes, and highly variable percent agreement between the two data sets among **biomes**. The necessity of long time series of earth surface information and the cost involved in producing global map products require an improved correspondence between products despite differing sensor systems and classification legends. GOFC-GOLD, IGOL, and GEOSS are international activities created to address (and eventually overcome) these problems.

See also: **Aerial Photography; AFRICOVER; Change Detection; Ecosystem Services; Hot Spot Identification; Land Degradation Assessment in Drylands (LADA) Project; Land Quality; Pixelizing the Social; Socializing the Pixel; TERRA-ASTER.**

Further Reading

Berounsky, Veronica, Environmental Monitoring in Rhode Island: A Synthesis, Developing Ecological Indicators: Vital Signs for Narragansett Bay and Coastal Ponds, Final Report [Online, December 2004], The University of Rhode Island Web Site, www.ci.uri.edu; DeFries, Ruth, Asner, Gregory, and Richard Houghton, eds., *Ecosystems and Land Use Change*, Washington, DC: American Geophysical Union, 2004; Giri, Chandra, Zhu, Zhiliang, and Bradley Reed, "A Comparative Analysis of the Global Land Cover 2000 and MODIS Land Cover Data Sets," *Remote Sensing of Environment* 94 (2005): 123–32; Intergovernmental Panel on Climate Change, IPCC Special Report on Land Use, Land-Use Change and Forestry [Online, December 2004], The IPCC Web Site, www.grida.no/climate/ipcc/land_use/007.htm; United Nations Framework Convention on Climate Change, International Panel on Climate Change, Good Practice Guidance for Land Use, Land-Use Change and Forestry [Online, December 2004], The IPCC Web Site, www.ipcc-nggip.iges.or.jp/public/gpglulucf/gpglulucf.htm; United States Environmental Protection Agency, EMAP Research Strategy: Environmental Monitoring and Assessment Program (EPA 620/R-02/002), Washington, DC: Environmental Protection Agency, 2002.

CHRISTIANE SCHMULLIUS

Monoculture. A land-use practice that involves the cropping of a single plant species in an area (including trees on plantations) year after year. Monocropping removes large amounts of specific nutrients from the soil and, in order not to decrease soil fertility, large amounts of artificial inputs such as mineral fertilizer need to be integrated into any monocultural agricultural production system, often with cascading negative effects upon the provision of **ecosystem services**—on freshwater and remote marine habitats even. Monocropping also increases the **vulnerability** of plants to disease, predation, and parasitism by providing abundant hosts or food for pest organisms, thus triggering increases in pest populations and reinforcing the input of biocides such as chemical pesticides with detrimental impacts upon native species. In general, monocultures lead to a decrease in the **biodiversity** of a given area, often involving simplifications of ecosystems and landscapes. This is also true for "tree farms" or monocultural forest plantations such as pines in the **Andes**, spruce in **west-central Europe** (notably Germany), and eucalyptus in **southern Africa**. In most of the cases, monocultures in the developing world date back to the impact of colonial rule on local land management. In Africa, for example, commercialization of agricultural commodities and the nature of **cash crops** themselves have been of major importance. Where

Monocultural cropping of Burley tobacco on a farm near Lexington in Fayette County, Kentucky, September 1940 (curing barn in the background).
PHOTO: Marion Post Wolcott. [Courtesy Library of Congress]

grown in single stands, and not intercropped or relay-cropped as in traditional gardens and fields, in particular **maize**, **cotton**, groundnuts, and **tobacco** offer poor ground cover, especially in the growing season where erosivity is highest. Monocultural specialization in the form of large-scale and high-input **land-use systems**, which have existed only for the last 50–100 years, is the current answer to the **consumption** needs of the world's population. It appears, though, that these production systems perpetuate and intensify earlier agricultural impacts such as **soil degradation** or pollution from fertilizer and pesticides, **soil erosion**, and habitat **conversion**. In addition, there are several social issues related to the consolidation of smaller farms into huge, often agro-industrial operations, with many interconnected environmental changes. Continuous, intensive monocropping may be productive and profitable in the short term—with monocultural specialization still being the dominant paradigm in today's **forestry** and **agriculture**—but the many negative impacts raise serious questions about the sustainability of "artificial forests" and high-input, intensive agriculture. Jason Clay poses the hypothesis that the externalities of monocultures are in no small part responsible for shifting **agricultural frontiers** around the world. Increasingly, methods of integrated nutrient and pest management become important in the discussion of more **sustainable land use** practices, and organic farming and polycultures are just components of these methods.

See also: **Agricultural Intensification; Agro-Industry; Biotechnology; Cascades; Coastal Zone; Colonization; Exotic Species; Land Concentration; Vulnerability.**

Further Reading

Blaikie, Piers M., and Harold Brookfield, *Land Degradation and Society*, London, New York: Routledge, 1987; Clay, Jason, *World Agriculture and the Environment: A Commodity-by-Commodity Guide to Impacts and Practices*, Washington, DC: Island Press, 2004.

HELMUT GEIST

Mountains. Upland environment covering a third of the world's land surface due to relatively high elevation above sea level or because of the presence of hilly or mountainous terrain. The land covers of these areas are affected therefore by the unique physical environmental conditions found at high elevations or with steep slopes. Temperatures, precipitation, and air pressure can be low enough to limit the growth and presence of living organisms. Soils are often poorly developed and shallow. There are numerous different ecological zones arrayed up the slopes of mountains, in addition to dramatic local and regional shifts with different exposures to sunlight and prevailing winds. These form environmental continua and can offer opportunities for complex systems of land use and natural resource extraction. For example, people have long exploited the numerous different ecological zones of tropical mountains through the management and **agriculture** of the landscape mosaics. This usage was also the case in many middle latitude mountainous regions in the past, although nowadays these lands are more valued for their scenic beauty and water resources. Future land-use changes are likely to be most affected by climate change, and the need to provide **access** within mountainous areas for transportation and the extraction of natural resources. The monitoring of these changes by remote sensing and the prediction of potential landscape conditions are necessary activities for the sustainable use of the highlands of the world.

Ecological Zonation, Complex Mosaics

High elevations have lower partial pressures of atmospheric gases than those lands found near sea level. Thus, gases such as carbon dioxide, needed for photosynthesis of plants, can be present in limiting amounts. The low air pressure also changes the temperature balances of the lower atmosphere and land surfaces because less heat is absorbed and maintained by gases. Nights are cool or cold, while temperatures can increase dramatically during the day due to incoming solar radiation, but cool off quickly in the shade or when evening comes. Many native plants have adaptations, such as low growth forms and hairy leaves, to cope with these daily temperature fluxes. Another consequence of energetic and thermal limitations is that mountain ecosystems have less **resilience**, or capacity to recover following a disturbance, than those found at lower elevations. The environmental conditions restrict the amount and rate of **regrowth**.

The vegetation types found in mountains are typically of a relatively small stature. There are dwarf forest types found near many altitudinal treelines, in addition to on top of mountain ridges with shallow soils. Up above the

Jura Mountains of France (left) and snowcapped Bernese Alps of Western Switzerland (right), October 10, 1994 (with Lake Geneva at the bottom of the image).

PHOTO: Space Shuttle. CREDITS: Image Science and Analysis Laboratory, NASA-Johnson Space Center, 18 Mar. 2005 (Earth from Space—Image Information) http://earth.jsc.nasa.gov/sseop.efs

treeline, a variety of herbaceous plants dominate landscapes, including meadows formed by grasses or sedges, plus forbs or shrubs with prostrate or cushion forms that hug the ground. At middle elevations, temperate mountains often are covered by evergreen conifer forests in places like western North America and New Zealand, while the European Alps, the **Himalayas**, and eastern North America have mixtures of coniferous and deciduous hardwood forests. Tropical mountains typically have diverse forests of evergreen hardwoods, although there are also conifer-dominated forests in Central American highlands and in some sites in **South America**, Africa, and the Pacific Ocean islands.

Local winds flow upslope during the day due to solar heating and thermal lifting. At night, the air parcels cool on the peaks and ridges, forcing cold downslope winds to form. These mountain and valley breezes are augmented in mountainous areas with glaciers and ice caps, which further cool the air and can form katabatic winds that flow downslope with gravity. In fact, even terrains that have no current permanent ice often show evidence of earlier cold conditions. Past glaciation, for example, left behind a landscape legacy of U-shaped valleys, ice-carved peaks and ridges, and valleys filled with moraines, erratic boulders, **wetlands**, and glacial lakes in mountains around the world.

"Aspect" refers to the compass direction of a slope, which is an important control on physical environmental conditions in rough terrain. For example, north-facing slopes in the Northern Hemisphere receive less direct solar radiation than do south-facing slopes (the opposite is the case in the Southern Hemisphere). The north-facing slopes will be cooler, photosynthetic rates will be lower, soil moisture will be higher, and fires will be less common. Often the **vegetation** and land cover vary with slope aspect, as a result. Another important feature is topographic shadow, a function of the angle of sun, which varies seasonally and with latitude, plus the height and slope angle of surrounding terrain. This aspect-controlled relationship can make for important differences in land cover from one side of a valley to another. Finally, prevailing winds that come from a particular direction can also leave an imprint on the landscape. Windward slopes face into the prevailing winds. They also tend to have abundant fog and sometimes heavy precipitation, while leeward slopes are sheltered, but may be dry and warm due to the heating and drying of descending air parcels, forming rain shadows.

The result is that mountains are complex mosaics, with ecological zonation varying with elevation, but with additional patchiness originating with aspect differences, exposure to sun and winds, and the persistence of legacies of past environmental conditions. Many mountains are geologically complex too, the result of past volcanism, compression and tension faults, plate subduction, and uplift. As soils develop on top of bedrock or deposited rock material, they likely will vary in their textures, chemistries, and degrees of edaphic development, depending on the original bedrock. Place-to-place differences in mountainous terrains are remarkable as a result, with different atmospheric, soil-related, geomorphic, hydrologic, and ecologic processes at work over short vertical and horizontal distances. In addition, natural disturbances, such as fires or landslides, alter those conditions by damaging or removing vegetation and altering topography. On a microscale, cryoturbation can alter soil and vegetation properties as ice repeatedly forms and melts in the top several centimeters of topsoil. The presence of permafrost at high elevations alters soils and can create bogs and other wetlands.

Biological diversity and the composition of plant and animal communities change with elevation and with the other sources of environmental heterogeneity. Because of the harshness of high-elevation environments, only species with unique adaptations will be found. Furthermore, many mountains or mountain chains are not connected to other high-elevation environments, creating island-like conditions in terms of their remoteness and inaccessibility. This isolation leads to genetic separation and, over long time periods, to the evolution of unique (endemic) species.

Land-Use Systems, Environmental Effects

Paradoxically, the highlands of tropical latitudes are among the most important landscapes in the world for the development of sustainable agricultural practices and the survival of ethnic groups. For example, it was in the mountains of the tropical Americas that many crops were domesticated one to six millennia ago. There are thousands of land races of potatoes in the tropical **Andes** of South America, with people mixing the genes of domesticated and wild species, and crafting variants adapted to local conditions and satisfying local tastes and needs. In places where water is limited, systems of irrigation farming were developed and communal labor arrangements were institutionalized to share that water and maintain flows. Individual fields are often small and scattered, a feature that takes advantage of the multiplicity of microclimates available and lessens the risk to crop loss due to natural disturbances and calamities caused by insect pests, crop diseases, and drought.

Pastoral systems in mountains are used to graze livestock in the high-elevation **grasslands**, on fallowed agricultural fields, and in shrublands opened by burning. In some cases, seasonality of precipitation requires the movement of livestock among different landscape locations as **pastures** change in quality and productivity. Because these kinds of land cover are found in almost any mountainous areas, the **land-use systems** present are typically multiple, with some people specializing as agriculturalists, shepherds, or traders, and others acting as generalists with lands or access to lands at all available elevations. Households organize labor and efforts to acquire needed foods and products, often through community-based arrangements of labor, land, and social ties. Land use can be ancient in some places, and so land cover as a result was anciently modified, for example, in some Mexican, Andean, African, and Asian landscapes. Other sites with extensive **forests** are being deforested currently.

Socioeconomic change is driving large shifts in tropical montane land use. **Urbanization** continues to draw many people away from rural areas, as health care, jobs, and education are more readily available in cities. Rural communities need to adapt to the loss of young people to the growing cities. Roads, put in to facilitate transportation, can also provoke unsustainable extraction of timber and wildlife from newly accessible areas, such as once-isolated cloud forests. When the cities are located in lowlands, then extraction of water and other natural resources from highland areas can be especially difficult to moderate, as the users of the resources are located far from the **environmental changes** caused by extraction. Some areas in the Andes, eastern Africa, and Asia have become tourist destinations.

The snow-capped Himalaya Mountains and the impressive high-elevation Tibetan Plateau were forced up by planetary plate tectonics, which have moved India into the Asian landmass at the rate of several centimeters each year. Over 45 million years, the resulting compressive forces have raised the tallest mountains in the world. These highlands are massive enough to affect the monsoons and global biogeochemical cycles. Their forests and mountain meadows are home to numerous endemic plants and animals. The climates that affect these habitats range from continental temperate climates in the northwest to tropical environments in the eastern Himalayas.

Land-use systems depend on rainfall amounts and on elevation, with farmers of **wheat** and barley at lower altitudes and herders of yaks and other livestock in higher environs. Although extensively and anciently utilized by people, the most abrupt recent **land-cover changes** are associated with the arrival of outsiders, especially tourists and the resulting demands placed on natural resources and protected areas. Future land-use changes will likely be driven by climate change and socioeconomic fluxes as ethnic, linguistic, and political groups search for ways to maintain some traditional values, while also allowing for needed changes.

Several of the land-use/land-cover changes noted for tropical mountain and Himalayan landscapes occurred in earlier years in temperate, middle-latitude mountains. For example, seasonal **transhumance** was at one time a defining characteristic of grazing systems that connected highlands to lowlands in **Europe** and in the **United States of America**. Ironically, the urbanization of these countries in the twentieth century meant that many highland areas were abandoned by their rural residents, permitting afforestation and a decrease in **soil erosion**. Abandoned agricultural fields regrew into forests, leaving only livestock grazing and timber management as important land-use systems over extensive areas. The environmental effects of cattle and timber harvesting practices are particularly contentious issues. Today, many high-elevation areas are most useful for their supply of water resources, used for drinking water and hydroelectric generation. The scenic areas are in great demand for short-term visits for recreation. Some anciently occupied areas in the European Alps are maintained as agricultural landscapes for their cultural, historical, and tourist values. Other temperate highland areas are being settled by former city dwellers interested in owning small tracts of land, for example, in the Rocky Mountains of western North America. Places downwind from industry and cities have been affected by acidic deposition and air pollution. Examples can be found in the mountains of California and the northeastern United States and in some mountainous areas in continental Europe.

Temperate mountains differ ecologically from tropical mountains because of their seasonal shifts in temperatures. Cold winters bring precipitation in the form of snow, and the types of vegetation present are often well correlated to the length of time the snow persists. In other words, the numbers of days the high-elevation sites are free from snow correspond approximately to the lengths of the growing season. Snowpack is also important hydrologically, as that is the source of water during the warm part of the year for mountain streams and rivers.

In many middle latitude mountains that are forested, humans have altered **fire** regimes, typically through suppression and fire prevention. These activities are necessary for safety and to protect buildings, but they run the risk of building up more fuel over time, leading inevitably to larger fires in the future. Prescribed burning is now attempted in places in the western United States and Europe, as are "let burn" policies, as long as no people or infrastructure are put in harm's way as a result. There is also a past legacy and current impact of mining in many places, often accompanied by degradation of water quality.

It is in the high mountains of the world where some of the most dramatic changes in land cover and shifting treelines are predicted to occur due to future global climate shifts. This concern exists because the plants and animals found there are already near their climatic limits for survival. Increased temperatures or atmospheric carbon dioxide would be predicted to increase photosynthetic rates. However, this increased productivity is often at the expense of dominance by the original, slow-growing plant species, which are replaced by lower-elevation species and by invasives. Nature reserves and protected areas need to be redesigned to prepare for altitudinal shifts in species and ecosystems. Precipitation is predicted to increase in some areas and decrease in others, so species responses will vary with their autoecology and with the specifics of changes in particular mountains.

In some places, snowpacks and glaciers provide environmental stability to people and their livelihoods, because they can count on meltwaters to supply irrigation needs and drinking water during warmer and drier times. Most of the world's mountain glaciers have negative mass balances and are retreating upslope. The loss of snow cover and permanent ice may require large shifts in human land use and water-resource utilization. For example, a decrease of hydrological supplies and increased seasonality might necessitate the design of larger reservoirs or more thrifty irrigation systems, or even force the development of alternative land-use systems.

Monitoring

Remote sensing of changes in land cover, for example the locations of treelines and snowlines, can permit **monitoring** of environmental change over large areas. This perspective is obviously of great practical benefit for use in the mountains of the world, which are by their nature hard to observe except from above. One important limitation on land-use/land-cover classifications is the influence of topographic shadow, which in some cases obscures up to a third of remotely sensed images. The recent development of digital elevation models permits better representation of complex mountain topography. Their use, combined with pre-classification image preparation, can reduce the information lost to shadow effects. Radar imagery can be used to isolate topographic features and even measure uplift. Some of the most sophisticated landscape computer modeling is now being done in order to predict shifts in ecological processes due to changes in physical environmental controls. Other computerized models are being used to predict when and where fires will occur and the likely landscape consequences of prescribed (or planned) burning. Natural hazards mapping and **modeling** are crucial for evaluating risks due to volcanic eruptions, earthquakes, floods, and debris flows. **Integrated assessment** models allow for the connecting of global and regional climate-change predictions to the likely landscape consequences. These and other geographical information science approaches will help predict future land-cover changes in mountains.

See also: **Atmosphere-Land Interlinkages; Biodiversity; Coffee; Community Involvement; Fallow; Forest Transition; Geographical Information System; Indigenous Knowledge; Land Abandonment; Mineral Extraction; Modification; Pastoral Mobility; Reforestation; Sustainable Land Use; Watershed.**

Further Reading

Barry, Roger G., *Mountain Weather and Climate,* 2nd ed., London: Routledge, 1991; Beniston, Martin, *Environmental Change in Mountains and Uplands,* London: Arnold,

2000; Booth, Douglas E., *Searching for Paradise: Economic Development and Environmental Change in the Mountain West*, Lanham, MD: Rowman and Littlefield, 2002; Funnell, Don, and Romola Parish, *Mountain Environments and Communities*, London: Routledge, 2001; Gerrard, A. J., *Mountain Environments: An Examination of the Physical Geography of Mountains*, Cambridge, MA: MIT Press, 1990; Godde, Pamela M., Price, Martin F., and Friedrich M. Zimmermann, eds., *Tourism and Development in Mountain Regions*, Wallingford, UK: CABI Publishing, 2000; Ives, Jack D., *Himalayan Perceptions: Environmental Change and the Well-Being of Mountain Peoples*, London: Routledge, 2004; Körner, Christian, *Alpine Plant Life: Functional Plant Ecology of High Mountain Ecosystems*, Berlin: Springer, 1999; Mountain Research Initiative [Online, February 2005], The MRI Web Site, mri.scnatweb.ch; Price, Martin F., ed., *Global Change in the Mountains*, New York: Parthenon, 1999; Shoumatoff, Nicholas, and Nina Schoumatoff, *The Alps: Europe's Mountain Heart*, Ann Arbor, MI: University of Michigan Press, 2001; The Mountain Institute [Online, August 2004], The TMI Web Site, www.mountain.org; Zheng, Du, Qingsong Zhang, and Shaohong Wu, *Mountain Geoecology and Sustainable Development of the Tibetan Plateau*, Dordrecht, NL: Kluwer Academic Publishers, 2000.

KENNETH R. YOUNG

Multiplier Effect. A basic economic concept that describes the indirect effects of an **investment** (or expenditure) in an economy as a function of the ways in which that initial investment circulates within the economy: the more it circulates, the greater the value to the economy, and the stronger the multiplier. Employment and income multipliers can be calculated and utilized to measure and predict the economic success of development strategies. Poor internal linkages and weak multipliers are often reasons for the limited success of export-oriented resource-based models of development, such as oil, gas, and metals extraction: in these settings large capital expenditures (on plant and equipment) typically produce weak developmental effects. Beyond its use as planning and evaluation tool, the value of the term lies in its emphasis on the internal structure of an economy, and the role this structure plays in magnifying the effects of an exogenous input. The term can be adapted to consider how the environmental effects of an external capital injection (foreign direct investment, remittance incomes, or central government spending, for example) are not limited to the injection itself, but are magnified (or mediated) by the ways in which it then circulates through the structure of the economy.

See also: **Exogenous Factor; Feedback; Mediating Factor; Mineral Extraction; Public Policy; Remittances Landscape.**

GAVIN BRIDGE

N

NASA Land-Cover/Land-Use Change (LCLUC) Program. Land-use/cover change research has become an important element of global change research programs on national as well as international levels, and is currently a separate element of the United States Climate Change Science Program, the scientific direction and strategy of which are coordinated by a U.S. interagency effort. Internationally, land-use/cover change research was on the agenda of the **Land-Use/Cover Change (LUCC) project** for one decade from 1995 onward, and is on the agenda of the newly developed **Global Land Project** (GLP) from 2005 onward. Following the LUCC and preceding the GLP efforts, the U.S. National Aeronautics and Space Administration (NASA) has supported several dozen projects under its Land-Cover/Land-Use Change (LCLUC) program to feed to the national and international research on land-change science. The NASA LCLUC program, which started as a combination of regional satellite-based studies, representative field-based process studies, and **modeling**, has grown into a large and multifaceted, interdisciplinary science program; and is linked to other NASA programs, such as Terrestrial Ecology, Terrestrial Hydrology, Water Cycle, Carbon Cycle, Ocean Biology, and Applications.

Land-change science is truly interdisciplinary. It requires an alliance of physical scientists (e.g., physical geographers, climatologists, ecologists, and hydrologists), remote-sensing scientists, and practitioners and social scientists (e.g., economists, demographers, human geographers, and anthropologists). Because land-change science breaks new ground in interdisciplinary research and the emergence of a new research paradigm ("integrated land-change science"), the NASA LCLUC program has made important contributions to methods on land-cover **change detection** and classification, on linking people and **pixels**, on the synergistic use of optical and radar satellite data, and in modeling the dynamic processes of land-use and **land-cover change**. The program aims to develop and use NASA **remote sensing** technologies to improve our understanding of human interactions with the environment, and thus provide a scientific foundation for understanding the sustainability, **vulnerability**, and **resilience** of human land use and terrestrial ecosystems. In doing so, a major goal of the LCLUC program is to further the understanding of the consequences of LCLUC on ecosystem goods and **ecosystem services**, the **carbon cycle**, the water cycles, and the management of natural resources.

The longer-term objectives of the LCLUC program are to develop the capability to perform repeated global inventories of land use and land cover from space to improve the scientific understanding of land-use/cover change processes and models necessary to simulate the processes taking place from local to global **scales**; to model and forecast land changes and their direct and indirect impacts, and evaluate the societal consequences of the observed and predicted changes; and to contribute to the establishment of the operational provision of land-use/cover-change data and information products, services, models, and tools for multiple users, for example, scientists, resource managers, and policymakers.

See also: **Change Detection; Pixelizing the Social; Socializing the Pixel.**

Further Reading

Gutman, Garik, Justice, Christopher, Sheffner, Ed, and Tom Loveland, "The NASA Land Cover and Land Use Change Program," in *Land Change Science: Observing, Monitoring and Understanding Trajectories of Change on the Earth's Surface*, eds. Garik Gutman, Anthony C. Janetos, Christopher O. Justice, Emilio F. Moran, John F. Mustard, Ronald R. Rindfuss, David Skole, B. L. Turner II, and Mark A. Cochrane, Dordrecht, NL: Kluwer Academic Publishers, 2004, pp. 17–29; U.S. National Aeronautics and Space Administration, Land-Cover/Land-Use Change Program [Online, December 2004], The NASA LCLUC Web Site, lcluc.gsfc.nasa.gov.

GARIK GUTMAN

National Institute of Public Health and the Environment. Conducts research into public health and environmental issues in the Netherlands, also operating as the Netherlands Environmental Assessment Agency (MNP in Dutch). Besides assessing implemented policy, the National Institute of Public Health and the Environment (RIVM) provides knowledge on the effectiveness and efficiency of prospective policy measures. It highlights potential bottlenecks and responds to current developments on environmental and nature issues. Both the present status of and future prospects for environment and nature policy are examined. Effects of different policy **scenarios** used in the assessments are supported by models developed by RIVM.

Effective exchange of information is needed for national and international policy development. RIVM provides information to the European Union and organizations such as the World Health Organization, the **United Nations Environment Programme**, the United Nations Economic Commission for Europe, the **Intergovernmental Panel on Climate Change**, and the World Bank. Many of these organizations have designated RIVM as a national focal point, which means that RIVM gathers and integrates information from the Netherlands, **Europe**, and/or the world and communicates this to the European Environmental Agency (EEA) and other international institutions. RIVM also hosts integrative global climate change models such as the **Integrated Model to Assess the Global Environment** (IMPEL), an international emission database (EDGAR), and the **History Database of the Global Environment**, a historical population and land-use/cover database of the past 300 years.

Today's MNP functions as the interface between science, policy, and politics, producing independent assessments on the quality of the environment for people, plants, and animals to advise national and international policymakers. MNP is one of four independent assessment agencies in the Netherlands. These agencies all have a role to play in giving substance to the World

Bank's People-Planet-Profit concept, with the Social and Cultural Planning Office of the Netherlands (SCP) dealing with "People," the Netherlands Bureau for Economic Policy Analysis (CPB), with "Profit," and MNP, along with the Netherlands Institute for Spatial Research (RPB), with the "Planet."

See also: **BIOME 300 Project.**

Further Reading

National Institute of Public Health and the Environment [Online, November 2004], The RIVM Web Site, http://www.rivm.nl; The Netherlands Environmental Assessment Agency [Online, September 2005], The MNP Web Site, http://www.mnp.nl.

KEES KLEIN GOLDEWIJK

Non-Timber Forest Products. Forest products not derived from logging but gained from collecting, such as fruits, nuts, medicines, woodcarvings, resins, and fibers. Based on data from the Peruvian Amazon, Charles Peters and colleagues claimed that especially impoverished forest dwellers could earn more money from tropical forests by collecting non-timber forest products (NTFPs) than from logging. This captured the imagination of those concerned with **conservation**, who reasoned that, if people could sell more NTFPs, they would be less inclined to clear the **forest**.

Using data from dozens of case studies in Africa, Asia, and **Latin America**, Manuel Ruiz-Pérez and colleagues found that most cases fell into the following three groups. First, especially in Asia, farmers manage NTFPs like crops, that is, they grow them in plantations or manage them intensively in forests (which are not the least disturbed, though). The families—usually not the poorest ones—specialize in the product and, indeed, get most of their income from it. They generally have secure **land tenure** and **access** to markets, and make a pretty good living without depleting their resources. Involved in the second group are mainly African farmers who tend to be poorer and collect their products from natural forests that are not intensively managed. They rely heavily on a number of different forest products just to barely survive, and often overharvest them. Forest products provide an important safety net for these people, but the future doesn't look bright. The third group of forest products represents a smaller share of farmers' income, but they allow this income to be more diversified. These cases fall between the first and the second group in terms of the farmers' incomes and revenue and the way they manage their resources. In summary, the **meta-analysis** suggests that there are not too many cases where selling products from largely unmanaged natural forests has helped save those forests or lift people out of **poverty**. Cultivating forest products can be a good business for better-off small farmers, while collecting products from natural forests clearly helps many people to survive.

See also: **Amazonia; Arc of Deforestation; Common-Pool Resources; Desertification; Economic Livelihood; Food Security; People at Risk.**

Further Reading

Peters, Charles M., Gentry, Alwyn H., and Robert O. Mendelsohn, "Valuation of an Amazonian Rainforest," *Nature* 339 (1989): 655–6; Ruiz-Pérez, Manuel, Belcher,

Brian, Achdiawan, Ramadhani, Alexiades, Miguel, Aubertin, Catherine, Caballero, Javier, Campbell, Bruce, et al., "Markets Drive the Specialization Strategies of Forest Peoples," *Ecology and Society* 9 (2004): [online] www.ecologyandsociety.org.

HELMUT GEIST

Normalized Difference Vegetation Index. A simple formula used in **remote sensing** studies to yield an estimate of the likelihood that **vegetation** was actively growing during a particular observation. The Normalized Difference Vegetation Index (NDVI) is computed by taking the ratio between reflected radiance values of the difference and sum from the red and near-infrared components of the electromagnetic spectrum and calculated using the formula NDVI = (near IR band – red band) / (near IR band + red band). The advantages of the NDVI are that it is easy to compute, a sensor needs to acquire data at red and near-infrared wavelengths, and it can be derived at a range of temporal and spatial resolutions from satellites such as **Landsat** Thematic Mapper, **IKONOS**, **advanced very high resolution radiometer**, **SPOT Vegetation**, and **medium resolution imaging spectrometer**. NDVI values can be correlated to parameters that can be measured in the field (e.g., leaf area index and **biomass**). They can also be compared over a time series of observations, as they are independent of illumination and viewing geometries, whereas atmospheric corrections of data are still required. The disadvantages of using NDVI values are that it is not a measurable geophysical product, and it has no clear meaning. Soil type can have a large influence on NDVI values. The NDVI has been extensively used in studies of **land-cover change**.

See also: **Change Detection.**

Further Reading

DeFries, Ruth S., and John R. G. Townshend, "NDVI-Derived Land Cover Classifications at a Global Scale," *International Journal of Remote Sensing* 15 (1994): 3567–86; Li, Zuotao, and Menas Kafatos, "Interannual Variability of Vegetation in the United States and Its Relation to El Niño/Southern Oscillation," *Remote Sensing of Environment* 71 (2000): 239–47.

KEVIN J. TANSEY

Nutrient Cycle. Movement, transformation, and storage of nutrients within and between the spheres of the earth system (biosphere, hydrosphere, lithosphere, and atmosphere). A nutrient is any element or compound that is necessary for or contributes to an organism's **metabolism**, growth, or other functioning. Depending on the relative amount of a nutrient species needed to maintain these functions, macro- and micronutrients are distinguished. The term "nutrient cycling" is in many aspects equivalent to the notion of biogeochemical cycling. However, biogeochemical cycles include the **carbon cycle** and are generally connoted with larger **scales** in space and time, as opposed to the nutrient cycle. The notion of nutrients commonly excludes carbon and water. However, the close link between climate/weather and nutrient cycles is obvious, because temperature and water availability are the main factors driving plant growth and decomposition of organic material, thus determining both accumulation and release of carbon and nutrients. The most important macronutrient elements are nitrogen, potassium, calcium, phosphorus, sulfur, and magnesium. The characteristics of a nutrient cycle are determined by the specific transformation processes between different nutrient compounds and by the transport processes of compounds within and between the spheres of the earth system.

Nutrient cycling in all terrestrial ecosystems, no matter whether they are natural (**forests**, rangelands, **grasslands**) or anthropogenic (agro-ecosystems, **pastures**, plantations), follows similar pathways; but nutrient pools, fluxes, and transformations may vary several orders of magnitude between systems. Both natural and managed ecosystems are open systems, which means that energy or matter can enter or leave them. The question of whether the inputs and outputs are balanced is closely related to the issue of **sustainable land use**. Natural systems tend to be almost balanced, that is, carbon and nutrient gains or losses are low and the ecosystems as such may persist over a long period of time (from a human perspective). Contrastingly, many managed systems tend to be highly unbalanced, either accumulating certain nutrients, a process that frequently occurs in irrigation farming, or in over-fertilized production systems; or losing nutrients by overexploitation of nutrient stocks (i.e., nutrient mining) or by inadequate management, which may promote losses via a number of pathways including leaching, erosion, volatilization, and others.

While the majority of principles underlying the processes and flows of the nutrient cycle are well understood, the impacts on land-use/cover change often are difficult to quantify. This fuzziness on the one hand originates from the variability as inherent property of ecosystems, and on the other hand is related to our imperfect knowledge of the specific quantities of nutrients cycling through the plethora of land-use and land-cover types. For many of the world's natural and managed ecosystems, no or only vague information about the respective nutrient cycles exists, or is not accessible. However, as all ecosystems exchange matter and energy across their boundaries, they all contribute to and influence global biogeochemical cycles. Gaseous or liquid and even particulate efflux of nutrients may be transported over large distances, and enter or pass through other ecosystems. In natural systems, large imbalances of nutrient fluxes often occur after disturbances of extreme climate conditions like drought, which may promote wind erosion; weather events like rainstorms, eventually causing soil losses or wind throws; or **fire**, which may convert large fractions of biomass-bound nutrients into ashes and leave the soil unprotected. Among managed ecosystems, especially intensively fertilized agro-ecosystems and high-density livestock systems tend to leak large amounts of nutrients into the atmosphere and/or surface and groundwater. Reactive compounds of nitrogen clearly dominate the nutrient fluxes within and between ecosystems. Accumulated global emissions of food production systems sum up to forty-three teragrams of ammonia and thirty-six teragrams of nitrous oxide (NO_x), including emissions from energy production. In the Northern Hemisphere, where the major emitters of nutrients reside, all downwind and downstream systems both marine and terrestrial receive considerable inputs of nitrogen. Thus, via far-distance transport and other mechanisms, land-use activities do affect the nutrient cycles of natural and managed ecosystems.

The high relevance of nutrient cycles for **modeling** land use and **land-cover change** is obvious, all the more as the spatiotemporal dynamics of nutrient-availability (among other factors) determine land use and vice versa (i.e., every land-use type determines a specific nutrient cycle). While the general importance of land-change research is widely accepted, the integration or application of models representing nutrient cycles is still rather exceptional, especially at large spatial scales (continental to global). In the realm of nutrient cycling, a large number of different concepts and models have been developed, ranging from simple nutrient balance approaches to numerous classes of process models of varying complexity. Many of these models are specialized to simulate either certain components of the earth system like soils or plants (crops or natural **vegetation**) or coupled systems of both. Contrasting to many recent models—such as the **LandSHIFT model**, **conversion of land use and its effects (CLUE) model**, and **integrated model to assess the global environment**(IMAGE)—most nutrient cycle models are not spatially explicit but point-based, which means they simulate, for example, a single plant or multiple plants, or a homogenous field or stand.

Nutrient Balance Models

Models based on nutrient balances are not an entirely new concept. Already in 1841 the chemist Justus von Liebig realized that the nutrients taken up from the soil by agricultural crops should be replenished. A number of approaches evolved over the years, in which most nutrient balance studies followed a fairly simple routine, starting by identification of the system boundaries and the key inputs and outputs to or from a variable number of sub-components. In the most simplistic approach, the system is considered a black box, in which nutrient depletion or enrichment of the system is simply a result of the difference between total nutrient input and total nutrient output. The more detailed compartment approach considers nutrient pools or compartments, connected by nutrient pathways and transfer rates. Estimation of flows of one or more nutrients is determined either through direct measurements or literature estimates or is based on transfer functions (e.g., for volatilization, deposition, leaching, erosion losses, and others). As nitrogen is the most limiting nutrient element with respect to plant growth and crop production, followed by phosphorus and potassium, most nutrient balance studies concentrate on these key elements. Initially, nutrient balances were applied to identify nutrient shortfalls and appropriate fertilization regimes and were carried out as simple accounting approaches of nutrients in agricultural or **land-use systems** in **Europe** and North America. More recently, nutrient balances are increasingly being used for other purposes, for example, to monitor and control levels of discharge in areas of high nitrate pollution, with the goal to limit adverse effects caused by high nutrient loads. Additionally, nutrient balance studies have been used from the farm level to continental scales in Africa and elsewhere to investigate changes in soil fertility (often declines) of a broad range of different land-use types. Furthermore, nutrient balances can be applied as indicators of sustainability with respect to soil fertility. In this context, balances are related to nutrient stocks in order to estimate the fraction of stock, which is used per cropping cycle to offset potentially negative balances.

Due to the straightforward approach, nutrient balances usually produce quick and indicative results about the present status of an ecosystem. Caused by the limited availability of measurement data, the assessment of pools and fluxes usually depends on a number of assumptions, models, or transfer functions, which may limit the precision or even bias the results, for example, if applied to large spatial scales, based on imperfect data. However, this statement also holds true for the process models discussed below. Nutrient balances provide a "snapshot" of the present status, a limitation that could partly be overcome by dynamically linking the balance to a land-use/cover-change model. Nutrient balances are black-box approaches in the sense that no processes or nutrient transformations are simulated explicitly. As a consequence, if boundary conditions and processes change over time, the validity of the model decreases, and a process model should be used instead of the balance approach.

Nutrient Cycle Process Models

Process models accounting for nutrient fluxes and transformations have been developed to simulate either soils or plants or both. Nutrient cycling in soils is dominated by the processes of decomposition of organic material, weathering of rocks, exchange processes involving clays and organic substances, and diffusive-advective transport of soluble compounds. These processes are driven by temperature, moisture, pH, and the availability of nutrients and organic material. Models simulating these processes use time steps between minutes and a month and between one and eight soil compartments of different physico-chemical properties, in up to ten soil layers. A similar bandwidth of models exists for the simulation of plant growth. These are driven either by temperature (using temperature optimum curves and growing degree days) or radiation (mimicking photosynthesis) or both, and constrained by water (e.g., using transpiration coefficients or empirical stress functions) and nutrient availability (e.g., by defining admissible ranges of C/N ratios). Carbon allocation schemes mostly are being made dependent on phenological stages. But as the nutrient availability to plants is strongly controlled by soil processes, only an integrated consideration of the plant and the soil system can provide an adequate representation of nutrient cycles. This integration is, for example, implemented in models such as DayCent, DNDC, EPIC, and WOFOST. However, several reasons limit the application of such process models in land-change studies: the high requirements of site-specific data (climate, soil, vegetation; and in the case of managed systems, additional information about management practices), the complex model structure with often intransparent parameter sensitivities, and the fact that many models are tested only for specific ecosystems or regions.

Implications for Modeling Land-Use/Cover Change

Land use and land-use changes affect input and output of nutrients in soils and vegetation, which can lead to a change in soil fertility, which in turn will affect **biomass** and crop production and finally **decision-making** on future land management. Thus, improved understanding of historical and present land-use and land-cover patterns and their effects on nutrient cycles can support the development of adaptation or mitigation strategies in alternative **scenarios** of land-use change. For this purpose, nutrient cycle models could be used in combination with land-use/cover change models to simulate the spatial distribution of various land-use types over different land units and to assess potential nutrient accumulations/depletions, as many environmental problems originate from disproportionate nutrient fluxes. Additionally, land-change models simulating land-use decisions would surely benefit from the **feedback** information of how land cover and land use are affecting the environment in terms of nutrient fluxes, and thus influencing future land-use decisions. Finally, more detailed information concerning the effects on adjacent ecosystems or other components of the earth system like water and atmosphere is needed by scientific disciplines simulating estuarine and marine ecosystems or the global climate. An exemplary coupling of land-use/cover change and nutrient cycle modeling on the regional scale is realized within the Patuxent landscape model, which has been developed by Robert Costanza and colleagues. On the global scale, such coupling has partly been carried out with IMAGE, and is on the way in the LandSHIFT model.

See also: **Agricultural Intensification; Atmosphere-Land Interlinkages; Century Ecosystem Model; Coastal Zone; Cumulative Change; Ecosystem Services; Eutrophication; Land Degradation; Soil Degradation; Soil Erosion; Systemic Change; Urbanization; Water-Land Interlinkages.**

Further Reading

Costanza, Robert, Voinov, Alexej, Boumans, R., Maxwell, Tomas, Villa, Ferdinando, Wainger, L., and Helena Voinov, "Integrated Ecological Economic Modeling of the Patuxent River Watershed, Maryland," *Ecological Monographs* 72 (2002): 203–31; Galloway, Jim, "The Global Nitrogen Cycle: Changes and Consequences," *Environmental Pollution* 102 (1998): 15–24; Melillo, Jerry M., Field, Christopher B., and Bedrich Moldan, eds., *Interactions of the Major Biogeochemical Cycles: Global Change and Human Impacts*, Washington, DC: Island Press, 2003; Parton, William. J., Holland, Elizabeth A., Del Grosso, Steve J., Hartman, Melanie D., Martin, Robin E., Mosier, Arvin R., Ojima, Dennis S., and David S. Schimel, "Generalized Model for NO_x and N_2O Emissions from Soils," *Journal of Geophysical Research* 106 (2001): 17, 403–19; Tiessen, Holm, ed., *Phosphorus in the Global Environment: Transfers, Cycles and Management*, Chichester, UK: Wiley, 1995; United Nations Food and Agriculture Organization, Assessment of Soil Nutrient Balance: Approaches and Methodologies [Online, October 2004], The FAO Web Site, www.fao.org; The University of Vermont, Gund Institute for Ecological Economics, PLM-Patuxent Landscape Model [Online, October 2004], The PLM Web Site, www.uvm.edu/giee/PLM; Vitousek, Peter M., Aber, John D., Howarth, Robert W., Likens, Gene E., Matson, Pamela A., Schindler, David W., Schlesinger, William H., and David G. Tilman, "Human Alteration of the Global Nitrogen Cycle: Causes and Consequences," *Issues in Ecology* 1 (1997): 1–15.

JÖRG A. PRIESS, AMARE HAILESELASSIE, AND MAIK HEISTERMANN

P

Paleo Perspectives. Evidence for changes in land use and land cover prior to the period of **industrialization**, which comes from a variety of sources. Documentary records of various kinds reach back as far as 5,000 years in Egypt and almost 4,000 years in **China**, though in most parts of the world they span a much shorter period. Evidence of past food plants comes from macroscopic remains found at archeological sites, for example in storage jars and pits, trash middens, floor sweepings, and coprolites. By far the most widespread indications of past land-cover change come from pollen analytical evidence.

Putting all these lines of evidence together, it is clear that the earliest evidence for plant **domestication** and incipient **agriculture** in the Near East predates the opening of the **Holocene** period. The cultivation of **rice** in East Asia began no later than 8,500 years ago. Pollen analysis documents the earliest discernible impact of human activities on land cover through **fire**, forest clearance, grazing, and cultivation. In **Europe**, there are relatively few signs of human impact before the spread of Neolithic people from around 8,000 years ago onward, though it seems quite likely that Mesolithic people had a crucial impact on marginal upland ecosystems as a result of the use of fire.

Several factors impede the development of quantitative estimates of pre-industrial human impacts on land cover. Even where a long history of human impact is well recognized, the tendency in recent years has been to discount this in the urge to use pollen analysis as a tool for climate reconstruction. Where the evidence for human impact is relatively easy to identify and evaluate, deriving quantitative estimates of **land-cover change** from pollen analytical evidence is a major challenge. This is even the case in the relatively simple situation where human activity replaces temperate deciduous **forest** with open **pasture** and cultivated land. It is much more difficult in many other types of **biome** where discrimination between forested and open landscapes is more difficult. Despite these difficulties, it is important to realize that in many long-settled areas of the world, evidence for extensive human impact over many millennia is incontrovertible. In **western Europe**, for example, there are many sites where pollen analytical evidence for **deforestation** and the creation of open landscapes during the late Bronze Age, Iron Age, and medieval times confirms a massive impact on terrestrial ecosystems.

There is an urgent need to re-evaluate Holocene pollen-analytical data in light of the growing interest in the possible impact of pre-industrial land-cover change on past climate, both directly through changes in **albedo** and moisture flux and indirectly though its impact on the **carbon cycle**. There is an equally urgent need to improve the basis for quantifying landscape openness from pollen diagrams. Only through such efforts will it be possible to establish the extent to which non-forest areas that were not managed through cultivation or pastoralism at the beginning of the eighteenth century, including extensive areas over which earlier human activities were subsequently discontinued, led to major shifts in ecosystem structure and function.

See also: **Environmental Change; Human Impact on Terrestrial Ecosystems (HITE) Initiative; Pristine Myth.**

Further Reading

Alverson, Keith, Bradley, Raymond S., and Thomas F. Pedersen, eds., *Palaeoclimate, Global Change and the Future*, Berlin: Springer Verlag, 2003; Oldfield, Frank, *Environmental Change: Key Issues in and Alternative Perspectives*, Cambridge, UK: Cambridge University Press, 2005.

FRANK OLDFIELD

Participatory Geographic Information System. The utilization of **geographical information system** (GIS) technology in the context of the needs and abilities of populations that will be involved with, and affected by, development and management decisions and projects. Participatory GIS is sometimes also termed "community-integrated GIS" or "public participation GIS." Participatory GIS distinguishes itself from traditional GIS through its being based on the assumption that stakeholders will significantly contribute to sustainable natural resource management by participating in the process of GIS application. For instance, participatory GIS can render processes of **decision-making** more transparent and accessible by producing maps that visualize development and management options. Participatory GIS emerged out of the need for new GIS methodologies that were not purely top-down and involving only GIS experts, but also the populations affected by decisions based on GIS applications. Therefore, the main objective of participatory GIS is to enable more democratic **land-use planning** processes. The challenges of participatory GIS, such as empowering weaker stakeholders and effectively integrating local knowledge, are still manifold. However, it has been successfully used, for example, in land-use planning with First Nations in **Canada**, in developing small-scale **watershed** management plans with local communities or in solving conflicts about **access** rights to natural resources in protected areas. The experience with participatory GIS so far seems to indicate that its success is influenced more by the social and political context in which GIS is used than by the technological constraints.

See also: **Arena of Land Conflict; Community Involvement; Conservation; Indigenous Knowledge; Participatory Land Management; Scenario.**

Further Reading

Abbot, Jo, Chambers, Robert, Dunn, Christine, Harris, Trevor, de Merode, Emmanuel, Porter, Gina, Townsend, Janet, and Daniel Weiner, "Participatory GIS: Opportunity or Oxymoron?" *International Institute for Environment and Development, PLA Notes* 33 (1998): 27–34.

SILVIA HOSTETTLER

Participatory Land Management. Encompasses a range of different research, development, and policy approaches that promote the involvement of local people in the management of their land and/or natural resources (**forests**, wildlife, **pastures**, agricultural fields, water, etc.), referred to more generally as participatory resource management. It emerged in the 1980s as an alternative to top-down, bureaucracy-heavy, and outsider- or state-enforced **conservation**-with-development planning, which was seen as failing on pragmatic and social justice grounds. Participatory land and resource management is proposed as an effective alternative because the participation of local community members will minimize conflicts that otherwise might arise as a result of introduced management rules; because local people often have detailed knowledge about their environment that could contribute to improved environmental management; and because local communities are often best situated to protect local resources and to enforce resource/land-use rules.

Participatory approaches to land management expanded in the 1990s with most rural development and conservation projects incorporating participatory components. This led to a proliferation in the use of related terms that refer to or suggest participatory processes, including devolution, decentralization, and empowerment, as well as management approaches that claim to be bottom-up, community-based, cooperative, and collective. Participatory research methodologies are pursued to incorporate **indigenous knowledge**, involve all stakeholders, and empower local communities. Despite the popularity of these terms, participation remains a complex process that is rarely complete in practice. A growing literature interrogates claims of **community involvement** or participation in land management projects by asking: How are communities defined? Who effectively participates in community projects? When do they participate—from research and planning to implementation to the **monitoring** and review stages of a program? How do people participate? Do they contribute knowledge and skills, or only voting power and labor? In asking these questions, critics acknowledge the need for participatory land management, but recognize that participation is a negotiated and contested process, the term itself dynamic and flexible. Moreover, environmental outcomes of participatory management are not straightforward, depending not only on project design but on the local socio-ecological contexts in which participatory land management programs are inserted.

See also: **Agrodiversity; Alternatives to Slash-and-Burn (ASB) Programme; Arena of Land Conflict; Institutions; Land Rights; Land-Use Policies; Participatory Geographic Information System; Population Dynamics; Project on People, Land Management and Ecosystem Conservation; Property; World Agroforestry Centre.**

Further Reading

Chambers, Robert, Pacey, Arnold, and Lori A. Thrupp, eds., *Farmers First: Farmer Innovation and Agricultural Research*, London: Intermediate Technology Productions, 1989; Jeffery, Roger, and Bhaskar Vira, eds., *Conflict and Cooperation in Participatory Natural Resource Management*, New York: Palgrave, 2001.

MARA JILL GOLDMAN, MATTHEW D. TURNER

Past Global Changes (PAGES) Project. A core project of the **International Geosphere-Biosphere Programme** since 1991 that supports research aimed at understanding the earth's past environment from a **paleo perspective** in order to make predictions for the future. The scope of interest of PAGES includes the physical climate system, biogeochemical cycles, ecosystem processes, **biodiversity**, and human dimensions on different time **scales** such as the Pleistocene, **Holocene**, last millennium, and the recent past. PAGES has five major research fields. Under focus 5 (past ecosystem processes and human-environment interactions), the project entails the **Human Impacts on Terrestrial Ecosystems (HITE)** initiative, Land Use and Climate Impacts on Fluvial Systems during the Period of Agriculture (LUCIFS), and Human Impact on Lake Ecosystems (LIMPACS). In March 2000, researchers from PAGES and the **Land-Use/Cover Change (LUCC) project** started a joint effort to reconstruct historical changes in permanent croplands at a global scale during the last 300 years, called the **BIOME 300 project**.

See also: **Environmental Change.**

Further Reading

Past Global Changes [Online, November 2004], The PAGES Web Site, www.pages.unibe.ch.

HELMUT GEIST

Pastoral Mobility. Refers to human movements directly associated with the grazing management of domestic livestock. It is most often used to refer to the long-distance movements conducted by herding or pastoral peoples around the world. While the mobility of herding peoples has long been seen as an archaic feature of primitive production systems, recent advances in the ecology of highly fluctuating rangeland systems (e.g., drylands and **mountains**) have characterized pastoral mobility as a necessary adaptation to climatic risk and an effective grazing management approach for variable environments both in terms of livestock production and the environment.

The mobility of herding peoples has long captured the imagination of outside observers. Whether negative or romantic in nature, extreme **perceptions** of the "wandering nomads" have hampered any real understanding of the role that mobility plays in the social life of pastoral peoples and in the ecology of their livestock and pastures. Mobility was seen by early researchers as a key feature of the material culture of herding peoples that distinguished them from others. Various herding groups were defined with respect to their mobility. Categories such as "nomadic," "semi-nomadic," "sedentary," and "settled" were used to categorize social groups ranging from communities to whole ethnic groups. With respect to land management, mobility was seen to produce an incentive structure not conducive to good land husbandry (pastoralist societies were seen not only as outwardly invasive socially but ecologically as well)—overgrazing local pastures before moving on. Early linkages of pastoral mobility with primitiveness, aggression, and poor land husbandry have proven to be highly persistent within national governments and the international development communities. An important political factor contributing to this persistence is the difficulty that mobility presents for national governments seeking to govern, control, and provide services to pastoral populations.

Longer-term studies of pastoral communities have shown that simple characterizations of the mobility of an ethnic group, community, and even household are highly misleading. Pastoral households show significant variation within and between years in their reliance on mobility to provide fodder to their animals. This variation reflects the particular household's resources and constraints, such as the number of animals it is managing; its **access** to herding labor; its access to local and distant water and **pasture**; and its reliance on other economic pursuits such as crop **agriculture**. In addition, broader factors that change over time, including climatic conditions, political insecurity, market opportunities, viability of social networks, cropland spread, and government actions, will affect when, where, and whether herders will move their livestock. Therefore, mobility should not be seen as some inherent cultural feature but as a term that encompasses a range of grazing management options that pastoral groups adopt in response to changing social and ecological situations. As a result of household and broader factors of change, pastoral mobility has declined significantly across wide areas of the world.

Resource managers working in dryland and mountain environments generally recognize the role that livestock mobility can play in both adjusting grazing pressures to forage availabilities at a regional scale, and producing temporal grazing patterns (graze-rest cycles) at the patch level that are more consistent with rangeland sustainability. In most areas of the world, livestock mobility can be maintained only through herding and pastoral mobility. As a result, there has been an upsurge of interest in reinvigorating pastoral mobility, particularly in dryland regions. These efforts have been plagued by a number of problems. First, there remains significant conceptual confusion around the term. What features of the grazing system are managers interested in promoting? Distance covered by livestock? Spatial dispersion of livestock? Distance covered between successive pastoral encampment points? Second, there exist many household level and broader factors that now make it difficult to reinvigorate pastoral mobility. Many pastoral peoples have come to rely on crop agriculture, sedentary ranching, and wage labor to supplement their pastoral livelihoods, which are restricted due to dwindling pasturelands available to their herds, and unfavorable government regulations. The movements of livestock and people are mediated by social networks across broad areas. Disruption of these networks will make it impossible for herders to socially navigate themselves across the pastoral landscape, even if they have the economic resources to do so. Reinvigoration of pastoral mobility will require a combination of state-level policy and local project initiatives working at different spatial scales.

See also: **Arena of Land Conflict; Cattle Ranching; Cultural Factors; Degradation Narrative; Desertification; Disequilibrium Dynamics; Great Plains; Transhumance.**

Further Reading

Fratkin, Elliot, Galvin, Kathleen A., and Eric A. Roth, eds., *African Pastoralist Systems: An Integrated Approach*, Boulder, CO: Lynne Reinner Publications, 1994; Niamir-Fuller,

Maryam, ed., *Managing Mobility in African Rangelands*, London: Intermediate Technology Publications, 1999; Scoones, Ian, ed., *Living with Uncertainty: New Directions in Pastoral Development in Africa*, London: Intermediate Technology Publications, 1994.

MATTHEW D. TURNER, MARA JILL GOLDMAN

Pasture. Landscapes that have a ground story in which grasses are the dominant **vegetation** life form. Definitions often include savannas, as well as the very sparsely grasslands of arid regions. **Grasslands** constitute one of the main land uses on earth, around 36×10^6 km^2, according to information from the **United Nations Food and Agriculture Organization**, including three major vegetation types. The first category is treeless grasslands, which may be temperate (the steppes of **Europe** and Asia) or tropical (Africa, **South America**, and Australia). The second type is savannas, the grasslands with a woody overstory, which often occur next to the treeless grasslands. The last category is the arid, shrubby grasslands. In Africa, South America, and Australia, these types tend to form a continuum of vegetation.

Given equivalent biophysical characteristics, grasslands are the exception rather than the rule, since long-lived woody vegetation (trees and shrub) will ultimately compete with the relatively short-lived grasses. In such contexts, the competitive advantage of woody species is significant, and grasslands tend

Cattle grazing in mountain pastures of Siskiyou County, California, July 1942.
PHOTO: Russell Lee. [Courtesy Library of Congress]

Fenced pasture on cut-over forest land in the Priest River Valley of Bonner County, Idaho, October 1939.

PHOTO: Dorothea Lange. [Courtesy Library of Congress]

to exist only in cases in which the dominance of woody species is suppressed. The factors that prevent dominance by woody plants are disturbance (fire), soil (physical and chemical properties), and temperatures (freezing). **Fires** are influential in the tropics and subtropics for suppressing woody plants. Grasslands are often located in areas with lower precipitation levels and under soil with specific characteristics (such as sandy soils or those subject to regular flooding) that inhibit woody and shrub vegetation. Finally, temperatures constitute a crucial factor that explains the distribution of grasslands at global **scales**.

Grasslands are exposed to **conversion** to croplands where temperatures and moisture availability provide conditions for agricultural expansion. In contrast, where the factors of soil or freezing temperatures are responsible for suppressing competition from woody species, grasslands remain devoted to pastoral uses. Nevertheless, in some environments such as the tropical regions, there is an increasing expansion of pasture to the detriment of

Overgrazed pasture with spreading gullies in the mountains of Bernanillo County, New Mexico, April 1940.
PHOTO: Russell Lee. [Courtesy Library of Congress]

forest. In both cases, grasslands and implanted pastures are exposed to **land degradation** as a result of overgrazing and lack of management practices. Many savannas and grasslands are under growing land-use pressure from grazing animals, cropping, and the extensive use of fire. The implanted pastures in the tropical and subtropical landscapes are exposed to the same degradation effects.

It has been documented that the most relevant **land-cover change** is tropical forest conversion to pasture for livestock production. For instance, in the Brazilian Amazon, around 65 million hectares had been deforested for agricultural uses by 2003. Pasture expansion is responsible for about 70 percent of forest conversion, while **fallow** and unutilized areas account for about one-fifth of the total deforested area, and one-tenth is likely under agricultural uses. There have been different explanations for the overwhelming expansion of pasture in tropical areas. The main explanation has stressed the influence of government policies that have provided incentives to cattle ranchers to settle down in frontier areas (such as tax holidays and subsidized credit). Although incentives have been removed over time, **cattle ranching** continues its expansion, mainly due to the development of infrastructure, and **access** to growing domestic and foreign markets that contribute to making beef cattle production a profitable activity.

In the Amazon, some grasses were introduced from Africa, principally *Brachiaria spp.* The overall stocking rate is less than one head per hectare, which is relatively low compared to other areas. Most cattle ranching in the Amazon is extensive, and there are no appropriate pasture management practices. The pastures are exposed to degradation after relatively short periods of use for cattle breeding, approximately ten to fifteen years. It is estimated that about half of the total pasture implanted might be degraded, supporting very low stocking rates. These areas show low plant cover, invasions by non-palatable native species, and often dense population with termite mounds. The main causes of pasture decline are inappropriate management practices, among them lack of maintenance and fertilization.

Intensification of cattle production in the Amazon is still incipient because of the large availability of cheap forestland and the high costs of pasture maintenance. It has been extensively argued that it could be cheaper to clearcut one hectare of forest to implant pasture than to recover the same hectare covered with degraded pasture. This trend might revert as obstacles for cattle ranching to expand over forest fringes increase (i.e., by enforcement of land-use regulations).

The solutions to pasture degradation are different, depending on both the causes and the stage of pasture degradation, which may range from a mere reduction of the pasture vegetation by the competence of some shrub species, to acute process of **soil degradation**, to reduction of vegetal **biomass**. The most commonly suggested solutions are soil fertilization, renovation of pasture, and the introduction of more intensive techniques of pasture management, using some sort of rotational systems over time and space, or interplanting pasture with some short-life crops. Furthermore, it has been suggested that pasture productivity could be enhanced by planting more diversified systems such as silvopastoral systems, allowing for some secondary succession. In some cases, however, degraded areas are abandoned and lead to forest **regrowth** to become the dominant vegetation.

Research focus on forest succession has found that forests regenerate vigorously in areas that were previously devoted to light agricultural use, and in which there was no complete removal of forest cover. Areas with moderately grazed pasture also are able to develop forest after abandonment, but biomass accumulation is lower than on light-use sites. In contrast, abandoned pastures subjected to heavy use have distinct patterns of succession because forest succession is scarce. Amazon forest can recover after large-scale pasture disturbance, and only **reforestation** is uncertain where land has been used too intensively for long periods. Some research done on soil nutrition concentration after pasture use has found that soil nutrition is independent of age and prior intensity of use in the pastures. There is some soil nutrition concentration when the pasture areas are abandoned, but then nutrient uptake by successional vegetation tends to reduce such soil nutrient stocks.

See also: **Agricultural Frontier; Agricultural Revolution; Alternatives to Slash-and-Burn (ASB) Programme; Amazonia; Brazilian Cerrado; Cumulative Change; Deforestation; Desertification; Disequilibrium Dynamics; Extensification; Great Plains; Hot Spots of Land-Cover Change; Land Abandonment; Land Rehabilitation; Pastoral Mobility; Savannization; Secondary Vegetation; Slash-and-Burn Agriculture; Tradeoffs; Transhumance; Tropical Dry Forest.**

Further Reading

Graetz, Dean, "Grasslands," in *Changes in Land Use and Land Cover: A Global Perspective,* eds. William B. Meyer and Billie L. Turner II, New York: Cambridge University Press, 1994, pp. 125–48; Uhl, Christopher, Buschbacher, Robert, and Emanuel A. S. Serrao, "Abandoned Pastures in Eastern Amazonia I: Patterns of Plant Succession," *The Journal of Ecology* 76 (1988): 663–81; Uhl, Christopher, Buschbacher, Robert, and Emanuel A. S. Serrao, "Abandoned Pastures in Eastern Amazonia II: Nutrient Stock in the Soil and Vegetation," *The Journal of Ecology* 76 (1988): 682–99.

PABLO PACHECO

Pattern Metrics. A set of quantitative indicators of landscape configuration, generated from thematic classifications and providing insight as to the patterns impacting landscape processes. Metrics measure many types of landscape configuration, including shape complexity, diversity, interspersion, proximity, and abundance. Pattern metrics are the implementation of the basic tenet of landscape ecology that understanding landscape processes requires not only knowing what is there but also knowing how it is configured and how, therefore, energy and materials flow in the landscape.

Landscape ecology premises that the **scale** at which landscapes are observed impacts the discernible patterns used to infer process; it further purports that those patterns comprising the landscape mosaic are key in understanding ecological processes by linking form to function. That is, the configuration of landscape elements both constrains and promotes the movement of materials and energy across the landscape. The landscape mosaic comprises three components: the matrix (background or dominant class), patches (contiguous cover of a class of interest having both edge and interior), and corridors (essentially interiorless, elongated, and relatively narrow patches that link or divide other landscape components). The scale of observation (and analysis) influences not only whether a landscape element is detectable but also whether it is defined as a patch, a corridor, or part of the matrix.

Once landscape elements are delimited into these configurational categories and attributed with their thematic class, pattern metrics can be generated. Pattern metrics can be run on either vector or raster data, and while some metric equations are the same between the two, others are different. Therefore when interpreting pattern metrics results, is it important to know the processing lineage. Vector-based metrics use a circle as perfect geometry for comparing shape departures. A circle is used because it represents the greatest area-to-perimeter ratio possible, and thus provides a baseline for comparison (that is not to suggest normatively that simple geometry is necessarily better than complex geometry). Using the raster data model pushes the perfect geometry to a square for comparative purposes. There is, however, an equally important consideration in the differences between pattern metrics run on vector versus raster data sets. Vector data sets typically represent landscapes more generally and smoothly than their raster counterparts, owing to the method by which that landscape was classified. Vector classifications are usually the results of manually digitizing off of a product (map, air photo, satellite image), where the analyst through interpretive skill is able to classify complex or mixed landscape classes. For example, if studying **agriculture** and digitizing agricultural areas off an air photo, the analyst would not digitize rows of soil separately from rows of plants in areas with row crop agriculture. With raster/automated processing, most classification treatments operate at the **pixel** level, resulting both in soil and **vegetation** being classified separately and a more pixelated product (depending upon the spatial grain). Unless smoothed or filtered first, raster products translate into more fragmented landscapes than does the manual/vector approach, and this difference can be observed in metrics.

While over fifty metrics have seen widespread use, most researchers now agree that there are many fewer (four to seven) classes of metrics that measure different components of landscape configuration. Additionally, while metrics are generated at the patch level, they can be reported at the class (thematic) and landscape levels as well via weighted averaging. Pattern metrics provide indications of shape complexity, abundance, interspersion, proximity, diversity, and, most recently, orientation. Not all metrics are supported at each level, and many require a minimum number of thematic classes or number of patches for the metric to be meaningful.

Several challenges in pattern metrics analysis remain. First, many parameters must be set before running metrics, such as how to define proximity (distance threshold, and in raster data also whether diagonal neighbors "count" or not) and edge versus interior (depth of edge). A slight change in these parameters (or the classification scheme) can produce dramatically different results. Second, the comparison of metrics results across time and/or areas is difficult since there are few absolute baselines in most pattern metrics. Holding number of classes, number of patches, and area size fairly constant can minimize artificial differences. Third, while the ecological importance of landscape pattern is clear, it is less clear what the ecological meaning of particular results is and when changes in the metrics represent ecologically significant change. In partial answer to the problems of comparison and ecological meaning, landscape ecologists have developed the neutral landscape method whereby real patterns and metrics can be compared to simulated (and therefore controlled) landscape pattern metrics.

See also: **Aerial Photography; Agent-Based Model; Auto-Correlation; Metabolism; Pattern to Process; Remote Sensing.**

Further Reading

Forman, Richard T. T., *Land Mosaics: The Ecology of Landscapes and Regions,* Cambridge, UK: Cambridge University Press, 1995; Frohn, Richard C., *Remote Sensing for Landscape Ecology: New Metric Indicators for Monitoring, Modeling, and Assessment of Ecosystems,* Boca Raton, FL: Lewis Publishers, 1998; Riitters, Kurt H., O'Neill, Robert V., Hunsaker, Carolyn T., Wickham, James D., Yankee, Dennis H., Timmins, S. P., Jones, K. Bruce, and B. L. Jackson, "A Factor-Analysis of Landscape Pattern and Structure Metrics," *Landscape Ecology* 10 1 (1995): 23; Turner, Monica G., *Landscape Ecology in Theory and Practice: Pattern and Process,* New York: Springer Press, 2001.

KELLEY A. CREWS-MEYER

Pattern to Process. A phrase that refers to the analysis of a spatial pattern of land cover in order to detect and understand the mechanisms that produced the pattern. "Process to pattern" is a related phrase that refers to the analysis of how the mechanisms of a land-change process manifest as a pattern on the landscape.

The fishbone pattern of **deforestation** in **Amazonia** is perhaps the best example of a distinct pattern of land change that has been linked to a particular **agricultural frontier** process. There is a main backbone of deforestation along the Transamazon Highway, and secondary ribs of deforestation along the secondary roads that are at regular intervals emanating from and perpendicular to the highway. This pattern of deforestation is caused by a sociological process of immigration to the forest via the highway and secondary roads.

There are numerous valid ways to characterize both a process and a pattern; therefore it is a challenge to link a particular process of land change with a particular pattern of land change. For example, Amazonian deforestation could be explained as a manifestation of the sum of the decisions of many individual farmers, each of whom is trying to maximize income for his or her family. Alternatively, Amazonian deforestation could be explained as a manifestation of the decisions of a much smaller number of regional planners who are encouraging immigration in order to alleviate unemployment in Brazilian cities. Both explanations lead to the fishbone pattern. Also, there are a variety of ways to measure a pattern, such as average size of patches, perimeter-to-area ratio of patches, etc. It is not immediately obvious which **pattern metrics** are best to characterize a particular pattern.

Linking pattern to process is especially important for scientists who create simulation models of land change. A model usually attempts to simulate the process of land change and then predicts the resulting pattern of land change. If there is a clear link between pattern and process, then the model can be validated by comparing the predicted patterns against reference data of land cover.

See also: **Cellular Automaton; Decision-Making; Land-Use History; Modeling; Pixelizing the Social; Socializing the Pixel; Transition Matrix; Validation.**

Further Reading

Liverman, Diana, Moran, Emilio, Rindfuss, Ronald, and Paul Stern, eds., *People and Pixels: Linking Remote Sensing and Social Science*, Washington, DC: National Academy Press, 1998; Nagendra, Harini, Munroe, Darla, and Jane Southworth, "From Pattern to Process: Landscape Fragmentation and the Analysis of Land Use/Land Cover Change," *Agriculture, Ecosystems and Environment* 101 (2004): 111–65; White, Roger, and Guy Engelen, "Cellular Automata and Fractal Urban Form: A Cellular Modelling Approach to the Evolution of Urban Land Use Patterns," *Environment and Planning A* 25 (1993): 1175–79.

ROBERT GILMORE PONTIUS JR.

Peatlands. One of the most important ecosystems in the world, representing 50–70 percent of **wetlands** on earth. They are also fundamental due to their vital functions in carbon, hydrological, and geochemical cycles. Although peatlands cover only 3 percent of the land and freshwater surface of the planet, they contain 30 percent of the carbon stored in soil and 10 percent of freshwater resources. It is estimated that the total carbon pool of wetlands exceeds that of the world's **forests**. They also support a very high diversity of species, many of which are unique for this ecosystem. Peatlands exist in varying climate conditions on different continents, from sea level to high-alpine conditions, and from tropical to boreal and Arctic zones. They occur everywhere on the globe, except where the climate is too dry or too cold to support plant life. In some countries they account for up to 30 percent of the land area.

However, peatlands have generally been seen as a valuable natural resource in the form of peat itself, which has many uses, including fuel, animal bedding, and growth substrate in **agriculture** and horticulture. Increased human demands place pressure on the use of these ecosystems, significantly changing the range and importance of the system's diverse functions, services, and resources. Due to growing needs of resources from peatlands, sustainable and wise use of these ecosystems is essential in order to assure that sufficient areas of peatlands remain to carry out their vital natural resources functions.

Definitions

Peatlands are areas with or without **vegetation** with a naturally accumulated peat layer at the surface. Peat is a brownish-black soil composed of sedentarily accumulated material, consisting of at least 30 percent dead organic material that results from carbon chemical biosynthesis. Inorganic soil particles are washed or blown into peatlands and also form part of the peat. The prominent characteristic of a peatland is the near-surface water table, as this makes peat accumulation possible—thus, it occurs in areas where precipitation/groundwater exceeds evapotranspiration/drainage. There are several types of peatlands, which are classified according to vegetation type and the source of water. A mire is a peatland where peat is currently being formed. The two main mire types are fens and bogs. A bog is a peatland that is raised above the surrounding landscape. It is fed mainly from rainwater. Some bogs support forests, but they are nutrient-poor. A fen is a peatland situated in depressions or on slopes and is fed from both rainwater and groundwater that are minerotrophic, acid or alkaline, and nutrient-rich or nutrient-poor.

Inventories

Peatland estimations vary substantially among sources. Presently, there is a lack of knowledge on the effective extent and precise location of peatlands, as different criteria are used to classify peatlands, depending on the scientific discipline.

Globally, it is estimated that 4 million km^2 are covered with peatlands, or approximately 3 percent of the planet. The largest known concentrations of peat are located in the Northern Hemisphere, being widespread within the boreal fringe, in Southeast Asia, and parts of the Amazon basin. Southern Hemisphere peatlands are generally smaller, sparsely distributed, and relatively shallow. Less than 1 percent of the world's peatlands are found in **South America**.

In the Northern Hemisphere, due to higher rainfall, peatlands occur over large areas of **Canada**, Alaska, Fennoscandia (Finland, Sweden, and Norway) and **Russia**, having an average thickness of 1.5 meters. It has been estimated that northern peatlands cover 346 million hectares and hold 455 billion metric tons of carbon, slightly less than the amount contained in all living organisms or in all atmospheric carbon dioxide.

In Southeast Asia, the total area of peatlands is estimated to be about 20 million hectares, which is approximately one-tenth of the entire extent of global peatland resource and at least 60 percent of the world's tropical peatlands. Most of the peatlands in Southeast Asia are found in Indonesia, which possesses over 70 percent of total peatland area in this region. Other major peatlands are located in Thailand, Malaysia, Vietnam, Brunei Darussalam, and the Philippines.

Functions

Peatlands are a particularly significant type of wetland and perform many vital functions as centers of hydrological functioning, as carbon sinks, and in terms of **biodiversity**. Peatland hydrological functions are more or less similar to those of wetlands. However, peat enhances normal wetland functions, making peatlands particularly efficient in water storage and highly effective as filters. Peat acts as a sponge, storing water for long periods of time and releasing it when necessary. Peat and peatland vegetation are also good filters, making peatlands highly effective in removing sediment, pollutants, and pathogens.

Peatlands are organic sinks, and it is estimated that a quarter of the earth's soil carbon is stored as peat. When life on earth began, there was more than ten times more carbon dioxide in the air than at present. Carbon-based organisms absorbed and used carbon dioxide and slowly reduced atmospheric carbon dioxide. When these organisms died, much of the carbon was locked up in the underground. The carbon was stored as peat in wetlands, and over geological time peat turned first to lignite and then coal. Carbon accumulation balances between inputs (**biomass** or primary production) and outputs (microbial activity that degrades organic matter) in redoxic conditions. This process occurs mainly in active mires.

Specific conditions of habitat in peatlands such as stresses (high acidity in bogs and nutrient-poor, redoxic conditions) call for adaptations of plants or animals to these ecosystems. Then, plant species require adaptations in physiology to deal with the toxic substances that originate under anaerobic conditions; as well as anatomy, such as aerenchyma that lead oxygen from the parts above ground to the root system; and growth form, including aerial roots.

Benefits and Values

Like all wetlands, peatlands provide a variety of goods and **ecosystem services**, both directly and indirectly, in the form of water supply and groundwater recharge, flood mitigation, energy, **forestry**, and fishery products. They also have a fundamental functional significance, the carbon stored in peat representing some 25 percent of the world soil carbon pool, which would contribute to global warming and climate disruption if released. However, some peatland uses and services are specifically predominant. In the Northern Hemisphere, peatlands are deep and extensive, which allows for their use as fuel. Peat has a low combustion temperature and low nitrogen oxide emissions and has been used traditionally for millennia in **Europe**, Fennoscandia, and Russia as a fuel, more recently for electricity generation. Industrial extraction of fuel peat has occurred in Europe since the nineteenth century with the invention of peat-working machines. Extensive drainage of bogs and fens for agricultural use in Europe began in the seventeenth century, although in some countries (like Sweden) cultivation of peat soils dates back to the early Iron Age. At present, half of all peat extracted is used in horticulture or for soil conditioning. Over 40 percent of the peat extracted in **South Africa** is used as a basis for horticultural substrates, and almost all Canadian peat is used for horticulture needs.

Threats to Peatlands

Since 1800, the total area of peatlands worldwide has been significantly reduced by approximately 10–20 percent due to climate change and human activities. All combined uses of peatland, through peat mining, drainage, and **conversion** to agriculture and forestry, lead to a global rate of loss of active peatlands estimated to be 0.1 percent per year. However, the losses have been much greater in Europe, where 52 percent of peatlands have now been lost. In some countries like the Netherlands, active peatlands that once covered 20 percent of the land surface are almost totally gone. Direct impacts on peatlands include drainage, land conversion, excavation, inundation, and visitor pressure. Peat extraction drastically reduces the water storage and filtering properties of the wetlands, impacts the carbon fixation function, and affects biodiversity negatively. Biologic or hydrologic **resilience** can occur only when peat is not industrially extracted. Drainage for agriculture and forestry still continue to be major causes affecting change in peatlands, especially in the tropics. **Forest degradation** through forest/peat fires, which have increased in tropical peatlands since the 1990s, caused destruction of these ecosystems in large areas in Southeast Asia, especially Indonesia, and released a huge amount of carbon dioxide to the atmosphere, potentially spurring global warming. Besides the increase of the **greenhouse gas** carbon dioxide (CO_2), indirect impacts on peatlands are a result of air pollution, water contamination, contraction through water removal, and infrastructure development.

Wise Use of Peatlands and Mires

The utilization of any natural resource should subscribe to the principles of sustainability. The effective utilization of peatlands involves a compromise among traditional practices, **conservation**, and tourism. Thus, peat extraction and some management actions are not compatible with undisturbed sites of high scientific and biodiversity value. Until recently, the conflict between conservationists and the peat industry seemed to be irreconcilable: preserving the natural heritage value and the bulk of sequestered carbon of peatlands will remain incompatible with any form of exploitation. However, in some cases limited peat extraction can increase biodiversity and may be sustainable in the long term. Therefore a wise use of peatlands and mires is at present encouraged. On an international level, the **Ramsar Convention on Wetlands** encourages the protection of peatlands worldwide (recommendation 6.1 of the 1996 Ramsar meeting). However, despite being the most extensive single wetland type in the world, less than 10 percent of the global peatland area is represented in the Ramsar list: wetlands of international importance, or "Ramsar sites" with significant peatland components, amount to 4.1 million out of 120 million hectares of protected wetlands; and of the 1,399 Ramsar sites, 443 are wetlands with significant peatland components (as of March 2004).

See also: **Atmosphere-Land Interlinkages; Biome; Carbon Cycle; Deforestation; GLOBWETLANDS; Modification; Paleo Perspectives; Water-Land Interlinkages.**

Further Reading

Global Environment Centre, Peat-Portal [Online, January 2005], The Peat-Portal Web Site, www.peat-portal.net; International Mire Conservation Group [Online, January 2005], The IMCG Web Site, www.imcg.net; Lappalainen, Eino, ed., *Global Peat Resources: Peatlands in Biosphere, Country Reports, Conservation and Use,* Jyväskylä, FL: International Peat Society, 1996.

LAURENCE HUBERT-MOY

People at Risk. Though at risk, people are not mere "victims" of **global environmental change** but have the ability to affect and shape how they live with a range of risks. Humans interact with nature and natural resources, producing negative and positive impacts. These interactions can operate at a local level (e.g., household) or they may be the product of larger-scale interactions (e.g., outcomes arising from development strategies, legislation, or economic systems). Understanding how global environmental change problems are conceptualized and how decisions, at various levels, including risks coupled to environmental change, are important research foci. Knowledge of the array of things that people do to better live with risk arising from global environmental changes is a key issue receiving heightened attention by scientists, governments, and wider civic society.

The impacts and risks associated with complex global environmental changes include ecological risks, climate risks, and risks to health and the environment. This complexity and breadth of issues in the environmental arena requires multiple approaches and perspectives. Traditionally, this field has been strongly influenced by scientists working in the biophysical science domain. Here, concerns include improving the understanding of the functioning of the earth as a complex system. With the strong shift toward sustainability and sustainable development or **sustainable land use**, however, perspectives are also increasingly being drawn from the social, economic, and political sciences, including questions of how terms such as risks and hazards are derived, framed, and constructed. Security risks associated with global environmental changes and the realm of issues linked to uncertainty, sustainability, and change are thus emerging as key themes allied to global environmental change research.

The notion of risk can be used and interpreted in many ways. Several academic and practitioner communities have been trying, for example, to articulate concepts that can assist both in improved understanding of the complex earth system and also to find ways to better "live" within such a system. Various terms have consequently been identified that are now embedded in the global environmental change discourse and literature, including terms such as risk, hazards, risk society, adaptive (co)-management, and risk reduction.

Global environmental change scientists like Carl Folke, Fikret Berkes, and Crawford Holling, among others, have been grappling with such terms as ecological risks and resilience, and have opened up spaces for expanded dialog between natural and social scientists working on issues related to environmental and other risks. Sociologists have also tried to frame understandings of risk from social and cultural perspectives, including examining the role of **institutions** and the cultural interpretations and practices enabling people to live with risks associated with global environmental change. Some of these approaches and uses of risk in the international science, policy, and development arena are explored more fully below.

Before detailing notions of risks and responses to risks, one needs to have some common reference point. Despite the absence of a common language to capture a universal term such as risk, it is usual to view risks as potential threats and dangers. Hence, the term risk is often used in association with the risk of something occurring (e.g., hazard or process) in a specific social setting (e.g., region at risk), as Roger Kasperson and colleagues use it, among others. Astrid von Kotze and Ailsa Holloway (1996, p. 5) talk about "elements at risk" and refer to communities, structures, services, or activities. They define risk as "the expected losses (lives lost, people injured, damage to property, and disruption of economic activity or livelihood) caused by a particular phenomenon." Consequently, "a societal element is said to be 'at risk' or 'vulnerable' when it is exposed to known hazards and is likely to be adversely affected by the impact of those hazards if and when they occur."

One of the key issues in living with risk to global environmental change is trying to understand the ways that systems (both ecological and human) generate, respond, and try to accommodate and manage risks. In the discussion that follows, several aspects will be examined that are linked to notions of societal risk in relation to global environmental change, including a risk in society, **vulnerability**, **resilience**, and building adaptive capacity.

Risk Society

Some of the most interesting analyses of risk in society have been those proposed by Anthony Giddens and Ulrich Beck. Rooted in discussions of modernity is the notion that the differentiation of risk is often viewed as occurring outside of society. Centralized institutions, such as national governments, are usually incapable of effectively responding to environmental risks because they are intricately bound up with these institutions. The need for reflexivity and improved understanding of risk then become critical in terms of how we approach and accommodate risks. In the words of Ulrich Beck (1992, p. 81), "environmental problems are not problems from our surroundings, but—in their origins and through their consequences—are thoroughly social problems, problems of people, their history, their living conditions, their relation to the world and reality, their social, cultural and political situations." Managing or adapting to risks therefore requires new approaches, because "what were once side-effects are challenging the core of our everyday assumptions," as Alan Irwin (2001, p. 57) puts it. He continues that "in line with this, science's 'monopoly on rationality' has begun to break down in the face of the new set of risks and challenges created by science itself."

The causes and responses to risks cannot therefore be identified and understood using only biophysical science and technological approaches. One also requires inputs that are understood from a "sociological" lens. Roger Kasperson and colleagues provide examples of how such concepts can be applied in various "regions at risk," including **Amazonia**, Eastern Sundaland, Nepal, **China**, and Mexico.

Risks are therefore the results of processes of active construction within particular contexts and settings. To effectively manage such risks, people need to be able to participate in the processes of **decision-making** (or as some call it, the "social discourse") that ultimately shape their lives and overall quality of life. One research area that is enabling us to better understand risk management in relation to global environmental change is determining vulnerability to a range of hazards and risks.

Vulnerability

Renewed debates on notions of vulnerability are increasingly trying to comprehend what factors generate vulnerability; including, for example, social, cultural, environmental, and political factors or complex combinations of these. This renewed attention around vulnerability has helped clarify the concepts of risks and disaster. Vulnerability, for example, has been described by Robert Chambers as being comprised of two dimensions or components: first, an internal dimension usually linked to adaptive capacity or coping; and second, an external dimension usually linked to a hazard or trigger that produces a shock or impact on an exposure unit (e.g., person, a farm, or city). In other words, as put forward by Omar Cardona (2004, p. 37), "vulnerability represents the physical, economic, political, or social susceptibility or predisposition of a community to damage in the case of the destabilizing phenomenon of natural and anthropogenic origin."

Vulnerability, however, is not a residual or impact of a disaster, and is therefore very difficult to measure. In a social context, vulnerability is not a uniform phenomenon and varies in time and space. Some people's coping capacity (internal component of vulnerability) in the very same area, and indeed in the very same house or shelter, may be markedly different to those in the same or similar setting or environment. Likewise the external dimensions may also vary. Moreover, external stresses may be in the form of single stress events or as part of multiple stresses. For example, in **southern Africa** those at risk to **climate impacts** are also often the resource poor who lack **access** to markets and information and are also vulnerable to a variety of health risks, such as **human immunodeficiency virus (HIV)/acquired immunodeficiency syndrome (AIDS)** and **malaria**.

Increasingly, those working at a very practical level, for example, humanitarian aid workers in various parts of the world (including southern Africa) attached to, for example, the World Food Programme or Save the Children Fund have to therefore find ways of measuring "differential" vulnerability and responses to risks to a number of triggering phenomena and ongoing processes such as HIV/AIDS, war, **poverty**, and drought. These factors are further complicated and compounded because they usually occur in various guises and at various rates of change (e.g., drought "creeps"). Vulnerability is hence referred to as being "chronic" or insidious. For improved understanding of such complex systems, a variety of social and physical science approaches are being developed and tested.

Resilience

Focusing only on vulnerability often tends to narrow the focus of concern to those negative aspects and processes related to risk. Increasingly, attention is also being directed toward more positive interpretations of risks and response to risks in both ecological and social systems and opportunities in relation to threats or risks, as done by Carl Folke and colleagues, for example.

Responses of ecosystems to global environmental change, moreover, are no longer viewed as being only linear and therefore easily controlled. Rather, natural and social systems are seen to be strongly coupled, interconnected systems that behave in non-linear ways and exhibit marked **thresholds** in their dynamics. Some researchers working with the "resilience alliance," for example, have been trying to understand and motivate for a more holistic understanding of resilience as a way to respond to risks. From their collective research, resilience is described as the magnitude of disturbance that can be tolerated before an ecosystem moves into a different state with a different set of controls. In a vulnerable system, even a small change may shift the ecosystem into an undesirable state with several consequences. This work has also helped those working on social dimensions of risk to better understand the functioning and interaction of various social networks and groups as ways of managing risks.

Enhancing Adaptive Capacity

One of the overriding concerns when trying to live with risks in coupled human-environment systems is to ultimately find ways to reduce negative outcomes to risks associated with global environmental change and to harness positive changes. These concerns have revived interest in trying to improve the resilience of systems, and enhancing social adaptive management and coping capacity to global environmental change. Enhancing adaptive capacity includes understanding the learning aspect of system behavior in response to disturbance and reflects the ability of social-ecological systems to cope with change without losing options for the future, as expressed by Carl Folke. Adaptive capacity is thus linked to a variety of features in systems, including biological and sociological diversity, such as **biodiversity** and **agrodiversity**. In resource-poor farming communities in many parts of Africa, for example, access to both on-farm and off-farm products (e.g., craft production and other forms of wage labor) has often been shown to be a necessary coping mechanism to help people withstand periods of shock and disruption of **economic livelihood**.

Managing and living with risks thus may mean new and innovative systems and institutions that may have to be designed and developed to enhance our ability to adapt to **environmental change**. One such research and advocacy area that reflects these needs and concerns is disaster risk reduction. Those interested in disaster management, for example, have moved the thinking in this field from reactive, event-driven responses when a risk materializes into a reality to a longer-term, proactive risk management and risk-reduction approach. Strong efforts to reduce the impacts and risks to disasters before the risks heighten into disasters are being profiled. Examples are the International Decade for Natural Disaster Reduction (IDNDR) publications and the International Strategy for Disaster Reduction (ISDR). Such a focus, however, requires interactions with various stakeholders and agents including development planners, government agencies, scientists, conservationists, non-governmental organizations (NGOs), and others. With this type of an approach, risks, threats to livelihoods, and possible hazards are usually identified and mapped using multidisciplinary approaches (e.g., the Sustainable Livelihoods Approach; Food Insecurity and Vulnerability Information and Mapping System [FIVIMS]—**United Nations Food and Agriculture Organization** [FAO]; and Risk Map—Save the Children Fund (SCF), United Kingdom). People and areas that are exposed to risks and/or threats are identified, using as a guide their ability to respond and adapt to threats and risks. Increasingly, however, those exposed to risks are in fact also exposed to daily risks as a result of poverty and not only to once-off events. Hence there is a need to now begin to better match development and disaster risk reduction efforts.

Living with risks requires innovative thinking. For those in developing countries, for example, there is a need to better match development and disaster risk reduction efforts. Whatever context, global environmental change cannot be ignored, and efforts to better live and adapt to changes require inputs of all, including non-governmental organizations, civic society, and scientists.

See also: **Agrodiversity; Arena of Land Conflict; Cumulative Change; Desertification; Food Security; Human Health; Systemic Change.**

Further Reading

Adger, Neil, "Social Capital, Collective Action, and Adaptation to Climate Change," *Economic Geography* 79 (2003): 387–404; Beck, Ulrich, *Risk Society: Towards a New Modernity,* London: Newbury Park, New Delhi: Sage, 1992; Beck, Ulrich, *World Risk Society,* Cambridge, UK: Polity, 1999; Berkes, Fikret, and Carl Folke, *Linking Social and Ecological Systems: Management Practices and Social Mechanisms for Building Resilience,* New York: Cambridge University Press, 1998; Blaikie, Piers, Cannon, Terry, Davis, Ian, and Ben Wisner, *At Risk: Natural Hazards, People's Vulnerability and Disasters,* London, New York: Routledge, 1994; Cardona, Omar D., "The Need for Rethinking the Concepts of Vulnerability and Risk from a Holistic Perspective: A Necessary Review and Criticism for Effective Risk Management," in *Mapping Vulnerability: Disasters, Development and People,* eds. Greg Bankoff, Georg Frerks, and Dorothea Hilhorst, London: Earthscan, 2004, pp. 37–51; Chambers, Robert, "Editorial Introduction: Vulnerability, Coping and Policy," *IDS Bulletin* 20 (1989): 1–7; Egeland, Jan, Living with Risk: A Global Review of Disaster Reduction Initiatives [Online, September 2004], International Strategy for Disaster Reduction Web Site, www.unisdr.org; Folke, Carl, Colding, Johan, and Fikret Berkes, "Building Resilience for Adaptive Capacity in Social-Ecological Systems," in *Navigating Social-Ecological Systems: Building Resilience for Complexity and Change,* eds. Fikret Berkes, Johan Colding, and Carl Folke, Cambridge, UK: Cambridge University

Press, 2002; Giddens, Anthony, *The Consequences of Modernity*, Cambridge, UK: Polity, 1990; Holling, Crawford (Buzz), "Understanding the Complexity of Economic, Ecological and Social Systems," *Ecosystems* 4 (2001): 390–405; Irwin, Alan, *Sociology and the Environment: A Critical Introduction to Society, Nature and Knowledge*, Cambridge, UK: Blackwell, Polity, 2001; Kasperson, Roger E., Kasperson, Jeanne X., and Billie L. Turner II, "Risk and Criticality: Trajectories of Regional Environmental Degradation," *Ambio* 28 (1999): 562–8; O'Riordan, Timothy, and Stephen Rayner, "Risk Management for Global Environmental Change," *Global Environmental Change* 1 (1991): 91–108; Research on Resilience in Social-Ecological Systems: A Basis for Sustainability [Online, September 2004], Resilience Alliance Web Site, www.resalliance.org; Von Kotze, Astrid, and Ailsa Holloway, *Reducing Risk: Participatory Learning Activities for Disaster Mitigation in Southern Africa*, Natal, South Africa: International Federation of the Red Cross and Red Crescent Societies, Department of Adult and Community Education, University of Natal, 1996; Walker, Henry A., "Understanding and Managing Risks to Health and Environment from Global Atmospheric Change: A Synthesis," *Human and Ecological Risk Assessment* 7 (2001): 1195–1209.

COLEEN H. VOGEL

Perception. Perception is the sum of all processes providing context and meaning to (sensorial) information from the environment, contributing to its awareness or apperception and allowing one to localize and recognize an object of perception (the "world of things" in its widest sense, which means objects as physical or social givens), to differentiate it from others, and to react to it (e.g., with language or with the body).

Environmental perception is about how an individual relates to his/her life space and makes sense of it. The life space does not only embrace the biotic and abiotic environment to provide the individual and his/her social group with all needed to survive, but also the space (including its components and the relation between these components) that has for them a social, political, cultural, or esthetic value. In the more general sense it is the space that, from an individual's perspective, has an effect on decision-making and where human actions have an impact; and where a mutual affectedness between humans and nature exists ("*Mitwelt*"). The perceived life space transcends the immediate local surroundings, encompassing to a different degree subnational, national, regional, and global dimensions. The increasing perception of a globalized environment is related to recent processes of economic, social, political, and cultural **globalization** processes and the concomitant increasing experience of a growing and accelerated interconnectedness of the world.

Environmental perception is mediated by worldviews; social, economic, and political structures; and by the immediate context of interaction between an individual and his/her environment. Factors such as knowledge (including ontologies) and skills, values, and norms; but also interpersonal influences, institutional settings, configurations of interest, technological possibilities, and constraints play an important role. It is also impossible to understand people's environmental perception without taking their history and the development, transformation, and perpetuation of their schemata of environmental knowledge and praxis, as well as their capacity to project themselves into the future, into account.

See also: **Cultural Factors; Forest Transition; Land Degradation; Landholding; Land-Use Policies.**

BARBARA GÖBEL

Pixel. A contraction of the phrase picture element. Pixels are the objects that compose satellite images and other raster-based digital data files. The term is most often used in association with **remote sensing** data products such as satellite imagery or digital **aerial photography**, but is sometimes used generically to refer to cells within raster data. Pixels arranged in a rectangular grid consisting of rows and columns of data form a complete image file. With passive remote sensing systems, pixel values represent a measurement of the ground surface reflectance (or brightness value) for a specific range of the electromagnetic spectrum. Different types of **vegetation** produce specific reflectances or spectral signatures. The pixel values for different bands of the electromagnetic spectrum can be combined and analyzed to produce classified land-cover maps using a variety of methods. Different satellite sensors are designed to detect different types of environmental features, with some sensors designed for terrestrial applications while others are designed for ocean or atmospheric applications. The spatial area covered by a single pixel is a product of the spatial resolution of the sensor. For example, the pixels in the panchromatic band of **IKONOS** data cover a 1 × 1 meter area on the ground. In contrast, pixels in some bands of **moderate resolution imaging spectroradiometer** imagery are 500 × 500 meters in size on the ground. Land-cover classification is complicated by the fact that at coarse spatial resolutions one pixel may include many different types of land cover on the ground.

See also: **Change Detection; Pattern Metrics; Pixelizing the Social; Thermal Band Analysis.**

Further Reading

Jensen, J. R., *Remote Sensing of the Environment: An Earth Resource Perspective*, Upper Saddle River, NJ: Prentice Hall, 2000; Liverman, Diana, Moran, Emilio F., Rindfuss, Ronald R., and Paul C. Stern, *People and Pixels: Linking Remote Sensing and Social Science*, Washington, DC: National Academy Press, 1998.

THOMAS P. EVANS

Pixelizing the Social. To "socialize the pixel" and to "pixelize the social" is to discern information embedded within spatial data on land-use/cover change that is directly relevant to the core themes of the land-change science agenda, by making **remote sensing** (RS) and **geographical information systems** (GIS) relevant to the social, political, and economic issues driving land-use/cover change. While a **pixel** is a unit of observation associated with satellite data on land cover, the term is used more broadly here to encompass all spatial analysis on land-use/cover change, even if the data are from other sources. Remote sensing, both data and image processing, and analysis through GIS have played crucial roles in the development of land-use/cover change models in both data creation and analysis.

During the course of the past ten years, significant progress has been made in acquiring spatial land-cover data sets from remotely sensed data, conceptualizing the basic geographic and environmental processes that are associated with land-use change. Numerous spatially explicit and heterogeneous land-use change models now exist, spurred by the vast amount of spatial land-use/cover data that are now available. The research agenda of the **Land-Use/Cover Change (LUCC) project** (and its associated groups) has demanded an interdisciplinary approach to **modeling**, with major contributions made by landscape ecologists, geographers, anthropologists, political scientists, economists, and demographers.

While the goal of both **socializing the pixel** and pixelizing the social is to better understand the spatial pattern of land-use/cover change, there are broad generalizations of the differences between the two, as envisioned here. To socialize the pixel is to develop land-change models that move from the RS imagery beyond its use in the natural sciences and toward the concerns of the social sciences per se. These models usually have as the unit of observation the satellite data pixel and include explanatory variables that can be "seen" from the remotely sensed data and calculated using GIS, such as distance measures; other spatial biophysical variables (e.g., soil, slope, elevation); and occasionally socio-economic "drivers," such as population or other sociodemographic characteristics, usually measured at some aggregate level, such as village or state.

To "pixelize the social" is to apply social science theories to spatial data in order to develop an underlying structural model that seeks to explain the human behavior that generates these patterns of land-use/cover change. These models usually take as the unit of observation the land under the control of an individual decision-maker and directly link the decision-maker to specific pixels. Of course, as research has progressed, and as more "socialization" has been done in the "socialize the pixel" realm and more "pixelization" has been done in the "pixelize the social" area, the boundaries between the two, rightly so, have become blurred.

The impacts of RS data and GIS to date have been strongest among the environmental and policy sciences because, respectively, space-based imagery observes the physical attributes of the biosphere, and various stakeholders and decision-makers require spatially explicit assessments of changes in the biosphere. While some areas of social science, specifically geography and regional science, as well as some specializations within economics such as urban economics, have always had a spatial focus; the majority of the human sciences have been slower to incorporate spatial issues as basic elements of research. However, land-use change is an inherently spatial process, so ignoring the spatial dimension in analysis is analogous to analyzing a dynamic process without knowing the chronological order of events. In order to explain and predict the spatial processes that result in land-use change, theoretical and empirical social science models must be developed to address where, when, and why these processes happen. In order to do this, improved spatial data, increased theoretical understandings of human behavior in space, and new methods to use these data and test these hypotheses are needed, that is, to pixelize the social.

During the past ten years there has been an increase in interest in spatial analysis within the broader social science community, as theoretical advances have been made in understanding human interactions across space, as spatial data sets of data of interest to social scientists have become more available. In addition to this increase in spatial data (remotely sensed data, such as satellite data, as well as other geo-referenced data), the advances in GIS and **global positioning systems** (GPS) availability and ease of use has increased the interest in spatial issues among economists. GIS software has a myriad of uses, including data visualization, organization, and integration across sources, as well as variable creation. Often spatial data for a study will come from a wide variety of sources as well as different **scales**, such as satellite data on land use, digital soil maps, tax assessment data on property values and structural characteristics, census data of different kinds, road networks, and school district boundaries. An important attribute of GIS is that it allows the linking of these data via their location in space to permit the assignment of the value of these data to the appropriate unit of observation in an analysis. GPS technology, which gives the coordinates of any location, allows a researcher to create a spatial database from "scratch" by linking household survey data with individual land-use management units. Finally, as spatial data and spatial data software applications have become more available, so too have the statistical tools to use these data appropriately. However, while there have been advances in each of these realms, further advances must be made in data generation and use, theoretical understanding of how individuals make choices concerning land use across space, and the methods used to develop models and test the hypotheses generated by both the data and theory, to truly pixelize the social.

Much of location-based spatial social science theory is derived from the monocentric city model. Briefly, the monocentric city model is an equilibrium model of urban spatial structure, where the distribution of land uses on a featureless plain around a central business district is the result of an equilibrium between a declining land price gradient and increasing transportation costs. This model results in concentric rings around the city made up of different land uses, with agricultural and **forest** lands as the residual land use beyond the developed land uses. For the most part, the only heterogeneity in the landscape is the distance to the central business district or some amenity or disamenity that is implicitly tied to this distance. In many situations, space matters in terms of complex spatial processes and must then be modeled as such at the appropriate scale of analysis. While some modeling approaches, such as **agent-based models**, can incorporate different spatial relationships and complexities, further advances are needed in the understanding of human behavior across space.

While many land-use change research projects use RS data for information on land use and land-use change, the issues discussed here do not involve the science of developing these data sets, but rather the creation of the necessary associated social science data to develop models of land-use change and how these data are linked with RS data. In many countries, different types of government census data sets exist that are at some level spatially explicit, but due to confidentiality issues, these data are aggregated to some level so that individuals cannot be identified. It is possible, through the use of GPS and household survey work, to create individual spatially explicit data that link individuals with specific parcels of land, but this is time-consuming and expensive. While any given piece of land is immobile in space, both the land manager and the boundaries of a parcel under an individual land manager can change over time, for example, as a farm is subdivided into residential lots, requiring longitudinal data to understand the land-use change process. An individual parcel can be under different land management units, such as owner and renter relationships, and individuals can control more than one piece of land. In addition, the unit of analysis is often at a different scale than other ancillary data, so decisions must be made on how to best combine diverse data sets to include in a land-use change model. Therefore, expertise in survey design, administration, and issues in modeling with such data are required.

When using spatial data, two related issues must be considered: how to use the data "creatively" and how to use the data "correctly." The former refers to developing ways of visualizing data to assist with the analysis as well as creating variables from spatial data that can be used in a model; the latter refers to issues of using these data in models. While many studies use data that are based on location, such as the impact of a new road on nearby forest clearing, a truly spatial study should contain a creative spatially explicit model of behavior or the interaction over space of the natural and human environments, or the explicit analysis of the spatial pattern of outcomes; or else some combinations of these and further development of these kinds of models are needed. The second modeling issue deals with the methodological concerns that follow from explicit consideration of spatial effects in such models, such as spatial **auto-correlation**. Finally, while theoretical conceptualization of land-use change processes lead to testable hypotheses, often the actual empirical specification of the model to be tested must take into consideration assorted limitations in data availability and quality, such as the use of proxies for variables of interest.

See also: **Georeferencing; Land Rent; Land-Use System; Pattern to Process; Property Thünen, Johann Heinrich von.**

Further Reading

Fox, Jefferson, Mishra, Vinod, Rindfuss, Ronald R., and Stephen J. Walsh, eds., *People and the Environment: Approaches for Linking Household and Community Surveys to Remote Sensing and GIS*, Amsterdam: Kluwer Academic Publishers, 2003; Geoghegan, Jacqueline, Pritchard, Lowell Jr., Ogneva-Himmelberger, Yelena, Chowdhury, Rinku Roy, Sanderson, Steven, and B. L. Turner II, " 'Socializing the Pixel' and 'Pixelizing the Social' in Land-Use and Land-Cover Change," in *People and Pixels: Linking Remote Sensing and Social Science*, eds., Diana Liverman, Emilio Moran, Ronald Rindfuss, and Paul Stern, Washington, DC: National Academy of Science Press, 1998, pp. 51–69; Nelson, Gerald C., and Jacqueline Geoghegan, "Modeling Deforestation and Land Use Change: Sparse Data Environments," *Agricultural Economics* 27 (2002):201–16.

JACQUELINE M. GEOGHEGAN

Population Dynamics. Demographic factors—including population size, growth, and density; fertility, mortality, and migration; and the age and sex composition of households—are known to be important factors influencing land use and cover change. Some early papers on **deforestation** and **desertification** focused on population factors as **driving forces** of land-cover change to the near exclusion of other variables. This may have been due in part to a proclivity among some researchers to identify population change as the single most important driver of all **environmental changes**. But it is likely that there were also practical reasons, such as the relative abundance of population data from censuses and surveys when compared to data on other drivers such as markets, government policies, agricultural practices, technology, roads, and settlements. Whatever the reasons, contemporary land-use/cover change research generally seeks a more balanced view of population relative to other drivers of land use and cover change.

In a **meta-analysis** of deforestation case studies, Helmut Geist and Éric Lambin found little support for the conclusion that population or other single factor causes such as shifting cultivation could explain the majority of tropical deforestation. Instead, they found that deforestation was driven by identifiable regional patterns of causal factor synergies, of which the most prominent were economic factors, **institutions**, and national policies. These, in turn, drove a range of **proximate causes**, including agricultural expansion, wood extraction, and infrastructure extension. Demographic factors such as natural increase or in-migration were explicitly mentioned in 61 percent of all case studies examined. Most of the explanatory power of population variables tends to be derived from interlinkages with other underlying forces. Many cases did not specify beyond broad notions of population pressure and growth, but those that did tended to identify in-migration more frequently than natural increase. Migration, especially in forest frontier areas, can lead to much more rapid increases in population than would be possible through natural increase (births minus deaths) alone.

Land degradation and desertification have also been attributed to population growth, especially in marginal environments that are characterized by constraints such as steep hillsides, poor soils, and drought-prone climates. Some have posited that increasing population densities in rural areas inevitably lead to land degradation and declining crop yields. Although there is evidence in parts of the world to support the notion of a downward spiral of population, **poverty**, and the environment, population size or density is rarely the sole contributing factor. Rather, as with deforestation, population variables tend to be part of a matrix of factors that include, in the case of land degradation, failed institutions, climate conditions, inherently poor soils, and lack of incentives for proper soil management (e.g., tenure insecurity and low market prices). Population displacements into marginal lands due to wars and civil conflicts, as has happened frequently in Africa and Central America, can also contribute significantly to land degradation.

There is much to be learned from regions where the downward spiral of **soil degradation** and poverty has not inevitably resulted from increasing population density. Case studies from the Machakos district in Kenya and in Chivi communal area in Zimbabwe suggest that improved soil productivity and more secure livelihoods can and do occur in a context of increased population density, and even benefit from it. The factors that can be important for averting land degradation include appropriate technologies for soil management adopted and propagated by local farmers (rather than imposed from outside); **access** to markets, road infrastructure, and development of local market towns for food processing; cash-cropping as opposed to purely **subsistence agriculture**; and development of local management capacity and skills through education and agricultural extension. Rather than blanket solutions, Ian Scoones urges more fine-tuned and people-centered development interventions in which the historical context and local specificity of needs are acknowledged. Such an approach recognizes local environmental knowledge, and sees the agricultural researcher as a facilitator rather than an expert prescribing solutions.

The **scale** of analysis is very important in terms of understanding the influence that population has on **land-cover changes**. If a researcher chooses to examine land-cover changes at the scale of the municipality simply because that is the level at which population data are reported, there is a potential to fall into an ecological fallacy. As an example of an ecological fallacy, say that population growth rates were found to be highly correlated with deforestation rates at the county level in a particular forest region. The researcher might conclude that population growth was a significant driver of deforestation. Yet, it may be that the population increased dramatically in urban areas contained within those counties, and that in fact the rural population, located where most of the deforestation occurred, remained relatively constant. This would be a form of spurious correlation; the real "culprit" might be something quite different, such as government policy or price mechanisms. Increasingly, spatially disaggregated population maps with the grid cell as the unit of analysis are being used to analyze land-cover changes. Examples include the Gridded Population of the World of the **Center for International Earth Science Information Network** and Oak Ridge National Laboratory's LandScan. These need to be interpreted with caution, since the underlying data still come from census administrative units.

Population size, density, or growth rate are not the only demographic factors of relevance to land-use and land-cover change. For example, household size and composition in **agricultural frontier** areas have been found to significantly affect land-use patterns, in particular the choice of more labor-intensive annual crops versus less-labor-intensive tree crops or livestock raising. Size and composition, in turn, are often related to the length of tenure. Security of **land tenure** has been found to affect fertility **decision-making** at the household level, with more secure tenure, *ceteris paribus*, resulting in lower fertility rates. This corroborates the theory that children are seen by rural households as a form of risk insurance and social security in highly uncertain environments.

Government policies can directly influence population dynamics in ways that lead to land-cover changes. Both Indonesia and Brazil have practiced various forms of resettlement to forest frontier areas. Some governments have deliberately encouraged migration of the rural landless as an alternative to less politically palatable **land reforms**.

See also: **Agricultural Intensification; Agrodiversity; Biodiversity; Boserup, Ester; Cash Crops; Center for the Study of Institutions, Population and Environmental Change;**

Consumption; Degradation Narrative; Economic Livelihood; Indigenous Knowledge; Malaria; Participatory Land Management; People at Risk; Population-Environment Research Network; South America; Urbanization.

Further Reading

Geist, Helmut J., and Éric F. Lambin, "Proximate Causes and Underlying Driving Forces of Tropical Deforestation," *BioScience* 52 (2002): 143–50; Scoones, Ian, "The Dynamics of Soil Fertility Change: Historical Perspectives on Environmental Transformation from Zimbabwe," *The Geographical Journal* 163 (1997): 161–9; Tiffen, Mary, Mortimore, Michael, and Francis Gichuki, *Population Growth and Environmental Recovery: Policy Lessons from Kenya* (IIED Gatekeeper Series no. 45), London: International Institute for Environment and Development, 1994.

ALEXANDER DE SHERBININ

Population-Environment Research Network. A project of the International Union for the Scientific Study of Population and the **International Human Dimensions Programme on Global Environmental Change** that seeks to advance academic research on **population dynamics** and the environment by promoting online scientific exchange among researchers from social and natural science disciplines worldwide. The Population-Environment Research Network (PERN) receives technical support from the Socioeconomic Data and Applications Center at the **Center for International Earth Science Information Network** of Columbia University, funded by the U.S. National Aeronautics and Space Administration.

Further Reading

The Population-Environment Research Network [Online, December 2004], The PERN Web Site, www.populationenvironmentresearch.org.

HELMUT GEIST

Poverty. Income below which a minimum nutritionally adequate diet plus essential non-food requirements are no longer affordable. Poverty is seen as both a potential cause and effect of land-use and **land-cover change**, although in each instance these assertions are contentious. Poverty, it is argued, can lead individuals, households, communities, and even states to make short-sighted land-management decisions. Degrading land resources, in turn, has an impact on the human livelihoods that depend on them, and these livelihoods are assumed to be those of the poor in many instances.

Defining Poverty

Poverty is a term that is open to multiple interpretations. It can be located at a number of spatial-social **scales**, from the individual, to the household, community, ethnic group, nation, and region. Poverty may be conceived of in relative terms (e.g., comparing households within a community, regions within a state, or states with one another). It may also be measured against a specific benchmark, such as an international poverty line. Finally, poverty may be assessed in a variety of ways, for example, in terms of monetary wealth or income, certain types of assets, or entitlements. The way in which poverty is conceptualized has an influence on who is defined as poor and how their interactions with the landscape are perceived.

Poverty as a Cause of Land Change

At the individual and household scale, it is argued that the poor are more likely to engage in environmentally deleterious behavior for at least two reasons. First, it is suggested that the poor (defined as households lacking resources) fail to maintain the natural resource base because they lack the means to invest in environmental **conservation** and remediation. As such, it is often argued that economic growth and environmental stewardship go hand in hand, as **economic growth** provides the resources necessary for sustainable environmental management. Second, it is asserted that the poor are more preoccupied with their survival in the present and therefore fail to appreciate higher-order amenities (such as environmental goods) and make natural resource management decisions that are short-sighted.

At the national scale, the theoretical framework of the environmental Kuznets curve has been used to suggest that there is a relationship between per capita income and a number of environmental measures. The model posits that countries in the early stages of development (and low per capita income) have a nominal impact on the environment. In the intermediate stages of development and income, it is suggested that countries often experience high rates of environmental degradation associated with rapid **industrialization**. Finally, the most developed and highest-income countries are supposed to have low levels of natural resource degradation given increasing **investments** in, and concern for, environmental amenities.

A number of studies have been undertaken (in all of the world's major regions) that at least complicate, and often contest, the narrative of poverty-induced environmental degradation. First, sustainable land management is not necessarily a capital-intensive process requiring monetary investment. In developing countries, many cultural ecologists have argued that traditional modes of land management are not necessarily inefficient or destructive. Furthermore, while industrial agricultural practices may lead to the higher production of certain crops, they often do so at an economic and environmental cost. In other words, the process of wealth creation may result in as much environmental destruction as it potentially remediates.

Second, some evidence regarding the famine behavior of extremely poor households in Africa suggests that they are often more concerned about the longer-term viability of their livelihoods than they are about shorter-term consumption (contradicting the notion that the poor are short-sighted). Furthermore, the concern for environmental amenities that is supposed to accompany wealth presumes that households and communities actually experience the environmental impacts of their production decisions. This increasingly is not the case in the most developed countries as, in an ever more globalized economy, sites of production and **consumption** are separated in space. There are burgeoning environmental justice and political ecology literatures describing how socially marginalized communities often bear the environmental costs of producing crops and goods for wealthier consumers in the other areas of the country or world. Wealthier households have little incentive to deal with the environmental costs of their consumption if they are unaware or spatially removed from such degradation.

Third, the trend hypothesized by the environmental Kuznets curve (and supported by a number of cross-national studies) often has as much to do with the export of dirty industries as it does with tighter environmental regulations and cleaner production. As such, analysis undertaken on a country-by-country basis may support the trend posited by the environmental Kuznets curve, whereas analysis undertaken at other scales may yield quite different results. This seeming contradiction, and the effects of varying analytical scales, is sometimes referred to as the modifiable areal unit problem.

It is not that poverty does not contribute to land-use and land-cover change in some instances (it clearly does), but that important definitional and scale issues, as well as approaches to technology and environmental accounting, influence characterizations of the poverty-land interface.

Land-Cover Change as a Cause of Poverty

Land-cover change has implications for land-dependent livelihoods. Declining productive capacity of the land may impoverish land-dependent livelihoods, just as increasing productive capacity may lead to enrichment. A variety of livelihood activities may be influenced by land-cover change. For example, decreased **forest** cover may lead to fewer opportunities for wild food collection, hunting, and some forest-dependent artisanal activities. Declining surface **biomass** may also alter the local hydrologic cycle, with implications for fisheries, flood plain **rice**, and groundwater availability (if there is a drop in the water table). Changing soil conditions have obvious implications for crop farming, but may also influence off-season grazing opportunities.

There is some debate regarding the degree to which poor (as opposed to rich) households, communities, and states are more susceptible to the adverse consequences of land-cover change. This debate about **vulnerability** revolves around at least two points. First is the degree to which the poor are more implicated in land-based livelihoods. Obviously, one's definition of poverty (and the social-spatial scale of this conception) influences such characterizations. In many rural communities, for example, the relatively wealthy actually control a bigger piece of the resource pie than the poor, which means that they may have more to lose from declines in surface biomass. In contrast, it may more convincingly be argued that the world's poorest countries frequently have more natural resource-based economies, and thus stand more to lose from the declining productive potential of the land. The second way in which the poor may be more sensitive to land-cover change concerns the ability of such households, communities, and states to cope with adverse **environmental change**. Here, the concern is that the poor have fewer reserves to fall back on in times of crisis.

Dynamic Adaptation

Very often, there is a circular process of dynamic adaptation. In many instances, this process has been depicted as a vicious cycle wherein poor households overtax the environment in order to survive, and this degraded environment further impoverishes the household. It is also quite possible, and well documented in several instances, that poor households will diversify into other activities or simply leave the landscape. In either case, the human burden on a particular landscape often declines.

See also: **Decision-Making; Degradation Narrative; Desertification; Driving Forces; Economic Livelihood; Human Immunodeficiency Virus (HIV)/Acquired Immunodeficiency Syndrome (AIDS); Land Degradation; Landholding; Mediating Factor; People at Risk; Population Dynamics; Resilience; Tradeoffs; Tragedy of Enclosure.**

Further Reading

Broad, Robin, "The Poor and the Environment: Friend or Foes?" *World Development* 22 (1994): 811–2; Gray, Leslie C., and William G. Moseley, "A Geographical Perspective on Poverty-Environment Interactions," *Geographical Journal* 171 (2005): 29–43; Moseley, William G., "African Evidence on the Relation of Poverty, Time Preference and the Environment," *Ecological Economics* 38 (2001): 317–26; Reardon, Thomas, and Scott A. Vosti, "Links Between Rural Poverty and the Environment in Developing Countries: Asset Categories and Investment Poverty," *World Development* 23 (1995): 1495–1506; Scherr, Sara, "A Downward Spiral? Research Evidence on the Relationship between Poverty and Natural Resource Degradation," *Food Policy* 25 (2000): 479–98; Swinton, Scott M., Escobar, German, and Thomas Reardon, "Poverty and Environment in Latin America: Concepts, Evidence and Policy Implications," *World Development* 31 (2003): 1865–72; Watts, Michael, "Poverty Gap," in *The Dictionary of Human Geography*, 4th ed., (reprinted), eds. Ron J. Johnston, Derek Gregory, Geraldine Pratt, and Michael Watts, Oxford, UK, Malden, MA: Blackwell Publishers Ltd., 2001, p. 627.

WILLIAM G. MOSELEY

Pristine Myth. The pervasive assumption that society's **modification** of nature is a relatively recent phenomenon—that remote or uninhabited lands are untouched or pristine. Humans have always altered the environments in which they live. Most areas on earth, even the hearts of **Amazonia** and the Australian outback, show a human imprint. Research that debunks the pristine myth demonstrates the history of human agency in modifying the land and contributes to basic understanding of nature-society relationships: it highlights the potential for nature to recover from anthropogenic disturbance, it shows that both ecosystems and human cultures exhibit dynamic change rather than remaining static, and it suggests that the path to more **sustainable land use** may require embracing humanized landscapes rather than the twentieth-century goal of "wilderness" preservation.

The pristine myth is often linked with what has been called *la leyenda verde* (the green legend) or the myth of the "ecological Indian," which asserts that indigenous groups embrace an environmental ethic that contrasts with their purportedly more environmentally destructive European colonizers. Commonly associated with both the Americas and Australia, scholars and lay people alike—perhaps due to a "nostalgic longing for the past and a simpler life"—often celebrate the idea that indigenous peoples are "in and of nature," living in balance with it, as noted by Shepard Krech. Complementing this view is the notion that the Americas, for example, were sparsely populated, and therefore the capacity for Amerindians to modify the land was low.

Instead of living in harmony with nature and treating it in a benign way, however, a wide range of evidence from a **paleo perspective** shows that indigenous groups in regions throughout the world modified the environment in profound ways. The degree of impact reflects the size and duration of occupation and use of an area. First, in the case of the Americas, there were far more people living in the New World than is widely known, with reasonable estimates lying somewhere between 43 and 65 million. Second, terraces and raised fields, the creation of anthropogenic soils (e.g., *terra preta* in the Amazon basin), canals, dams, and other components of irrigation farming, and the use of **fire** to produce and maintain productive landscapes, are among the many examples of sophisticated and intensive production systems found throughout the Americas before the Columbian encounter. Finally, these production systems are linked to many of the same kinds of environmental problems witnessed today: large-scale **deforestation**; **soil erosion** and nutrient loss; **wetland** sedimentation; weed and pest invasion; and water stress.

Debunking the pristine myth is more than a historical exercise. That the landscapes in precolonial regions were socially constructed holds implications for how humans relate to nature today. Debate over whether pristine nature exists reflects conflict between those who argue that humans are a part of wild nature, and those who embrace a wilderness-civilization dichotomy. Those who celebrate humanized landscapes see lessons for how to foster sustainable development.

First, the perpetuation of the pristine myth coincides with the idea of *terra nullius* (land belonging to no one), the two ideas mutually reinforcing each other and justifying the **colonization** of thinly settled lands in the Americas and Australia. In Mexico, for example, throughout colonial history and continuing into the twentieth century, the two concepts have been used to justify the colonization and control of frontier locations. Despite the presence of indigenous populations, Porfirio Díaz (the president of Mexico from 1876–1880 and 1884–1911) declared the bulk of the country's tropical forests to be *terrenos baldíos* (unoccupied or vacant lands), which he quickly subdivided and awarded to colonists or international logging companies. Despite evidence of past human impact as well as the ongoing occupation and use of forests, for over 100 years nature **conservation** has focused primarily on using reserves to exclude people and to protect pockets of "wilderness." Uncovering flaws in the assumptions of *terra nullius* and pristine nature exposes the injustice of taking away lands from indigenous groups in the past and present.

Second, research on the pristine myth not only identifies the agency of humans to alter the earth, but also the **resilience** of nature and its capacity to recover from significant anthropogenic shocks. Caution is needed, therefore, in referring to "environmental destruction." The most widely cited examples of environmental recovery are from the New World. The first is in the Maya lowlands of Mesoamerica, which were almost completely deforested during the Classic Maya period, but returned to mature forest after the civilization's collapse around AD 800–900. Second, with a 90–95 percent depopulation of the Americas after Europeans and their Old World diseases arrived, environmental conditions "improved" (e.g., forest cover expanded) as cultural features (e.g., raised fields and terraces) were abandoned. While the scale and circumstances of the depopulation in both cases are unique, there are multiple examples in more recent periods of environmental recovery on smaller scales, and without associated extreme population shifts.

Third, by showing how societies worldwide have modified and adapted to nature, detailed regional environmental histories not only reflect the dynamism of ecosystems, but also the evolution of cultures. That neither is static draws into question attempts to preserve, fixed in time, a snapshot of a particular historical period. By showing that aboriginal groups treated nature in far from a benign way, present-day environmentalists are forced to question calls for a return to a "simpler" or more traditional way of interacting with nature. While the pace and the scale of human impact over the last 200 years is unprecedented, cultures throughout history have exploited nature for their own self-interest.

Finally, with the vast majority of nature worldwide either inhabited or being actively used by people, research that questions the pristine myth implicitly endorses the notion that society today needs to embrace humanized landscapes. The modern meaning of wilderness, which encompasses sublime landscapes where nature rules and humans are absent, may be obstructing rather than advancing sustainable development. If humanized landscapes are the norm, then a solution to environmental problems that creates barriers between people and nature is problematic. The solution to environmental problems is not to put a box around select pockets of "wilderness" while continuing with business as usual elsewhere. In both the **United States of America** and Chile, for example, the bulk of wilderness areas under protection are remote and with low population densities, not in those landscapes that are most vulnerable to human modification. In short, the focus on wilderness preservation may distract from the need to change nature-society relationships in society writ large. There is evidence, however, that the influence of the pristine myth may be declining in the Americas. In the 1980s and 1990s, there was a dramatic expansion across **Latin America** not only in protected areas but also in a celebration of humanized landscapes as part of nature conservation.

But not all are so quick to celebrate the social construction of nature. In stark contrast to the sustainable development movement, many environmentalists see the embrace of humanized landscapes as justifying continued human impact on the land. Many point out that while landscapes such as the forests of the Maya lowlands may have recovered from past use, the world today is more populous, more interconnected, and has a much greater capacity to alter the land through ever more efficient technologies; therefore, it is unlikely that the kinds of shifts in human-environment conditions that took place in the past will occur in the foreseeable future.

Other critics argue that the debate over the character of the pre-Columbian American landscape has often focused on the polar assertions that the continent was either a pristine landscape or a humanized one. This polarization detracts from a reasoned assessment of the impact of the First Americans on the land. As such, it hampers efforts to establish ecological restoration and conservation plans.

In the case of North America, Thomas Vale argues that the pre-Columbian landscape was neither entirely pristine nor intensely humanized. Rather, the landscape was a mosaic of pristine wilderness and differentially humanized patches. There were intensely humanized farmed areas along the East Coast; more moderately modified plains; the West Coast was altered to varying degrees, primarily by anthropogenic fire; and there were large tracts of wilderness (or nearly so) in the Southwest and at high elevations.

Thomas Vale identifies intensity, spatial, and temporal ambiguity in the pristine myth debate. First, the intensity of human impact in a settlement and its associated agricultural fields was profound, while a forest where gathering and hunting occurred was much less so. Second, human settlements in the western United States were highly uneven in their distribution across the landscape, often concentrated along streams. Third, some human impacts, such as hunting, may have had a relatively short-lasting impact on the land when compared with **agriculture**. Thomas Vale argues, therefore, that the American landscape was a mosaic, ranging from intensely humanized to largely pristine wilderness.

See also: **Agricultural Frontier; Degradation Narrative; Driving Forces; Great Plains; Land-Cover Change; Land-Use History; Proximate Causes.**

Further Reading

Cronon, William, ed., *Uncommon Ground: Toward Reinventing Nature*, New York: W. W. Norton & Co., 1995; Denevan, William M., "The Pristine Myth: The Landscapes of the Americans in 1492," *Annals of the Association of American Geographers* 82 (1992): 369–85; Head, Lesley, *Second Nature: The History and Implications of Australia as Aboriginal Landscape*, Syracuse, NY: Syracuse University Press, 2000; Oelschlaeger, Max, *The Idea of Wilderness: From Prehistory to the Age of Ecology*, New Haven, CT: Yale University Press, 1991; Sluyter, Andrew, *Colonialism and Landscape: Postcolonial Theory and Applications*, Landham, MD: Rowman & Littlefield Publishers, 2002; Vale, Thomas, ed., *Fire, Native Peoples and the Natural Landscape*, Washington, DC: Island Press, 2002; Whitmore, Thomas M., and B. L. Turner II, "Landscapes of Cultivation in Mesoamerica on the Eve of Conquest," *Annals of the Association of American Geographers* 82 (1992): 402–25.

PETER KLEPEIS, PAUL LARIS

Private Property. Refers to a form of **property** in which a single actor enjoys extensive use and control rights on a resource (ownership). All rights associated with a resource are bundled in the hands of an actor. Private property has received much attention in theory and policy on land use, motivating **land privatization** programs and **land reforms** in many countries. The prominence of the concept derives from its constitutive role in political liberalism and capitalist economies at the level of theory and practice. Private property is thought to promote political participation and efficient resource use. Private property on land is far from common, however. **Land rights** around the world remain subject to extensive control by states and other collective entities. For example, land rights pertaining to **common-pool resources** often take the form of common property regimes.

Private property and land use mutually influence each other. Historically, the intensification of **agriculture** has been associated with the emergence of private property. For example, agricultural intensification in eighteenth-century

Private property owners who have found a little coal on their land in roadside West Virginia, mining and selling it along the highway in September 1938.
PHOTO: Marion Post Wolcott. [Courtesy Library of Congress]

Europe went along with the enclosure of common lands. In the contemporary world, programs of land privatization and land registration are major drivers of land-use change. There is empirical evidence from Africa and Thailand that land registration has contributed to **agricultural intensification** and **investments** in **soil conservation** in some regions. Forestland allocation has been connected with the rapid expansion of smallholder tree plantations in Vietnam, re-greening previously barren hills. At the same time, policy promoting private property has also caused **land degradation**. Privatization has been linked with overgrazing and range degradation in **China** and Central Asia. Land privatization in central and eastern Europe may have allowed agricultural intensification in some regions, but it has also contributed to environmental degradation.

Land use influences the nature of property. **Ester Boserup** has hypothesized that changes in agricultural practices driven by rising population densities lead to the emergence of private property in land. As land becomes scarce, individual land users expand the bundles of rights to land, eventually achieving private property. At a more empirical level, research has shown that settlers in frontier regions often gain private property in land by being the first to clear a particular parcel of land. In Africa, tree planting strengthens private claims on land, even though they may not amount to complete private property. In Vietnam, terracing and other investments in land strengthen individual land rights, which may be relatively restricted otherwise.

See also: **Cadastre; Institutions; Land-Use Policies; Public Policies.**

Further Reading

Boserup, Ester, *The Conditions of Agricultural Growth*, New York: Aldine, 1965; Macpherson, C. B., ed., *Property: Mainstream and Critical Positions*, Toronto: University of Toronto Press, 1978.

THOMAS SIKOR, JOHANNES STAHL

Probit/Logit Model. Probit and logit models are members of a wider class of non-linear probability models that associate a set of explanatory variables with the probability of observing a dependent variable that is discrete. It is common to model land-cover types, for example, forest and non-forest (cleared land), as possible outcomes that are distinct from one another.

A linear probability model is used when the dependent variable takes on a continuous value as a function of the independent variables and their associated parameters. For example, one could predict the level of agricultural output in tons as a function of land, labor, and capital inputs. A linear model is less useful at predicting discrete breaks in the data. For example, **forest** may be cleared for agricultural uses when its opportunity cost exceeds the benefit in keeping the land in forest. One could use a linear dependent variable formulation of the probability of land **conversion** (i.e., anything beyond 0.50 could be cleared; below 0.50 left in forest), but there is no guarantee that predicted values will be well-behaved; that is, fall in the probability range from 0 to 1.

Thus, non-linear transformations to represent probability of a particular event or outcome are the norm. Logit models are derived from a logistic function, or the probability of a particular outcome relative to all other alternatives. Probit models are derived from the normal distribution, with the outcome taking a particular value (i.e., 1 = forest; 0 = cleared) depending on some **threshold** determined by the linear combination of independent variables and their associated parameters. The simplest formulations of both models represent a binary framework (1,0), but multinomial expansions exist for both models; that is, multiple categories of a single dependent variable.

One challenge for land-use/cover change research is the extreme likelihood of spatial and/or temporal non-stationarity; land-change processes are often spatially or temporally dependent. These problems are of course compounded by other violations of ideal conditions (heteroskedasticity, non-ormality, etc.). However, corrections for both temporal and/or spatial dependence to nonlinear probability models are cumbersome. Logistic models by nature do not have an error structure; that is, explicit unmodeled or autonomous variation. Thus, it is generally impossible to correct for spatial or temporal **auto-correlation**. Probit models, because they are derived from the normal distribution, have a defined error structure, but correspondingly, they are much more sensitive to misspecification, which could lead to bias and inefficiency of the estimated parameters.

See also: **Change Detection; Continuous Data; Discrete Data; Modeling; Hot Spot Identification.**

Further Reading

Fleming, Mark M., "Techniques for Estimating Spatially Dependent Discrete Choice Models," in *Advances in Spatial Econometrics*, eds. Luc Anselin, Raymond Florax, and Serge Rey, Heidelberg, GE: Springer 2004, pp. 145–68; Long, J. Scott, *Regression Models for Categorical and Limited Dependent Variables* (Advanced Quantitative Techniques in the Social Sciences, vol. 7), Thousand Oaks, CA: Sage Publications, 1997; Mertens, Benoît, and Éric F. Lambin, "Land-Cover Change Trajectories in Southern Cameroon," *Annals of the Association of American Geographers* 90 (2000): 467–94.

DARLA K. MUNROE

Project on People, Land Management and Ecosystem Conservation. A collaborative effort among scientists from across the developing world to develop sustainable and participatory approaches to **conservation**, especially of biodiversity, within small farmers' agricultural systems and in participation with the farmers. The Project on People, Land Management and Ecosystem Conservation (PLEC) has been developed since 1993 by the United Nations University (UNU), and received funding from 1998 to 2002 from the Global Environmental Facility (GEF), managed by the World Bank. The project is executed by UNU through a network of five locally based clusters that have been established in West Africa (Ghana, Guinea), East Africa (Kenya, Tanzania, Uganda), the Asia-Pacific (**China**, Thailand, Papua New Guinea), and America (Brazil, Peru, Mexico, Jamaica). About 80 percent of all participants are from these countries. Scientists from Australia, **Japan**, the **United States of America**, and the United Kingdom are also involved, with **Harold Brookfield** being a driving force behind the network. PLEC uniquely provides both for South-to-South cooperation and South-to-North twinning arrangements. The specific objectives are to establish historical and baseline comparative information on **agrodiversity** and **biodiversity** at the landscape level; to develop participatory and sustainable models of biodiversity management based on farmers' technologies and knowledge within agricultural systems at the community and small-area levels; to recommend approaches and policies for sustainable agrodiversity management to key government decision-makers, farmers, and field practitioners; and to establish national and regional networks for capacity strengthening within participating institutions, and to carry forward the aims of PLEC. The core of PLEC's work is in its demonstration site villages. Here, PLEC becomes the farmers' own enterprise, and scientists are the facilitators, not the instructors. The scientists identify and demonstrate farmers' practices that are environmentally, socially, and financially sustainable, and which sustain biodiversity. They help farmers in achieving their own conservationist goals. Collaborating farmers manage varied biophysical conditions, growing a range of crops and using biodiversity with discretion. The PLEC approach differs from mainstream agricultural research at experiment stations under controlled conditions. By integrating locally developed knowledge of soils, climate, and other physical factors with scientific assessments of their quality in relation to crop production, a set of sustainable agricultural technologies can be devised so that agricultural diversity is maintained. The participatory process will eventually enhance farmers' and local communities' ability to adapt to **environmental change** as well as to social and economic change. Since 2002, PLEC has focused on mainstreaming lessons it has learned into national and international policies and training **institutions**.

See also: **Alternatives to Slash-and-Burn (ASB) Programme; Community Involvement; Consultative Group on International Agricultural Research; Indigenous Knowledge; Land-Use Policies; Participatory Land Management; World Agroforestry Centre.**

Further Reading

Project on People, Land Management and Ecosystem Conservation [Online, November 2004], The PLEC Web Site, www.unu.edu/env/plec.

HELMUT GEIST

Property. Refers to social relationships between persons with respect to goods of material or symbolic value. Analysis of property involves attention to various kinds of social actors recognized to take part in property relationships; property objects, being material and cultural goods considered as valuable; and types of relationships, often expressed in terms of rights and obligations, including their temporal dimensions. For example, an analysis of property on a particular tree examines the products associated with the tree (leaves, fruits, timber, etc.), the actors making claims on the tree (male loggers, female gatherers, herders, etc.), and the types of rights and duties making up the relationships among actors with regard to the tree products (use for **consumption**, collection for sale, etc.). Property is a fundamental concept for the study of **land-use systems** because property affects land-use practices, and because land-use practices influence property relationships. **Private property**, in particular, has been at the core of a longstanding debate about the relationship between **land rights** and **land change as a forcing function in global environmental change**. But property also has other functions beyond production, such as the protection of social stability and political participation.

The rights constituting property relationships are commonly divided into use rights and control rights. Use rights include rights of cultivation, passage, grazing, and collection, among others. Control rights refer to the rights of excluding other claims, transfer the object, make major alterations to it, and sell the object. Edella Schlager and Elinor Ostrom suggest a simple classification scheme into five types of natural resource rights. The right of **access** entitles its holder to walk on to a resource system. The right of withdrawal allows its holder to enjoy resource products. The right of management includes the right to regulate use patterns within a local group of resource users and the right to transform the resource by making improvements and investments. The right of exclusion is about determining who will have an access right, and how that right may be transferred. The right of alienation refers to selling or leasing of the rights of management and exclusion. One can speak of ownership, or private property, if one actor holds all five types of rights concurrently. Ownership is far from universal, as many resource users do not have the rights of alienation. Their "bundles of rights" are limited to the rights of access, use, management, and, perhaps, exclusion.

The distinction between use and control rights informs a customarily used classification of property regimes into four broad types. One speaks of private property when a single actor enjoys extensive use and control rights on a resource. Under state property, it is the state that exercises significant use and control rights, although it may grant limited use rights to other actors. Common property refers to a situation in which some collective entity, such as a village community, holds control rights and grants use rights to its members. Open access, in turn, occurs when many actors assert use rights in the absence of control rights. This classification is a highly simplified representation of actual property relations, however. Property relations are in reality much more varied, being characterized by complex combinations of use and control rights among multiple actors.

Property exists simultaneously at various layers of social organization. Notions of property are fundamental to cultural norms and social values. Property is a key object in laws and regulations enacted by states, religious entities, international treaties, and other forms of social organization. Property exists at the level of actual social relationships, being embedded in wider social relationships, such as kinship, religious, or political affiliations. Finally, property relationships become manifest in social practices, that is, in the interactions of various types of actors with concrete property objects and rights. By way of concrete practices, the actors reproduce and modify actual social relationships, laws and regulations, and cultural norms in material and symbolic struggles. Property relations at the different layers are interconnected, although property at one layer is analytically distinct from property at another layer. In addition, common are discrepancies between the forms property assumes at the different layers. The effects of legal property reforms on land use, therefore, depend on the degree to which they are translated into actual social relationships and practices.

The relations between property and land use are mediated by broader access relations. The concept of access, in its sociological interpretation, describes the ability of actors to derive benefits from a resource. It is broader than the property concept, as property focuses on rights-based access. In contrast, access to a resource includes attention to other structural and relational mechanisms influencing the benefits derived from rights. These include access to technology, capital, markets, labor, and knowledge. For example, many actors beyond wood collectors themselves enjoy access to benefits derived from charcoal. They include transporters moving charcoal to the city and retailers selling it to urban customers. The relations between property and land use, therefore, are mediated by numerous other factors, as research on the effects of land rights and registration has demonstrated.

See also: **Cadastre; Common-Pool Resources; Institutions; Mediating Factor; Transhumance.**

Further Reading

Benda-Beckmann, Franz von, and Keebet von Benda-Beckmann, "A Functional Analysis of Property Rights, with Special Reference to Indonesia," in *Property Rights and Economic Development: Land and Natural Resources in Southeast Asia and Oceania*, eds., Toon van Meijl and Franz von Benda-Beckmann, London, New York: Kegan Paul

International, 1999, pp. 15–65; Ribot, Jesse C., and Nancy L. Peluso, "A Theory of Access," *Rural Sociology* 68 (2003): 153–81; Schlager, Edella, and Elinor Ostrom, "Property-Rights Regimes and Natural Resources: A Conceptual Analysis," *Land Economics* 68 (1992): 249–62.

THOMAS SIKOR, JOHANNES STAHL

Proximate Causes. Proximate causes are the direct or immediate human activities that change land use or land cover. While underlying **driving forces** provide the impulse and context for change, proximate causes relate to the mechanisms of change or the immediate reasons for it. Distinguishing features include closeness (i.e., proximity) to the site of change, and the evident role of human agents in making and implementing decisions to carry out change. These human agents are the channels or media through which the underlying driving forces operate to effect change at particular places and times. They "reflect" the underlying driving forces, according to William Meyer and B. L. Turner II, or give expression to them.

The conventional separation into proximate causes and underlying driving forces implies a hierarchical structuring of the factors determining change in land use and land cover (see figure below).

While this model can be helpful in providing a framework for analysis, it is, like all models, a simplified version of reality. It provides only limited space for factors such as the development of road networks. "Infrastructural development" is shown as a "proximate cause" in the figure, but arguably it has a status and role that are different from those of agricultural expansion or wood extraction. Well-known examples (e.g., from Rondônia in Brazil) have demonstrated an association between the extension of the road network

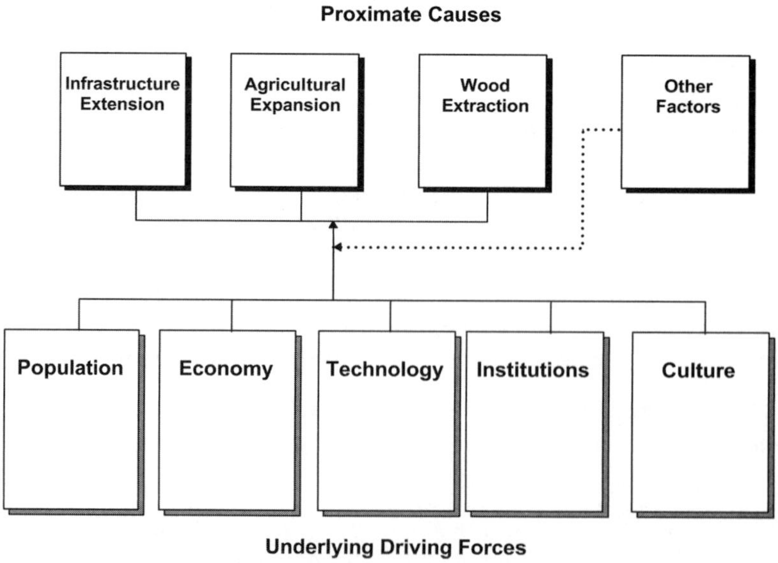

Underlying driving forces and proximate causes.
SOURCE: Simplified after Geist and Lambin (2002)

and forest clearing for pioneer agriculture. In this instance, economic development is manifested in the growth of a road network that permits penetration of the **forest** and hence forest clearance for **agriculture**. In practice, the influences operating on land use and land cover form a complex web: separation into a two-tier structure is somewhat arbitrary.

The framework of underlying driving forces and proximate causes has been used in research on tropical **deforestation** since the early 1990s, but certain parallels can be recognized with "political economy" and behavioral approaches employed, for example, in work on agricultural change in developed countries from the 1970s.

The explicit notion of "proximate causes" is associated most closely with tropical deforestation. Proximate causes in terms of agricultural expansion and wood extraction are frequently cited, but both of these have many variant forms and they are not necessarily mutually exclusive. Down through history, the former has been the main cause of deforestation. Wood extraction on its own would not necessarily lead to the permanent removal of the forest, unless activities such as livestock grazing or **environmental change** such as soil deterioration also occurred.

Agricultural Expansion

Underlying driving forces of population growth and economic development can lead to increasing demand for food and hence expansion of the agricultural area. Usually such expansion is at the expense of the forest. In short, a plausible link can be perceived between population growth and deforestation. The relationship, however, is not deterministic: an alternative response to increasing demand for food could be **agricultural intensification** in existing areas. Also agricultural expansion can take various forms, by no means all of which are closely related to a simple growth in demand for food.

Helmut Geist and Éric Lambin reviewed 152 case studies of deforestation as reported in ninety-five journal papers, and analyzed them in terms of conclusions about causation. Overall, agricultural expansion was found to be by far the most frequent proximate cause of deforestation: they found that it applied in 96 percent of cases. Different authors, using different approaches, have reached similar conclusions about the primacy of agricultural expansion.

Shifting cultivation is a form of agriculture that has often been blamed for, or implicated in, deforestation. The term, however, is variously defined and interpreted, and confusion has arisen in particular between "traditional" shifting cultivation, usually practiced by indigenous peoples as **swidden cultivation** or "swidden fallow" farming, and "colonist" shifting cultivation or **slash-and-burn agriculture**. The former, if practiced in long cycles, does not necessarily result in deforestation, and indeed in some parts of the world people and forests have co-existed in this way for millennia. A mosaic of cleared patches, successional stages, and mature forest can exist in perpetuity. If the cycles shorten, through population growth or some loss of land available to the shifting cultivators, then the mosaic is truncated and full recovery to mature forest cannot be maintained. In other words, traditional shifting cultivation is likely to be a proximate cause of deforestation only if it is de-stabilized through population growth or curtailment of **access**. Geist and

Lambin identify traditional shifting cultivation as a proximate cause in 30 percent of the cases they have reviewed. The **United Nations Food and Agriculture Organization**'s (FAO) Global Forest Resources Assessment 2000 suggests that the contribution of shifting cultivation in terms of area deforested is much less.

Permanent agriculture is a more frequently reported proximate cause of tropical deforestation. Again, however, different forms exist. Divisions are made into large-scale and small-scale farming (or commercial farming and **subsistence agriculture**) and also into pastoral farming and cropping. Clear continental contrasts seem to exist in terms of relative importance of large- and small-scale agriculture: in Africa, "small scale" is far more important than "large scale," while the reverse is true in **Latin America**. There are, however, other possible divisions, including those involving formal land-settlement schemes as opposed to informal settlement. Variations exist both in the underlying drivers and the characteristics of the resulting deforestation.

Forest clearance for **cattle ranching** and soybean production is closely associated with Brazil. Some of the most dramatic images of tropical deforestation relate to these end uses, and as a result it is possible that their relative importance as proximate causes in the tropical zone as a whole has been exaggerated. Geist and Lambin report that only 6 percent of the deforestation cases they examined in Asia and 16 percent in Africa involved cattle ranching, compared with 82 percent in Latin America. This clearly illustrates the essential spatial variability of the causes of deforestation and thus helps to explain why simple, convincing explanations for deforestation have been so elusive. The case of Brazil, however, highlights the role of economic factors as underlying driving forces. The integration of **Amazonia** into the national and international economy has triggered an episode of change in land cover, in the same way that the integration of parts of India into the world economy in the nineteenth century led to the replacement of forest by tea plantations.

Agriculture has also expanded into the forest in some areas as a result of planned land settlement. Examples include the Bolivian lowlands, and areas such as Sumatra and Kalimantan in Indonesia where government-sponsored "transmigration" schemes have settled small farmers in previously unoccupied areas. According to the analysis of Geist and Lambin, this form of **colonization** occurred in around 40 percent of cases, but was much less frequent in Africa than in Asia and Latin America.

If agricultural expansion is the major proximate cause of deforestation, it might be assumed that agricultural retrenchment would also result in change in land use and/or land cover. To some extent this has happened, for example in central Europe and New England. While this change can sometimes be "active" (e.g., involving decisions to establish forest plantations on land released from agriculture), it can also be passive. While a human decision may be taken to discontinue agriculture, no such decision is necessarily involved in relation to forest invasion. The change can therefore be regarded as more "passive," and arguably the proximate causes of it are more complicated.

Wood Extraction

The harvesting of timber from the forest at a rate greater than that of natural increment depletes the forest and is an obvious candidate as proximate cause of deforestation. Logging in general and clearcutting in particular have become symbols, in the eyes of many, of environmental or **land degradation**. Forest exploitation for timber has sometimes been on a "cut and run" basis: in Southeast Asia, for example, a series of countries were affected successively in the second half of the twentieth century, and the result was a reduced forest area and a depleted forest resource. A similar episode occurred in the **United States of America** in the nineteenth century, giving rise to fears of a timber famine. Yet wood extraction need not necessarily result in a change in land cover, and in any case it seems that it rarely acts alone to reduce forest areas.

Wood extraction is reported by Geist and Lambin to have been involved in two-thirds of the deforestation cases they examined, but the frequency ranged from nearly 90 percent in Asia to little over half in Latin America. They subdivide wood extraction into "commercial" and "fuelwood," the former (for trade) being almost twice as important numerically as the latter (for domestic use). Again, the continental pattern shows strong contrasts; this ratio is inverted in the case of Africa.

The underlying driving forces differ between the two segments, with economic development presumably being a major driver of commercial extraction and population growth more directly significant in relation to extraction as fuelwood. It can probably also be concluded that the ecological impact of these two types of extraction varies according to forest type.

Commercial extraction of timber on a large scale requires roads (or railways). The construction of a communications network increases the accessibility of the forest and permits its penetration by small settlers. They may then proceed to clear further areas of forest for agriculture. In other words, wood extraction and agricultural expansion may combine to bring about removal of the forest in a synergistic way. This illustrates the general point that combinations of proximate causes, as well as of underlying driving forces, are usually more significant than single factors. According to Geist and Lambin's analysis, single-factor causation applied in fewer than 10 percent of cases, with the remainder involving two or more causal factors. They use the notion of tandems, such as "infrastructure-agriculture" or "logging-agriculture," to convey the message that combinations are more important than single causes.

Infrastructural Expansion

Serious problems of analytical logic are encountered when the role of infrastructure such as roads is considered. Infrastructural development permits or facilitates agricultural expansion and wood extraction, but it is debatable whether it can be viewed as a "proximate" cause. It can certainly be a permissive and/or intermediate causal factor, but it is questionable whether it can be regarded as the same category as agricultural expansion or wood extraction. Nevertheless, "transport extension" is reported by Geist and Lambin to have applied in nearly two-thirds of deforestation cases overall, and over three-quarters of those in Latin America. They also confirmed that it was very frequently involved in combination with agricultural expansion and wood extraction.

This factor in recent times has been closely associated with road building, but on a longer time scale, the advent of cheap forms of shipping was probably of major significance. Forests have tended to disappear more readily in accessible coastal lowlands than in the interiors of land masses. This pattern may be related to patterns of **land quality** such as soil fertility, but accessibility and ease of export have almost certainly also been significant.

Proximate Causes and Deforestation

Combination and contingency are key themes. The clear conclusion from the work of Geist and Lambin is that multiple causation is far more common than single-factor causation. Less immediately obvious but equally significant is the notion that the role of a process such as agricultural expansion, in terms of deforestation, depends on the setting. This setting includes both locational aspects or **initial conditions**, in relation to transport networks, and biophysical and social characteristics. In short, some forest types are more easily removed than others, and some social climates are more likely to be associated with small-scale agricultural expansion, for example, than others. A further complication is that **feedback** influences can and do link proximate causes and underlying driving forces: neither is fixed or independent.

"Proximate Causes" as an Investigative Instrument

Thus far, the hierarchical framework of underlying driving forces and proximate causes has been used mainly in analyses of tropical deforestation. A critique of that framework can readily reveal limitations of a logical and/or operational nature even in relation to that type of change in land cover. As yet, experience with the explicit application of the framework in other types of change in land use and land cover is limited, and therefore a general evaluation of the approach cannot be reached. It may be, however, that adaptation of this framework, or the development of an alternative one, may be required for the analysis of changes other than tropical deforestation. Modification or adaptation may be needed even if understanding of tropical deforestation is to deepen.

Leaving aside factors such as climatic change, most alterations to land use and land cover are the result of human **decision-making**. It follows that an understanding of them requires an understanding of the context of the decision. Various schematizations of land-use decision environments have been attempted, but most of them have been more useful as pedagogic tools than as research instruments. Most of them involve two or more levels, such as the behavior of the individual decision-maker and the "environment" in which (s)he operates. Attempts have been made to advance understanding of land-use change both at the "behavioral" level and that of prevailing regional political economy. The linking of the two levels is logical and desirable, but it has proved difficult to achieve in practice, and it raises wider philosophical issues of structure and agency. The significance of human agents is sometimes acknowledged in work on tropical deforestation, but their full incorporation into the "underlying proximate" framework has not yet occurred. Even if that "immediate" level were added, and the framework further modified to accommodate intermediate levels, problems would remain. The simplicity of the framework would be compromised and linearity of the underlying cause-effect model sustained.

In recent years, actor-network theory has attracted much attention in the social sciences. It avoids the problem of a hierarchical structuring of causality, and its use in land-use analyses can circumvent intractable problems of whether infrastructure, for example, is a different category of factor from wood extraction. It also provides the means of avoiding issues such as the question of whether there are fundamental differences between active and passive factors. On the other hand, while actor-network theory can help in promoting an understanding of a system, it does not yield results akin to those of the statistical analyses that have flowed from most work on land-use/cover change in recent years. It may therefore be less likely to produce results that can effectively inform policies relating to issues such as tropical deforestation.

There is nothing inherent in models of the "underlying driver/proximate causes" type that logically restricts their application to situations of linearity or of independence of variables. Indeed in relation to the latter, work with them has emphasized tandems or other combinations. Nevertheless, they do not necessarily point to the significance of non-linear relationships and **thresholds** of change, and it remains to be seen whether they can be adapted to cope with reversals of trends such as that associated with the **forest transition**.

The understanding of change in land use and land cover raises epistemological and well as practical issues. The quest for "explanation" that can be expressed in apparently simple (and preferably) numerical form has at best been only partly successful. It has been located within a context that can perhaps be described as traditional and reductionist—within a belief that effects can be explained by causes and that causes can be isolated and that they exist, in chain-like form, at different levels. The alternative, some would claim, is the despair that comes from acknowledging that the complexity of the real world is irreducible and inexplicable. The notion of "proximate causes" clearly lies in the first position rather than in the second: it is a notion that is problematic, for a variety of reasons, but to say that it has limitations is not to deny its helpfulness in contributing to a fuller understanding of change.

See also: **Agent-Based Model; Anthropocene; Biodiversity; Biomass; Degradation Narrative; Desertification; Exotic Species; Extensification; Forest Anomalies; Mediating Factor; Population Dynamics; Pristine Myth; Secondary Vegetation; Turnover; Vulnerability.**

Further Reading

Geist, Helmut J., and Lambin, Éric F., "Proximate Causes and Underlying Driving Forces of Tropical Deforestation," *BioScience* 52 (2002): 143–50; Meyer, William B., and B. L. Turner II, eds., *Changes in Land Use and Land Cover: A Global Perspective*, Cambridge, UK: Cambridge University Press, 1994; Meyer, William B., and B. L. Turner II, "Human Population Growth and Global Land-Use/Cover Change," *Annual Review of Ecology and Systematics* 23 (1992): 39–61.

ALEXANDER S. MATHER

Public Policy. A course of action plan for that dimension of human activity that is regarded as requiring governmental or societal regulation or intervention. As Thomas Dye once summarized, public policy is focused on "what governments do, why they do it, and what difference it makes." It is concerned with how public issues and problems such as the impacts of land-use/cover change come to be defined and constructed, and how they are placed on the political and policy agenda. According to Stuart Nagel, the study of "the nature, causes, and effects of public policies" requires frameworks, metaphors, and models for thorough policy analysis. The clear distinction between the public and the private began to blur from the late nineteenth century onward. Almost all areas of social life that political economists have regarded as private were penetrated by public policies, among them sectors as diversified as housing, agricultural **colonization**, rural welfare, conservation, and urban planning.

The closeness of public policy to issues of land-use/cover change is a reminder that the use of land is a highly political activity or **arena of land conflict**. By implication, especially **land-use planning** or **land-use policies** in *strictu sensu* indicate a belief that determination of land use should not be left to chance, to individual landowners, or to the market alone. Governments have the duty to address those aspects of society and economy that individuals cannot handle in an appropriate, society-embracing way. Although **private property** is one sacred cow in modern societies, it is the private market that has yielded often undesirable and sometimes disastrous results, among them loss of **biodiversity**, **land degradation**, air and water pollution, and **urban sprawl**, reinforced by free riding behavior of individuals. The seed of modern public policy is that society should aim to improve the techniques of governing a capitalist system by making the environment less vulnerable. An intelligent management is needed for guaranteeing more efficiency in the public sector, like the promotion of intergovernmental cooperation to better manage global **common-pool resources**.

One of the consequences of governments taking responsibility for reconciling public and private interests was the development of bureaucracy as a more rational form of organization. Max Weber underlined that the growth of bureaucracy was due to the process of an encompassing "rationalization" in industrial society and that a bureaucrat who characterized a rational, neutral professional ethics would—different from the politician—best answer the purpose to serve the public interest. During the last decades of the twentieth century, this distinction between the roles and ethics of a politician and a bureaucrat has decreased, as argued by representatives of the "new right" such as Friedrich Hayek and Milton Friedman, who refer to the failures of public policy caused, among other reasons, by the blur of the public and private sector and a disappearance of a clear distinction between the sphere of the market versus the state.

Whereas it has been the state as prime actor designing and implementing public environmental policies, **globalization** resulted in the renaissance of local policies becoming more and more important in a world in which cities and regions competed for investors on the one hand and consumers on the other. The strategic management of places is becoming a major task for local policymakers who need to strongly promote their region or city. Local policymakers know best how to promote their locality in the optimum fashion. Federal policies offer an important and useful framework for land management, but the main impetus for efficient and **sustainable land use** and management is expected to derive from local policies. Due to globalization, the strategies and roles of governments at the regional, state, and federal levels have greatly changed during the last two decades, resulting in a reinvigoration of local policymaking.

During the last years, public policy analysis has increasingly been based on comparative research relying on methods such as statistical analysis of several case studies across countries or a more focused comparison of a specific policy area, for example, environmental or entrepreneurship policy, within a selected number of comparable countries. Policy analysis is challenged by complex interactive processes and mechanisms related to policymaking that are often intertwined and not clearly separated from each other. Public policy analysis has to take into account processes such as elections, or roles such as of politicians, parties, and interest groups who have a major impact on policymaking. The complexity of policy analysis increases if subliminal context variables are taken into consideration, such as business environment, access to information, or freedom of expression, which affect policymaking to a high degree. In this regard, public policy-related **cultural factors** such as public attitudes, values, and beliefs rank high among the many pressures or underlying **driving forces** of land change, both local and regional and acting in various combinations in different geographical locations of the globe. For example, among factors associated with tropical **deforestation** are public unconcern and missing basic values with respect to the environment, in particular dominant frontier mentality, prevailing attitudes of nation-building, modernization and development, and low public morale. Likewise, and in terms of **landholding**, medium- and large-scale landholders' main motivation very often is to achieve the largest economic benefits through their better connection to markets, usage of industrial inputs, and hired labor force. In these groups, land-use **decision-making**, and particularly deforestation, are contingent upon the profit they will obtain from expanding arable land for commercial crops or **pasture** for beef cattle production. They will, therefore, deforest more in cases in which the prices of commodities are higher than their production costs. The latter is likely to happen when they benefit from prices of support, subsidized and cheap credit, preferential export taxes, favorable exchange rates for export commodities, and low price of inputs. Especially the medium- and large-scale enterprises have in diverse contexts benefited from policies that have for different reasons tended to provide them additional institutional rents (i.e., cheap credit, access to roads, etc.). On the other hand, next to same policies have negatively biased small farmers and limited them in expanding their **access** to assets. Nonetheless, an example of how public policy can enhance low deforestation, in a larger region titled the **arc of deforestation**, stems from **Amazonia**. As in most other states of Brazil, government policy and discussions in the state of Acre have often centered around a developmental strategy involving forest products. Acre, however, shows a contrasting reality because the dominance of extractive economies in this area is based primarily on rubber extraction, and rubbertappers resisted cattle ranchers' invasions and lobbied the federal government for the creation of extractive reserves in the late 1980s and early 1990s. It was **non-timber forest products** extraction that has proved to be more environmentally sound, also tending to favor social equity. Nevertheless, pressures on deforestation persist due to the implantation of some areas in pasture.

For land-use **tradeoffs** in developing countries, Steve Vosti and Thomas Reardon have proposed analyses to be organized around a "critical triangle" of public policy goals that include productivity growth, **poverty** alleviation, and sustainable resource use. Some other policy analysts extended this approach to include **human health**, and a collection by David Lee and Christopher Barrett includes a large number of examples of approaches to quantitative policy analysis of tradeoffs among the "critical triangle" for cases of **agricultural intensification**. In terms of the sociopolitical properties of land use at the national-scale level, in particular **China** is in a crucial moment of its history in finding a balance among **food security**, **economic growth**, and ecological **conservation**.

One of the biggest challenges the world has ever faced is climate change. Its potentially destructive potential has forced the world community into a multilateral process of negotiations, to prevent at least the most disastrous consequences and to slow down the speed of climate change. Global warming and policy issues are increasingly intertwined. The controversies among national and local policymakers all over the world circle around new or at least re-designed international trade agreements, questions of national sovereignty versus global governance, and ideological debates about future strategies for the promotion of economic growth. In this context, the role of public policies has increased to enforce control over the interests of the private sector, as well as governments affecting **global environmental change** in a negative way. Only by the common efforts of the world community and a global policy-driven strategy will the challenge of climatic change be deactivated. Environmental groups supported by sympathizers, for example, in foundations and in the federal government, aim to restrict and phase out the use of fossil fuels as sources of energy. Such measures would reduce emissions of **greenhouse gases** into the atmosphere. International climate policy is based on the **Kyoto Protocol**, which calls on industrialized nations to carry out, within one decade, drastic cuts in greenhouse gas emissions that stem mainly from the burning of fossil fuels. In November 2004, the Russian Duma ratified the protocol, so that the accord was finally brought into force. As of November 2004, the agreement had been ratified by 127 countries representing 61 percent of emissions, which is a success based on an enduring and determined climate policy. The United States, although a signatory to the protocol, has neither ratified nor withdrawn from the protocol.

See also: **Agenda 21; Carson, Rachel Louise; Community Involvement; Consumption; Decentralization; Economic Restructuring; Integrated Assessment; Investments; Land-Use System; Multiplier Effect; Property; Reforestation; Scenario; Swidden Cultivation; Tragedy of Enclosure; United Nations Convention to Combat Desertification; Vulnerability.**

Further Reading

Audretsch, David B., Grimm, Heike, and Charles Wessner, eds. *Local Heroes in the Global Village: Globalization and New Entrepreneurship Policies*, New York: Springer Science & Media Inc., 2005; Dye, Thomas R., *What Governments Do, Why They Do It, What Difference*

It Makes, Tuscaloosa: University of Alabama, 1976; Friedman, Milton, *Capitalism and Freedom,* Chicago: University of Chicago Press, 1969; Hayek, Friedrich A., *The Road to Serfdom,* Chicago: University of Chicago Press, 1994; Lee, David R., and Christopher B. Barrett, eds., *Tradeoffs or Synergies? Agricultural Intensification, Economic Development and the Environment,* Wallingford, UK: CAB International, 2001; Nagel, Stuart S., *The Policy Studies Handbook,* Lexington, MA: Lexington Books, 1975; Parsons, Wayne, *Public Policy: An Introduction to the Theory and Practice of Policy Analysis,* Cheltenham, UK: Edward Elgar, 1995; Platt, Rutherford H., *Land Use and Society: Geography, Law, and Public Policy,* Washington, DC: Island Press, 2004; Rausser, Gordon C., and Rachael E. Goodhue, "Public Policy: Its Many Analytical Dimensions," in *Handbook of Agricultural Economics, Volume 2B: Agriculture and Food Policy* (Handbook in Economics no. 18), eds. Bruce L. Gardner and Gordon C. Rausser, Amsterdam: Elsevier Science, 2002, pp. 2057–102; Repetto, R., and M. Gillis, *Public Policies and the Misuse of Forest Resources,* Cambridge, UK: Cambridge University Press, 1988; Vosti, Steve A., and Thomas Reardon, eds., *Sustainability, Growth, and Poverty Alleviation: A Policy and Agroecological Perspective,* Baltimore, MD: Johns Hopkins University Press, 1997; Weber, Max, *The Protestant Ethic and the Spirit of Capitalism,* New York: Charles Scribner's Sons, 1930.

HEIKE GRIMM

R

Ramsar Convention on Wetlands. An abbreviation of the "Convention on Wetlands of International Importance Especially as Waterfowl Habitat," which is an intergovernmental treaty adopted in February 1971 in the Iranian city of Ramsar. The convention came into force in 1975 and provided a global framework for the **conservation** of **wetlands**. In December 2004, the Ramsar convention was ratified by 143 contracting parties, and 1,399 wetlands have been designated for inclusion in the "Wetlands of International Importance" list, covering about 1.2 million square kilometers. Originally, the convention emphasized the conservation and wise use of wetlands to provide habitat for waterbirds, but all aspects of conservation and sustainable use are now considered with a broadened perspective. However, the designation of "Wetlands of International Importance" is often conflicting and debatable, as preliminary identification of national priorities in each country is required, as well as international agreements in the case of transfrontier wetlands, shared water systems, and species.

See also: **Coastal Zone; GLOBWETLANDS; Millennium Ecosystem Assessment.**

Further Reading

The Ramsar Convention Secretariat [Online, December 2004], The Ramsar Convention on Wetlands Web Site, www.ramsar.org.

LAURENCE HUBERT-MOY

Reforestation. Forest **regrowth** that constitutes an increasing fraction of the earth's cover and performs a crucial role in mitigating the harmful effects of **global environmental change**. There is thus a critical need to better understand the biophysical and human driving forces associated with reforestation. The term **forest transition** has been widely used to describe changes in forest cover with increased economic development and **industrialization**. The environmental Kuznets curve, on which much of this discussion is based, predicts a rapid increase in **deforestation** during the initial period of economic development for a nation, followed by a turnaround period of net reforestation in economically developed nations. There is increasing evidence now that this trend is not confined to developing countries. A range of socio-economic and institutional **driving forces** have been associated with reforestation, including demographic transitions, **agricultural intensification**, **urbanization**, global collapses in certain agricultural market prices (such as of **coffee**), shifts toward non-wood energy sources, restrictive national policies, **community involvement**, and increased cultural awareness. Factors such as **land-use history** and soil fertility significantly impact the dynamics of reforestation, with the rate of regrowth much faster in areas of shifting cultivation compared to intensively farmed areas and in the fertile tropics compared to the temperate zone. It is important to recognize that the observed rates of forest regrowth are often much lower than the rates of deforestation in preceding periods. Reforested areas can take several decades to return to **biodiversity** levels approximating those in the original old-growth forests, and may never reach these levels in areas where there is large-scale planting of non-native species. **Scale** is a crucial factor impacting the perceived trajectory of change, with reforestation frequently reported as occurring in smaller abandoned clearings, or on tree lots planted by local people to alleviate problems caused by increasing scarcity of wood. Nevertheless, where properly managed, reforestation has multiple beneficial impacts on global environmental change, providing crucial environmental or **ecosystem services** such as **carbon sequestration**, **watershed** protection, habitat for endangered species, and support for **forest**-dependent communities, indicating that there is a pressing need to identify and support the factors associated with the increasing expansion of secondary forest cover across the world.

See also: **Conservation; Degradation Narrative; Exotic Species; Fallow; Forestry; Green Wall Project; Hot Spots of Land-Cover Change; Kyoto Protocol; Land Rehabilitation; Land-Use Policies; Mediating Factor; Property; Public Policy; Secondary Vegetation.**

Further Reading

Fairhead, James, and Melissa Leach, "False Forest History, Complicit Social Analysis: Rethinking Some West African Environmental Narratives," *World Development* 23 (1995): 1023–35; Moran, Emilio F., Brondizio, Eduardo S., Tucker, Joanna M., da Silva-Forsberg, Maria Clara, McCracken, Stephen, and Italo Falesi, "Effects of Soil Fertility and Land-Use on Forest Succession in Amazônia," *Forest Ecology and Management* 139 (2000): 93–108; Nagendra, Harini, Southworth, Jane, and Catherine Tucker, "Accessibility as a Determinant of Landscape Transformation in Western Honduras: Linking Pattern and Process," *Landscape Ecology* 18 (2003): 141–58.

HARINI NAGENDRA

Regrowth. Conversion of farmland on slopes to **forest** or **grassland** for the aim of restoring **vegetation** and restraining ecological deterioration in **China**, for example. After serious flooding disasters happened to the reaches of the Changjiang and Yellow rivers, the government of China began to realize the urgent necessity to protect and improve the **watershed** environment of the nation's two largest rivers. It is not only a large ecological project, but also a socio-economic project implemented all around western China from the year 2000 onward, covering eleven provinces. As a principle, land with a slope between 15 and 25 degrees should be returned to forests and grassland. It is estimated that the total regressed area is 9.5 million hectares, accounting for 24.9 percent of the total cultivated land in western China. In the mountainous and plateau areas of western China, a tremendous amount of forestland and grassland was impoldered in the past to create arable land to support the growing population. However, water and soil on the land slopes are seriously lost when the slopes are cultivated. The main policy of regrowth is the **conversion** of farmland to forestland (grassland), planting trees, and thus sealing mountain passes. Regrowth will play an important role in the control of environmental deterioration and protection of the environment in China. It has been well implemented, but it is still necessary to keep enough cultivated land for **food security** and to avoid negative impacts such as food shortages.

See also: **Conservation; Decentralization; Fallow; Forest Anomalies; Green Wall Project; Himalayas; Land Degradation; Reforestation; Soil Erosion; Tradeoffs.**

QI LU

Remittances Landscape. An emergent landscape that is driven by the investment of remittances. People migrate in order to ensure their own survival and to improve the economic well-being of their families. Once in their new host countries, migrants often begin to send money home to provide for the basic needs of their families or to support community development projects. In the past ten years, these migrant financial flows, known as remittances, have doubled worldwide.

Remittances can be divided into private and collective remittances. Private remittances are funds that are sent directly from an individual migrant to a family member in his or her country of origin. Collective remittances are sent by a group of migrants to their community of origin to support community and development projects or to contribute to disaster-related recovery efforts. Collective remittances make up approximately 1 percent of total remittances.

According to the World Bank, remittances to eighty countries amounted to nearly $60 billion (in U.S. dollars) in 2000, which is 6 billion more than total net foreign aid from countries of the Organization for Economic Cooperation and Development (OECD). The largest flow of remittances goes to **Latin America** and the Caribbean. Migrant remittances to Mexico in 1996 amounted to $4.22 billion, or fourteen times the total sum of net foreign aid received. Other countries where remittances amount to at least three to nine times the amount of net foreign aid include India, Morocco, Sudan, Croatia, the Dominican Republic, Colombia, and Brazil. In many countries, remittances have replaced agricultural exports to a large extent. Income from **agriculture** as a percentage of the gross national product (GNP) dropped from 21 percent in 1980 to 18 percent in 1997, and remittances as a percentage of GNP rose from 3 percent to 15 percent during the same period. In most South Asian countries, and particularly in Bangladesh, Sri Lanka, India, and Pakistan, migration and remittances are becoming so important that they have become a key focus in development planning.

Many studies reveal a growing flow of remittances from urban to rural areas. Urban migrants from rural households apparently prefer to invest in rural lifestyles and preserve or improve the quality of their rural households rather than to invest in urban areas to which they migrate. However, the implications for land use, land cover, and landscapes of this inflow of capital and the medium- to long-term loss of labor are disputed. The prevailing thought is that remittances are used to cover basic needs but are not invested in agriculture. On the other hand, there are numerous studies from Latin

America, West Africa, South Asia, and elsewhere that show that migration and remittances are also used to improve agricultural or **land-use systems**.

The way remittances are invested depends on the particular situation of the family receiving the money. Nonetheless, in general the largest part of remittances (an estimated 80 percent) are used for **consumption** and household needs (food, clothing, health care, education, etc.). Investment in land, livestock, agricultural inputs, and in-house construction and maintenance is also relatively common but secondary to daily needs. Still less is used for business **investments**, savings, and loan repayments. Agricultural investment—for instance the purchase of improved inputs such as fertilizer and the expansion of irrigation—is more likely to occur in situations where political/economic incentives and environmental conditions are favorable, but typically only after consumption demands that immediately improve family welfare are met. The key factor deciding the way remittances are invested is insecurity; therefore the type of insecurity affects the type of investment.

Despite the impressive increase in remittances worldwide, and the substantial amount of research on migration, there is relatively little knowledge of the impact of remittances on landscapes. At present, no commonly used definition of remittances landscapes seems to exist. One possible criterion that can be applied in search of a definition is the following: a certain type of landscape is being transformed into another type mainly due to the investment of remittances.

So far, the transformations of at least two types of landscape due to the investment of remittances have been identified. The first transformation is that of agricultural landscape to a peri-urban landscape: investment of remittances in house construction replacing traditional dwellings by large, brick and cement homes. Many migrants decide to leave their country in the hope of earning enough money in another country to be able to build a house in their community of origin. This can lead to the emergence of peri-urban landscapes of "cultivated real estate," meaning that land is valued as a safe investment, and a place to demonstrate one's success with a large home, but not as an agricultural investment. This type of remittances landscape dominated by large ostentatious houses to demonstrate achievement and to have a home for retirement has also been described as a "landscape of conspicuous retirement." In a few migrant sending areas, this transformation has included the partial abandonment of cultivated land because remittances substitute agricultural income and are sufficiently large to allow households to abandon an **economic livelihood** based on agriculture, for example, in Bolivia, Ecuador, Mexico, Philippines, Thailand, and India. In the African context, even though remittances are also invested in housing, meeting basic needs continues to be the central preoccupation. Nevertheless, some migrants are able to build homes for their retirement, thereby encroaching on farming land around cities, contributing to the conversion of an agricultural to a peri-urban landscape in Nigeria and Morocco, for example. The second transformation is that of **forest** and agricultural landscape to a pastoral landscape, that is, investment of remittances in livestock to transform forested and agricultural land into **pasture**. The conversion of previously cultivated fields into degraded pasture is also included in this category, with examples stemming from Ecuador, Mexico, the Dominican Republic, and Brazil.

Another type of remittances landscape has been identified, namely the transformation of landscapes affected by natural disasters and civil wars to reconstructed landscapes. Especially in Central America but also in some African countries, much of the remittances sent back by migrants have helped to rebuild countries after civil war or natural disasters in Honduras, El Salvador, Nicaragua, Lebanon, and Somalia, for example. However, reconstructed landscapes are peri-urban, urban, agricultural, or pastoral landscapes that are "restored" to the original type of landscape. As the type of landscape does not change due to the investment of remittances, reconstructed landscapes are not treated here as a third type of remittances landscape. Furthermore, remittances that are invested into agricultural improvements in an agricultural landscape do not change the basic type of landscape and are therefore not included in this typology.

A tentative definition of remittances landscape could thus be the emergence of a certain type of landscape that is driven by the investment of remittances. However, how exactly land cover will be affected by the various ways remittances are invested cannot be generally predicted. The investment of remittances, and therefore the transformation of land use, land cover, and landscapes, will largely be influenced by the cultural, socio-economic, political, and environmental context from which migrants leave and to which they remit. If these conditions are positive, then it is more likely that remittances are invested into agricultural production. Thus, a predominantly agricultural landscape is maintained as opposed to an emerging peri-urban remittances landscape where previously cultivated land may be abandoned and/or turned into real estate. The emergence of a pastoral landscape due to the investment of remittances into livestock can potentially lead to large-scale changes in land cover, in particular through **deforestation**.

See also: **Agricultural Intensification; Decision-Making; Land Abandonment; Land Degradation; Land Rehabilitation; Mediating Factor; Multiplier Effect; Population Dynamics; Urban-Rural Fringe.**

Further Reading

Anonymous, "The Longest Journey: A Survey of Migration," *Economist* November 2 (2002): 1–16; Durand, Jorge, and Douglas Massey, "Mexican Migration to the United States: A Critical Review," *Latin American Research Review* 27 (1992): 3–42; International Institute for Environment and Development, Inquiry on Migration and Development [Online, August 2004], The IIED Web Site, www.iied.org; Jokisch, Brad, "Migration and Agricultural Change: The Case of Smallholder Agriculture in Highland Ecuador," *Human Ecology* 30 (2002): 523–50; McDowell, Christopher, and Arjan de Haan, Migration and Sustainable Livelihoods: A Critical Review of the Literature [Online, August 2004], The IDS Working Paper Web Site, www.ids.ac.uk; Nii Addy, David, Wijkström, Boris, and Colleen Thouez, Migrant Remittances: Country of Origin Experiences—Strategies, Policies, Challenges and Concerns [Online, August 2004], The International Migration Policy Programme Web Site, www.livelihoods.org; Sander, Cerstin, and Samuel Maimbo Munzele, Migrant Labor Remittances in Africa: Reducing Obstacles to Developmental Contributions [Online, August 2004], The World Bank Africa Region Working Paper Series Web Site, www.worldbank.org; Tiemoko, Richmond, Migration, Return and Socio-Economic Change in West Africa: The Role of the Family

[Online, August 2004], The Sussex Centre for Migration Working Paper Web Site, www.sussex.ac.uk.

SILVIA HOSTETTLER

Remote Sensing. The best practical way of spatially explicit **monitoring** of land-use and land-cover changes on a regional to global scale is from space. It is urgently needed to take data from the sequence of individual moderate- and high-resolution sensors and develop long-term climate data records of sufficient quality for the study of global change. The challenge in creating these records is to account for the evolving instrument design and performance since the 1970s, and necessitates dynamic product continuity.

The foremost observational requirement for land-use/cover change research is to extend the thirty-two-year record of high-resolution **Landsat** observations, which is the indispensable foundation for global land-cover inventories. Availability of the global thirty-meter spatial resolution data that have been collected from Landsat-7 increases the potential for detailed studies of landscape change on regional and global **scales**, a task that only a few years ago was not possible for some regions of the world. The long-term record of high-resolution observations will be continued through the Landsat Data Continuity Mission, which is being developed by the U.S. National Aeronautics and Space Administration (NASA) and the **United States Geological Survey (USGS) Program**. It is important that the data quality and availability from this mission be adequate for land-use/cover-change research, and emphasis needs to be given to instrument calibration, data quality assessment, data acquisition strategy, and data policy. It is also critical that high-resolution observations (ca. thirty meters) become part of the operational environmental satellite suite, enabling data continuity without gaps in the data record.

While high-resolution satellite data are used to map and quantify **land-cover change** at the regional scale, moderate resolution data are used to classify and characterize land cover at the global scale. The TERRA and AQUA satellite products, especially from the **moderate resolution imaging spectroradiometer** (MODIS), extend the twenty-three-year moderate-resolution data record from the **advanced very high resolution radiometer** (AVHRR), providing global land-cover products, including continuous fields of tree cover and burned area. During the last several years there has been an increasing emphasis on quality assessment and **validation** for MODIS satellite products, which sets an important precedence for future missions. The newly developed MODIS-based system is being used to detect changes in land cover at spatial resolutions of 250 meters, allowing the monitoring of potential **hot spots of land-cover change** over the globe. The continuity of these moderate-resolution data will be sustained by the future National Polar Orbiting Environmental Satellite System (NPOESS) mission, which will have on board the "MODIS-like" Visible Infrared Imaging Radiometer System (VIIRS). Quality of VIIRS observations and products must be assured to enable the measurement suite developed for the MODIS time series to be continued.

In addition to the longest continuous series of observations from Landsat and the AVHRR-MODIS-VIIRS line, there are other very useful optical coarse-to-moderate resolution data that are used for monitoring land surface. The International Satellite Cloud Climatology Project (ISCCP) global data set contains data on surface reflectances and temperatures. It has been based on compilation of all available geostationary satellite observations during the past two decades and could be used for analyzing global surface changes on a coarse spatial resolution. Data from the U.S. geostationary satellites (GOES) covering the Western Hemisphere have been extensively used for monitoring **fires** over the Americas. The above data sets are available at NASA data archives and/or from the National Climate Data Center (NCDC). Night-lights data from the Optical Line Scanner (OLS) on-board the DMSP satellite have been effectively used in monitoring tropical fires and **urbanization**. These data have been collected at the National Geophysical Data Center (NGDC). Another source of global land optical observations during the last few years is the Multiangle Imaging Spectro-Radiometer (MISR) on-board TERRA satellite providing angular reflectance measurements, which have been successfully utilized to characterize surface bi-directional properties and derive landscape composition.

During the last few years, NASA has been investing in hyperspatial resolution data (fewer than ten meters) for mapping land-cover change with data from the commercially operated **IKONOS** system. Results have demonstrated the utility of these data for mapping logging and **forest degradation** and validating high-resolution interpretations. While hyperspatial resolution data are becoming available for local uses, there remain significant hurdles to their adoption, including access to data, cost, and knowledge of the use of geospatial information.

In the New Millennium EO-1 mission, NASA has launched technology demonstration projects to explore the utility of high-resolution thermal (ASTER) and hyperspectral (Hyperion) measurements. The latter provides insight into vegetation cover composition and soil type discrimination.

Lidar technology—the active optical observations—is useful for fine-scale mapping of urban areas and provides elevation data that can be applied directly to existing/traditional applications. Lidar observations also provide information on **vegetation** structure that cannot be directly derived from passive optical data. NASA recently launched ICESat with the Geoscience Laser Altimiter System (GLAS) on board. Potential for using GLAS data for land-cover change studies is being investigated.

In addition to optical measurements, microwave observations have been widely utilized for mapping changes at the land surface. Radar measurements are especially useful in the areas of persistent cloudiness. Among the most useful applications with active microwave observations from space borne radars is monitoring flood events. **Biomass** estimation and classification of **forest**-cover types have produced encouraging results with available radar data.

The **NASA Land-Cover/Land-Use Change (LCLUC) program** has a goal to develop the capability to perform repeated global inventories of land use and land cover from space to identify changes occurring at the land surface. Priority for **change detection** under NASA LCLUC has been given to areas of the world that have been undergoing the most rapid changes and where stresses from human activities are likely to increase most rapidly.

Emphasis has been placed on the exploitation of satellite remote sensing data as the best source of information on the spatial distribution of land cover and rates of landscape change on regional, continental, and global scales. Time series of satellite data described above are analyzed in the LCLUC program to provide a consistent record of global land-cover change; characterize the end states of land-cover modification in regions of high population density; and study the impacts of spatial patterns and past history of land-use conversion on carbon and water cycles, and ecosystem changes—for example, changes in biodiversity, interaction of land-cover change with climate, and the sustainability of ecological services.

See also: **AFRICOVER; Airborne Light Detection and Ranging; BIOME 300 Project; Center for International Earth Science Information Network; Continuous Data; Desertification; Discrete Data; Earth Resources Observation Systems (EROS) Data Center; ENVISAT ASAR; European Remote Sensing (ERS-1/-2) Satellites; Forest Anomalies; Global Burnt Area 2000; Global Land Cover Map of the Year 2000; Global Observation of Forest and Land Cover Dynamics; Global Positioning System; Global Terrestrial Observing System; GLOBWETLANDS; Hot Spot Identification; Land Cover Classification System; Land-Use History; Normalized Difference Vegetation Index; Pattern Metrics; Pixel; SaHEL Land Cover; Socializing the Pixel; TERRA-ASTER; Thermal Band Analysis; Tropical Ecosystem Environment Observations by Satellite (TREES) Project.**

Further Reading

Gutman, Garik, Janetos, Anthony C., Justice, Chris O., Moran, Emilio F., Mustard, John F., Rindfuss, Ronald R., Skole, David, Turner, B. L., II, and Mark Cochran, eds., *Land Change Science: Observing, Monitoring and Understanding Trajectories of Change on the Earth's Surface (Remote Sensing and Digital Image Processing Series 6)*, Dordrecht, NL: Kluwer Academic Publishers, 2004; Rogan, J., "Remote Sensing Data and Methods in Land Cover Change Detection: A Review and Implication for Operational Large Area Monitoring," *Progress in Physical Geography* (in press).

GARIK GUTMAN

Resilience. Capacity for renewal and persistence in a dynamic environment facing disruption and change. In 1973, Crawford Holling first defined the concepts of resilience and stability. Stability is the tendency of a system to return to an equilibrium following a disturbance. "Ecological resilience" measures the amount of disturbance that an ecosystem can absorb before it changes into an alternative stability domain. Crawford Holling distinguished it from the "engineering resilience" that is the ability to return to a steady state following a perturbation. The ecological resilience paradigm includes research on alternative states or transitions. "Social resilience" implies that not only the relevant ecosystem, but also the corresponding human and management systems are able to absorb perturbations such that the system as a whole remains able to sustain (on average) a reasonable flow of benefits over time. "Social resilience" is then the human ability to cope with and adapt to social and **environmental change**, mediated through appropriate institutions. It remains unclear, however, whether resilient ecosystems enable resilient communities. Using the structure proposed by the Resilience Alliance, the concept of resilience in a broad sense has three defining characteristics. It is a measure of the amount of change the system can undergo and still retain the same controls on function and structure, the degree to which the system is capable of self-organization, and the community's ability to build and increase its capacity for learning and adaptation.

The resilience concept is interlinked with the concepts of sustainability and **vulnerability**. **Sustainable land use** exists if the use and management of resources do not lead to the loss of future options. The quality in a resource that renders its use sustainable is its resilience. Vulnerability is the negative side of resilience: when a social or ecological system loses resilience, it becomes vulnerable to change that previously could be absorbed. In a resilient system, change has the potential to create opportunity for development, novelty, and innovation. In a vulnerable system, even small changes may be ravaging. **Land degradation**, to put it very simply, is loss of resilience.

The degree to which the social-ecological system can build and increase the capacity for learning, adaptation, and responding in a manner that doesn't constrain future opportunities is a central aspect of resilience. In the face of the inevitability of changes (but with uncertain impacts), the resilience of the system demands evolving and adaptive strategies. Short-term responses to changes in land-based activities, which are identified as coping mechanisms, are one component of this adaptive capacity. The second component is related to cultural and ecological adaptations and represents long-term adaptive strategies. Coping mechanisms often take the form of emergency responses in abnormal seasons or years. Adaptive strategies, on the other hand, are the ways in which individuals, households, and communities change their productive activities and modify local rules and **institutions** to secure livelihoods. Compilation and analysis of case studies that illustrate adaptive management by local or indigenous people (a process of adjusting management actions and directions, as may be appropriate) are two goals of the Resilience Alliance.

See also: **Agrodiversity; Biodiversity; Disequilibrium Dynamics; Economic Growth; Economic Livelihood; Land-Use System; Mediating Factor; People at Risk; Pristine Myth; Threshold; Tradeoffs; Turnover.**

Further Reading

Adger, Neil W., "Social and Ecological Resilience: Are They Related?" *Progress in Human Geography* 24 (2000): 347–64; Berkes, Fikret, and Dyanna Jolly, "Adapting to Climate Change: Social-Ecological Resilience in a Canadian Western Arctic Community," *Conservation Ecology* 5 (2001): The Ecology and Society Web Site, www.ecologyandsociety.org/vol5/iss2/art18/; Folke, Carl, *Social-Ecological Resilience and Behavioural Responses* (Beijer Discussion Paper 155), Stockholm, SW: The Beijer Institute, 2002; Gunderson, Lance H., Holling, Crawford S., Pritchard, Lowell, Jr., and Garry D. Peterson, "Resilience," in *The Encyclopedia of Global Environmental Change, Vol. 2: The Earth System: Biological and Ecological Dimensions of Global Environmental Change*, eds. Harold A. Mooney and Josep G. Canadell, Baffins Lane, Chichester, UK: John Wiley & Sons, 2002, pp. 530–1; Resilience Alliance, Research on Resilience in Social-Ecological Systems: A Basis for Sustainability [Online, January 2005], The Resilience Alliance Web Site, www.resalliance.org.

NATHALIE STEPHENNE

Rice. An ancient grain, with *Oryza sativa* being by far the most important cultivated rice species worldwide. The origins of rice have been much debated, but the precise time and place of its first development remains unknown. Wild species of rice have been found over a broad area extending south and east from India and south **China** across continental Asia and insular Southeast Asia.

Rice terraces carved out of the hillsides hundreds of years ago by Ifuago farmers on the Philippine Islands, photo taken between 1890 and 1923. PHOTO: Marion Post Wolcott. [Courtesy Library of Congress]

Rice is presently cultivated in nearly all parts of the world where the soil and climate are favorable. It is the most important **food crop** for human beings and the primary staple for more than half of the world's population. Asia accounts for about 90 percent of production and **consumption** of rice. There are numerous ways of using the crop, including the straw.

Rice is the dominant food crop in the humid subtropics (76 percent of the area under food grains), humid tropics (75 percent), and subhumid tropics (51 percent). It is an important crop in the subhumid subtropics (36 percent) and semiarid tropics (19 percent), but is insignificant in the semiarid and cool subtropics.

There are about 120,000 varieties of cultivated rice. The most important cultivar groups are indica and japonica (or sinica) rice. Indica rice accounts for more than 75 percent of global trade and is grown mostly in tropical and subtropical regions. Japonica rice, typically grown in regions with cooler climates, accounts for about 12 percent of global rice trade. Other types of rice are aromatic rice, primarily jasmine from Thailand and basmati from India and Pakistan; and glutinous rice, grown mostly in Southeast Asia and used in desserts and ceremonial dishes.

Cultivated rice is generally considered a semiaquatic annual grass, although in the tropics it can survive as a perennial. Physical requirements such as warm temperatures and available water determine its geographical distribution. Rice is grown as irrigated rice, rainfed lowland and upland rice, and flood-prone rice. Irrigated rice accounts for 54 percent of the harvested rice area and contributes 76 percent of global rice production. It is concentrated mostly in the humid and subhumid subtropics and humid tropics.

World rice production in 2004 exceeded 600 million tons and has almost tripled in the last four decades. The most important rice-producing countries are China and India, with about 51 percent of the total rice production. Other important countries are Indonesia, Bangladesh, Thailand, and Vietnam.

Demands for rice have steadily increased, mainly due to population growth. In the past centuries, increasing demand for rice was met mainly through increase in production area. In the last four decades, rice's cultivated area further increased by about 30 percent to around 150 million hectares, which is more than 10 percent of total global arable land. However, the increase in rice production area has gradually declined during the last twenty-five years largely due to competition with other land-use types and for environmental reasons. In some countries such as China, land use for rice production has even decreased in recent decades. Growing demands for rice are increasingly met by higher productivity.

Global rice yields more than doubled from 1961 to 2004 and have almost reached 4 t ha^{-1}. These yields are mainly a result of the **Green Revolution** and associated progress in breeding for new varieties and improved crop management, including extensive use of synthetic fertilizers and agrochemicals. Rates of yield increases and presently obtained yields largely vary among countries and regions. The highest yields are obtained for irrigated rice, for example, in China, Indonesia, and Vietnam, among others. High cropping intensity and extensive use of agrochemicals affect **human health** and the environment.

An additional matter of concern is the contribution of intensive rice production to global warming through the release of methane (CH_4) into the atmosphere. Methane, one of the principal **greenhouse gases**, accounts for 15–20 percent of current radiative forcing. Its concentration has more than doubled over the past 200 years. Rice paddy soils are one of the major anthropogenic CH_4 sources. It has been estimated that the 60 percent increase in rice production required by the year 2020 to meet the demands of an increased population may increase CH_4 production by up to 50 percent. In contrast, the **Intergovernmental Panel on Climate Change** (IPCC) has recommended immediate reductions of 8 percent in anthropogenic emissions of CH_4 to stabilize atmospheric concentrations at current levels.

However, rice production is not only contributing to but also affected by climate change. Rice is a C3 plant with a pathway of carbon fixation typically more responsive to CO_2 increase than for C4 plants such as **maize**. Experimental and modeling studies suggest increases in productivity of about 20–25 percent if present CO_2 concentration doubles. However, projected increases in temperature will reduce rice yields so that the combined CO_2 and temperature effects will tend to compensate, which, however, will vary depending on the specific growing conditions and crop management.

The population of Asia is growing at 1.8 percent per year, until the middle of the next century, which will increase the demand for rice by about 70 percent over the next thirty years. Accordingly, rice production has to be raised by another 70 percent over this time period to sustain **food security**. Since expansion of rice-growing areas is largely limited, further increases in yields appear to be the only option to meet the growing demand for rice. However, rice yields in the high-yielding irrigated regions tend to approach a ceiling. Also, most rice is grown in small farms with limited **investment** capital to intensify production. The environmental and socio-economic consequences associated with any attempts to further increase the supply of rice are largely unclear.

See also: **Agricultural Intensification; Biotechnology; Cash Crops; Coastal Zone; Domestication; Cumulative Change; Environmental Change; Land Privatization; Land Rights; Malaria; Paleo Perspectives; Population Dynamics; Turnover; Wetlands.**

Further Reading

Kimball, Bruce A., Kobayashi, Kazuhiko, and Marco Bindi, "Responses of Agricultural Crops to Free-Air CO_2 Enrichment," *Advances in Agronomy* 77 (2002): 293–368; Matthews, Robin, and Reiner Wassmann, "Modelling the Impacts of Climate Change and Methane Emission Reductions on Rice Production: A Review," *European Journal of Agronomy* 19 (2003): 573–98; Parry, Martin L., Rosenzweig, Cynthia, Iglesias, Ana, Livermore, Matthew, and Günther Fischer, "Effects of Climate Change on Global Food Production under SRES Emissions and Socio-economic Scenarios," *Global Environmental Change* 14 (2004): 53–67; Ramankutty, Navin, and Jonathan A. Foley, "Estimating Historical Changes in Global Land Cover: Croplands from 1700 to 1992," *Global Biogeochemical Cycles* 13 (1999): 997–1027; Riceweb [Online, February 2005], The Rice Web Site, www.riceweb.org; United Nations Food and Agriculture Organization [Online, February 2005], The FAO Web Site, www.fao.org; Vergara, B. S., and S. K. de Datta, "*Oryza sativa L.*," in *Plant Resources of South-East Asia Handbook 10: Cereals*, eds. Gerard J. H. Grubben and Soetjipto Partohardjono, Leiden, NL: Backhuys Publishers, 1996, pp. 106–15.

FRANK EWERT

River Basin. Part of the land surface of the earth from which rain and snow drain into a river and which eventually flows into a lake or an ocean. It is also called a "drainage basin" or "catchment area." River basins are defined by natural hydrological boundaries that are usually **mountain** ridges between two neighboring basins. Basins can be divided into **watersheds** or areas of land around a smaller river or stream. In a river basin, all the water, including the wastewater produced by human beings, drains to a large river.

River basins absorb and channel all the runoff from snowmelt and rainfall. All the watersheds within a basin constitute a coherent system through the linking of the river system. By this function, a river basin forms a critical link and an integrator among the hydrologic, ecological, and socio-economic components of a river basin. It also supplies the resources we need and regulates the environment we live in if it is managed in a sustainable way.

Within each river basin, water travels over the surface and across **forest**, farm fields, **grassland**, villages, and city streets, or it seeps into the soil and travels as groundwater on its way to the main river channel. Land cover varies in its evaporation, transpiration, and water retention properties, and then has different hydrologic functions. Change in land use and land cover can have important impacts on the hydrology of river basins.

As water moves downstream, it erodes soil on slopes and carries and redeposits gravel, sand, and silt. This function of rivers makes them a major

An elevated view of Potomac River and Tidal Basin in Washington, D.C., ca. 1920–1950.
PHOTO: Theodor Horydczak. [Courtesy Library of Congress]

force of landscape **modification**. The upper reaches of a river basin are dominated by erosional landforms. Land use in this part is usually at risk of water and **soil erosion**. Sedimentary landforms like alluvial plains dominate on the lower part of a river basin, and land use in this part is usually at risk of flooding when the seasonal discharge exceeds the capacity of the river drainage system.

Water also transfers chemicals, excess nutrients, organic matter, and bacteria. The nutrients carried by river are usually the main resources for aquatic (marine or lacustrine) ecosystems. These functions of rivers make river basins into coupling means between hydrologic and biogeochemical cycles and between land and ocean ecosystems.

As gathering grounds, river basins play an important role in the ocean-air-land-water cycle. Perhaps the most important resources that river basins provide for human beings are fresh drinking water. When wisely managed, river basins can also provide **access** to food, hydropower, building materials, medicines, and recreational opportunities. By acting as natural "filters" and "sponges," well-managed basins play a vital role in water purification, water retention, and regulation of flood peaks. In many parts of the world, seasonal flooding remains the key to maintaining fertility for grazing and **agriculture**.

The function of river basins as a provider of resources and regulator of the environment has long been recognized and taken advantage of by human beings. Probably the greatest achievements and foci of river basin management during the past two centuries have been engineering works for drinking water supply, irrigation farming, hydroelectricity, sanitation systems, and flood-reduction purposes. This tightly focused engineering approach to river basin management, however, has put river basins under threat. Construction of dams and dykes, excessive pumping of groundwater aquifers, long-distance transfers of water from one basin to another, and drainage of **wetlands** disrupt natural hydrologic, biogeochemical, and ecological systems, resulting in mounting conflicts among both human communities and sectors, and a degradation of overall human welfare in the medium to long term.

The present situation of environment and development calls for a more holistic and integrated river basin management. A new **perception** of river basins, namely, river basins as not only water drainage systems but also part of the whole land surface of the globe, is needed with the new approach. Thus, the integrated river basin management should coordinate **conservation**; management; and development of water, land, ecosystem, and related resources across both sectors and human communities.

See also: **Coastal Zone; Land-Ocean Interactions in the Coastal Zone (LOICZ) Project; Siltation; Water-Land Interlinkages.**

Further Reading

Calder, Ian R., *The Blue Revolution: Land Use and Integrated Water Resources Management*, London: Earthscan, 1999; Calder, Ian R., "Hydrologic Effects of Land-Use Change," in *Handbook of Hydrology*, ed. David R. Maidment, New York: McGraw-Hill, 1993, pp. 13.1–13.50; Chorley, Richard J., and Barbara A. Kennedy, *Physical Geography: A Systems Approach*, London: Prentice-Hall, 1971; Global Water Partnership, *Integrated Water*

Resources Management (Global Water Partnership Technical Advisory Committee Background Papers no. 4), Stockholm, SW: Global Water Partnership, 2000.

XIUBIN LI

Russia. The largest country in the world, comprising 13.1 percent of the earth's land surface. Russia has about one-quarter of the territory of 1.7 billion hectares (of which 1.63 billion hectares are covered by soil and vegetation) in **Europe**, and the rest (behind the Ural Mountains) forms Asian Russia. Russian land is a natural resource of global importance both environmentally and economically. Globally, this land contains the largest deposits of natural gas, diamonds, and wood, and it occupies the second place in oil, coal, and other important natural resources. Russia has 23 percent of the world's **forests** and more than half of the world's bogs (**peatlands** cover about one-fifth of the country's area). The largest wild area of the planet comprises more than two-thirds of the country's land: about 26 percent of the last frontier forests are in Russia. Russian land plays an important role in global biogeochemical cycles: about 350 Pg organic carbon (C) (about 17 percent of the global) are accumulated in the top one-meter layer of Russian soils and bogs, and 40 Pg carbon are stored in living **biomass** (or 2.45 kg C m^{-2} for land covered by **vegetation**). On average, net primary production is estimated to be 4.36 Pg C yr^{-1} or 267 g C $m^{-2} yr^{-1}$, with a clear zonal gradient from arctic desert and tundra (120 g C $m^{-2} yr^{-1}$) to temperate forests and steppe (530 g C $m^{-2} yr^{-1}$). During the recent decades, Russian terrestrial ecosystems (mostly forests and **wetlands**) served as a net carbon sink of 400–700 Tg C yr^{-1}.

The major part of the country falls under a temperate to cold climate, and the zone of cold coniferous forests comprises 70 percent of total land. More than 60 percent of Russian territory is covered by permafrost; **mountains**, desert, and semi-desert regions together occupy almost half of the country's land, and arable lands cover only 7.3 percent. Agricultural lands, particularly in densely populated regions, show negative anthropogenic impacts up to a critical state and cannot be restored in a natural way, that is, only with special ameliorative activities. The area of critical ecological situations in Russia exceeds 250 million hectares or 15 percent of the country's territory. The current condition has been formed historically, but mostly it is a heritage of the seventy-year Soviet period, substantially aggravated by insufficient policies during the last fifteen years of the transition to a market economy.

The most dramatic climate change over the globe is expected to occur in the high latitudes of Russia. In spite of probably increasing productivity of terrestrial biota, negative consequences of climate change (in particular, thawing permafrost and accelerating regimes of natural disturbances) are likely to be more tangible.

Current Land Use and Land Cover

The two-dimensional official Russian land-use and land-cover classification includes seven land-use categories (distribution of land by destination and administration). In millions of hectares in 2002, they are the following: **agriculture**, 400.7; settlements, 18.9; industry, transport, etc., 17.2; specially protected areas, 34.2; forest fund, 1,103.1; water fund, 27.8; and unused land reserves, 107.9; each of the above categories includes different land-cover classes.

Agricultural land (i.e., territories, which are systematically used for production of agricultural goods) comprises 220.9 million hectares, including 123.5 million hectares of arable land. Russia has 0.85 hectares of arable land per capita (less than in **Canada** and Australia) and is seventh in the world in the amount of arable land weighted by potential biological production. Due to estimates by the Russian State Committee on Land Policy, the condition of agricultural lands is not satisfactory. The processes of active degradation of soil cover (increasing **soil erosion**, negative transformation of soil water regime, desertification, contamination, paludification, dehumification and decline of humus horizon, secondary carbonization, losses of nutrients, etc.) are typical for land of almost all densely populated regions. The area of the agricultural and arable lands under erosion threat comprises 64 and 71 percent, respectively. More than 50 million hectares of agricultural land (including 35 million hectares of arable land) suffer from water and wind erosion, basically in regions of steppe and semi-desert zones. It leads to annual decreasing of humus content on arable land at 0.7 t ha^{-1}. About 45 percent of arable land has an insufficient amount of humus, phosphorous (23 percent), and potassium (9 percent). High deficit and negative balance of nutrients are typical for agricultural land as a whole [e.g., the balance of NPK in 1997 comprised –97 kilograms per hectare (+21.5 – 118.5)]. **Desertification** is observed in twenty-eight administrative regions out of a total of eighty-two. The first desert in Europe has arisen and is dramatically increasing in Kalmikia and along the lower reaches of the Volga River. Paludification affects 30–50 percent of agricultural lands in the humid zone of European Russia. Areas of salty soil are increasing in many regions, in particular on irrigated land. Many regions suffer from contamination of soil by heavy metals, sulfur, radionuclides, and other industrial impacts. For instance, about 40,000 different accidents with spills of oil took place in western Siberia during only three years (1995–1997). Average production of agriculture is low and measured by 2000 (t ha^{-1}): grain crops, 1.5; sugarbeets, 18.5; sunflower, 0.83; potatoes, 9.7; vegetables, 14.9; etc. Agricultural production decreased by 10–30 percent during the last 10–15 years. The current **land reform** (started in 1991), which aims at the dismantling of the Soviet system of state properties, still did not present substantial results. About 265,000 agricultural farms on 14.4 million hectares (or 0.8 percent of land area of the country) accounted for 2.2 percent of the total agricultural production in 2002. In contrast, subsidiary plots (about 20 million plots on 6.2 million hectares, with an average size of 0.38 hectares) generate 60 percent of all agricultural production, namely 92 percent of potatoes, 77 percent of vegetables, and 59 percent of meat. In 1999, the productivity of each hectare of cultivated land on private plots was 23.5 times that of agricultural enterprises and thirty times that of farms.

Forestland including closed forests and temporarily unforested areas, clearcuts, and burnt areas makes up 51.6 percent of the country's land.

According to the State Forest Account in 2003, closed forests cover 776.1 million hectares or 45.4 percent of the total land area. Close to one-fifth of the forests are classified as protected forests, 7.6 percent are destined mostly for ecological services with prohibited or restricted industrial logging, and 69.4 percent are destined for timber extraction. Due to large areas of protected and low-productive forests, only 45 percent of Russian forests are industrial logging zones. The absolute majority (about 95 percent) of Russian forests are in the **boreal zone**, especially coniferous species (70 percent) dominated by larch, pine, spruce, fir, and stone pine (Russian cedar). Soft-leaved deciduous forests cover 17 percent, mostly secondary birch and aspen. Relatively small areas (2.5 percent) are covered by valuable hardwood deciduous trees. The rest (about 10 percent) mostly presents dwarf pine and other shrubs, in regions where tall trees are not able to grow due to severe climatic conditions. The growing stock of Russian forests (i.e., the total amount of stem wood of all living trees) was 88.3 billion cubic meters in 2003.

Large areas of treeless bogs (154.2 million hectares) are mostly found in peatlands and on peat soils. Surface water (71.7 million hectares) includes 120,000 rivers with a total length of 2.3 million kilometers, and about 2 million freshwater and salty lakes. Special protected areas, corresponding to categories I–II of the World Conservation Union (IUCN) and totaling 34 million hectares, include ninety-five state natural reserves; thirty-five national and natural parks; eleven federal preserves; and other territories that have special environmental, scientific, cultural, esthetic, and sanitation functions. Roads cover 7.9 million hectares, built-up areas and land under construction is 5.5 million hectares, and 1.1 million hectares fall under land that has been destroyed by industrial activity.

Legislatively, about 800 million hectares are indicated as resident regions for indigenous people of the north. Of this area, northern deer **pastures** cover about 300 million hectares. Previous and current policies led to the development of degraded and contaminated large territories there. In all northern pastures, 65 percent are degraded to medium to strong degrees. The amount of fodder from lichen pastures decreased during the last fifty years more than twofold. With more than 30 million hectares, the European north is the most heavily polluted area. Increasing heat contamination on permafrost, and processes of thermocarst and solifluction are widely distributed. Altogether, more than 3 million hectares in central European Russia and in the central Ural Mountains are contaminated by radionuclides due to the Chernobyl nuclear accident in 1986.

Land-Use History

During the last 2,000 years, major features and current conditions of land cover in Russia were formed by gradually increasing and regionally specific anthropogenic impacts resulting in decreasing forest cover. In central and southern regions of European Russia, a special system of agricultural land use (under which forest land was cleared and used for crops for three to ten years, following a recovery of the land for the following forty to eighty years during which **secondary vegetation** encroached upon these areas) led to increasing fragmentation of landscapes and eventually formed large open territories. In the north, human-induced burning of vast areas, aiming at development and improvement of northern deer pastures, also caused a substantial decrease of forest cover. This practice resulted in the 150–200 kilometer southward shift of the treeline and likely formed the current forest tundra zone. By 1700, forest covered about 50 percent of European Russia. Land cover of Asian Russia was regulated by rather stable regimes of natural disturbances, and forest likely covered 65–70 percent of the land there. The documented history of land-use/cover changes in European Russia started in the 1700s and was put on a solid legislative basis after the *Manifesto on General Land-surveying* (1765). During some 200 years (1696–1888), forested areas decreased significantly for all major parts of European Russia. Agricultural areas increased by 127 million hectares (from 7 percent to 27.3 percent). During the first half of the nineteenth century, 150,000–200,000 hectares of forests were cleared annually; during 1860–1890, this rose to about 900,000 hectares per year, while only fewer than 30,000 hectares were planted annually. From 1700 to 1914 (the year of minimal forest cover in European Russia during the last 300 years), the forest cover decreased from 52.7 percent to 35.2 percent. About 67 million hectares of closed forests were converted to other land-use categories during 1700–1910. The reform of 1861, which ended serfdom in Russia, had a pronounced effect on land use. Peasants received 40 million hectares of lands. Arable land increased in 1860–1887 from 88.8 million to 117.2 *dessiatinas* (=2.7 acres), and the average yield increased from twenty-nine *pods* (=16.38 kilograms) in 1861–1870 to thirty-nine *pods* in 1890–1891. **Colonization** of western Siberia and the Far East, which started in the seventeenth century, substantially intensified after the construction of the Trans-Siberian Railroad (1980s) and in particular after Piotr Stolypin's reform (1906–1917). The latter aimed at developing a wide layer of independent agricultural producers, and substantially triggered the migration of population to new lands further east. In 1917, Russia produced 20 percent of the world's grain (the Soviet Union produced only 12 percent in the 1990s). The October Revolution (1917) outlawed **private property**, which then became state property. The history of **agriculture** in the Soviet Union was heterogeneous—from "military communism" (1918–1921) through an era of market-oriented "New Economic Policy" (1922–1927) to collectivization (1927–1937; 99.4 percent of all cultivated land was in *kolkhozes* by 1937). In the following period (up to the 1990s), the Soviet Union generated a rather mechanized collective agricultural production; but simultaneously it was a period of discrimination against the rural population, and humiliation and annihilation of the peasants. In spite of many attempts to introduce sustainable agriculture (e.g., Stalin's so-called plan of transformation of nature, 1948–1953, and a large-scale developing of virgin lands in the 1960s), the Soviet period did not solve the issue of national **food security** and left agricultural lands in an unsatisfactory condition.

Reliable and detailed information on the dynamics of land-use/cover change in Russia is available for the period since 1961, when the first forest inventory was done for the entire country. In spite of the wide occurrence of natural and human-induced disturbances and about 17 billion cubic meters of wood that was harvested in Russia from 1950 to 2000, the area of closed forests increased to 70.5 million hectares (10 percent to the area of 1961) during 1961–2003, basically due to the high natural regeneration capacity of Russian boreal forests, suppression of wild fire (the area of burnt forests decreased from 57.6 million to about 30 million hectares), and **reforestation** activities (in 2003, 18.6 million hectares of planted forests were accounted for). During 1960–1990, annual timber harvest increased from 350 to 450 million cubic meters. The intrinsic features of the Soviet forest industry were the use of unsustainable logging technologies and machinery, and substantial losses of harvested wood. The harvest was provided in the most productive, valuable species and accessible forests. It resulted in the substantial decrease of forest quality. During the last four decades, the area of secondary birch and aspen forests increased by 13 percent, and the area of mature coniferous forests decreased by about one-third.

The severe economic and social decline in territories of the former Soviet Union has dramatically impacted land-use/cover dynamics during the last fifteen years. In agriculture, it caused increasing **land abandonment**, that is, encroachment of trees, shrubs, and **grasslands** (more than 35 million hectares from 1988 to 2003, or an estimated 25 percent of all land under cultivation). In many regions, abandoned land is covered by weeds and becomes a source of crop diseases and pests. In the summer of 1999, thirty Russian regions suffered from locust invasions for the first time during the last fifty years. As a new tendency, the human formation of "narcotic landscapes" is observed in some regions of Asian Russia on degraded and abandoned agricultural lands. Application of mineral fertilizers decreased from eighty-eight kilograms per hectare in 1990 to fifteen kilograms per hectare in 1999 (about thirty-seven times less than in the Netherlands), and organic fertilizers from 3.5 t to 0.9 t ha^{-1}. Any systems measures destined toward improvement of the quality of agricultural lands have not been implemented during the last fifteen years. **Land quality** has been decreasing: for example, during 1990 to 1996, the area of acid soils increased by 14 percent, and the area of soil with low humus content by 5 percent. The forest sector is on a substantial decline. During the last decade, the annual amount of harvested wood covered 120–150 million cubic meters, while the annual allowable cut, that is, the many-year norm of sustainable harvest, is estimated to be 500 million cubic meters. However, harvest mostly decreased in remote regions, and harvested areas are concentrated in regions with developed infrastructure. The amount of harvested wood increased in many dense populated areas. High prices for fuel have caused increased tree harvesting in forest deficit regions. According to estimates of non-governmental organizations, illegal logging is 20–30 percent of the officially reported numbers. The last decade was the warmest in the Russian territories for all the documented history of climate records in Russia. It caused an unprecedented increase in natural disturbances: satellites estimated the average annual area of vegetation **fires** in Russia to be about 10 million hectares in 1997–2003, with half to two-thirds of the fires having occurred on forestland. Outbreaks of dangerous insects were reported on about 10 million hectares in 2001–2002, and a substantial part of these outbreaks happened to be in the northern part of the country, where outbreaks have never been observed in the past.

Climate change might dramatically impact the condition of ecosystems and land cover throughout Russia. Over the twenty-first century thus far, the predicted increase of mean annual temperature varies from +4 to +9°C for different models, scenarios, and regions of Russia. The relative increase of precipitation will likely be smaller than that of warming. It will support desiccation of landscapes in many vast regions of Asian Russia, and will likely provoke a dramatic acceleration of natural disturbances (fire and insects' outbreaks). Melting of permafrost will cause the physical destruction of northern landscapes due to thermokarst and solifluction, thus threatening built-up structures. A substantial increase of anthropogenic pressure on fragile ecosystems at high latitudes can be expected due to dilative processes of exploration and extraction of natural resources. Taken together, all these influences might negatively impact the hydrological regime of vast territories, leading to steppization of landscapes and "green desertification" (i.e., the long-term replacement of forests by grasslands and wetlands). In order to provide the conditions for a transition to **sustainable land use**, Russia needs to develop and implement an anticipatory strategy of adaptation of lands to, and mitigation of the negative consequences of, global change.

See also: **Carbon Cycle; Conservation; Extensification; Hot Spots of Land-Cover Change; Industrial Revolution; Land Rehabilitation; Mineral Extraction; Nutrient Cycle; Salinization; Taiga; Transhumance.**

Further Reading

Kraninka, Olga N., Sun, Guoqing, Shugart, Herman H., Kharuk, Vyacheslav, Kasischke, Eric, Bergen, Kathleen M., Masek, Jeffrey G., Cohen, Warren B., and Doug R. Oetter, "Northern Eurasia: Remote Sensing of Boreal Forest in Selected Regions," in *Land Change Science: Observing, Monitoring and Understanding Trajectories of Change on the Earth's Surface* (Remote Sensing and Digital Image Processing Series 6), eds. Gutman, Garik, Janetos, Anthony C., Justice, Chris O., Moran, Emilio F., Mustard, John F., Rindfuss, Ronald R., Skole, David, Turner, Billie L. II, and Mark A. Cochrane, Dordrecht, NL, Boston, London: Kluwer Academic Publishers, 2004, pp. 123–38; Milanova, Elena V., and Maria Sennikova, Land Use and Land Cover Changes in Russia [Online, March 2005], The Encyclopedia of Life Support Systems Web Site, www.eolss.net; Shvidenko, Anatoly, and Sten Nilsson, "Synthesis of the Impact of Russian Forests on the Global Carbon Budget for 1961–1998," *Tellus* 55B (2003): 391–415.

ANATOLY SHVIDENKO

S

Sahel Land Cover. A characterization of land-use and land-cover trends for ecological regions in the Sahel of West Africa, carried out by the **Earth Resources Observation Systems (EROS) Data Center** of the **United States Geological Survey (USGS) program**, working in close collaboration with the AGRHYMET Regional Center in Niamey, Niger. SaHEL land cover is about mapping and quantifying changes in natural resources through land use and **land-cover changes** in nine countries: Burkina Faso, Cape Verde, Chad, the Gambia, Guinea-Bissau, Mali, Mauritania, Niger, and Senegal. The objective is to characterize land-change trends using historical and current satellite imagery and supporting ground information from three points in time (1965, using declassified CORONA photography; 1985; and 2000 using **Landsat** Thematic Mapper data), and to better understand the socioeconomic and biophysical **driving forces** of regional land-use and land-cover change. For this reason, the USGS-AGRHYMET partnership includes the Institut du Sahel in Bamako, Mali, to include issues related to socioeconomic trends and policy research as they relate to understanding the driving factors of **environmental change**.

See also: **Aerial Photography; BIOME 300 Project; History Database of the Global Environment; Land-Use History; Proximate Causes; Remote Sensing; Sudan-Sahel.**

Further Reading

United States Geological Survey, Science for a Changing World [Online, March 2005], The USGS Web Site, www.usgs.gov.

THOMAS R. LOVELAND

Sahelian Land Use (SALU) Model. A dynamic simulation model designed to represent regional-scale land-use change processes in the African Sahel region over a time scale of around thirty years backward and forward. Identified from an in-depth literature review of local case studies from the **Sudan-Sahel**, the dominant **driving forces** and processes of land-use change in the region are represented by a combination of simple equilibrium equations and knowledge rules. The model assumes that three land uses are generating the basic resources for the population: natural **vegetation** areas from which fuelwood is extracted, cropland that produces food for subsistence and market needs, and pastoral land that feeds livestock.

In this model, the **exogenous factors** or variables (rural and urban human population, livestock population, rainfall, and cereals imports) drive yearly changes in land-use allocation. For any given year, the land-use demand is calculated under the assumption of equilibrium between the production and

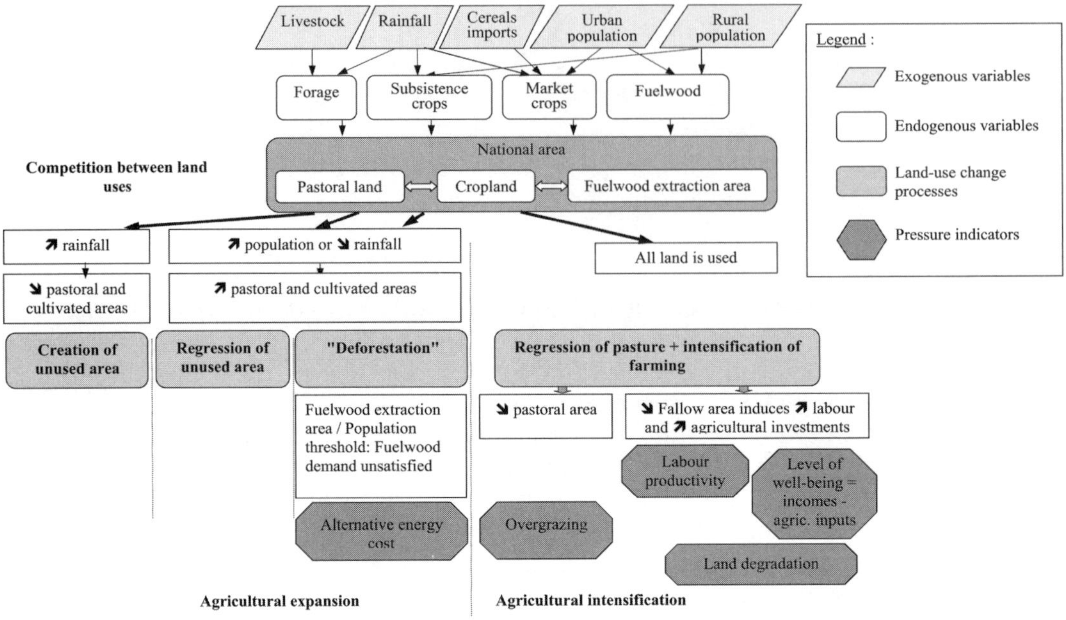

Structure of the Sahelian Land Use Model.

consumption of resources. Thus, the supply of food and energy resources derived from the areas allocated to the different land uses must satisfy the demand for these resources by the human and animal populations, given the exploitation technologies used at a given time. The model simulates two processes of land-use change: agricultural expansion at the most extensive technological level, followed by **agricultural intensification**. Driven by changes in human and animal populations, agricultural expansion leads to **deforestation** or to a regression of pastoral land. Once the expansion of cropland and pastoral land has occupied all unused land, and when fuelwood extraction areas have reached their minimal area, additional demand for **food crops** results mainly in agricultural intensification. This mostly takes place as a shortening of the **fallow** cycle, compensated by the use of labor and agricultural inputs to maintain soil fertility. The model also generates several endogenous pressure indicators, which are symptoms of land-use change and can be used to identify sustainability **thresholds** and dynamic **feedbacks**.

See also: **Cash Crops; Endogenous Factor; Extensification; Modeling; Pasture; Subsistence Agriculture.**

Further Reading

Stephenne, Nathalie, and Éric F. Lambin, "A Dynamic Simulation Model of Land-Use Changes in Sudano-Sahelian Countries of Africa (SALU)," *Agriculture, Ecosystems & Environment* 85 (2001): 145–61.

NATHALIE STEPHENNE

Salinization. The accumulation of water-soluble salts such as sodium, calcium, magnesium, potassium, sulfate, and chloride within the soil profile.

Salinization occurs naturally in areas characterized by a net upward flux of soil water, that is, mostly in semiarid and arid regions. Land-use/cover change can enhance or cause salinization of soils. However, the major cause of human-induced salinization is improper management of irrigated soils, sometimes called secondary salinization or irrigation salinization. This is considered a major contributor to **soil degradation** in some parts of the world. A related problem is alkalization.

Being a product of chemical weathering, salts are abundant in soils, surface runoff, and ground water. Under humid climate conditions, salts are washed out of soils during the initial phase of soil formation and carried off by streams to the oceans. Therefore, saline soils providing ecological niches are found in humid areas only under very specific conditions, that is, close to the oceans, in areas with saline marine sediments or where saline groundwater reaches the surface. Under semiarid to arid conditions, saline soils occur naturally in low-lying areas and depressions receiving additional water and solutes from the catchment area. As the water evaporates, the solutes are left behind and accumulate. Higher grounds are usually not affected, because salts are either not mobilized or drained. The rise in salt concentrations in the soil decreases the osmotic potential of the soil and diminishes the plants' ability to take up water. This results in drought damages to plants long before salt concentrations reach a toxic level.

Salinization is an immanent climate-induced hazard for irrigation farming in semiarid to arid environments, irrespective of the political or economic system. Its major cause is improper design and management of irrigation schemes (irrigation salinization). This includes the excessive application of water due to inefficient water distribution systems that eventually cause the groundwater table to rise within the reach of capillary rise, the use of brackish or saline surface or groundwater, and the neglect of proper drainage systems. Another cause of salinization is the removal of **vegetation** in drylands, again causing a rise of groundwater and enhancing evaporation and accumulation of salts in the upper soil horizons.

Although flushing of soils is a potential method of regenerating saline soils, it is considered economically unfeasible and ecologically doubtful. Therefore, salinization has to be considered an irreversible process of soil degradation. Drawing together different estimates from the 1990s, the Global Environment Outlook (GEO 3 report) states that over 30 percent of all irrigated land (2.55 10^6 km^2) is affected by, and about 10 percent severely degraded from, salinization, leading to the abandonment of nearly 4 percent of the total irrigated area annually. As the total irrigated area continues to rise, this implicates a strong pressure on other land resources.

See also: **Agricultural Intensification; Coastal Zone; Cotton; Desertification; Environmental Change; Land Rehabilitation; Watershed; Water-Land Interlinkages; Wetlands.**

Further Reading

Umali, Dina L., *Irrigation Induced Salinity: A Growing Problem for Development and the Environment* (World Bank Technical Paper 215), Washington, DC: The World Bank, 1993; United Nations Environment Programme [Online, March 2005), http://www.unep.org/geo/geo3/], *Global Environment Outlook 3: Past, Present and Future Perspectives,*

London: Earthscan, 2002; United States Salinity Laboratory [Online, March 2005], The USSL Web Site, www.ussl.ars.usda.gov.

JUSSI BAADE

Sauer, Carl Ortwin (1889–1975). American geographer and professor at the University of California–Berkeley (where he built a distinguished graduate school) who is widely credited for reinvigorating cultural and historical geographies and unifying physical and human geographies in his time. Seeing geography as inseparable from human history inasmuch as the earth is profoundly affected by humanity, he creatively worked the concepts of anthropology, archeology, and sociology into geography through studies of the human occupation of a specific region over intervals of historic time. Enduring influence of his cultural landscape studies, as manifest in such notions as "cultural agency," "cultural hearth," and "historically accumulative effects" remains a countercurrent to the prevalent view of seeing "clocklike" processes of land-use/cover change amenable to computer **modeling**.

See also: **Domestication.**

Further Reading

University of Colorado–Boulder, Department of Geography, Geographers in the Web [Online, December 2004], The Carl Ortwin Sauer Web Site, www.colorado.edu/geography/giw/sauer-co/sauer-co.html.

KE CHEN

Savannization. The process whereby a particular land cover is converted to savanna, either naturally or as a result of human activity. The word savanna is used to describe a landscape with a continuous herbaceous layer, composed of grasses and forbs, with scattered woody vegetation that includes trees and shrubs. However, the exact composition of the herbaceous layer and the degree of woody vegetation cover varies across savannas. Although both natural and anthropogenic savannas exist today, the origin of specific savannas is difficult to pinpoint in many cases. The process of savannization has occurred throughout the world, converting forested areas or areas along a forest-savanna into savanna. This process has been highly debated in many areas where savannas may be of natural origin but are sustained by human activities, including **deforestation** and regular burning of the landscape. Savannization also refers to a purposive management strategy in arid and semiarid areas.

Ecosystems across the world have exhibited change in **vegetation** composition, configuration, and health due to multiple forcing factors, many of which are human in origin, whether directly (e.g., deforestation, **agricultural extensification**) or indirectly (e.g., via increased competition for water, impacts on local and regional climate). Savannas occupy 20 percent of the earth's surface and show marked change in vegetation composition, function, and **biodiversity**. Savannas are typically defined by two components: a continuous grassy or herbaceous layer and a discontinuous woody component. Many of the differences in traditional and newer models of tree/grass ratios in savannas have been researched of late, which is important, as tree density is often used as a surrogate for determining whether an area is **grassland**, savanna, or woodland. The transition of a given area into one of these land-cover types has been hypothesized to be potentially controlled by physical factors (e.g., temperature, precipitation, soil texture), biologically meaningful factors (such as plant-available moisture and available nutrients), and disturbance factors (fire, flood, trampling, and herbivory/grazing). These ecosystems and their ecological fluctuation or stability are considered important because of their high levels of biodiversity (faunal as well as floral) and socioeconomic benefits. Globally, savannas and grasslands contribute approximately one-third of net primary productivity.

Savannas typically have a climate consisting of alternating wet and dry seasons. The prominent dry season results in dormant grasses that become susceptible to **fire** in contrast to a wet season during the year. These areas include tropical Africa, Asia, Australia, and both North and **South America**.

Savannization has been offered in some cases as a tool for combating **desertification**; in this case savannization occurs by turning deserts into savanna by dryland-adapted afforestation of rangelands. This process helps to control erosion by promoting greater ground cover and changing water pathways, whether the water source is irrigation or rainfall, and has been successfully implemented in Israel. Soil **conservation** and concomitant water protection result in sustainability of greater vegetative land cover that is also of higher forage quality. Grazing, then, is more productive and allows for fire suppression through controlled grazing. It is hypothesized that impacting the water budget through changing water storage and overall evapotranspiration promote a positive **feedback** between available moisture and subsequent microclimate conditions, though these relationships have not been exhaustively modeled to date. Thus, savannization has two differing connotations: as a management strategy for fighting desertification (converting desert to savanna) and as a seminatural process involving either woody encroachment into grasslands or woody mortality in woodlands.

Desavannization, the process of converting savanna into another ecosystem type, is also occurring around the world. Both North American and African savannas face impending woody encroachment, whereby the woody species density eventually exceeds the level associated with savannas and moves into that of woodland. Desavannization may also occur when woody species are systematically eliminated, leaving primarily grasses and forbs, turning the savanna into a grassland.

Most savannas operate in a climatic context (precipitation and temperature) whereby very small changes can trigger wholesale **conversion** into or out of woodlands and grasslands. These shifts, documented in both North America and Africa, may also occur in response to disturbance regimes. The two primary disturbance regimes affecting savannization and desavannization are fire and grazing. Fire and/or grazing convert woodlands to savanna, and can also convert savannas to grasslands if frequent or severe enough. Fire suppression and grazing limitations have the opposite effect, allowing grasslands to segue into savanna and savanna to convert to woodland. **Land-use policies** concerning fencelines and grazing management or fire suppression therefore have immediate, direct, and substantial impacts on savannas and their rates of areal growth or decline.

While fire and grazing have negative and positive short-term effects on aboveground productivity, composition, and diversity, their independent and interactive effects on woody plants are fundamental to production and maintenance of savannas and grasslands. However, the effects of season of disturbance have rarely been examined and as such are especially inviting for analysis via an intra-annually rich satellite time series. Similar to fire, grazing effects on herbaceous plant species composition can be dependent on season. However, depending on grazing intensity, this effect can be less than that caused by fire, selectivity of grazers, or climate.

A third disturbance regime, flooding, has also been found to impact savannas in Africa and South America. For example, in the Okavango Delta of Botswana, floodwaters from the Angolan highlands each year spread through the low-energy river system into nearby grasslands, savannas, and woodlands.

As flooding extent increases over several years, savanna can be lost to grasslands or gained in woodlands due to woody species mortality. When flooding extent is lower than normal for several consecutive years, savannization occurs in the grassland areas while desavannization occurs in the more distal woodlands.

Studies of seasonal effects of disturbance, particularly fire, are fraught with difficulties. The causes of post-disturbance response of a plant community may be due to direct (interruption or impact on phenological phase) or indirect (**modification** of conditions in post-disturbance environment) mechanisms and are difficult to identify and therefore model. A coarse prediction of post-disturbance compositional outcomes may be achieved by allocating plant species to guilds, also known as functional groups. However, this method may be flawed due to both the variation within guild and the variation of species phenology year to year and place to place. Furthermore, any direct or indirect effects of disturbance may be obscured by both the variation in response time of some components of the vegetation and the possible overriding effect of climate on this ecosystem. This suggests that any examination of plant community responses to ecological disturbance must have high replication across time and space to extract local geographic and short- and long-term effects, again suggesting a gap that an earth **remote sensing** approach could potentially help fill.

The vast majority of the literature on savannization and desavannization tends to treat those processes as either wholesale sweeping change or species substitution. Little of the work considers the spatial nature of the conversion into or out of savanna ecosystems. That is, the spatial process by which savannas morph has not been sufficiently studied. For example, few studies of savannization employ **pattern metrics** to better describe the configuration of savanna landscape elements. When savannization occurs, does it move along a frontier, advancing into its new territory at the shared ecotone or edge? Or does savannization occur more through percolation and subsequent patchy **colonization** of woody species, fragmenting the former ecosystem until it reaches a new background or dominant land-cover type of savanna? The answers lie in comparing the way savanna species (particularly woody species) propagate into new areas, and the global inventory of savanna woody species is vast.

There is some controversy as to whether defining savannas (as separate from woodlands or grasslands) can be performed solely on the basis of woody species density (whether stem count or canopy closure). Furthermore, disentangling anthropogenic effects from natural effects has proven difficult. For example, lightning-induced fires may be natural in origin but start fires that burn at higher temperatures with greater fuel loads because of human fire suppression. However, recent work comparing savannas from Australia, Africa, and North America suggests that more than many other ecosystem types, savannas may in fact be incredibly similar across widely varying conditions. The implication for global savanna studies is that the processes at work on differing savanna species around the world may in fact function very similarly, making research in savannization applicable to savanna landscapes worldwide. As the threat of global warming and population-driven fire frequency increase, even portions of **Amazonia** have been identified as likely targets for natural savannization in the next century, adding to the increase in the world's savanna ecosystems.

See also: **Degradation Narrative.**

Further Reading

Archer, Steve, Boutton, Thomas W., and Kathy A. Hibbard, "Trees in Grasslands: Biogeochemical Consequences of Woody Plant Expansion," in *Global Biogeochemical Cycles in the Climate System*, eds. Ernst-Detlef Schulze, Martin Heimann, Sandy Harrison, Elisabeth Holland, Jonathan Lloyd, Ian C. Prentice, and David Schimel, San Diego, CA: Elsevier Academic Press, 2001; Belsky, A. Joy, "Tree/Grass Ratios in East African Savannas: A Comparison of Existing Models," *Journal of Biogeography* 17 (1990): 483–9; Bourlière, François, ed., *Ecosystems of the World 13: Tropical Savannas*, New York: Elsevier, 1983; Scholes, Robert J., and Steve R. Archer, "Tree-Grass Interactions in Savannas," *Annual Review of Ecology and Systematics* 28 (1997): 517–44.

KELLEY A. CREWS-MEYER, AMY L. NORMAN

Scale. Scale is the spatial, temporal, quantitative, or analytic dimension used to measure and study objects and processes. Two important components of scale are extent and resolution. Extent is the magnitude of a dimension used in measuring, whereas resolution is the precision used in this measurement. For example, if a map consists of square **pixels**, then the amount of land portrayed by the collection of pixels determines the extent, and the size of the individual pixels determines the map's resolution. In this case, the length of the side of the pixel is the measurement of the map's resolution, which is also known as the grain.

Categorical scale is the level of detail of the categories in an analysis. For example, in classification of land cover, a hierarchy may be defined in which many detailed categories can be aggregated into fewer coarser categories, which can be aggregated further into even fewer, more general categories.

The word scale frequently causes confusion because it has many meanings. The most common source of confusion derives from the fact that some scientists use simple descriptors such as "small" and "large," while they simultaneously fail to specify which version of the definition of scale they are using. If a scientist uses the word scale to mean "extent," then "small scale" means a small study area. If a scientist uses scale to mean the ratio between the map and the ground (e.g., 1/1,000), then "small scale" means a large study area for a given size of paper map. If a scientist uses the word scale to mean resolution or categorical detail, then the size of the scale is independent of extent. There would be less confusion if scientists used the adjective "fine" to describe the small grain of highly detailed resolutions and "coarse" to refer to large grains of less-detailed resolutions. The adjectives "small" and "large" should be reserved to describe extent.

The operational scale is the scale at which a certain process (e.g., a land use decision) manifests on the landscape. Land use and cover changes are the results of many interacting processes. Each of these processes operates over a range of scales in space and time. Land-use decisions are made at different levels of social organization (e.g., the individual or community level). Level refers to level of organization in a hierarchically organized system and is characterized by its rank ordering in the hierarchical system. Other examples of levels include the organism, ecosystem, landscape, social system, nation, and globe. If the operational scale does not correspond with the scale of the data, then the scientist might not be able to determine the correct processes. **Land-use systems** rarely produce a single scale that can be regarded as correct or optimal for measurement and prediction, because landscapes usually demonstrate many interacting processes. An optimal scale of analysis for a specific data set might exist where predictability is highest; however, that scale will not necessarily be the same for other analyses. Therefore, it is desirable to determine the scale of the analysis from a careful examination of the phenomenon, rather than using *a priori* scales of observation.

The interaction of processes at various scales and the mismatch between operational and data scales are common. These can cause results of empirical analysis of land-use change to be sensitive to the selection of both extent and resolution. In cases where an investigation attempts to use data that are readily available, the extent and resolution of the analysis are sometimes determined by factors that are independent of the phenomenon. For example, if the phenomenon of interest is **deforestation**, then the available data are frequently in the form of satellite images, which are available in pre-determined extents and resolutions that are functions of the path and sophistication of the satellite used to create the image. If the scientist adopts the satellite image as the study area, then both the extent and the resolution are dictated by the format of the available data, not the phenomenon. Scientists are developing tools to measure the degree to which changes in scale influence the analysis of the phenomenon. This challenge is closely related to geography's famous and unsolved "Modifiable Areal Unit Problem," which states that statistical results are sensitive to changes in the areal unit. Numerous examples in land-use change research have indicated that the analysis of the relation between land-use change and its **driving forces** should not be restricted to a single scale. Multiscale and multilevel techniques are better able to unravel the complex interactions and **feedbacks** over scales that determine the functioning of the land-use system. Scale remains one of the major challenges of geographic analysis, and is consistently ranked as a top research priority by the University Consortium on Geographic Information Science.

See also: **Auto-Correlation; Environmental Change; Integrated Assessment; Land Cover Classification System; Land-Use History; Mediating Factor; Millennium Ecosystem Assessment; Pattern Metrics; Pixelizing the Social; Remote Sensing; Socializing the Pixel; Validation; Vulnerability.**

Further Reading

Gibson, Clarke C., Ostrom, Elinor, and Toy-Kyeong Anh, "The Concept of Scale and the Human Dimensions of Global Change: A Survey," *Ecological Economics* 32 (2000): 217–39; Openshaw, Stan, *The Modifiable Areal Unit Problem*, Norwich, UK: GeoBooks, 1984; Pontius, Robert G., Jr., "Statistical Methods to Partition Effects of Quantity and Location During Comparison of Categorical Maps at Multiple Resolutions," *Photogrammetric Engineering & Remote Sensing* 68 (2002): 1041–9; Verburg, Peter H., and You Qi Chen, "Multi-Scale Characterization of Land-Use Patterns in China," *Ecosystems* 3 (2000): 369–85; Verburg, Peter H., de Groot, Wouter T., and A. (Tom) Veldkamp, "Methodology for Multi-Scale Land-Use Change Modelling: Concepts and Challenges," in *Global Environmental Change and Land Use*, eds. A. J. (Han) Dolman, A. (Jan) Verhagen, and C. A. (Ilse) Rovers, Dordrecht, NL, Boston, London: Kluwer Academic Publishers, 2003, pp. 17–51.

ROBERT GILMORE PONTIUS JR., PETER H. VERBURG

Scenario. A hypothetical, possible future pathway that describes dynamic processes and represents sequences of events over a period of time, including causally related states, **driving forces**, events, consequences, and actions. The word scenario comes from the dramatic arts. In the theater, it is an outline of the plot; for a movie, a scenario sets forth details relevant to the plot. Its formal intellectual roots trace back to the Manhattan Project six decades ago when world-renowned nuclear physicists explored and evaluated the possibility that the energy buildup from a full-scale explosion of the hydrogen bomb might ignite a devastating deuterium reaction in the skies and the oceans. In the 1950s, Herman Kahn and Anthony Weiner at the Rand Corporation used the concept of scenarios in a series of strategic studies for military planning purposes. In the corporate world, scenarios were refined at Royal Dutch/Shell by Pierre Wack in the 1970s and 1980s, and Shell became a leader of the scenario approach to business planning. The word scenario came to the attention of the general public in 1972 with the publishing of *The Limits to Growth* by Dennis Meadows and colleagues. Today, scenario development is used in a variety of different contexts ranging from political decision-making, to business planning, to local community management, and to global environmental understanding.

Garry Peterson and colleagues provide a good first point of entry to define a scenario. They propose to position scenarios along two axes: the degree of uncertainty and the degree to which a system can be controlled. As depicted in the figure, scenarios are useful in systems where uncertainty is high and controllability low. Most definitions of scenarios agree with this assumption. However, the highest level of agreement is on what a scenario is not.

A scenario is not a prediction, understood as the best possible estimate of future developments; neither is a scenario a forecast, the best estimate from a particular method or model. Last, a scenario is also not a projection, as it depends on assumptions about drivers and leads to "what if" statements.

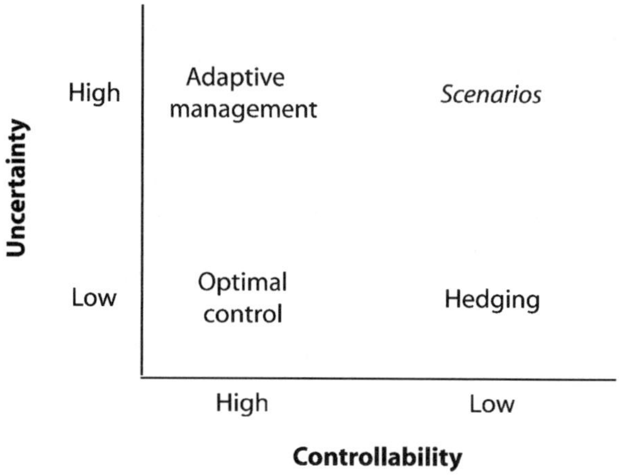

Degree of controllability and uncertainty of the system being studied. Adapted from Petersen et al. (2003).

A first definition of scenarios states that they are hypothetical sequences of events constructed for the purpose of focusing attention on causal processes and decision-points. Alternatively, scenarios can be defined as plausible, challenging, and relevant stories about how the future might unfold that can be told in both words and numbers, as done by the Millennium Ecosystem Assessment.

A large diversity of types of scenarios has been developed, and various attempts to classify them have been undertaken. For example, Philip Van Notten and colleagues propose a classification based on degree of complexity, exploration versus decision support, and degree of formalization. Other useful classification criteria include forecasting versus backcasting, or whether participatory methods are employed. Within the framework of land-use/cover-related research, it is particularly the division between qualitative and quantitative scenarios that is of interest.

Qualitative scenarios are usually in the form of narrative storylines and involve stakeholders during the development. Stakeholders can be involved in two fundamentally different manners: in-depth interviews and group sessions, and workshops. The advantages of interviews are larger sample size, larger reproducibility and replicability, and thus arguably a higher "scientific" value, although the latter can be disputed. Nevertheless, there are important advantages of a workshop setting that outweigh the possible drawbacks. A group setting enables stakeholders to create a common vision; it can be a stimulus for processes of social interaction and learning; it is the ideal setting for development of complex, multidisciplinary scenarios; and it might enable viewpoints that might not have been discovered in individual interviews. Given that participatory processes are at the heart of the development of qualitative scenarios, an important goal is usually the process itself. Stimulating discussion among stakeholders and initiating long-term participation in the **decision-making** process are often as important as the actual results of the exercise. The final output is usually in the form of a storyline, although similar methods are increasingly being employed for qualitative model building, of which the output can be semi-quantitative, for example.

Quantitative scenarios are almost always created to serve as inputs into **modeling**. They therefore rarely serve as products of their own. Of particular interest are area-based scenarios that are linked to spatially explicit models. Key questions are less related to how the future might unfold and more to where these changes will take place. Typical elements of area-based scenarios are the location of national parks, migration, or the construction of new roads. The focus of the model and therefore of the scenarios is on the distribution of land use. The availability of data can potentially limit the scope and richness of quantitative scenarios.

A particularly pressing issue is establishing the link between qualitative outputs from employing participatory methods and quantitative (i.e., data-demanding) models. Many of the recent (global) scenario efforts—such as those of the **Intergovernmental Panel on Climate Change**, the **United Nations Environment Programme**, and the **Millennium Ecosystem Assessment**—developed narrative storylines that were subsequently quantified and linked to models. However, in most documented cases, the link between the stories and models remains rather rudimentary, and both products are developed in parallel but separate exercises. Among the novel methods that have been employed is the use of **agent-based models** that can be directly parameterized by stakeholders.

Another type of scenario that has increasingly been developed in recent years is the multiscale scenario. VISIONS and MedAction are good examples of how multiscale scenarios can be developed using stakeholders and how decision support systems (DSS) can be used to quantify them. The Millennium Ecosystem Assessment has advocated a multiscale approach as well in its conceptual framework that serves as the guidelines for a large number (more than thirty) of sub-global assessments. The ultimate goal is to link a multitude of local studies through a meso-level assessment, which includes scenarios, to four global storylines.

Historically, land-use (change) scenarios are mostly developed to serve as inputs for land-use change models. Therefore, most land-use scenarios are quantitative rather than qualitative, although recently participatory methods are increasingly being employed to develop qualitative scenarios. As most currently available scenarios are linked to a model, they logically follow the framework of that model. The architecture of many land-use models consists of two parts: a large-scale, non-spatial part where total quantities of each land-use type are projected for a number of scenarios; and a small-scale spatially explicit part where those quantities are subsequently allocated. Land-use scenarios therefore often have a place-based component.

Projections of large-scale developments are usually related to processes that are slower and act over large areas like climate change, macro-economic developments, or institutional factors like the expansion of the European Union. Depending on the spatial extent of the study area, these non-spatial aspects can be related to quantitative storylines. In most land-use models, these large-scale developments are a direct input into the allocation part of the model.

The future changes of all dynamic drivers of land-use change that are spatially explicit are part of any land-use scenario. Typical examples are the construction of a new road—linked to accessibility, the institutionalization of a national park, or migration patterns—linked to population density. Often, these place-based assumed changes will heavily influence resulting land-use patterns.

Increasingly, scenarios are being recognized as one of the best tools in participatory approaches, like stakeholder workshops and focus groups.

Discussion on possible futures is an excellent way to avoid conflict, increase awareness, and start a process of mutual learning. It is therefore to be expected that the role of qualitative scenario development will increase in the coming decade. At the same time, DSSs have rapidly become easier to use, also by laypeople. Top-down, bottom-up iterative procedures linking locally developed narratives and DSSs that can be discussed during series of stakeholder workshops could be a great step forward.

See also: **Change Detection; Integrated Model to Assess the Global Environment; Land-Use Planning; Participatory Geographical Information System; Socializing the Pixel.**

Further Reading

Alcamo, Joseph, Kok, Kasper, Busch, Gerald, Priess, Jörg, Eíckhout, Bas, Rounsevell, Mark, Rothman, Dale, and Maik Heistermann, "Searching for the Future of Land: Scenarios from the Local to Global Scale," in *Land Use and Land Cover Change: Land Processes, Global Impacts*, eds. Éric F. Lambin and Helmut Geist, Berlin: Springer, 2006; Kahn, Herman, and Anthony Weiner, *The Year 2000: A Framework for Speculation on the Next Thirty-Three Years*, New York: McMillan, 1967; Meadows, Donella H., Meadows, Dennis L., Randers, Jørgen, and William W. Behrens III, *The Limits to Growth: A Report for the Club of Rome's Project on the Predicament of Mankind*, New York: Universe Books, 1972; Peterson, Garry D., Cumming, Graeme S., and Stephen R. Carpenter, "Scenario Planning: A Tool for Conservation in an Uncertain World," *Conservation Biology* 17 (2003): 358–66; Rotmans, Jan, Van Asselt, Marjolein B. A., Anastasi, Chris, Greeuw, Sandra C. H., Mellors, Joanne, Peters, Simone, Rothman, Dale S., and Nicole Rijkens-Klomp, "Visions for a Sustainable Europe," *Futures* 32 (2000): 809–31; Wack, Pierre, "Scenarios: Uncharted Waters Ahead," *Harvard Business Review* 63 (1985): 72–89; Xiang, Wei-Ning, and Keith C. Clarke, "The Use of Scenarios in Land-Use Planning," *Environment and Planning B: Planning and Design* 30 (2003): 885–909.

KASPER KOK

Secondary Vegetation. A transition state in which vegetation is in the process of recovering to a more mature state of primary vegetation following natural or anthropogenic disturbances. Much of the world's **vegetation** has been altered by a variety of anthropogenic forces. Natural events such as forest **fire**, landslides, and flooding also cause significant **modification** to vegetated landscapes, after which vegetation begins to recover. Following these types of disturbances, vegetative cover goes through a process of recovery to a more mature condition.

Secondary vegetation differs from primary vegetation in terms of species composition and vegetative structure. An early state of succession usually consists of hardy species that survive well under harsh conditions. These pioneer species can gradually change the ecological conditions that make the environment suitable to other species unable to initially colonize disturbed areas. For example, following forest clear-cutting, the arrival of new pioneer species can provide enough shade to allow the growth of species adapted to low sunlight conditions.

Some literature refers to secondary vegetation as a condition that specifically follows human disturbance while other literature includes both human and natural disturbances as precursors to secondary vegetation. The more inclusive definition has been used here. Specific examples of secondary vegetation include forest **regrowth** following agricultural or **land abandonment**, catastrophic fire events, or clearcutting. In addition, secondary vegetation is a key component of the cycle of **swidden cultivation**, in which secondary growth is a phase in a cycle of agricultural production. Thus, the term secondary vegetation applies even though the system will never actually reach a mature vegetated state.

Landscapes modified by human actions such as timber harvesting, pasturing, and **slash-and-burn agriculture** can revert to a more natural vegetated state after abandonment, but this process takes many years. The intensity and duration of disturbances affect the rate of vegetation recovery. An important part of the growth of secondary vegetation is the presence of a seed bank from which new vegetation growth can develop. Areas that have been under a disturbance regime such as cattle **pasture** for many years may have lost the seed bank necessary for the regrowth of tree species and thus take many decades to return to what can be considered a mature vegetated state.

It should be noted that a primary vegetation state is not a static endpoint target. Even in mature vegetated areas, species composition can change within local areas due to natural events such as tree falls. A large contiguous area of **forest** can be considered in a mature state, but within the larger forest there are local areas with different species compositions and ecological conditions. Thus, the **scale** and context of a vegetated area are important in considering whether the term secondary vegetation is applicable. Forests subject to small-scale disturbances such as natural tree falls or low-intensity selective cutting are generally not considered secondary vegetation.

See also: **Amazonia; Biomass; Boreal Zone; Driving Forces; Ecological Colonization; Fallow; Land-Use History; Pristine Myth; Proximate Causes; Reforestation.**

Further Reading

Chokkalingam, Unna, and Wil De Jong, "Secondary Forest: A Working Definition and Typology," *International Forestry Review* 3 (2001): 19–26.

THOMAS P. EVANS

Siltation. The accumulation of fine primary or flocculated particles (clay, silt, sand, and organic matter) in ditches, rivers, lakes, and other water bodies. Siltation is a natural subprocess of sediment transport occurring when flow velocity of water carrying suspended sediment is reduced below a particle or floc size depending on **threshold**. Alluvial plains, **wetlands**, and many landforms in the **coastal zone** result from siltation. Land-cover and land-use changes and alternations of watercourses (e.g., building of locks and dams) have a strong impact on rates of instream and downstream sediment transport and siltation.

A major concern is enhanced siltation of waterways, estuaries, and reservoirs due to **deforestation** and **soil erosion** in the catchment areas of water bodies. This threatens aquatic habitats; reduces water depth and flow capacity in waterways, eventually hindering navigation and increasing the flooding hazard; and diminishes storage capacity of flood control, fresh water, and hydropower reservoirs. The World Commission on Dams estimates that

Aswan High Dam in southern Egypt, March 2001 (before completion of the dam in 1970, 110 million tons of silt was deposited by the annual flood of the Nile, which is now trapped behind the dam).

PHOTO: Space Shuttle. CREDITS: Image Science and Analysis Laboratory, NASA-Johnson Space Center, 18 Mar. 2005 (Earth from Space—Image Information). http://earth.jsc.nasa.gov/sseop.efs

0.5–1 percent of the world's freshwater storage capacity in small and large reservoirs is lost to siltation annually. Land **conservation** measures reducing runoff, soil erosion, and sediment yield to watercourses help to reduce siltation of water bodies in the long term. Short-term measures to mitigate siltation include dredging and flushing of waterways and reservoirs.

See also: **Land Degradation; River Basin; Water-Land Interlinkages; Watershed.**

Further Reading

International Sedimentation Initiative [Online, March 2005], The ISI Web Site, portal.unesco.org; World Commission on Dams, Dams and Development: A New Framework for Decision-Makers [Online, March 2005], The WCD Web Site, www.dams.org/report.

JUSSI BAADE

Slash-and-Burn Agriculture. A colloquial as well as a pejorative term often referring to the method of clearing and preparing land that is common among swidden cultivators. Slash-and-burn can also refer to the practice of using **fire** simply to clear **forests** for permanent cultivation, **pasture**, or further development. This practice was employed on a massive scale during the European settlement of the frontiers of the eastern **United States of America**.

Some scientists argue that sustainable **swidden cultivation** is being replaced by unbalanced, unsustainable slash-and-burn agriculture because the population of swidden cultivators has expanded so rapidly that their traditional

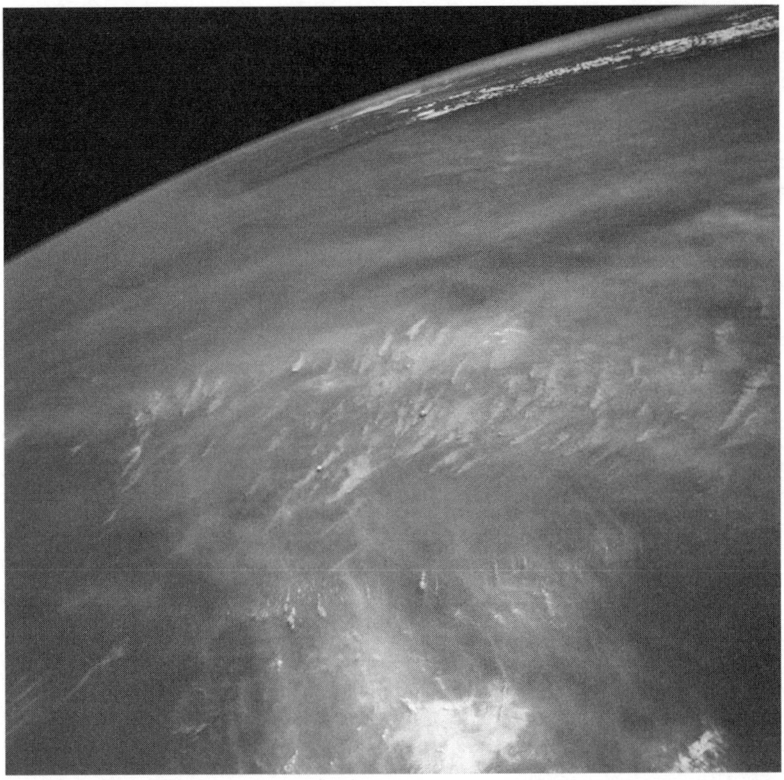

Major Smoke palls due to fires for land clearing by slash-and-burn practices in the humid tropical forest in the state of Rondônia, western Brazil, in September 1984.

PHOTO: Space Shuttle. CREDITS: Image Science and Analysis Laboratory, NASA-Johnson Space Center, 18 Mar. 2005 (Earth from Space—Image Information). http://earth.jsc.nasa.gov/sseop.efs

practices can no longer support their numbers. Under these circumstances, traditional cultivators reduce the length of **fallow**, leading to unsustainable slash-and-burn agriculture.

Norman Myers tries to distinguish between what he calls "shifting" and "shifted" cultivators. The first are traditional peoples who practice **sustainable land use** with a sufficiently long fallow period for soil fertility to regenerate. The second, the shifted cultivators, are small-scale farmers who find themselves squeezed out of traditional farmlands and feel they have no alternative but to pick up machete and matchbox and head for the forests as the last unoccupied lands available. Norman Myers estimate these peoples, who might be called slash-and-burn agriculturalists, as numbering 200–600 million.

Christine Padoch and colleagues point out that the problem with this distinction is that identifying particular forms of swidden as ideal (swidden cultivators) and others as problematic (shifted cultivators or slash-and-burn agriculturalists) is that many site- and situation-specific variations from those considered to be ideal may often be misinterpreted as shortcomings in technical expertise or agricultural performance.

See also: **Agricultural Frontier; Agricultural Intensification; Alternatives to Slash-and-Burn Programme; Colonization; Degradation Narrative; Extensification; Land-Use History; Secondary Vegetation; Tragedy of Enclosure.**

Further Reading

Fox, Jefferson, "How Blaming 'Slash and Burn' Farmers Is Deforesting Mainland Southeast Asia," *Asia Pacific Issues* 47 (2000): 1–9; Myers, Norman, "Tropical Forests: The Main Deforestation Fronts," *Environmental Conservation* 20 (1993): 9–16; Padoch, Christine, Harwell, Emily, and Adi Susanto, "Swidden, Sawah, and In-Between: Agricultural Transformation in Borneo," *Human Ecology* 26 (1998): 3–20.

JEFFERSON FOX

Socializing the Pixel. To "socialize the pixel" and to "pixelize the social" are to discern information embedded within spatial data on land-use/cover change that is directly relevant to the core themes of the land-change science agenda, by making **remote sensing** (RS) and **geographical information systems** (GIS) relevant to the social, political, and economic issues driving land-use/cover change. While pixel refers to a unit of observation associated with satellite data on land cover, the term is used more broadly here to encompass all spatial analysis on land-use/cover change, even if the data are from other sources. Remote sensing, both data and image processing, and analysis through GIS have played crucial roles in the development of land-use/cover change models in both data creation and analysis.

During the course of the past ten years, significant progress has been made in acquiring spatial land-cover data sets from remotely sensed data, conceptualizing the basic geographic and environmental processes that are associated with land-use change. Numerous spatially explicit and heterogeneous land-use change models now exist, spurred by the vast amount of spatial land-use/cover data that are now available. The research agenda of the **Land-Use/Cover Change (LUCC) project** (and its associated groups) has demanded an interdisciplinary approach to **modeling**, with major contributions made by landscape ecologists, geographers, anthropologists, political scientists, economists, and demographers.

While the goal of both "socializing the pixel" and "**pixelizing the social**" is to better understand the spatial pattern of land-use/cover change, there are broad generalizations of the differences between the two, as envisioned here. To "socialize the pixel" is to develop land-change models that move from the RS imagery beyond its use in the natural sciences and toward the concerns of the social sciences per se. These models usually have as the unit of observation the satellite data **pixel** and include explanatory variables that can be "seen" from the remotely sensed data and calculated using GIS, such as distance measures, other spatial biophysical variables (e.g., soil, slope, elevation); and occasionally socio-economic "drivers," such as population or other sociodemographic characteristics, usually measured at some aggregate level, such as village or state.

To "pixelize the social" is to apply social science theories to spatial data in order to develop an underlying structural model that seeks to explain the human behavior that generates these patterns of land-use/cover change.

These models usually take as the unit of observation the land under the control of an individual decision-maker and directly link the decision-maker to specific pixels. Of course, as research has progressed, and as more "socialization" has been done in the "socialize the pixel" realm and more "pixelization" has been done in the "pixelize the social" area, the boundaries between the two, rightly so, have become blurred.

Landscape ecologists were early developers of spatially explicit models of land-use change used to predict changes in spatial patterns of the landscape. The early models were simple grid-based models that merely calculated the percent change of each land-cover type during a time period and predicted future changes by assuming that these proportionate changes remained constant over time. More sophisticated models were then developed that estimated these changes as a function of other explanatory variables and not just simply a function of previous land-use changes. While many of these models had sophisticated treatment of ecological relationships that affected or were a result of land-use/cover change, they were very simple with respect to human behavior.

Subsequent models developed by geographers and other social scientists included some variables that captured some aspects of human behavior. For example, distance to urban centers, and variables that reflect the biophysical heterogeneity of the landscape have been commonly included for economic reasons. Distance to urban centers matters because the accessibility to markets affects transportation costs of getting agricultural goods to markets; whereas biophysical features of the landscape, for example certain soils being preferred for agricultural use, will affect the choices of land managers. Typically, these models estimate the effects of such explanatory variables as slope, soil, elevation, aspect, location, and population measures on different types of land-use/cover change. The results of these estimated models have often been intuitive. For example, in models that focus on **deforestation**, the effects of the explanatory variables are often as expected: the higher the elevation, the smaller the probability of deforestation; the further a pixel is from the road, the less likelihood of deforestation; the closer a pixel is to a market or a village, the greater the probability of deforestation.

However, there are issues of aggregation and **scale** with linking these diverse data sets. For example, in a single model, the size of the unit of observation, the satellite pixel, will often be of a different scale than the data for the explanatory variables. Slope and elevation data are often derived from a digital elevation model that can be at a different scale, and soils can come from maps at an entirely different scale, while any socio-economic census data will be at some other governmental entity scale, such as state or town, the size of which is often determined by population levels. Therefore, decisions must be made in how to best link these data for modeling purposes. For example, a model that includes socio-economic census data must somehow allocate those values of those variables to the individual pixels, such as assigning all pixels in a particular village the same value for population density, even though, in reality, not all pixels are likely impacted equally by the individuals living in the village. As a result, the estimated effect of such variables will only be the average effect on land-use change for the scale at which the variable is measured.

These estimated models are often subsequently used to simulate the effect of different **scenarios** on land-use change, such as to simulate the effect on future land-use change of a moratorium on logging or road building or the effects of **urbanization**, abolition of national parks, extension of national parks, **soil erosion**, or crop disease at certain elevations. However, for simulation or predictive purposes, all models are limited by the variables that are included in them and how closely these variables are linked to individual **decision-making** on land-use change. For example, in some models, the only simulations that can be performed are changes that are imposed on the explicit features of the landscape, such as building a new road or prohibiting certain land uses, as there are no explicit variables directly linked to human behavior. As a result, the impact of an agricultural policy change, such as a subsidy change on a farmer's decision to farm his land, cannot be predicted. The only way to simulate such a policy with these models would be to assume the land-use decision of the farmer in response to the policy.

Another consideration of the "socializing the pixel" models is that the unit of analysis is either an individual pixel or some aggregation of landscape units, rather than the individual decision-maker. For this reason, having information on the boundaries of individually owned land parcels, rather than just the boundaries between two dissimilar land-use pixels, is greatly preferred. For example, individuals owning large land parcels may react differently to a policy than those with small land parcels. Distinguishing the effects of a policy change among large and small landowners is important and only possible if ownership boundaries are known.

See also: **Cadastre; Driving Forces; Forest Anomalies; Landholding; Land-Use System; Pattern to Process; Property.**

Further Reading

Geoghegan, Jacqueline, Pritchard, Lowell, Jr., Ogneva-Himmelberger, Yelena, Chowdhury, Rinku Roy, Sanderson, Steven, and B. L. Turner II, "'Socializing the Pixel' and 'Pixelizing the Social' in Land-Use and Land-Cover Change," in *People and Pixels: Linking Remote Sensing and Social Science*, eds. Diana Liverman, Emilio Moran, Ronald Rindfuss, and Paul Stern, Washington, DC: National Academy of Science Press, 1998, pp. 51–69; Irwin, Elena G., and Jacqueline Geoghegan, "Theory, Data, Methods: Developing Spatially-Explicit Economic Models of Land Use Change," *Agriculture, Ecosystems and Environment* 84 (2001): 7–24; Verburg, Peter H., Kok, Kasper, Pontius, Robert Gilmore, Jr., and A. (Tom) Veldkamp, "Modelling Land Use and Land Cover Change," in *Land Use and Land Cover Change: Local Processes, Global Impacts*, eds. Éric Lambin and Helmut Geist, Berlin: Springer, 2006.

JACQUELINE M. GEOGHEGAN

Soil Degradation. The loss or deterioration of soil properties and soil ecological and socio-economic functions due to direct or indirect human interventions (sometimes more narrowly defined as the reduction of soil productive capacity due to human action). Natural changes of soil properties and soil functions are called soil formation. Soil degradation is often used as an indicator for **land degradation**, and sometimes soil degradation and land degradation are used synonymously. Because of the close relationship between soils and land cover, it is a major concern under changing land use and land cover.

Situated at the interface of the lithosphere, atmosphere, biosphere, hydrosphere, and anthroposphere, soils play a vital role in supporting life and human well-being. Soils filter and store water, plants root in soils providing nutrients and water, soils sequester carbon, and most human activities take place on soils. Under natural conditions, soils are in a dynamic equilibrium with climate and **vegetation**. In most cases, human utilization of soils includes the removal of the natural vegetation, resulting in a deterioration of the dynamic equilibrium and ecological soil functions while fulfilling socioeconomic functions. This seems to be an unavoidable conflict of interests.

Processes contributing to soil degradation include a wide range of often-inadvertent events ranging from local contamination from leakages to the worldwide erosion of soils accompanying some agricultural management practices. Frequently, processes contributing to soil degradation are related to each other in positive **feedback** mechanisms. **Soil erosion** and soil organic matter depletion are a good example.

The Global Assessment of Soil Degradation (GLASOD), up to date the only consistent global evaluation of human-induced soil degradation, distinguishes four main types of soil degradation: water erosion, wind erosion, chemical deterioration (including loss of nutrients and organic matter, **salinization**, acidification, and pollution), and physical deterioration (including compaction, sealing and crusting, waterlogging, and subsidence of organic soils). Taking into account the degree and the relative extent of degradation, severity of soil degradation was assessed on a national level by expert judgment.

On the global **scale**, 15 percent of all soils accounted for (130×10^6 km^2) were found to be degraded. Strongly and extremely degraded soils considered irreclaimable cover an area of 3×10^6 km^2, mostly in Africa and Asia. The most important processes are soil erosion by water and by wind, causing degradation on 56 percent and 28 percent of the degraded soils, respectively. Chemical deterioration affects 12 percent and physical deterioration 4 percent of the degraded area. The three most important causative factors are overgrazing, **deforestation**, and agricultural mismanagement, which are held responsible for soil degradation on 92 percent of the affected area. Soil degradation of the remaining area is due to overexploitation of the vegetation cover for domestic use and bio-industrial activities.

See also: **Anthropocene; Carbon Sequestration; Cotton; Degradation Narrative; Desertification; Modification; Monoculture; Pasture.**

Further Reading

EEA (European Environment Agency), Environment in the European Union at the Turn of the Century (Chapter 9), Copenhagen: EEA, 1999 [Online, March 2005], http://reports.eea.eu.int/92-9157-202-0/en; Natural Resources Conservation Service, World Soil Resources [Online, March 2005], The NRSC Soils Web Site, soils.usda.gov/use/worldsoils; Oldeman, L. R., Hakkeling, R. T. A., and W. G. Sombroek, *World Map of the Status of Human-Induced Soil Degradation: Global Assessment of Soil Degradation (GLASOD)*, 2nd ed., Wageningen, NL: International Soil Reference and Information Center, Nairobi, Kenya: United Nations Environment Programme, 1991.

JUSSI BAADE

Soil Erosion. The accelerated displacement of soil or parent material by natural drivers, that is, water, wind, and gravitational processes, following

Soil erosion in croplands and remaining chimney of abandoned farm house in Greene County, Georgia, April 1939.
PHOTO: Marion Post Wolcott. [Courtesy Library of Congress]

Soil erosion gullies on farm land in Caswell County, North Carolina, October 1940.
PHOTO: Marion Post Wolcott. [Courtesy Library of Congress]

human interference with the landscape. The natural displacement of soil or parent material lowering hills and mountains and filling up basins since millions of years ago is referred to as (geological) erosion or denudation. Soil erosion is a major driver of **soil degradation**, contributes to the siltation and pollution of water bodies, and might impair air quality. Climate, soil properties, and topography determine the type of, and **vulnerability** to, soil erosion. However, land use, land cover, and land management are the prime factors controlling the extent and severity of soil erosion.

Soil erosion is a complex phenomenon resulting from human interference with landscape properties that control the evolution of landforms. Interacting in a complex manner, climate, soil properties, and topography set the scene for soil erosion to determine the (prevailing) type of soil erosion, that is, soil erosion by water or by wind and the vulnerability of a landscape to soil erosion. For example, sloping topography is a prerequisite for soil erosion by water, but not for soil erosion by wind; the steeper a slope, the higher the risk of severe soil erosion by water. Sandy soils are very susceptible to soil erosion by wind, but less so to soil erosion by water. Humid climate conditions hinder soil erosion by wind, while semiarid areas are subject to soil erosion by both wind and water. The third type, soil erosion by gravitational processes, is often associated with soil erosion by water, like shallow land sliding at the edge of gullies formed by water.

Climate, soil properties, and topography set the scene, but human land use and land management cause soil erosion and control its extent and severity. Vegetation cover and soil management are crucial. Any disturbance of the vegetation cover protecting the soil against the shear stress exerted by wind and flowing water or the impact of raindrops detaching soil particles can trigger soil erosion. Deterioration of soil properties (which usually accompanies the disturbance of the vegetation cover), like compaction or loosening of the soil surface, reduction of soil organic matter content, and soil infiltration and water storage capacity, increases the susceptibility to soil erosion.

This implies that a wide range of direct and indirect human activities can cause soil erosion. The most common and fateful ones are **deforestation**, overgrazing, and tillage-based **agriculture**. However, in fragile environments even the trampling of foot paths can result in soil erosion. Therefore, soil erosion has to be considered an inevitable consequence of human utilization of the earth's surface, at least to some extent.

Soil erosion results in onsite and offsite damages. Onsite damages include the loss of minerogenic material, soil organic matter, and nutrients, contributing to an often-gradual degradation of soil depth and soil fertility. Ultimately, this degradation might reach a point where the land-use practice causing the soil erosion is brought to an end. Strongly eroded and abandoned field sites presently either under **forest** cover, as in some sites in the northern loess zone of **Europe**, or already bare of soil as in some places in the Mediterranean, highlight the ultimate onsite consequences of soil erosion. Offsite damages result from the deposition of material mobilized by soil erosion. The covering of adjacent fields or roads with blown or washed-out sediment, the **siltation** of waterways and reservoirs, and the input of nutrients and pesticides into aquatic habitats are issues of concern.

The fact that **vegetation** cover and soil management are crucial for soil erosion implies the opportunity to control soil erosion. Basically, activities to mitigate soil erosion are characterized by either reducing disturbance of plant cover and soil properties or even improving plant cover, both in time and space (e.g., **fallow** plant cover, fertilization of grassed areas); and strengthening soil structural stability and soil hydraulic conductivity by reducing or even giving up tillage operations. More technical and more costly solutions include terracing of sloping fields, implementing subsurface drainage, and treating fields with soil stabilizing chemicals. The basic principles have been well known for decades. However, the utilization of soil-conserving land-management measures often meets resistance due to traditional, practical, and economic reservations. However, discussions of **carbon sequestration** in soils might create a new momentum for the application of soil **conservation** measures, which not only increase soil organic matter content, but also help to mitigate the soil erosion problem.

Soil erosion is a problem controlled by several factors acting at the field scale and might therefore vary from field to field. This makes it rather difficult to assess soil erosion on smaller scales, that is, regions, nations, continents, or the whole world. Often, models are used to estimate the soil erosion risk based on climate, soil properties, and topography. However, the disadvantage is that land use and land management, that is, the factors determining the actual severity of soil erosion, are not taken into account. The Global Assessment of Soil Degradation (GLASOD) estimated the global dimension of soil erosion based on expert judgments from all major countries. According to GLASOD, soil erosion by water (including mass movements) and by wind (including overblowing as an offsite effect) are the most important processes causing soil degradation: 56 percent of areas with degraded soils are deteriorated from soil erosion by water and another 28 percent from soil erosion by wind. The results have been criticized for several reasons; and taking into account the considerable changes in land use over the past decade, the data are certainly in danger of becoming obsolete. However, the GLASOD assessment remains to date the only consistent global evaluation of the actual status of soil erosion.

See also: **Agricultural Revolution; Cash Crops; Cassava; Degradation Narrative; Desertification; Environmental Change; Hot Spots of Land-Cover Change; Land Degradation; Land-Use System; Monoculture; Nutrient Cycle; Water-Land Interlinkages; Watershed.**

Further Reading

Favis-Mortlock, Dave, Soil Erosion Site [Online, March 2005], The Soil Erosion Web Site, www.soilerosion.net; Morgan, Roy, *Soil Erosion and Conservation*, 3rd ed., Oxford, UK: Blackwell, 2005; Oldeman, L. R., Hakkeling, R. T. A., and W. G. Sombroek, *World Map of the Status of Human-Induced Soil Degradation, Global Assessment of Soil Degradation* (GLASOD), 2nd ed., Wageningen: International Soil Reference and Information Center (ISRIC), and Nairobi, KE: United Nations Environment Programme (UNEP), 1991 [Online, March 2005], www.isric.nl/Docs/Glasod.zip.

JUSSI BAADE

South Africa. Lying at the southern tip of Africa, the present-day landscape of South Africa reflects the country's colonial history and particularly the racial policies pursued during the twentieth century. Like the rest of **southern Africa**, hunter-gatherer hominids have been present in the area for over a million years, and pastoralism and cultivation have been important land uses for about 2,000 years.

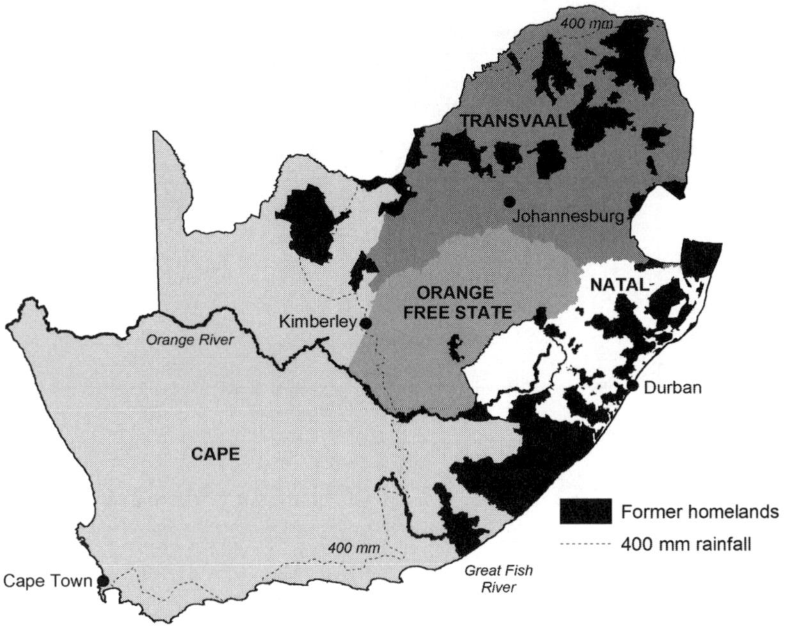

The pre-1944 political units of South Africa.

Rainfall in South Africa decreases from east to west, forming a dry western half (<400 mm per year), and a wetter eastern half and southern coastal plain (400–1500 mm p.a.) (see figure titled "The pre-1944 political units of South Africa"). Most of the region receives rainfall in summer, except for the extreme southwest, which receives winter rainfall. The dry western region is covered by shrublands (the succulent and the Nama Karoo) and the fine-leafed acacia savannas of the Kalahari region. The wetter eastern half consists of **grasslands** and either broad-leafed savanna on sandy infertile soils, or acacia savannas on more fertile clay soils. The southwestern tip was originally covered by sclerophyll thicket, known as *fynbos*; both the eastern grasslands and the *fynbos* are now substantially transformed by **agriculture**. Small, isolated patches of **forest** occur along the **mountain** escarpment in the southern and eastern parts of the country.

By 1900, approximately 3.5 percent of the total land area of South Africa was cultivated; 96.2 percent was covered by natural **vegetation**, and 0.3 percent consisted of urban settlements (see figure titled "Changes in land use"). At the end of the twentieth century, 85 percent of South Africa was still covered by natural vegetation, while 14 percent had been converted to planted crops and 1.3 percent to urban land uses. Approximately 1.3 million hectares (8 percent of the cultivated area) was under irrigation. The untransformed area in 2000 consisted of 5 percent formally protected land, while 75 percent was used for activities such as livestock ranching and, increasingly, nature-based tourism ventures. Approximately 5 percent of South Africa was covered by natural vegetation but in a degraded state. About 68 percent of the country consisted of freehold land, 14 percent was communal tenure, and 18 percent was owned by the state. South Africa is currently a net food

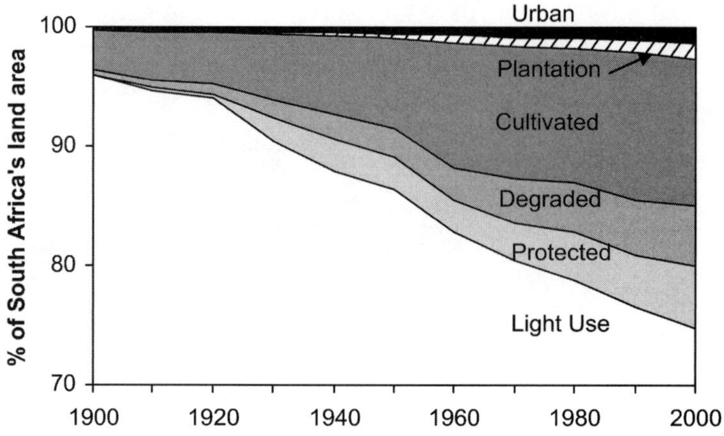

	Light use	Protected	Degraded	Cultivated	Plantation	Urban
1900	95.94	0	0.41	3.37	0.02	0.26
1910	94.65	0.32	0.62	3.97	0.1	0.34
1920	94.02	0.32	0.85	4.29	0.16	0.37
1930	90.51	1.88	1.42	5.52	0.25	0.42
1940	87.87	2.7	2.06	6.56	0.31	0.5
1950	86.39	2.72	2.41	7.52	0.37	0.6
1960	82.8	2.79	2.68	10.37	0.69	0.67
1970	80.52	3.02	3.81	11	0.81	0.83
1980	78.76	4.03	4.26	11.14	0.94	0.87
1990	76.55	4.39	4.64	12.33	1.01	1.08
2000	74.76	5.3	4.95	12.23	1.47	1.29

Changes in land use over the twentieth century

exporter in most years, and produces a wide range of food, timber and fiber, and horticultural, cosmetic, and medicinal crops.

European Settlement (1652–1910)

During this period, land-cover impacts in South Africa closely tracked the growth and spread of the human population. The Dutch East India Company set up a trading post in 1652 in present-day Cape Town to provide fresh food for the company's ships that rounded the Cape on their way to East Africa and Asia. The settlement expanded over the next 150 years, and by the early 1800s, approximately 26,000 European farmers had settled the wetter region along the southern coast between Cape Town and the Great Fish River (see first figure). These immigrants engaged primarily in **subsistence agriculture** and produced little for export. During this time there were also dispersed pockets of subsistence cultivation in the interior east of the 400-millimeter rainfall line, which had existed for up to 1,500 years, in the areas settled by various Bantu-speaking African tribes. The largest settlements were those of the Zulu and Xhosa people along the east coast, and the Sotho, Ndebele, and Venda people in the northeastern regions of present-day South Africa. To the west of the 400-millimeter rainfall line (the limit of cultivation) were the sparsely distributed Khoikhoi (also called the San), who were hunter-gatherers and pastoralists.

Upheavals in the political, economic, and social structures of the Bantu-speaking communities in the first half of the nineteenth century led to widespread migrations and dramatically altered ownership and land-use patterns. The period is known as the *mfecane* or *difaqane* and was brought about by a combination of factors including population growth, depletion of natural resources, and devastating drought and famine. The slow southward migration of the Bantu people in Africa over the preceding 2,500 years had reached its southern limit, and pressure for land increased. The introduction of **maize** (*Zea mays*) from the Americas in the eighteenth century via the Portuguese in Mozambique had enabled a growth in the population, as maize produced more food per unit land or labor than the main indigenous staples sorghum and millet. However, maize requires more water than local grains, and the onset of a severe ten-year drought starting in about 1800 had devastating impacts as people fought for supplies of grain and cattle. Large numbers of people were displaced, and changes in military organization took place. Warfare continued over the following decades and led to the establishment of a number of Bantu states with large armies, the most successful and aggressive being that of the Zulu kingdom. To secure foodstuffs and protect themselves from the Zulu marauders, several defensive states formed in the 1820s and 1830s, most notably the Swazi and Sotho kingdoms, which form the core of the modern-day countries of Swaziland and Lesotho.

Between 1795 and 1806, the British captured the Cape Colony from the Dutch. The British attempted to alleviate the land problems of the expanding European population in the southwestern Cape by sending imperial armies to fight against the Xhosa in the region of the Great Fish River, expelling most of the African population in that area by about 1820. In order to stabilize the frontier, about 5,000 British immigrants assisted by the British government were settled on smallholder farms in the region at this time. British rule and the increasing number of British settlers, however, engendered growing hostility from the Boer community (farmers primarily of Dutch, German, and French descent). Beginning in 1836, about 12,000 people, Boer families together with their servants, left the eastern Cape frontier and traveled by ox-wagon into the high-lying interior. This movement, known as the "Great Trek," led to the eventual establishment of two republics (the Orange Free State in the 1840s and the South African Republic, later known as the Transvaal, in the 1850s) in the northeast of present-day South Africa (see first figure). The economic mainstay of the Boer republics during the first thirty years of their existence was wildlife, and especially a trade in hides and ivory. The abundant herds of large ungulates were eliminated by 1870. By this time approximately 50,000 migrants of European descent had settled in the two republics, practically all living in rural areas and, like the African people among whom they settled, making a living from small-scale subsistence agriculture, extensive pastoralism, and hunting. During the period from 1800 to 1865, the total European population increased twelve-fold from about 20,000 to nearly 250,000 (180,000 in the Cape Colony, 18,000 in Natal, and 50,000 in the Boer republics).

The diffusion of British and Boer farmers into the interior had two major impacts. First, the Europeans had firearms and shot large numbers of indigenous mammals, so that the animals' populations were substantially reduced.

Herbivores, and particularly the megaherbivores, play an important role in ecosystem processes. Their removal, together with altered **fire** regimes, has been blamed for the "bush encroachment" problem that has rendered an estimated 2.5 percent of the savanna region of South Africa uneconomical for livestock ranching. Second, modern technologies, particularly the establishment of boreholes in the late nineteenth century, enabled farmers to introduce large numbers of livestock into previously inhospitable parts of South Africa, such as the arid karoo area. The marked decline in livestock numbers in many of these areas in the early twentieth century, following an initial peak, has been taken as evidence of **land degradation** in these environments due to early overstocking.

The discovery of diamonds in 1867 and large deposits of gold in 1886 initiated the shift to an economy based on manufacturing and service industries. Mineral discoveries attracted large numbers of European immigrants and African migrants in the late nineteenth century. The diamond mines of Kimberly brought about a doubling of the European population in the Cape Colony between 1865 and 1900, to reach a total of 400,000. Similarly, in fewer than fifteen years, the gold mines in Johannesburg attracted a population of 75,000 Europeans and 100,000 African laborers, mostly from the Portuguese-ruled areas of Mozambique. The growth of the urban population necessitated the establishment of surplus-producing commercial agriculture and the first large-scale commercial timber plantations to provide support timber for the mines. The discovery of gold in the South African republic was the major factor leading to the South African War (1899–1902), in which the British captured the Boer republics, and eventually led to the formation of the Union of South Africa in 1910, establishing the country's present-day boundaries. The **urbanization** of thousands of Africans led to the establishment of large multiracial communities in the new industrial cities, competition between the African and European populations for jobs, and increasing bargaining strength on the part of the Africans that was seen as a threat to the supply of cheap labor for the mines. These threats prompted the introduction of the first pass law in the Cape Colony in 1872, limiting the mobility of and economic opportunities for Africans. This formed the foundation for subsequent discriminatory laws that had far-reaching effects on the South African society, economy, and environment throughout the twentieth century.

The Period of White Dominance (1910–1994)

Following the unification of South Africa in 1910, several pieces of racially discriminatory legislation were introduced, the most important of which was the Natives Land Act of 1913. This act restricted land ownership for Africans (4 million people; two-thirds of South Africa's population) to 7.5 percent of the country's land area, and only on a customary rather than a freehold basis. Although this was increased to 13 percent of South Africa in 1936, the legislation had significant impacts, and the effects on land-cover are still clearly visible today (see image). The population densities in these "homeland" areas were five to ten times greater than in the rural areas in the rest of South Africa. At least 1.5 million people were forcibly removed from white-owned areas and resettled in the homelands over the course of the twentieth century. Most of the population were peasant farmers, but the small **landholdings**; insecure tenure; and the lack of capital and **access** to technology, advice, and markets made it very difficult to make a living from farming. When apartheid ended in 1994, the average African farm was 5.2 hectares in size, while the average white-owned farm was 1,300 hectares. The result was that most of the economically active sector of the population worked as poorly paid migrant laborers in the cities of "white" South Africa, leaving the elderly and children to manage the land. The consequent shortage of able-bodied labor in the homelands, together with the lack of capital for agricultural inputs and conservation, led to deteriorating land productivity in many of these areas. Gathering wood, dung, and crop residues to provide fuel for cooking and heating further reduced soil fertility. Large numbers of livestock, kept primarily as a form of capital investment rather than for meat, were a significant factor in the denudation of vegetation in these areas. The so-called "betterment" schemes, implemented from the 1940s to the 1970s, aimed to address land degradation in these areas, but largely failed as they did not address the underlying socio-political problems.

Land degradation has also been an issue of significant concern in white-owned commercial farming areas since the late nineteenth century. Efforts

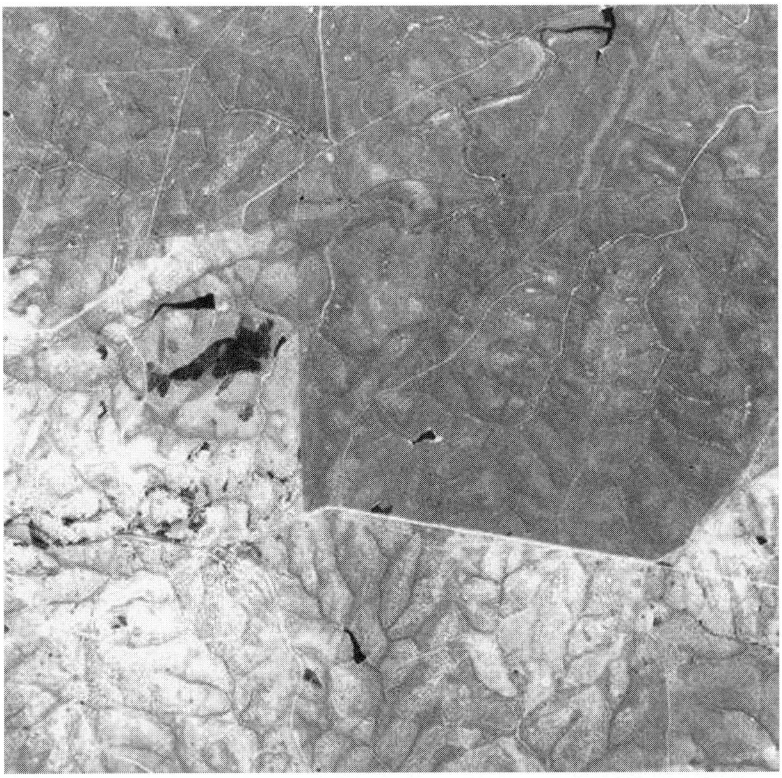

Land sat image showing the Krüger National Park (north) and denuded former homeland area (south).

CREDITS: Tobias Landmann, Council for Scientific and Industrial Research, South Africa to combat land degradation took the form of government subsidies for conservation works, legislation, agricultural extension, and farmer study groups, and were generally successful at improving the **conservation** status of these areas. The effects of drought and **soil erosion** received particular emphasis. The practice of corralling domestic livestock at night was identified as a major factor leading to soil and vegetation degradation, and consequently a program of exterminating jackals and other predators was initiated in the early twentieth century. Cheap fencing material was provided to create paddocks for rotational grazing management, and water points were provided so that livestock did not have to walk long distances. Particularly influential was the work of John Acocks, who believed that the margins of the Karoo shrubland had been expanding into the grasslands since colonial times as a result of overgrazing by commercial small-stock farmers. The expanding Karoo hypothesis, subsequently found to be unsupported by evidence, heavily influenced policy development until the 1980s, specifically interventions such as the Stock Reduction Scheme (1969–1978) and the National Grazing Strategy (1985). Soil conservation and veld management acts were introduced in the mid-1900s and required landowners to, among other things, obtain permission for the cultivation of virgin soil and steep slopes, to prevent waterlogging and **salinization**, to restore eroded land, to control the number and type of livestock kept on the land, and to control wild fires and invader plants.

The government's racial policies led to growing international isolation from the 1950s until the end of apartheid in 1994. Consequently, agricultural policies increasingly emphasized food self-sufficiency. White farmers received privileges such as loans from the government's Land Bank, labor law protection, and crop subsidies. Marketing boards were established to stabilize the production of many crops, and various drought subsidies and rebates existed to keep white farms economically viable. These policies were partly responsible for the 4-million-hectare (3 percent of South Africa's land area) increase in cultivated area that occurred between 1955 and 1965. Large increases in crop yields of white-owned commercial farms occurred in the 1960s, resulting from the introduction of higher-yield cultivars, pesticides, fertilizers, and irrigation. In contrast, the productivity in the homeland areas throughout the twentieth century stayed at the same level it was in the early 1900s.

The New Democracy (1994–Present)

The end of apartheid brought the need to reverse the discriminatory laws around **land rights** and **land tenure**. Two **land reform** programs have been instituted: a legally driven restitution process to return land that was confiscated in terms of the 1913 and 1936 Land Acts; and a market-based redistribution process to significantly increase the share of South Africa under freehold tenure by Africans. The marketing boards have been phased out, and agricultural produce is now freely traded. Currency fluctuations are a significant factor impacting the economic viability of especially export agriculture. The total area under cultivation is unlikely to expand much due to the climatic limitations of the country. On the other hand, nature-based tourism is growing annually at approximately five times the rate of the agricultural sector, resulting in a significant shift out of pastoral land uses and into tourism ventures, changing the policy emphasis in the country. According to the United Nations medium-variant projections, the total population of South Africa is expected to stay at approximately the current size over the next half-century, but become increasingly urbanized. This is likely to bring an increasing demand for protein as people grow wealthier, which is likely to be met by increasing use of feedlots in livestock production.

See also: **Cattle Ranching; Colonization; Degradation Narrative; Green Revolution; Mineral Extraction; Spatial Diffusion.**

Further Reading

Acocks, John P. H., "Veld Types of South Africa," *Memoirs of the Botanical Survey of South Africa* 8 (1953): 1–128; Biggs, Reinette, and Robert J. Scholes, "Historical Changes in Natural Capital in South Africa: An Approach Using Changes in Biodiversity," in *Restoring Natural Capital in the 21st Century: Emerging Countries of the South*, eds. Sue J. Milton and James Aranson, Washington, DC: Island Press, 2005; Biggs, Reinette, and Robert J. Scholes, "Land-Cover Changes in South Africa 1911–1993," *South African Journal of Science* 98 (2002): 420–4; Byrnes, Rita M., ed., *South Africa: A Country Study* [Online, September 2004], Federal Research Division, United States Library of Congress, Country Study Web Site, http://countrystudies.us/south-africa; Dean, W. Richard J., and Ian A. W. MacDonald, "Historical Changes in Stocking Rates of Domestic Livestock as a Measure of Semi-Arid and Arid Rangeland Degradation in the Cape Province, South Africa," *Journal of Arid Environments* 26 (1994): 281–98; Fairbanks, Dean H. K., Thompson, Mark W., Vink, Dineke E., Newby, Terence S., van den Berg, Hendrik M., and David A. Everard, "The South African Land-Cover Characteristics Database: A Synopsis of the Landscape," *South African Journal of Science* 96 (2000): 69–82; Grossman, David, and Mark V. Gandar, "Land Transformation in South African Savanna Regions," *South African Geographical Journal* 71 (1989): 38–45; Hoffman, M. Timm, and Ally Ashwell, *Nature Divided: Land Degradation in South Africa*, Cape Town, South Africa: University of Cape Town Press, 2001.

REINETTE (OONSIE) BIGGS, ROBERT J. SCHOLES

South America. This continent covers an area of approximately 17.8 million km^2; the west of South America is dominated by the **Andes** Mountains, which form the continental divide between the Atlantic and Pacific oceans, and a coastal belt extending from Tierra del Fuego to Panama; while the east is dominated by the Amazon (5.87 million km^2), Orinoco (966,800 km^2), and Paraguay-Paraná (3.1 million km^2) **river basins**, which drain 55 percent of the continent, and the Guiana and Brazilian highlands, and smaller **watersheds** in the north and in southern Argentina, which flow into the Caribbean and South Atlantic, respectively.

Extending from approximately 11° north to 56° south, South America's climate, soils, and **vegetation** are diverse. Tropical humid, dryland, mid-latitude, polar, and **mountain** climate zones are represented on the continent. The main climatic influences are the Intertropical Convergence Zone, the Southern Hemisphere air masses, and the orographic effect of the Andes. Natural vegetation and non-irrigated agricultural land uses are in dynamic equilibrium with climate. Short-term perturbations, particularly the El-Niño Southern Oscillation (ENSO), affect normal climate patterns.

The **Global Land Cover Map of the Year 2000** map for South America indicates that a little over 75 percent of South America was covered by natural vegetation with varying levels of disturbance. Four types of **forests** cover 46.4 percent of the land area (humid tropical, 35.5 percent; dry tropical, 8.3 percent; flooded tropical, 1.5 percent; and temperate, 1.1 percent), **grasslands** and shrublands cover 18.1 percent, and steppe vegetation covers 6.9 percent. South America is globally important in terms of **biodiversity**. Five of the world's twenty-five biodiversity hotspots—the Atlantic forest, **Brazilian Cerrado**, central Chile, Choco-Darien-western Ecuador, and tropical Andes—cover 945,130 km^2 (5.3 percent) of the continent, of which 24.5 percent is protected. The expansion of protected areas for **conservation** has been an important driving force in land-use change during the 1980s and 1990s.

Migration and urban population growth have been important land-use change drivers in the region since the eighteenth century. Population reconstructions for the **BIOME 300 project** show low densities ($<$5 persons/km^2) during the eighteenth century with exception areas around Rio de Janeiro, Santiago de Chile, and Lima. During the nineteenth century, these cities provided foci for population growth. The high population densities around Lima pushed northward along the Pacific coast and southward into the Andes. By 1900, high population densities existed from Lima to northern Venezuela along the coast and in the Andes (particularly major cities such as Bogotá) and southward to Cochabamba. High population densities existed in the northeast Brazil drought triangle (Alagoas, Paraíba, and Pernambuco states) by 1850, and by 1900 high population densities extended from Maranhão (northeast Brazil) along the Atlantic coast to Buenos Aires province (Argentina). The early spatial distributions of population are reflected in the contemporary distribution of people. Low densities ($<$5 persons/km^2) in much of **Amazonia** are surrounded by an arc of high population density that extends from eastern Venezuela through Colombia, Ecuador, Peru, Bolivia, northern Argentina to Buenos Aires, and then northward along the Atlantic coast to northeast Brazil. A second belt of relatively dense population extends from Buenos Aires through Córdoba and Mendoza provinces to central Chile.

Approximately 266 million South Americans people live in urban areas, and urban populations are particularly high in Argentina, Chile, Uruguay, and Venezuela. **Urbanization** is an important land-use change driver, even though urban areas account for only about 0.1 percent (11,400 km^2) of the land area. Large, rapidly growing urban areas (e.g., Buenos Aires, Lima, Rio de Janeiro, and São Paulo) continue to be locally significant drivers through **urban sprawl**, by the demands exerted on rural areas in terms of goods and services, and by attracting rural out-migrants.

Migration to South America, particularly from **Europe** and **Japan**, from the mid-nineteenth to early twentieth centuries, underpinned much of the continent's population growth. Many migrants settled in the **Southern Cone region**, and the area under cropland expanded considerably. The subtropical grasslands (pampas) of Argentina and Uruguay were converted to large farms from the mid-nineteenth century to rear cattle (for meat and hides) and cereals for export. The Argentine pampas now mainly comprise highly mechanized commercial farms, while in Uruguay the pastoral economy is stagnating. Patagonia saw development of large sheep estancias at the same time at the expense of steppe vegetation. The grasslands and shrublands of interior and northern Argentina have been converted more recently for sugarcane, citrus, and soybean cultivation. Large estancias producing beef, **maize**, sugarcane, and **cotton** characterized southern and eastern Paraguay from the mid-nineteenth century, but recently soybean cultivation has expanded and the dry thorn shrublands and woodlands of northern Paraguay have become areas of **colonization**. According to the **United Nations Environment Programme**, most of the continent's drylands were undergoing **desertification** at the end of the twentieth century.

Southern Brazil also witnessed much immigration in the mid-nineteenth and early twentieth centuries, and the region's sub-tropical and temperate forests (the Atlantic forest) were converted for **cattle ranching**, **coffee**, and soybeans. The loss of the Atlantic forest extended southward into Paraguay, Uruguay, and Argentina, and northward along the Brazilian coast. The staggering decline in **tropical humid forests** (16 million hectares between 1990 and 1997, a rate of 0.38 percent per annum) is generally a more recent phenomenon. The majority of these forests in South America are found in the Amazon basin, which covers the Brazilian Legal Amazon and parts of Bolivia, Peru, Ecuador, Colombia, and Venezuela. Significant land-use conversions to these sparsely populated forests began about 150 years ago in **Amazonia** with a succession of boom-and-bust cycles based on **non-timber forest products** (most notably rubber), and demands for land from mining, timber extraction, and livestock. However, most land-use conversion has taken place in the last thirty to fifty years, and its underlying driving forces have been the development of regional economic markets (e.g., MERCOSUR), integration into global economic markets, human **population dynamics** (especially the need to resettle the rural and urban poor), and infrastructure development as part of national development planning. Much of this has taken place in an **arc of deforestation** in eastern and southern Brazil; in western Amazonia in Bolivia, Peru, Ecuador, and Colombia; and in southern Venezuela. Major drivers have been legal and illegal timber extraction, **mineral extraction**, hydroelectric power generation, the formation of cattle ranches, and the expansion of croplands. The conversion of forests to **pastures** has been a major land-use change in Brazil since 1960s, and continues to be so even though government subsidies have been withdrawn and conversion has proven unsustainable in the medium-term. Many pastures are being converted to soybean cultivation. In western Amazonia, there have been efforts to settle people at the western margins of the Amazonian lowlands, along the Carretera Marginal de la Selva. A number of deforestation hot spots occur in this area, some of which are coca production zones. Western Amazonia and foothills of the Andes from Lake Maracaibo to northern Argentina are major hydrocarbon production zones. Oil and gas exploitation have led to localized **deforestation** and provided **access** to settlers. Much of the montane tropical forest of the Andes is protected for nature conservation, but mid-altitude forests in Colombia and Peru have been extensively cleared for coffee. The temperate forests and alpine ecosystems of the southern Andes have been conserved mainly for watershed protection and/or tourism or converted during the twentieth century to commercial forests with introduced species such as Monterrey pine (*Pinus radiata*). In central Argentina and Chile, significant amounts of land have been converted to viticulture.

See also: **Agricultural Frontier; Agricultural Intensification; Economic Growth; Exotic Species; Hot Spots of Land-Cover Change; Land Degradation; Malaria; Poverty; Tropical Dry Forest.**

Further Reading

Denevan, William, *Cultivated Landscapes of Native Amazonia and the Andes*, Oxford, UK: Oxford University Press, 2001; Eva, Hugh D., Belward, Alan S., De Miranda, Evaristo E., Di Bella, Carlos M., Gond, Valéry, Huber, Otto, Jones, Simon, Sgenzaroli, Matteo, and Fritz Steffen, "A Land Cover Map of South America," *Global Change Biology* 10 (2004): 731–44.

ANDREW MILLINGTON

Southeast Asia, Mainland. Mainland Southeast Asia includes the nations of Cambodia, Laos, Myanmar, Thailand, and Vietnam. The region harbors an immense wealth of natural resources, including globally important stocks of **forests** and biological diversity as well as many major river systems. It is also home to a rich heritage of indigenous cultures. Little is known about land-use and **land-cover change** in the region before the early 1800s when the British began to map the boundaries between their newly acquired territories in Myanmar and Siam. John Richards and Elizabeth Flint documented land-use and land-cover change in mainland Southeast Asia between 1880 and 1980. They concluded that agricultural expansion and, to a lesser extent commercial logging, caused substantial land-use change in the region during this century. The greatest single demand for land **conversion** came from pioneer peasant farmers clearing land for wet rice paddies, as the sparsely inhabited river deltas of the region became **rice** granaries. During this period, temporary crops such as **maize** and **cassava** increased by 330 percent in area to become 17 percent of the total land areas. Permanent crops such as rubber increased by a factor of 6.5 during the century but still represented only 1.5 percent of the total land area in 1980.

Over the last five decades, countries that make up the region have been under vastly different economic and political regimes; and these differences influence land use and land cover in the region today. Thailand has had an open market and democratic government. Cambodia, Laos, and Vietnam are socialist states that have differed in the timing and ways in which they opened their economies to world markets and provided private usufruct rights to natural resources. Myanmar remains a closed economic and political system. Cultural differences in the ways in which different ethnic groups use lands, their customary trading practices, and their relationships with other groups affect how they use land and their responsiveness to different government polices and market pressures.

Forest Cover

Forestry statistics in this region are notoriously unreliable and differ by country, time collected, and collection methods. In order to ensure that the data are as comparable as possible, this discussion is limited to statistics of the **United Nations Food and Agriculture Organization** (FAO) for 1990 and 2000. These data (see table titled "Forest cover") suggest that Cambodia,

Forest cover in 1990 and 2000 (total and as percent of national land area)

	1990	2000				
	Forest Area (ha × 1,000 and %)	Natural Forest (ha × 1,000)	Annual Change (ha × 1,000)	Total Forest (ha × 1,000 and %)	Annual Change (ha × 1,000)	Annual Change (ha × 1,000)
Cambodia	9,896 / 56	9,245	90	9,335 / 53	−56	−0.6
Laos	13,088 / 57	12,507	54	12,561 / 54	−53	−0.4
Myanmar	39,588 / 60	33,598	821	34,419 / 52	−517	−1.4
Thailand	15,886 / 31	9,842	4,920	14,762 / 29	−112	−0.7
Vietnam	9,303 / 29	8,108	1,711	9,819 / 30	52	0.5
Total	**87,761 / 44**	**73,300**	**7,596**	**80,896 / 40**	**−686**	**−0.8**

Source: FAO 2001.

Laos, and Myanmar maintain approximately 50 percent forest cover. Annual rates of forest loss in these countries during this period ranged from 0.4 percent in Laos to 1.4 percent in Myanmar, with the greatest annual reduction of forest cover occurring in Myanmar. In 2000, Thailand and Vietnam maintained approximately 30 percent forest cover, with Thailand losing approximately 0.7 percent annually and Vietnam gaining approximately 0.5 percent annually. In recent years, Thailand and Vietnam have undertaken strong efforts to reverse the rapid rates of **deforestation** of previous years. This has occurred through both natural regeneration of previously logged areas as well as the establishment of forest plantations (FAO statistics include rubber as forest plantations).

Croplands

Extensification of cultivated areas continued in various parts of the region throughout the last five decades although most alluvial lowland areas had already long been occupied. In 1961, rice accounted for nearly two-thirds of all arable land in the region (29.3 million hectares). By 2002, this had increased to three-quarters of all arable land (36.5 million hectares), ranging from 131 percent in Vietnam (because of double-cropping) to 53 percent in Cambodia (see table titled "Changes in arable and rice land").

Northeast and eastern Thailand were **hot spots of land-cover change** over the last half-century. Harald Uhlig reported a 47 percent decrease in forest cover in this area and even 75 percent loss in the provinces of Chon Buri and Rayong as spontaneous settlers cleared roughly four to five million hectares of previously little-used but not necessary forested uplands. During this period the proportion of cultivated land under rice in the northeast fell from 96 to 76 percent; the total number of farming holdings increased by over 36 percent; and the total cultivated area increased by 53 percent as maize, kenaf (*Hibiscus sabdariffa*), and cassava were planted. This expansion was driven by export prices as well as numerous loopholes in legal codes and the political difficulty of enforcing their provisions; and inadequacies in survey and land registration permitted continued expansion.

Government-sponsored land settlement schemes were not as large as those found in Indonesia and the Philippines but did occur. In Thailand some

Changes in arable and rice land between 1962 and 2002

	Arable Land 1961 (ha × 1,000)	Arable Land 2002 (ha × 1,000)	Rice Land 1962 (ha × 1,000)	Rice Land 2002 (ha × 1,000)	Rice as Percent of Arable Land 1962	Rice as Percent of Arable Land 2002
Cambodia	2,840	3,700	2,286	1,966	0.80	0.53
Laos	640	800	590	690	0.92	0.86
Myanmar	9,900	9,550	4,654	6,200	0.47	0.65
Thailand	10,400	16,800	6,540	9,990	0.63	0.59
Vietnam	5,550	5,700	4,889	7,485	0.88	1.31
Total	**29,330**	**36,550**	**18,959**	**26,331**	**0.65**	**0.72**

Source: IRRI 2004.

241,000 settler families were placed on about 700,000 hectares under various programs. In Vietnam, state-sponsored resettlement of lowland Kinh settlers in the central highlands caused extensive alienation of lands that minority peoples formally occupied. After 1975, large tracts of forestland were incorporated into state farms and state enterprises on which the newly arrived Kinh were settled. More recently, waves of spontaneous migrants, often themselves ethnic minorities (Tay, Nung) from land-scarce provinces in the northern **mountains**, have displaced local minorities from better lands located near roads. Intensification has affected different agricultural or **land-use systems** in different ways and to different degrees. Wet-rice systems, whether irrigated or rainfed, have undergone the highest degree of **agricultural intensification**. Irrigation is the major method of intensifying production and has generally encouraged demand for labor and capital outputs. Irrigated rice accounted for about 26 and 32 percent of the land area in Myanmar, Thailand, and Vietnam in 1976 and 1996, respectively. In 1996 irrigated rice represented 20 percent of all rice land in Thailand and 51 percent of Vietnam (see table titled "Irrigated rice land").

The greatest expansion in rice production has been in Vietnam, especially in the southern delta region following the abandonment of land collectivization in 1989 and the introduction of free markets. Vietnam now exports 1–1.5 million tons of rice per year and ranks at the world's third-largest exporter. Average yields in most southern provinces have reached six tons per hectare per season, against a level of 1.5 tons per hectares before decollectivization. On the Plain of Reeds, previously a region of low-yielding floating rice production, intensification has seen rice replaced by maize grown for sale as animal feed, as livestock raising has expanded markedly.

In Thailand, supplementary wet season irrigation and drainage have been implemented on a large scale since 1960. The expansion of irrigation farming, and in the Chao Phraya Delta improved drainage, has typically been accompanied by the use of double cropping, high-yielding varieties, mechanization of tillage, the use of fertilizers, insecticides, and other pesticides. In the Chiang Mai valley in northern Thailand, triple cropping began in the late 1960s. By 1980, two-fifths of the land was triple cropped in glutinous rice (main season), rice or vegetables, and non-glutinous rice rotation.

This was possible due to reliable water supplies, partly from tube wells; and the use of high-yielding varieties for the second and third crops, together with heavy fertilizer use, raising main-crop yields to 5.5 tons per hectare.

Irrigated rice land between 1980 and 1996 (total and as percent of all rice land)

	Irrigated Rice 1980 (ha × 1,000) and (%) Rice Land	Irrigated Rice 1996 (ha × 1,000) and (%) Rice Land
Cambodia	** / **	** / **
Laos	** / **	** / **
Myanmar	873 / 18%	1,535 / 27%
Thailand	2,006 / 22%	1,894 / 20%
Vietnam	2,250 / 40%	3,572 / 51%
Total	**5,129 / 26%**	**7,001 / 32%**

Source: IRRI 2004.

Rainfed cropping systems have become more intensive by increasing the number of crops obtained from the same fields and the yields of single-crop systems. In northeast Thailand, for example, where rice grows from August to December, it has been mixed with crops such as sesame, peanuts, cassava, **tobacco**, and vegetables. In Vietnam, rice-rice rotations give lower net returns than rice-vegetable (eggplant, cowpea, squash) rotations because of high material and labor inputs used for rice, and so cowpeas or peanuts have been added to rice-rice rainfed rotations.

In upland areas of mainland Southeast Asia, **swidden cultivation** has been the dominant land-use practice for at least a millennium. Jefferson Fox and John Vogler summarize eight sites where they documented changes in land cover in upland areas across the region over the last fifty years. They found that at the level of secondary forest **regrowth** after swidden cultivation, land cover remained stable across the region throughout this fifty-year period. This level of abstraction, however, hid major changes that occurred within the sub-categories of **secondary vegetation** as well as in the size and number of land-cover fragments. These studies suggested that the land-use change that occurred was due to intensification of swidden cultivation and the introduction of **cash crops**.

Such intensification is marked by reduced **fallow** to the extent that in some areas cultivation is now semi-permanent. In the Mae Chaem and Mae Sarieng **watersheds** of northern Thailand, the fallow period of the former long fallow swidden cultivation systems of the Karen and Lua people has now been reduced to five years or less. However, because without external inputs the productivity of upland rice falls drastically, many farmers have switched to non-rice crops such as soybeans, maize, cabbage, and other cash crops, accompanied by heavy use of chemical fertilizers, herbicides, and pesticides. In northern Vietnam, the amount of land devoted to swidden cultivation decreased in the 1990s as farmers devoted more labor and capital to permanent agricultural fields.

Cash cropping has been promoted throughout the region, but particularly in Thailand and Vietnam. Cash crops include temperate vegetables such as tomatoes, carrots, green beans, potatoes, and cabbage, as well as cut flowers, and subtropical fruits such as persimmon, apricot, plum, lychee, and longan. These crops exploit the uplands' biophysical advantage—its temperate climate—for off-season and high-value commercial crops for lowland markets and export.

Agribusiness companies are having increased influence on farming systems in many parts of the region. These companies are interested in purchasing, and sometimes processing, agricultural products for both domestic and export markets. In recent years there has been enormous expansion of various forms of **contract farming**, offering farmers alternative ways to earn income. For example, in Chiang Mai province, there are many canning and food processing plants that subcontract farmers to deliver products that meet particular grade specifications. Inputs are prescribed and grid-pricing systems of payments used for crops such as soybeans, green beans, and bamboo shoots. Farmers in some areas of Thailand and Vietnam have begun to allocate a substantial part of their land to plantation crops such as sugarcane. Others have allocated land to **coffee** on contract to township sugar mills and coffee producers.

Although intensification has been a dominant process in recent decades and although the agricultural workforce continues to grow, areas of **disintensification** as well as areas of production stability can be found. Disintensification does not imply active withdrawal of labor from agriculture so much as the use of off-farm, non-agricultural work to supplement subsistence. This has involved some swidden cultivators but many more lowlanders. This process has been particularly strong in Thailand although it has occurred elsewhere as well. In Thailand, few farmers rely solely on **agriculture** to meet their needs. Rural households are dividing their time between farm and non-farm activities, constructing livelihoods that are increasingly hybrid, both spatially and sectorally. Whether these changes mean that rural people will eventually become permanent city dwellers or not is not yet clear. They may choose to live in rural areas and to split their time between farm and non-farm activities.

See also: **Agro-Industry; Biodiversity; Colonization; Conservation; Forestry; Hot Spot Identification; Land Reform; Land Rights; Malaria; River Basins; Subsistence Agriculture.**

Further Reading

Fox, Jefferson, and John Vogler, "Land-Use and Land-Cover Change in Montane Mainland Southeast Asia," *Environmental Management* 36 (2005): 394–403; Hill, Ronald D., "Stasis and Change in Forty Years of Southeast Asian Agriculture," *Singapore Journal of Tropical Geography* 19 (1998): 1–25; International Rice Research Institute, World Rice Statistics [Online, December 2004], The IRRI Web Site, www.irri.org; Rerkasem, Kanok, "Uplands Land Use," in *Social Challenges for the Mekong Region*, eds. Mingsarn Kaosa-ard and John Dore, Chiang Mai, Thailand: Social Research Institute, Chiang Mai University, 2003, pp. 323–46; Richards, John F., and Elizabeth P. Flint, "A Century of Land-Use Change in South and Southeast Asia," in *Effects of Land Use Change on Atmospheric CO_2 Concentrations: South and Southeast Asia as a Case Study*, ed. Virginia H. Dale, New York: Springer Verlag, 1993, pp. 15–68; Rigg, Jonathan, and Sakunee Nattapoolwat, "Embracing the Global in Thailand: Activism and Pragmatism in an Era of Deagrarianization," *World Development* 29 (2001): 945–60; Uhlig, Harald, "Spontaneous and Planned Settlement in Southeast Asia," in *Agricultural Expansion and Pioneer Settlements in the Humid Tropics*, eds. Walther Manshard and William B. Morgan, Tokyo: United Nations University Press, 1988, pp. 7–43; United Nations Food and Agriculture Organization, Global Forest Resources Assessment 2000, Rome: FAO, 2001.

JEFFERSON FOX

Southeast Europe. Encompassing the whole Balkan Peninsula excluding Greece, post-communist Southeast Europe (SEE) is the least-developed region in **Europe**. This was caused by the heritage of a century-long and successive occupation by the Ottoman Empire (fourteenth to nineteenth century); and then by its inconvenient geo-economic position; and, after World War II, by the impact of a rigidly planned communist economy based on the exhaustion of all natural and human resources. Since the disintegration of Yugoslavia in the early 1990s, SEE has comprised seven countries, with a total area of 613,000 km^2, that is, larger than France and slightly smaller than Texas (about 6 percent of Europe). Four countries are post-Yugoslavian republics (Slovenia is considered part of **east-central Europe**). Among them, the most populous country is Serbia and Monte Negro (10 million inhabitants), and the least populous is Macedonia (2 million). Bosnia and Herzegovina, and Croatia each have about 4.5 million inhabitants. Within all SEE, the most populous country is Romania (22 million), with Bulgaria and Albania holding populations of 8 and 3 million inhabitants, respectively. In 2002, SEE had about 55 million inhabitants, which is 8.5 percent of the European population (with a population density of about eighty-eight persons square kilometers, i.e., far less than east-central Europe).

All countries are maritime except Macedonia. Romania and Bulgaria lie at the coast of the Black Sea, separated by the Danube River. The Danube enters the Black Sea through an ecologically valuable marshy delta in the boundary territory between Romania and Ukraine.

Geologically, SEE belongs to the Balkan region/peninsula. The Sava River and the lower Danube create its northern boundary; the western border is created by the indented island-rich seacoast of the Adriatic Sea, and the shoreline of the Black Sea forms the eastern border. SEE is geologically one of the youngest regions in Europe, as a part of the tertiary Alpine-Himalayan folder zone (with earthquakes being typical there). The surface is predominately mountainous. Plains and lowlands comprise only about 20 percent of the total area. Large lowlands are situated in eastern Croatia, in central-northern Serbia (surrounding Beograd and Vojvodina) and in Romania and Bulgaria along the Danube River. The Dinaric Alps stretch along the Adriatic coastline, and therefore their western slopes are very sheer. Only in Albania and at part of the Croatian seashore can one find coastal lowlands. Large karst areas are typical. **Mountains** of the Alpine-Carpathian system and the Transylvanian Plateau form the natural backbones of Romania and Bulgaria. The Danubian lowland fills the north of Bulgaria; and the south holds the highest ranges of the Balkans (Rodopi, Rila, and Pirin at 2,900 meters). In the Dinaridy region, stony rendzins, litosols, and so-called *terra rossa* prevail, with kambisols in lower parts.

In terms of climate, SEE lies at the dividing line between the temperate and Mediterranean subtropical climates. The majority of the SEE region belongs to the temperate climate zone. The inland area belongs to a zone of warm fluvial continental climate (Cfb type) with high temperature amplitudes throughout the year. Its opening to the north causes an inflow of cooler air from the Panonian basin in east-central Europe. A Mediterranean etesian climate with dry summers is typical for the Adriatic coast. It enables intensive recreational uses and large areas of wine growing.

In inland parts of the SEE, **forests** cover about one-third of the territory, and forest types vary with altitude from oak to beech with mixtures of maple and hornbeam (up to 1,800 meters in the south). This mountainous landscape is typical for its so-called medieval Walachian **colonization**—with dispersed farms at mountainous **pastures** (cattle, mutton), which survived until recent times. At the highest level, forests are made up of conifers, pine (mainly in Bosnia and Herzegovina), and spruce (in Serbia). The treeline ranges from 1,800 to 2,100 meters.

Original oak forests in the Mediterranean region were almost entirely extinguished by logging, especially during the rule of the medieval Republic of Venice, then by grazing (goats, sheep), and by **fire**. Oak forests were replaced by a macchia type of **secondary vegetation** (laurel, myrtle, frygana, and others), followed by evergreen oak forests in a few spots. **Deforestation** led to significant **soil erosion** and denudation, especially on steep slopes of the Dalmatian coast and its islands, where rocky ground of gray limestone emerged even (e.g., "polje," with lakes in southern Serbia). There, **reforestation** is impossible. The Balkan landscapes as well as that of the Apennine peninsula represent examples of the longest human impact on landscape configuration in Europe.

At the dawn of the **Industrial Revolution**, traditional agriculture, cattle grazing, and handicrafts dominated in the central and southern regions, whereas factories emerged in the north, and early capitalistic relationships developed already in the eighteenth century. In 1804, today's Croatia, as well as the former Hungarian Transylvania (from 1920 onward, part of Romania), were under the rule of the Habsburg Monarchy. Especially in the case of Croatia, this "historical start" over the rest of SEE has an imprint until nowadays. **Agriculture** and industry were most developed in Croatia and Serbia at the end of the nineteenth century (including Slovenia). Between the 1870s and 1912, almost all countries that were under Ottoman rule gained partial and consequently full independence, with Romania and Bulgaria (due to Russian pressure upon the Ottoman Empire) exerting control over the Bosporus and Dardanelle straits. After World War I, the kingdom of Serbs, Croatians, and Slovenes was established; and, after World War II, changed into a federal state called Yugoslavia, which gradually collapsed after 1990.

A remarkable agrarian character remained in the whole of SEE. In the Serbian lowland region of Banat, amelioration of land transformed marshlands into arable lands in the eighteenth century, triggering an intensive cultivation of cereal and **cash crops**. At the end of the nineteenth century, Croatia and Serbia were the most economically advanced regions. However, still after World War I, Yugoslavia remained an undeveloped agrarian country with subtle industry and less-developed agriculture—in comparison with Czechia, Hungary, and Slovenia in east-central Europe, for example. The same applies to Romania. However, due to extensive cereal cultivation in the fertile Walachian lowland, Romania became the fifth-largest producer of **maize** and the ninth-largest producer of **wheat** in the world before World War II.

After World War I, land ownership in the entire SEE was atomized into a big number of small farms and small number of great estates. The collectivization of agriculture took place from late 1940s until the late 1950s. A lot of cooperatives and state farms were established. The organizational structure

East Central and Southeast Europe: Land Use 1961–1990–2002 in Percent

	1961		1990		2002	
Land category	ECE	SEE	ECE	SEE	ECE	SEE
Arable land	49.2	35.0	45.3	33.0	43.2	31.4
Permanent cultures	1.8	2.6	1.8	2.7	1.6	2.3
Permanent grassland	14.0	19.9	12.9	21.3	13.1	19.7
Agricultural land	65.0	57.5	60.0	57.0	56.8	53.4
Forests-woodland	25.2	31.6	28.0	31.9	29.6	31.7
Other areas	9.8	10.9	12.0	11.1	12.5	14.9
TOTAL	100.0	100.0	100.0	100.0	100.0	100.0

Source: Statistical database of the UN Food and Agricultural Organization (http://www.fao.org).

of agriculture from the communist period partially survived after 1990. In the late 1990s, almost no new and capitalist-style corporate farms existed. Individual small farms continued cultivation on 50–60 percent of the agricultural land, in Albania even on 80 percent. Socialistic **industrialization** deepened the regional differences between urbanized regions and countryside. It also reinforced the deep differences among former Yugoslavian states, having endangered the integrity of Yugoslavia from 1980 onward. An agricultural crisis set in, especially in poor mountainous regions, with small farms not able to survive in many cases, so that depopulation of these regions followed. In the fertile Walachian lowland of Romania, the relatively important mining of oil and natural gas had a bad environmental impact.

In 2003, the average gross domestic product (GDP) (purchasing power parity per capita) was about the half of that of east-central Europe (i.e., about $6,390). These differences are reflected also in the average share of agriculture in total GDP, which is about four times higher than that of east-central Europe: 12 percent. Together with the low technological and organizational level of agriculture, these data indicate a prevailing rural character. The share of agricultural land in the total territory in 2002 was highest in Romania (62 percent) and Croatia (56 percent), and lowest in Albania (40 percent). Similar trends apply to arable land. Croatia and Macedonia (28 percent and 25 percent, respectively) hold the highest share of permanent **grassland**. The landscapes of SEE are largely wooded, with huge regional differences in the type of forests, though. Bosnia and Herzegovina and Macedonia hold the largest forest cover (44 and 35 percent, respectively), with Albania at 34 percent, Bulgaria at 33 percent, Romania with 27 percent, and Serbia and Montenegro having 28 percent.

The above table presents overall data of different developments in land use in SEE for the last thirty years (including the influence of transformation through capitalism). It is remarkable that east-central Europe has a much higher share of arable land than SEE. On the other hand, SEE shows a higher proportion of permanent **grassland**, is much more forested, and shows also a higher share of other areas (e.g., lakes and denudated rocky areas). Trends of land-use changes after 1989 in SEE differ a little from the development in east-central Europe. The area of arable land decreased significantly because mechanisms of **land rent** started to work again, and as a consequence, less-fertile soils have been abandoned and mostly transformed to permanent grasslands. The share of permanent grassland increased during 1961–1991, but decreased in 1991–2002. No changes occurred in forests and woodlands. In contrast, "other areas" increased considerably after 1990, probably as a consequence of **soil erosion** and **land abandonment** due to a large decrease of arable land.

See also: **Agricultural Revolution; Economic Restructuring; Land Reform; Mineral Extraction.**

Further Reading

Jelavich, Barbara, *History of the Balkans*, Cambridge, UK, New York: Cambridge University Press, 1983; Kann, Robert A., *A History of the Habsburg Empire, 1526–1918*, Berkeley, CA: University of California Press, 1974; Magocsi, Paul R., *Historical Atlas of East-Central Europe*, Seattle, WA, London: University of Washington Press, 1993; Sugar, Peter F., *Southeastern Europe under the Ottoman Rule*, 1354–1818, Seattle, WA: University of Washington Press, 1977; United Nations Food and Agricultural Organization, Statistical Database [Online, December 2004], The FAO Web Site, www.fao.org; U.S. Central Intelligence Agency, The World Factbook [Online, December 2004], The CIA Publication Web Site, www.cia.gov/cia/publications; Zagoroff, Slavcho, and Stanley P. King, eds., *The Agrarian Economy of the Danubian Countries, 1933–1945*, Stanford, CA: Stanford University Press, 1955.

LEOŠ JELEČEK

Southern Africa. The region south of approximately 15° S, covering the modern countries of Namibia, Botswana, Zimbabwe, southern Mozambique, **South Africa**, Lesotho, and Swaziland. Southern Africa has been influenced by anatomically modern humans for about 200,000 years, and their hominid ancestors for over a million years. For most of the period, human impacts on the environment were limited to the effects of hunting, gathering, and the use of fire. About 2,000 years ago, Bantu agropastoralists arrived in southern Africa, bringing domesticated livestock, crops, and metal-working technology, which substantially increased human pressures on the landscape. The arrival of the Europeans about 400 years ago and their settlement of the region's interior over the past 150 years coincided with the industrial era and formed the basis for extensive transformation of the region during the twentieth century.

Almost 90 percent of southern Africa is covered by natural **vegetation**, although only about 8 percent of the region lies in formally protected areas. Most of the non-protected area is used for livestock grazing, although a growing fraction is being converted to nature-based tourism activities. Cultivation presently accounts for approximately 9 percent of the land area, and urban land uses for less than 2 percent. About half the region is covered by savanna; the remainder consists primarily of arid shrublands in the southwest and **grassland** in the southeast (see figure titled "The natural vegetation of southern Africa"). Most of the region has summer rainfall with a single peak between October and April; the exception is the extreme southwest, which experiences winter rainfall. Namibia, Botswana, and the western interior of

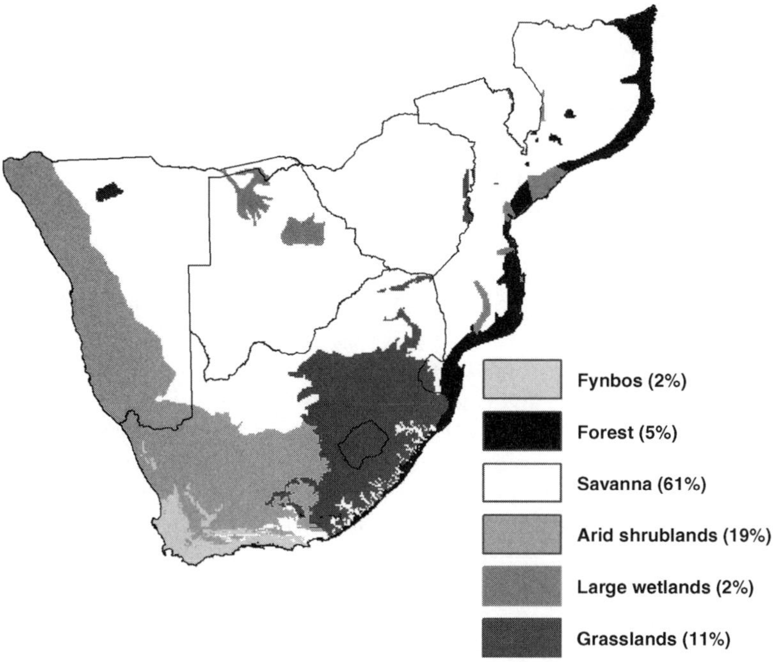

The natural vegetation of southern Africa, still covering almost 90 percent of the area, and a primary determinant of land use in the region.

South Africa are hot and arid; Zimbabwe, southern Mozambique, and the eastern parts of South Africa mostly have a semiarid or dry subhumid climate.

Early Human Influence

The San, also known as the Khoikhoi, bushmen, and a variety of other names, have inhabited southern Africa for at least 30,000 years. They were hunter-gatherers whose impact on the land cover was qualitatively little different than that of ancestral humans. **Fire** is thought to have been used by hominids for more than 1.5 million years for cooking and warmth. The San also used fire for hunting: large mammals were attracted to purposely burning specific pieces of land at certain times. The San harvested plants such as the marula (*Sclerocarya birrea*), !nara melon (*Acanthosicyos horridus*), rooibos (*Aspalathus linearis*), and buchu (*Agathosma betulina*), which continue to be highly valued today and in some cases are now grown commercially for regional and international markets. The pre-historic population density of the San is unknown, but probably low. The total hunter-gatherer population at the time of European arrival in the mid-seventeenth century has been estimated at about 50,000 people, translating to a density of about five people per 100 square kilometers.

The agropastoral Bantu-speaking people arrived from the north over a period of hundreds of years, starting about 2,500 years ago and reaching the southern Cape in present-day South Africa perhaps 1,000 years ago. They brought with them sheep and cattle, which had been domesticated in Asia, and millet, domesticated in North and West Africa. They also knew how to work metal, particularly iron, and make ceramic pots. As a result, their capacity and need to hunt large mammals (including carnivores) and to cut trees for fuel, building timber, and cultivated lands was much greater than that of the San, whom they progressively displaced into the more marginal arid southwestern parts of the region. With the arrival of the Bantu, a portion of the wild herbivore population was replaced by domestic livestock, and small areas of natural vegetation were cleared for cultivation. Bantu population densities in the pre-colonial period are estimated at 0.5–5 people per square kilometer, that is, at least an order of magnitude greater than those of the San.

Africa seems to be unique in that the growing human population did not lead to a rapid demise of large wildlife populations, but co-existed with large numbers and a wide diversity of other mammals. In pre-colonial times, people may have had a controlling influence on the populations of slow-breeding megaherbivores such as elephants, which in turn can have significant local impacts on vegetation. People certainly had the capacity to hunt elephants successfully, and a trade in ivory has existed for at least a millennium on the east coast of Africa.

Hunting, pastoralism, cultivation, and other human activities in the pre-colonial era had consequences for land cover, certainly locally, and probably also regionally. For example, the Tswana people on the edges of the Kalahari Desert lived in settlements of several tens of thousands of people each, covering a hundred hectares or more of stone walling. The oral history, records of early European explorers, and archaeological evidence agree that the settlements would each last a few decades before moving to another location. The reason for moving may have included depletion of grazing and wood in the vicinity of the town, depletion of the water supply, and/or a buildup of pathogens.

European Settlement and the Current Era

The first permanent European settlement in southern Africa was established in 1652 in the area of present-day Cape Town, but significant settlement of the interior took place only after 1835. The technology and political trade links that the Europeans brought and the start of the industrial era provided the basis for the largest-scale transformation of the region's land cover to date, taking place over the course of the twentieth century. A measure of the **land-cover change** that has occurred is that the regional population at the time the European settlers arrived in the mid-seventeenth century was probably no more than a few million people, with a similar number of livestock, and cultivating an area of under a million hectares. In 2000, the population was approximately 80 million, with 75 million livestock (cattle, sheep, and goats) and 35 million hectares of cultivated land. The following figure gives an estimate of the location and extent of areas presently under different forms of land use.

Pastoralism

Today, livestock production is the land use that occupies the largest fraction of southern Africa, and domestic livestock (especially cattle) are the

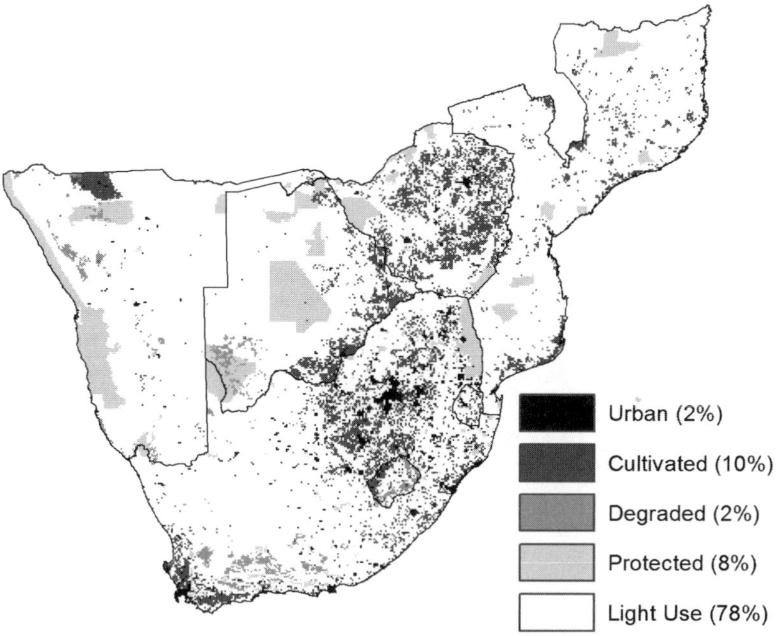

The major land uses in southern Africa. Most of the untransformed "light use" areas are used for livestock grazing and, under certain circumstances, may become degraded.

largest consumers of vegetation in the region. While livestock have been present throughout southern Africa for approximately 2,000 years, the numbers and breeds changed significantly during the twentieth century. Unlike in many parts of the world, livestock production in southern Africa seldom involves a deliberate transformation of the land cover. Livestock are almost exclusively raised on natural vegetation, taking the place of large indigenous herbivores. Pastoralism in southern Africa therefore generally has a fairly low impact on the indigenous flora and fauna. Only if livestock numbers are very high relative to capacity of the natural vegetation to support them does a change in land cover take place. In these cases, the natural vegetation persists, but in a degraded form—typically, vegetation production (i.e., biomass and cover) is reduced to about half of the non-degraded state, often accompanied by significant changes in the dominant species of plants, and sometimes by a structural change to greater or less dominance by woody plants, and accelerated loss of soil.

The major increase in livestock numbers in southern Africa followed **industrialization** toward the end of the nineteenth century, aided by the drilling of boreholes to provide surface water, immunization against disease, and feed supplementation. The first significant changes in livestock numbers took place in South Africa with a "boom" in merino sheep, introduced in the Karoo and eastern Cape areas in about 1870. Their numbers grew to a peak and then declined by almost half in the 1920s, possibly because the vegetation cover was permanently altered from a grassy to a shrubbier, less palatable form. A European fashion demand for feathers and the **domestication** of ostriches led to an ostrich population boom in the southern Cape during the early twentieth century. This boom was revived in the late twentieth century, but this time driven by a demand for ostrich meat. In the drier parts of the region (receiving below 450 millimeters of mean annual rainfall), sheep and goats are the dominant livestock. Donkeys are important draft animals in poorer communities throughout the region, but make up a small proportion of the domestic livestock biomass.

Cattle are mainly kept in the wetter (greater than 450 millimeter annual rainfall) savanna and grassland areas of the region. In southern Africa, cattle are not kept only for meat: especially in traditional societies living on communal lands, cattle are an important way of accumulating assets and demonstrating wealth and status. Cattle numbers in southern Africa increased more gradually than those of sheep and goats, reaching approximately their current levels in the 1960s. In Botswana, the Kalahari Desert region, which had no permanent water, was opened to **cattle ranching** by the drilling of deep boreholes during 1970–2000. The low-lying parts of Zimbabwe, Mozambique, and northeastern South Africa were unfavorable for cattle due to the presence of the tsetse fly, which carries the *Trypanosoma* protozoans that cause *nagana* in cattle and sleeping sickness in humans. The tsetse fly was eradicated from most of this area after the invention of DDT, from about 1940 onward, so that cattle ranching is now an important land use in these areas.

Deforestation and Land Degradation

Deforestation in its narrow sense has never been an extensive process in southern Africa, because less than 1 percent of the land surface supported tall, closed forest, even in pre-colonial times. The settler demand for large-dimension timber for ships, buildings, wagons, and later mine and railway infrastructure quickly depleted the local sources (mainly along the southern and eastern coast of South Africa), which had halved in extent by 1850.

Deforestation in the broader sense, that is, reduction in woody cover of the landscape (most of southern Africa is partly wooded), had been going on at a gradual rate for millennia. For example, it is calculated that one *kraal* (a pen for corralling cattle, built by every household and necessary to protect the livestock from predators) consumes 150 tons of wood. During the twentieth century, two divergent land-cover patterns emerged with respect to woody cover in areas with a savanna climate (a hot, wet summer and a warm, dry winter; about 60 percent of southern Africa). Those lands occupied by European cattle ranchers showed a marked increase in woody plant biomass, often to the point of becoming uneconomical for ranching use. This phenomenon, known as bush encroachment, has also been observed following European settlement in Australia, Texas, and Argentina, and has been attributed partly to changes in the prevailing fire regimes. Both the intensity and frequency of fires are believed to have been reduced, resulting in a rapid increase in shrub and tree cover, which further suppressed fires.

On the African communal lands, on the other hand, there was a marked loss in tree cover over the twentieth century. By the 1950s, land ownership by Africans in South Africa, Zimbabwe, Namibia, and Mozambique was restricted by law to a minor part of the land area (13 percent in South Africa, the so-called homelands; about 45 percent of Zimbabwe, the "Tribal Trust

Lands"; and 27 percent of Namibia). **Poverty**, high densities of people and livestock, and a breakdown in traditional **institutions** of land management led to widespread clearing, overgrazing, and **soil erosion**. The consequences on land cover were dramatic and will remain visible for decades after the repeal of the land-ownership laws in the 1980s and 1990s. Large areas of the "homelands" are today effectively peri-urban settlements or **remittances landscapes**, economically sustained by social and wage remittances from urban areas.

Fires collectively burn an area of about 3.7 million km^2 per year in the region of Africa south of the equator (a third of the land area, mostly in savannas and grasslands). At the present time, more than three-quarters of fires in southern Africa are ignited by people, mostly in order to manage ecosystems for grazing, honey, and agriculture. While fires may lead to substantial changes in land cover for a period of days to weeks, and in the very long term (centuries) have shaped entire landscapes, in the annual to decadal time frame they are typically "included disturbances" that do not change the land cover. Fires are a natural and required feature of southern Africa's ecology. Most savannas and grasslands become unusable for grazing if fire is excluded. Unless the total amount of **biomass** burned changes over time, there is no net emission of carbon dioxide—the carbon dioxide emitted by wildfires is reabsorbed within weeks to years by the regrowth of the vegetation. Humans have probably been the principal source of ignition of wildfires in Africa throughout the past million years, and there is little evidence that the overall extent of fires in southern Africa has changed substantially over this period.

Cultivation

Taken over the full extent of southern Africa, crop cultivation is the land use that has had the biggest impact on the terrestrial biota of the region. Most of the cultivated land is in South Africa (36 percent of total cultivated area in the region) and Zimbabwe (29 percent) in large commercial farms (although in Zimbabwe these are currently being reclaimed by the government as part of its land redistribution policy).

Large-scale cultivation in southern Africa started in the early twentieth century. In South Africa, the area under cultivation grew steadily from the end of the South African (Anglo-Boer) War in 1902. A major phase of expansion occurred in the 1960s that was related to the introduction of fertilizers, new cultivars (particularly of **maize**), mechanization, and irrigation. The area under cultivation reached a plateau in the 1970s, when most of the economically arable land was under cultivation, and has declined slightly since agricultural subsidies were reduced. Cultivation in Zimbabwe began to expand from the subsistence level in the 1920s. The major expansion in the 1960s was driven by an influx of white settlers and new agricultural technologies: **tobacco**, hybrid maize, and agricultural lime in particular. Redistribution of this land to smaller-scale African farmers with few technical and financial resources, starting in the 1990s, will have as-yet-unknown land-cover consequences.

The cultivated area in the other countries in southern Africa is currently small in comparison and is largely limited by climatic and soil constraints.

Subsistence agriculture, while not occupying extensive areas, plays an extremely important role in the livelihoods of impoverished rural communities. At the start of the twenty-first century, most of the economically arable land in southern Africa was under cultivation, and a range of **scenarios** shows little further transformation. The major remaining potential for agricultural transformation lies in Angola, Zambia, and northern Mozambique, north of the area covered in this article.

Beginning in the early twentieth century, and accelerating after 1950, substantial areas (currently 1.5 million hectares in South Africa and 0.15 million hectares each in Swaziland and Zimbabwe) of moist highland grassland were converted to exotic timber plantations (eucalyptus, pine, and Australian *acacia* species) for the provision of mine-support timber, lumber, poles, tannin, and paper pulp. Further expansion was slowed in the 1990s by recognition of the deleterious consequences of plantations on **biodiversity** and streamflow in planted **watersheds**.

See also: **African Trypanosomiasis; Colonization; Conservation; Green Revolution; Hot Spots of Land-Cover Change; Land Rights; People at Risk.**

Further Reading

Biggs, Reinette, and Robert J. Scholes, "Land-Cover Changes in South Africa 1911–1993," *South African Journal of Science* 98 (2002): 420–4; Dean, W. Richard J., and Ian A. W. MacDonald, "Historical Changes in Stocking Rates of Domestic Livestock as a Measure of Semi-Arid and Arid Rangeland Degradation in the Cape Province, South Africa," *Journal of Arid Environments* 26 (1994): 281–98; Geist, Helmut J., "Causes and Pathways of Land Change in Southern Africa during the Past 300 Years," *Erdkunde* 56 (2002): 144–56; Hoffman, M. Timm, "Human Impacts on Vegetation," in *Vegetation of Southern Africa*, eds. Richard M. Cowling, David M. Richardson, and Shirley M. Pierce, Cambridge, UK: Cambridge University Press, 1997, pp. 507–34; Scholes, Robert J., and Reinette Biggs, eds., *Ecosystem Services in Southern Africa: A Regional Assessment*, Pretoria, South Africa: Council for Scientific and Industrial Research, 2004, available online at www.maweb.org/en/subglobal.safma.aspx.

REINETTE (OONSIE) BIGGS, ROBERT J. SCHOLES

Southern Cone Region. Southern part of **South America**, which includes Argentina, Chile, and Uruguay, and consists of subtropical and temperate ecosystems (see figure). The southern half of this region, from approximately 40° to 55° S in Argentina, is known as Patagonia. While geographically isolated, the principal land uses in the Southern Cone are linked directly to global markets in agricultural and forest products. Regional land changes affect the viability of local production systems, such as sheep ranching, and a range of environmental services, including **biodiversity** and the **carbon cycle**. Despite these connections to regional and earth systems, the Southern Cone has received relatively little attention from global change and sustainability science. Lacking are integrated studies of the relationships between identified land changes and climatic, local biophysical, political-economic, social, and cultural conditions.

Significant European **colonization** of southern South America did not occur until the nineteenth century, late in global terms. Today it remains a

relatively sparsely settled region of the world; however, its temperate **forests** and **grasslands**, the two land covers emphasized here, have been dramatically altered by human action. Throughout the Southern Cone there has been extensive forest clearing and thinning, and the introduction of non-native flora and fauna (e.g., via forest plantations). Intensive sheep and cattle ranching have led to **soil erosion**, changes in species composition, and falling productivity in the region's native grasslands. The range of biophysical, socioeconomic, and political factors driving these kinds of land changes in other parts of the world are represented in the Southern Cone as well. Also represented are societal attempts to slow human impact, most notably through conservation reserves.

Forest Change

Grass prairies dominate the central and southern parts of Uruguay, and the northern part of the country is covered by subtropical forest and shrub formations, although only 5.72 percent of Uruguay's land area is forested. These forests are being actively managed, however, and there was a 4 percent loss of total forest cover between 1990 and 2000. Almost 90 percent of Uruguay is devoted to **agriculture** or **cattle ranching**.

As of 2000, Argentina had 34,648,000 hectares of forest (12.7 percent of the country's land area), with 926,000 hectares in plantations. According to data of the **United Nations Food and Agriculture Organization** (FAO), there was a net loss of 285,000 hectares of forest at an annual rate of change of –0.8 percent from 1990–2000. Forest change in the country is not a new problem. In the late nineteenth century, the government sold large blocks of land in the country's northern Chaco region to timber companies and cattle barons. Subsequently, most of the forests there were clearcut and now remain as desert scrub.

In 2000, Chile had 15,536,000 hectares of forest (20.7 percent of the country's land area), with 2,017,000 hectares in plantations. Forest-cover change shows a loss of 20,000 hectares from 1990–2000, at an annual rate of –0.1 percent, according to FAO data. Analyzing Chilean forest change in more depth gives a sense of the kinds of forest change dynamics witnessed in the larger region.

Human impact on Chilean forests is most severe in the coastal and lowland zones of the south-central part of the country (35.6°–42.3° S). Endangering the forests is a mix of **proximate causes**, including uncontrolled logging, plantation forestry (usually using introduced species of pine and eucalyptus), farming, the rearing of livestock, invasive species, **fire**, and agro-industrial pollution. There is continuing road building, such as the controversial Southern Coastal Highway between Valdivia and Puerto Montt, which has been rerouted to help protect coastal rainforest. Plantation forestry was initiated in 1940 and has been subsidized by the government since 1974 (Forest Law 701), which has led to some 2 million hectares of *Pinus radiata* being planted between 36° and 39° S latitude. Now less than 45 percent of forest cover in Chile is mature native forest.

Among the underlying **driving forces** of forest change in Chile are policies enacted under the dictatorship of Augusto Pinochet (1973–1990). As part of a suite of neoliberal policies, Pinochet targeted native forests and the intensive exploitation of natural resources nationwide. This focus led to incentives for establishing forest plantations as well as the implementation of a policy in 1985 to sell native forestland, much of which was quite cheap compared to industrial forests elsewhere in the world, and that attracted the attention of multinational corporations. At the same time, the Chilean government's oversight of the logging industry has been weak, with lack of enforcement, **monitoring**, and appropriate fines. Government policy that facilitated logging combined with market forces in 1987 when there was a boom internationally in demand for Chilean wood chips and pulp. The result was the large-scale conversion of native forests to plantations of **exotic species**, principally by multinational corporations, which led to what has been described as the "tragedy of the Chilean forest."

The prominent degradation of forests in south-central Chile hides some more positive trends, however. First, many industries in the country are becoming certified by ISO 14,001 (requiring an environmental management system) and the Forest Stewardship Council. Second, the Ley Marco Environmental Framework Law of Chile, which was created in 1994 and put into effect in 1997, mandates a system of environmental impact evaluation (SEIA). Akin to the environmental impact assessment component of the U.S. National Environmental Policy Act (NEPA), SEIA requires comprehensive assessment of the environmental impacts of development projects. Third, **forestry** shows signs of movement toward more sustainable methods. For example, government regulations in both Chile (since 1974) and Argentina (more recently) require a shelterwood system of forest management (regionally called a *corte de protección* or protection cut). Southern beech (*Nothofagus* spp.), which dominates the forests of Patagonia, has patchy cycles of regeneration due to shade intolerance and the natural mortality of trees; a shelterwood system protects the overstory and mirrors natural disturbance and regeneration processes.

Finally, as is the case for all of **Latin America**, there is a growing embrace of nature **conservation** in the Southern Cone. Many of the national parks in Argentina and Chile were established because of the desire to establish a presence along the long-disputed international border between the two countries rather than for nature conservation. But in Chile, decree-law no. 18,362 created the national public system of protected areas (SNASPE) in 1984, which seeks to protect ecosystem and evolutionary processes as primary goals. One half of the area of Chile's southernmost Region XII is under protection, and 40 percent of the forests in the Argentinean **Andes** of Patagonia are under public protection as well. These protected areas do not include private reserves, such as Estancia Yendegaia in Tierra del Fuego. These figures are somewhat misleading, however, given that most of the protected areas in Patagonia are remote, have low species richness and endemism, and often lie above 600 meters in altitude, consisting of ice and unvegetated terrain.

There is also controversy about some of the nature conservation areas. For example, American Douglas Tompkins has purchased extensive landholdings in Chile, including the 320,000-hectare Pumalin Park in the country's Palena province, which he established as a large private nature reserve. Many Chileans resent that the reserve, owned by a foreign national and extending from the Argentinean border to Chile's Pacific coast, cuts the country in half. In another example of controversy, on the site of a failed logging project by a U.S.-based corporation, the investment company Goldman Sachs announced in September 2004 the establishment of a 280,000-hectare nature reserve in Tierra del Fuego, Chile. Environmentalists are celebrating the protection of the area's forest; however, the decade-long battle over what was, by most accounts, one of the most innovative sustainable logging initiatives to date highlights the difficulty in balancing the environment and development in the region.

Grassland Change

For the most part, too far south for intensive agricultural production, Patagonia's native grasslands have been the site of sheep ranching since the late nineteenth century. Sheep numbers peaked in the mid-twentieth century, after which a combination of market shifts and degraded rangelands resulted in a steady decline. Shrub encroachment and **soil degradation** are widespread phenomena in arid and semiarid rangelands around the world. **Land degradation** affects most of Patagonia and is linked primarily to overstocking of the rangelands. The landscape has been altered so dramatically that it is difficult to determine pre-European composition and characteristics of the vegetation. In the Patagonian steppe, shrubs provide relatively poor forage for livestock compared with the region's perennial grasses; shrub steppe is found to have more severe erosion and higher water loss than in grass dominated areas; and the poor soil characteristics of the degraded shrub steppe may inhibit the reestablishment of perennial grasses, making attempts to reverse degradation processes difficult.

Despite Patagonia being the most threatened landscape in Argentina and Chile, there are no reserves in either country to protect grasslands. David Aagesen identifies four key human driving forces in the degradation of Patagonian rangelands: ranchers' lack of concern about soil conditions, skewed **land tenure** that results in small landholding size (which affects economies of scale and rangeland management strategies) and public lands where there tends to be more severe change, the lack of a diversified regional economy, and top-down and largely ineffective government policy. Human impact on grasslands has been exacerbated by both the strong winds that dominate the region and also by dry periods in the twentieth century. Fire history has also shaped current forest and shrubland patterns across Patagonia.

Intensive Agriculture and Ranching

The humid subtropical zone of eastern Argentina and Uruguay has moderate temperatures and precipitation throughout the year. Land covers range from prairie in the pampas of Argentina to a mixed forest-grassland cover in Uruguay. The pampas, along with the western Chaco region of Argentina and the Central Valley of Chile, are the most prominent agricultural zone in the Southern Cone. Often described as having unsurpassed agricultural potential, the pampas of Argentina are an extensive fertile plain that extends in an arc of hundreds of kilometers around Buenos Aires. Given its rich and deep topsoil, and a geographic location that historically has been more accessible to international markets than the rest of the Southern Cone, starting in the last quarter of the nineteenth century the region was initially devoted almost entirely to **pasture**, grain cereals, and **maize**. After a bust in the livestock industry in the 1910s, other agricultural uses, such as **cotton** and fruits, were added.

Once the site of over 300 grasses and other forage plants, development of the area involved widespread replacement of these native species with introduced pasture grasses, alfalfa, and feed grain. By the 1950s, overstocking of the rangelands and other poor land management practices, wind and water erosion, and the region's irregular climate (drought affects more profoundly the introduced grasses and crops than it does native varieties) had resulted in severe land degradation, such as soil erosion and weed expansion. The unequal distribution of land tenure meant that smallholders were pushed to more marginal lands in the western pampas. More recently, continuous cultivation and double cropping have also led to soil degradation. Despite these changes, however, most of the pampas in Argentina are not severely degraded. But after forests were cleared from the western edge of the Chaco, an agricultural boom occurred in the 1960s that focused initially on grains, beans, and oil seeds. Subsequently, drought, wind erosion, and poor land management practices resulted in severe land degradation.

The Central Valley of Chile (approximately 32°–38° S) has a Mediterranean climate and some of the richest soils in South America. The region, much of it irrigated, is the country's agricultural heartland; there is widespread mining activity, in particular, exploiting the area's extensive copper reserves; and the zone is highly urbanized. Twentieth-century land change in the area is profound.

See also: **Agricultural Intensification; Desertification; Forest Degradation; Mineral Extraction; Urbanization.**

Further Reading

Aagesen, David, "Crisis and Conservation at the End of the World: Sheep Ranching in Argentine Patagonia," *Environmental Conservation* 27 (2000): 208–15; Armesto, Juan J., Tozzi, Ricardo, Smith-Ramirez, C., and Mary T. K. Arroyo, "Conservation Targets in South American Temperate Forests," *Science* 282 (1998): 1271–2; Food and Agriculture Organization of the United Nations, Global Forest Resources Assessment 2000, FAO Forestry Paper no. 140, Rome [Online, April 2004], The FAO Web Site, www.fao.org; Gustavo Zarrilli, Adrián, "Capitalism, Ecology, and Agrarian Expansion in the Pampean Region, 1890–1950," *Environmental History* 6 (2001): 561–83; Neira, Eduardo, Verscheure, Hernán, and Carmen Revenga, *Chile's Frontier Forests: Conserving a Global Treasure*, Washington, DC: World Resources Institute: Comité Nacional Pro Defensa de la Fauna y Flora, and the Universidad Austral de Chile, 2002; Parizek, Bernardo, Rostagno, Cesar M., and Roberto Sottini, "Soil Erosion as Affected by Shrub Encroachment in Northeastern Patagonia," *Journal of Range Management* 55 (2002): 43–8; Pauchard, Anibal, and Pablo Villarroel, "Protected Areas in Chile: History, Current Status, and Challenges," *Natural Areas Journal* 22 (2002): 318–30; Sawers, Larry, "Income Distribution and Environmental Degradation in the Argentine Interior," *Latin American Research Review* 35 (2000): 3–33; Veblen, Thomas T., Kitzberger, Thomas, Villalba, Ricardo, and Joseph Donnegan, "Fire History in Northern Patagonia: The Roles of Humans and Climatic Variation," *Ecological Monographs* 69 (1999): 47–67.

PETER KLEPEIS

Spatial Diffusion. The movement and dispersion of phenomena across geographic space, and through time, from an initial location to a wider area. The diffusing phenomena can include people (migration), ideas (innovation), goods, services, urban development, species, or diseases. Diffusion reflects human interaction in space and time, and diffusion processes contribute to the development of spatial structure and land-use patterns. There are two basic types of spatial diffusion: relocation diffusion, where an object no longer remains in the originating area, for example, the migration of people; and expansion diffusion, where an object passes between areas, but its influence remains in the originating area, for example, the movement of information. Expansion diffusion further subdivides into contagious diffusion, which depends on direct contact between people and is, therefore, a function of proximity or distance, for example, the transmission of contagious diseases; and hierarchic diffusion, in which movement occurs through sequences of classes or hierarchies, for example, transmission of ideas through social networks.

The diffusion of innovation is of particular interest in land-use studies because it can affect the ways in which land-use decisions are made and the rates of land-use change. The diffusion of innovation often follows a logistic curve, which reflects an initially rapid growth rate in the number of adopters of an innovation, after which the rate of adoption slows with time as the number of adopters reaches saturation. The presence of diffusion processes can often be observed in geographic space through the typical wave patterns they produce. The wave height may, for example, reflect the number of new adopters of an innovation, the number of migrating people, or the incidences of a disease. Diffusion waves change character with distance from the time and point of origin.

Torsten Hägerstrand developed perhaps one of the earliest and best-known models of spatial diffusion. The Hägerstrand model was used to simulate the diffusion of knowledge about a new **pasture** management technique in an agricultural community in Sweden. The model represented the spread of information between farmers through personal contact (contagious diffusion). Distance was treated explicitly as the factor controlling the frequency of contacts between farmers and therefore the rate and location of new adoption. Farmers in close proximity had a higher probability of contact than farmers further apart, although these contacts were also driven by a random element implemented using Monte Carlo simulation techniques. Each farmer could theoretically have his own contact-distance relationship (depending on the range and frequency of his movements); however, in practice a mean information field was used to represent the probability of contact between all farmers. The model was further developed to favor the movement of knowledge through farmers with larger farms (i.e., hierarchic diffusion). The Hägerstrand model was very important in land-use science as it demonstrated how land-use change could be understood as a spatial dynamic process driven by the diffusion of knowledge. The model was the precursor of many present-day land-use models based on **cellular automata** and agent-based approaches.

See also: **Agent-Based Model; Decision-Making; Economic Restructuring; Integrated Model; Modeling.**

Further Reading

Hägerstrand, Torsten, *Innovation Diffusion as a Spatial Process*, Chicago: The University of Chicago Press, 1967.

MARK ROUNSEVELL

Species Extinction. The disappearance of an organism or naturally occurring population that can interbreed only with others of its species (microorganisms, fungi, plants, and animals), as a result of hunting, natural disaster, climate change, overspecialization, unsuccessful competition for resources, introduction of **exotic species**, and/or human-caused stresses associated with the **proximate causes** and underlying **driving forces** of land-use/cover change, including the use of **fire**. Extinction of species has occurred since the beginning of life on earth. In fact, most of the organisms that have ever existed are now extinct: the species currently living amount to 2–3 percent of those that have ever lived; all others have become extinct, typically within about 10 million years of their first appearance. Five significant extinction events occurred, including the extinction of megafauna at the end of the Pleistocene era, but these events collectively seem to have ended no more than 5–10 percent of the species that ever lived. The current estimate for the total number of species possibly lies in the 5–15 million range, with a best guess of about 7 million. However, the species at risk now represent an unusually high proportion of all those that ever lived. It appears that a recent pulse of extinction started during the late Quaternary period. Evidence indicates that a massive (sixth) extinction event, driven by human activities, has been underway for some 40,000 years. Human activities have greatly accelerated the rate of species extinction far beyond the natural rate. There is agreement that the average species has a life span of about 5–10 million years, and that the rate of extinction is 1–0.1 species per million species-years. For the past 300 years, recorded extinctions for a few groups of organisms reveal rates at least several hundred times the rate expected on the basis of geological record. Rodolfo Dirzo and Peter Raven estimate that extinction rates in the next fifty years are some 1,000 times higher than the background rate of 1.0. Historically, overhunting has been the most consistent explanation of the megafaunal extinctions of the past 40,000 years. In addition, "first-contact extinctions" of species over the last 1,000 years are numerous on islands such as Madagascar and New Zealand, that is, extinction correlated with the arrival of humans including the introduction of alien invasive species. Today's major causes of threat are habitat loss and degradation related to agricultural activities (plantations, crops, and livestock farming), extraction activities (logging, harvesting, mining, and fishing), and the development of infrastructure (human settlements, industry, roads, dams, and power plants and lines). Direct exploitation such as hunting (also trading and collecting) is the current driver second in importance.

The rate of extinction had and still has an important influence on the world's **biodiversity**. This is true despite the fact that only a very few groups of organisms are well-enough known to be assessed for extinction. For example, some nineteen of each twenty tropical moist **forest** species are unknown to science at present, so that the effects of ongoing forest burning result in tremendous biodiversity losses. The International Union for Conservation of Nature and Natural Resources (IUCN) lists about 11,200 threatened species facing a high risk of extinction in the near future resulting from human activities. Although this is a small number relative to the total number of species (i.e., less than 1 percent), it includes 24 percent and 12 percent of all mammals and birds, respectively. For plants, the IUCN number represents only 2–3 percent of the known species, but individual country evidence suggests far more extinctions. In the **United States of America**, for example, it is estimated that about 33 percent of the plant species native to the country are threatened with extinction, including 24 percent of the conifers (most seriously threatened, though, are 69 percent of the freshwater mussels and 51 percent of the crayfish). Therefore, Rodolfo Dirzo and Peter Raven propose that a reasonable *interim* estimate would be that, at present, a third of the plant species of the world are threatened. As for birds and mammals, they provide a global picture of extinction as follows: at least 500 (but probably closer to 600) out of 1,192 threatened bird species and some 565 of the 1,137 threatened species of mammals will go extinct in the next fifty years, due to habitat loss and fragmentation, and mainly in the tropical forest zone. They conclude that, at any event, more than a third of the existing species on earth could disappear with the destruction of tropical forests only, and that it is reasonable to envision the loss of two-thirds of the species on earth by the end of the twenty-first century.

See also: **Agrodiversity; Biodiversity Novelties; Carson, Rachel Louise; Conservation; Cumulative Change; Deforestation; Ecological Colonization; Forests; Marsh, George Perkins; Paleo Perspectives; Tropical Dry Forest; Tropical Humid Forest; Turnover.**

Further Reading

Dirzo, Rodolfo, and Peter H. Raven, "Global State of Biodiversity and Loss," *Annual Review of Environment and Resources* 28 (2003): 137–67; International Union for Conservation of Nature and Natural Resources (IUCN), Species Survival Commission [Online, September 2004], Red List of Threatened Species Web Site, www.redlist.org.

HELMUT GEIST

SPOT Vegetation. A series of satellites originally developed by the French Space Agency (CNES), the first being launched in 1986 and the latest (SPOT-5) launched in 2002. The latest two platforms carry the Vegetation sensor. The SPOT Vegetation system has four spectral bands at visible and infrared wavelengths with a resolution of one kilometer. The Vegetation Programme was developed by a number of partners for the **monitoring** of land-surface parameters with daily frequency at global scale. The Vegetation system will complement the existing high-spatial resolution capabilities of the SPOT series. Data from SPOT Vegetation has been used to map the distribution of burnt areas at the global scale, assess the impact of natural disasters, and map land cover and **land-cover change**.

See also: **Advanced Very High Resolution Radiometer; Global Burnt Area 2000; Global Land Cover Map of the Year 2000; GLOBCARBON; Normalized Difference Vegetation Index; Remote Sensing; Tropical Ecosystem Environment Observations by Satellite (TREES) Project; Vegetation.**

Further Reading

French Space Agency (CNES) [Online, August 2004], The SPOT Vegetation Programme Web Site, www.spot-vegetation.com; Lupo, Fred, Reginster, Isabella, and Éric F. Lambin, "Monitoring Land-Cover Changes in West Africa with SPOT VEGETATION: Impact of Natural Disasters in 1998–1999," *International Journal of Remote Sensing* 22 (2001): 2633–9; Mayaux, Philippe, Bartholomé, Etienne, Fritz, Steffen, and Alan Belward, "A New Land-Cover Map of Africa for the Year 2000," *Journal of Biogeography* 31 (2004): 861–77.

KEVIN J. TANSEY

Subsistence Agriculture. Agricultural production that is oriented toward home **consumption** and characterized by low application of external production factors and low output per unit of land, in contrast to market-oriented or commercial agriculture. The main farm inputs in subsistence agriculture (or subsistence farming) are family labor and land, while capital **investment** is minimal to nonexistent. Subsistence farming relies on the cultivation of **food crops** to fulfill the immediate nutrition demands of the household. The amount of farm output determines the amount a family can consume in pure subsistence agriculture. However, the term subsistence agriculture involves several ambiguities. This stems from the fact that farm households are producers and consumers at the same time, in contrast to conventional economic theory where firms produce goods by maximizing their profits, while households supply labor and consume the products. Consequently, subsistence-oriented farm households are economic entities, which make both production and consumption decisions. The objective of subsistence land users is not profit maximization but the maximization of subjective utility, which includes **tradeoffs** in limited labor allocation between the necessary satisfaction of consumption requirements and the achievement of leisure time. Pure subsistence farmers are completely outside of market exchanges and produce exclusively for home consumption, while a pure commercial farmer produces solely for sale on markets. The term subsistence agriculture is not clearcut and implies a certain threshold along a gradient between zero market integration and a 100 percent share of marketed agricultural production. The proportion of production devoted to a household's home consumption is termed the "degree of subsistence" of a farm household. Often, a 50 percent threshold is used. Historically, most farmers were engaged in subsistence-oriented agricultural production. This is still the case in the least-developed countries (LDCs), but pure subsistence farming is becoming increasingly rare and most farmers today produce some marketable surplus. The degree of subsistence of farming households shows substantial geographic variations, but is especially high in LDCs. Subsistence agriculture also increased after the collapse of socialism in the transition countries of **east-central Europe**.

The use of only family labor in combination with traditional farming tools limits the amount of land a farm household can cultivate. As cultivated fields on sloping land tend to lose fertility under permanent cultivation, the most economic method with the highest return per unit of labor is shifting or **swidden cultivation** to restore the fertility of plots under **fallow**. The expansion of areas devoted to subsistence production in permanent and shifting agricultural cultivation is identified as one of the major causes of tropical **deforestation**. Population growth through natural growth or inmigration often leads to this expansion of cropped areas in traditional subsistence-based communities. Increasing market orientation and intensification of subsistence-based economies brought about by economic development may be a major factor contributing to land-use and **land-cover changes** in the future.

See also: **Agricultural Frontier; Agriculture; Cassava; Decision-Making; Human Immunodeficiency Virus (HIV)/Acquired Immunodeficiency Syndrome (AIDS); Landholding; Land-Use System; Population Dynamics; Proximate Causes.**

Further Reading

Chayanov, Alexander V., *The Theory of Peasant Economy*, Madison, WI: University of Wisconsin Press, 1986; Ellis, Frank, *Peasant Economics*, Cambridge, UK: Cambridge University Press, 1993; Upton, Martin, *The Economics of Tropical Farming Systems*, Cambridge, UK: Cambridge University Press, 1996; Singh, Inderjit, Squire, Lyn, and John Strauss, *Agricultural Household Models: Extensions, Applications, and Policy*, Baltimore, MD: Johns Hopkins University Press, 1986.

DANIEL MÜLLER

Suburbanization. The decentralization of cities caused by outward movement of households and firms to the urban periphery. The process of suburbanization has been taking place for well over a century and has occurred in cities worldwide. However, it accelerated markedly in the latter half of the twentieth century and has proceeded faster and farther in the **United States of America** than in any other country.

As Alex Anas, Richard Arnott, and Kenneth Small detail, suburbanization is fundamentally tied to changes in transportation and communication systems. At the beginning of the nineteenth century in pre-industrialized cities, intra-urban transport took place mainly by horse and wagon or by foot. People lived where they worked, and there was a clear demarcation line between the dense city and sparsely settled countryside. Within cities, the rich and poor lived together, although the rich, for the most part, outbid the poor for the most central and convenient locations. The industrial age led to massive rural-to-urban migration as newly built factories and mills attracted people from the countryside and spurred urban agglomerations. In the United States, the advent of the railroad in the 1850s and then the electric streetcar in the 1870s brought about the first real suburbs as increasing numbers of upper- and middle-income commuters moved out along railcar and then streetcar lines. The result was a "hub-and-spoke" pattern of development: a compact production-oriented core and radial streetcar lines with residences concentrated around stations located along these lines. The "streetcar suburbs" marked the first real emergence of land-use patterns characterized by a clear separation between residential and commercial/industrial uses. The separation of land uses was further reinforced by local zoning laws that became common in the United States by the 1920s.

By far the most influential force underlying suburbanization has been the automobile. Assembly-line production beginning in 1908 made the automobile increasingly affordable for households, and trucks gradually replaced the horse-drawn vehicles as the preferred intra-city transport for businesses. As a result, businesses and households were able to take advantage of lower land values further away from the city center; residential suburbs developed at the edges of cities while businesses moved further from the central city core as business districts expanded. In the United States, the federal government initiated support for road building in the 1920s that by the 1970s had led to an extensive network of high-speed, interstate highways and secondary roads.

After World War II, the increase in automobile ownership and roads led to unprecedented suburban growth in the United States. The share of metropolitan population living in central cities declined from 64 percent in 1948 to 39 percent in 1990. Spurred by rising household incomes, abundant and accessible rural land, and the relatively lower suburban housing prices, many city residents flocked to the suburbs. Higher crime rates and racial tensions in cities, exacerbated by forced desegregation of schools in the 1950s, led to "white flight" from the cities and the emergence of homogenous, segregated suburban jurisdictions comprising higher-income households that offered higher-quality schools and services. A vicious cycle of central city decline and suburban growth resulted as more and more middle class families left the city for better suburban neighborhoods, the tax base of central cities eroded further, and **poverty** became concentrated in the central cities.

In addition to highway and road-building subsidies, other federal policies have fueled suburbanization in the United States. For example, federal income tax policies allow taxpayers to deduct mortgage interest and **property** taxes from taxable income, which creates a substantial subsidy for homeownership. Because land for new residential development is located at the urban fringe, this subsidy has fostered even greater suburbanization.

The suburbanization of employment in the United States also accelerated markedly after World War II. The percent of manufacturing located in U.S. central cities declined from 67 percent in 1948 to 45 percent in 1990. Corresponding percentages of wholesale and retail trade declined from 92 to 49 percent and from 75 to 48 percent respectively between these years. Intracity and intercity trucking, commuting via automobile, and the switch from multistory to single-story plants that exploited new horizontal production technologies (e.g., assembly-line production) led manufacturing to locate to suburban areas where land was cheaper and circumferential highways provided easy access to interstate highways. Advances in information technologies in the 1980s and 1990s (including fax machines, e-mail, the Internet, and teleconferencing) further eroded the benefits of physical proximity and have allowed firms to be increasingly footloose.

Suburbanization has also occurred outside of the United States during the post-World War II era, although to a lesser degree. Cross-country comparisons of urban **decentralization** indicate that cities in other developed countries, including **Japan**, **Canada**, and Germany, are relatively less suburbanized. Likewise, a number of cities in developing countries have become more decentralized over time. Interestingly, some evidence suggests that, while central-city densities are much higher in non-U.S. cities, the rate of decentralization of some non-U.S. cities has been comparable to that of U.S. cities. It is difficult to isolate the factors that contribute to the differences in suburbanization between the United States and other countries. Scholars have emphasized a variety of factors, including the greater abundance of land in the United States, a greater reliance on automobiles versus public transportation, a more extensive system of intra-metropolitan highways, less government protection of rural land, greater suburban fiscal autonomy, higher crime rates in central cities, and greater ethnic and racial diversity in the United States.

Urban decentralization has grown more complex in recent decades. Rather than uniform decentralization, cities have evolved into polycentric forms characterized by a number of suburban subcenters in addition to the traditional urban center. Agglomeration economies in production, for example, from manufacturers that share the same input supplier or from office firms that are supported by the same nearby hotel and restaurants, have led to clusters of suburban employment and, in some cases, the emergence of "edge cities"—new concentrations of office and retail activities outside core areas of metropolitan areas. The emergence of edge cities, along with continued road building and advances in communications technologies, has pushed the urban periphery out even further into rural areas. In many developed countries, suburbanization has given way to exurbanization—development of peri-urban regions that are beyond the suburbs, but still within the commutershed of an urban area.

See also: **Industrialization; Leapfrogging; Public Policy; Urban Sprawl; Urbanization; Urban-Rural Fringe.**

Further Reading

Anas, Alex, Richard Arnott, and Kenneth A. Small, "Urban Spatial Structure," *Journal of Economic Literature* 36 (1998): 1426–64; Jackson, Kenneth T., *Crabgrass Frontier: The Suburbanization of the United States,* New York: Oxford University Press, 1985; Mieszkowski, Peter, and Edwin S. Mills, "The Causes of Metropolitan Suburbanization," *Journal of Economic Perspectives* 7 (1993): 135–47; O'Sullivan, Arthur, *Urban Economics,* 5th ed., New York: McGraw Hill, 2004.

ELENA G. IRWIN

Sudan-Sahel. Bioclimatic region lying south of the Sahara Desert and stretching from the Atlantic Ocean in the west through Sudan in the east. The Sudan-Sahel is a semiarid tropical region experiencing high temperatures and strongly seasonal rainfall (most all rain falling from May through September). The northern and southern boundaries of the Sahel and the Sudan regions have been generally defined by long-term mean annual rainfall (LMAR). Climate in the region is driven by the movement of the Intertropical Convergence Zone. Annual rainfall increases as one moves south from the desert edge into the Sudan-Sahel with a corresponding increase in the length of the summer rainy season from one to four months. The boundary between the Sahel and the Sahara Desert to the north has been variously defined as 100–200 millimeters LMAR. To the south, the boundary between the Sahel and the Sudan has been variously defined as the 500–600 millimeter LMAR. The Sudan borders the Guinean bioclimatic zone to the south at a variously defined boundary of 800–1000 millimeter LMAR.

Vegetation in the Sudan-Sahel grades from **grassland** steppe in the northern Sahel to open savanna in the southern Sudan. Herbacous vegetation in the Sahel is dominated by short-cycle annual grasses and forbs, the species composition of which changes from year to year based on the magnitude and distribution of summer rains. Perennial grasses increase in prevalence in the southern Sudan. Vegetative productivity in both zones is limited not only by sparse and highly seasonal rainfall but by low soil fertility. Due to a combination of underlying sedimentary rock formations and earlier historical periods of higher humidity and weathering in the region, soils are acidic, low in organic matter content and exchange capacity, and with particle size distributions dominated by fine sands. Plant growth is thus multiply constrained in the region, with the availability of nitrogen, phosphorus, and moisture seen as the major constraints. The relative importance of these limiting factors changes seasonally and year-to-year within any particular location, but the importance of moisture availability is generally higher in locations of the northern Sahel while soil fertility is more likely to be the dominant constraint in the southern Sudan. The Sudan-Sahel is a region with multiple constraints to vegetative productivity but one where the relative importance of these constraints varies along a sharp north-south gradient.

The Sudan-Sahel is where many African precolonial empires developed (e.g., Nubia, Ghana, Mali, Songhay, Segou, Kanem-Bornu, Haussaland, Sokoto, Toukouleur, and Maasina). While these larger polities developed for different reasons and sustained themselves in different ways, one general reason for the succession of empires in the region was the economic surplus gained through the control of regional trade. North-south regional trade networks in cattle, ivory, gold, slaves, cola nuts, salt, etc. crossed through the Sudan-Sahel region. These goods were produced at different points along the steep north-south ecological gradient from the rainforest to the south of the region extending to the desert north. Ecological specialization of production across a relatively short north-south distance was a major stimulus for the development of this trade in the region that played a role in the formation of precolonial states and cultural identities.

Agricultural strategies have developed in response to this environment. Major grain crops in the region are millet and sorghum (with some floodplain **rice**). Millet is grown as a rainfed crop in all of the region except the northern half of the Sahel zone (less than 300 millimeters LMAR). Sorghum, while rare in the Sahel (cultivated on low-lying fine-grained soils), becomes more common in the Sudan zone. Farmers' agricultural strategies reflect the multiple constraints that crop agriculture faces in the region. Risk of crop failure due to insufficient moisture during a portion of the growing season is a major issue particularly as one moves north into the Sahel zone. Where arable land is not scarce, farmers will often manage widely dispersed fields, only committing the necessary labor (for weeding, etc.) to bring a field to harvest for those fields that have received sufficient rainfall. Farmers have historically managed the wide variety of millet and sorghum cultivars that require different-length growing seasons and with different resistances to drought. Farmers will often be required to reseed their fields as crops fail during the beginning of the rainy season—utilizing different cultivars as the remaining growing season shortens.

When moisture is sufficient, soil fertility quickly becomes a major limiting factor in crop **agriculture**. While inorganic fertilizers have become more prevalent in the region over the past several decades, there remain significant economic constraints to their widespread use. Historically, farmers have relied on a combination of legume intercropping (e.g., cowpea, *Faidherbia albida*), long-term **fallow** systems (outfields) and nutrient applications as organic matter amendments (infields). While green manure systems exist, the major form in which nutrients are harvested from a wider area and applied to cropped fields has been through animal grazing and manure application. It is the Sudan-Sahel region where systems of manure contracting between livestock herders and farmers are most well documented.

Livestock producers in the region have long relied on seasonal mobility to reduce their exposure to drought risk and to take advantage of the different forage opportunities across the north-south ecological gradient. Seasonal movements to the northern Sahel during the rainy season and back to agricultural areas in the southern Sahel and Sudan during the dry season are commonly referred to as **transhumance**. Long-distance transhumance has worked to reduce the risk of crop damage in the agricultural zone while presenting regional livestock to the sparser but more nutritious forage in the northern Sahel. These systems of long-distance transhumance have been seriously eroded over the past thirty years as drought-induced decapitalization of the herds' "pastoralists" have caused many to rely on crop agriculture; rebel activity in the north has increased risks; and new livestock owners have preferred to manage their livestock within the agricultural zones during the rainy season. Therefore, **pastoral mobility** has generally become more circumscribed in the region. Notable exceptions are those cases where mobility has increased due to local shortages of **pastures**.

Ethnic and caste identities have long been associated with different occupational specializations. Production strategies of Sudano-Sahelian households are multifaceted, often extending across categories such as pastoralist, farmer, city dweller, etc. Still, there remain dominant economic activities associated with different ethnic identities. For example, a wide range of ethnic groups (e.g., Wolof, Malinke, Bambara, Songhay, Mossi, Haussa, etc.) are seen as dryland farmers of the region. Likewise, there are a number of different ethnic groups whose identities are tied to livestock husbandry (e.g., Fulani, Touareg, Toubou, Dinka). In all cases, actual productive activities are broader and more diverse than those considered prototypic or ideal for a particular ethnic or caste group.

The countries of Senegal, Mali, Burkina Faso, Niger, Chad, and Sudan are countries that contain substantial areas of both Sahelian and Sudanian zones. Except for Sudan, these countries were part of the French West African colonial system from the late nineteenth century to the early 1960s. The French, through their system of taxation, encouraged the production of **cash crops** such as **cotton** and groundnut. With the expansion of cash cropping and an acceleration of population growth, cropped area expanded rapidly in the 1950s and 1960s—a historic period of relatively high rainfall. During this period, cropped area expanded northward onto pastoral rangelands. With the drought of 1968–1973, this period of expansion came to an abrupt halt, leaving those who had invested their labor into cash crop production, as during the early colonial period, particularly vulnerable.

Since the 1968–1973 drought, rainfall in the Sudan-Sahel region has remained below its long-term average. This represents one of the best examples of a seemingly persistent shift in rainfall regime in the world—leading to speculation as to its relationship to global climate change or more regional biogeophysical **feedbacks**. Reduced vegetative productivity and rapidly increasing human populations have raised concerns about anthropogenic environmental degradation in the region. In fact, the Sudan-Sahel has been commonly cited as one of the best examples of widespread **desertification**. In addition, it served as a prominent example for linking **poverty** and **land degradation** and as such contributed to the development of the concept of sustainable development. Contemporary assessments have greatly reduced the prior estimates of anthropogenic environmental degradation in the region.

This dry period has severely affected people living in the Sudan-Sahel. Rural peoples have continued their efforts to reduce their exposure to risk by pursuing non-agricultural occupations (for some family members) and moving their production activities to the southern Sudanian and even Guinean zones where rainfall is more consistent. As a result, there has been a southerly shift of cropping and livestock presence and an acceleration of rural-to-urban migration over the last thirty years. **Conservation** and development funding for the Sahel have declined significantly since the 1980s and are now focused on economic and ecological stabilization through **decentralization** of **decision-making**. The most common forms of rural development activity in the zone include poverty banks, gardening projects, and programs to improve the management of village lands. In the Sudan, there has been greater private **investment** in agriculture with an expansion of cash crop production in some areas, particularly cotton (e.g., the Sikasso region of Mali).

See also: **Modification; People at Risk; Sahel Land Cover; Sahelian Land Use (SALU) Model; Vulnerability.**

Further Reading

Gritzner, Jeff, *The West African Sahel: Human Agency and Environmental Change*, Geography Research Paper no. 226, Chicago: The University of Chicago, 1988; Le Houérou, Henri N., *The Grazing Land Ecosystems of the African Sahel*, Berlin, New York: Springer Verlag, 1989; Mortimore, Michael, *Adapting to Drought: Farmers, Famines and Desertification in West Africa*, Cambridge, UK: Cambridge University Press, 1989.

MATTHEW D. TURNER

Sustainable Land Use. The management of land for agricultural and other purposes, usually through the integration of ecological with socio-economic and political principles, in order to ensure continuous satisfaction of human needs for present and future generations without comprising the ability of future generations or populations in other locations to meet their needs (intra- and intergenerational equity). According to Euro Beinat and Peter Nijkamp, at least three dimensions of sustainable land use can be distinguished. First, the husbandry dimension relates to the durability, exploitability, and continuity of natural resources over a long time horizon. Related actions, for example, are the application of crop-rotation systems—rather than **monocultures**—the careful use of scarce natural resources, and the rehabilitation of degraded land. Second, the interdependence dimension implies aspects of **land fragmentation**, land segmentation, and relations among different types of land uses. In contrast to **agro-industry**, for example, traditional farming is more often based upon interdependence according to which the farm and the surrounding (semi)natural areas achieve a flowing equilibrium based on interaction and mutual system resilience. Third, with concepts such as option value or existence values, the ethics dimension refers to obligations toward future generations: sustainable land use is given if the use and management of resources do not lead to the loss of future options, with **resilience** as a **land quality** feature that renders land use sustainable.

From a wide array of scientific evidence, Éric Lambin distilled three general conditions or components of human-environment interactions, involving multiple—that is, temporal, historical, social, economic, socio-political, cultural, technological, financial, and institutional—dimensions each. These conditions are shown to control the transition toward sustainable land use, that is, success or failure (including collapse) in land use and management. The first component is information on the state of the environment. It relates to the understanding by decision-makers of **land degradation** and of alternative management practices, as driven by knowledge, information, and communication. The second component is motivation to sustainably manage the environment, and it relates to the sources of the behavior of agents. The third component is the capacity to implement sustainable management, and it relates to the provision of appropriate physical, technical, and institutional infrastructure necessary for sustainable land management.

See also: **Agrodiversity; Alternatives to Slash-and-Burn (ASB) Programme; Community Involvement; Decision-Making; Endogenous Factor; Forest Transition; Indigenous Knowledge; Institutions; Land Rehabilitation; Land-Use Planning; Land-Use Transition; Public Policy; Tradeoffs.**

Further Reading

Beinat, Euro, and Peter Nijkamp, "Land Use Planning," in *Encyclopedia of Global Change: Environmental Change and Human Society*, vol. 2 (J–Z), eds. Andrew S. Goudie and David J. Cuff, Oxford, UK, New York: Oxford University Press: 2002, pp. 27–33; Lambin, Éric F., "Conditions for a Transition Towards a Sustainable Use of Environmental Resources," *Global Environmental Change* 15 (3) (2005): 177–80; Turner, B. L., "The Sustainability Principle in Global Agendas: Implications for Understanding Land-Use/Cover Change," *Geographical Journal* 163 (1997): 133–40.

HELMUT GEIST

Swidden Cultivation. Also called shifting cultivation and derogatively **slash-and-burn agriculture**, swidden cultivation is among the oldest, most complex, and multifaceted forms of **agriculture** in the world. Estimates of the amount of land cultivated worldwide by swidden cultivators vary from 2.9 billion hectares to roughly half of the land area in the tropics.

Swidden is a system in which **vegetation** felled in patches of **forest** during the dry season is burned before the onset of the rainy season to open the site and release nutrients. The cleared fields are cultivated and harvested for one or more years, and then left to lie **fallow** for varying periods to allow secondary forest to regrow. Indigenous farmers manage the system in ways that integrate production from both cultivated fields and diverse secondary forests, including everything from grass and bushes in the early stages, to young open-canopy trees, to mature closed-canopy tree communities. Clifford Geertz described a swidden as "[a] natural forest . . . transformed into a harvestable forest," and others have suggested that vast areas considered primary or virgin forest are really late-secondary forests on lands previously cleared for swiddens. **Harold Brookfield** and Christine Padoch argue that swidden cultivation is not one system but many hundreds or thousands of systems. Some cultivators rotate some of their fields but also practice permanent cultivation on part of their land. Other swidden farmers create complex agroforests in which every plant has its uses, and they modify the landscape by making small soil-retention terraces on steep slopes.

Opposition to swidden cultivation is based mainly on misconceptions of how the system works, especially misunderstanding of its effects on soil as compared to the effects on soil of permanent agriculture. Swidden cultivation involves only temporary use of forestlands, not permanent loss, as is the case with permanent agriculture, human settlements, mining, and dam-building. In fact, swidden has a significant advantage over permanent agriculture: the fallow periods allow soil to stabilize and give forest vegetation an opportunity to regrow, providing a home for a variety of life forms. Because of its fallow period, swidden agriculture promotes both greater **carbon sequestration** and **biodiversity** conservation than permanent agriculture. Failure to understand the swidden agricultural system and its effects on forest regeneration has led scientists to overestimate the amount of "**deforestation**" that has occurred. Failure to understand successional vegetation has also led to government policies, mostly failures, for settling swidden agriculturists.

See also: **Agricultural Intensification; Agroforestry; Alternatives to Slash-and-Burn (ASB) Programme; Conservation; Degradation Narrative; Pristine Myth; Public Policy; Regrowth; Secondary Vegetation.**

Further Reading

Brookfield, Harold, and Christine Padoch, "Appreciating Agrodiversity: A Look at the Dynamism and Diversity of Indigenous Farming Practices," *Environment* 36 (1994): 7–11; Brown, Sandra, and Ariel E. Lugo, "Tropical Secondary Forests," *Journal of Tropical Ecology* 6 (1990): 1–32; Geertz, Cliford, *Agricultural Involution: The Processes of Ecological Change in Indonesia*, Berkeley, CA: University of California Press, 1963; Thrupp, Lori Ann, Hecht, Susanna B., and John O. Browder, *The Diversity and Dynamics of Shifting Cultivation: Myths, Realities, and Policy Implications*, Washington, DC: World Resources Institute, 1997.

JEFFERSON FOX

Syndrome. Fundamental high-level cause of land-use/cover change. Summarizing a large number of case studies under the umbrella of the **Land-Use/Cover Change (LUCC) project**, Éric Lambin and colleagues found that land-use change was driven by a combination of the following five syndromes or high-level causes: resource scarcity leading to an increase in the pressure of production on resources; changing opportunities created by markets; outside

Typology of land-change syndromes

	Resource scarcity causing a pressure of production on resources	Changing opportunities created by markets	Outside policy intervention	Loss of adaptive capacity, increased vulnerability	Changes in social organization, in resource access, and in attitudes
Slow	Natural population growth and division of land parcels; Domestic life cycles leading to changes in labor availability; Loss of land productivity on sensitive areas following excessive or inappropriate use; Failure to restore or to maintain protective works of environmental resources; Heavy surplus extraction away from the land manager;	Increase in commercialization and agroindustrialization; Improvement in accessibility through road construction; Changes in market prices for inputs or outputs (e.g., erosion of prices of primary production, unfavorable global or urban-rural terms of trade); Off-farm wages and employment opportunities;	Economic development programs; Perverse subsidies, policy-induced price distortions and fiscal incentives; Frontier development (e.g., for geopolitical reasons or to promote interest groups); Poor governance and corruption; Insecurity in land tenure;	Impoverishment (e.g., creeping household debts, no access to credit, lack of alternative income sources; weak buffering capacity); Breakdown of informal social security networks; Dependence on external resources or on assistance; Social discrimination (ethnic minorities, women, lower classes people, or caste members);	Changes in institutions governing access to resources by different land managers (e.g., shift from communal to private rights, tenure, holdings, and titles); Growth of urban aspirations; Breakdown of extended family; Growth of individualism and materialism; Lack of public education and poor information flow on the environment;
Fast	Spontaneous migration, forced population displacement, refugees; Decrease in land availability due to encroachment by other land uses, e.g., natural reserves (i.e., tragedy of enclosure);	Capital investments; Changes in national or global macro-economic and trade conditions leading to changes in prices (e.g., surge in energy prices, global financial crisis); New technologies for intensification of resource use;	Rapid policy changes (e.g., devaluation); Government instability; War;	Internal conflicts; Illness (e.g., HIV); Risks associated with natural hazards (e.g., leading to a crop failure, loss of resource or loss of productive capacity);	Loss of entitlements to environmental resources (e.g., expropriation for large-scale agriculture, large dams, forestry projects, tourism, and wildlife conservation), leading to an ecological marginalization of poors;

Lambin, Geist, and Lepers (2003).

policy intervention; loss of adaptive capacity and increased **vulnerability**; and changes in social organization, in resource **access**, and in attitudes. Some of the syndromes are experienced as constraints, forcing local land managers into degradation, innovation, or displacement pathways. Other fundamental causes are associated with the seizure of new opportunities by land managers seeking to realize diverse aspirations. Each of these syndromes can apply as a slow evolutionary process that changes incrementally at the time scale of decades or more, or as a fast change that is abrupt and occurs as a perturbation that affects human-environment systems suddenly (see table titled "Typology of land-change syndromes"). Only a combination of several causes, with synergetic interactions, is likely to drive a region into a critical trajectory. Some of the syndromes leading to land-use change are predominantly **endogenous factors**, such as resource scarcity, increased vulnerability, and changes in social organization; while others, such as changing market opportunities and policy intervention, are mostly **exogenous factors**, though mediated by some local factors.

The basis of the syndrome approach is the insight from a large array of case studies that not all causes and all levels of organization are equally important. For any given human-environment system, a limited number of causes are essential to predict the general trend in land use. The syndrome approach dates back to the pioneering work of Gerhard Petschel-Held and colleagues at the Potsdam Institute for Climate Impact Research (PIK) in Germany. PIK has described archetypical, dynamic, coevolutionary patterns of human-environment interactions; and a taxonomy of syndromes links processes of **land degradation** to both changes over time and status of state variables. The approach is applied at the intermediate functional **scales** that reflect processes taking place from the household level up to the international level. For example, the "overexploitation syndrome" represents the natural and social processes governing the extraction of biological resources through unsustainable industrial logging activities or other forms of resource use. The typology of PIK syndromes reflects expert opinion based on local-scale examples, and the approach aims at a high level of generality in the description of mechanisms of environmental degradation.

See also: **Cumulative Change; Driving Forces; Human Immunodeficiency Virus (HIV)/ Acquired Immunodeficiency Syndrome (AIDS); Mediating Factor; Proximate Causes; Tragedy of Enclosure.**

Further Reading

Lambin, Éric F., Geist, Helmut J., and Erika Lepers, "Dynamics of Land Use and Cover Change in Tropical and Subtropical Regions," *Annual Review of Environment and Resources* 28 (2003): 205–41; Petschel-Held, Gerhard, Lüdeke, Matthias K. B., and Fritz Reusswig, "Actors, Structures and Environments: A Comparative and Transdisciplinary View on Regional Case Studies of Global Environmental Change," in *Coping with Changing Environments: Social Dimensions of Endangered Ecosystems in the Developing World*, eds. Beate Lohnert and Helmut Geist, Aldershot, UK: Ashgate, 1999, pp. 255–93.

HELMUT GEIST

Systemic Change. A characteristic of biogeochemical cycling in which change in any part or at any location in the earth system can create significant consequences throughout the system. The structure and function of the earth system, including the biosphere or that part of the system in which life can exist, is maintained by the cycling of biogeochemicals such as carbon, nitrogen, and water among the oceans, land, and atmosphere. This cycling, for the most part, is globally fluid; changes in the ocean, land, and atmospheric stocks or disruption in the cycling among these stocks, if of sufficient magnitude, affect the earth system at large. The capacity of humankind to affect directly and sufficiently most biogeochemical cycles to trigger systemic change in the earth system is recent, a product of advances in fossil fuel and synthetics technology over the past 300 years, although the systemic consequences of **deforestation** may have begun much earlier. By the end of the twentieth century, the production and **consumption** activities of humankind matched nature in terms of affecting critical biogeochemical cycles. Perhaps the most notable example is the human impact on the **carbon cycle**. The burning of fossil fuels primarily in highly developed economies and, to a lesser extent, global deforestation, has released terrestrial-based stocks of carbon, generating unprecedented, rapid buildup of carbon dioxide (CO_2) in the atmosphere, raising concerns about changes in global climates.

See also: **Atmosphere-Land Interlinkages; Change Detection; Cumulative Change; Economic Growth; Nutrient Cycle; Water-Land Interlinkages.**

Further Reading

Kasperson, Roger E., Turner, B. L., II, Meyer, William B., Dow, Kirstin, Golding, Dominic, Kasperson, Jeanne X., Mitchell, Robert C., and Samuel J. Ratick, "Two Types of Global Environmental Change: Definitional and Spatial-Scale Issues in Their Human Dimensions," *Global Environmental Change: Human and Policy Dimensions* 1 (1990): 14–22; Steffen, Will, Sanderson, Angelina, Tyson, Peter, Jäger, Jill, Matson, Pamela, Moore, Berrien III, Oldfield, Frank, et al., eds., *Global Change and the Earth System: A Planet under Pressure,* Berlin: Springer Verlag, 2004.

B. L. TURNER II

T

Taiga. The northern **forest** belt in Eurasia and North America, now often used synonymously in both continents for boreal forest (to which *Boreas*, the Greek mythological god of northern winds, gave his name). The Russian term тайга for "forest" is of Altaic origin for "little twigs." The taiga consists of a nearly continuous circumpolar belt of coniferous trees that covers about 17 percent of the earth's land surface, between 50° and 60° northern latitudes (about 1.5 billion hectares or 3.7 billion acres) and one-third of all forested land, thus representing the single largest terrestrial **biome** and wilderness of the planet. The region is of great importance because of its size, **climate impacts**, and **biodiversity**.

The taiga belongs to the cold continental climate zone with long, severe winters (up to six months with mean temperatures below freezing) and short summers (50–100 frost-free days) with high temperatures. Mean annual precipitation is 400–500 millimeters, mostly as summer rains, but low evaporation rates make this a humid climate.

The boreal forest communities of North America and Eurasia display a number of ecological similarities. For this reason, Taylor Ricket and colleagues assign them into their only circumpolar (of fourteen) biome, the boreal forest/taiga biome. Overlying formerly glaciated areas and partially permafrost locations on both continents, the forest is a mosaic of successional and subclimax plant communities sensitive to varying environmental conditions. The forests are dominated by dark evergreen conifers including *Pinus* (pine), *Abies* (fir), *Picea* (spruce), and *Larix* (larch or tamarack), and *Thuja* (arborvitae) in North America. Hardwoods include paper birch (*Betula papyrifera*) and quaking aspen (*Populus tremuloides*). Shrubs include willows and members of the heath family (*Ericaceae*). Ground cover is dominated by mosses and lichens. Large tracts are formed by recurring disturbances, such as forest fires and insect infestations. **Fires** have occurred more frequently in the past decades than they would if lightning had been the only cause. The Russian Forest Service estimates that 90 percent of fire events are caused by people.

Northward, the boreal forest merges into the circumpolar tundra biome. This transition zone represents the climatically sensitive northern timberline. Further latitudinal differentiation is undertaken into "sparse northern taiga" in northern Eurasia and "lichen woodland" in **Canada**, respectively. This open coniferous forest is followed southward by "middle and southern taiga" and "closed forest," respectively. In the **United States of America**, the southern border—temperate broadleaf deciduous forest—is dominated by white pine (*Pinus strobus*), sugar maple (*Acer saccharum*), and American beech (*Fagus americanus*). Only in northern Eurasia does there exist a strong longitudinal distinction: the western taiga, with less severe climate conditions, and the eastern taiga beyond river Yenisey (at 90° E) with extreme continentality. In the western section, dense forests of spruce and fir ("dark taiga") on moister locations alternate with pine, shrubs, and grasses on drier soils. Extreme frost resistance enables the deciduous larch to be dominant in the east ("light taiga").

Fur-bearing predators like the lynx (*Felis lynx*) and various members of the weasel family (e.g., wolverine, fisher, pine marten, mink, ermine, and sable) belong to the characteristic fauna. Large herbivores are more closely associated with successional stages where there is more nutritious browse available and include elk or wapiti (*Cervus elaphus*, known as red deer in **Europe**) and moose (*Alces alces*, known as elk in Europe). Among birds, insect-eaters like wood warblers are migratory and leave after the breeding season.

Many indigenous people live in boreal forests, including the Dene, Inuit, Cree, and Athabaskans of North America; the Saami of Scandinavia; the Ainu of northern **Japan**; and the Nenets, Yakut, Udege, and Altaisk of Siberia. In recent times, settlement of peoples from outside the **boreal zone** has taken place, and the exploitation of the resources is now posing a threat to the fragile ecosystem and its native peoples.

Agricultural land use occurs in Canada and mainly in the European part of northern Eurasia. The taiga remains the world's largest timber reserve and carbon stock despite huge stretches of swamp and peat bog. Large-scale industrial **forestry** is by far the most important threat affecting boreal forests today. In Scandinavia, large-scale exploitation of the forests has transformed virtually all forestland into intensively managed **secondary vegetation**. In Siberia and parts of Alaska and Canada, primary forest still exists, but forest demand from **China**, Japan, and Mongolia transforms the taiga into a "pulp factory." Other pressures include oil and gas exploration, road building, and mining.

See also: **Hot Spots of Land-Cover Change; Peatland; Russia; Wetlands.**

Further Reading

Ricketts, Taylor, Dinerstein, Eric, Olson, David, Loucks, Colby, Eichbaum, William, DellaSala, Dominick, Kavanagh, Kevin et al., *Terrestrial Ecoregions of North America: A Conservation Assessment*, Washington, DC: Island Press, 1999; Taiga Rescue Network, The Boreal Forest [Online, December 2004], The Taiga Web Site, www.taigarescue.org; Yaroshenko, Alexey, Potapov, Peter, and Svetlana Turubanova, *The Last Intact Forest Landscapes of Northern European Russia*, Moscow: Greenpeace Russia, 2001.

CHRISTIANE SCHMULLIUS

Technological-Scientific Revolution. According to Jaroslav Purš, the technological-scientific revolution (TSR) is characterized by the use of electric power to drive machines, by combustion engines, the development of heavy chemistry, the introduction of improved machines and technological-chemical processes in a number of major industries, the beginning of automation, production belt systems, and a more extensive use of scientific knowledge in production practice, including **agriculture**. The TSR, following the **Industrial Revolution**, had three phases: the first from the 1870s to the turn of the century, the second from then until the 1930s, and the third lasting until the end of the 1940s. In parts of the literature, the term "Second Industrial Revolution" is used as a synonymous term. TSR also indicates a period in which industrial hegemony moved from Britain to Germany and the **United States of America**.

After culmination of the **Agricultural Revolution**, the feeding of the rapidly increasing non-agricultural population in the growing cities required an increase of agricultural production efficiency instead of the expansion of cultivated land area. The only solution was a more intensive exploitation of inherited land-use structure. This has been carried out through capital **investments** into the most fertile plots by mechanization, deep tillage, use of chemicals, crop rotation, etc.

Not all elements of TSR (in industry) could be directly related to agriculture, because of the special conditions for circuit of capital depending on natural processes of organic growth, physico-geographical conditions, and limited area of land suitable for cultivation. The TSR in agriculture then meant the complete introduction of free crop rotation systems, mechanization by extensive use of combustion engines, electricity as a new means of energy, use of artificial fertilizers produced in industry, land reclamations (ameliorations), and the use of the tractor as the "universal machine in agriculture."

Its first phase was typical for the rapid introduction of more sophisticated agricultural machines and tools (yet driven by animal power and only at big estates, very rarely also by steam engines); by slow increase of synthetic fertilizer use, the acceleration of land improvements (including rejoining of field patterns); application of scientific methods in plant and animal production; and by the founding of agricultural high schools, research stations, and farmer cooperatives (credit, storage, purchasing, marketing, etc.). Such innovations are named the **Green Revolution** in parts of the literature.

The second phase meant the quantitative and qualitative boom of all features of the first phase, and mainly their growing penetration into small farming. The larger use of chemicals started together with deep tilling after World War I to play the decisive role in reviving and improving the natural fertility of land and in protection of plants against insects and diseases. The principal role was played by the internal combustion engine and electricity, which brought fundamental change in the energy balance in agriculture: the move from solar energy to fossil fuel energy for pulling or driving the drilling, sewing, grain threshing, and cleaning machines; and in animal production for preparing of feed (cutters, grinders, pumping water, centrifuging milk, etc.). Safe electric lighting enabled the night labor to work in peak seasons and spread the indoor breeding of livestock. The tractor as a universal tractive and motive machine, comparable to the steam engine in the Industrial Revolution, was suitable mainly in plant production because of its "areal" use character, and at large farms replaced horse power and cattle. It also made it possible for machines of greater weight and higher performance to be used in the fields, for deeper and faster tilling, and for faster transportation in agriculture.

These processes made a large part of the human labor force available for industry. The number of horses used could be reduced, and permanent

grassland that produced fodder could now be used for production of other plants. The TSR in agriculture needed a much larger amount of capital, because it was re-oriented toward **agricultural intensification**. The land-use trends at the beginning of the twentieth century changed dramatically: generally, the acreage of arable land decreased from this time until today; and permanent grasslands decreased at even higher rates, all to the detriment of **forest** and other areas or land uses such as roads, mining, and built-up areas.

See also: **Agro-Industry; Boserup, Ester; Carson, Rachel Louise; Driving Forces; Economic Restructuring; Extensification; Land Reform; Land Rehabilitation; Land Rent; Proximate Causes; Urbanization.**

Further Reading

Jeleček, Leoš, "Changes in the Production and Techniques in the Agriculture of Bohemia 1870–1945," in *Agriculture in the Industrial State*, eds. Michael A. Havinden and Edward J. T. Collins, Reading, UK: University of Reading, Rural History Centre, 1995, pp. 126–45; O'Brien, Patrick, and Prados Leandro de la Escosura, "Agricultural Productivity and European Industrialization, 1890–1980," *Economic History Review* XLV (1992): 60–92; Purš, Jaroslav, "Complex Revolution of the Modern Age and Industrial Revolution," *Historica* XIX (1980): 135–70; Thompson, Francis M. L., "The Second Agricultural Revolution, 1815–1880," *Economic History Review* 21 (1968): 62–73; Zanden, J. L. van, "The First Green Revolution: The Growth of Production and Productivity in European Agriculture, 1870–1914," *Economic History Review* XLIV (1991): 215–39.

LEOŠ JELEČEK

TERRA-ASTER. An Advanced Spaceborne Thermal Emission and Reflection Radiometer (ASTER) on-board the Earth Observing System (EOS) TERRA satellite of the National Aeronautics and Space Administration (NASA), launched in the year 2000. ASTER acquires spectral information at fourteen different wavelengths from 0.52 to 11.65 μm (visible and near-infrared to thermal infrared). The spatial resolution of the data set varies between fifteen and ninety meters. The ASTER detectors are pointable in the across-track direction, which enables detailed terrain models to be produced. ASTER data has been used to investigate the spatial and temporal distribution of evapotranspiration rates, for **monitoring** glaciers, and to study thermal properties of the urban environment.

See also: **Remote Sensing.**

Further Reading

National Aerospace Agency, Earth Observing System TERRA [Online, August 2004], The Advanced Spaceborne Thermal Emission and Reflection Radiometer (ASTER) Web Site, terra.nasa.gov/About/ASTER.

KEVIN J. TANSEY

Thermal Band Analysis. All objects with a temperature above absolute zero emit energy. Within land-use/cover-change analyses, thermal band (3–14 μm) satellite data measure the emission of energy from the earth's surface. This information, when converted into temperatures, can be used to link directly to other processes (e.g., micrometeorological). Satellite-based thermal analysis offers an accurate and viable means of **monitoring** different environments and estimating spatial aspects of their energy budget. Land surface processes are of paramount importance for the redistribution of energy and moisture between the land and the atmosphere. These exchanges of radiative, moisture, and heat fluxes affect biosphere development and physical living conditions at the earth's surface. The thermal band allows for a continuous representation of land cover and may be as important as reflectance-based **vegetation** indices used in much of today's research using **remote sensing**.

See also: **Advanced Very High Resolution Radiometer; Along Track Scanning Radiometers (ATSR) World Fire Atlas; Atmosphere-Land Interlinkages; Continuous Data; Pixel.**

Further Reading

Southworth, Jane, "An Assessment of Landsat TM Band 6 Thermal Data for Analyzing Land Cover in Tropical Dry Forests," *International Journal of Remote Sensing* 25 (2004): 689–706.

JANE SOUTHWORTH

Threshold. A point where a system is unstable and may switch abruptly between alternative stable states or equilibria. Thresholds may occur in all kinds of systems: ecological, social, or both. When a system moves from one equilibrium to another, this is often denoted as transition. Stable systems respond smoothly to gradual changes; near a threshold rapid, often unpredictable changes in the system can occur. Small changes in conditions may then result in qualitatively different trajectories. Consider the following example of a socio-ecological system. Woodlands and grassy open landscapes can under certain climatic/soil conditions be alternative stable states. Landscapes can be kept open by herbivores (domesticated or not) or **fires** even under conditions where woodlands, once established, are stable, because herbivores cannot destroy adult trees and shading reduces grass cover so that fires cannot spread. Increasing human use can result in a loss of **resilience** of woodlands until the threshold is reached—then the system flips into the **grassland** state.

See also: **Biomass; Canada; Change Detection; Disequilibrium Dynamics; Environmental Change; Land-Use Transition; People at Risk; Tradeoffs; Vulnerability.**

Further Reading

Berkes, Fikret, and Carl Folke, eds., *Linking Social and Ecological Systems,* Cambridge, UK: Cambridge University Press, 1998; Muradian, Roldan, "Ecological Thresholds: A Survey," *Ecological Economics* 38 (2001): 7–24; Scheffer, Marten, Carpenter, Steve, Foley, Jonathan A., Folke, Carl, and Brian Walker, "Catastrophic Shifts in Ecosystems," *Nature* 413 (2001): 591–6.

HELMUT HABERL

Thünen, Johann Heinrich von (1783–1850). Landowner near Rostock, Germany, who is noted for his work on the relationship between location and land use. Interested in how the economics of land use were influenced by distance to market, he published his method of analysis in *The Isolated State*, in 1826. His analysis focused on transport costs. Net profit depended on the market price of a commodity and its production cost plus transport cost. Transport cost for any one commodity, and hence profitability, depended on distance to market. Different commodities had different characteristics in relation to transport: the profitability of bulky, low-value commodities would decline more steeply with increasing distance than would that of compact, high-value commodities. The relative profitability of different commodities would therefore vary with distance from market. This, in turn, would translate into concentric zones or rings of different commodity production (i.e., land uses) arranged around the central market, sometimes referred to as a von Thünen pattern. Thünen's analysis was based on several assumptions—that profit maximization was the objective, that land quality was uniform, that a single market existed, and that transport costs varied with distance. When these assumptions are relaxed, the simplicity and regularity of the model break down, although the logic of the method of analysis survives. Thünian approaches have subsequently been applied at a variety of **scales**, including the global scale (with a central market in northwest **Europe**) as well as that of the village and farm. They have also been applied to study **deforestation** processes and to intra-urban land uses. Revisionists have attempted to adapt the static model to dynamic situations and in particular to regions with expanding cities.

See also: **Agent-Based Model; Cash Crops; Decision-Making; Economic Growth; Food Crops; Land Rent; Land-Use System; Modeling.**

Further Reading

Grotewold, Andreas A., "Von Thünen in Retrospect," *Economic Geography* 35 (1959): 346–55; New School University, The History of Economic Thought [Online, December 2004], The Johann Heinrich von Thünen Web Site, http://cepa.newschool.edu/het/profiles/thunen.htm; Thünen, Johann Heinrich von, *Von Thünen's Isolated State: An English Translation of Der Isolierte Staat* (trans. C. M. Wartenburg, ed. P. Hall), Oxford and New York: Pergamon, 1966.

ALEXANDER S. MATHER

Tobacco. The world's most widely grown non-food crop (in more than 120 countries) with a fairly low share of less than 1 percent of all arable land, though, and a globally widespread legal drug the use of which is one of the largest causes of premature death and the most common preventable cause of death in the world. Since about the mid-1960s, a global shift of tobacco production has occurred. Compared with 1700, when nearly the entire world production of tobacco was concentrated in Brazil, parts of the Caribbean, and in the Chesapeake colonies of New England (where most tobacco was grown and harvested by slave labor), with the breakdown of colonial rule from the mid-nineteenth century onward, tobacco farming spread nearly all over the world. Farming is extremely labor-intensive (100 times that of wheat), and the number of women and child laborers has increased with a production shift to developing countries. There, contract farming under the

Tobacco field in Windsor Locks, Connecticut, September 1937.
PHOTO: Edwin Locke. [Courtesy Library of Congress]

Flue-cured (Virginia) tobacco barn on Irwinville Farms in the Rural Resettlement Administration Project at Irwinville in Irwin County, Georgia, June 1936.
PHOTO: Carl Mydans. [Courtesy Library of Congress]

supervision of multinational cigarette or leaf buyer corporations is the current major trend. It has been described by Jason Clay as follows: tobacco companies act as banks, extending credit to farmers at the beginning of the season (for seed, fertilizer, pesticides, and technical support); in return, farmers agree to sell their crop to the company at harvest; company inspectors make regular visits to the farms to make sure their guidelines are followed; when the harvest is in and the time comes to pay the farmer, the leaf buyers determine the leaf grade and the price; as a common result, they often end up paying producers less than the original loans.

At present, most of the quantities entering world trade originate from zones of the developing world where frost-free days, a sufficient long dry season (allowing for harvesting and curing the crop), and low-cost conditions of production are optimal. By 2000, around four-fifths of the more than 120 growing countries, or the equivalent of around 90 percent of all global land under tobacco, had become located in developing countries. **China**, India, and Brazil produce more than half of the world's tobacco, with China and India being the largest producers as well as consumers (in both instances, most of the tobacco consumed is grown locally). The principal importing countries are **Russia**, followed closely by Germany, the **United States of America**, the Netherlands, the United Kingdom, and **Japan**. In developed market economies, tobacco production is increasingly being phased out, and tobacco use has stabilized (actually expected over the next century in **Europe** and the United States). Driven by these trends, mainly (and large-scale anti-smoking campaigns as well), multinational cigarette companies have applied **corporate strategies** to increasingly shift their manufacturing capacities into low- and middle-income countries where most of today's tobacco is grown and where new and large consumer markets are expected to develop (due to large population size, low rates of smoking among women, increasing disposable incomes available to large numbers of people for **consumption**, and poor law enforcement in terms of tobacco control policies). As corporations such as Altria, Japan Tobacco, Philip Morris, British-American Tobacco, and Imperial Tobacco build new factories; enter into joint ventures; and buy formerly state-owned factories; at the same time international leaf dealers have followed. These are companies that source tobacco around the world, ensure quality control, and then sell to manufacturers. These buyers set up a closed chain of leaf procurement and processing facilities near the factories. The world leaf market is currently dominated by three U.S.-based companies: Dimon, Standard Commercial, and Universal, working with the major tobacco manufacturers to determine which countries will produce how much tobacco leaf and what kind. In a constant drive to increase profits and reap benefits from **economic liberalization**, the companies regularly shift production from one country to another (in the **miombo** zone, for example, from Zimbabwe to Uganda, Tanzania, and Zambia, currently), based on the concessions that they can negotiate with the local governments. This happens irrespective of the impact this may have on local growers, local ecosystems, or national economies.

It has been estimated that tobacco use currently kills 4 million people per year, and that by 2030 it will kill 10 million people annually, with 7 million of these deaths in low- and middle-income countries of the developing world. The global shift of tobacco production coincides with a global shift in tobacco use, and two patterns overlap as follows: first, educated and prosperous people increasingly abandon smoking, while the practice becomes more concentrated among the poor in most societies; and, second, smoke products have undergone a change from dark, naturally cured tobaccos (used predominantly by men) to bright, artificially cured, and highly addictive tobaccos (increasingly used by women and teenagers). In particular, American-blend type cigarettes such as Marlboro bear a globally high prestige, and Virginia or flue-cured tobacco is the major component of these highly addictive cigarettes.

In growing areas of the developing world, tobacco is commonly grown more in highlands than lowlands, and more in drylands (under semiarid to semihumid climates) than in humid ecosystems. The share of tobacco in all arable land is highest in producer countries of subtropical East Asia and in **southern Africa**. Less than half of the global land under tobacco is grown with naturally cured varieties, that is, using natural variations in temperature and humidity to dry up the leaves (air- and sun-curing). The majority of global land under tobacco is cropped with artificially cured varieties, that is, those using heat from external sources such as wood and coal (fire- and flue-curing). While green tobacco leaves are directly exposed to heat and smoke in fire-curing, flue-curing requires the leaves to be dried by radiant heat from flues or pipes connected to a furnace (see photo titled "Flue-cured (Virginia) "tobacco" barn). In Tanzania, for example, the use of wood as firewood and polewood (for curing barn construction) is widespread among flue-cured Virginia (about 80 percent) and dark fire-cured tobacco (about 20 percent of total national production in the 1990s). Experiments with coal in the artificial curing of flue tobacco failed, and no cost-effective alternatives to wood have so far been introduced on a large scale (which is also valid for most African producer countries, at least). Less than 10 percent of the Virginia farmers in the country are known to use wood from their own woodfuel plantings, and most of the wood has been taken from open, accessible forests and woodlands, thus contributing to **deforestation** at an estimated 11,000 hectares annually. In Tanzania, curing of tobacco is the second largest consumer of wood after the domestic sector, producing 4.36 billion cubic meters of carbon dioxide and 238 million cubic meters of carbon monoxide per season. Local case studies point to the fact that in areas where tobacco is grown, the tobacco sector is the leading commercial consumer of wood and forest products, accounting for up to 20 percent of all regional consumption. Globally in the 1990s, an estimated twenty (stacked) cubic meters of wood were used to produce one ton of tobacco in those areas where the energy came from wood. This translates into 200,000 hectares of forests and woodlands that were removed by tobacco farming each year during the 1990s, accounting for nearly 5 percent of total net deforestation in the respective producer countries of the developing world. Tobacco-related deforestation happens in about seventy producer countries and, in particular, serious, high, and medium degrees have been estimated for half of these countries.

Tobacco is grown in **monoculture** and contributes to the simplification of rural landscapes and ecosystems through **agricultural intensification**. When tobacco is grown on the same land repeatedly with minimal rotation with other crops, there is a tendency for the soil to become exhausted and for crop pests to become endemic. This is why monocultural tobacco cropping requires ever-increasing inputs of pesticides and chemical fertilizer. Furthermore, the tobacco plant is unique in that it depletes major soil nutrients such as nitrogen, phosphorus, and potassium at five times the rate of **food**

crops, with especially potassium needed and absorbed in uniquely large amounts. This is due to the design of the crop as a smoke product reaching out for high levels of nicotine enrichment. The uptake of nitrogen is vital for growth, especially the green color and nicotine content; phosphorus is essential for nutrition, especially the promotion of ripening; and potassium has a positive influence on leaf combustibility and the (bright) color of flue-cured leaf. Two agricultural practices, "topping" and "desuckering," are carried out to manipulate natural plant growth before ripening and harvest in order to deliver higher (nicotine) yields (by about 10–15 percent). Topping is the manual removal of inflorescences, done with most cigarette and cigar tobaccos. It forces all nutrients not to go into seed but into leaf production instead, making the upper leaves grow longer, wider, thicker, and darker in color. It further stimulates root growth and slows down ripening rate, thus draining more nutrients from the soil. However, as an unintended consequence, topping stimulates the growth of suckers, which—if not removed at least once a week—would cause more harm than the flower would have done if left. Desuckering is the manual removal of suckers, occasionally supported by chemicals, and needs to be carried out several times, because as soon as suckers are removed, yet more suckers are stimulated to grow. It has been estimated that in air-cured burley tobacco in Malawi, about 1 percent of yield is lost for every day suckers are left unremoved on the plant. Thousands of additives and pesticide residues in a cigarette (an estimated 4,000 in total) not only directly impact the health of smokers, but their application also exacts a toll on agro-ecosystems and the livelihood system of tobacco farming societies. Apart from excessive pesticide usage, one environmental health problem peculiar to tobacco work is green tobacco sickness, with nearly half of all tobacco workers experiencing it at least one time during the tobacco harvest. Green tobacco sickness is a form of nicotine poisoning through dermal absorption (resulting from workers brushing against wet tobacco leaves), and the symptoms include nausea and vomiting, headache, dizziness, blood pressure fluctuation, and abdominal cramping. Because of the toxic chemicals in tobacco, any waste that cannot be used to make other, more valuable tobacco products is used to manufacture insecticides.

The potential to improve tobacco growing is poor for three major reasons: most production has shifted to growing countries where there is less ability to influence production practices; multinational leaf buyers usually have little interest in sustainable production; and few alternatives exist in many developing countries to wood-based energy, especially in Africa. The recommended focus of **public policy** is therefore on curbing the global tobacco epidemic by stopping smoking, but not making production sustainable. Indeed, global regulatory actions were triggered in the late 1990s, and as per February 27, 2005, an international tobacco control convention came into force, with the international standard now to be translated into national law by those countries that signed and ratified.

See also: **Agro-Industry; Cash Crops; Economic Restructuring; Human Health.**

Further Reading

Brown, Valerie J., "Tobacco's Profit, Workers' Loss?" *Environmental Health Perspectives* 111 (2003): 284–7; Clay, Jason, *World Agriculture and the Environment: A*

Commodity-by-Commodity Guide to Impacts and Practices, Washington, DC: Island Press, 2004; Framework Convention on Tobacco Control [Online, March 2005], The FCTC Web Site, www.fctc.org; Geist, Helmut, "Global Assessment of Deforestation Related to Tobacco Farming," *Tobacco Control* 8 (1999): 18–28; Geist, Helmut, "Soil Mining and Societal Responses: The Case of Tobacco in Eastern Miombo Highlands," in *Coping with Changing Environments: Social Dimensions of Endangered Ecosystems in the Developing World,* eds. Beate Lohnert and Helmut Geist, Aldershot, UK, Brookfield, VA: Ashgate, 1999, pp. 119–48; Geist, Helmut, "Transforming the Fringe: Tobacco-Related Wood Usage and Its Environmental Implications," in *Environment and Marginality in Geographical Space: Issues of Land Use, Territorial Marginalization and Development in the New Millennium,* eds. Roser Majoral, Heiko Jussila, and Fernanda Delgado-Cravidão, Aldershot, UK, Burlington, VA: Ashgate, 2000, pp. 87–118; Jacobs, Rowena, Gale, H. Frederick, Capehart, Thomas C., Zhang, Ping, and Prabhat Jha, "The Supply-Side Effects of Tobacco-Control Policies," in *Tobacco Control in Developing Countries,* eds. Prabhat Jha and Frank J. Chaloupka, New York: Oxford University Press, 2000, pp. 311–41.

HELMUT GEIST

Tradeoffs. A central feature of rational choice among alternatives when resources are scarce, which almost always is the case. Indeed, "making tradeoffs" is one way of referring to inevitable choices in a world of binding resource constraints. Even private choices (within the realm of a single household, farm, or firm) involve problematic tradeoffs. These are not just hard choices, but choices that can be difficult to analyze formally and to quantify; such as those involving natural resources, risk, and uncertainty. One important example is the impact of current management decisions for a particular plot of land on its future productivity.

There also is the prospect—indeed the likelihood—of significant offsite effects from decisions about land-use and **land-cover change** and, in turn, of landscape, larger **scale**, and cross-scale interactions of these effects in space and time. These offsite effects are called environmental or production externalities by economists and are an important class of what others refer to as **ecosystem services**. Pollination and water pollution are two classic examples, respectively, of positive and negative externalities.

Private and social rates of tradeoff differ whenever there are externalities. In this situation, society could do better than the sum total of independent private decisions. Indeed, private decisions alone may produce unsustainable or even disastrous collective results. Because externalities so often are associated with land use and cover change, the focus of analysis appropriately has been on broader challenges of social choice and public policy concerns, including issues of efficiency of economic incentives, human welfare (including health), and environmental impacts of different land-use alternatives, as well as equity of outcomes across people (winners and losers) at a single point in time and across generations.

The potential number of dimensions that could be considered in this social choice problem is vast. Practical attempts at policy analysis, by necessity, have to make pragmatic choices about what to include in the assessment of tradeoffs. For the important case of land-use tradeoffs in developing countries, Steve Vosti and Thomas Reardon have proposed that analyses be organized around a "critical triangle" of public policy goals: productivity growth,

poverty alleviation, and sustainable resource use. Charles Crissman and colleagues extended this policy analysis approach to include **human health**. A collection edited by David Lee and Christopher Barrett includes a large number of examples of approaches to quantitative policy analysis of tradeoffs among the "critical triangle" of **public policy** goals for the case of **agricultural intensification**.

Quantifying tradeoffs starts with assessment of what really is possible. The dimensions of the "possibilities set" to be considered must span multiple policy goals in terms of the attributes of the various goods and services involved. Of particular interest is the "possibilities frontier," which comprises the envelope of efficient outcomes from the set of technically feasible combinations of factors of production (natural, manufactured, financial, social, and human). The shape of this possibilities frontier is the rate of transformation between efficient outcomes or, in more general terms, the rate of tradeoff. "Efficient outcomes" is used here in Pareto's sense: these are states of the world in which it is not possible to produce additional amounts of one good or service without reducing production of another. However, particularly in the case of many developing countries, outcomes may fall far short of possibilities. For such apparent "win-win" opportunities, though, it is important to realize that the necessary collective action or policy reform to achieve a better result for society also entails real costs. (Moreover, what is "good" or "bad" from a broad social perspective is a non-trivial valuation problem that will be taken up below).

Despite a growing body of empirical work, much remains to be done to establish empirical patterns and regularities in the shape of tradeoffs curves. While some relationships among policy goals may be represented by convex possibilities sets with tradeoff curves that are smooth and continuous, there is persuasive qualitative reasoning and a growing body of quantitative results suggesting that non-convexities (discontinuities and **thresholds**) and hysteresis (path dependence) are the norm, particularly for natural resources and environmental aspects of the social choice problem. Indeed, whether or not these tradeoff relationships are smooth is a central feature distinguishing neoclassical and ecological approaches to environmental economics. Others argue that non-convexities and thresholds also should be taken more seriously in development economics. Moreover, in a dynamic context, the rates and patterns of technological change that influence the shape of the tradeoffs curves are mediated by **institutions** and policies, not just biophysical or engineering relationships. The specific shape of these tradeoffs is a fundamental determining factor regarding policy options and the opportunity costs, efficiency, and equity of possible outcomes for society. Thus, approaches that can help refine our empirical understanding have a great deal of practical importance as well as scientific interest.

Two ongoing initiatives for integrated assessment of land-use alternatives illustrate the range of fruitful (yet contrasting) empirical strategies as well as some important common elements. The tradeoff analysis (TOA) method is an example of a more formal approach that combines spatially referenced biophysical and economic data with crop growth models and econometric-process simulation models and biophysical models. By avoiding the "representative farm" concept, a major advantage of this econometric and spatially explicit approach it that the TOA method can represent the heterogeneity of individual land-use decisions made on farms and aggregate their consequences within a landscape context. These advantages, and the additional rigor of the approach, do come at a cost, however, in terms of required data and specialized statistical and modeling skills. By way of comparison, the matrix approach of the **Alternatives to Slash-and-Burn (ASB) Programme** is much simpler in its data requirements and analytical setup, emphasizing replicability and comparability across sites in the search for empirical patterns. The ASB matrix approach uses chronosequences to identify representative land uses that then become the focus of measurement and analysis at benchmark sites across the humid tropics. The rows of the ASB matrix are specific land uses, ranging from natural **forests** and forest extraction activities to agroforests, simple tree-based systems, rotational bush **fallows**, continuous annual cropping, and **pastures/grasslands**. The columns are indicators of specific goals from global perspectives (**biodiversity** and carbon stocks), national policymakers' perspectives (**economic growth**, employment) and local perspectives (agronomic sustainability, adoptability by smallholders). This approach has made it possible to undertake cross-continental comparisons of patterns in tradeoffs from data collected in the **tropical humid forest** margins in the Amazon, Congo basin, and Southeast Asia. The ASB matrix approach is useful for identifying big (order of magnitude) differences and is a useful framework for testing a wide range of indicators for specific policy goals, but is not as sophisticated spatially or statistically as TOA. An additional practical advantage of the ASB matrix is the relative ease of expanding the scope of the comparative analysis of land-use alternatives. Data for additional rows (land-use alternatives) or columns (indicators) can be incorporated without affecting the other elements of the matrix.

Similarities in the TOA and ASB approaches to measuring tradeoffs are at least as important as their methodological differences. Most importantly, each provides a proven framework for planning, coordinating, and reporting multidisciplinary, integrated assessments of policy-relevant tradeoffs associated with land use and cover change. Each also is valuable as a tool for *ex ante* impact assessment. Both engage policymakers and other key stakeholders and likely users of the analysis at the outset in order to identify policy-relevant criteria and indicators for empirical assessment. Each takes a modular multidisciplinary approach to measurement of specific indicators, so that the necessary depth of expertise can be incorporated into the **integrated assessment**. Both also take a multiscale approach, building up from land-use decisions at the farm level as units of analysis but placing these plot-level activities within a broader environmental context.

Perhaps the most interesting similarity between the ASB and TOA approaches concerns the fundamental question of valuation of alternatives. While each approach uses indicators derived from market prices when these are available, neither employs contingent valuation techniques for non-market services, which predominate for environmental issues. Why not use non-market techniques in an effort to aggregate across alternatives and to identify a "socially optimum" point on the tradeoff curve? First, there is no consensus about the utility or reliability of these non-market valuation techniques for the really big issues (e.g., **species extinction**) resulting from land-cover change, particularly in a developing country context. Second, land-use and land-cover change often involve conflicts of interest among diverse stakeholder groups (ranging, for example, from farmers to upstream and downstream communities, national policymakers, and international **conservation** organizations). These groups and others each may well have their own distinct values for alternative outcomes. In a static sense, then, the social optimum probably does not exist. Both ASB and TOA engage with policymakers and other elements of society as a means of discovering social valuation through participation in complex, dynamic political processes. Linking science directly with policy and political processes to support negotiation among different groups is likely to be especially important for complex, interlinked problems where there are conflicting interest groups and uncertainties about alternatives.

See also: **Arena of Land Conflict; Carbon Cycle; Deforestation; Driving Forces; Forest Degradation; Hot Spot Identification; Land-Use Policies; Mediating Factor; Millennium Ecosystem Assessment; Regrowth; Resilience; Sustainable Land Use; Threshold; Vulnerability; Watershed.**

Further Reading

Antle, John M., and Susan M. Capalbo, "Econometric-Process Models for Integrated Assessment of Agricultural Production Systems," *American Journal of Agricultural Economics* 83 (2001): 389–401; Crissman, Charles C., Antle, John M., and Susan M. Capalbo, eds., *Economic, Environmental, and Health Tradeoffs in Agriculture: Pesticides and the Sustainability of Andean Potato Production,* Boston: Kluwer Academic Publishers, 1998; Dasgupta, Partha, and Karl-Goran Mahler, eds., *The Economics of Non-Convex Ecosystems,* Dordrecht, NL: Kluwer Academic Publishers, 2004; Lee, David R., and Barrett, Chris B., eds., *Tradeoffs or Synergies? Agricultural Intensification, Economic Development and the Environment,* Wallingford, UK: CAB International, 2001; Montana State University [Online, November 2004], The Tradeoff Analysis Project Web Site, www.tradeoffs.montana.edu; Stoorvogel, J., Antle, John M., Crissman, Charles C., and W. Bowen, "The Tradeoff Analysis Model: Integrated Bio-Physical and Economic Modelling of Agricultural Production Systems," *Agricultural Systems* 80 (2004): 43–66; Tomich, Thomas P., van Noordwijk, Meine, Vosti, Steve A., and J. Witcover, "Agricultural Development *with* Rainforest Conservation: Methods for Seeking *Best Bet* Alternatives to Slash-and-Burn, with Applications to Brazil and Indonesia," *Agricultural Economics* 19 (1998): 159–74; Tomich, Thomas P., Cattaneo, A., Chater, S., Geist, Helmut J., Gockowski, J., Kaimowitz, David, Lambin, Éric F., et al., "Balancing Agricultural Development and Environmental Objectives: Assessing Tradeoffs in the Humid Tropics," in *Slash and Burn: The Search for Alternatives,* eds. Cheryl A. Palm, Steve A. Vosti, P. A. Sanchez, and P. J. Ericksen, Columbia University Press, 2005; Van Noordwijk, Meine, Tomich, Thomas P., de Foresta, H., and G. Michon, "To Segregate or to Integrate? The Question of Balance between Production and Biodiversity Conservation in Complex Agroforestry Systems," *Agroforestry Today* 9 (1997): 6–9; Van Noordwijk, Meine, Tomich, Thomas P., and Bruno Verbist, "Negotiation Support Models for Integrated Natural Resource Management in Tropical Forest Margins [Online, November 2004], The Conservation Ecology Web Site, www.consecol.org/vol5/iss2/art21; Vosti, Steve A., and Thomas Reardon, T., eds., *Sustainability, Growth, and Poverty Alleviation: A Policy and Agroecological Perspective,* Baltimore, MD: Johns Hopkins University Press, 1997.

THOMAS P. TOMICH

Tragedy of the Commons. First articulated by biologist Garrett Hardin in 1968 as a situation in which a group of self-interested individuals exploit a commonly owned resource to a level beyond that which is socially optimal.

The logic of the tragedy of the commons potentially applies to a range of open access/**common-pool resources**, including fisheries, pasture, unpolluted waterways, and communal **forests**. Such resources are rivalrous (one individual's use of the resources diminishes the benefits available to another user) but non-excludable (**access** to the resource by specific individuals cannot be limited).

Garrett Hardin uses the example of a commonly owned **pasture** in which individuals decide how many of their own cattle to graze on the pasture. Individuals capture the benefits from cattle that they graze, but the costs of this grazing—degradation of the pasture—are shared among all users. When deciding how many cattle to graze, Garrett Hardin assumes that individuals account only for their private benefits and costs. (The difference between the private and social costs of an individual's action is often referred to as an external cost.) Because external costs are not accounted for by individuals, at the individual level the benefits of adding an additional cow to the pasture may exceed the individual's cost, although the social benefits of the additional cow will be below its social cost. The result is that too many cattle are grazed, resulting in **land degradation** beyond a socially optimal level. Garrett Hardin's theoretical arguments were formalized by economist Scott Gordon to demonstrate socially sub-optimal exploitation of an open access fishery by profit-maximizing harvesters. Both authors' models are based on the assumption that individuals act independently to maximize private benefits. Under this theoretical model, privatization of access rights to the common-pool resource can create incentives for socially optimal levels of resource exploitation.

Political scientist Elinor Ostrom has demonstrated that in many common-pool resource regimes, individuals may coordinate to develop a set of institutional rules that lead to lower levels of resource degradation. Case study and experimental research have identified conditions under which successful rules are more likely to develop. These include the ability to establish and enforce clear rules that restrict conditions of resource use, development of rules that allocate access in proportion to access costs, participation in governance by those affected by the rules, and the ability to monitor resource use and impose graduated sanctions for rule violations. If these conditions hold, common-pool resource regimes may be as successful as private-**property** regimes in maintaining a socially optimal level of the natural resource.

See also: **Community Involvement; Decision-Making; Institutions; Land Privatization; Land Rights; Participatory Land Management; Public Policy.**

Further Reading

Gordon, H. Scott, "The Economic Theory of a Common-Property Resource," *The Journal of Political Economy* 62 (1954): 124–42; Hardin, Garrett, "The Tragedy of the Commons," *Science* 162 (1968): 1243–8; Ostrom, Elinor, *Governing the Commons: The Evolution of Institutions for Collective Action*, New York: Cambridge University Press, 1990.

DAWN CASSANDRA PARKER

Tragedy of Enclosure. Coupled social and environmental consequences of land encroachment and demarcation using boundaries, usually associated with changes in **land rights** and **land-use systems**. In developing nations, the state often acts in conjunction with businesses and multilateral institutions to deny small farmers or grassroots actors access to **common-pool resources** hitherto managed by them through local **institutions**. In effect, common property regimes are taken over by the state for large-scale commercial exploitation, either by its own agencies or by allied business interests using the legal-political powers of the state. Raymond Bryant and Sinéad Bailey point to notable cases or **proximate causes** such as the creation of extensive networks of reserved forests, national parks and "government lands" or "nationalized" territories in many parts of Asia, Africa, and **Latin America**, including areas of large-scale logging, **mineral extraction**, **cattle ranching**, production of **cash crops**, and dam construction. They point out that political ecology has been instrumental in stressing that post-colonial "development" was typically associated with the dissolution of many grassroots or local institutional arrangements that had so far governed the commons. The second point noted by Raymond Bryant and Sinéad Bailey is that the tragedy of enclosure served to further marginalize poor local land managers in the measure that their access to environmental resources essential for their **economic livelihoods** was restricted or denied. Not only was **access** to common resources ended, but these actors were often forced into a situation whereby they had to work ecologically marginal lands elsewhere in order to survive. The end result was that marginalization became a defining trait for most impoverished local groups and individuals as they were displaced from newly created **forest** reserves, national parks, or "development" projects. Within the broad category of impoverished and marginalized people, some actors are usually worse affected than others, notably women and indigenous minorities. The term **economic restructuring**, though unspecific with regard to geographic and historical context, insists that multiple mechanisms or processes of corporate, technological, and institutional change be considered together as a totality (rather than as separate variables) to drive local and regional environmental change associated with the enclosure of the commons. For example, the restructuring of developing economies following the adoption of **economic liberalization** policies includes a "property-regime effect" whereby the institutions of **property** through which land and resources are managed are progressively privatized and the social and environmental "commons" enclosed. The so-called Washington Consensus of liberalization and privatization continues to produce **land change as a forcing function in global environmental change** through the creation of private, exchangeable rights to resources such as land, water, or plant and animal life, which embeds resources in a distinctive management regime, one that prioritizes land-use practices that maximize the production of exchange values rather than use value. Éric Lambin and colleagues found that forced population displacement in conjunction with a decrease in land availability due to encroachment by other land uses (such as natural reserves) features "resource scarcity causing pressure of production on resources" as a fundamental high-level cause or **syndrome** of land change. It is associated here with rapid **land-cover change**, while the overall syndrome also includes slow rates of land change associated with heavy surplus extraction away from the land manager, for example. Summarizing a large number of case studies, the resource scarcity and production pressure syndrome was found to be just one out of five syndromes, and it appears as if only a combination of several causes, with synergetic interactions, is likely to drive a region into a critical trajectory of population displacement and **land degradation**.

See also: **Agro-Industry; Conservation; Corporate Strategies; Driving Forces; Land Concentration; Land Rent; People at Risk; Poverty; Private Property; Public Policy; Yellowstone Model.**

Further Reading

Bryant, Raymond L., and Sinéad Bailey, *Third World Political Ecology*, London, New York: Routledge, 1997; Geist, Helmut, "Exploring the Entry Points for Political Ecology in the International Research Agenda on Global Environmental Change," *Zeitschrift für Wirtschaftsgeographie* 43 (1999): 158–68; Glennie, Paul, "Enclosure," in *The Dictionary of Human Geography*, 4th ed. (reprinted), eds. Ron J. Johnston, Derek Gregory, Geraldine Pratt, and Michael Watts, Oxford, UK, Malden, MA: Blackwell Publishers Ltd., 2001, p. 206; Kates, Robert W., and Viola Haarman, *Poor People and Threatened Environments: Global Overviews, Country Comparisons, and Local Studies* (Research Report 91–20), Providence, RI: Brown University, The Alan Shawn Feinstein World Hunger Program, 1991.

HELMUT GEIST

Transhumance. Transhumance involves the seasonal movement of livestock herds in order to make the best use of resources and to minimize risk. All pastoralist systems juggle the basic production constraints of forage, water, minerals, and disease in a wider context of alternative possible economic activities, changing markets, and shifting security. Many operate in arid or semiarid environments, characterized by variability and unpredictability of both rainfall and primary production, and most pastoralist strategies involve movement and patch use. Transhumance is found in most pastoralist systems, such as sub-Saharan Africa; the Mediterranean including southern **Europe**, the Near East, and the Maghreb; the European, for example, Swiss Alps, Pyrenees; Middle East (Kirghizstan, Turkmenistan, and Afghanistan); Central Asia (Mongolia and Siberia); and New World systems, e.g., Andean and Australian.

Where key resources are predictable, transhumance involves regular movements between wet/dry seasons or winter/summer **pastures**. Arid and variable systems involve more frequent, less predictable movements in response to patchy rainfall and forage. Movements may be influenced by raiding and violent conflict, making certain areas too hazardous to use. Depending on the livestock species involved and on local conditions, transhumance may cover anything from a few kilometers (e.g., European Alps) to hundreds (e.g., Fulani in the inland delta of the Niger) or even thousands of kilometers (Saharan Tuareg camel pastoralists). Movements may be altitudinal (e.g., Swiss alpine transhumance), exploit relatively subtle differences between floodplain and higher land (e.g., the Sudd in southern Sudan), or use a latitudinal system (Fulani groups moving among Sahelian, Sudanian, and Guinean zones).

Transhumant movements may be undertaken primarily to manage herd conditions through making best use of forage, water, and mineral availability; and transhumant animals often have significantly better health and

Winter feeding on cattle ranch in Madison County, Montana, March 1942.
PHOTO: John Vachon. [Courtesy Library of Congress]

production. Different livestock species and/or herds may follow different transhumant patterns (as among the Turkana) or exploit the same area in sequence (as for the inland delta of the Niger) as quality and quantity of forage varies with growth stage and with availability of crop residues.

Many pastoralist transhumance systems have developed to minimize exposure to disease risk while optimizing forage use. Fulani cattle transhumance is often timed to leave higher-risk parts of the range as the rains come and vector populations increase, moving into drier rangelands with little vector-borne disease, to exploit the temporary wet season flush of annual grasses. As the rains pass and the seasonal pastures are used up, herds return to wetter areas with more woody **vegetation**, perennial pastures, and crop residues. These areas have potentially high vector populations and disease transmission, but the risk decreases as vector populations decline in the dry season.

Animals in good condition can withstand a degree of disease challenge, and low levels of exposure stimulate a degree of immunity.

Other considerations enter into transhumance decisions. Fulani families in northern Burkina Faso minimize crop damage by keeping the herds away from growing crops after the harvest, when they may move animals to pasture on crop residues as part of a manuring contract with settled farmers. Other Fulani groups move near markets to sell pastoral produce. Transhumance may facilitate participation in social gatherings.

Among systems and households concentrating on pastoralism, the whole family may move with the animals. Families may join together to travel and camp (Turkana); seasonally, large groups of families may congregate (e.g., Tuareg at the *cure salée*). In more agropastoralist groups, one or more young adult males may take the family herd(s) on transhumance, depending on household demography and the broader network of livelihoods strategies, or cattle may be entrusted to a transhumant herder while the household stays put.

Transhumant systems depend on rights of tenure, **access**, and through-passage between point-centered, key limiting common pool resources (water, pasture, minerals) around which the radius of control fluctuates according to conditions. Access has customarily been negotiated by others in need, through token gifts, labor provision, trade, relations of intermarriage, fostering and adoption, and substantive payments in cash or in kind. Negotiated access often operates customarily in a graded way through orders of precedence. Access by users with successively lower priority claims becomes decreasingly likely as conditions deteriorate. The social **institutions** necessary to manage these transhumant systems are inherently flexible, with fuzzy social and spatial boundaries, unlike "modern" westernized **land tenure** systems based on imported European legal codes and imposing total exclusion. Currently, in addition to the restrictions imposed by land being set aside as protected areas, **land privatization** is increasingly threatening transhumant mobility, disrupting through-routes and barring pastoralists from key resources given over to cultivation.

See also: **Andes; Common-Pool Resources; Degradation Narrative; Disequilibrium Dynamics; Grassland; Land Rights; Pastoral Mobility; Property; Sudan-Sahel.**

Further Reading

Adamu, X., and Kirk-Greene, B., eds., *Pastoralists of the West African Savanna*, Manchester, UK: Manchester University Press, International African Institute, 1986; McCabe, J. T., *Cattle Bring Us to Our Enemies: Turkana Ecology, History, and Raiding in a Disequilibrium System*, Ann Arbor, MI: University of Michigan Press, 2004; Niamir-Fuller, M., ed., *Managing Mobility in African Rangelands: The Legitimization of Transhumance*, London: IT Publications, Rome: Food and Agricultural Organization of the United Nations, 1999.

KATHERINE HOMEWOOD

Transition Matrix. A table of numbers arranged in rows and columns that shows the amount of land that transitions from one category to another category between two points in time. Typically, the rows of the matrix show the categories at the initial time and the columns show the categories at the subsequent time. The entry in row *i* and column *j* of the matrix gives the area of land that transitions from category *i* at the initial time to category *j* at the subsequent time. The entries on the diagonal give the area of the map that persists between the two points in time. It is standard practice to append a "totals column" on the right that gives the total areas for each category at the initial time, and a "totals row" on the bottom that gives the total areas for each category at the subsequent time.

When the categories at the initial time are the same as the categories at the subsequent time, it is helpful to create an additional "losses column" on the right that gives the total area lost for each category, and a "gains row" on the bottom that gives the total areas gained for each category. It can be advantageous to express the areas in the matrix as a proportion of the study area by dividing all the entries in the matrix by the total area in the study. The creation of the transition matrix is one of the most common first steps in analyzing change among land categories between two points in time.

One can analyze the numbers in the transition matrix at various levels of detail, depending on the particular question. At the simplest level of analysis, one compares the numbers in the totals column to the corresponding numbers in the totals row in order to compute the net change for each category, which is the change in quantity of area. At a slightly more sophisticated level of analysis, one computes the change in location for each category (i.e., swap), which is formed by pairing as many gaining areas as possible with losing areas. In this manner, a scientist can budget the total change on a map in terms of two pairs of components: gains and losses, and net and swap. In order to examine transitions among individual categories, one must examine the entries within the matrix. The off-diagonal entries can be analyzed at a variety of levels of detail. At the simplest level, scientists consider the size of the entries, in which case the larger numbers are usually more important. In order to detect a systematic transition, each off-diagonal entry should be analyzed with respect to the relative size of all the categories.

The transition matrix is the foundation of Markov chain analysis, which is a common statistical tool. Scientists can use Markov chain analysis to extrapolate land change into the future, where the category of a **pixel** has a certain probability of transitioning to a different category at each time step based on the information in a probability matrix. The probability matrix derives directly form the transition matrix. If the probabilities from one time interval are the same as the probabilities of other intervals, then the probabilities are said to be stationary, in which case it is reasonable to use them to extrapolate over various ranges of time.

There are now methods to compute the transition matrix for cases where each unit of observation (e.g., pixel) is soft-classified, which means that it has simultaneous partial membership in multiple categories. Therefore, the entire transition matrix can be computed and analyzed at multiple **scales**.

See also: **Change Detection; Land-Use Transitions; Pattern to Process; Probit/Logit Model.**

Further Reading

Baltzer, Heiko, "Markov Chain Models for Vegetation Dynamics," *Ecological Modeling* 126 (2000): 139–54; Pontius, Robert Gilmore, Jr., Shusas, Emily, and Menzie

McEachern, "Detecting Important Categorical Land Changes while Accounting for Persistence," *Agriculture, Ecosystems and Environment* 101 (2004): 251–68; Pontius, Robert Gilmore, Jr., "Statistical Methods to Partition Effects of Quantity and Location during Comparison of Categorical Maps at Multiple Resolutions," *Photogrammetric Engineering and Remote Sensing* 68 (2002): 1041–9.

PONTIUS, ROBERT GILMORE, JR.

Tropical Dry Forest. The preferred zone for **agriculture** and human settlement worldwide, because environmental and biotic stresses are the lowest compared to other tropical life zones, and other life zones are too cold, hot, dry, or wet. When contrasted against wet or **tropical humid forests**, tropical dry forests (TDFs) cannot be defined as the last frontier but the first frontier of economic development. In fact, early farming communities and the earliest civilizations in Mesoamerica, for example, developed in the tropical dry forest regions, and there is a strong possibility that today's major tropical crops and food animals originated from tropical dry forest ecosystems. Worldwide, 42 percent of tropical forests are tropical dry forests.

Various political and economic factors have increased anthropogenic stresses in these ecosystems, leading to severe disturbances and widespread clearing. Currently, a significant gap exists in terms of TDF research when compared with tropical moist or rainforests where, for many complex political and institutional reasons, international funding has been more prominent. Since 1945, a total of 2,300 research articles have been produced from tropical environments, of which (in 2004) only 10 percent related in some form to tropical dry forests, according to the Science Citation Index. Efforts aimed at generating information regarding TDF are scattered and limited to a few sites in the Americas and elsewhere such as in the tropical dry forests of India. No TDF multidisciplinary and comparative studies exist for the Americas and other regions in Africa and Asia.

TDFs have less species diversity than the wet forests, though the former one has more endemic species that the latter one. It is estimated that the Chamela's tropical dry forest region on the Pacific coast of Mexico is probably one of the sites with the highest number of endemic tree species in Mesoamerica. Tropical dry forests serve as home to rare and charismatic creatures as jaguars, howler monkeys, scarlet macaws, toucans, and tapirs. It is estimated that when the Spaniards arrived in Mesoamerica (the area covered by southern Mexico and Central America), TDF covered as much extension as the tropical wet forest, yet today it is estimated that only about 2 percent of all the tropical dry forest in the region may be considered intact. For the Guanacaste Conservation Area in Costa Rica alone, new biodiversity inventories have found in the order of 120,000 species of insects, 60,000 species of fungi, and 9,000 species of plants. Therefore, when **conversion** to agriculture/**pasture** land takes place in tropical dry forest regions, the risk of losing species is high, even in small/localized areas.

Our understanding of the human and biophysical dimensions of TDF change and their cumulative effects is still in its early stages of academic discovery, since more emphasis has been placed on tropical evergreen forests, especially the Amazon basin. Therefore, there is a need for continuous and systematic efforts to understand and integrate TDF phenological patterns and social/anthropogenic mechanisms at three basic levels: in the context of conservation biology, the context of land-use and **land-cover change** that are taking place on this rich **agricultural frontier**, and in the context of local and national development policies that contribute to degradation of tropical dry forests.

Accurate forest-cover assessments are important to define natural resource management strategies and policies for **conservation**. It is through information about the location, the extent, and the status of forest areas that threats to **biodiversity** from **deforestation** fronts ("hot spots") can be identified and solutions can be evaluated and implemented. The understanding of the causes and consequences of land-cover change in dry forest environments and its impacts on the environment through time is key to predicting possible negative effects on biological resources. In addition, their cascading effects across the many components of a functional ecosystem have negative impacts on the services they provide to mankind. In fact, our current understanding of the cumulative effects of human-dominated landscapes in tropical dry forest is strongly linked to a clear understanding of the **proximate causes** and underlying socio-economic, biophysical, and **driving forces** of land-use and land-cover change. These forces acting at different **scales** (from the international sphere to the unit of the production level) contribute to regional environmental deterioration trends that in many cases cannot be fully understood because of the lack of sound spatial, socio-economic, and ecological databases that have been built in the context of an interdisciplinary approach. It is through the combination of these data sets that the cumulative effects of land-use/cover change in a given environment can be soundly evaluated and remediation/control policies can be implemented.

See also: **Hot Spots of Land-Cover Change; Forest; Species Extinction.**

Further Reading

Kalacska, Margaret, Sánchez-Azofeifa, G. Arturo, Calvo, Julio, Rivard, Benoît, Quesada, Mauricio, and Daniel Janzen, "Species Composition, Similarity and Diversity in Three Successional Stages of a Tropical Dry Forest," *Forest Ecology and Management* 200 (2004): 227–47; Sagar, R., Raghubanshi, A. S., and J. S. Singh, "Tree Species Composition, Dispersion and Diversity along a Disturbance Gradient in a Dry Tropical Forest Region of India," *Forest Ecology and Management* 183 (2003): 61–71; Janzen, Daniel H., "Tropical Dry Forests: The Most Endangered Major Tropical Ecosystem," in *Biodiversity*, ed. Edward O. Wilson, Washington, DC: National Academy Press, 1988, pp. 130–7.

ARTURO SÁNCHEZ-AZOFEIFA

Tropical Ecosystem Environment Observations by Satellite (TREES) Project. A research program managed by the **Joint Research Centre (JRC) of the European Commission**, in close cooperation with the Commission's Environment Directorate-General, and local partners in developing countries, and dedicated to the development of **forest**-cover assessment throughout the humid tropics. Initiated in the early 1990s, the project used the global imaging capabilities of a number of earth-observing satellites, including Europe's SPOT 4 and the **European remote sensing (ERS-1/-2) satellites** to provide the latest information on the state of the world's **tropical humid forests**. Indeed, in spite of the importance of the tropical forests, our knowledge concerning their rates of change remains limited. The project made use of an extensive set of **remote sensing** satellite imagery. The initial objectives of the TREES project were to develop techniques for global tropical forest mapping, to develop techniques for **monitoring** active deforestation areas, and to set up a comprehensive tropical forest information system. The ultimate goal is to establish an operational observing system that can detect and identify changes in the tropical forest cover of the world.

The first phase of this research program (TREES-I) has allowed the creation of the most complete, up-to-date set of maps available that document the distribution of the world's remaining humid tropical forests. These maps provide an unprecedented view of one of the most important **biomes** on the planet.

The primary objectives of the second phase of this research program (TREES-II) were to produce relevant information, more accurate than currently available, on the state of the humid tropical forest ecosystems from new remote sensing-based approaches and to analyze this information in terms of **deforestation** and **forest degradation** trends. TREES-II was completed in 2002.

TREES-II's results clearly show that deforestation in the humid tropics is still a major global environmental issue. The project provides the most accurate, consistent figures on rates of deforestation throughout the humid tropics currently available. Between 1990 and 1997 a staggering 5.8 million hectares of humid tropical forest were lost each year. This is an area twice the size of Maryland or Belgium. A further 2.3 million hectares per year of forest are detected as highly degraded—becoming increasing fragmented, heavily logged, and/or burnt. Although the statistics document the trends up to 1997, the most recent maps from the project (around year 2000) provide no grounds to believe that this situation was improving in the recent years.

The maps, information on forest-cover status, and rates of change are based on uniform, independent, and repeatable methods. These new data have already reduced uncertainties in dealing with carbon sink issues, and they provide accurate baseline views of this hugely valuable global resource and help in planning strategies for effective **conservation** of its biological diversity.

TREES, endorsed by the international **Land-Use/Cover Change (LUCC) project**, clearly demonstrates the important role of sound scientific evidence to support policy. The close collaboration with local partners in developing countries and international governmental or non-governmental organizations combined with state-of-the-art analysis of satellite imagery has proved a powerful combination.

See also: **Biodiversity; Carbon Cycle; Forests; SPOT Vegetation.**

Further Reading

Achard, Frédéric, Eva, Hugh, Stibig, Hans-Jürgen, Mayaux, Philippe, Galego, Francisco, Richards, Timothy, and Jean-Paul Malingreau, "Determination of Deforestation Rates of the World's Humid Tropical Forests," *Science* 297 (2002): 999–1002;

Joint Research Centre of the European Commission, Institute for Environment and Sustainability, Global Vegetation Monitoring Unit [Online, November 2004], The Tropical Ecosystem Environment Observations by Satellite Web Site, www-gvm.jrc.it/tem/ProductsArchive.htm.

FRÉDÉRIC ACHARD

Tropical Humid Forest. Major forest type of the world that has year-round abundant precipitation (typically more than 1,600 millimeters annually), occurs either near the equator or on low-latitude windward coasts, and which is currently undergoing rapid changes as humans increasingly extract timber resources, expand and intensify agricultural areas, and migrate into previously unsettled areas. The evergreen and seasonal forests of the tropical humid bioclimatic zone, also called rainforests, differ from **tropical dry forests** (and woodlands) in that the latter occur in drier areas farther from the equator, holding broad-leaved deciduous trees. The domain of the dry seasonal forests of continental Southeast Asia is intermixed with the humid forests domain, while the separation between these two domains is clearer in Africa and Latin America. Yet in spite of the importance of tropical rainforests, our knowledge concerning their distribution and rates of change remains surprisingly limited.

Monitoring systems based on **remote sensing** provide information on **land-cover change** (including cover type and biophysical parameters) and disturbance due to **fire**. By using the repeat coverage of the earth observation (EO) satellites, it has been possible to monitor forest-cover change in an operational mode since the early 1980s. In the following, summary findings are presented originating from the Forest Resources Assessment program of the **United Nations Food and Agriculture Organization** (FAO), initiated in the early 1980s, and of the **Tropical Ecosystem Environment Observations by Satellite (TREES) Project**, estimating tropical forest remaining cover and its rate of change for the 1990s.

FAO Forest Resources Assessment Programme

The FAO Forest Resources Assessment (FRA) program provides, in principle, two estimates for the full tropics separately through two methodological approaches: the country survey (CS), which is based on the compilation and standardization of national data and the estimates of which are generally considered as reference figures for the tropics; and the remote sensing survey (RSS), which provides statistical estimates at a continental level, derived from forest-cover change maps (interpretation of thirty-meter resolution satellite imagery). Internal inconsistencies have been highlighted for the FRA CS estimates, which might be due to the difficulties of standardizing country level data obtained from official intergovernmental processes. On the other hand, it is well accepted that a sample can provide reliable estimates of **deforestation** if the sample size is sufficiently large. The FRA RSS uses a random sample covering 10 percent of the whole tropical belt, designed for forest cover estimation.

For both surveys (CS and RSS), a 10 percent canopy cover threshold is used as the minimum definition of forest in tropical countries. The FRA RSS findings indicate that the world's tropical forests covered 1.57 billion hectares in 2000, about 12 percent of the world's land area. The net change in tropical forest area was –8.6 million hectares per year, representing the difference between a deforestation rate of 9.2 million hectares per year of natural forests and an expansion of 0.6 million hectares per year of natural forests and forest plantations. The reported FAO CS rates of change in forest area appear larger than the estimates from satellite-based studies. Most of the world's forest losses were in the tropics. The rate of net change was slightly lower in the 1990s compared to the 1980s, due to a higher estimated rate of forest expansion in the 1990s. Country studies point at **land rights** as a common main determinant behind deforestation, and the RSS indicates that direct **conversions** of **forests** to permanent **agriculture** were more prominent than shifting agriculture in forest change processes.

Estimates of tropical forest area and forest-cover change during the 1990's from FAO and TREES surveys

	Forest area 2000 (10^6 ha)		Annual forest area change1990–2000	
Continent	TREES Domain	Non-TREES Domain	TREES Domain	Non-TREES Domain
Southeast Asia	260		−2.3	
Africa	191	292	−0.7	−1.5
Latin America	646	121	−2.3	−1.9
Tropics		**1,510**		**−8.7**

Note: "TREES domain" refers to: (a) the humid tropical forest biome of Latin America excluding both Mexico and the Atlantic forests of Brazil, (b) the humid tropical forest biome of Africa, and (c) the tropical forest biome of Southeast Asia and India, excluding only the dry biome of India. "Non-TREES domain" is defined as all remaining tropical forests, i.e., dry forests of Latin America, Africa and India, and all Mexico forests and Atlantic forests of Brazil.

TREES' Monitoring of Humid Tropical Forests

The TREES project examines **hot spots of land-cover change**, namely deforestation with fine spatial resolution (thirty meters) satellite imagery. TREES uses a stratified systematic sample covering 6.5 percent of the humid tropical domain, designed for forest cover change estimation. The project provides the most accurate, consistent figures on rates of deforestation throughout the humid tropics currently available. The TREES and FRA RSS estimates can be considered complementary, as they are both based on a statistical sampling strategy using satellite imagery and have been designed to give continental estimates. Both methods provide measurement of tropical forest-cover change in a uniform, independent, and repeatable manner, and moreover are consistent with each other (see table titled "Estimates of tropical forest").

The results of the TREES project clearly show that deforestation in the humid tropics is still a major global environmental issue. Analysis is now complete and shows that in 1990 there were some 1.15 billion hectares of humid tropical forest. Between 1990 and 1997 a staggering 5.8 million hectares of humid tropical forest were lost each year. This is an area twice the size of Belgium or Maryland. A further 2.3 million hectares per year of forest are detected as highly degraded—becoming increasing fragmented, heavily logged, and/or burnt. Although the statistics document the trends up to 1997, the most recent maps from the project (from 1999 and 2000) provide no grounds to believe that this situation is improving.

Although a global phenomenon, the spatial detail and ability to compare different regions of the world provided by TREES reveals considerable variation around the world. An annual global deforestation rate of 0.5 percent is alarming enough, but the project shows that Southeast Asia is losing its tropical forests twice as fast as anywhere else. While the regional forest change rate for **Latin America** is relatively low, the overall gross deforestation is a dramatic 2.5 million hectares per year, the same as that of Southeast Asia. Furthermore, this change is confined to several "hot spots," where rates can reach at least 3 percent loss per year.

See also: **Agricultural Frontier; Alternatives to Slash-and-Burn (ASB) Programme; Biome; Forest Degradation.**

Further Reading

Achard, Frédéric, Eva, Hugh, Stibig, Hans-Jürgen, Mayaux, Philippe, Galego, Francisco, Richards, Timothy, and Jean-Paul Malingreau, "Determination of Deforestation Rates of the World's Humid Tropical Forests," *Science* 297 (2002): 999–1002; DeFries, Ruth, Houghton, Richard A., Hansen, Matthews, Field, Chris, Skole, David, and John Townshend, "Carbon Emissions from Tropical Deforestation and Regrowth based on Satellite Observations from 1980s and 90s," *Proceedings of the National Academy of Science* 99 (2002): 14,256–61; Food and Agriculture Organization of the United Nations, *The Global Forest Resources Assessment 2000* (FRA 2000): Main Report, Rome: FAO, 2001; Harcourt, Caroline S., and Jeffrey A. Sayer, *The Conservation Atlas of Tropical Forests: The Americas*, London: Macmillan, 1996.

FRÉDÉRIC ACHARD, HUGH EVA, PHILIPPE MAYAUX, HANS-JÜRGEN STIBIG

Turnover. Amount of time it takes to replace the **biomass** of a population or ecosystem, measured as the ratio of stock (mass, e.g., standing crop of biomass) and flow (mass per time unit, e.g., net primary production) and expressed in time units (e.g., per year) or a synonym: for example, transit time or residence time. Habitat ecologists sometimes use the term in the sense of turnover of species, indicating the time required for a complete change in species composition by means of **ecological colonization** and **species extinction**. Ecosystems are characterized by typical rates of biomass turnover. Turnover time is probably related to the resistance and resilience of ecosystems. Systems with fast turnover may be displaced from their state easily, but recover quickly, that is, their resistance is low but their **resilience** high. For example, an annual **grassland** with a turnover of one year recovers quickly after a **fire**. By contrast, a system with a slow turnover is not easily changed completely, but if it is, it recovers very slowly if at all, that is, its resistance is high and its resilience low (e.g., a **forest**). Land use fundamentally alters ecosystem patterns and processes. As a consequence, the turnover of biomass is usually accelerated considerably. Societies show preferences for annual plants such as cereals, and very often replace systems with larger biomass stocks such as forests and savannas. Moreover, as **colonization** aims at increasing useful biomass output (e.g., net primary productivity of harvestable goods), stocks are reduced due to juvenilization effects (e.g., reduction of stand age in forests). The consequences of this acceleration of turnover for ecosystem functioning and patterns have received little, if any, scrutiny and are therefore at present largely unknown.

See also: **Agriculture; Agrodiversity; Biodiversity; Disequilibrium Dynamics; Domestication; Forestry; Human Appropriation of Net Primary Production; Proximate Causes.**

Further Reading

Haberl, Helmut, Erb, Karlheinz, Krausmann, Fridolin, Loibl, Wolfgang, Schulz, Niels B., and Helga Weisz, "Changes in Ecosystem Processes Induced by Land Use: Human Appropriation of Net Primary Production and Its Influence on Standing Crop in Austria," *Global Biogeochemical Cycles* 15 (2001): 929–42; Ricklefs, Robert E., and Gray L Miller, *Ecology*, 4th ed., New York: Freeman and Company, 1999.

KARLHEINZ ERB

U

United Nations Convention on Biological Diversity. Presented at the United Nations Conference on Environment and Development (UNCED) in Rio de Janeiro in 1992 (Earth Summit), the convention defines biological diversity as the variability among living organisms from all sources, including terrestrial, marine, and other aquatic ecosystems and the ecological complexes of which they are a part. Thus, it encompasses diversity within species, between species, and of ecosystems. The Convention on Biological Diversity (CBD) came into effect in 1993, and a summary of the key elements that highlight and characterize the **biodiversity** problem is provided in the preamble:

1. The importance of biological diversity for maintaining life-sustaining systems of the biosphere
2. The concern that biodiversity is being significantly affected by certain human activities
3. The awareness that **conservation** and sustainable use of biodiversity are critical in meeting the food, health, and other needs of the growing world population
4. The awareness that **access** to and sharing of both genetic resources and technologies are essential.

However, the CBD failed to address the "theft" of biological resources that developing countries claim has occurred since colonialism. In contrast, developed countries have argued that biodiversity is part of the "global commons" for the benefit of humankind, regardless of where it is located.

See also: **Agenda 21; Agrodiversity; Biodiversity Novelties; DIVERSITAS; Domestication; Exotic Species; Millennium Ecosystem Assessment; Proximate Causes; Species Extinction; Tragedy of the Commons.**

Further Reading

McConnell, F., "The Convention on Biodiversity," in *The Way Forward: Beyond Agenda 21*, ed. Felix Dodds, London: Earthscan, 1997, pp. 47–54; United Nations Environment Programme, Secretariat of the Convention on Biological Diversity [Online, September 2004], The Convention on Biological Diversity Web Site, www.biodiv.org.

HELMUT GEIST

United Nations Convention to Combat Desertification. An international agreement negotiated by the **United Nations Environment Programme** that addresses **desertification**. Through today, the United Nations

Convention to Combat Desertification (UNCCD) provides the most authoritative definition of the issue at stake, stating that desertification means **land degradation** in arid, semiarid, and dry sub-humid areas resulting from various factors, including climatic variations and human activities. In 1977, the United Nations Conference on Desertification (UNCOD) adopted the Plan of Action to Combat Desertification (PACD), and the question of how to tackle desertification was still a major concern for the United Nations Conference on Environment and Development (UNCED) in Rio de Janeiro in 1992. UNCED supported a new, integrated approach to the problem, emphasizing actions to promote sustainable development at the community level. It also called on the United Nations (UN) General Assembly to establish a procedure for the preparation of an agreement to combat desertification, particularly in Africa. In December 1992, the UN General Assembly agreed and adopted resolution 47/188. The convention was adopted in June 1994, and came into force on December 26, 1996, that is, ninety days after the fiftieth ratification was received. The Conference of the Parties (COP) is the convention's supreme governing body, and in 2003 over 180 countries were parties. Not denying the fact that many people are failing to thrive and, indeed, survive in drylands, the realization of UNCCD is a major causal element of why desertification continues to be perceived as a global environmental change issue despite weak empirical data on the global **scale**. Following UNCOD from the 1970s onward, many governments of countries with large dryland zones, but also international development and relief agencies, have been using the coupled notion of desiccation and **poverty** to rescue an ideology of authoritarian intervention in rural **land-use systems** and to claim rights to stewardship over resources previously outside their control. Despite the fact, for example, that the landscape in Ivory Coast is becoming more wooded and not desertified, the desertification and **degradation narrative** is present in the National Environmental Action Plan. Still, there is an explicit focus of intervention measures on Africa, and the full text of UNCCD consequently reads as the "United Nations Convention to Combat Desertification in Those Countries Experiencing Serious Drought and/or Desertification, Particularly in Africa." This is directly tied to the political and managerial outcomes of the Rio conference rather than based on empirical evidence, which shows current and rapid **hot spots of land-cover change** in the drylands of Central Asia rather than in Africa. Therefore, political ecologists like Neil Adger and colleagues recognize a "desertification crisis rhetoric," since the theme was added to the global agenda at UNCED due to persistence from African countries and because the **United States of America** unexpectedly supported the African position. It is believed that the support came in response to the environmental criticism the United States had received for failing to support the **United Nations Convention on Biological Diversity** and for dragging its feet in the preparations for the UN Framework Convention on Climate Change. Others interpret the concession of U.S. support for UNCCD as part of a **tradeoff** between developing and developed countries insofar as the European Union together with the United States expected African states to be supportive regarding the remaining issues in the Rio declaration.

See also: **Community Involvement; Institutions; Land Degradation Assessment in Drylands (LADA) Project; Millennium Ecosystem Assessment; Public Policy; Tragedy of Enclosure.**

Further Reading

Adger, Neil W., Benjaminsen, Tor A., Brown, Katrine, and Hanne Svarstad, "Advancing a Political Ecology of Global Environmental Discourses," *Development and Change 32* (2001): 681–715; Geist, Helmut, *The Causes and Progression of Desertification* (Ashgate Studies in Environmental Policy and Practice), Aldershot, UK, Burlington, VT: Ashgate, 2005; United Nations Convention to Combat Desertification [Online, February 2005], The UNCCD Web Site, www.unccd.int.

HELMUT GEIST

United Nations Environment Programme. A specialized agency of the United Nations that was created in 1972 in response to a recommendation from the United Nations (UN) Conference on the Human Environment in Stockholm, Sweden. Located in Nairobi, Kenya, the United Nations Environment Programme (UNEP) coordinates the environmental activities of UN agencies, works to increase national capacity for **conservation** or environmental protection in member states, and oversees environmental **monitoring** and assessment activities under the Earthwatch program (such as the Global Environmental Monitoring System, GEMS) and the gathering of scientific information (such as through the **Millennium Ecosystem Assessment**). In 2004, UNEP and the **United Nations Food and Agriculture Organization** (FAO), based on the recommendations of **Agenda 21**, responded to the needs for standardized land-cover data by initiating the **Global Land Cover Network**. For example, the global data set of the **Global Land Cover Map of the Year 2000** has been published by UNEP in association with the FAO and the **Joint Research Centre (JRC) of the European Commission**. The database has been adopted by Millennium Ecosystem Assessment as the baseline for 2000 and is being used by UNEP's World Conservation Monitoring Centre in its report on the state of the world's protected areas. In general terms, UNEP works to increase the ability of the UN system to address environmental problems such as **desertification**, and, more specifically, assists in negotiations of international agreements and provides support to them, such as the **United Nations Convention to Combat Desertification** and the **United Nations Convention on Biological Diversity**. Nearly half of the multilateral environmental agreements concluded since the organization's creation have been under UNEP auspices. UNEP provides secretariat services to a number of these agreements and acts as an implementing agency for projects carried out under the Global Environmental Facility and the Montreal Protocol Multilateral Fund. In addition, UNEP has worked to set guidelines for various environmental concerns such as the management of **common-pool resources** and environmental impact assessments.

See also: **AFRICOVER; Global Terrestrial Observing System; GLOBCOVER; Investments; Land Cover Classification System; Land Degradation; Scenario.**

Further Reading

United Nations Environment Programme, Environment for Development [Online, January 2005], The UNEP Web Site, www.unep.org.

HELMUT GEIST

United Nations Food and Agriculture Organization. One of the largest specialized agencies of the United Nations (UN), both in membership by states and its staff, which has multiple roles in providing services to governments through the compilation of data and a variety of information and support services in the fields of **agriculture**, agricultural commodities, fisheries, **forestry**, and **food security**. Located in Rome, Italy, the UN Food and Agriculture Organization (FAO) had a predecessor entity in the International Institute of Agriculture in Rome and was created at the same time as the United Nations (in 1945). The FAO, drawing upon its networks and technical staff, is also a source of advice to governments on agricultural policy and planning as well as administrative and legal planning. Through its field program, it gives assistance to a large number of developing countries. At the time of its creation, the primary focus of the FAO was on the improvement of the production and **consumption** of food and other agricultural products. This was fully in line with the concepts and strategies of the **Green Revolution**. In the course of the 1980s and 1990s, the FAO was criticized for being top-heavy, centralist, and out of touch with the mainstream of creative thinking and analysis. Thus, the FAO was not a crucial factor in the considerations of the follow-up to the major UN conferences of the 1990s, which are the basis of today's international environment-development agenda. Only since about the World Food Summit in 1996, the FAO now has created the necessary preconditions for reinventing itself as an effective player in the multilateral system. For example, **AFRICOVER** was developed as a project in response to a number of national requests for assistance to develop reliable and georeferenced information on natural resources. Its interpretation and classification are based on the FAO **Land Cover Classification System**, which was conceived by an international working group and was developed by several FAO services, with the support of a large number of national and international agencies. Likewise, the global data set of the **Global Land Cover Map of the Year 2000** has been published by the FAO in association with the **Joint Research Centre (JRC) of the European Commission** and the **United Nations Environment Programme**. The database has been adopted by **Millennium Ecosystem Assessment** as the baseline for 2000.

See also: **Global Land Cover Network; Global Terrestrial Observing System; Land-Use Planning; Land-Use Policies; Proximate Causes.**

Further Reading

Food and Agriculture Organization of the United Nations, Helping to Build a World without Hunger [Online, January 2005], The FAO Web Site, www.fao.org.

HELMUT GEIST

United States of America. The land-cover history of the United States is largely a result of the nation's mid-latitude location; physical resources; the federal government's role in land dispersal, reservation, and management; the dominant technologies at the time of settlement, and society's demand for land-based resources. Initial **land-cover changes** subsequently were modified by increased population, **globalization** of the economy, maturation of

Wheat farming area near Goldendale in Klickitat County, Washington, in August 1939 (with abandoned farm in foreground and occupied farm beyond).
PHOTO: Dorothea Lange. [Courtesy Library of Congress]

the nation's settlement system, and development of new technologies that have resulted in the exploitation of different resources and **modification** of old ones. Regions with poor innate suitability, which are unable to respond to increases in the intensity of land use that is associated with greater technical inputs, were either bypassed by settlers and businesses or temporarily exploited and then allowed to revert to a "natural" land cover.

The Natural Resource Base

The major natural land-cover patterns were the result of temperature, precipitation, elevation gradients, and the flow of energy and moisture from the Pacific Ocean, Gulf of Mexico, and Atlantic that primarily followed the flow of the mid-latitude westerly winds. In general, temperatures decline south to north and precipitation declines east to west. These gradients are modified by three north-to-south mountain systems: the Appalachians in the east, Rocky Mountains in the interior west, and Pacific ranges along and near the west coast. These **mountains** block and channel air masses from the west and allow air from the south and north to penetrate to the continent's interior. The general land-cover result was moist forests east of the Mississippi River and on the windward slopes of mountains; **grasslands** between the

Mississippi River and Rocky Mountains; and arid land grasses and shrubs from the Rocky Mountains to the Sierra Nevada and Cascade Mountain ranges.

Presettlement

The land cover of the United States prior to European settlement is subject to debate. The noted biogeographer A. Küchler mapped potential natural vegetation of the lands now representing the conterminous forty-eight United States. Potential natural vegetation was defined as the **vegetation** that would cover the land if there were no disturbances from man or nature. Essentially, A. Küchler's map represents climax vegetation, an ecological condition that was unlikely in many areas due to natural disturbances and land uses associated with indigenous people. Still, the map provides a starting point for understanding U.S. land change: prior to European settlement, 50 percent of the conterminous United States was forested, grasslands covered 24 percent, shrublands covered 18 percent, and open woodlands (grassland-forest savannas) covered nearly 8 percent of the land.

Clearly, vegetation at the time of settlement had been modified by both indigenous people and by natural events. For example, hurricanes and other wind storms frequently damaged southeastern and northeastern forests, creating gaps and modifying successional status. In the west, **fires** changed the composition of forests and created a varied quilt of different seral stages. There is also evidence that indigenous populations and economies were sufficient to have a significant local to regional impact on the natural vegetation of the United States. Indigenous populations were relatively high in 1500 and then declined following the spread of infectious diseases. As a result, the impact of indigenous people due to land uses associated with their farming, grazing, hunting, and urban activities was likely to be highest in 1500 and then was reduced.

Development Frontier

The primary development frontier in the United States was east to west with secondary frontiers moving south to north from New Spain (Mexico) during the seventeenth and eighteenth centuries and a mining frontier that moved east from the California gold fields through the mountainous west during the second half of the nineteenth century. The most important land-cover changes were the result of U.S. expansion westward from the original thirteen states. The historical geography of U.S. land-cover changes follows four epochs of change that were the result of the time of initial land cover; settlement patterns; the predominant technologies of the time, especially transportation and energy technologies; the resources that attracted settlement; and the prevailing government policies associated with land.

Epoch I: 1620–1820

The forests east of the Mississippi River and south of the Great Lakes were settled by farmers who cleared their land with axes. Preferred areas of farming often included fields abandoned by indigenous people. The primary settlement motivation was to acquire farmland, and the primary resource was good soil. Most land was converted to fields and **pasture** with some small, developed areas in villages and incipient cities. In the South, global demands for **cotton** and **tobacco** were a **driving force**, while the northern crops of **wheat** and corn were produced for the domestic grain and meat markets. Before 1776, the British tried to prevent settlement west of the Appalachians, which confined virtually all land **conversion** to the eastern areas. After the American Revolution, land that was not within the boundaries of a state was considered to be in the public domain until statehood. The primary federal government role was to survey and sell the land to farmers and speculators as quickly as possible.

Epoch II: 1820–1870

Between 1820 and 1870, land conversion accelerated because of the nation's increased population and the application of **Industrial Revolution** technology to land conversion. The application of steam power to shipping and early railroads increased the magnitude of land conversion. The nation's forests were converted rapidly as the nation's larger population demanded additional farmland, wood for heating and construction, and fuel for steam engines, which consumed prodigious quantities of wood. Most forest conversion was located near water bodies, because waterpower sites were the primary industrial energy source and the inland waterway system of canals and rivers supported most long-distance transportation. The primary federal role continued to be the surveying and disposition of land as quickly as possible.

Epoch III: 1870–1920

By 1870, the settlement frontier had reached the forests of the northern states and the grasslands of the **Great Plains**. Improvements in steam technology, the availability of cheap steel, and commercial development of electricity allowed new resources to be developed. Coal became the dominant energy source, and areas with it had rapid land conversion. The demands for wood increased because of the nation's growing population, settlement of the Great Plains where lumber had to be imported from forested regions, and use of wood for mine supports and railroad ties.

The northern forests were commercially harvested and abandoned when forestry companies chose to relinquish land ownership and move to western mountains that had areas of virgin forests. The eastern half of the grassland interior was converted from natural grass to cropland and pastures, while the arid western half became a mosaic of dryland **wheat** fields and natural rangeland depending upon soil quality and crop prices. Farther west, the arid, intermountain west remained in natural grass and shrubs except where irrigation farming was available.

The band of needle leaf evergreen trees that stretched from Maine to Minnesota and along the slopes of the Appalachian Mountains and Cumberland Plateau began to be commercially logged. Where there was adequate soil, farmers moved in as the forestry companies left, but in the north the soil was often infertile and secondary forests began to grow, and in the south the slopes were often so steep that after a few seasons of farming, most of the good soil had eroded. By 1920, virtually all of the old-growth forest of the eastern United States had been harvested. Drainage of **wetlands** and conversion to farmland was significant in the glaciated portions of the Midwest.

In response to public concerns about environmental **conservation** and timber shortages, and settler reluctance to acquire lands that could not be profitably farmed, the federal government began to reserve lands to create national forests, national parks, and federal grazing lands. All of these were in the mountainous and arid west. These lands had minimal land conversion from their original cover, even when commercial **forestry** and grazing were permitted. After 1900, the federal government began to purchase private lands in the hilly areas of the east where farming had failed or forests had been cut and not yet regrown.

Epoch IV: 1920–Present

By 1920, the U.S. settlement pattern and infrastructure system was established. One of the two major conversions was the return of trees to cleared land that had been cleared for farms or commercial forestry. This occurred throughout the states that bordered the Gulf of Mexico, Atlantic Ocean, and Great Lakes as well as substantial parts of the Appalachian and Ozark highlands. Florida, the exception, had massive drainage and forest clearance during this epoch. The second major transformation was the growth in urban areas. After 1920, the population of the nation grew rapidly, and the automobile allowed a lower-density settlement pattern. Meanwhile, the general increase in per capita income led to larger residential lots and houses. The result was that more land per capita was converted to developed land covers (e.g., houses, residential lots, streets and highways, and parking lots), and the nation's developed area increased faster than the population.

During the 1930s, the federal government purchased degraded farmland in much of the Great Plains and converted the land to national grasslands. By the end of the twentieth century, the federal government owned about 28 percent of the United States. With the exception of urban offices and military bases, most of this land has the same land cover that it did in 1620.

During the 1950s and 1960s, the nation made a series of substantial one-time land conversions to transportation, reservoirs, and **urbanization**. The construction of the federal interstate highway system converted more than 700,000 hectares to highways and intersections that often became development nodes leading to further conversion to developed covers. States and cities also were responsible for substantial street and highway development. Construction of reservoirs had begun during the Industrial Revolution, but approximately 60 percent of the nation's reservoir storage capacity was constructed during the 1950s and 1960s. Most of these are located in the humid East, but most of the large ones are west of the Mississippi River. After 1930, up to 2 million farm ponds were constructed, with the greatest concentration in the eastern Great Plains and Southeast. Most of these are less than one hectare in size. Before 1980, significant areas of hardwood wetlands in the Mississippi alluvial plain and other river bottoms in the South were drained and converted to cropland.

After 1970, the nation's population growth, and associated conversion to developed covers, slowed in the traditional growth core of the northeastern and midwestern states and shifted to the warmer southern states, especially from California to Florida and north to North Carolina. Secondary population growth occurred in rural areas within commuting distance of cities and in amenity regions associated with oceans, lakes and reservoirs, mountains, and warm areas.

The Major Transformations

Visually, the primary land-cover transition was the decline in **forest** from 50 percent to 29 percent of the nation's land cover. Partially compensating for the loss of trees to farmland and developed areas were trees in areas historically devoid of forests. These trees were generally the result of fire prevention, the planting of farm windbreaks, and settlements of all sizes from hamlets to conurbations where trees were planted. Virtually all of the nation's humid and sub-humid grassland had been converted to crops and pasture covers. West of the 100th meridian, areas with good soil and **access** to water were typically converted to irrigated cropland.

Contemporary Land Cover

The documentation of the contemporary land cover of the United States is somewhat incomplete. A 2000 land-cover map of the nation is being developed and will provide the most complete land-cover description of all fifty states ever produced. A 1992 map of the land cover of the conterminous United States (excluding Alaska and Hawaii) reveals that forests cover over 29 percent of the land, cropland and pasture now occupy 26 percent, shrublands cover 18 percent, and grasslands cover 16 percent. The developed footprint of the conterminous United States is just over 2 percent of the area. Wetlands cover 4 percent, barren lands comprise 2 percent, and fresh water covers nearly 2 percent. The types and extent of land cover in the conterminous United States reflect the resource potential based on the inherent biotic and abiotic features of different regions of the country; and the influence of transportation systems, government policy, and shifting social and economic conditions.

See also: **Agricultural Revolution; Change Detection; Colonization; Corporate Strategies; Desertification; Economic Growth; Economic Restructuring; Extensification; Food Crops; Hot Spots of Land-Cover Change; Industrialization; Land Degradation; LANDFIRE Project; Land Reform; Land Rights; Land Tenure; Land-Use History; Leapfrogging; Mineral Extraction; Pristine Myth; Proximate Causes; Public Policy; Reforestation; River Basin; Suburbanization; Technological-Scientific Revolution; United States Geological Survey (USGS) Program; Urban-Rural Fringe; Urban Sprawl; Tobacco; Yellowstone Model.**

Further Reading

Borchert, John R., "American Metropolitan Evolution," *Geographical Review* 57 (1967): 301–32; Conzen, Michael, ed., *The Making of the American Landscape,* Boston: Unwin Hyman, 1990; Denevan, William M., "The Pristine Myth: The Landscape of the Americas in 1492," *Annals of the Association of American Geographers* 82 (1992): 369–85; Dilsaver, Lary M., and Craig E. Colten, eds., *The American Environment: Interpretations of Past Geographies,* Lanham, MD: Rowman and Littlefield Publishers, Inc., 1992; Graf, William L., "Dam Nation: A Geographic Census of American Dams and Their Large-Scale Hydrologic Impacts," *Water Resources Research* 35 (1999): 1305–11; Hart, John

Fraser, "Half a Century of Cropland Change," *The Geographical Review* 91 (2001): 525–43; Hart, John Fraser, "Loss and Abandonment of Cleared Farm Land in the Eastern United States," *Annals of the Association of American Geographers* 58 (1968): 417–40; Küchler, A. W., *Potential Natural Vegetation of the Conterminous United States* (Special Publication no. 36), Washington, DC: American Geographical Society, 1964; McIlwraith, Thomas F., and Edward K. Muller, eds., *North America: The Historical Geography of a Changing Continent*, 2nd ed., New York: Rowman & Littlefield Publishers, Inc., 2001; Prince, Hugh, *Wetlands of the American Midwest: A Historical Geography of Changing Attitudes*, Chicago: University of Chicago Press, 1997; Ramankutty, Navin, and Jonathan A. Foley, "Estimating Historical Changes in Land Cover: North American Croplands from 1850 to 1992," *Global Ecology and Biogeography* 8 (1999): 381–96; Vogelmann, James E., Howard, S. M., Yang, L., Larson, C. R., Wylie, B. K., and N. Van Driel, "Completion of the 1990's National Land Cover Data Set for the Conterminous United States from Landsat Thematic Mapper Data and Ancillary Data Sources," *Photogrammetric Engineering and Remote Sensing* 67 (2001): 650–62; Whitney, Gordon G., *From Coastal Wilderness to Fruited Plain: A History of Environmental Change in Temperate North America 1500 to the Present*, Cambridge, UK: Cambridge University Press, 1994; Williams, Michael, *Americans and Their Forests: A Historical Geography*, Cambridge, UK: Cambridge University Press, 1989.

DARRELL NAPTON, THOMAS R. LOVELAND

United States Geological Survey (USGS) Program. The U.S. Geological Survey (USGS) plays a significant land-cover research and development role. A noteworthy contribution occurred in the early 1970s when James Anderson, USGS chief geographer, led the development of "A Land Use and Land Cover Classification System for Use with Remote Sensor Data." This system established a hierarchical classification scheme and intellectual framework for using the then-new **Landsat** imagery, as well as **aerial photography**, for mapping land use and land cover. Each level of the classification hierarchy was matched with the information potential of different **scales** of remotely sensed data (i.e., Level I and Landsat multispectral scanner data, and Level II and high altitude aerial photography). The framework developed by James Anderson and colleagues was used to produce a twenty-one-class national land-cover map based on manual interpretation of 1970s aerial photographs and classified images. The map, often referred to as LUDA (land-use data analysis), was the first comprehensive land-cover map of the **United States of America**.

The USGS continues to produce land-cover maps of the United States. In 2000, the National Land Cover Database (NLCD) was completed for the conterminous United States. The NLCD was produced using a modified Anderson classification scheme, circa-1992 Landsat Thematic Mapper (TM) imagery, ancillary data sets, and computer-assisted classification methods. Development of a new NLCD covering all fifty states began in 2001 using circa-2000 Landsat 7 Enhanced Thematic Mapper data and improved classification methods that were designed to improve land-cover map accuracy and consistency.

Two additional nationwide land-cover mapping activities are underway in the USGS. The Gap Analysis Project produces natural **vegetation** maps using the same 1992 and 2000 Landsat data used to produce the NLCD maps.

The **LANDFIRE project** is an effort to produce detailed vegetation maps corresponding to wildfire fuels. Like NLCD and the Gap Analysis Project, LANDFIRE uses the same circa-2000 Landsat data. The three 2000 USGS national land-cover and vegetation-mapping projects (NLCD, Gap Analysis, and LANDFIRE) share data and are closely coordinated in order to ensure map consistency.

The USGS has been a leader in global-scale land-cover mapping. It led the development of the first validated **IGBP-DIS 1-km global land cover data set**. USGS land-cover specialists continue to investigate large-area land-cover topics and are contributors to the **Global Land Cover Map of the Year 2000** (GLC-2000).

USGS land-cover research spans a wide range of topics and spatial and temporal scales. Studies of contemporary U.S. land-cover trends (1972–2000) and paleo-ecological investigations of historical vegetation ranges are also underway. Research on large-area mapping and accuracy assessment strategies is part of the USGS core land-cover research program.

See also: **Center for International Earth Science Information Network; Earth Resources Observation Systems (EROS) Data Center; Remote Sensing; Sahel Land Cover.**

Further Reading

Anderson, James R., Hardy, E. E., Roach, J. T., and R. E. Witmer, *A Land Use and Land Cover Classification System for Use with Remote Sensor Data,* U.S. Geological Survey Professional Paper 964, Reston, VA: U.S. Geological Survey, 1976; United States Geological Survey, Earth Resources Observation Systems (EROS) Data Center, Land Cover Characterization Program [Online, November 2004], The USGS Land Cover Web Site, landcover.usgs.gov.

THOMAS R. LOVELAND

Urban Sprawl. A catch-all phrase that typically refers to low-density urban development that may also be characterized by a lack of contiguity, compactness, or centrality. No universally agreed-upon definition of sprawl exists. While some define sprawl as anything negative associated with urban form and others define it in terms of market failures, the most useful definitions are in terms of quantifiable aspects of an urban land-use pattern. This approach then allows further investigation of trends, causes, and consequences of this pattern. The literature ranges from very simple operational definitions, such as population per urban acre, to multidimensional definitions that seek to capture a variety of pattern aspects, including density, contiguity, concentration, centrality, mixed use, and proximity. Measurement of sprawl is limited by data availability and complicated by the dynamic process of land-use change, which implies that sprawl is not a discrete outcome, but rather a matter of degree and that sprawl may be a transient phenomenon.

Many of the factors that are hypothesized to cause sprawl are the same factors that have been linked to **suburbanization**, including cheap transportation dominated by automobiles and trucks; central-city fiscal and social problems; advances in information technologies; rising household incomes; and lifestyle preferences for new houses, large lots, and open space. A variety of external costs are commonly associated with sprawl, including overconsumption of natural resource lands, increased traffic congestion, and the added costs of providing more roads, utility lines, and other public services.

See also: **Extensification; Land Privatization; Leapfrogging; Urbanization; Urban-Rural Fringe.**

Further Reading

Breuckner, Jan, "Urban Sprawl: Diagnosis and Remedies," *International Regional Science Review* 23 (2000): 160–71; Galster, George, Hanson, Royce, Ratcliffe, Michael R., Wolman, Harold, Coleman, Stephen, and Jason Freihage, "Wrestling Sprawl to the Ground: Defining and Measuring an Elusive Concept," *Housing Policy Debate* 12 (2001): 681–717.

ELENA G. IRWIN

Urbanization. Urban built-up areas make up a small proportion of global land covers—2 to 3 percent by most estimates—yet their influence on surrounding landscapes is profound. This is because urban areas are centers of population and economic activity that draw on their hinterlands for renewable and nonrenewable resources such as water, food, energy, and raw materials for manufacturing. The urban-rural linkage thus becomes a critical element in understanding how urban areas impact land-use and **land-cover change**.

Coming up with a consistent definition of what constitutes an urban land-cover type is not as simple as it might at first appear. There is in fact a

Skyline of mid-New York in 1931.
PHOTO: William Frange. [Courtesy Library of Congress]

gradient from wild areas to rural agricultural lands to peri-urban areas to suburban and then urban. Traditionally researchers treated as urban anything that fell within a city's administrative borders or that met a certain population density threshold. However, using administrative boundaries is clearly deficient, since in many cases urban built-up lands spill over a city's limits, and in other instances city boundaries may have been drawn in such a way as to include large tracts of relatively undeveloped land. In recent years, **remote sensing** scientists, demographers, and other researchers have sought to develop more consistent definitions for what constitutes "urban" land cover.

The Global Urban-Rural Mapping Project (GRUMP) of the **Center for International Earth Science Information Network** utilized a combination of data sources—including the Gridded Population of the World, nighttime lights, and point estimates for city population—to develop a globally consistent urban extents database. Utilizing these data, it is possible to identify the patterns in geographic location of urban areas, including the kinds of land-cover classes in which urban areas tend to be located. Globally, the urban areas tend to be located in **coastal zones**, where 10.2 percent of the land is urban, followed by cultivated areas (6.8 percent), islands (4.7 percent), and in close proximity to inland waters (3.2 percent). Drylands, forests, and mountainous areas are all under 2 percent urban land cover. According to the study, 46.2 percent of the world's population resides in urban areas, and 40 percent of the urban population resides in cities under 500,000 in size, with a quarter residing in cities between 1 and 5 million in size. Some 6.7 percent of the world's population resides in "megacities" of 10 million or more. These megacities are generally located in the coastal zone, on islands, or near inland waters.

The processes and patterns of urbanization are different in developed and developing countries. In most developed countries, large cities arose during the **Industrial Revolution** as economies of scale were generated by the concentration of capital and labor in urban areas. Mills would locate near sources of raw materials and cheap power (rivers or coal deposits), and people flocked to the mills from the countryside, creating urban agglomerations. In 1950, seven of the ten largest cities were in the industrialized world. In the latter part of the twentieth century, the economies of major cities became less dependent on heavy industry and more tied to the service sector. In the Americas, **suburbanization**, reliance on cars rather than public transportation, and the decentralization of jobs led to progressive low-density development on the fringes. In **Europe** and **Japan**, the same phenomenon has occurred though to a lesser extent due to shortages of land, the protection of farmland, and greater concentration of development near public transportation.

Developing country cities have grown very rapidly in the past fifty years such that today, seven out of the ten largest urban agglomerations are in the developing world, with Mexico City, Mumbai, and Saõ Paulo following after Tokyo as the second, third, and fourth largest cities in the world. While industrial activity played a significant role in the development of Asian and Latin American cities (less so in Africa), the recent explosion of growth is also related to a combination of rapid population growth, lack of economic opportunities in rural areas, rural landlessness, political instability, and droughts and other natural disasters. This, coupled with a lack of **land-use planning**, weak zoning, and insufficient urban infrastructure (water and sanitation), has led to a deteriorating quality of life in many developing country urban areas, especially in slum or squatter settlements. As the poorest residents often reside in previously undeveloped lands on steep slopes or in low-lying flood-prone areas, they have an elevated risk of being affected by natural hazards.

These problems point to a major difference in the environmental outcomes of urbanization in developed and developing countries. Over a period of 150 years, urban managers in industrialized nations have developed mechanisms to mitigate the impact of household and industrial **consumption** and wastes on surrounding ecosystems through water and sanitation systems, sewerage treatment, automobile and industrial air pollutant emissions standards, environmental impact assessments for new development, zoning codes, and waste disposal regulations. Even so, many developed countries face problems of "brownfields"—former industrial zones that are difficult to redevelop due to buildup of hazardous wastes—as well as periodic air and water quality emergencies. Therefore, in light of rapid urban growth and a comparative lack of resources, it is not surprising that urban managers in developing countries struggle to meet the water, sanitation, and waste disposal needs of city dwellers.

From an ecological perspective, the intensity of the transformation of the landscape in urban areas has certain repercussions for ecosystem processes such as **nutrient cycling** and hydrology. These have not been as well studied as ecology in natural areas, largely because scientists have tended to see these areas as human-transformed landscapes of little ecological interest. This is beginning to change, and there are new studies examining urban ecosystems, including long-term ecological research sites in the U.S. cities of Baltimore, Maryland, and Phoenix, Arizona. Generally, urban areas are places of intensified nutrient cycling, as nutrients such as plant and animal food products are shipped to cities and consumed, and resulting wastes often end up in rivers, lakes, and coastal zones, affecting aquatic ecosystems. Urban areas also have a dramatic impact on hydrology from two perspectives. On the one hand, impervious surfaces result in more dramatic spikes of runoff during stormwater discharge, affecting flooding in downstream areas. On the other, water demand in urban areas often results in the diversion of water from distant areas, which affects ecosystems in both the source (rural) and destination (urban) areas.

Urban areas can also have significant impact on local microclimates through such phenomena as the urban heat island effect. Urban heat islands are created by the relatively low **albedos** (reflectivity) of urban land surfaces, and the relative absence of vegetative cover. **Vegetation** absorbs energy in the visible near infrared band of the spectrum, converting it to plant growth through photosynthesis. By contrast, paved surfaces absorb the sun's energy and re-radiate it after the sun has set. Research in Atlanta, Georgia (United States) found that the growing season was increased by several days as a result of the heat island effect. Also, the elevated air temperatures in Atlanta create local convection that results in thunderstorms in downwind areas.

As stated above, just because urban areas make up only a small proportion of global land area does not mean that they can be discounted as an insignificant contributor to land-use and land-cover change. On the contrary, if one traces urban-rural linkages, it becomes clear that the "**ecological footprint**" of urban areas is several orders of magnitude larger than the urbanized area itself. Urban centers rely on their hinterlands for water, food, fuel, and natural resources of all types. One study estimated that the cities of Baltic Europe appropriate for their resource consumption and waste assimilation an area of forest, agricultural, marine, and wetland ecosystems that is 565–1,130 times larger than the area of the cities themselves.

Cities also expropriate water from large distances. In some cases, such as in the New York City water supply districts, this may have the effect of conserving large areas of land that might otherwise be developed. In others, such as in Los Angeles, the expropriation may have a negative impact on the ecology of natural water bodies such as Mono Lake and the Colorado River, where much of the water is drained before it crosses the Mexican border.

Urban area energy demands may also have significant impacts on land cover in distant areas. Many rural agricultural areas and natural ecosystems have been submerged under the reservoirs created by large hydroelectric dams, such as the Three Gorges Dam in **China** and the Aswan High Dam in Egypt. At 500 kilometers in length, Lake Nasser, formed by the Aswan High Dam, is the largest artificial lake in the world. Mining, oil drilling, and other land-based energy extraction activities make up small but important portions of global land cover. Many African cities are still heavily dependent on **biomass** fuels such as wood and charcoal for household heating and cooking, which can have a significant impact on **deforestation** in surrounding areas.

A significant impact of urban areas on the hinterlands is the need for sinks (dumping sites) for the great volumes of waste they produce. Urban wastes can be solid, liquid, or gaseous. Problems of solid waste disposal have become increasingly problematic as landfills have reached their maximum capacities. Cities generally have few solutions other than to truck their waste at great cost to distant landfills, or to incinerate it. Owing to shortages of space for landfills, European and a number of North American cities have increasingly opted for incineration. Problems arise from emissions of toxic fumes, especially dioxins, lead, and mercury, and then the disposal of toxic ash.

Atmospheric pollutants can have a significant impact on land use and land cover. Carl Folke and colleagues estimate that the carbon dioxide emissions of major cities alone would consume 95 percent of the carbon absorptive capacity of the world's forests. Another major problem is air pollutant impacts on natural vegetation and crops. Acid rain has been a documented problem that damages **forest** lands in North America and eastern Europe. William Chameides and colleagues examine the impact of ground-level ozone on crop production in three large "metro-agro-plexes"—areas of intense agricultural and industrial production in North America, northern Europe, and China/Japan—and find that it may be reducing global crop production by a few percent. This impact is likely to intensify due to projected increases in nitrogen oxides, an important precursor to ozone.

Urbanization pathways lead to different impacts on rural landscapes in the developed and developing worlds. In the developed world, large-scale urban agglomerations and extended peri-urban settlements fragment the landscapes of such large areas that various ecosystem processes are threatened. Ecosystem fragmentation, however, in peri-urban areas may be offset by urban-led demands for **conservation** and recreational land uses.

See also: **Climate Impacts; Economic Restructuring; Extensification; Hot Spots of Land-Cover Change; Human Immunodeficiency Virus (HIV)/Acquired Immunodeficiency Syndrome (AIDS); Industrialization; International Human Dimensions Programme on Global Environmental Change; Investments; Land Reform; Land Rent; Leapfrogging; Malaria; Mineral Extraction; People at Risk; Population Dynamics; Reforestation; Suburbanization; Urban-Rural Fringe; Urban Sprawl.**

Further Reading

Chameides, William L., Kasibhatla, Prasad S., Yienger, James, and Hiram Levy, "Growth of Continental-Scale Metro-Agro-Plexes, Regional Ozone Pollution, and World Food Production," *Science* 264 (1994): 74–7; Elvidge, Christopher D., Sutton, Paul C., Wagner, Thomas W., Ryzner, Rhonda, Vogelmann, James E., Goetz, Scott J., Smith, Andrew J., et al., "Urbanization," in *Land Change Science: Observing, Monitoring and Understanding Trajectories of Change on the Earth's Surface* (Remote Sensing and Digital Image Processing Series no. 6), eds. Garik Gutman, Anthony C. Janetos, Chris O. Justice, Emilio F. Moran, John F. Mustard, Ronald R. Rindfuss, David Skole, B. L. Turner, II, and Mark Cochrane, Dordrecht, NL, Boston, and London: Kluwer Academic Publishers, 2004, pp. 315–28; de Sherbinin, Alex, Balk, Deborah, Yager, Karina, Jaiteh, Malanding, Pozzi, Francesca, Giri, Chandra, and Antoinette Wannebo, Social Science Applications of Remote Sensing: A CIESIN Thematic Guide [Online, December 2004], The CIESIN Web Site, sedac.ciesin.columbia.edu/tg; Folke, Carl, Jansson, Asa, Larsson, Jonas, and Robert Costanza, "Ecosystem Appropriation by Cities," *Ambio* 26 (1997): 167–72; Urbanization and Global Environmental Change [Online October 2005], The UGEC Web Site, co.hunter.cuny.edu/%7Eshodges/UGEC/index.htm.

ALEXANDER DE SHERBININ

Urban-Rural Fringe. Semi-rural areas that lie beyond built-up, urbanized areas, but that fall within the influence of urban areas. These areas consist of a mix of urban and rural functions that are spatially interspersed. A variety of terms have been used to describe these areas, including *peri-urban, exurban, urban fringe, rural-urban fringe, urban-rural fringe, urban periphery, extended metropolitan regions,* and *desakota* (meaning literally a village city).

While urban-rural settlement patterns vary by country, region, and time period, they are often characterized as low-density, scattered urban development that consists of largely residential and sometimes industrial land uses; existing rural towns often with new development on the periphery; and intervening rural land uses. Due to the low density and fragmented nature of this development, the impact on population redistribution to these areas is far less than the impact on land **conversion**.

Population growth in urban-rural areas—a process sometimes referred to as "exurbanization" or "counterurbanization"—is largely the result of **urbanization** pressures. In some ways, urban-rural fringe areas are the "new suburbs" associated with expanding urban areas. However, they differ from their suburban counterparts in ways other than just their location vis-à-vis the urban core. Rural economic activities, such as **agriculture** and other natural resource-based industries, are still a substantial contributor to the local economy. A mix of rural residents, for example, farmers and residents of rural towns and new residents moving outward from urban areas live in these areas. Government controls, including land-use controls and locally provided public services, are often much less extensive than in the built-up urban areas.

Developed countries witnessed a "rural renaissance" or "population turnaround" in the 1970s—faster population growth in nonmetropolitan versus metropolitan areas. While these trends waned in the 1980s and appeared to reverse themselves in certain countries, the 1990s saw renewed growth in semi-rural areas of many developed countries. Analyses of longer-term trends support the hypothesis that counterurbanization has been a dominant trend for the past several decades in many developed countries, including **Canada**, the **United States of America**, Great Britain, and other countries in **Europe**.

Similar trends in some developing countries have been documented as well. For example, Victor Sit and Chun Yang studied recent growth patterns in the Pearl River delta of **China** since the economic reforms of 1978 and found that the predominant growth pattern has been a more equal level of urbanization across the region coupled with a declining importance of the primate regional city. Even in sub-Saharan Africa, where economic growth is largely absent, increased demand for new, higher-quality residential development has spurred peri-urban growth in some metropolitan areas.

Low transportation costs, due to the lost cost of automobiles and an extensive roads network; new manufacturing technologies; advances in communications technologies; and rising household incomes are some of the primary **driving forces** that have lessened the advantages of urban agglomeration and led to urban-rural fringe growth. In addition, increased demand for environmental amenities and the **economic restructuring** of rural areas of the United States, Canada, and other developed countries have transformed some rural areas into recreation and tourism-based places with high-valued environmental amenities. In China and other Asian countries, economic **globalization** has led to growth that favors smaller cities and semi-rural areas due to production technologies that require large tracts of land and use information technologies that reduce the need for urban proximity. Peri-urban residential growth in some developing countries has also been tied to increasing incomes of a growing professional class and remittances from migrants.

See also: **Economic Growth; Leapfrogging; Remittances Landscape; Suburbanization; Urban Sprawl.**

Further Reading

Briggs, John, and Ian E. A. Yeboah, "Structural Adjustment and the Contemporary Sub-Saharan African City," *Area* 33 (2001): 18–26; Champion, Anthony G., "Urban and Regional Demographic Trends in the Developed World," *Urban Studies* 29 (1992): 461–82; Daniels, Tom, *When City and Country Collide*, Washington, DC: Island Press, 1999; Nelson, Arthur C., "Characterizing Exurbia," *Journal of Planning Literature* 6 (1992): 350–68; Sit, Victor F. S., and Chun Yang, "Foreign Investment-Induced Exo-Urbanization in the Pearl River Delta, China," *Urban Studies* 34 (1997): 647–77.

ELENA G. IRWIN

V–Y

Validation. A process that measures the consistency between theory and data, which is used frequently to test models (see figure titled "The position of validation"). If a theory is consistent with accurate and comprehensive data, then the theory is said to be valid. In the context of **modeling**, Edward Rykiel Jr. defines validation as "a demonstration that a model within its domain of applicability possesses a satisfactory range of accuracy consistent with the intended application of the model." In order to test the validity of any particular theory, the theory must be translated into a form such that it can be compared directly to data. For example, if a scientist wants to test the theory that agricultural market forces drive land change, then the scientist could express this theory in terms of a model that generates a prediction in the form of a map. Validation is the process of measuring the agreement between the model prediction and independent data. If there is a good match between the model prediction and independent data, then the method used to make the prediction is said to be valid. Scientists must respect rigorous definitions of "model prediction," "independent data," and "good match."

In order to distinguish between model prediction and independent data, it is crucial to distinguish between model calibration and model validation. Calibration is the process of creating a model such that it is consistent with some subset of the entire available data set. It is essential that the available data set be split into two separate subsets, called calibration data and validation data. The modeler uses the calibration data and theory to create the model. The validation data are reserved for the validation procedure and must not be used to help to create the model. After the model is calibrated using the calibration data, the model generates a prediction that is then compared to the validation data. The word "prediction" refers to the fact that the model writes the output before the model sees the validation data. There are two common techniques to separate calibration data from validation data. The first method is separation through time, where calibration data consist of information at or before some specific time, and validation data consist of information after that time. Under this first arrangement, the validation procedure tests whether the model is valid as it predicts across time. The second common method is separation through space, where calibration data are from one location and the validation data are from some other location. Under this second arrangement, the validation procedure tests whether the model is valid as it extrapolates across space.

Much of the confusion about validation derives from lack of distinction between goodness-of-fit of calibration and goodness-of-fit of validation. The former refers to how well a model can fit a mathematical form to existing data during the calibration phase, where the data are used to help define

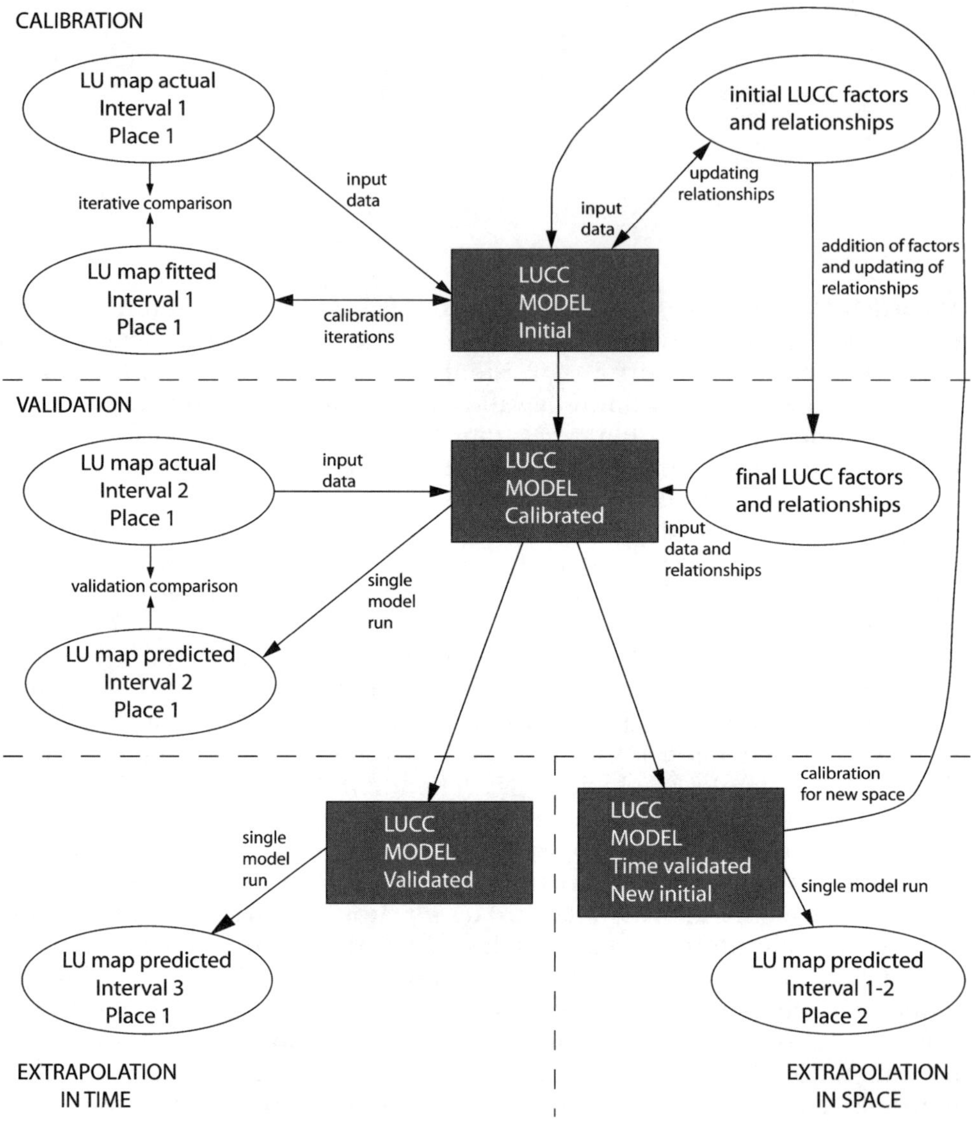

The position of validation in testing models.

the mathematical form and the parameters. For example, an R-squared value in regression is a measure of goodness-of-fit of calibration, if all the data are used to create the regression line. This type of calibration is referred to as model fitting, not as model prediction. Model prediction occurs when the scientist uses the fitted model to attempt to predict the validation data. When scientists use the terms "modeled" and "simulated," it is not clear whether they mean "fitted" or "predicted"; therefore, it is not clear whether the goodness-of-fit refers to calibration or validation. An elaborate fitting procedure can attain a high goodness-of-fit for calibration; however, such a fit does not necessarily indicate that the model will have a good fit for validation. High-speed computers can induce modelers to over-fit a model, so the model becomes calibrated to both the signal and the noise in the calibration data; so the fit of the calibration can be high. When an over-fitted model extrapolates to data that were not used for calibration, the resulting measure of goodness-of-fit of validation is likely to be less than the goodness-of-fit of calibration. In order to assure an appropriate measure of a model's predictive power, one must separate the information used for calibration from the information for validation. This can be difficult for places where availability of information is limited and for models that demand substantial data for calibration.

The validation exercise assumes that the scientist already has knowledge of the time and place to validate the model, because proper validation requires accurate validation data. However, usually the major purpose of the model is to understand a phenomenon for a time or place that is unknown, for which data do not exist. Therefore, in order to capitalize on the validation exercise, the scientist should translate the measurement of validation into a statement about confidence in a model as it extrapolates into times and places that are unknown. Extrapolation is the process whereby a model makes predictions beyond the extent of the calibration and validation data. If there is a good fit between the model prediction and the validation data, then the model might be able to make accurate extrapolations to other spatial and temporal extents. If the characteristics and mechanisms of the other spatial and temporal extents are similar to the characteristics and mechanisms that existed during the calibration and validation phases, then the model should be able to extrapolate to other extents with a level of accuracy similar to the performance in the validation. If the model fails to attain a good fit in the validation, then one should have little confidence in the model's ability to extrapolate accurately to other extents. A high goodness-of-fit of validation is a necessary condition for a scientist to have confidence in a model's predictions to other spatial and temporal extents. If the mechanisms that the model simulates are consistent during the calibration, validation, and extrapolation phases, then a high goodness-of-fit in the validation phase is also a sufficient condition for a scientist to have confidence in a model's extrapolations to other spatial and temporal extents. Obviously, we will not know *a priori* whether the mechanisms during the calibration and validation phases will continue into the extrapolation, because we can never know something until we have empirical data. Alas, if empirical data existed, there would not be much reason to make a model prediction. Ultimately, separation of calibration from validation, and objective measurement of validation are important techniques that a scientist should use in order to measure the level of trust one should have in a model as it is used to help one learn about what one does not yet know.

There are no agreed criteria among scientists concerning either the method or the level of what is considered a "good match" between validation data and a model prediction. A reasonable minimum criterion would be that the agreement between the validation data and the prediction from a scientist's model should be better than the agreement between the validation data and the prediction from a null model. The prediction from a null model is the naive prediction that one would make if one were not to create any analytical model. For example, suppose the purpose of the model were to predict the change in a land-cover map from an initial time to a subsequent time. The calibration data consist of information at or before the initial time. The validation of the model would compare the independent validation map of the subsequent time to the prediction map of the subsequent time from the scientist's model. The prediction map of the subsequent time from the null model would be exactly the map of the initial time. In other words, the null model would predict no change between the initial time and the subsequent time. In many cases, the agreement between the independent validation map and the prediction map from a null model can be large, because the only disagreement between those two maps is the change between the two points in time, which can be small relative to the study area. It is not advisable to use a map of randomness as the prediction of the null model, because a naive prediction would be pure persistence rather than randomness.

Validation of a land-change model usually requires the comparison between a validation map and a prediction map. Hence, there is a need to measure objectively the agreement between any pair of maps in order to determine which among the many possible prediction maps has the closest agreement with the validation map. There are numerous mathematical methods to compare the patterns in maps. The methods can be separated into those that apply to a real variable and those that apply to a categorical variable.

The usual first step in comparing two maps of a real variable is to plot the values in a Y versus X space where the vertical axis shows the prediction values and the horizontal axis shows the validation data. Each point in the plot denotes an element on the map, such as a **pixel** or a polygon. The closer the points are to the Y = X line, the better the fit between the validation data and the prediction of the scientist's model. This fit should be compared to the fit between the validation data and the prediction of the null model. Two standard measurements of how tightly the values are clustered around the Y = X line are the root mean square error (RMSE) and the mean absolute error (MAE).

A common first step in computing the agreement between a validation map and a prediction map of a categorical variable is to calculate the percentage of pixels classified correctly. This statistic should be compared to the percentage correct in the comparison between the validation data and the prediction from the null model. The percentage correct is the most common statistic reported because it is simple to compute and is relatively intuitive to interpret. The disadvantage is that it fails to capture patterns that are immediately obvious to the human eye. There are many other statistics available that compute **pattern metrics** to compare the patterns in two maps. Whatever statistic is chosen, the scientist should be cognizant of two important components in the comparison of the pattern between two maps of a common categorical variable: comparison in terms of quantity of each category and comparison in terms of location of each category. Comparison in terms of quantity considers whether the proportion of each category on one map is similar to the proportion of the corresponding category on the other map. Comparison in terms of location considers whether the position of each category on one map is similar to the position of the corresponding category on the other map. Robert Gilmore Pontius Jr. describes how to budget the agreement and disagreement for these two components.

In some cases, the validation data are of a categorical variable, and the prediction from the model is of a real variable. For example, the validation data may consist of pixels in which 0 indicates land persistence and 1 indicates land change, whereas the prediction from the model is a real number that indicates probability of land change. The scientist would like to know whether high probabilities of land change are concentrated at locations of true land change according to the Boolean validation map. For this case, there are a variety of statistical measures of agreement. One of the most appropriate ones is the relative operating characteristic (ROC).

None of the statistical measurements above match perfectly the human eye's ability to recognize patterns. Therefore, all of the methods of statistical map comparison should be complemented by a visual assessment to determine whether the selected statistic is measuring the characteristic that the scientist thinks is important. However, visual assessment can be influenced dramatically by subjective aspects of map production, such as selection of the color palette. Therefore, objective statistical measurement is essential to maintaining scientific rigor.

Any measurement to compare two maps can be extremely sensitive to the **scale** of the analysis. Therefore, it is advisable to compute the measurement at various scales to examine the degree to which the results are sensitive to changes in scale. Increasingly, scientists are creating methods to examine how results are sensitive to scale. Robert Gilmore Pontius Jr. shows how to compute the components of agreement and disagreement in terms of quantity and location at multiple resolutions. He and colleagues describe how to compute the ROC for multiple resolutions and how to compute the entire cross-tabulation matrix at multiple resolutions. They further show how to measure the value of information at nested levels of stratification structure, and how changes in categorical scale influence the measurement of map agreement, while Kasper Kok and colleagues perform their validation at multiple scales. Multiple-scale analysis allows the statistics to conform more closely to the patterns and clusters that the human eye sees.

Multiple-scale validation is also important because it allows the scientist to see whether the model makes predictions at scales that are relevant to the purpose of the model. Many models are calibrated and make predictions at fine resolutions that match the available data. For example, satellite data are commonly available in the form of thirty-meter pixels. However, the relevant questions concerning land change may occur at coarser resolutions. For example, many global climate models operate at scales of 1 degree longitude by 1 degree latitude. Scientists are sometimes reluctant to change the scale of the raw data because any adjustments introduce additional artifacts into the data. Multiple resolution validation allows a scientist to see whether the thirty-meter resolution model performs sufficiently accurately at a resolution that is relevant to the purpose of the model.

When there is lack of fit between the data and the prediction, then the scientist must consider the quality of both the model and the data. The scientist should improve the quality of whichever is worse. If the data quality is high, then the scientist will create a better model by changing the model to make it more consistent with the data. If the data quality is low, then the scientist should focus on getting better data; otherwise the scientist might be induced to make an inaccurate model by redesigning the model to be consistent with inaccurate data.

There is tremendous desire on the part of some scientists to label a model as valid or not. It is not particularly useful to attempt to crown a model as valid or to condemn a model as invalid. It is more useful to state carefully the degree to which a model is valid. Validation should measure the performance of a model in a manner that enables the scientist to know the level of trust that one should put in the model. Useful validation should also give the modeler information necessary to improve the model.

See also: **Agent-Based Model; Change Detection; Remote Sensing.**

Further Reading

Hagen, Alex, "Fuzzy Set Approach to Assessing Similarity of Categorical Maps," *International Journal of Geographic Information Science* 17 (2003): 235–49; Jolliffe, Ian T., and David B. Stephenson, eds., *Forecast Verification*, West Sussex, UK: Wiley, 2003; Kok, Kasper, Farrow, Andrew, Veldkamp, Tom A., and Peter Verburg, "The Need for Multi-Scale Validation in Spatial Land Use Models," *Agriculture, Ecosystems and Environment* 85 (2001): 223–38; Oreskes, Naomi, Shrader-Frechette, Kristin, and Kenneth Belitz, "Verification, Validation, and Confirmation of Numerical Models in the Earth," *Science* 263 (1994): 641–46; Pontius, Robert Gilmore, Jr., "Quantification Error Versus Location Error in Comparison of Categorical Maps," *Photogrammetric Engineering & Remote Sensing* 66 (2000): 1011–66; Pontius, Robert Gilmore, Jr., "Statistical Methods to Partition Effects of Quantity and Location during Comparison of Categorical Maps at Multiple Resolutions," *Photogrammetric Engineering & Remote Sensing* 68 (2002): 1041–49; Pontius, Robert Gilmore, Jr., Agrawal, Aditya, and Diana Huffaker, "Estimating the Uncertainty of Land-Cover Extrapolations while Constructing a Raster Map from Tabular Data," *Journal of Geographical Systems* 5 (2003): 253–73; Pontius, Robert Gilmore, Jr., and Kiran Batchu, "Using the Relative Operating Characteristic to Quantify Certainty in Prediction of Location of Land Cover Change in India," *Transactions in GIS* 7 (2003): 467–84; Pontius, Robert Gilmore, Jr., Huffaker, Diana, and Kevin Denman, "Useful Techniques of Validation for Spatially-Explicit Land-Change Models," *Ecological Modelling* 179 (2005): 445–61; Pontius, Robert Gilmore, Jr., and Jeffrey Malanson, "Comparison of the Structure and Accuracy of Two Land Change Models," *International Journal of Geographical Information Science* 19 (2004): 243–65; Pontius, Robert Gilmore, Jr., and Nicholas Malizia, "Effect of Category Aggregation on Map Comparison," *Lecture Notes in Computer Science* 3234 (2004): 251–68; Pontius, Robert Gilmore, Jr., and Laura Schneider, "Land-Use Change Model Validation by a ROC Method for the Ipswich Watershed, Massachusetts, USA," *Agriculture, Ecosystems & Environment* 85 (2001): 239–48; Pontius, Robert Gilmore, Jr., and Beth Suedmeyer, "Components of Agreement in Categorical Maps at Multiple Resolutions," in *Remote Sensing and GIS Accuracy Assessment*, eds. Ross S. Lunetta and John G. Lyon, Boca Raton, FL: CRC Press, 2004, pp. 233–51; Ritters, Kurt H., O'Neill, Robert V., Hunsaker, Carolyn T., Wickham, James D., Yankee, Dennis H., Timmins, Sidey P., Jones, K. Bruce, and Barbara L. Jackson, "A Factor Analysis of Landscape Pattern and Structure Metrics," *Landscape Ecology* 10 (1995): 23–39; Rykiel Jr., Edward J., "Testing Ecological Models: The Meaning of Validation," *Ecological Modelling* 90 (1996) 229–44.

ROBERT GILMORE PONTIUS JR., KASPER KOK

Vegetation. Green mantle of plants that covers the land surface, consisting of assemblages of plant species that, depending on the prevailing ecological conditions of each site or region, range from a small number of species in a simple spatial and structural arrangement, to very complex and highly diverse arrangements. Vegetation is conspicuous; it is absent only in areas that are too dry or cold to sustain it. The flora of a region are the complete range of plant species found there. The world's natural or semi-natural vegetation includes about 400,000 plant species.

The kinds of organisms constituting vegetation are those belonging to the *Plantae* kingdom, of which vascular plants are the most prominent. Vascular plants are those that contain vessels, functioning as a support structure and as a means of distributing the water and nutrients uptaken by the roots to the various plant tissues. Vascular plants can range from giant trees to tiny herbs and include vegetative forms such as woody shrubs, vines, and tall herbs with non-woody stems. Some trees may live for up to a thousand years, while annual herbs may live for only a few months and produce seed to establish future generations. Reproduction can be by seed or by vegetative means. Vascular plants include angiosperms, gymnosperms, ferns, and a few other groups. Angiosperms are, by far, the group with the larger number of species, which reflects their evolutionary success, and include trees, shrubs, and herbs. Gymnosperms are composed of tree and shrub species, whereas ferns are mostly herbaceous.

Each vegetation type is characterized by the growth form of its dominant plants (the largest, most abundant, characteristic plants). Examples of growth forms are annual herbs, broadleaf evergreen trees, or drought deciduous shrubs. Vegetation is also characterized by the architecture of its canopy layers, which together with the prevailing growth forms contributes to its

Desert vegetation in Yuma County, Arizona, March 1942.
PHOTO: Russell Lee. [Courtesy Library of Congress]

outer appearance, or physiognomy. A vegetation type that extends over a large region is called a formation. The prevailing climatic conditions determine the length of the growing season, which corresponds to that part of the year when the biophysical processes that make plants grow and reproduce are active; canopy phenology or the timing of leaf production and loss generally tracks this period. The length of the growing season and the total amount of foliage displayed determine the productivity of the system. Perennial plant species that never lose their leaves are called evergreens, whereas those that lose their leaves by the end of the growing season are called deciduous.

Plants need a substrate, from which they uptake mineral nutrients and water. They also need light and temperatures within their range of tolerance (which may vary widely). Biotic interactions, such as competition or consumption by herbivores, constitute additional constraints to their establishment, growth, and reproduction. Vegetation type distribution and abundance are controlled by the interaction of many biotic and abiotic environmental factors, such as the length of the dry season, soil nitrogen availability, or degree of shading from the canopy of taller plants. For each plant species there are favorable ranges of intensities of each factor, marginal ranges where unfavorable conditions cause stress, and ranges of intolerance along environmental gradients. The interaction of these gradients determines the potential ecological range of each vegetation type.

Temporal change is ubiquitous in vegetation. Regular changes, correlated with seasons, include the sequence of annual or other periodic variations that are part of the normal cycle of growth. These phenological changes do not correspond to abrupt changes in species composition. A different type of change is that resulting from sequential replacement of the dominant set of plant species by a different set. These sequential replacements, known as plant succession, correspond to the unfolding of the life history of those plants having propagules available at the site. In theory, grasses establish first, and then longer-lived shrub species dominate, followed by several types of trees, culminating in an equilibrium community dominated by shade-tolerant species. However, for a given set of environmental conditions the sequence of establishments is a function of initial floristic composition, differential dispersal, growth, and survival. The occurrence of disturbances, such as land clearing and **fire**, can interfere with succession, promoting alternative successional pathways and maintaining vegetation at a given successional stage. For example, savanna systems in sub-Saharan Africa are maintained by the periodic burning of woody vegetation, which would otherwise establish and become dominant.

Given appropriate availability of water, temperature, and light, plants synthesize their food materials through photosynthesis: the process by which chlorophyll traps light energy to produce sugars from carbon dioxide and water. Carbon dioxide is drawn from the atmosphere through pores in leaves and stems—the stomata—which are open when water is available for transpiration, and are shut otherwise, halting the process. Transpiration is the process by which leaf temperature is decreased, as a large amount of energy is required to convert liquid water to vapor, resulting in evaporative cooling. Plant green tissues are thus directly involved in transfers of energy, carbon, and water between the atmosphere and terrestrial ecosystems. The primary products of photosynthesis are converted into carbohydrates, fats, and proteins through other plant biochemical processes, and the surplus that is not consumed in plant biochemical activities (respiration) is accumulated in seeds, fruits, and vegetative tissues. Mass accumulation by photosynthesis in ecosystems is called primary production. **Biomass** accumulation also depends on the ability of some plant species to fixate nitrogen in their roots by a symbiotic association with specific bacteria. Nitrogen is an essential nutrient needed by all plants in relatively large quantities and is not available from the bedrock underlying soil formation. Once the nitrogen-fixing plants incorporate atmospheric nitrogen into their own system, it can be made available to other plants through decay. The leaves, stems, and trunks that constitute plant canopies also play a physical role in the energy budget of their ecosystems because, depending on their architecture and density, canopies intercept light and water, also affecting wind patterns. The water trapped in plant canopies is evaporated back into the atmosphere and thus is not used to replenish soil reservoirs; similarly in some tropical forest ecosystems almost no photosynthetically active light reaches the ground. Vegetation has a major role in ecological and biogeochemical processes: it modulates **atmosphere-land interlinkages**, produces the energy that feeds all organisms in the trophic chain, protects and improves soil quality, and is the matrix where many other terrestrial species live. Photosynthesis has been identified as the most important chemical process in the biosphere, because it provides energy used by all living organisms. It also supplied the energy now stored in the fossil fuels used in industrial societies.

Given its structural, functional, and ecological roles, vegetation is commonly used to denominate terrestrial ecosystems or **biomes**, in association with climate. For example, one broad classification of the world biomes is that of tropical forests, temperate forests, cold forests, forest/shrub/grassland combinations, **grassland**, tundra, and sparse or no vegetation.

Even though **climate impacts** on vegetation have been known for a long time, awareness of vegetative **feedbacks** on climate is recent. In addition to evaporation and transpiration of water, reflection and absorption of incoming radiation, and transfer of momentum to/from the lower boundary of the atmosphere, recent studies have shown that the relative amounts of vegetation and the spatial patterns of vegetated/bare areas can affect rainfall patterns. Vegetation also plays a central role in the **carbon cycle**. Several studies have shown that in the last two decades there has been a net carbon dioxide uptake in the northern and tropical biomes (despite tropical deforestation). This is a consequence of anthropogenic activities that increase atmospheric carbon dioxide concentrations, which in turn drive higher photosynthetic activity and net primary productivity in some terrestrial ecosystems. The role of vegetation as a carbon sink, however, depends on the transfer of carbon to forms with long residence time (e.g., wood), which in turn depends on appropriate land-management practices. Vegetation is also a source of carbon, and eventually all carbon fixed by photosynthesis is returned to the atmosphere. However, the return rate can vary from almost instantaneous (**deforestation** and biomass-burning) to slow (heterotrophic respiration by decomposers). The net carbon dioxide release due to land-use change during the 1980s (mainly tropical deforestation and **conversion** of natural vegetation to **agriculture**) has been estimated as 0.6–2.5 Pg C/yr.

Humans have used and manipulated vegetation cover since ancient times. Activities such as harvesting, hunting, sheltering, food growing, and **cattle ranching** involve various, ever more sophisticated processes of land clearing and intervention. There is virtually no vegetation type on earth that has not been somewhat influenced by anthropogenic activities, and in many parts of the developed world there are no remains of the potential natural vegetation types. Concomitantly, increases of population density in developing countries are associated with **agricultural intensification** and occupation of increasingly hazardous areas. Clearing of natural and semi-natural vegetation increases the risk of problems associated with **land degradation**, such as **soil erosion** and nutrient depletion from soils, which may determine a decrease in the capability of the environment to support adequate levels of food production while hampering other environmentally dependent mechanisms of raising income.

Vegetation has always been central in land-cover mapping. Its role in ecosystems and its ability to function as an indicator of the prevailing environmental conditions determined that different land-cover and land-use classification schemes always used vegetation as a main reference (presence/absence, condition, type, or use). The most traditional classification schemes rely on vegetation physiognomy or on its use by humans. For example, **forest** is a vegetation type dominated by tall woody plants (above six meters), and scrub is dominated by woody plants less than six meters tall, whereas classes such as irrigated agriculture or dryland agriculture use related classes. More recently, the use of **remote sensing** for mapping and **monitoring** land cover led to the development of new land-cover classification systems, such as the **Land Cover Classification System** (LCCS) used by the **United Nations Food and Agriculture Organization** (FAO).

See also: **Boreal Zone; Coastal Zone; Conservation; Disequilibrium Dynamics; Ecological Colonization; Fallow; Food Security; Holocene; Human Appropriation of Net Primary Production; Mountains; Nutrient Cycle; Pasture; Regrowth; Savannization; Secondary Vegetation; Taiga; Tropical Humid Forest; Tropical Dry Forest; Wetlands.**

Further Reading

Aber, John D., and Jerry M. Melillo, *Terrestrial Ecosystems*, 2nd ed., Oxford, UK: Academic Press, 2001; Breckle, Siegmar-Walter, *Walter's Vegetation of the Earth*, Berlin: Springer, 2001.

MARIA JOSÉ P. DE VASCONCELOS

Vulnerability. The extent to which changes both within and outside an entity—acting alone or in combination—predispose it to harms over time. A concept considered capable of bringing global change research to focus, vulnerability is a subject of diverse interpretations bandied about that are fueled by differences from many sources. A diversity of terminology has been enlisted to define and clarify its polyphony of meaning and implications (evidenced by its plural form "vulnerabilities") and its proper operationalization (including measurement), ranging from vocabularies from hazard, risk, and **food security** literatures where the study of vulnerability finds its root, to terms from systems ecology. Vulnerability defined in these terms varies widely, including a synonym for **poverty**, marginalization, fragility, or brittleness; a function of different combinations of exposure, sensitivity, resistance/robustness, irreversibility, coping, and so on; a relative state along a continuum from **resilience** to susceptibility; and an error term to account for residual consequences after adaptation has been taken. Some of these terms tend to illuminate and strain against more precise definitions at once, owing simultaneously to semantic baggages from mother disciplines, positivist overtones, and their equally (if not more) complex meanings. Consequently, the very term "vulnerability" itself has become vulnerable to misuse, even abuse.

Ongoing debate surrounding vulnerability points to a wide variety of contributing factors operating at different **scales** and in different contexts, which may be woven into different "explanatory frameworks" falling along several analytical axes. These variant frameworks have significant bearing on alternative answers to three interrelated questions: what (who) is vulnerable, and to what? How does a vulnerable state come about? Where are the solutions?

The "what" question highlights differences in understanding the connections between social and biophysical dimensions of vulnerability, and the extent to which exposure alone can be used to reliably measure vulnerability. The problem of specifying what is vulnerable often arises when we consider the variegated forms vulnerable conditions take: degradation, marginalization, deprivation, instability, etc. "Vulnerable" in one condition does not necessarily mean vulnerable in the other. Although vulnerability is frequently used to describe attributes of different systems at different levels, its mechanisms and implications in each setting are usually different. Vulnerability of parts does not necessarily translate into vulnerability of the whole, and vice versa (e.g., artificial flooding in flood control). The reliability of using exposure (meaning increased contact with environmental hazards) as a surrogate indicator of vulnerabilities also needs to be considered here. In a given place, vulnerability to one type of hazard does not mean vulnerability to the other; then how about multi-hazards? Besides, exposure unit is difficult to define due to the notorious boundary problem that impedes separation between **endogenous factors** and extraneous or **exogenous factors**. Cascades of relations often render upstream causes of exposure difficult to trace. Moreover, the biophysical and social dimensions of vulnerability rarely correspond in a given place as the concept of exposure might suggest: the most vulnerable people do not necessarily live in the most vulnerable regions.

The "how" question underscores the local and historical configuration of vulnerability and its structural shaping forces. One definitional approach is to take exposure to a harmful event as given and search for explanations of differential consequences. In this vein, social vulnerability is seen to have two sides: an external side of shocks and stresses to which a system is subject (exposure) and an internal side—a lack of coping capability (a function of resistance, and the rate of recovery) that is largely determined by socio-economic structure anchored in **property** relations (e.g., architecture of entitlements). Those focusing on "social construction of vulnerability" even take exposure formally out of the definition of vulnerability. Some structural explanations emphasize configuration of contextual factors of a geographic place in shaping the **syndrome** of vulnerabilities, while others see vulnerability of complexity as intrinsic to technological societies; yet others, seeing the need to distinguish complexity (large number of fully distinct elements and their spatial distribution) and complicatedness (many interlocking but structurally similar parts), attribute vulnerability to the very fabric of modern industrial societies that, in contrast to the evolutionary tendency of biological systems to become complex, have replaced structural diversity and complexity with structurally similar parts (e.g., **monocultures**) arranged in very complicated patterns. It is this complicatedness that builds industrial efficiency that also creates vulnerability to failure in that the more centralized structures have the ability to transfer both positive and negative repercussions over border areas and lower the margins of tolerance. When temporal dimension is brought in, adaptability is emphasized, where two research paths diverge: one projective, interpreting vulnerability as a residual variable to be measured by the adverse consequences remaining after the processes of adaptation has taken place; another reconstructive, attempting to use historical analogs to profile local constraints that limit the capacity to adapt. The distinction between chronical and current vulnerability as well as the notion of "windows of vulnerability" are all relevant here; the latter sees heightened vulnerability arising from conjunction of circumstances. Special note should be made about the interactions among **institutions**, nature, and society with the ultimate "pathology" unfolding in the forms of institutional rigidity, ecological brittleness (lower margin of tolerance), and resulting societal dependencies and vulnerabilities.

The "where" question brings to the fore the differences in the identification of "winners and losers," the values placed on the differential consequences and their **tradeoffs**, the distributions of responsibilities based on appraisal of these distributed consequences, and the level of uncertainties inherent to all of the above. Clearly, the burdens of vulnerabilities are not distributed evenly across places and social scales, hence the notion of differential vulnerabilities. Tangible or intangible "harms" is a value-laden term that means different things to different people, as studies of discourse of social vulnerability illustrate. Also, harms that are certain to happen are different from potential harms to be avoided. Irreversible loss of land and culture causes much more harm to people displaced from a dam building than the potential harm avoided by the dam. For an individual household, vulnerability to job loss is very different from vulnerability to loss of entitlement to a social safety net. Lessening vulnerability of a social system is not necessarily equivalent to the enhanced level of security for the social groupings and the ecological systems residing within its boundary, whose delineation is a matter of uncertainties. For technological infrastructures, equal amounts of protection from hazardous events do not ensure equal capacities to recover nor equal distribution of benefits for different groups of end users, contingent upon such systemic attributes as criticality, substitutability, connectedness, and dependency. Difficulties of stratifying degrees of vulnerabilities across social and spatial scale are exacerbated by the often-neglected normative dimension of vulnerability. Reducing vulnerability to changes does not mean maintaining status quo, or defending a vulnerable (read as rigid, unfair, etc.) establishment, in the same sense that the sustainability (of biosphere) should not be used to justify sustainability of everything. In addition, redundancy and diversity/diversification (in both the cultural and biological sense) that are seen as conducive to sustainability have to be achieved at the expense of efficiency. All these issues will enter into policy debate and bear on the critical question of "for whom or for what are the priorities in designing responses?"

The three questions posed above all point to the importance of scale of observation and the level of analysis in understanding vulnerability in its finer grain. Both need to be fine-tuned to correspond to the appropriate level where policy leverages can be applied to achieve their goals. A rarely clarified question is how vulnerability is linked with sustainability. Arguably, vulnerability is a variable state essentially intrinsic to all evolving, complex systems, while sustainability is not. Tough tradeoffs must be made between higher and lower levels, and between global and local interests in sustainability transition, often meaning (if not prescribing) increased level of vulnerability of different systemic units to changes at different junctures.

See also: **Biodiversity; Cascades; Driving Forces; Feedback; Hot Spot Identification; Human Immunodeficiency Virus (HIV)/Acquired Immunodeficiency Syndrome (AIDS); People at Risk; Proximate Causes; Threshold.**

Further Reading

Turner, B. L., II, Kasperson, Roger E., Matson, Pamela A., McCarthy, James J., Corell, Robert W., Christensen, Lindsey, Eckley, Noelle, et al., "A Framework for Vulnerability Analysis in Sustainability Science," *Proceedings of the National Academy of Sciences of the United States* 100 (2003): 8074–9.

KE CHEN

Water-Land Interlinkages. On its transit overland from the **watershed** to the **coastal zone**, water is exposed to the properties of the terrestrial surface, which, in conjunction with land-use and **land-cover change**, are important determinants for both water quantity (i.e., sufficient supply of freshwater to support human and natural systems) and water quality (i.e., suitability of supply for an intended use). It is well recognized that direct land changes such as **deforestation** and **reforestation**, cropland change (such as **agricultural intensification**), **mineral extraction**, and **urbanization**—but also indirect effects stemming from underlying **driving forces**—hold considerable potential to significantly modify or even disrupt hydrological cycles.

Overview

The recent intensification of agriculture, in particular, has its largest impacts on both freshwater and marine ecosystems, which are and continue to be greatly eutrophied by high rates of nitrogen and phosphorus release from agricultural fields. Aquatic nutrient **eutrophication** triggers the growth of blue-green algae that renders water unpalatable and increases the growth of weeds. It increases the turbidity of water and links to several other types of impacts: it can lead to shifts in the structure of food chains, including fish killings, outbreaks of nuisance species, loss of **biodiversity**, and negative

Flooding of low-lying areas by extreme high tides at Holland Cliffs Shore on mid-Patuxent River, Maryland, in fall 1993.

PHOTO: Mary Hollinger. CREDITS: NOAA Central Library and the Office of the CIO/HPCC (America's coastline collection).

health impacts. For example, water shortages already exist in many regions of the world (with more than one billion people without adequate drinking water, and thus consequences), and 90 percent of the infectious diseases in developing countries are transmitted from polluted water.

It has early been estimated that, on a global scale, withdrawals (i.e., water removed from a source and used for human needs) from streams, rivers, and aquifers, combined with instream flow requirements (altogether totaling 6,780 cubic kilometers per year), already account for more than 50 percent of total accessible runoff. Claims have arisen that the human impact on the terrestrial water cycle during the last fifty years (actually dating back to about 4,000 years with water engineering in association with cropland expansion) has likely exceeded natural forcings of continental aquatic systems in many parts of the world.

Several interconnected properties intervene into the relationships shaping the impact of land-use/cover change upon freshwater hydrology. First, the life history of water on a landscape encompasses its appearance already in precipitation until its exit to the ocean. Indeed, any land-use decision very often turns out to be a water decision, be it the **conversion** of coastal marshes into **agriculture**, settlements or industrial zones, or the **modification** of farmland through the development of freshwater resources for irrigation farming. In a broader earth system perspective, responses to these influences are already discernible to reverberate through the hydrological cycle, which go well beyond the direct human appropriation of freshwater and coastal zones for drinking, agriculture, and industry.

Various **tradeoffs** need to be addressed between the potential benefits of land-use/cover change and the potentially negative consequences upon the hydrological cycle. Clearly, agricultural ecosystems have become incredibly good at producing food, and it has been the irrigation of croplands that

Surrounded by desert and steppe, the Aral Sea (Kazakhstan, Uzbekistan) is still the fourth largest lake in the world (since the development of agricultural irrigation projects in the late 1950s, the sea has shrunk drastically and two islands can be seen emerging in July 1985).

PHOTO: Space Shuttle. CREDITS: Image Science and Analysis Laboratory, NASA-Johnson Space Center, 18 Mar. 2005 (Earth from Space—Image Information). http://earth.jsc.nasa.gov/sseop.efs

contributed enormously to **food security**. Currently, around 40 percent of all agricultural production comes from irrigated areas, and global food production foreseeably becomes largely dependent on artificial irrigation systems. On the other hand, current agricultural practices involve deliberately maintaining ecosystems in a highly simplified, disturbed, and nutrient-rich state—with threats to biodiversity and the supply of water for food production. Especially in irrigated dryland zones of the world, the negative hydrological consequences of land-use/cover change are most pressing. They illustrate modern or industrial society's capacity to transform large coupled human-environment systems rapidly such as in the case of the Aral Sea basin (see image above), or the collapse even of ancient societies. In any case, and not only in drylands, the increased yields of food production have environmental costs that cannot be ignored, especially if the rates of nitrogen and phosphorus triple, and the amount of irrigated land doubles in the coming decades. Not included in tradeoff considerations so far are the large social costs associated with social disruptions due to dam construction (and population displacement) and other large-scale water infrastructure constructions, particularly along transnational rivers in dryland zones of the world.

Many insights into the hydrological consequences of land-use/cover change stem from the experimental manipulation of land cover at rather small spatial, observable scales such as research plots, hill slopes, and small catchment areas (e.g., 100–1,000 hectares). These manipulations prove that human activities can modify or disrupt interception losses by different plant species, soil infiltration, storm runoff, water yields, flood peaks, evapotranspiration rates, concentration of water-quality constituents, snow accumulation, and snow melting. However, extrapolating from such studies to larger systems such as **river basins** is confounded by the diversity of **land-use systems** as well as hydrological systems.

There is an unequal distribution of water resources or hydrological systems, for which the demand likewise varies greatly. Regions in which water quantity is particularly stressed by human demands are western North America; areas bordering the Sahara Desert; the Arabian peninsula; and several densely populated zones in Asia, namely India, Pakistan, and northeastern **China**.

The dominance of different processes changes at different **scales**. For example, land-use/cover changes in the upstream of a catchment may have a different impact on hydrology than changes downstream; and processes of interception, infiltration, and storage dominate at the plot scale; while channel processes assume a greater role with increasing catchment size. Only few examples exist for controlled long-term studies of the impacts of permanent land conversions at multiple scales such as **forest** to agriculture or agriculture to urban cover.

The impact of land-use/cover change also varies in terms of time scale. On the scale of catchments, changes usually occur at irregular-time intervals, while crop planting, drainage for afforestation, and other changes at the plot scale occur fairly regularly. Furthermore, river and lake quality can be restored in quite a short time, while destroyed biodiversity can take several thousands of years to recover to the original condition. Especially in the case of sediments, even after a "beneficial" land-use change, there will be still enough sediment in the system from prior human-induced or natural erosion that would lead to increased sediment loads in the rivers. In most of the cases, the time taken to contaminate a system is only a fraction of the time required to later clean up the same system. Remedies against high loadings of pathogens as a result of large population centers, for example, can be effective within less than one year, while eutrophication and micro-pollutants may contaminate the system for up to 100 years. Agrochemicals, in particular, may have a large impact over a long period of time. Also, mining and other sources of suspended load may have an impact on the ecosystem over many years.

Identifying a linkage between land-use and hydrological change often implies isolating the impact of biophysical forces and climatic variability, in particular. There is some reason to assume that climate change and long-term climatic fluctuations are particularly inherent to extreme weather events such as flooding and droughts. Likewise, human interventions at the micro-scale appear easily possible and are well documented, but the change in flood peaks, sediment load, and base flow at the large scale becomes much stronger when dominated by natural processes. Also, distinguishing the impact of land-use/cover change on hydrology from the impact of climatic variability is more difficult at the catchment scale than at the plot level.

Forest Zones

In **forest** zones, land clearance through deforestation—with the subsequent loss of the surface organic layer and decline in soil organic matter—increases overland stream flow through decreased evapotranspiration as well as raindrop detachment of soil particles, sheet erosion, rill erosion, gullying, and downstream sedimentation, though the latter processes are often episodic. The hydrological consequences following deforestation, however, are highly variable, and depend upon a wide array of host factors. At the watershed scale, forest clearing generally results in a significant increase in annual water yield, but generalization across different streamflow response measures remains difficult. On deforested slopes in particularly steep terrain, rates of erosion are maximized if the terrain is subsequently subjected to intensive cropping, **fire**, or both. Barren wastelands, unable to support any **vegetation**, can be the end result of this process, and the foothills north of Mexico City are an example of this. In forest zones where either revegetation occurs rapidly or **secondary vegetation** and **regrowth** are part of the land-use cycle, the effects of forest clearing on hydrological cycles are transient and less extreme. It has been noted though that geological conditions can often override the effects of land-use/cover change.

Reforestation (or afforestation) is seen to help reverse the hydrological responses to deforestation. Complete reversal of trends, however, depends upon the restoration of both vegetation and soil properties that had been characteristic of native forests within a particular climate on a particular parent material. In temperate climates, for example, hydrological processes are restored quite gradually, given the slow speed of soil development there. It is estimated that old-growth forests help to completely recover water flows to original levels only after about 150 years past disturbance. There appears to be a clear link between forests and the quality of water, a much more sporadic link between forests and the availability of water quantity, and a variable link only between forests and the constancy of flow at the catchment level.

At the scale of river basins, modeling studies for the Amazon basin suggest that complete conversion of the rainforest to degraded **pasture** would cause annually a decrease of 26 percent in mean precipitation, 30 percent in evapotranspiration, and 18 percent in runoff; while other studies have produced similar, if less dramatic, responses to deforestation. It has been shown that changes in riparian forest and vegetation have a major impact on the instream biota as well as on the pollutants entering the stream or river stretch.

There is little scientific evidence for the largest, most damaging flood events being caused by deforestation at the global scale. Studies in **South America**, **South Africa**, and in the **Himalayas** indicate that the increase in infiltration capacity of forested lands over non-forested lands is insufficient to influence major downstream flooding events. Rather, the intensity, amount, and spatial distribution of rainfall appear to be the key elements determining the extent and magnitude of damage caused by such disasters, with local geology, land use, and topography being important, concomitant factors.

Croplands

Agriculture is the largest consumer of water by humans worldwide (nearly 85 percent of total human consumptive use), and land-use practices associated with agricultural intensification have been identified to exert an impact on hydrological cycles, especially where improper cultivation techniques were applied in environments with high natural variability. In the **United States of America**—where per capita withdrawals were about 1,700 cubic meters in 1995—crop irrigation accounts for more than 40 percent of the withdrawn freshwater on average, but the figure is as high as 85 percent in California, where agriculture accounts for only 3 percent of the state's economic production. In Africa—where reported water uses range from approximately 600–800 m^3/person/year in Egypt, Libya, the Sudan, and some other countries to under 20 m^3/person/year in the poorest countries of the continent—as much as 90 percent (or more) of reported water uses go to agriculture. As a matter of fact, large countries that produce grain in **monocultures**—such as **Canada**, the United States, Argentina, and Australia—all have significantly higher per capita water uses than average. In another group of large countries with extensive irrigation—such as India and China—agricultural water use numbers approximately 90 percent.

The central arguments for water-land interlinkages in croplands are that, first, those agricultural practices that can retard **soil erosion** are practices that will also increase infiltration, thus reducing and delaying surface runoff (eroded soils absorb 87 percent less water through infiltration than uneroded soils); and, second, that most problems such as **salinization** are not problems at all with rainfed crops, because the soils are naturally washed away, for instance. In the U.S. southern **Great Plains**, for example, the paucity of farm-level conservation strategies—combined with a period of extreme drought and the Great Depression—caused the Dust Bowl of the 1930s, during which two to twelve inches of topsoil were removed by wind and water erosion and hundreds of thousands of farming households were deprived of their **economic livelihood**. Most of the region eventually recovered from the Dust Bowl episode. For example, changes in streamflows and sediment yields were observed and simulated for several watersheds in Texas, and both observational data and modeling results indicated a significant decline in erosion and **siltation** (reservoir sedimentation) during the period 1910–1984—attributable to the combined effects of conversion from rural to urban land, changes in agricultural crops (i.e., replacement of **cotton** by **wheat** and sorghum), and the implementation of soil and water **conservation** measures beginning in the 1940s (i.e., terracing, contour plowing, strip cropping, and no-till cropping).

In the high-intensity agricultural production zones across the world, the use of commercial fertilizers—containing primarily nitrogen, phosphorus, and potassium—bear several effects on water quality, extending into stream chemistry across both watershed zones and coastal systems. Indeed, there is a direct and quantitative link between the amounts of nitrogen in the major rivers of the world and the magnitude of agricultural nitrogen inputs to their watersheds. In particular, small but severely disturbed agricultural areas such as animal feedlots export about 100 times more nitrogen and phosphorus than other types of agricultural land use. One of the major (unintended) consequences of land-use intensification has been the contamination of shallow groundwater with nitrates, especially due to an explosive growth in fertilizer application following the end of World War II. Nitrate concentrations of 10–20 mg NO_3-N/liter are frequently observed in most shallow aquifers of agricultural areas in North America, for example. Due to the tendency of nitrate concentrations greater than ten milligrams to cause methemoglobinemia in infants and to form carcinogenic nitrosamines in the human intestine, these waters are undrinkable. In the United States, about 40 percent of water is deemed unfit for drinking or recreational use due to contamination by microorganisms, pesticides, and fertilizers; and more than 76 million Americans are infected (and 5,000 die per year) as a result of pathogenic *Escherichia coli* (E. coli) and related foodborne pathogens. In recent decades, more U.S. livestock production systems have moved closer to urban areas, contaminating water and food with manure there.

It has been noted that the effect of decreasing forest cover and increasing agriculture is not linear. As agriculture expands to more than 70 percent of land cover, the transformation of the last remaining normal landscape traps for NO_3 to agriculture—such as **wetlands** and riparian forests—makes NO_3 concentrations in streams rise exponentially. For example, in an estimation of the biogeochemical effects of land-use change in the Choptank basin of the Chesapeake Bay area in the eastern United States over the last 150 years, it was found that conversion of forest to agriculture in the eighteenth and nineteenth centuries increased nitrogen and phosphorus by a factor of two, but application of fertilizers in the twentieth century resulted in about a fivefold increase in nitrogen. Modeling results suggest that not all NO_3 in the groundwater may appear in the base flow, with hydric watershed soils driving denitrification by as much as 80 percent below the expected concentrations based on land use.

Drylands

Irrigation farming, or the consumptive and nonrecoverable use of water by irrigated crops, is a major component of the water balance at many scales, and the availability of freshwater is a key factor for intensification and expansion of agriculture in drylands, in particular. The hydraulic control on previously natural river systems—with damming, water extraction, and redirection of flows as the most important engineering works—alters the behavior of rivers immediately and exerts an array of impacts reverberating through the coupled land-water system.

Soil salinization is one of the problems associated with irrigation farming. It has been estimated that about half of all existing irrigated soils worldwide are adversely affected by salinization, and that the amount of world agricultural land destroyed by salinized soils is 10 million hectares. Another problem associated with crop irrigation is waterlogging, which means that—in the absence of adequate drainage—water levels rise in the upper soil levels, including the plant root zone, and crop growth is impaired. Such irrigated fields are sometimes called "wet deserts" because they are rendered unproductive. In India, for example, waterlogging adversely affects 8.5 million hectares of cropland and results in the loss of as much as 2 million hectares to grain every year.

In most developed countries, the total area of irrigated land has been relatively stable over the last decade, but irrigation has remained a major offstream use of both surface and groundwater resources. In the western United States, for example, groundwater withdrawals for irrigation have been among the most significant impacts of agricultural activities, contributing to dramatic increases in evapotranspiration, excessive declines in water tables, surface subsidence, and soil salinization. Following recoverage from the Dust Bowl episode, changing regional and national economies have promoted substantial pumping of groundwater for crop irrigation. Globally, groundwater aquifers provide an estimated 23 percent of water per year available for sustainable use—with approximately 60 percent of the water intended for crop irrigation never reaching the crop due to water losses during pumping and transport—while the United States relies disproportionately on water pumped from aquifers (i.e., 65 percent). The capacity of the large Ogalla aquifer, for example, which underlies parts of Nebraska, South Dakota, Colorado, Kansas, Oklahoma, New Mexico, and Texas, has decreased 33 percent since about 1950, and water withdrawal is three times faster than its recharge rate. This continues to trigger decreasing water levels and subsequent abandonment of agricultural land, thus raising new concerns about the long-term sustainability of cultivation in the wider region.

The abandonment of irrigated agricultural land—due to soil and water degradation (but also as a consequence of shifting priorities in water allocation)—has been identified as a common feature of land-use irrigation zones of arid and semiarid regions across the world. In the United States, about 150,000 hectares of agricultural land have already been abandoned solely because of high pumping costs. Despite **land stability** in irrigation farming, depleted freshwater resources and land-use legacies have been playing together in the western United States at some locations in producing an abundance of opportunistic shrubs and non-native annual plants as well as lower species diversity than lands that were never cultivated.

In contrast, developing countries that hold large dryland zones—such as Central Asia or Middle Eastern countries along the Euphrates River—have been experiencing an explosive growth in irrigation over the last decade, often driven by huge government-funded water engineering projects. As in developed countries, groundwater withdrawal, declining water tables, and subsequent **land abandonment** are common. In the agriculturally productive Chenaran plain in northeastern Iran, for example, the water table has been declining by 2.8 meters annually since the late 1990s, likewise in Guanajuato in Mexico by as much as 3.3 meters per year. In countries like Turkey and Syria, a pattern of reallocation of land and water use has been typical, that is, irrigated lands along river bottoms and floodplains got abandoned for upland sites, due to a myriad of factors including groundwater depletion, reallocation for surface water, salinization, and waterlogging. In Central Asia, the impacts of land-use/cover change upon the hydrological cycle in low-lying ecosystems over the last 300 years are linked to a typical pathway of water and **land-use transition**. The transition more or less paraphrases those impacts that are typical for the transformation from a predominantly rural mode to a largely industrialized mode of farming and society.

Large hydraulic structures for river regulation are often designed to support the extension of irrigation farming—besides their many other purposes, e.g., for electricity generation, domestic water supply, and flood control. Examples are dams, reservoirs, diversions, levees, artificial channels, and detention/retention ponds. Only 23 percent of the flow in 139 of the largest rivers in the Northern Hemisphere remains unaffected by reservoirs, and the equivalent of 40 percent of the total global runoff to the oceans is intercepted by large dams. There is growing evidence that land-use changes associated with the introduction of hydraulic control of especially large river systems in drylands contribute to rapid water degradation, disruption of hydrological cycles, and the partly irreversible collapse even of regional human-environment systems such as the Aral Sea basin through processes of **desertification**. There is further evidence that some of the responses reverberate through the coupled land-water cycle, strongly impacting hydrological conditions through changes in the partitioning of incoming solar radiation between evapotranspiration and sensible heat, which in turn affect the amount of water that runs off into riverine systems or infiltrates into soil.

In the Lake Chad basin of northern Africa, for example, long-term decreases in lake area, lake level, and river discharges were primarily attributed to climatic variations (rainfall has been declining since the 1960s), but increases in water losses from rapidly growing irrigation explained a large proportion of the variation. Likewise, a decline in potential evapotranspiration in a regional-scale water development project of arid, southeastern Turkey could be attributed to increasing irrigated land area, playing together with decreasing wind speed and increasing atmospheric humidity. From a **paleo perspective**, records for both regions prove that pumping for crop irrigation means the mining of fossil water, because groundwater reservoirs have not been fully recharged for thousands of years. Worldwide, the Nile and Syr-Darya rivers are among the most heavily regulated rivers, partly demonstrating drastic changes in coupled land-water cycles. The Nile, after the erection of the Aswan High Dam in 1968, shows reduced overall discharge, truncated peak flows, higher lower flows, and a seasonal shift in the timing of the natural hydrograph. In the case of the Syr-Darya River, the progressive losses of discharge are associated with expanded water use for irrigation and the contraction of the Aral Sea. In the Aral Sea basin, water from the two rivers entering the inland sea was diverted from the 1960s onward, and led to an enormous decrease in the area and volume of the sea over the following thirty-five years, that is, from 6.7×10^4 km^2 to 3.2×10^4 km^2 and 1,064 km^2 to 310 km^2, respectively. Most studies anticipate its complete disappearance within the next twenty-five years. Associated with the hydrological consequences of land-use/cover change in the sea basin have been an array of ecological influences reverberating in the coupled land-water-atmosphere complex (e.g., surface temperature and local climate changes) as well as an array of social influences on economic livelihoods (e.g., decline in agricultural productivity due to salinization and water logging) and human health.

Owing to impoundments, still waters (i.e., the standing stock of river channel water) have increased worldwide by more than 700 percent relative to the natural state. The consequences of such water aging for associated material transport are to trap a substantial proportion of the incoming suspended sediments and to modify the concentration of dissolved components of nitrogen, phosphorus, and silicon. It has been estimated that the current registered 45,000 largest reservoirs (i.e., greater than 0.5 km^3 storage capacity) trap nearly 30 percent of global sediment flux destined for the ocean, and that this estimate of sediment retention rises further with inclusion of about 800,000 smaller impoundments—and, of course, with continued dam construction.

Urban Zones

The ramifications of land transformation toward urban cover and its impact on the hydrological cycle have been qualitatively described for various stages of urbanization. They include decreases in transpiration from loss of vegetation, decreases in infiltration due to decreased perviousness associated with urban development (streets, roofs, sidewalks, parking lots, etc.), increases in storm runoff volumes, increases in flood peaks, declines in water quality from discharges of sanitary wastes to local streams and rivers, and reductions in baseflow.

Both empirical analyses and modeling results suggest that urbanization and **suburbanization** increase stream flow through increased runoff (and thus flood potential), but also decrease water quality when the amount of impervious surface in a watershed exceeds 10–15 percent of the total land cover. In catchments, the degree of impervious areas in catchments is often directly related to the size of floods. Several studies point to the fact that increased nitrate-nitrogen exports across river basins (such as that of the Mississippi River) can be associated with the amount of developed land there. Agricultural intensification and the disposal of human waste in septic systems are clearly the principal causes of elevated NO_3 in groundwater, and contribute to the eutrophication of aquatic ecosystems, besides various consequences for **human health**.

See also: **Atmosphere-Land Interlinkages; Climate Impacts; Decision-Making; Degradation Narrative; Ecosystem Services; Environmental Change; LandSHIFT Model; Nutrient Cycle; People at Risk; Vulnerability.**

Further Reading

Eshleman, Keith N., "Hydrological Consequences of Land Use Change: A Review of the State-of-the-Science," in *Ecosystems and Land Use Change* (Geophysical Monograph no. 153), eds. Ruth DeFries, Gregory Asner, and Richard Houghton, Washington, DC: American Geophysical Union, 2004, pp. 13–29; Gleick, Peter H., "Water Use," *Annual Review of Environment and Resources* 28 (2003): 275–314; Kabat, Pavel, Claussen, Martin, Dirmeyer, Paul A., Gash, John H. C., Bravo de Guenni, Lelys, Meybeck, Michel, Pielke Sr., Roger A., Vörösmarty, Charles J., Hutjes, Ronald W. A., and Sabine Lütkemeier, eds., *Vegetation, Water, Humans and the Climate: A New Perspective on an Interactive System*, Berlin: Springer, 2004; Meybeck, Michel, and Charles Vörösmarty, "Human-Driven Changes to Continental Aquatic Systems," in *Global Change and the Earth System: A Planet under Pressure* (The IGBP Series), eds. Will Steffen, Angelina Sanderson, Peter D. Tyson, Jill Jäger, Pamela A. Matson, Berrien Moore III, Frank Oldfield et al., Berlin: Springer, 2004, pp. 112–3; Mustard, John F., and Thomas R. Fisher, "Land Use and Hydrology," in *Land Change Science: Observing, Monitoring and Understanding Trajectories of Change on the Earth's Surface* (Remote Sensing and Digital Image Processing no. 6), eds. Garik Gutman, Anthony C. Janetos, Christopher O. Justice, Emilio M. Moran, John F. Mustard, Ronald R. Rindfuss, David Skole, B.L. Turner II, and Mark A. Cochrane,

Dordrecht, NL: Kluwer Academic Publishers, 2004, pp. 257–76; Pimentel, David, Berger, Bonnie, Filiberto, David, Newton, Michelle, Wolfe, Benjamin, Karabinakis, Elizabeth, Clark, Steven, Poon, Elain, Abbett, Elizabeth, and Sudha Nandagopal, "Water Resources: Agricultural and Environmental Issues," *BioScience* 54 (2004): 909–18.

HELMUT GEIST

Watershed. Boundary defining an area where all water flows into a single drainage basin. In Britain, a watershed traditionally refers to the divide between catchments, where water falling on one side of the divide will flow into a particular **river basin**, and water on the other side will flow into a different river basin. A typical watershed divide is generally located along a **mountain** range or topographical high point. In North America, watershed commonly refers to the area within this boundary, and is defined as the total area where all the water falling within that land flows into the same drainage basin, or the region contributing to the supply of a waterbody. In this latter sense, watershed is used interchangeably with "catchment" or "drainage basin." A watershed can be defined at a range of **scales**, depending on the resolution of interest. The largest watersheds, for example, are the global watersheds that separate land areas draining into the Atlantic, Pacific, and Indian oceans. These watersheds can be further subdivided into continental watersheds separating drainage into various oceans. Watersheds can be defined at progressively finer scales, such as the drainage area for a single river or stream.

As noted by David Thomas and Andrew Goudie, the watershed is the basic unit of landscape hydrology. While watersheds are usually drawn based on map elevation contours, actual drainage is likely to differ due to disparities between surface and subsurface flow. The hydrological cycle acting within a watershed unit drives processes of runoff, **soil erosion**, sedimentation, and evapotranspiration. These processes are influenced by land use within a watershed unit. Dense **vegetation**, for example, will increase water infiltration and decrease runoff. Removal of vegetation, or the presence of urban areas, is associated with increased overland flow and increased risk of **soil erosion** and flooding. Similarly, watershed topology drives climate, waterflow, and sediment transport, and thus influences the natural and potential land covers and land uses for a given catchment region.

Characteristics and processes of land cover and land use are driven, to a large extent, by hydrological systems; this makes the watershed an important unit of study for land-use/cover analyses. As noted by William Cunningham and colleagues, a watershed is a natural boundary that provides an excellent basis for analysis of land-cover processes; land and water systems can be integrated easily, downstream effects can be conceptualized, and the watershed provides an excellent unit for management planning. The nested nature of watersheds (i.e., the ability to select a well-defined watershed unit at a variety of scales) means that various resolutions of land-cover processes can be assessed, understood, and integrated using the same conceptual unit of analysis. Dennis Soden and Brent Steel add that watersheds are relatively well-defined, rational boundaries that can be easily understood. As a result, the watershed has emerged as a popular unit of analysis for land-cover and land-use research, management, and planning.

See also: **Eutrophication; Land-Use Planning; Mineral Extraction; Mountains; Reforestation; Regrowth; Tradeoffs; Water-Land Interlinkages.**

Further Reading

Cunningham, William P., Ball, Terence, Cooper, Terence, Gorham, Eville, Hepworth, Malcolm T., and Alfred A. Marcus, *Environmental Encyclopedia*, Detroit, MI: Gale Research Inc., 1994; Snelder, Ton H., and Barry J. F. Biggs, "Multiscale River Environment Classification for Water Resources Management," *Journal of the American Water Resources Association* 38 (2002): 1225–39; Soden, Dennis L., and Brent Steel, *Handbook of Global Environmental Policy and Administration*, New York: Marcel Dekker, 1999; Thomas, David S. G., and Andrew Goudie, eds., *The Dictionary of Physical Geography*, 3rd ed., Oxford, UK: Blackwell Publishing, 2000.

LEA BERRANG FORD

West-Central Europe. Western part of a region that is often conceived of as central Europe in a geographic context, comprising the countries of Austria, Belgium, Germany, Luxembourg, the Netherlands, and Switzerland. The region falls within the ecological zone of temperate nemoral climate and is thus characterized by warm summers that hold most of the annual precipitation, and by cool, but not extremely cold, winters. In the coastal lowlands in the north, particularly Belgium, the Netherlands, and northern Germany, the climate is milder and more humid; toward the east the climate is becoming slightly continental. Accordingly, temperate deciduous and mixed **forests** constitute the zonal potential natural **vegetation**. Deviations with respect to climate and potential vegetation occur mainly in the southern part of west-central Europe, which is dominated by the Alps, which reach an elevation above 4,000 meters. With respect to soils, podzols prevail in the northern part, while cambisols and luvisols dominate most of the rest of the region.

West-central Europe covers an area of 550,000 km^2 and accounts for about 11 percent of the total European territory (excluding **Russia** and the Ukraine). In 2000, the total population of west-central Europe was about 124 million, which is about one-quarter of the European total. The west-central European countries are among the economically most advanced, highly industrialized countries in the world. Taken together, they produced about one-third of the total European gross domestic product (GDP) in 2001. The share of urban population in west-central Europe amounts to 85 percent, significantly above the European average (75 percent). West-central Europe includes some of the most densely populated regions in **Europe**. A belt of highly urbanized and industrialized land stretches from Belgium and the Netherlands (average population density above 350 persons/km^2) through the German industrial zone of the *Ruhrgebiet* to southern Germany, and continues south of the Alps into northern Italy.

Land Use and Landscape Patterns

The west-central European landscape is heterogeneous and includes large urban/industrial areas, regions of very intensive and specialized agricultural production, and extensively used landscapes dominated by traditional mixed agriculture, woodlands, and **pastures**. The area used for human settlement, transport infrastructure, and industrial production occupies approximately 15 percent of the total land area, but may exceed 30 percent in the densely populated regions in the north. About half of this urban fabric is actually built-up land covered by artificial surface. Significant areas, predominantly in eastern Germany and the *Ruhrgebiet*, are characterized by vast, often abandoned open pit coal mines and the corresponding industries. Almost half of the total area of west-central Europe is agricultural land. More than 60 percent of this area is used for rainfed crop production; the rest is **grassland**. Agricultural yields are 30–40 percent above the European average due to favorable conditions and highly intensive production. Roughly 30 percent of the total area is covered with woodlands and forests. The wooded areas are concentrated in the hilly and mountainous regions in the southern and central part of the region, while the northern lowlands are largely deforested.

The northern part of west-central Europe is characterized by fertile—in its western parts highly urbanized—coastal lowlands stretching from Belgium and the Netherlands into northeastern Germany. A significant share of this area, particularly in the Netherlands, was gained by coastal land reclamation and drainage of **peatland** during the last centuries. Cattle farming on grasslands prevails in the marshier areas of the **coastal zone**, whereas large areas further inland and eastward are devoted to intensive high-input crop production, mostly grain and rapeseed, but also root crops such as sugarbeets. Especially the sparsely populated eastern part of Germany is characterized by huge farm operations and an agro-industrial landscape that is a heritage of collective farming of the former German Democratic Republic. The central part of west-central Europe features low **mountain** ranges with mixed farming and woodlands, but also extensive fertile lowlands with intensive crop production. The southern part is dominated by the mountainous regions of the Alps. Mont Blanc in the French Alps is Europe's highest peak and reaches an elevation of 4,807 meters. The Alps are comparatively sparsely populated (an average of 75 inhabitants/km^2) and cover a total territory of approximately 192,000 square kilometers, with the major parts situated in Austria (30 percent), northern Italy (27 percent), southeastern France (21 percent), and Switzerland (14 percent). Major land-use types in the Alpine regions are grassland-based cattle farming and **forestry**. Population and infrastructure are concentrated in the major valleys. Non-urban areas in these valleys are mostly used for intensive, mixed agriculture. The Alps are one of Europe's main tourist regions (370 million overnight stays per year). Alpine tourism is an important economic factor in the region and contributes roughly 10–15 percent to total GDP in Austria and Switzerland. However, the impact of tourism, and particularly winter tourism, on Alpine ecosystems and land use there is considerable. Among the most crucial environmental issues related to alpine mass tourism are the large volume of traffic, construction activities, waste and sewage generation, mechanical production of artificial snow, and skiing.

LAND-USE HISTORY

Europe has a long history of intensive agricultural land use. As early as in the fifth century BC, west-central Europe had lost a large part of its forest cover to land clearing for **agriculture**. The expansion of agricultural area in this region reached its all-time high during the thirteenth century when medieval land clearing had reduced forest cover to less than 20 percent. Overpopulation of favored areas and large-scale **deforestation** contributed to considerable **soil erosion** and deterioration of agricultural land in the Middle Ages. In combination with deteriorating climate and a number of years with particularly bad harvests, these **land degradation** tendencies resulted in a downward trend in population that culminated around 1350 when the bubonic plague (Black Death) wiped out 25–35 percent of west-central Europe's population. This drastic and sudden population decline resulted in the large-scale abandonment of human settlements, and consequently triggered major, comparatively rapid changes in land use and land cover. Cropland and grain production receded, and the land was increasingly used for grazing. Forests re-advanced into the fields, and woodland area more than doubled. In the time period between 1450 and the beginning of **industrialization**, the agricultural area gradually expanded again as population and food demand grew. By 1800, woodland was again reduced to less than 30 percent and remained only to the minimum extent that it was necessary to provide for timber and firewood, respectively on habitats due to soil conditions or morphology not worth the effort of cultivation.

At the eve of industrialization, most of west-central Europe's agricultural land was farmed in cereals (e.g., rye), mostly in three-course rotation with **fallow**. In most parts, animals were only subsidiary to provide draft power and manure. Animal-rearing with subordinate crop farming only prevailed in the alpine regions in the south and in the marshy lowlands in some parts of northern west-central Europe. Agriculture was most advanced in Belgium and the Netherlands, where root crops and leguminous fodder crops were introduced early into the crop rotation and allowed for more livestock and higher yields. Beginning in the eighteenth century, this modern type of agriculture gradually expanded all over central Europe and allowed for a doubling of agricultural output during the nineteenth century.

After World War II, the most significant driver of land-use change in west-central Europe was **agricultural intensification**. Supported by agricultural policies, including massive national and European Union subsidies, large-scale mechanization and the use of artificial fertilizers and pesticides were introduced within only two decades. This resulted in, among other things, highly specialized agro-industrial landscapes in large parts of central Europe and multiplied output and productivity. The consequences were increasing farm and field size, a reduction of crop diversity, the introduction of **monocultures**, and the spatial separation of arable farming and livestock husbandry. Agriculture intensified on the most productive lands, while less-favored areas were increasingly abandoned. One of the outcomes of these processes was a steady decrease of the area of agricultural land and a gradual process of **reforestation** in many central European regions, for example, in the Alps. The total area of agricultural land, above all grassland, has decreased by roughly 15 percent in central Europe during the last fifty years.

In the mid-1980s, the subsidized agricultural overproduction became a severe economic burden. Instead of its former focus on fostering production, agricultural policy and subsidization schemes were reformed to include a number of measures to encourage farmers to adopt more extensive and less-polluting farming practices. As a consequence, inputs such as artificial fertilizer began to level off and even decrease. A significant proportion of the region's cropland (10–15 percent) was taken out of production and managed within programs for temporary or permanent set-aside land. Organic farming experienced a significant growth, and the number of certified organic farms in west-central Europe has almost tripled since 1993. More than 1.2 million hectares or approximately 5 percent of the total agricultural area was farmed organically in 2002, the share being highest in Austria (9 percent) and Switzerland (7 percent). Another measure to reduce grain production with significant spatial impact was the promotion of the cultivation of so-called "energy crops," which caused the share of rapeseed and sunflower of total cropland to increase from less than 5 percent in the 1980s to 20 percent in 2002.

Crucial Current Issues

At present, hardly any natural or undisturbed areas can still be found in the central European landscape. Cultural landscapes in which traditional small-scale farming remained in place are currently considered to be among the most diverse and valuable landscapes. This includes, above all, regions in the mountainous and pre-alpine regions of Austria, Germany, and Switzerland, where richly structured agrarian landscapes with mixed farming remained; these were regions with high-alpine grassland farming or extensively used heath lands in the north. These diverse landscapes still cover approximately 5–15 percent of central European territory and are regarded as being of high cultural and ecological value. However, these regions are particularly threatened either by intensification or **land abandonment** and reforestation.

Among the most intensively discussed issues in central European land use are the high intensity of agricultural land use, which results in negative environmental impacts; the increasing reforestation and depopulation in marginally productive regions; the adverse effects of mass tourism on alpine landscapes; and the persistent growth of areas covered by settlements, industry, and transport infrastructure. Land consumed by urbanization and infrastructure is ever increasing. For example, in Germany each year about 0.11 percent of the total area (5 m^2/inhabitant/year) is converted into built-up land.

See also: **Agricultural Revolution; Agrodiversity; East-Central Europe; Land Rights; Land-Use Policies; Mineral Extraction; Urbanization; Western Europe.**

Further Reading

Brouwer, Floor M., Thomas, A. J., and M. J. Chadwick, *Land Use Changes in Europe: Processes of Change, Environmental Transformations and Future Patterns*, Dordrecht, NL: Kluwer Academic Publishers, 1991; European Environmental Agency [Online, August 2004], The EEA Web Site, www.eea.eu.int/products; International Commission for the Protection of the Alps [Online, August 2004], The CIPRA Web Site, www.cipra.org; Pounds, Norman J. G., *An Historical Geography of Europe*, Cambridge, UK: Cambridge University Press, 1990; United Nations Food and Agricultural Organization [Online, August 2004], The FAO Web Site, www.fao.org.

FRIDOLIN KRAUSMANN

Western Europe. Countries of the pre-2004 European Union (EU) with the addition of Iceland, Norway, and Switzerland. This definition is used by the **United Nations Food and Agriculture Organization** (FAO) in compiling its land-use statistics, but it is to some extent arbitrary. The notion of western Europe has been fluid, and its limits are variously defined. In much of the second half of the twentieth century, it was delineated with reference to "Eastern Europe," with the boundary along the Iron Curtain. Prior to the Cold War, it would have been used to describe a small area including France, Ireland, and the United Kingdom (UK) together with the Low Countries (Belgium, Luxembourg, and the Netherlands). Even this narrower definition, however, would have depended on the treatment afforded to Norway and the Iberian countries of Spain and Portugal, and to the recognition of "northern" and "southern" Europe.

Most of the land cover of western Europe has been modified by human activity. Few if any areas are free from human influences. Even the high **mountains** have been affected to some degree by air pollution from power stations and other sources. Over most of the lowlands and intermediate altitudes, the original **vegetation** has been completely replaced by crops or severely modified by the grazing of farm animals.

Land Cover

In the absence of humans, most of western Europe would have been clothed in **forest**. Little remains of this forest in its natural condition. In the maritime lands of Ireland, Scotland, and Denmark, the forest extent had fallen to under 10 percent by modern times. Once removed, **regrowth** or forest regeneration in the moist and exposed climate was difficult, especially if peat formed on deforested ground. In the more continental interior, the forest survived more extensively, but not without human **modification**. According to land-use statistics of international agencies, the relative extent of forest "undisturbed by man" ranges from zero (in several countries) to a maximum of 17 percent in Sweden: parts of today's forests have been planted, while most of the remainder has been affected by some form of management. Perhaps the fate and state of the forest epitomize the degree of human impact on land cover. Over sizable areas, "artificial" land cover, in the form of built-up areas with sealed soil surfaces, has been created; and over much larger areas, near-**monocultures** of intensively produced crops have resulted in drastic reductions in **biodiversity**.

CORINE land cover (CLC) information from the **Coordination of Information on the Environment (CORINE) database** exists for much of the region. Based primarily on satellite imagery, the CORINE project quantifies land cover at a scale of 1:100,000, using a classification system of forty-four categories that can be grouped hierarchically into larger classes. Maintained by the European Environment Agency, the database can provide aggregations for various spatial units such as countries, EU statistical areas, biogeographical provinces, and river catchments. Gaps in coverage mean that a breakdown for the whole of western Europe as here defined cannot be achieved, but it is clear that semi-natural habitats are restricted to a very small proportion of the land area.

The first table (below) summarizes the pattern of variation in land cover across two transects, from maritime Ireland to continental Austria, and from the southwest to the northeast.

In each of the countries included in the table, the proportion of artificial surfaces is under 10 percent. Though western Europe is usually regarded as heavily urbanized, the urban area is still small and in all except a few countries is very low. The relative extent of arable land is also small. Even when "complexes" and permanent crops are included, it remains under 50 percent in each of the countries. "Forest and transitional woodland" forms the most variable category. It covers more than three-quarters of the land area of Finland but less than one-tenth in Ireland and the Netherlands. One of the most distinctive features of the maritime periphery is the low forest extent compared with the continental interior: this extent, however, is much greater than it was a hundred years ago.

Trends

The dynamics of land cover and land use in western Europe are complex in detail, and different trends are discernible at different time **scales**. Over the long sweep of history, the forest has retreated and agricultural land has expanded. Over shorter time scales, different trends have dominated. Over the last 100–200 years, a **forest transition** has occurred and forests have expanded significantly, while the agricultural area has contracted. The second table in this section summarizes FAO data for western Europe: these data are associated with some problems, and indeed FAO discontinued the publication of statistics on forests and woodland in this form after 1995. The table is included here simply as an indicative summary of trends: (total) agricultural area and arable area contracted by 14 and 16 percent respectively between 1961 and 2001, while the forest area increased by 12 percent. The third table in this section focuses on the 1990s and includes more reliable forest data.

Land cover across western European transects

	Ireland	France	Austria	Portugal	The Netherlands	Finland
Artificial surfaces	1.2	3.8	1.8	1.4	9.5	0.5
Arable land	5.9	25.8	13.6	17.7	20.9	6.3
Vineyards, fruit trees, and olive groves	0	2.6	0.7	13.9	0.3	0.1
Pastures	56.7	14.7	11.0	0.1	31.0	0
Complexes etc.^	6.3	19.3	10.9	13.9	16.8	0.4
Forest and transitional woodland	6.1	27.3	44.4	41.3*	8.0	78.5
Natural grass and moors	6.8	3.9	8.8	8.5	2.0	1.9
Bogs	14.3	0.3	0.2	0.3	1.1	2.6
Bare and water surfaces	3.1	2.4	8.7	2.9	10.3	9.8

^ Complex cultivation patterns, and land principally under cultivation but with significant areas of cultivate land;
* including agroforestry.
Source: Based on Corine Land Cover 90 Version 12/2000, European Environment Agency, Copenhagen.

One of the biggest changes in western European land use and landscapes in the twentieth century has been forest expansion, although that process has operated more strikingly in the peripheries of individual countries than in their cores. In recent years, this process has been especially marked in the maritime periphery, as the third table indicates. Forest extents on the maritime fringes are low, but the rate of expansion is high. In countries in the maritime fringe, most of the expansion is through forest plantations. In the more continental parts of the region, it is mostly through natural regeneration. This process is especially marked in alpine regions, where the retreat or **disintensification** of agriculture has been followed by the spread of woodland over alpine **pastures**.

Trends in agricultural and forest areas in western Europe (ha × 1,000)

	Agricultural Land	Arable Land	Forest and Woodland
1961	170,012	89,021	108,957
1971	162,215	81,617	116,612
1981	157,767	80,193	120,538
1991	150,561	78,548	122,437
2001	145,755	75,151	*

*Statistics discontinued after 1995.
Source: Based on FAO FAOSTATS.

Land is coming out of agriculture especially on the more extensive margins, though some is also being lost through **urbanization** at the intensive margin. There is some evidence that arable land is becoming increasingly concentrated in the areas with more favorable soils and climate, such as the Paris basin.

These trends of forest expansion and agricultural contraction are accompanied by other less evident but equally significant processes. For many centuries, the forest resource in western Europe dwindled in volume as well as in area. In recent times, however, this process has been reversed. Timber removals are now equivalent to less than two-thirds of the annual increment, with the result that the volume of standing timber is increasing. While the goal of sustainability might indicate that it is desirable that removals not exceed growth, the aging of the forest may bring its own problems.

Trends within agriculture are different. Here the dominant process has been **agricultural/intensification** of the more productive areas. This intensification has led to dramatic improvements in yields, but these have been achieved at an environmental cost. Semi-natural habitats have been converted to cropland as drainage has been improved, field boundaries have been removed, and grazings plowed up. The result has been a loss of biodiversity and an opening and simplification of the ecosystem and landscape. The

Change in forest and agricultural areas in western Europe in the 1990s (%)

	Ireland	France	Austria	Portugal	The Netherlands	Finland
Forest cover (annual)	3.0	0.4	0.2	1.7	0.3	n.s.
Agricultural area (decennial)	−21.9	−2.8	−3.1	+4.5	−2.9	−7.7

Forest cover: annual change 1990–2000; n.s. not significant; agricultural area: change 1990–2000.
Sources: Forest cover: FAO 2001; Global Forest Resources Assessment 2000, Main Report; *FAO Forestry Paper 140*; Rome: FAO; Agricultural area: FAOSTAT.

process has probably been most severe in the core areas of eastern England, northern France, and the Low Countries. It is now slowing and perhaps in some cases being reversed: net loss of hedge length in England, for example, stabilized in the 1990s, and habitat re-creation has been carried out in some areas under agri-environmental programs.

Driving Forces

Historically, demography or human **population dynamics** was a key driver. High population densities provided a ready market for food and encouraged agricultural expansion and intensification. Over the last 100–150 years, however, there has been a partial decoupling of demographic trends from land-use change. On the one hand, transport improvements, internationalization, and **globalization** have ensured that local populations are not dependent on local food production. On the other hand, most but not all countries in western Europe have sought to shield their domestic agriculture from the effects of cheap imports, and thus the extent of decoupling has been less than it might otherwise have been. The recent stabilization of population might be expected to lead to a lessening of dynamism of land use, but again the relationship has been complex. While urban populations have stabilized, urban areas are still expanding, in part because of increasing numbers of households. Even if the urban area is still relatively very small, the local environmental effects can be large, especially in the more heavily urbanized areas such as the Randstat area of Holland and the London area, where formerly separate settlements now merge together in **urban sprawl**.

A weakening of the effects of traditional drivers such as population growth would in any case now be expected, because the influences of these drivers are now mediated through complex sets of **land-use policies** and plans. These are especially prominent in two spheres—the regulation of the expansion of urban land uses over rural land, and support for agriculture. In the absence of such plans and policies, the character of land cover and use in western Europe would be significantly different.

The Role of the State

The significance of government intervention in land use is unmistakable. In the nineteenth century, the state became a powerful entity across most of western Europe, and since then it has involved itself in a widening range of areas and activities that relate to land use. These include rural welfare, resource **conservation** and strategic interests, and environmental management. The supra-state, in the form of the European Union, has of course also become a prominent actor, associated in particular with **agriculture**, but increasingly also involved in other areas.

The EU's Common Agricultural Policy (CAP) has been a major influence, but national policies in Iceland, Norway, and Switzerland have had comparable effects. In essence, agricultural policies weakened the direct relationship between producer and consumer of food. For reasons of national self-sufficiency and rural welfare, farm prices have been supported at levels well above those of the world market. Production has accordingly been higher than it would have been in the absence of these policies, as have intensity of management and extent of area. The influence of wars and food shortages shaped interventionist policies, but arguably they were maintained long after the specter of scarcities had receded. The inevitable consequence has been a tendency toward overproduction of food relative to regional demand, and the high cost of production of that food meant that it could be exported only if subsidized. By the 1990s, the existence of a problem had been recognized, but an acceptable solution was elusive. A push to decouple support measures from production began tentatively during the 1990s, with a more radical step in 2004. In place of production-related subsidies, farmers will receive direct support. As long as they manage their land in accordance with defined environmental criteria, they will receive payments regardless of their volume and type of production. This change in approach could have consequences in terms of land use and land cover that are both profound and spatially variable. One **scenario** is that agricultural production will decline in the more marginal areas, while continuing to increase through further intensification in the areas that are environmentally more favorable. On the other hand, environmental regulation, imposed by the state in response to changing societal demands, may limit the spatial concentration to which such a trend could lead. Already, pollution loadings from intensive livestock units and fertilizer applications are high in some areas.

Another manifestation of the role of the state is its attempts to curb urban expansion. Greenbelt and similar policies have been pursued for more than half a century. Their effects, however, have been partial: urban expansion has continued apace. In theory, the increasing emphasis now placed on brownfield development should help to curb expansion. Nevertheless, with an apparent problem of overproduction of food, it will be difficult to prevent the **conversion** of further areas of agricultural land into urban uses.

Overview

The role of land in relation to economy and society in western Europe has been transformed. Today, land-based industries account for only tiny proportions of national gross domestic products (GDPs) and employment. The historical role of land in producing food and wood has weakened: memories of physical scarcities are fast receding. The role of land in providing a pleasant environment and the setting for pleasant lifestyles has grown as the productive role has weakened. Increasingly, farms are being purchased as residential rather than productive units. In time, this trend is likely to have its own impact on land use and landscape. The effect will probably be to hasten the rate of agricultural contraction. It may not accelerate the rate of forest expansion, but it will probably lead to new patterns and styles of tree planting, perhaps reminiscent of those of country estates of the past. In other words, a trend of agricultural area may be sustained as the environmental Kuznets curve suggests, along with an inverse-Kuznets curve of forest area. As yet, however, there is little indication that the top of a Kuznets-type curve of urban area has been reached. Perhaps it is unfortunate that the scope for complementarity between the agricultural and urban trends is so limited. Historically, cities grew up in fertile plains: subsequent expansion inevitably eats into the stock of fertile agricultural land that is most likely to be the scene of further intensification.

See also: **Agricultural Revolution, Consumption; East-Central Europe; Economic Restructuring; Europe; Forestry; Hot Spots of Land-Cover Change; Peatlands; Public Policy; Reforestation; West-Central Europe.**

Further Reading

Kronert, Reinhold, Baudry, Jacques, Bowler, Ian, and Annette Reenberg, eds., *Land Use Changes and Their Environmental Impact in Rural Areas in Europe,* New York: Parthenon, 1999; Kuusela, Kullervo, *Forest Resources in Europe 1950–1990,* Cambridge, UK: Cambridge University Press, 1994; Tracey, Michael, *Government and Agriculture in Western Europe 1880–1980,* New York: Harvester Wheatsheaf, 1989; United Nations Economic Commission for Europe and Food and Agriculture Organization, *Forest Resources of Europe, CIS, North America, Australia, Japan and New Zealand,* New York: United Nations, 2000.

ALEXANDER S. MATHER

Wetlands. Since about the second half of the twentieth century, a substitute for terms such as "swamp", "marsh", and "bog". Wetlands include a wide variety of habitat types, such as **peatlands**, mires, fens, and mangroves. Most wetlands are natural, but some wetlands are artificial, such as farm ponds, reservoirs, sewage farms, gravel pits, and salt pans. Once regarded as unhealthy areas or wastelands of little value, wetlands are being recognized as one of the most biologically productive of all ecosystems. The quite recent awareness of wetlands' functional role and their increasing disappearance in farming landscapes have involved the need to inventory them for **conservation** or restoration purposes. However, the inventories of wetlands are very often debated, insofar as the definitions of these ecosystems are numerous and the interests of users and managers for these areas are greatly conflicting.

Marsh vegetation on the mid-Patuxent River, Maryland, September 1996.
PHOTO: Mary Hollinger. CREDITS: NOAA Central Library and the Office of the CIO/HPCC (America's Coastlines Collection).

Definitions

Most wetlands are transition areas between terrestrial and aquatic ecosystems. There is no single, universal, indisputable definition for the term wetland because the gradation between totally dry and totally wet environments is continuous but also because the reasons for defining wetland vary. This explains why wetlands have been defined in various ways, and a multitude of definitions have been proposed.

L. Cowardin and colleagues define wetlands as transitional lands between terrestrial and aquatic systems where the water table is usually at or near the surface, or the land is covered by shallow water. With regard to this classification, a wetland must have one or more of the following attributes: at least periodically, the land supports predominantly hydrophytes; the substrate is predominantly undrained hydric soil; and the substrate is nonsoil and is saturated with water or covered by shallow water at a given time during the season of growth each year.

In this classification, the limit between wetland and upland is defined according to the three following boundaries: the boundary between land with predominantly hydrophytic cover and land with predominantly mesophytic or xerophytic cover; the boundary between predominantly hydric soil and predominantly nonhydric soil; or in the case of wetlands without vegetation or soil, the boundary between flooded or saturated land at a given time each year and land that is not in these hydrologic conditions. The limits between wetland and deepwater systems are also defined. Although the boundary between wetland and deepwater systems is important for inventory purposes, it is unusually at issue in conflicts of standardization and is not mentioned completely in the regulatory definitions of wetland.

The classification by L. Cowardin and colleagues relies on the use of soils, **vegetation**, and hydrology factors for wetland identification. It introduces the concepts of hydrophytes and hydric soils and of predominance both for hydrophytes or undrained hydric soils in wetland areas. In addition, this classification includes some areas that lack vascular plants or soils. These concepts have been later developed in wetland delineation manuals. For example, the U.S. Army Corps of Engineers defines wetlands as areas that are inundated or saturated by surface or groundwater at a frequency and duration sufficient to support a prevalence of vegetation typically adapted for life in saturated soil conditions.

In most current approaches to the mapping and assessment of wetlands, the identification variables used to determine wetland presence are hydrophytic vegetation, hydric soil, and hydrodynamics. The concept of wetland comprises a number of characteristics, including the elevation of the water table with regard to the ground surface, waterlogging, duration of surface water, soil types that form under temporarily or permanently saturated conditions, and various kinds of plants and animals that are adapted to a wet environment. In other terms, wetland is defined as land where the water table is at, near, or above the land surface long enough each year to allow hydric soils formation and to support hydrophytes' growth. The single characteristic that is common to all wetlands is the presence of more soil moisture than necessary to support the growth of most plants. In certain types of wetlands, soils and vegetation are poorly developed or even absent because of important and frequent fluctuations of water levels, turbidity rates, or very high concentration levels of salts or other substances in the water or substrate.

The use of scientific principles for wetland definition contributes to stabilize both the discrimination of wetlands from the other ecosystems and the determination of wetland boundaries for the purposes of inventory, assessment, and management. It also helps to rationalize the application of regulations. In addition, legal implications need to be considered in the application of any wetland definition, because specifying the exact boundary of a wetland that present and projected future users will accept is not an easy task in these frequently conflicting and sometimes controversial areas.

Wetland Functions

Wetlands perform many vital functions, including three main categories: hydrological, biogeochemical, and ecological functions. Hydrologic functions concern regulation of water regime through water storage, flood mitigation, groundwater recharge, and discharge. These functions reduce the amplitude of flooding peaks downstream and maintain basic flow rates by buffering flow distributions. Biogeochemical functions encompass the transformation and cycling of elements, the accumulation of inorganic sediments and peats, and the retention and removal of dissolved substances from surface waters. These functions improve water quality, retain nutrients and other elements, and affect aquatic and atmospheric chemistry. In particular, wetlands play an important role in nitrogen and carbon storage. They can have high rates of denitrification and nitrogen fixation, contributing to control non-point source pollution. Wetlands are major natural sources of reduced gases such as methane and sulfur compounds. Wetlands act also as sediment sinks, in particular in floodplains. But they also contribute, with **rice** fields, to be the largest emitters of methane to the atmosphere due to the anoxic conditions occurring in their flooded soils and their high primary production, totaling roughly 40 percent of the methane input to the troposphere. Ecological functions include habitat maintenance and food web support. This latter includes maintenance of wetland plant communities that provide food and habitat for waterfowl and other animals, thus maintaining biodiversity. Some wetlands are among the most productive environments of the world.

Benefits and Values

Wetlands provide ecological services and economic benefits such as water supply, food supply, biodiversity, and recreation. Most of the sources used to inventory wetlands provide some information on their values and benefits. However, this information frequently takes the form of a summarization of the **biodiversity** values and human use, with few quantitative or economic data included. The exceptions are the productivity of artificial wetlands, rice paddies, Salinas, and fish ponds. The benefits and values of all wetlands for biodiversity and human uses are often highlighted at a global scale. Information is particularly detailed for mangroves, where values and benefits include flood reduction, coastal protection, sediment accumulation, and fish nurseries. Similar descriptions are available for **peatlands**. Protected areas are also valued by people for various reasons, including **conservation**, fishing, and tourism. For example, in **Europe** there is a focus on the values of protected areas, especially as feeding or breeding habitat for birds.

Wetland Inventories

Wetlands are distributed worldwide, from tropical to Arctic countries, but the real extent of wetland is not yet precisely determined. Numerous detailed and precise wetland inventories are conducted on a local scale, and national inventories are developed in a number of countries, either through national wetland policies or implementation of wetland priorities within national environmental action plans and national biodiversity strategies. The global areal extent of wetlands at the earth surface has been estimated by several institutions, but they are uncertain because it is difficult to identify and classify wetlands on a global scale. For example, the World Conservation Monitoring Centre has estimated that they roughly cover some 5.7 millions km^2, which is about 6 percent of the earth's land surface. Based on the wetland inventory in each region of the **Ramsar Convention on Wetlands**, supplemented by a review of regional and international inventories, a global review of wetland resources and priorities for wetland inventory by C. Finlayson and A. Spiers estimated that the area of wetlands worldwide ranges from 7.48 million km^2—the "best" minimum global estimates—to 12.79 million km^2, which corresponds to the amount of the regional minimum estimates. These variations highlight major discrepancies in the areal estimates, which can be attributed to many factors, such as differences in the definition of wetlands, gaps and inaccuracies in the information, the techniques used to collect and interpret the basic data, and the scale of the analyses.

Wetland areas are often underestimated for a number of reasons. Some of the main omissions are due to lack of **access** (e.g., Siberian or Brazilian wetlands), or to their dispersion. For example, small riparian wetlands are generally not inventoried because they are inserted in agricultural landscapes in the bottomlands of the headwater catchments or **watersheds**, scattered in the rural landscape. They are often neglected in national and regional wetland inventories although they strongly influence hydrology, water quality, and biodiversity. Some inventories are carried out by groups with narrow topical interests like habitat maintenance, thus excluding parts of wetland areas. In forested areas, wetlands are also underestimated and difficult to determine, especially in those areas where the water table is below the surface.

Moreover, the areal extent of wetlands is being modified as a result of land-use changes. As a result, once a large inventory is set up, the areal distribution must be monitored and revised periodically while existing ones are updated and new local or regional inventories are created.

Globally, there is very little information available or even attempts made to evaluate the loss of wetlands on a systematic basis, except in North America and **western Europe**. In the **United States of America**, most of the wetland loss occurred during the first half of the twentieth century. Between the first phases of European **colonization** and the 1980s, about half of the wetland area in the conterminous states was converted to other land uses, mainly by drainage, for agricultural purposes. Federal intervention by various acts of swampland and ambitious drainage projects of the U.S. Army Corps of

Engineers (USACE) promoted wetland conversion, and this resulted in the creation of productive croplands that today constitute an important part of the agricultural resource base of the United States. **Conversion** has been distributed very unevenly. The conversion, which had exceeded 80 percent in a number of states by the end of the 1980s, was highest where conversion was both feasible and advantageous, or where wetlands were of limited extent but were located on the most favorable areas for agriculture and population growth. In Europe, loss of wetlands dramatically increased from the second half of the century with **agricultural intensification**. The loss of wetlands worldwide has been estimated at 50 percent of those that existed in the beginning of the twentieth century. This evaluation underestimates the amount of wetland loss, because it includes inland wetlands and possibly mangroves, but not marine wetlands and large estuaries. Since the 1950s, tropical and sub-tropical wetlands, particularly peatlands, swamp forests, and mangroves, have increasingly been converted to cultivated areas.

The driving factors of variations in areal extent of wetlands have both natural and anthropogenic origins, but human factors are predominant, with **agriculture** considered the principal cause of wetland loss worldwide. By 1985, it was estimated that 56–65 percent of available wetland had been drained for intensive agriculture in North America and Europe, 27 percent in Asia, 6 percent in **South America**, and 2 percent in Africa. These figures are probably underestimated, as in the last thirty years **deforestation** has strongly increased the loss of wetlands, mostly in developing countries.

Moreover, the loss of wetlands is the most visible impact of human use on these ecosystems, but parts of existing ones have been degraded, leading to the loss or, at least, to the decrease of their functional efficiency. The Ramsar site database provides a regularly updated but still uneven and incomplete analysis of threats to wetlands. It indicates that 84 percent of Ramsar-listed wetlands had been eliminated or were threatened by ecological change. The most widespread threats were drainage for agriculture, pollution, settlements and **urbanization**, and hunting.

Consequences of Wetland Loss and Degradation

Loss and degradation of wetlands cause important and sometimes irreversible damage to the provision of **ecosystem services**. For example, the issue of water allocation and distribution is linked to the rate and extent of wetland loss and degradation. Planning actions on the rivers like regulation and on associated natural waterbodies and wetlands have impacts on water supply in diminishing underground water reserves and increasing **salinization**, but also on biodiversity in decreasing fish stocks because of degraded habitat and impeded migration.

The conversion of wetlands for agriculture by drainage and transformation of these areas into **grasslands** and/or dry cropland and urban areas leads to major changes in hydrology, vegetation, soil characteristics, and therefore biogeochemical cycles. When wetlands are totally drained, they are excluded from wetland inventories, but it is necessary to determine the areal extent of lost wetlands, because the influence of human activity can be assessed and these areas can be seen as "potential" wetlands that can, under certain circumstances, be rehabilitated and become functional again.

Wetland degradation is more subtle, because it remains as wetland, but is functionally altered, which is not always easy to assess. For example, grazing without hydrologic control results in changes in vegetation and consequent changes in biodiversity. Partial water management for the purpose of reducing the frequency and extent of flooding either for agricultural purposes or fishing leads to wetland degradation of biogeochemical functions and possibly removes carbon in important quantities through grazing and/or peat harvesting.

Wetland areas are influenced by onsite factors like drainage, but also by offsite or upland areas where deforestation or intensive cultivation generally lead to enhanced runoff, **soil erosion**, sedimentation, and increasing pollutant fluxes to aquatic systems. From the perspective of land use, wetlands are drinking-water and irrigation-water sources, and a direct source of food and grazing grounds. Uplands surrounding wetlands must be taken into account because of their controls on hydrology and nutrient supply, which affect wetland functions.

For existing wetland inventories, which are often incomplete, attention must be given to the inventory of priority wetland habitats, focusing on those with the greatest risk of destruction and degradation, and those for which there is little or no information. It appears necessary to consider wetlands that have not been included in wetland inventories, but also to improve knowledge of wetlands that have been better mapped than others, but where serious inconsistencies and few areal estimates exist; more comprehensive inventory is required in that case. Priority wetland habitats designated in the review by C. Finlayson and A. Spiers are the following:

1. Sea grasses in South Asia, the South Pacific, South America, and some parts of Africa that are under increasing threat from pollution, coastal development, destructive fishing practices, recreational use, etc.
2. Salt marshes and coastal flats, because they are under increasing threat worldwide, particularly in Africa, Asia, and Oceania due to increasing coastal development; mangroves, to better determine their loss that is occurring at a dramatic rate in many parts of Africa, Southeast Asia, and Oceania through deforestation, land reclamation, and development for aquaculture.
3. Arid-zone wetlands, which are important when considering the increasing population pressures and water supply, although poorly inventoried.
4. Peatlands, which are well mapped in comparison with other wetland types, are threatened by drainage for agriculture and afforestation in Europe, Asia, and North America in particular, despite their importance as a global carbon sink and economic resource, and are poorly known in tropical regions such as Southeast Asia.
5. Small riparian wetlands can be included in that summary, as they are generally not inventoried despite their important functional role.

Besides natural wetlands, artificial wetlands that can partially compensate for the loss and degradation of natural wetlands have to be taken into account in inventories of priority wetland habitats: reservoirs, dams, salinas, paddies, and aquaculture ponds are increasingly important in many regions, particularly in Asia, Africa, and the neotropics, where they can provide habitat for wildlife and many values and benefits.

In developing countries, land-use changes are increasingly eliminating wetlands. In addition, given the expected increase in human population, mostly in developing countries, the pressure to convert wetlands is expected to increase even more with growing food requirements. Demographic pressure and related agricultural and urban land use could be the most important factors affecting wetland distribution changes in the future and could dominate some of the expected changes in sea level, temperature, and carbon dioxide concentration variations associated with possible climate changes.

In response to the extent and rate of wetland loss and degradation, numerous scientific and conservation programs involved in wetlands currently recommend rehabilitation of degraded wetlands whenever possible so that their structure and functions can be restored. They also outline the need for stronger protective measures for still-remaining functional wetlands. The definition of priority conservation actions should ensure a wise use of these areas.

See also: **Atmosphere-Land Interlinkages; Carbon Cycle; Coastal Zone; Cumulative Change; Extensification; GLOBWETLANDS; Land Degradation; Land Rehabilitation; Malaria; Modification; Monitoring; River Basin; Taiga; Water-Land Interlinkages.**

Further Reading

Birkett, C., et al., Global Wetland Distribution and Functional Characterization: Trace Gases and the Hydrologic Cycle [Online, January 2005], The Global Analysis and Integrated Modeling Web Site, gaim.unh.edu; Cowardin, L. M., Carter, V., Golet, F., and E. LaRoe, Classification of Wetlands and Deepwater Habitats of the United States [Online, January 2005], The U.S. Fish and Wildlife Service Web Site, www.nwi.fws.gov; Finlayson, C. M., and A. G. Spiers, eds., Global Review of Wetland Resources and Priorities for Wetland Inventory [Online, January 2005], The Wetlands Web Site, www.wetlands.org; Mitsch, W., and J. Gosselink, *Wetlands*, New York: Van Nostrand-Reinhold, 1993; National Research Council, Committee of Wetlands, Wetlands: Characteristics and Boundaries [Online, January 2005], The National Academy Press Web Site, www.nap.edu; U.S. Army Corp of Engineers, *Wetlands Delineation Manual* (Technical Report Y-87-1), Washington, DC: Department of the Army, 1987.

LAURENCE HUBERT-MOY

Wheat. Next to **rice**, wheat is the most important cereal worldwide and provides almost 20 percent of all human food energy. Wheat (or *triticum*) is used for making bread, pastries, crackers, biscuits, macaroni, spaghetti, bulgur, couscous, and breakfast food, as a brewing ingredient, and for alcohol production and livestock feed.

Domestication of wheat started around 7500–6500 BC in the Middle East. In prehistoric times, wheat was grown in Greece, Persia, Egypt, and throughout **Europe**. It spread into **China** and southwest Asia and was finally introduced by the Spanish to the New World in the sixteenth century.

Mule-drawn combine in wheat field of Walla Walla County, Washington, July 1941.
PHOTO: Russell Lee. [Courtesy Library of Congress]

Wheat is adapted to a wide range of environments. It is grown from the equator to the Arctic Circle and from sea level up to 4,500 meters high. Bread and durum wheat are the most important wheats and are grown on about 90 percent and 10 percent, respectively, of the total area cultivated with wheat. Durum wheat is less frost-resistant and requires warmer temperatures than many bread wheats and is grown mainly in Mediterranean environments. Bread wheat is more variable with about 25,000 varieties grown across a wide range of conditions worldwide.

Wheat can resist frost; the forms that have the ability to survive low temperatures in winter are known as winter wheat. However, about two-thirds of the total area devoted to wheat (i.e., 220 million hectares in 2004) is cultivated with spring wheat. As opposed to spring wheat, winter wheat requires vernalization at cold temperatures to reach the generative stage. For both spring and winter wheat, phasic development depends on temperature and, as for most cultivars, also on day length. The degree of sensitivity to cold temperatures and day length varies with cultivar and is an important parameter characterizing the agro-ecological adaptation. Wheat has a C_3 photosynthetic pathway. Optimum growth temperatures are between 10 and 24°C. Growth stops at above 35°C, and temperatures above 40°C induce irreversible heat stresses. Unlike spring wheat, most winter wheat is grown under rainfed conditions.

Historic changes in wheat area and production are closely related to population growth and economic development. During the last three centuries, the demand for food steadily increased with the rising number of people and the industrial development in many countries. Expansion of cropland including wheat land was for a long time the most effective way to increase food production but has eventually reached the limits set by available land area, other land uses, and environmental reasons. Global wheat harvested area has more than doubled within the last hundred years and reached its maximum extension around the 1980s. A reduction in wheat-growing area has been observed in recent decades due to abandonment of cropland in mainly developed countries. Presently, about 15 percent of all arable land is used for wheat production. Most wheat is grown in the European Union, China, India, and the **United States of America**, accounting together for more than 55 percent of the global wheat production. Other important wheat-growing countries are Australia, **Canada**, and **Russia**. In the first half of the twentieth century, wheat was produced mainly in developed countries; but during the second half of the century, production gradually shifted toward developing countries, which presently produce almost 50 percent of all wheat.

In recent decades, raising productivity per unit land has become the most important factor in meeting the growing demand for food. Average global wheat yields were about 2.9 t ha^{-1} in 2004 and have increased by almost 200 percent since 1961. Large differences among countries, however, are evident, which is due mainly to differences in climatic and socio-economic conditions. High yields are obtained in countries of the European Union: they often exceed 8 t ha^{-1}, for example, in the Netherlands, United Kingdom, Germany, and France. Yields have increased with a high rate since the **Green Revolution** in the 1960s. Advances in technology through improved varieties and crop management, including the use of fertilizers and biocides, have nearly tripled wheat yields in many countries within a few decades. Increasing wheat yields has in some regions, for example the European Union, resulted in overproduction and abandonment of cropland. However, supply is still insufficient for many countries in the developing world, and further increase in the demand for wheat due to population and **economic growth** is expected for the future. Since cropland expansion is limited, increases in production have to be realized via higher yields through further development of production technologies. The long-term effects on the environment of intensive crop production and the use of genetically modified organisms remain unclear but are of increasing concern.

Future productivity of wheat will also depend on changes in climatic conditions and atmospheric carbon dioxide (CO_2) concentration as projected by the **Intergovernmental Panel on Climate Change**. Experimental evidence and **modeling** studies suggest that doubling the present CO_2 concentration may increase wheat productivity by 20–30 percent, which may ameliorate some of the negative effects of increasing temperature and reduced rainfall. Effects of climate changes are projected to vary depending on the region. Wheat may extend into colder regions in the North and South due to increasing temperatures and disappear from warmer regions in the South due to increasing stresses from drought and heat. However, relationships that will determine future changes in yields and the spatial distribution of wheat are complex and not well understood. Impacts of climate change will also depend on the **vulnerability** and more specifically on the adaptive capacity of the agricultural systems, which again will depend on the socio-economic conditions of the farms, regions, and countries.

See also: **Agricultural Intensification; Agricultural Revolution; Cash Crops; Consumption; Driving Forces; Economic Restructuring; Food Crops; Great Plains; Land Abandonment; Land Degradation; Mountains; Population Dynamics; Turnover.**

Further Reading

Ewert, Frank, Rounsevell, Mark D. A., Reginster, Isabelle, Metzger, Marc, and Rik Leemans, "Future Scenarios of Agricultural Land Use in Europe: I: Estimating Changes in Productivity," *Agriculture, Ecosystems & Environment* 107 (2005): 101–16; Parry, Martin L., Rosenzweig, Cynthia, Iglesias, Ana, Livermore, Matthew, and Günther Fischer, "Effects of Climate Change on Global Food Production under SRES Emissions and Socio-Economic Scenarios," *Global Environmental Change* 14 (2004): 53–67; Ramankutty, Navin, and Jonathan A. Foley, "Estimating Historical Changes in Global Land Cover: Croplands from 1700 to 1992," *Global Biogeochemical Cycles* 13 (1999): 997–1027; Rounsevell, Mark D. A., Ewert, Frank, Reginster, Isabelle, Leemans, Rik, and Timothy R. Carter, "Future Scenarios of Agricultural Land Use in Europe: II: Projecting Changes in Cropland and Grassland," *Agriculture, Ecosystems & Environment* 107 (2005): 117–35; United Nations Food and Agriculture Organization [Online, February 2005], The FAO Web Site, www.fao.org; Van Ginkel, M., and R. L. Villareal, "Triticum L," in *Plant Resources of South-East Asia Handbook 10: Cereals*, eds. G. J. H. Grubben and Soetjipto Partohardjono, Prosea Foundation, 1996, pp. 137–43.

FRANK EWERT

World Agroforestry Centre. One of the international agricultural research institutions supported by the **Consultative Group on International Agricultural Research**, also known as the International Centre for Research in Agroforestry (ICRAF). Taking a broad view of the economic, social, cultural, and environmental roles of trees on farms and at the landscape/**watershed** level, ICRAF seeks to improve livelihoods for poor people in developing countries while enhancing sustainability of land use, local ecosystems, and the global environment. The Centre's headquarters are in Nairobi, Kenya. Nearly 200 professional staff work in twenty-nine developing countries within seven regional programs in Africa, Asia, and **Latin America**. ICRAF works in partnership with farmers, national scientists, policymakers, and others to conduct **agroforestry** research and development, strengthen the capacity of partners, communicate useful information about agroforestry, and provide scientific leadership in the field of integrated natural resource management.

See also: **Agrodiversity; Alternatives to Slash-and-Burn Programme; Biodiversity; Ecosystem Services; Forestry; Land Rehabilitation; Land-Use Policies; Natural Resources; Participatory Land Management; Poverty; Sustainable Land Use; Water-Land Interlinkages.**

Further Reading

World Agroforestry Centre, Transforming Lives and Landscapes [Online, November 2004], The World Agroforestry Centre Web Site, www.worldagroforestrycentre.org.

THOMAS P. TOMICH

Yellowstone Model. Legislation to create the world's first national park—Yellowstone—in Wyoming was passed on March 1, 1872. It was the starting point for national governments to start annexing and managing "natural" areas for various objectives (e.g., nature protection and recreation), and has loomed large in the psyche of conservationists and park administrators ever since. Yellowstone was promoted as a wilderness: an area untouched by humans, an area where nature remained undisturbed. The wilderness concept that has evolved in North America is deeply entrenched (e.g., in 1964, the **United States of America** passed the Wilderness Act) but is not as strong elsewhere. The strict definition of wilderness as a place untouched by people is strongly disputed on the basis of archaeological, biological, and paleoecological evidence. Retrospectives on the establishment of Yellowstone National Park illustrate that the area was considered undisturbed because Indians were ignored entirely. In one sense, then, the Yellowstone Model of **conservation** comprises large areas of wilderness being annexed by national governments, usually to protect nature and/or to encourage recreation. When the Yellowstone Model is applied to the design of a nature reserve, there is often a strong assumption of exclusion. People are evicted from the land they live on, or their **access** to natural resources (e.g., fuel and medicinal plants collection) is restricted. In the worst-case **scenarios**, areas are controlled by guards, they may be fenced off, and access is granted only to scientists and/or tourists. This practice is also colloquially known as "fortress conservation" or the "fences-and-fines approach." National parks were established using the Yellowstone Model in the first part of the twentieth century throughout the world (e.g., Southern [now Nahuel Huapi, Argentina], 1902; Udjon Kulun [Indonesia], 1915; and Krüger [**South Africa**], 1926). But the Yellowstone Model has led to other types of protected areas (PAs)—not simply large areas that exclude indigenous people—being established. The International Union for Conservation of Nature and Natural Resources (IUCN) currently recognizes seven PA categories. Therefore, while strict protectionism is one clear outcome of the Yellowstone Model, the growth in PAs has been another. At the Fifth World Parks Congress (Durban, South Africa, 2003), IUCN reported 102,102 PAs covering 18.8 million square kilometers. This equates to 11.5 percent of the earth's surface under some kind of protection, an area greater than that under permanent arable crops. Statistics show that the number and area of PAs is growing exponentially. The land-use and **land-cover change** implications of the Yellowstone Model are found within and outside park boundaries. Access regulations in many types of protected areas can allow relatively small areas in parks to recover to near-natural land covers. However, the communities that have been evicted from the parks or have had their access restricted often settle around park perimeters. The areas around their settlements often become degraded, and illegal access to natural resources within the PAs continues.

See also: **Arena of Land Conflict; Public Policy; Tragedy of Enclosure.**

Further Reading

Gomez Pompa, A., and A. Kaus, "Taming the Wilderness Myth," *BioScience* 42 (1992): 271–9; Stevens, Stan, "The Legacy of Yellowstone," in *Conservation through Cultural Survival*, ed. Stan Stevens, Washington, DC: Island Press, 1997, pp. 13–32.

ANDREW MILLINGTON

Selected Bibliography

Abrol, Yash P., Satpal Sangwan, and Mithilesh K. Tiwari, eds. *Land Use: Historical Perspectives: Focus on Indo-Gangetic Plains.* New Delhi, India: Allied Publishers, 2002.

Bičík, Ivan, Pavel Chromý, Vit Jančá, and Helena Janů, eds. *Land Use/Land Cover Changes in the Period of Globalization.* Prague, Czech Republic: Charles University Faculty of Science, International Geographical Union Commission on Land-Use and Land-Cover Change, 2002.

Blaikie, Piers M., and Harold Brookfield. *Land Degradation and Society.* London, New York: Routledge, 1987.

Bridge, Gavin. "Contested Terrain: Mining and the Environment," *Annual Review of Energy and Resources* 29 (2004): 205–59.

Bryant, Raymond L., and Sinéad Bailey. *Third World Political Ecology.* London, New York: Routledge, 1997.

DeFries, Ruth, Gregory Asner, and Richard Houghton, eds. *Ecosystems and Land Use Change.* Washington, DC: American Geophysical Union, 2004.

Foley, Jonathan A., DeFries, Ruth, Asner, Gregory P., Barford, Carol, Bonan, Gordon, Carpenter, Stephen R., Chapin, F. Stuart, Coe, Michael T., Daily, Gretchen C., Gibbs, Holly K., et al. "Global Consequences of Land Use." Science 309 (2005): 570–74.

Fox, Jefferson, Ronald Rindfuss, Stephen Walsh, and Vinod Mishra, eds. *People and the Environment: Approaches for Linking Household and Community Surveys to Remote Sensing and GIS.* Dordrecht, NL: Kluwer, 2003.

Geist, Helmut. *The Causes and Progression of Desertification.* Aldershot, UK, Burlington, VT: Ashgate, 2005.

Geist, Helmut J., and Éric F. Lambin. "Proximate Causes and Underlying Driving Forces of Tropical Deforestation." *BioScience* 52 (2002): 143–50.

Gutman, Garik, Anthony Janetos, Christopher Justice, Emilio Moran, John Mustard, Ronald Rindfuss, David Skole, B. L. Turner II, and Mark Cochrane, eds. *Land Change Science: Observing, Monitoring and Understanding Trajectories of Change on the Earth's Surface* (Remote Sensing and Digital Image Processing Series 6). Dordrecht, NL: Kluwer, 2004.

Haberl, Helmut, Matthis Wackernagel, and T. Wrbka, eds. "Land Use and Sustainability Indicators." *Land Use Policy* 21 (2004).

Himiyama, Yukio, Alexander Mather, Ivan Bičík, and Elena V. Milanova, eds. *Land Use/Cover Changes in Selected Regions in the World: Volumes I–IV.* Asahikawa, JP: International Geographical Union Commission on Land-Use and Land-Cover Change, 2001, 2002, 2005.

Johnston, R. J., Peter J. Taylor, and Michael Watts, eds. *Geographies of Global Change: Remapping the World in the Late Twentieth Century.* Oxford, UK: Blackwell, 2002.

Kasperson, Jeanne X., Roger E. Kasperson, and B. L. Turner II, eds. *Regions at Risk: Comparisons of Threatened Environments.* Tokyo: United Nations University Press, 1995.

Lambin, Éric, Baulies, Xavier, Bockstael, Nancy, Fischer, Günther, Krug, Thelma, Leemans, Rik, Moran, Emilio, et al. *Land-Use and Land-Cover Change (LUCC): Implementation Strategy* (IGBP Report 48, IHDP Report 10). Stockholm, SW: International Geosphere-Biosphere Programme, and Bonn, GE: International Human Dimensions of Global Environmental Change Programme, 1999.

Lambin, Éric, and Helmut Geist, eds. *Land Use and Land Cover Change: Local Processes, Global Impacts.* Berlin: Springer, 2006.

Lambin, Éric, Geist, Helmut, and Erika Lepers. "Dynamics of Land Use and Cover

Change in Tropical and Subtropical Regions." *Annual Review of Environment and Resources* 28 (2003): 205–41.

Lambin, Éric F., Turner, B. L. II, Geist, Helmut J., Agbola, Samuel B., Angelsen, Arild, Bruce, John W., Coomes, Oliver, et al. "The Causes of Land-Use and Land-Cover Change: Moving beyond the Myths." *Global Environmental Change* 11 (2001): 261–9.

Lepers, Erika, Lambin, Éric F., Janetos, Anthony C., DeFries, Ruth, Achard, Frédéric, Ramankutty, Navin, and Robert J. Scholes. "A Synthesis of Information on Rapid Land-Cover Change for the Period 1981–2000." *BioScience* 55 (2005): 115–24.

Liverman, Diana, Emilio Moran, Ronald Rindfuss, and Paul Stern, eds. *People and Pixels: Linking Remote Sensing and Social Science.* Washington, DC: National Academy Press, 1998.

Mather, Alexander S. "The Forest Transition." *Area* 24 (1992): 227–37.

Mather, Alexander S. "Forest Transition Theory and the Reforesting of Scotland." *Scottish Geographical Journal* 120 (2004): 83–98.

Meyer, William B., and B. L. Turner II, eds. *Changes in Land Use and Land Cover: A Global Perspective.* Cambridge, UK, New York, Melbourne, AU: Press Syndicate of the University of Cambridge, 1994.

Moran, Emilio F., and Elinor Ostrom, eds. *Seeing the Forest and the Trees: Human-Environment Interactions in Forest Ecosystems.* Cambridge, MA: MIT Press, 2005.

Morris, Dick, Freeland, Joanna, Hinchliffe, Steve, and Sandy Smith. *Changing Environments.* Chichester, UK: John Wiley & Sons, 2003.

Oldfield, Frank, *Environmental Change: Key Issues in and Alternative Perspectives,* Cambridge, UK: Cambridge University Press, 2005.

Neil, Roberts, ed. *The Changing Global Environment.* Oxford, UK: Blackwell, 1994.

Parker, Dawn C., Manson, Steve M., Janssen, M. A., Hoffmann, M. J., and Peter Deadman. "Multi-Agent System Models for the Simulation of Land-Use and Land-Cover Change: A Review." *Annals of the Association of American Geographers* 93 (2003): 316–40.

Ramankutty, Navin, Foley, Jonathan A., and N. J. Olejniczak. "People on the Land: Changes in Global Population and Croplands during the 20th Century." *Ambio* 31 (2002): 251–7.

Reynolds, James, and Mark Stafford Smith, eds. *Global Desertification: Do Humans Cause Deserts?* (Dahlem Workshop Report 88). Berlin: Dahlem University Press, 2002.

Rudel, Thomas K, Coomes, Oliver T., Moran Emilio, Achard, Frédéric, Angelsen, Arild, Xu, Jianchu, and Éric F. Lambin. "Forest Transitions: Towards a Global Understanding of Land Use Change." *Global Environmental Change* 15 (2005): 23–31.

Steffen, Will, Sanderson, Angelina, Tyson, Peter, Jäger, Jill, Matson, Pamela, Moore, Berrien III, Oldfield, Frank, et al. *Global Change and the Earth System: A Planet under Pressure.* Berlin: Springer, 2004.

Tri-Academy Panel. "Population and Land Use in India, China, and the United States: Context, Observations, and Findings." In *Growing Populations, Changing Landscapes: Studies from India, China, and the United States,* edited by Indian National Science Academy, Chinese Academy of Sciences, and U.S. National Academy of Sciences. Washington, DC: National Academy Press, 2001, pp. 9–72.

Turner, B. L., II. "Toward Integrated Land-Change Science: Advances in 1.5 Decades of Sustained International Research on Land-Use and Land-Cover Change." In *Challenges of a Changing Earth: Proceedings of the Global Change Open Science Conference, Amsterdam, The Netherlands, 10–13 July 2001* (The IGBP Series), edited by Will Steffen, Jill Jäger, David Carson, and Clare Bradshaw. Berlin: Springer, 2002, pp. 21–26.

Turner, B. L., II, Skole, David, Sanderson, Steven, Fischer, Günther, Fresco, Louise, and Rik Leemans. *Land-Use and Land-Cover Change Science/Research Plan* (IGBP Report 35, HDP Report 7). Stockholm, SW: International Geosphere-Biosphere Programme, and Geneva: Human Dimensions of Global Environmental Change Programme, 1995.

United Nations Environment Programme: One Planet, Many People: Atlas of Our Changing Environment. Nairobi: UNEP, 2005.

Veldkamp, Tomas, and Éric Lambin, eds. "Predicting Land-Use Change." *Agriculture Ecosystems and Environment* 85 (2001).

Verburg, Peter H., and Tom Veldkamp, eds. "Spatial Modeling to Explore Land Use Dynamics." *International Journal of Geographic Information Science* 19 (2005): 99–102.

Walsh, Stephen, and Kelley Crews-Meyer, eds. *Remote Sensing and GIS Applications for Linking People, Place, and Policy*. Dordrecht, NL: Kluwer, 2002.

Wisner, Ben, Blaikie, Piers M., Cannon, Terry, and Ian Davis. *At Risk: Natural Hazards, People's Vulnerability, and Disasters*. 7th ed. London: Routledge, 2003.

Index

Note: Page numbers in bold type indicate main encyclopedia entries.

List of Contributors

(in alphabetical order; * = co-authored entry; † = introductory essay)

Achard, Frédéric (Dr.)
Joint Research Centre of the European Commission
Institute for Environment and Sustainability
Italy
1. Hot Spots of Land-Cover Change*
2. Joint Research Centre (JRC) of the European Commission
3. Kyoto Protocol*
4. Tropical Ecosystem Environment Observations by Satellite (TREES) Project
5. Tropical Humid Forest*

Arneth, Almut (Dr.)
Lund University
Department of Physical Geography and Ecosystems Analysis
Sweden
1. Atmosphere-Land Interlinkages
2. Integrated Land Ecosystem-Atmosphere Processes Study
3. International Geosphere-Biosphere Programme

Baade, Jussi (Prof.)
Friedrich Schiller University Jena
Department of Geography
Germany
1. Carbon Sequestration
2. Salinization
3. Siltation
4. Soil Degradation
5. Soil Erosion

Bartholomé, Etienne (Dr.)
Joint Research Centre of the European Commission
Institute for Environment and Sustainability
Italy
1. Global Land Cover Map of the Year 2000*

Belward, Alan (Dr.)
Joint Research Centre of the European Commission
Institute for Environment and Sustainability
Italy
1. Global Land Cover Map of the Year 2000*

Berrang Ford, Lea
University of Guelph
Department of Population Medicine
Canada
1. African Trypanosomiasis
2. Human Immunodeficiency Virus (HIV)/Acquired Immunodeficiency Syndrome (AIDS)
3. Malaria
4. Watershed

Biggs, Reinette (Oonsie) (Dr.)
University of Wisconsin, Madison
Center for Limnology
United States of America
1. South Africa*
2. Southern Africa*

Bridge, Gavin (Prof.)
Syracuse University
Department of Geography
United States of America
1. Corporate Strategies
2. Economic Liberalization
3. Economic Restructuring
4. Investments
5. Mineral Extraction
6. Multiplier Effect

Brouwer, Floor (Prof.)
Agricultural Economics Research Institute
The Netherlands
1. Agriculture
2. Land Abandonment

Brown, Daniel G. (Prof.)
University of Michigan
School of Natural Resources and Environment
United States of America
1. Agent-Based Model
2. Auto-Correlation

Busch, Gerald (Dr.)
University of Kassel
Center for Environmental Systems Research
Germany
1. Carbon Cycle
2. Land Degradation

Chase, Thomas N. (Prof.)
University of Colorado
Cooperative Institute for Research in Environmental Sciences and Department of Geography
United States of America
1. Climate Impacts*

Chen, Ke (Dr.)
Stockholm Environment Institute
SUSTF International Secretariat
Sweden
1. Extensification
2. George Perkins Marsh Institute
3. Land Tenure
4. Land-Use System
5. Sauer, Carl Ortwin
6. Vulnerability

Crews-Meyer, Kelley A. (Prof.)
University of Texas
Department of Geography and the Environment
United States of America
1. Access*
2. Airborne Light Detection and Ranging*
3. Change Detection
4. Global Positioning System*
5. Pattern Metrics
6. Savannization*

Cunfer, Geoff (Prof.)
University of Saskatchewan
Department of History
Canada
1. Great Plains

de Sherbinin, Alexander (Dr.)
Columbia University
Center for International Earth Science Information Network
United States of America
1. Center for International Earth Science Information Network
2. Consumption
3. Population Dynamics
4. Urbanization

de Vasconcelos, Maria José P. (Prof.)
Tropical Research Institute
Cartography Centre
Portugal
1. Cellular Automaton
2. Fire
3. Vegetation

de Vries, Bert J. M. (Prof.)
Netherlands Environmental Assessment Agency
Global Sustainability and Climate
The Netherlands
1. Integrated Model to Assess the Global Environment*

Eickhout, Bas (Dr.)
Netherlands Environmental Assessment Agency
Global Sustainability and Climate
The Netherlands
1. Integrated Model to Assess the Global Environment*

Erb, Karlheinz (Dr.)
Klagenfurt University
Institute of Social Ecology
Austria
1. Biomass
2. Colonization
3. Ecological Colonization
4. Ecological Footprint
5. Turnover

Eva, Hugh (Dr.)
Joint Research Centre of the European Commission
Institute for Environment and Sustainability
Italy
1. Global Land Cover Map of the Year 2000*

2. Hot Spots of Land-Cover Change*
3. Tropical Humid Forest*

Evans, Thomas P. (Prof.)
Indiana University
Department of Geography
United States of America
1. Center for the Study of Institutions, Population and Environmental Change
2. Decision-Making
3. Pixel
4. Secondary Vegetation

Ewert, Frank (Dr.)
Wageningen University
Plant Production Systems
The Netherlands
1. Green Revolution
2. Maize
3. Rice
4. Wheat

Fox, Jefferson (Dr.)
East-West Center
Program on Environment
United States of America
1. Fallow
2. Slash-and-Burn Agriculture
3. Southeast Asia, Mainland
4. Swidden Cultivation

Freitas, Helena (Prof.)
University of Coimbra
Department of Botany
Portugal
1. Biodiversity
2. Eutrophication
3. Grassland

Geist, Helmut (Prof.)
University of Aberdeen
Department of Geography
United Kingdom
1. *Agriculture, Ecosystems & Environment*
2. Agrodiversity
3. Agroforestry
4. Agro-Industry*
5. *Ambio, A Journal of the Human Environment*
6. Anthropological Center for Training and Research on Global Environmental Change
7. Biodiversity Novelties
8. Biome
9. BIOME 300 Project
10. *BioScience*
11. Blaikie, Piers
12. Boserup, Ester
13. Brookfield, Harold Chilingworth
14. Carson, Rachel Louise
15. Cassava
16. Center for International Forestry Research
17. Contract Farming
18. Conversion
19. Cultural Factors*
20. Deforestation
21. Desertification*
22. DIVERSITAS
23. Domestication
24. *Ecology and Society*
25. Endogenous Factor
26. *Environment*
27. Exogenous Factor
28. Exotic Species
29. Feedback
30. Forest Degradation
31. *Frontiers in Ecology and the Environment*
32. *Global Environmental Change*
33. Global Environmental Change and Food Systems (GECAFS) Project
34. Human Health
35. IKONOS
36. Initial Conditions
37. International Food Policy Research Institute
38. *Journal of Arid Environments*
39. *Journal of Land Use Science*
40. *Land Degradation & Development*
41. Land Degradation Assessment in Drylands (LADA) Project
42. Land Quality
43. Land Stability
44. Land-Ocean Interactions in the Coastal Zone (LOICZ) Project
45. Landsat
46. Land-Use Legacy
47. Land-Use Transition
48. Land-Use/Cover Change (LUCC) Project
49. Large-Scale Biosphere-Atmosphere Experiment in Amazonia
50. Maathai, Wangari M.
51. Marsh, George Perkins
52. Mediating Factor
53. Meta-Analysis

54. Millennium Ecosystem Assessment
55. Miombo
56. Mode of Interaction
57. Moderate Resolution Imaging Spectroradiometer
58. Modification
59. Monoculture
60. Non-Timber Forest Products
61. Past Global Changes (PAGES) Project
62. Population-Environment Research Network
63. Project on People, Land Management and Ecosystem Conservation
64. Species Extinction
65. Sustainable Land Use
66. Syndrome
67. Tobacco
68. Tragedy of Enclosure
69. United Nations Convention on Biological Diversity
70. United Nations Convention to Combat Desertification
71. United Nations Environment Programme
72. United Nations Food and Agriculture Organization
73. Water-Land Interlinkages

Geoghegan, Jacqueline M. (Prof.)
Clark University
Department of Economics
United States of America
1. Economic Growth
2. Pixelizing the Social
3. Socializing the Pixel

Göbel, Barbara (Dr.)
International Human Dimensions Programme on Global Environmental Change
Germany
1. Earth System Science Partnership
2. Globalization
3. International Human Dimensions Programme on Global Environmental Change
4. Perception

Goldman, Mara Jill
University of Wisconsin
Department of Geography
United States of America
1. Participatory Land Management*
2. Pastoral Mobility*

Grégoire, Jean-Marie (Dr.)
Joint Research Centre of the European Commission
Institute for Environment and Sustainability
Italy
1. Global Land Cover Map of the Year 2000*

Grimm, Heike (Dr.)
University of Erfurt
School of Public Policy
Germany
1. Agro-Industry*
2. Cultural Factors*
3. Public Policy

Gutman, Garik (Dr.)
NASA Land-Cover/Land-Use Change (LCLUC) Program
United States of America
1. NASA Land-Cover/Land-Use Change (LCLUC) Program
2. Remote Sensing

Haberl, Helmut (Prof.)
Klagenfurt University
Institute of Social Ecology
Austria
1. Cascades
2. Human Appropriation of Net Primary Production
3. Metabolism
4. Threshold

Haileselassie, Amare (Dr.)
Institute of Soil Science and Forest Nutrition
George August University
Germany
1. Nutrient Cycle*

Heistermann, Maik
University of Kassel
Center for Environmental Systems Research
Germany
1. LandSHIFT Model*
2. Nutrient Cycle*

Himiyama, Yukio (Prof.)
Hokkaido University of Education
Japan

1. International Geographical Union (IGU) Commission on Land-Use and Land-Cover Change
2. Japan

Homewood, Katherine (Prof.)
University College London
Department of Anthropology
United Kingdom
1. Disequilibrium Dynamics
2. Transhumance

Hostettler, Silvia
Ecole Polytechnique Fédérale de Lausanne
Laboratory of Urban Sociology
Switzerland
1. Participatory Geographic Information System
2. Remittances Landscape

Hubert-Moy, Laurence (Prof.)
University of Rennes 2
Department of Geography
France
1. Peatlands
2. Ramsar Convention on Wetlands
3. Wetlands

Hwang, Manik (Prof.)
Seoul National University
Department of Geography Education
South Korea
1. Coastal Zone

Irwin, Elena G. (Prof.)
Ohio State University
Department of Agricultural, Environmental, and Development Economics
United States of America
1. Leapfrogging
2. Suburbanization
3. Urban Sprawl
4. Urban-Rural Fringe

Jeleček, Leoš (Prof.)
Charles University
Department of Social Geography and Regional Development
Czech Republic
1. Agricultural Revolution
2. Cadastre
3. East-Central Europe
4. Industrial Revolution
5. Land Reform
6. Land Rent
7. Southeast Europe
8. Technological-Scientific Revolution

Jepson, Wendy (Prof.)
Texas A&M University
Department of Geography
United States of America
1. Biotechnology
2. Brazilian Cerrado

Klein Goldewijk, Kees (Prof.)
Netherlands Environmental Assessment Agency
Global Sustainability and Climate
The Netherlands
1. Center for Sustainability and the Global Environment
2. History Database of the Global Environment
3. Land-Use Change in a Global, Historical Perspective†
4. National Institute of Public Health and the Environment

Klepeis, Peter (Prof.)
Colgate University
Department of Geography
United States of America
1. Degradation Narrative*
2. Land-Use History*
3. Pristine Myth*
4. Southern Cone Region

Kok, Kasper (Dr.)
Wageningen University and Research Center
Department of Environmental Sciences
The Netherlands
1. Integrated Assessment
2. Scenario
3. Validation*

Krausmann, Fridolin (Dr.)
Klagenfurt University
Institute of Social Ecology
Austria
1. Industrialization
2. West-Central Europe

Laris, Paul (Prof.)
California State University
Department of Geography
United States of America
1. Degradation Narrative*
2. Land-Use History*
3. Pristine Myth*

Lawrence, Peter J. (Dr.)
University of Colorado
Cooperative Institute for Research in Environmental Sciences and Department of Geography
United States of America
1. Climate Impacts*

Li, Xiubin (Prof.)
Chinese Academy of Sciences
Institute of Geographic Sciences and Natural Resources Research
China
1. China
2. River Basin

Linke, Sophia
United Nations Food and Agriculture Organization
Environment and Natural Resources Service
Italy
1. Land Cover Classification System*

Loveland, Thomas R. (Prof.)
South Dakota State University
United States of America
1. Earth Resources Observation Systems (EROS) Data Center
2. IGBP-DIS Global 1-Km Land Cover Data Set
3. LANDFIRE Project
4. Sahel Land Cover
5. United States of America*
6. United States Geological Survey (USGS) Program

Lu, Qi (Prof.)
Chinese Academy of Sciences
Institute of Geographic Sciences and Natural Resources Research
China
1. Green Wall Project
2. Land Rehabilitation
3. Regrowth

Mather, Alexander S. (Prof.)
University of Aberdeen
Department of Geography
United Kingdom
1. Driving Forces
2. Europe
3. Forest
4. Forest Transition
5. Forestry
6. *Land Use Policy*
7. Land-Use Policies
8. Proximate Causes
9. Thünen, Johann Heinrich von
10. Western Europe

Mayaux, Philippe (Dr.)
Joint Research Centre of the European Commission
Institute for Environment and Sustainability
Italy
1. Global Land Cover Map of the Year 2000*
2. Hot Spots of Land-Cover Change*
3. Tropical Humid Forest*

Millington, Andrew (Prof.)
Texas A&M University
Department of Geography
United States of America
1. Agenda 21
2. Conservation
3. Land-Cover Change†
4. South America
5. Yellowstone Model

Mollicone, Danilo (Dr.)
Joint Research Centre of the European Commission
Institute for Environment and Sustainability
Italy
1. Kyoto Protocol*

Morais, João M. F. (Prof.)
International Geosphere-Biosphere Programme
Sweden
1. Global Land Project

Moseley, William G. (Prof.)
Macalester College
Department of Geography
United States of America

1. Cotton
2. Food Security
3. Poverty

Müller, Daniel (Dr.)
Humboldt University
Institute for Agricultural Economics and Social Sciences
Germany
1. Aerial Photography
2. Cash Crops
3. Food Crops
4. Georeferencing
5. Hot Spot Identification*
6. Subsistence Agriculture

Munroe, Darla K. (Prof.)
Ohio State University
Department of Geography
United States of America
1. Hot Spot Identification*
2. Probit/Logit Model

Nagendra, Harini (Dr.)
Indiana University
Center for the Study of Institutions, Population, and Environmental Change
United States of America
1. Community Involvement
2. Decentralization
3. Forest Anomalies
4. Himalayas
5. Reforestation

Napton, Darrell (Prof.)
South Dakota State University
Department of Geography
United States of America
1. United States of America*

Neuenschwander, Amy L.
University of Texas
Department of Geography and the Environment
United States of America
1. Airborne Light Detection and Ranging*
2. Global Positioning System*

Norman, Amy L.
University of North Carolina
Carolina Population Center
United States of America
1. Access*
2. Savannization*

Oldfield, Frank (Prof.)
University of Liverpool
Department of Geography
United Kingdom
1. Anthropocene
2. Environmental Change
3. Holocene
4. Human Impact on Terrestrial Ecosystems (HITE) Initiative
5. Paleo Perspectives

Pacheco, Pablo (Dr.)
Center for International Forestry Research
Embrapa Amazônia Oriental
Brazil
1. Agricultural Frontier
2. Amazonia
3. Arc of Deforestation
4. Cattle Ranching
5. Economic Livelihood
6. Land Concentration
7. Land Fragmentation
8. Landholding
9. Pasture

Parker, Dawn Cassandra (Prof.)
George Mason University
Department of Geography
United States of America
1. Ecosystem Services
2. Integrated Model
3. Tragedy of the Commons

Parton, William J., Jr. (Dr.)
Colorado State University
Natural Resource Ecology Laboratory
United States of America
1. Century Ecosystem Model

Pischke, Frederik
United Nations Food and Agriculture Organization
Environment and Natural Resources Service
Italy
1. Global Land Cover Network*

Pontius, Robert Gilmore, Jr. (Prof.)
Clark University
Department of International Development, Community, and Environment

United States of America
1. Pattern to Process
2. Scale*
3. Transition Matrix
4. Validation*

Priess, Jörg A. (Prof.)
University of Kassel
Center for Environmental Systems Research
Germany
1. Agricultural Intensification
2. Disintensification
3. LandSHIFT Model*
4. Nutrient Cycle*

Robbins, Paul (Prof.)
University of Arizona
Department of Geography and Regional Development
United States of America
1. Institutions

Rounsevell, Mark (Prof.)
University of Louvain
Department of Geography
Belgium
1. Global Environmental Change
2. Greenhouse Gases
3. Intergovernmental Panel on Climate Change
4. Spatial Diffusion

Sánchez-Azofeifa, Arturo (Prof.)
University of Alberta
Department of Earth and Atmospheric Sciences
Canada
1. Canada
2. Tropical Dry Forest

Schmullius, Christiane (Prof.)
Friedrich Schiller University
Institute of Geography
Germany
1. Boreal Zone
2. ENVISAT ASAR
3. European Remote Sensing (ERS-1/-2) Satellites
4. Monitoring
5. Taiga

Scholes, Robert J. (Prof.)
CSIR Division of Water, Environment and Forest Technologies
South Africa
1. Global Observation of Forest and Land Cover Dynamics
2. Global Terrestrial Observing System
3. South Africa*
4. Southern Africa*

Sessa, Reuben (Dr.)
United Nations Food and Agriculture Organization
Environment and Natural Resources Service
Italy
1. AFRICOVER
2. Global Land Cover Network*
3. Land Cover Classification System*

Shvidenko, Anatoly (Prof.)
International Institute for Applied Systems Analysis
Forestry Program
Austria
1. International Institute for Applied Systems Analysis
2. Russia

Sikor, Thomas (Dr.)
Humboldt University
Institute for Agricultural Economics and Social Sciences
Germany
1. Common-Pool Resources
2. Land Privatization
3. Land Rights
4. Private Property*
5. Property*

Southworth, Jane (Prof.)
University of Florida
Department of Geography and Land Use and Environmental Change Institute
United States of America
1. Coffee
2. Continuous Data
3. Discrete Data
4. Thermal Band Analysis

Stahl, Johannes
Humboldt University
Institute for Agricultural Economics and Social Sciences
Germany
1. Private Property*
2. Property*

Stephenne, Nathalie (Dr.)
Joint Research Centre of the European Commission
Italy
1. Geographical Information System
2. Indigenous Knowledge
3. Land Evaluation
4. Land-Use Planning
5. Resilience
6. Sahelian Land Use (SALU) Model

Stibig, Hans-Jürgen (Dr.)
Joint Research Centre of the European Commission
Institute for Environment and Sustainability
Italy
1. Global Land Cover Map of the Year 2000*
2. Hot Spots of Land-Cover Change*
3. Tropical Humid Forest*

Tansey, Kevin J. (Dr.)
University of Leicester
Department of Geography
United Kingdom
1. Advanced Very High Resolution Radiometer
2. Albedo
3. Along Track Scanning Radiometers (ATSR) World Fire Atlas
4. Coordination of Information on the Environment (CORINE) Database
5. Global Burnt Area 2000
6. GLOBCARBON
7. GLOBCOVER
8. GLOBICE
9. GLOBSCAR
10. GLOBWETLANDS
11. JERS Radar Mosaics
12. Medium Resolution Imaging Spectrometer
13. Normalized Difference Vegetation Index
14. SPOT Vegetation
15. TERRA-ASTER

Tomich, Thomas P. (Dr.)
World Agroforestry Centre
Alternatives to Slash-and-Burn Programme
Kenya
1. Alternatives to Slash-and-Burn (ASB) Programme
2. Consultative Group on International Agricultural Research
3. Tradeoffs
4. World Agroforestry Centre

Turner, B. L., II (Prof.)
Clark University
George Perkins Marsh Institute
United States of America
1. Cumulative Change
2. Land Change as a Forcing Function in Global Environmental Change†
3. Systemic Change

Turner, Matthew D. (Prof.)
University of Wisconsin
Department of Geography
United States of America
1. Arena of Land Conflict
2. Desertification*
3. Participatory Land Management*
4. Pastoral Mobility*
5. Sudan-Sahel

Verburg, Peter H. (Prof.)
Wageningen University and Research Center
Department of Environmental Sciences
The Netherlands
1. Conversion of Land Use and Its Effects (CLUE) Model
2. Modeling
3. Scale*

Vogel, Coleen H. (Prof.)
University of the Witwatersrand
School of Geography, Archaeology and Environmental Studies
South Africa
1. People at Risk

Young, Kenneth R. (Prof.)
University of Texas
Department of Geography
United States of America
1. Andes
2. Latin America
3. Mountains

About the Editor

Helmut Geist is professor of human-environment at the University of Aberdeen in Scotland, United Kingdom. From 2000 until 2005, he was Executive Director of the IGBP-IHDP Land-Use/Cover Change (LUCC) project at LUCC's International Project Office at the University of Louvain in Louvain-la-Neuve, Belgium. His research interests include political ecology, sustainable land use, land-change theory, and human-driven environmental changes such as land degradation, deforestation, desertification, and urbanization. Based at universities in the United Kingdom, Belgium, Germany, and the United States, he was involved in the World Bank's Human Development Network on tobacco control (1998), in reviews of the European Commission's desertification research (2003), and in the United Nations Millennium Ecosystem Assessment (2004).